Biological Control

Biological Control

ROY G. VAN DRIESCHE
Department of Entomology, University of Massachusetts, Amherst, MA

THOMAS S. BELLOWS, JR.
Department of Entomology, University of California, Riverside, CA

KLUWER ACADEMIC PUBLISHERS
BOSTON/DORDRECHT/LONDON

Distributors for North, Central and South America:
Kluwer Academic Publishers
101 Philip Drive
Assinippi Park
Norwell, Massachusetts 02061 USA
Telephone (781) 871-6600
Fax (781) 871-6528
E-Mail <kluwer@wkap.com>

Distributors for all other countries:
Kluwer Academic Publishers Group
Distribution Centre
Post Office Box 322
3300 AH Dordrecht, THE NETHERLANDS
Telephone 31 78 6392 392
Fax 31 78 6546 474
E-Mail services@wkap.nl>

Electronic Services <http://www.wkap.nl>

Library of Congress Cataloging-in-Publication

Van Driesche, R.G.
Biological control / Roy G. Van Driesche and Thomas S. Bellows, Jr.
p. cm.
Includes bibliographical references and index.

ISBN-13:978-1-4612-8490-1 e-ISBN-13:978-1-4613-1157-7
DOI:10.1007/978-1-4613-1157-7

1. Pests--Biological control. 2. Pollination by animals. 3. Pollination by insects. 4. Angiosperms--Evolution. I. Bellows, T.S. II. Title
SB975.V375 1996 95-31557
632'.96--dc20 CIP

Softcover reprint of the hardcover 1st edition 1996

Third Printing 2001 by Kluwer Academic Publishers

Cover photos courtesy of: J.A. MacDonald, P. Kenmore, and J.K. Clark.
Cover design: Robert Freese

This printing is a digital duplication of the original edition.

Printed on acid-free paper.

Contents

PREFACE

Nearly twenty years have passed since the publishing of a broadly-based textbook on biological control (Huffaker and Messinger 1976). In the interim, other works on biological control have been either briefer treatments (DeBach and Rosen 1991), or collections of essays on selected topics (Waage and Greathead 1986; Mackauer and Ehler 1990). Our text has been written to fill what we believe is a need for a well-integrated, broadly-based text of appropriate length and degree of technical detail for teaching a one semester upper level course in biological control. We have attempted to focus on principles and concepts, rather than on biological control of particular taxa or biological control by particular kinds of natural enemies. Therefore, for example, the reader will find the material on biological control of weeds integrated with biological control of insects and mites into chapters on principles, techniques, and applications rather than presented separately. Only biological control of plant pathogens is addressed independently, an appraoch made necessary by the many special features and concepts in plant pathogen biological control.

In addition to essential material on such expected topics as natural enemy introduction, taxonomy of natural enemies, and the history of the discipline, we have sought to emphasize several areas we view as of special importance. In particular, we have emphasized natural enemy evaluation (Chapter 13) as a key feature of biological control studies. We have developed new treatments on conservation of natural enemies (Chapter 7) and biological control of environmental pests (Chapter 21). We have also explored topics of a broader nature, such as the philosophy of integration of biological control in pest management systems (Chapter 14), the role of extension in teaching biological control to farmers (Chapter 19), and the effects of government policies on the degree to which biological control is employed (Chapter 20).

This text has been written to encourage training of a new generation of biological control scientists committed to both the understanding of biological control and its safe use to solve pest problems. Biological control, while enjoying a century of practical use, is a critically important means of addressing current pest problems in both crop protection and the preservation of natural systems. It is our hope that this text will contribute to the training of scientists who achieve tomorrow's successes in biological control.

Preparation of this book benefitted in many ways from interactions with and contributions from colleagues. The book's early development and outline was aided by discussions with J. Waage and P. Kenmore. Reviews of the book in part or whole, were contributed by F. Bigler, C. Campbell, R. Charudattan, W. Coli, J. Coulson, E. Delfosse, J. Elkinton, T. Fisher, H. Frank, L. Gilkeson, R. Goeden, D. Headrick, M. Hoddle, M. Hoffmann, D. Hogg, M. Johnson, H. Kaya, J. van Lenteren, N. Lepla, J. Lewis, C. McCoy, D. Meyerdirk, O. Minkenberg, R. Prokopy, M. Samways, M. Schauff, J. Sutton, B. Vinson, J. Waage, and M. Wilson. M. Hassell assisted in composing portions of Chapter 18. Assistance in checking the names and authors of species in the text was provided by C. Breidenbaugh, C. Lombardi, C. Meisenbacher, and B. Orr. J. Rewa prepared many revisions of the typescript. Our thanks also go to Greg Payne, at Chapman & Hall, who has been unwavering in his patience and support, and to those who encouraged our early interest in entomology and biological control, P. Ritcher, J. Lattin, E. Dickason, S. Knapp, J. Owens, W. Whitford, M. Hassell, and our colleagues at the University of Massachusetts and the University of California. We also extend a special thanks to our wives and families for their patience and support in this project.

R. G. Van Driesche
T. S. Bellows

Biological Control

SECTION

I

Origins and Scope of Biological Control

In Chapters 1 and 2, the origins and scope of biological control are considered. In Chapter 1, the ecological origins of pests are discussed, and biological control is defined. To provide students of biological control a view of the origins of the discipline, the historical development of the concepts and practices of biological control are briefly presented. Precedent-setting past cases are illustrated, together with some recent important successes. The historical record of biological control with respect to establishment and pest suppression rates of projects is reviewed, together with literature sources for the historical record of biological control activities in particular countries or regions.

In Chapter 2, the range of organisms that have been targets of biological control projects is discussed, as well as the kinds of organisms that have been employed as agents of biological control. The principal methods by which biological control may be implemented (conservation, introduction, augmentation) are defined and discussed. These are considered in depth in Chapters 8 through 12 in Section III.

CHAPTER

1

Pest Origins, Pesticides, and the History of Biological Control

HUMAN NEEDS AND THE ORIGINS OF PESTS

The human population is large and still expanding. To gain more farmland, native ecosystems are being rapidly converted to human use, destroying forests, soil, and native plants and animals. To produce sufficient food, commercial and subsistence farming systems must be highly productive, but sustainable and nonpolluting. However, to preserve the world for the future, space must be left for wild animals and wild places. To do both of these things is the great challenge of the early twenty-first century. Part of the solution to this problem is biological control, the foundation on which sustainable, nonpolluting pest control for tomorrow's farms must be built.

Where do pests come from? Some pests are created because of how we grow crops. Crop defenses such as toughness or repellent compounds are decreased by selection. Crops are grown in large patches, with uniform planting and harvesting schedules. These practices have potential to reduce plant defenses against herbivores. Other pests arise because movement of organisms around the world creates new species combinations. Sometimes, local herbivores adopt crop species that are introduced into a country. In South Africa for example, 68% of 188 arthropod pests of 14 introduced crops are native species not previously in contact with the crops they now attack (Dennill and Moran 1989). In other cases, pest herbivores that attack crops accompany crop cultivars as stowaways when cultivars are moved to new locations. The natural enemies that suppress the herbivores, however, are likely to be left behind (Fig. 1.1), allowing herbivore populations to thrive in an enemy-free environment, reaching high, pest densities. Since 1850, the number of nonnative insect species in the United States has expanded over tenfold (Fig. 1.2). Of these, over 200 have become severe pests, and over 500 are lesser pests (Fig. 1.3). Origins of the adventive arthropods in the United States are discussed by Sailer (1983). In the United States, adventive species makes up 39% of the insect pests on crops, 27% of the insect pests of forests, 7% of the animal pests, 31% of the plant pathogens of vegetables, 73% of the weeds of cultivated crops, and 41% of the weeds of pastures (Pimentel 1993). Origins of many weeds are similar to those of pest arthropods in that many are adventive species no longer suppressed by specific herbivores. Unlike arthropods, however, many pest weeds are species deliberately introduced for various reasons (Foy et al. 1983).

Figure 1.1 Invasions of some adventive species are rooted in commerce, both as unintentional stowaways during shipment (as in the case of plant products [A]), or as intentionally introduced species with unrecognized harmful potential, (as in the case of the giant African snail, *Achatina fulica* Bowditch, sold in some locations as a pet [B]). (Photographs courtesy of USDA-APHIS.)

Consideration of the origins of pests suggests that solutions for our pest problems lie in the modification of agriculture systems (where possible without loss of productivity) to conserve natural enemies of crop pests and in the reconnecting of herbivores with their natural enemies, where these have become separated. Methods to achieve these ends and to protect native species and ecosystems from the effects of aggressive adventive species, via conservation, augmentation, and introduction of natural enemies, are the subjects of this book.

PROBLEMS WITH PESTICIDES

Chemical pesticides are commonly used for the control of many pests. In contrast to biological control methods, use of such pesticides does not require information on the ecological origins of pests. Pest suppression is achieved temporarily by killing (or for plant pathogens, preventing the growth of) as many members of the pest population as possible through repeated applications of chemical products, as needed. Worldwide pesticide use has increased twelvefold since the early 1950s (Fig. 1.4), and costs paid by farmers in the United States for pesticides increased sixfold between 1951 and 1976 (Eichers 1981). Chemical pesticides have proved effective in many cases, particularly in controlling weeds and plant diseases, but have been ineffective

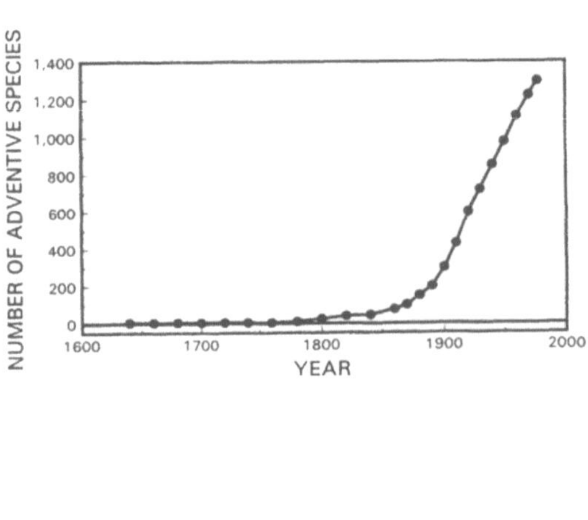

Figure 1.2 Total number of adventive insect species in the United States from 1640 to 1977 (after Sailer 1978).

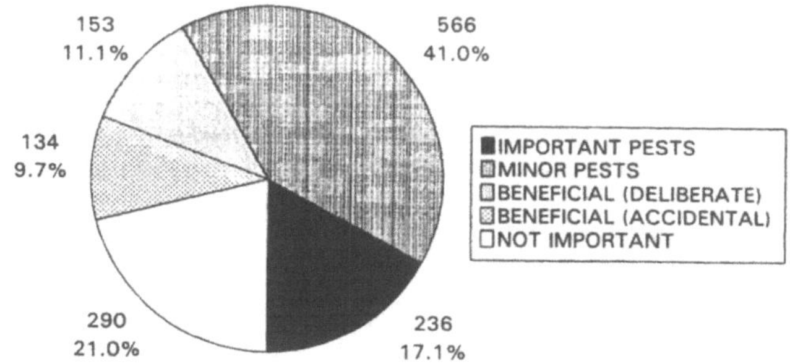

Figure 1.3 Numbers of adventive insect species in the United States in various categories in 1977 (after Sailer 1978).

in others, for example, controlling cotton insects in many countries (Bottrell and Adkisson 1977). Some of the problems associated with using pesticides include failure of pest control, contamination of the environment, and damage to human health. Concern over these issues has prompted some countries to seek to reduce pesticide use.

Pesticides create pest control problems when they fail to control the target pest or when they create new pests. Resistance to pesticides is the main way in which pesticide use can lead to pest control failure. Resistance develops in pest populations through the differential survival of members of the pest population that best detoxify or avoid exposure to the pesticide. Over several generations, pest populations may develop that can no longer easily be killed by one or more pesticides. Since 1945, the number of insects, weeds, and, most recently, plant pathogens that have become resistant to pesticides has increased dramatically (Brent 1987) (Fig. 7.5).

Another way in which pesticide use can foster outbreaks of pests is the destruction of the target pest's natural enemies (Trichilo and Wilson 1993). Many pests and potential pests are held in check, at least partially, by various predacious, parasitic, pathogenic, or antagonistic species. Most pesticides kill natural enemies, which often are more sensitive to the pesticide

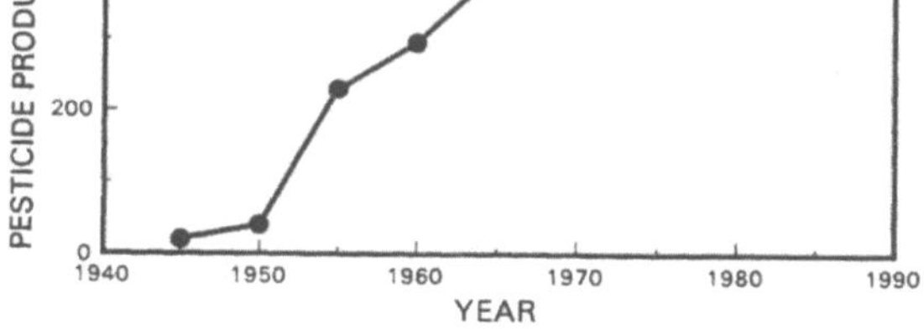

Figure 1.4 Pesticide production in the United States (after Pimentel 1991).

than is the pest itself (Croft 1990). Insecticides, for example, kill beneficial insects. Acaricides kill predacious mites. Fungicides suppress growth of fungi that protect plants against plant diseases and also reduce fungi that kill pest arthropods. When these natural enemies are destroyed, the pest individuals that remain after the pesticide application survive longer, causing their numbers to rebound to yet higher levels. Outbreaks of rice brown planthopper, *Nilaparvata lugens* (Stål), in Asia, for example, have been shown to be greatly increased by the destruction of spiders and other natural enemies by pesticides (Heinrichs et al. 1982).

Pesticides also create new pests when natural enemies of ordinarily harmless species (or minor pests) are destroyed by chemicals (Kerns and Gaylor 1993). These "secondary pests" then reach higher densities than normal and begin to cause economic damage. They are then considered pests by farmers and become targets of pesticides themselves. Spider mites on many crops are examples of species that have become pests because of the use of pesticides. The leaf miner *Phyllonorycter crataegella* (Clemens) in apple (*Malus pumilla* Miller) is a pest because its parasitoids are suppressed by pesticides used in apple orchards for control of other pests (Maier 1982).

In areas in which pesticide use is common, small quantities of some pesticides may move out of the treated area and end up in soil, water, birds, and other wild animals, as well as being found in the crop itself. The consequences of these residues varies from none to serious. Herbicides and soil fumigants have contaminated groundwater supplies; hydrochlorinated pesticides such as DDT have caused regional extinctions of some raptorial bird species; and the more toxic organophosphate pesticides have caused human poisonings (see Graham 1970; Metcalf 1980; Dempster 1987; Newton 1988). Accidental poisonings and contamination are especially likely when farmers do not understand the toxicity of the materials they are using, when they cannot read product instructions, or do not have the necessary protective equipment. As farmers and the public at large gain increased knowledge of these problems, the demand for other forms of pest control, in particular biological control, increases.

Biological control, however, is not simple. Its successful use requires highly trained specialists to conduct research on natural enemies, and well-informed extension workers and farmers to implement their use.

DEFINITION OF BIOLOGICAL PEST CONTROL

The scope of biological control—what kinds of targets it may be used against, what kinds of natural enemies exist to aid in biological control, and what methods work for employing them—are discussed in detail in Chapter 2. Here, we define the process and set the limits of what the remaining chapters will address. Biological control is the use of parasitoid, predator, pathogen, antagonist, or competitor populations to suppress a pest population, making it less abundant and thus less damaging than it would otherwise be. Insects, mites, weeds, plant diseases, and vertebrates all may be targets of biological control. Biological control may be the result of purposeful actions by man or may result from the unassisted action of natural forces. Biological control may be employed either for suppression of crop or forest pests, or for restoration of natural systems affected by adventive (nonnative) pests. Not all nonchemical control is biological control. Plant breeding, cultural control, and use of semiochemicals, if directly intended to influence the pest, are not biological control. These same approaches, however, may in some cases have a role to play in biological control if they are directed not at the pest but at conserving or enhancing its natural enemies. Breeding plants to directly resist pests, for example, by being toxic or having some other pest-suppressive features, is not biological control. But breeding plants to make them better sites for parasitoids or predators to search and attack pests is biological control. Chemicals extracted from plants or microbes and used for pest control are not biological controls. Biological control is a population-level process in which one species' population lowers the numbers of another species by mechanisms such as predation, parasitism, pathogenicity, or competition. In this book, we discuss principles of biological control broadly with examples drawn from a variety of target taxa; consequently, we do not address biological control of particular taxa (insects, weeds, and so on) in separate chapters, except for biological control of plant pathogens. Plant pathogens involve some considerable differences from biological control of other taxa and requires separate treatment.

Some Terminology

Because movement of organisms to new locations (with consequent loss of natural enemies) is important to many aspects of biological control, it is often necessary to describe the origins of organisms. The terms (native, exotic, endemic, immigrant, introduced) commonly employed in such descriptions, however, present some difficulties because of conflicts in their meanings or unintended implications. If, for example, we refer to a pest that has arrived to a new location as an "introduced pest" (as is commonly done), we prompt the question "Who introduced it?", when in many cases no one did.

To resolve questions of terminology, in this text we follow Frank and McCoy (1990) and use the following terminology:

(1) **indigenous (or native)** organisms are those organisms in a specified area, that arose evolutionarily in their current taxa in that location

(2) **precinctive** organisms are the subset of the indigenous organisms of a given location that occur nowhere else

(3) **adventive** organisms are those species in a specified location that did not evolve there, but arrived there from elsewhere (i.e., the opposite of native)

(4) **immigrants** are those adventive organisms in a specified location that arrived there without the deliberate, purposeful aid of man. (This group includes both actively dispersing organisms, and ones arriving as stowaways on plants or other commodities moved by man.)

(5) **introduced** organisms, which are those brought to a location by the conscious choice of man (i.e., food crop species, ornamental or forage plants, pets and domestic animals, biological control agents).

In this text we distinguish between **parasites** and **parasitoids** (see Chapter 2 for definitions), but employ **parasitism** and **parasitized** when referring to the action of either parasites or parasitoids.

HISTORY OF BIOLOGICAL CONTROL

Like all human advances, the development of biological control followed no master plan, but surged or stagnated at the whim of insights, luck, personal endeavor and, in more recent decades, institutional momentum. The history of the development of the method is really seamless, but for discussion can be divided into four parts.

Early Observations and Development of Key Concepts

In this section we trace the earliest intuitive uses of natural enemies. We discuss the development of the concepts of predation, parasitism, and disease in invertebrates; weed control agents; and antagonists and parasites of plant pathogens. We point out the first suggestions and attempts for practical use of such agents. This brief presentation has been summarized from an excellent account of the topic by DeBach and Rosen (1991) to which the reader is referred for more detail.

Insect Predators. Before the formal development of natural history as a science in western Europe during the Renaissance period, farmers in other parts of the world were already making use of some types of predacious arthropods. In both China and Yemen, ant colonies were moved between sites for control of pests in tree crops (citrus [*Citrus* spp.] and dates [*Phoenix dactylifera Linnaeus*]) (Coulson et al. 1982). Also, in China, spiders were manipulated for pest control (Sparks et al. 1982). These practices, dating back several thousand years, were developed by farmers through direct observation of these predators, which are large enough to be visible and whose life cycles are easily understood. In Europe, one of the first written proposals to use predacious insects for pest control was made in 1752 by the taxonomist Carl Linnaeus who remarked that "Every insect has its predator which follows and destroys it. Such predatory insects should be caught and used for disinfesting crop-plants" (see Hörstadius 1974). By the early 1800s, naturalists such as Erasmus Darwin and various American entomologists were suggesting that predators such as syrphid flies and coccinellid beetles should be employed to combat aphids in greenhouses and certain outdoor crops (Kirby and Spence 1815). The first international movement of a predacious invertebrate was made in 1873 by the American entomologist C.V. Riley, who sent the mite *Tyroglyphus phylloxerae* Riley and Planchon to France to combat the grape phylloxera (*Daktulosphaira vitifolii* [Fitch]). The mite established but unfortunately had no practical value.

Insect Parasitism. In contrast to insect predation, which was easily understood (at least in concept) by comparison to predacious mammals, insect parasitism was more difficult to correctly understand. In fact, early observers (for example, Aldrovandi in 1602) who witnessed parasitoids emerging from butterfly larvae misunderstood the process as a transformation in which the parasitoids were another stage of the larva produced by a type of metamorphosis (Fig. 1.5) (Bodenheimer 1931). The first person to publish a correct interpretation of insect parasitism was the British physician Martin Lister who in 1685 noted that the ichneumon wasps emerging from caterpillars were the result of eggs laid in the caterpillars by adult female ichneumons. In 1700, Antoni van Leeuvenhoek, the Dutch microscopist, correctly interpreted parasitism of aphids by a species of *Aphidius* wasp.

Following these initial observations, many other workers studied other parasitoids and described their biologies. The nineteenth century saw an explosive increase in the number of scientific works produced on the taxonomy, biology, and ecology of insect parasitoids and predators by scientists such as M.M. Spinola, J.W. Dalman, J.L.C. Gravenhorst, J.O. Westwood, Francis Walker, C. Rondani, A. Förster, J.T.C. Ratzeburg, and many others (DeBach and Rosen 1991). This immense new body of knowledge provided the practical tools needed to begin to turn concepts about biological control into a usable technology.

The first suggestion to move parasitoids between countries for pest control was made in 1855 by the New York entomologist Asa Fitch, who proposed to import parasitoids of the wheat midge, *Sitodiplosis mosellana* (Géhin), from Europe. No importations, however, were made. Not until almost 30 years later, in 1883, was the first parasitoid species, *Cotesia glomerata* (Linnaeus) (= *Apanteles glomeratus* Linnaeus), successfully moved between continents (from England to the United States) and established where released (Riley 1885).

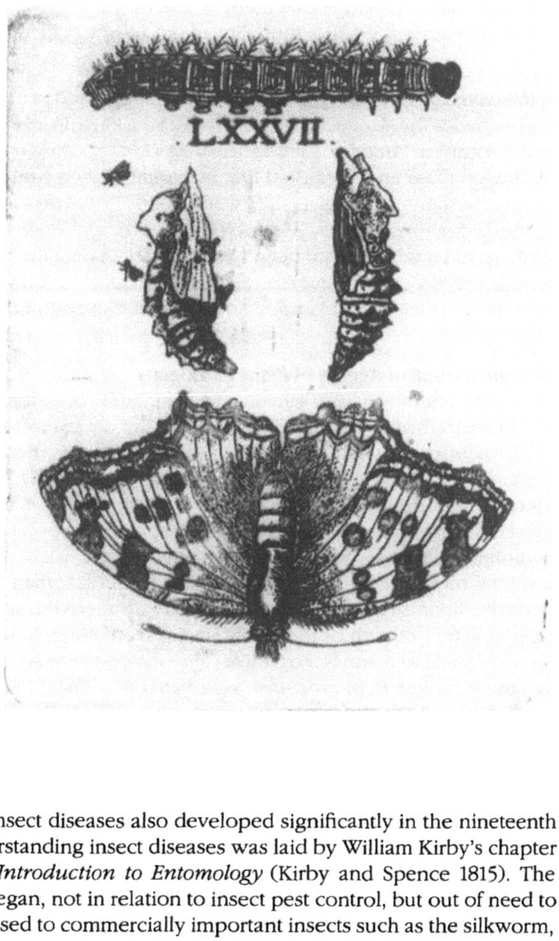

Figure 1.5 A woodcut depicting parasitoid wasps emerging from a lepidopteran pupa (after Goedaert 1662). (Photograph by Bancroft Library, University of California, Berkeley.)

Insect Diseases. Knowledge of insect diseases also developed significantly in the nineteenth century. The foundation for understanding insect diseases was laid by William Kirby's chapter on "Diseases of Insects" in *An Introduction to Entomology* (Kirby and Spence 1815). The understanding of insect disease began, not in relation to insect pest control, but out of need to control damage from diseases caused to commercially important insects such as the silkworm, *Bombyx mori* (Linnaeus).

Agostino Bassi of Lodi, Italy, was the first to demonstrate experimentally the infectious nature of insect disease in his study of the white muscardine diseases (caused by the fungus *Beauveria bassiana* [Balsamo] Vuillemin) of silkworm in 1835. Further work on silkworm diseases was done in 1865–70 by Louis Pasteur in France which correctly identified both vertical and horizontal transmission routes for infection (for an explanation of terms see Chapter 16 on the biology of pathogens).

The first suggestion to use insect pathogens for pest control was made by Bassi in 1836 when he proposed that liquids from putrefied cadavers of diseased insects could be mixed with water and sprayed on foliage to kill insects. The first field trials of pathogens, however, were not conducted until 1884, when the Russian entomologist Elie Metchnikoff developed a facility to mass produce spores of the pathogenic fungus *Metarhizium anisopliae* (Metchnikoff) Sorokin and used them in field tests against larvae of the sugar-beet curculio, *Cleonus punctiventris* (Germar), causing 55–80% mortality of larvae in the field.

Arthropods for Control of Weeds. The first suggestion of using insects to control weedy plants was made by the American entomologist Asa Fitch in 1855, who noted that some European weeds found in America such as toadflax (*Linaria vulgaris* Miller) had no American insects feeding on them and suggested that importing insects from Europe might solve the problem. The first actual use of insects for weed control occurred in 1863 in southern India when an imported cochineal insect, *Dactylopius ceylonicus* (Green), was moved into the region from northern India where it had been observed decimating the pest cactus *Opuntia vulgaris* Miller (Goeden 1978). The first international movement of a biological weed control agent happened soon afterwards when this same species was moved to Sri Lanka.

Biological Control Agents of Plant Pathogens. Sanford (1926) was one of the first to recognize competition between saprophytic and pathogenic organisms for nutrients at the site of initial infection as a form of biological control of plant disease. He observed that potato scab (caused by *Streptomyces scabies* [Thaxter] Waksman and Henrici) could be reduced by adding organic matter (grass clippings) to soil. He speculated, correctly, that this was due to antagonistic effects from saprophytic organisms. Early attempts to use augmentative applications of desirable competitive organisms were, however, often unsuccessful, and it is now recognized that environmental conditions are crucial in affecting the balance between species, either in the soil or on the phylloplane (leaf surface) (Faull 1988; Blakeman 1988). Massive doses of inoculum gradually diminish unless habitat conditions are altered so as to confer a selective advantage on the population growth of the antagonist. Nevertheless, in some circumstances, the approach has been made to work. For example, inoculation of conifer stumps after timber felling with the antagonist fungus *Phanerochaete gigantea* (Fries: Fries) Rattan et al. suppressed the ability of the pathogen *Heterobasidion annosum* (Fries: Fries) Brefeld to attack stumps, with a subsequent lower rate of infection of new trees at the site (Rishbeth 1963).

Biological control of plant diseases can occur via several distinct mechanisms, including competition for nutrients between a pathogen and a harmless species, parasitism, and production of antibiotics (Faull 1988). Concepts of plant disease biological control have been articulated by Baker (1985), Cook and Baker (1983), and Campbell (1989). Relative to other forms of biological control, control of plant diseases is new and has a briefer history.

First Successes: Impetus for Creation of Institutions

By the 1880s, enough knowledge had accumulated concerning natural enemies of arthropods and weeds for their practical use. In this section, we discuss a series of successful and highly visible biological control projects that impressed the societies and governments involved and, in several locations, led to the creation of formal institutions charged with carrying out more biological control projects. These institutions provided greater continuity and increased resources, leading to an expansion in the number of biological control projects undertaken. While many projects could be cited (see DeBach and Rosen 1991 and Clausen 1978 for a fuller listing), those discussed here forcefully demonstrate the value of the method.

Whether or not a given project was catalytic in causing the formation of a biological control institution was influenced by several factors. (See Sawyer 1990 for a history of the unsuccessful effort to develop strong biological control institutions within the United States Department of

Agriculture in the 1888–1951 period.) First, the project needed to provide dramatically visible control. The pest had to be easy to see and the control had to happen rapidly after natural enemy releases began, so the two were easy to connect as cause and effect. Second, the pest itself had to be very serious and well known. Biological control of pests that had only recently invaded an area and still had a limited distribution—however large their potential for damage might have been—seems to have made less of an impression on society than control of pests that were widespread and already causing important losses. Third, the society in which the project took place had to have suitable resources and governmental structure to be receptive and able to react to the success and create institutions to conduct further work.

Introduction of Predacious Arthropods: The Cottony Cushion Scale and Vedalia Beetle in California (U.S.A.). In about 1868, a pest of citrus, the cottony cushion scale, *Icerya purchasi* Maskell, was found in California. By 1886, this pest was on the verge of destroying the citrus industry in southern California. Chemical measures were tried (cyanide fumigation of trees) and had only limited effect. Production was being severely reduced. In or about 1887, it was learned that the native home of the scale was Australia, where *I. purchasi* was not considered a pest. In the next two years, two efforts were made to secure natural enemies from Australia, one well known and one less well-recognized. The famous story is that Albert Koebele, a representative of the United States Department of Agriculture, went to Australia in 1888 and collected a parasitic fly, *Cryptochetum iceryae* (Williston), and a predacious coccinellid, *Rodolia* (formerly *Vedalia*) *cardinalis* (Mulsant), since known as the vedalia beetle (Fig. 1.6). The less famous story is that W.G. Klee, a California State Inspector of Fruit Pests, through correspondence with

Figure 1.6 Introduction of natural enemies. Importation of the Vedalia beetle (*Rodolia cardinalis* [Mulsant]) into California led to control of the cottony cushion scale, *Icerya purchasi* Maskell. (Photograph by M. Badgley.)

an entomologist in Australia, Frazer Crawford, imported the same parasitic fly in 1887. Of these two insects, the beetle became the center of attention in southern California, since it provided immediate and dramatic control. The vedalia beetle multiplied profusely on scale-infested trees and was soon widely distributed by growers. Within two years, the beetle controlled the pest throughout the state. The fly also established and eventually became the dominant control agent for the pest in the coastal areas of the state.

The cottony cushion scale project had far-reaching effects on biological control by demonstrating that arthropod predators could be manipulated in ways that solved serious problems and produced large economic benefits. Because the economic benefit from the work was so dramatically obvious to the influential citrus growers in the state, it stimulated the official pursuit of biological control by state government, leading ultimately to a long series of projects by entomologists employed by the University of California. In addition, contacts between entomologists from California and entomologists in Australia (during collection of the beetle) and in countries such as Chile (one of many countries to which the beetle was later sent) stimulated interest in introductions of natural enemies in many countries (Caltagirone and Doutt 1989). Such early introductions of natural enemies required methods to preserve small colonies of natural enemies for extended periods; modern transportation has greatly shortened the travel times required to ship natural enemies, which can be transported in 2–5 days between most locations, without need for rearing colonies (Fig. 1.7).

Introduction of Parasitic Arthropods: The Sugarcane Leafhopper in Hawaii (U.S.A.). In 1900, a pest of sugarcane (*Saccharum officinarum* Linnaeus), the leafhopper *Perkinsiella saccharicida* Kirkaldy, was discovered in the Hawaiian Islands. By 1903, sugar yields were being depressed by the pest. It was learned that the species could be found in Queensland, Australia, and caused no important damage there. Inspired by the Vedalia project in California, the entomologists R.C.L. Perkins and A. Koebele went to Australia to search for natural enemies. Between 1904 and 1916, six species of egg parasitoids were imported into Hawaii, of which the most effective turned out to be *Anagrus optabilis* (Perkins). These six parasitoids established and together significantly reduced the problem. Later, an egg predator, *Tytthus mundulus* (Breddin), was also imported and this predator, in combination with the parasitoids, provided complete control. This project, and others, demonstrated the role of parasitoids in control of pest arthropods and stimulated the development of a very active state program that currently makes Hawaii an important center for biological control.

Introduction of Arthropod Herbivores for Weed Control: Prickly Pear and Cactoblastis Moth in Australia. In the 1800s, several species of ornamental cacti were introduced into Australia. Some species, especially the prickly pear species *Opuntia inermis* de Candolle and *Opuntia stricta* (Haworth), spread rapidly in forest and grazing lands such that, by 1925, 24 million hectares of land were densely infested and rendered economically valueless. In 1920, the Commonwealth Prickly Pear Board was formed to send entomologists to South America to search for insects that attacked the weed, and to establish quarantine laboratories in Australia to facilitate the testing of candidate agents. Ultimately, some 50 species of insects were collected and sent to Australia. (See pp. 51–55 of Wilson [1960] for details of this project.) Twelve species established and exerted some control prior to importation of the agent that ultimately solved the problem, the moth *Cactoblastis cactorum* (Bergroth) (and its associated plant pathogens). This agent was collected in 1925 in Argentina and colonized in the field in Australia in 1926. By

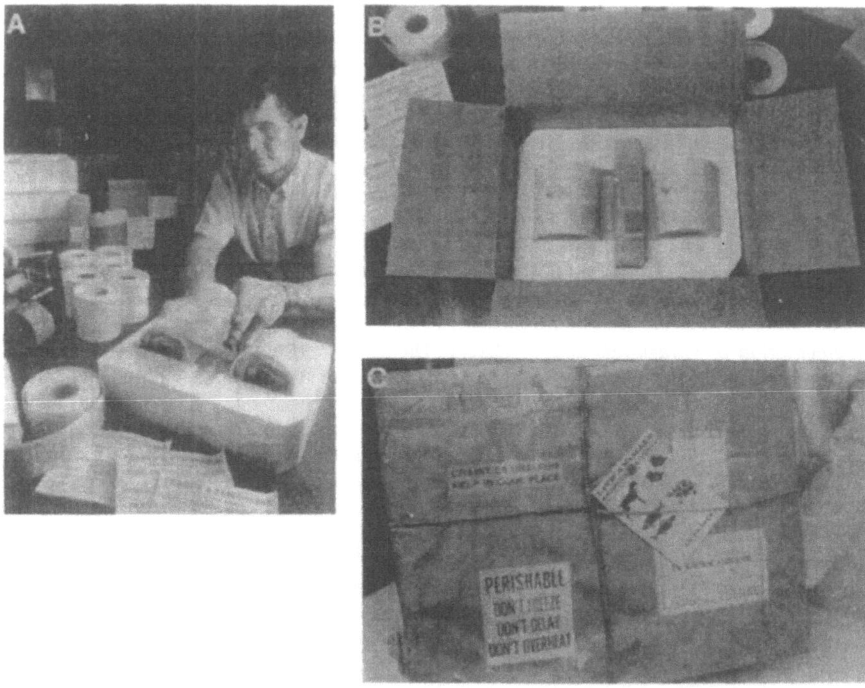

Figure 1.7 A container for international shipment of natural enemies; note use of Styrofoam insulation and pack of artificial ice (in center) for cooling (foam top of box removed for photograph) (A,B); assembled package and associated shipping labels (C). (Photographs courtesy of USDA/BIRL, M. Heppner [A], S.R. Bauer [B], and R.M. Hendrickson [C].)

1930–32, a general collapse of cactus stands at the original release sites had occurred, and the moth quickly spread and controlled the cacti throughout the infested area. The dramatic collapse of such large plants, so rapidly and over such a large area exceeded all expectations. The project so deeply impressed the Australian government that strong biological control institutions were created that today make Australia one of the world leaders in the biological control of weeds. This project, together with the control in the 1950s of Klamath weed, *Hypericum perforatum* Linnaeus, in California (Huffaker and Kennett 1959) convincingly demonstrated the practical ability of imported insects to suppress adventive weeds.

Introduction of Arthropod Pathogens: The Rhinoceros Beetle and Its Baculovirus. The first intentional, successful use of an introduced arthropod pathogen for pest control occurred in 1967. The rhinoceros beetle, *Oryctes rhinoceros* (Linnaeus), is an important pest of coconut (*Cocos nucifera* Linnaeus) and oil palms. The beetle is native to southeast Asia and during the present century spread to many Pacific islands. In 1963, a baculovirus of the larvae and adults was discovered in Malaysia and in 1967 the virus was introduced into Western Samoa, where it controlled the pest (Bedford 1980, 1986). The virus was subsequently moved to other islands with similar success. In the Maldives (in the Indian Ocean), for example, the percentage of damaged palms fell from 40–60% before introduction to about 10% afterwards (Zelazny et al. 1990, 1992). One of the reasons for the success of this virus is that infected adult beetles remain alive for many weeks and act as vectors, introducing the virus to larvae living in breeding sites. This project demonstrated the potential for pathogens to be successful as biological control agents through introduction, however, successful movements of arthropod pathogens have been rare, and consequently biological control work with pathogens has focused on the augmentative use of artificially produced and applied pathogens. (Methods of use of natural enemies—introduction, conservation, augmentation—are discussed in Chapter 2).

Augmentation of Parasitoids and Predators: The Greenhouse Whitefly and Two-Spotted Spider mite. Unlike the preceding examples, all of which are based on the intentional international movement of natural enemies, their liberation, and subsequent spread as self-sustaining populations, in glasshouse crops another approach was needed. Because glasshouse crops are partially isolated from the outside environment, natural enemies often have to be introduced in each cropping cycle. Among the most important pests of protected cultivation are the greenhouse whitefly, *Trialeurodes vaporariorum* (Westwood), and the two-spotted spider mite, *Tetranychus urticae* Koch.

The first step toward the development of a biological control system for pests in glasshouse crops occurred in 1926 when a grower in England noted dark-colored (parasitized) whiteflies in his glasshouse and called them to the attention of the entomologist E.R. Speyer. The responsible parasitoid, *Encarsia formosa* Gahan, was subsequently found to provide control of the pest and commercial production was begun (Speyer 1927; Hussey 1985). Production lapsed in 1949 due to competition from newly developed pesticides and did not resume again until 1967. While several groups were involved in the early 1970s in the production of natural enemies for use in glasshouses, a critical event was the decision in 1965 by a Mr. Koppert in Holland to begin to produce beneficial predator mites and insects for greenhouse use. His business developed a number of improved methods for rearing, sorting, and shipping both *Encarsia formosa* and predacious mites. Several large insect-rearing firms now supply a variety of beneficial arthropods to growers for augmentative releases in glasshouses.

Whitefly control with parasitoids required that all other pests in glasshouse crops also be controlled with natural enemies or chemicals compatible with *Encarsia formosa*. Another important pest on tomatoes (*Lycopersicon lycopersicum* [Linnaeus] Karsten ex Farwell) and cucumbers (*Cucumis sativus* Linnaeus) was the two-spotted spider mite. While several species of natural enemies were considered for control of this pest, the species that ultimately proved the most effective was discovered in 1960 on a shipment of orchids sent from Chile to Germany (Bravenboer and Dosse 1962). This species, *Phytoseiulus persimilis* Athias-Henriot, proved voracious and easy to rear and is now widely used in greenhouses (Fig. 1.8).

Extensive research was required to develop the methods to rear these natural enemies and to use them effectively (Hussey 1985). These agents made the commercial application of biological control in glasshouses possible. By 1994, *Phytoseiulus persimilis* was being applied on over 8000 ha of glasshouse crops annually and *Encarsia formosa* on nearly 5000 ha (van Lenteren 1995), demonstrating that predacious and parasitic arthropods can be reared efficiently to provide cost-effective biological control. This success provided the incentive for the development of the commercial insectary business throughout much of the world.

Figure 1.8 Augmentation of natural enemies. Seasonal augmentative releases of the predacious mite *Phytoseiulus persimilis* Athias-Henriot provides control of phytophagous mites in glasshouses. (Photograph courtesy of M. Badgley.)

Augmentation of Arthropod Pathogens: Bacillus thuringiensis Berliner. Early in this century, a bacterial disease of larvae of the flour moth *Anagasta kuehniella* (Zeller) was noted by Berliner (1911) and the causative agent described as *Bacillus thuringiensis.* Tests were conducted on various Lepidoptera and, by 1938, the first commercial preparation based on this organism was marketed under the name Sporeinet in France (Jacobs 1951). Many strains of this organism were isolated from various hosts, but prior to the 1980s commercial products were limited in their activity to larvae of certain Lepidoptera. Isolates since have been obtained that are effective against many other taxa, including Diptera and Coleoptera (van Essen and Hembree 1980; Herrnstadt et al. 1987). The effectiveness and selectivity of this pathogen are based on its production of one or more of several toxins that paralyze and kill hosts, with certain toxins affecting some host groups and not others. This pathogen represents the earliest commercially successful use of a microbial pathogen as a formulated pesticide (Fig. 1.9). The toxins of this bacterium are stomach poisons that kill only certain groups of insects, permitting the development of integrated pest management systems in which natural enemies are conserved while selectively suppressing particular pests. The success of this product has been important in stimulating the development of commercial interest in the location, production, and use of microbes as pest control products. Information about this pathogen and its use are summarized by Entwistle et al. (1993).

Conservation of Native Natural Enemies: The Rice Brown Planthopper. The rice brown planthopper, *Nilaparvata lugens*, is an occasional pest of rice in many parts of Asia. In the 1960s and 1970s, cultural changes in rice (*Oryza sativa* Linnaeus) varieties, fertilization, and cropping practices occurred which encouraged farmers to use pesticides to suppress this minor pest. The pesticides destroyed spiders and other generalist natural enemies and led to a cycle of more

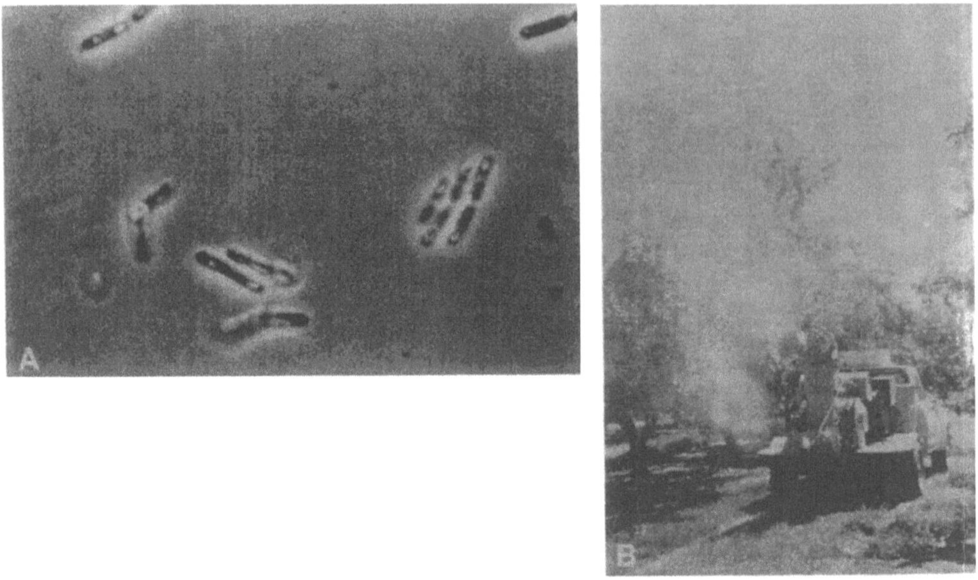

Figure 1.9 Use of pathogens as biopesticides. The bacterium *Bacillus thuringiensis* Berliner (A) can be cultured and applied to crops (B) for control of certain pests, such as the navel orangeworm, *Amyelois transitella* (Walker). (Photograph courtesy of W. F. Burke [A] and P. V. Vail [B].)

frequent and more devastating pest outbreaks, as demonstrated experimentally by Heinrichs et al. (1982). The cycle was broken when farmers were taught to understand the importance of spiders and other predators and to avoid routine pesticide applications (Fig. 1.10). Following such an educational campaign in Indonesia, reduced pesticide use on rice was accompanied by stable or increased rice yields. This major biological control success demonstrated the importance of native natural enemies and the value of extension systems that educate farmers as to the merits of biological control and the risks of pesticide use (Kenmore 1988). This project has influenced agricultural policy in developing countries and international aid organizations, placing emphasis on conserving native biological control agents, rather than encouraging pesticide use.

Recent Successes

Documenting the long list of successful biological control projects is not our purpose. However, discussions of some recent highly successful cases of biological control help illustrate that biological control has not been superseded by pesticides, biotechnology, or other technological developments, and is still an effective solution to many important pest control problems. We have chosen four projects to illustrate some of the recent successes of biological control through the introduction of new natural enemies. Each of these projects was highly effective and economically valuable to the countries involved.

Figure 1.10 Conservation of natural enemies. Field training of farmers in Narshingdi, Bangladesh, in recognition of natural enemies of pests of rice. (Photograph courtesy of P. Kenmore, FAO.)

Alfalfa Weevil. The alfalfa weevil, *Hypera postica* (Gyllenhal), invaded the eastern part of the United States in about 1945 (Day 1981). By the mid-1960s it had spread throughout most of the alfalfa-producing areas in the eastern half of the country. Larval densities were so high that most alfalfa (*Medicago sativa* Linnaeus) growers regularly made one or more pesticide applications per year to suppress damage. Because alfalfa is grown on nearly one million ha in the northeastern United States, pesticide use was extensive, especially in the key dairy states of New York and Wisconsin. Starting in 1959, the United States Department of Agriculture introduced a series of parasitoids of this pest, collected in Europe (Day 1981). These suppressed the pest to such low levels that farmers were able to reduce their use of pesticides against the pest by over 73% (Day 1981). The most effective species were two ichneumonid larval parasitoids in the genus *Bathyplectes* and a parasitoid of the adult weevil, *Microctonus aethiopoides* Loan. This project, over a twenty year period, cost about one million dollars, but by 1968 paid back annual benefits of over $8,000,000 (Day 1981), demonstrating that introduction of natural enemies remains a cost effective approach to solving insect problems.

Cassava Mealybug in Africa. The cassava mealybug, *Phenacoccus manihoti* Matile-Ferrero, invaded West Africa from South America in about 1973. By the early 1980s, it was found in much of tropical Africa and was spreading up to 300 km per year (Fig. 1.11), greatly reducing yields of cassava (*Manihot esculenta* Crantz), the basic subsistence food of nearly 200 million people. A parasitoid, *Epidinocarsis lopezi* (De Santis), of the mealybug was collected in South America after the mealybug was correctly identified and finally located in Paraguay. This parasitoid was introduced into Africa and proved to be highly effective (Norgaard 1988; Neuenschwander et al. 1989) (Fig. 3.9). A massive rearing and release program was conducted, and the pest was soon under biological control in nearly all parts of its range in Africa, and a basic food crop was again productive and its yields secure. More than any other recent project, this success has caused international development agencies to seriously consider using biological control based on the introduction of new species of natural enemies in their projects.

Floating Fern in Papua New Guinea. *Salvinia molesta* D.S. Mitchell is a floating fern from Brazil that was introduced into many tropical countries in the early part of the twentieth century

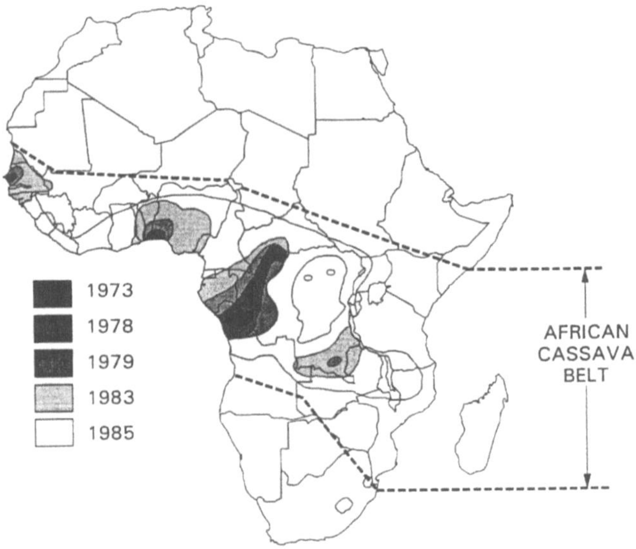

Figure 1.11 Spread of the cassava mealybug (*Phenacoccus manihoti* Matile-Ferrero) in Africa following its invasion from South America (after Herren 1987).

as an aquarium plant and botanical curiosity. The plant reached Papua New Guinea in the early 1970s and rapidly spread over the surface of the Sepik River and associated oxbow lakes. Mats of weed formed up to 1 m thick, preventing boats from moving freely along the river and isolating villages (Mitchell et al. 1980). In the 1960s and 1970s, there were efforts to locate natural enemies of the weed in South America because of problems in other regions. However, little was achieved because mis-identification of the plant species led to the collection of ineffective herbivores from related plants. After the weed was correctly identified and its native range discovered in 1978, more appropriate natural enemies were obtained. The weevil *Cyrtobagous salviniae* Calder and Sands proved highly effective in controlling the weed (Thomas and Room 1986; Room 1990) (Figs. 8.1, 8.2). Promising control programs have since been initiated in several additional countries including India and Namibia (Thomas and Room 1986; Forno 1987). This project achieved its objective of permanently ridding large areas of a damaging aquatic weed at moderate cost where other forms of control would have been totally infeasible.

Narrow-Leaf Skeletonweed in Australian Wheat. Skeletonweed, *Chondrilla juncea* Linnaeus, is an immigrant weed infesting dryland wheat (*Triticum aestivum* Linnaeus) areas in Australia. Three genetic types exist (narrow, broad, and intermediate leaf). The narrow-leaf form (the most abundant type) was controlled by a rust pathogen, *Puccinia chondrillina* Bubak and Sydow, imported to Australia from Italy. Later, a second rust strain was introduced for control of the intermediate-leaf form (Cullen 1978, 1985; Hasan and Wapshere 1973; Hasan

1981; Watson 1991). This project was the first successful use of an introduced plant pathogen for the control of an immigrant pest, and the first successful biological control of a weed in an annual crop.

THE HISTORICAL RECORD OF BIOLOGICAL CONTROL EFFORTS

Success Rates of Biological Control Projects Employing Natural Enemy Introduction

The historical record of past biological control projects is important because it documents the frequency of past success. This record can be evaluated with respect to particular cases (which agent has been most successful for the control of a particular pest), or for rates of success across many projects. The former is an important source of specific information, useful when a new location faces a problem dealt with successfully in the past in another area. The latter provides perspective on the probability of suppressing the target pest of any particular new project. The world database on biological weed control through natural enemy introduction is given by Julien (1992), who summarizes releases by agent and target, with notes on establishment and effects on the target plant. Clausen (1978) summaries worldwide introductions of parasitoids and predators for arthropod targets up through 1968, and Luck (1981) provides a more up-to-date listing of introductions of parasitoids for arthropod control. A worldwide, comprehensive database for predators and parasitoids for arthropod targets (BIOCAT) is maintained by the Commonwealth Agricultural Bureaux in the United Kingdom (Greathead and Greathead 1992). Hall et al. (1980) analyzed data in Clausen (1978) and found that 16% of all biological control projects resulted in complete control of the target pest in the project's location and an additional 42% resulted in partial control. In considering analyses of success rates, it is important to consider whether rates are calculated on a per project (involving potentially many agents) or per agent basis (which necessarily is lower than the rate per project). The figures of Hall et al. (1980) indicate that biological control, while not guaranteed to solve any one particular problem, is on average an effective strategy. Since benefit/cost ratios of biological control projects are large (3:1 to 100 or more:1) project success rates of 16% or more are sufficient for national programs of biological control to be highly cost effective as a form of social investment of public funds (Bellows 1993).

The historical record of projects of biological control through augmentation or conservation have not been summarized, although Frank and McCoy (1994) provide a list of 49 biological control agents imported into Florida (U.S.A.) by commercial organizations.

Regional Summaries of Past Projects

The historical records of biological control introductions in various countries have been summarized and provide an important source of information. These include: for insects worldwide (Clausen 1978; Luck 1981), for weeds worldwide (Goeden 1978; Julien 1992), the Pacific region (Waterhouse and Norris 1987), New Zealand (Cameron et al. 1989), Australia (Wilson 1960), southeast Asia and the Pacific region (Rao et al. 1971), Thailand (Napompeth 1990), Malaysia (Ghee 1990), Taiwan (Chen and Chiu 1986), South Korea (Choi and Lee 1990), Hawaii (Funasaki et al. 1988a), western and southern Europe (Greathead 1976), the British Commonwealth Caribbean and Bermuda (Cock 1985), Latin America (Altieri et al. 1989), sub-Saharan Africa (Greathead 1971), South Africa (Hoffmann 1991), Israel (Argov and Rossler 1988), Canada (Anon. 1971; Kelleher and Hulme 1984); and western North America (Nechols et al. 1995).

BIOLOGICAL CONTROL AND INTEGRATED PEST MANAGEMENT

Biological control projects may exist in their own right, independent of other control efforts. This is particularly true of pests in uncultivated areas, aquatic weeds, rangeland weeds, pests of ornamental plants, and forest insects. Many crop pests, however, are members of pest complexes all of which must be controlled simultaneously for effective crop production. In such cases, integration of potentially conflicting controls (biological control and chemical controls, but also perhaps biological control and plant breeding or agronomic practices) must be achieved. This subject is developed in Chapter 14. Basic questions in such integration are which controls are essential and which are optional. In the period following 1945, when pesticide use was extremely popular, chemical controls came to be considered the basic controls and biological controls were viewed as secondary or unnecessary. These biological controls were expected to play some (usually minor) role, if they could function despite chemical applications. If the biological control agents were unable to function in crops receiving pesticide applications, it was taken as proof that biological control agents were unimportant.

In recent decades (1980s to present), a change has gradually taken place with respect to this perception of priorities. The concept of integrated pest management systems based on biological control (as opposed to merely including some biological control) has arisen. Under this concept, biological control agents are seen as essential and of first priority in building pest control systems. Unlike in the past, biological control-based pest management does not permit the inclusion of control practices that destroy the biological control basis for pest suppression in the crop. Efforts to develop pest control systems based on biological control are in their early stages in many systems and well developed in others and represent the current evolution of biological control implementation. In this book, we explore both biological control through natural enemy introductions and the building of biological control-based integrated pest management systems by a range of methods, including introductions, augmentation, and conservation.

CHAPTER

2

Kinds of Biological Control Targets, Agents, and Methods

TARGETS OF BIOLOGICAL CONTROL

Biological control was first used to control insects, mites, and weeds (DeBach 1964a; Huffaker and Messenger 1976; Clausen 1978; Waterhouse and Norris 1987; Cameron et al. 1989; Julien 1992). Application of the method broadened with time and other invertebrates, plant pathogens, and even some vertebrates are now considered as likely targets (Cook and Baker 1983; Ross and Tittensor 1986; Campbell 1989; Madeiros 1990; Madsen 1990; Singleton and McCallum 1990; Stirling 1991).

Insects

Pest insects have been the most common type of organism against which biological control has been employed (Laing and Hamai 1976). Worldwide, over 543 species of insects have been targets of more than 1200 programs of biological control introductions (Greathead and Greathead 1992) and more are targeted through programs of natural enemy conservation and augmentation. These have included members of most of the important herbivorous orders, Homoptera, Diptera, Hymenoptera, Coleoptera, Lepidoptera, as well as smaller numbers in other groups. Homoptera has been the order against which biological control through introduction has been successfully employed most frequently (Greathead 1986a). This pattern of successful use of biological control is caused by the high frequency with which scales, aphids, and whiteflies move internationally on plants in trade (due to their small size and inconspicuousness), the large number of species in these groups which are important pests, and the frequency with which parasitoids and predators are significant factors in restraining the densities of such insects.

Mites

Several families of mites have been targets of biological control efforts. These include rust mites of the family Eriophyidae (Gruys 1982; Abou-Awad and El-Banhawy 1986), tarsonemid mites (Huffaker and Kennett 1956), and, most frequently, Tetranychidae, the spider mites (McMurtry 1982). Actions have included introduction of mite predators (principally Phytoseiidae and

Coccinellidae), conservation of native mite predators, and augmentative release of reared mite predators (McMurtry 1982).

Other Invertebrates

After insects and mites, snails have been the invertebrate group against which biological control efforts have been most frequently directed. Snails of concern have been either herbivorous species which damage crops or medically important snails which are intermediate hosts for pathogens which cause diseases of humans or domestic animals. Among the crop pests, the principal concern has been with edible species such as the giant African snail, *Achatina fulica* Bowdich, which has been spread to many areas by deliberate introduction for use as food (Waterhouse and Norris 1987). Of the medically important snails, those which are intermediate hosts for schistosomiasis have been of greatest concern (Greathead 1980; Pointier and McCullough 1989; Madsen 1990).

Biological control efforts against other types of invertebrates have been very rare. Clausen (1978) records the introduction of an egg parasitoid of a poisonous spider, *Latrodectus mactans* (Fabricius), to Hawaii. The parasitic fly *Pelidnoptera nigripennis* (Fabricius) has been imported into Australia for release against the adventive millipede *Ommatoiulus moreletii* (Lucas) (Bailey 1989).

Weeds

Plants in many taxonomic groups have become pest weeds in a variety of habitats, including forest, agricultural and rangeland, and native ecosystems, both terrestrial and aquatic. At least 116 species of plants in 34 families have been targets for biological control (Table 2.1) through introduction of invertebrate herbivores or plant pathogens (Julien 1992). About half (47%) of the weed species involved have been in the three families of Asteraceae, Cactaceae, and Mimosaceae. Other families, however, have contained individual species of great economic

TABLE 2.1 Taxonomic Range of Weeds Against which Biological Control Has Been Attempted via Introduction of Invertebrates

Plant Family	No. of Weed Species	Plant Family	No. of Weed Species
Amaranthaceae	1	Hydrocharitaceae	1
Anacardiaceae	1	Lamiaceae	1
Araceae	1	Loranthaceae	1
Asclepiadaceae	1	Malvaceae	2
Asteraceae	32	Melastomataceae	2
Boraginaceae	2	Mimosaceae	9
Cactaceae	22	Myricaceae	1
Caesalpiniaceae	1	Passifloraceae	1
Caryophyllaceae	1	Polygonaceae	3
Clusiaceae	2	Pontederiaceae	1
Chenopodiaceae	2	Proteaceae	2
Convolvulaceae	1	Rosaceae	4
Cuscutaceae	4	Salviniaceae	1
Cyperaceae	1	Scrophoulariaceae	3
Ehretiaceae	1	Solanaceae	1
Euphorbiaceae	2	Verbanceae	2
Fabaceae	4	Zygophyllaceae	2

importance and have been the focus of intensive efforts; among these are the Clusiaceae (*Hypericum perforatum*, St. John's wort), Salviniaceae (*Salvinia molesta*, water fern), and Verbenaceae (*Lantana camara* Linnaeus, lantana). Grasses have not been a group against which biological control has been commonly applied, but some needs and opportunities exist for biological control of pest grass species (Wapshere 1990).

Plant Diseases

Antagonist organisms have been used to prevent or suppress some plant diseases. Many plant pathogens are potentially affected by biological control agents (Cook and Baker 1983; Campbell 1989). Examples of plant pathogens against which antagonists have been directed include species of *Agrobacterium*, *Fusarium*, *Heterbasidion*, *Pythium*, *Erwinia*, *Pseudomonas*, *Sclerotinia*, *Rhizoctonia*, and *Cryphonectria* (Schroth and Hancock 1985; Campbell 1989). Lindow (1985a) discusses antagonists for foliar pathogens.

Vertebrates

Feral populations of vertebrates, such as rats, pigs, goats, sheep, rabbits and opossums, are important pests damaging grazing lands, forests, and nature conservation in many areas (see Chapter 21). Many of these species, however, also are desirable species in other contexts. Biological control efforts targeted at vertebrates must use agents that are sufficiently specific to protect other vertebrates. Such projects can only be undertaken in locations where conflicts between the need to control feral populations and protect domestic populations of the same species either do not exist, or have been judged in favor of the control efforts.

In addition to the introduction of pathogens with narrow host ranges or enhancement of habitats for native predators, genetic methods for vertebrate control have been developed. These are based on the use of genetically-modified vertebrate pathogens, such as the myxoma virus of rabbits, to cause infected females to develop antibodies against the species' sperm, preventing conception (Deeker 1992; Barlow 1994).

KINDS OF BIOLOGICAL CONTROL AGENTS

Natural enemies, the agents used in biological control, are the fundamental resource with which biological control success is achieved. Agents come from an array of taxonomic groups and have diverse biological and populational properties. These characteristics play a large role in the success or failure associated with the use of any particular group of natural enemies and a detailed appreciation of the biologies of many different natural enemy groups is of great value.

Insect Parasitoids

Parasitoids are arthropods that kill their hosts (unlike true parasites such as fleas or tapeworms) and which are able to complete their development on a single host (unlike predators which generally must consume several prey to complete their development) (Doutt 1959; Askew 1971; Vinson 1976; Vinson and Iwantsch 1980a; Waage and Greathead 1986; Godfray 1994). Parasitoids have been the most common type of natural enemy introduced for biological control of insects (Hall and Ehler 1979; Greathead 1986a). Most parasitoids that have been used in biological control are in the orders Hymenoptera (Fig. 2.1) and, to a lesser degree, Diptera.

Figure 2.1 The braconid parasitoid *Aphidius matricariae* Haliday parasitizing its aphid host *Diuraphis noxia* (Mordvilko). (Photograph by M. Badgley.)

While use has been made of parasitoids in at least 26 families, certain groups stand out as having more species employed in biological control projects than others. The most frequently used groups in the Hymenoptera have been the Braconidae and Ichneumonidae in the Ichneumonoidea, and the Eulophidae, Pteromalidae, Encrytidae, and Aphelinidae in the Chalcidoidea (Table 2.2). In the Diptera the most frequently employed group has been the Tachinidae (Greathead 1986a). Parasitoids are also found in the insect orders Strepsiptera and Coleoptera (some members of such families as the Staphylinidae, Meloidae, Rhipiphoridae), although parasitism is not typical of the Coleoptera (Askew 1971).

Predators of Arthropods and Other Invertebrates

The predacious habit is extremely widespread among insects and can be found in most orders and a large number of families (Fig. 2.2) (Hagen et al. 1976a; Borror et al. 1989). Predacious insects have been introduced for control of immigrant pests, and native predators are of major importance in the suppression of both native and immigrant herbivores. Groups of predators most frequently recognized as significant for pest suppression in agriculture and forestry include some 32 families (Table 2.3). Of these, the Anthocoridae, Pentatomidae, Reduviidae, Carabidae, Coccinellidae, Staphylinidae, Chrysopidae, Cecidomyiidae, Syrphidae, and Formicidae are among the predators most commonly found preying on pest species in crops. Spiders are virtually all predacious (Foelix 1982) and while usually not specialized as to prey species, do show habitat specialization. The role of spider complexes in suppressing groups of pests in crops and other habitats has become better recognized in recent years (Clarke and Grant 1968; Mansour et al. 1980; Riechert and Lockley 1984; Nyffeler and Benz 1987; Bishop and Riechert 1990). Spider mites, an important agricultural pest group, have no parasitoids and are held in check to a large degree by predators. These include predacious thrips, some coccinellid beetles and, most importantly, mites in the family Phytoseiidae. Slugs and

TABLE 2.2 Number of Times Parasitoids Have Been Established in Biological Control Introduction Programs in Families from which at Least One Species Has Been Released[1]

	No. Parasitoid		No.	No.	No. Effective
	Genera	Species	Pest Species	Occasions	Control Cases
Hymenoptera					
Evanoidea					
Stephanidae	1	1	1	1	—
Ichneumonoidea					
Braconidae	23	66	59	158	53
Ichneumonidae	30	45	28	72	22
Proctotrupoidea					
Proctotrupidae	—	—	—	—	—
Scelionidae	4	12	11	23	6
Platygasteridae	4	7	6	12	5
Diapriidae	—	—	—	—	—
Cynipoidea	3	4	4	7	1
Chalcidoidea					
Trichogrammatidae	2	12	12	24	—
Eulophidae	21	36	47	72	23
Mymaridae	4	7	9	15	9
Chalcididae	2	3	2	4	—
Eurytomidae	—	—	—	—	—
Torymidae	1	1	1	2	—
Pteromalidae	15	26	221	49	17
Encyrtidae	34	61	401	132	53
Aphelinidae	13	59	32	185	90
Eupelmidae	2	4	3	4	—
Bethyloidea					
Bethylidae	2	2	2	3	—
Dryinidae	2	2	1	2	—
Scolioidea	3	13	10	21	3
Diptera					
Pygotidae	—	—	—	—	—
Cryptochetidae	1	2	2	5	5
Tachinidae	27	30	27	69	35
Muscidae (*Acridomyia*)	—	—	—	—	—
Sarcophagidae	—	—	—	—	—
Strepsiptera	—	—	—	—	—
Totals	194	393	2741*	860	216*

*Totals are reduced because more than one parasitoid has often been established on the same host in a single country.
[1]Greathead (1986a).

snails are attacked by predacious snails, sciomyzid flies (whose larvae find and kill one or several snails during their development), and some carabid beetles.

Vertebrate predators which attack insect pests are diverse and include insectivorous birds, small mammals, lizards, amphibians, and fish, some of which have been used in the past as agents of biological control (Davis et al. 1976). While birds and mammals are generally not used as agents for introduction against immigrant pests, indigenous species are believed to be important sources of mortality for some pests, particularly in stable environments such as forests (Bruns 1960; Bellows et al. 1982a; Campbell and Torgersen 1983; Nuessly and Goeden 1984; Atlegrim 1989; Crawford and Jennings 1989; Higashiura 1989; Zhi-Qiang Zhang 1992).

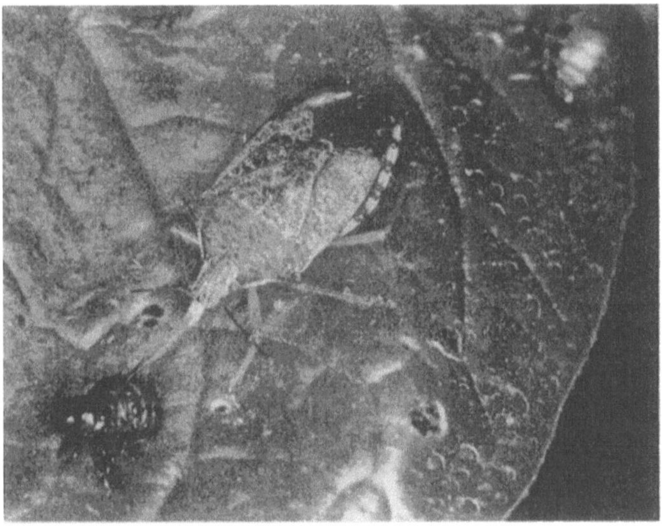

Figure 2.2 The pentatomid *Podisus maculiventris* (Say) attacking a larva of the chrysomelid *Leptinotarsa decemlineata* (Say). (Photograph courtesy of D. Ferro.)

Table 2.3 Some Important Families of Predacious Arthropods

Thysanoptera (thrips)	Cybocephalidae
Aeolothripidae	Staphylinidae (rove beetles)
Phloeothripidae	Neuroptera (nerve-winged insects)
Thripidae	Chrysopidae (green lacewings)
Hemiptera (true bugs)	Hemerobiidae (brown lacewings)
Anthocoridae (minute pirate bugs)	Diptera (true flies)
Gerridae (water striders)	Cecidomyiidae (gall flies)
Miridae (plant bugs)	Chamaemyiidae (aphid flies)
Nabidae (damsel bugs)	Sciomyzidae (marsh flies)
Pentatomidae (stink bugs)	Syrphidae (flower flies)
Reduviidae (assassin bugs)	Hymenoptera
Veliidae (broad shouldered water striders)	Formicidae (ants)
Phasmatidae (ambush bugs)	Vespidae (yellow jackets and paperwasps)
Coleoptera (beetles)	Sphecidae (hunting wasps)
Carabidae (ground beetles)	Acari (mites)
Cicindelidae (tiger beetles)	Phytoseiidae
Dytiscidae (predacious water beetles)	Stigmaeidae
Cleridae (checkered beetles)	Hemisarcoptidae
Coccinellidae (lady bird beetles)	Araneae (spiders)

Fish, in contrast, have mainly been utilized through augmentative releases into water bodies for the suppression of mosquito larvae (Miura et al. 1984).

Pathogens and Predators of Vertebrates

When sufficiently selective agents can be identified for safe use, pathogens are sometimes employed to suppress pest vertebrates. Agents that would be unacceptable in settled areas may be acceptable on oceanic islands where no human settlements exist to create conflict between suppression of feral populations of the species (seen as pests) and husbandry of domestic populations of the same species (seen as economically valuable resources). Thus, the release of the feline panleucopaenia pathogen for control of feral house cats on unsettled islands which are home to seabird colonies (van Rensburg et al. 1987) is acceptable, but in other areas, use of this pathogen would be unacceptable. Other vertebrate pathogens employed for biological control include the myxoma virus of rabbits (Fenner and Ratcliffe 1965; Ross and Tittensor 1986) and some helminths that attack rodents (Singleton and McCallum 1990). Interest also exists in the possible use of venereal diseases to suppress population growth rates of feral goats on uninhabited oceanic islands where goats cause extensive damage to native vegetation (Dobson 1988).

Vertebrate species that are predators of vertebrates are rarely sufficiently specific to allow their safe introduction to areas outside of their natural range. However, some native vertebrates have been successfully enhanced through habitat modification or provision of nesting boxes, as for example the barn owl, *Tyto alba* Linnaeus, for control of rats on oil palm plantations in Malaysia (Madeiros 1990; Mohd 1990) (Fig. 2.3).

Pathogens and Nematodes Attacking Arthropods

Insect pathogens include a range of bacteria, viruses, fungi, and protozoa (Brady 1981; Miller et al. 1983; Maramorosch and Sherman 1985; Moore et al. 1987; Burge 1988; Tanada and Kaya 1993). Natural epizootics of certain of these agents are important sources of mortality for some species (Fuxa and Tanada 1987). In addition, some pathogens have been formulated and commercially marketed as insecticides (Cherwonogrodzky 1980; Falcon 1985).

Bacteria. Of the various pathogen groups, bacteria have been most successfully brought into commercial use. Three species of spore-forming bacteria in the genus *Bacillus* are currently used for the control of several groups of pests: *Bacillus popilliae* Dutky (larvae of scarabaeid Coleoptera), *Bacillus thuringiensis* (Lambert and Peferoen 1992), including var. *kurstaki* (larvae of Lepidoptera), var. *israelensis* (larvae of Diptera, in several families), and var. *tenebrionis* (larvae of chrysomelid Coleoptera), and *Bacillus sphaericus* Neide (larvae of culicid Diptera) (Falcon 1985; Lüthy 1986; Osborne et al. 1990). Bacteria are more amenable to commercial use than viruses because many species can be grown in fermentation media and often do not require the expensive culturing of live insect hosts as do the viruses. Most emphasis has been placed on *B. thuringiensis*, of which about 30 subspecies and more than 700 strains have been isolated. Some *Bacillus thuringiensis* products contain both live bacteria and associated toxic proteins; others contain no live material, but only the toxic chemical proteins the bacteria have produced in culture. Products containing no live spores were initially used for application in regions in Asia where silkworm culture was important and where live bacterial products were viewed as a potential risk (Aizawa 1987).

Figure 2.3 The barn owl, *Tyto alba* Linnaeus, a predacious vertebrate. (Photography by M. Badgley.)

Viruses. At least sixteen families of viruses have been found to be pathogens of insects (Entwistle 1983; Moore et al. 1987; Tanada and Kaya 1993). Of these, members of the Baculoviridae (Fig. 2.4) have been the focus of attempted commercial use because they frequently cause lethal infections and are only known to cause disease in insects (Payne 1986). This family contains the nuclear polyhedrosis and granulosis viruses. The nonoccluded viruses, formerly placed in the Baculoviridae, are now unclassified. The biology of this family has been reviewed by Granados and Federici (1986). Few viruses have been successfully marketed as commercial products because of high production costs and narrow host specificity. Existing products are mostly ones supported at public expense (Falcon 1976, 1985; Morris 1980).

Fungi. Most fungi that attack insects are in the family Entomophthoraceae within the subdivision Zygomycotina or in the Deuteromycotina (species known only from asexual forms) (Brady 1981; Zimmermann 1986). The higher classification of fungi varies between authors. In this text, we follow the system outlined by Ainsworth (1971) and Ainsworth et al. (1973). Fungal epidemics occur periodically and can cause high levels of mortality in the affected arthropod populations (see Goh et al. 1989). The fungus *Zoophthora radicans* (Brefeld) Batko from Israel was introduced to Australia to aid in the suppression of the lucerne aphid, *Therioaphis trifolii* (Monell) f. *maculata* (Milner et al. 1982). Other fungi have been of interest for the development of commercial biological pesticides (Ferron 1978; Gillespie 1988). To date, successful development of mycoinsecticides has been extremely limited, because of narrow host range and high humidity requirements for germination or sporulation (Moore and Prior 1993). Fungal pesti-

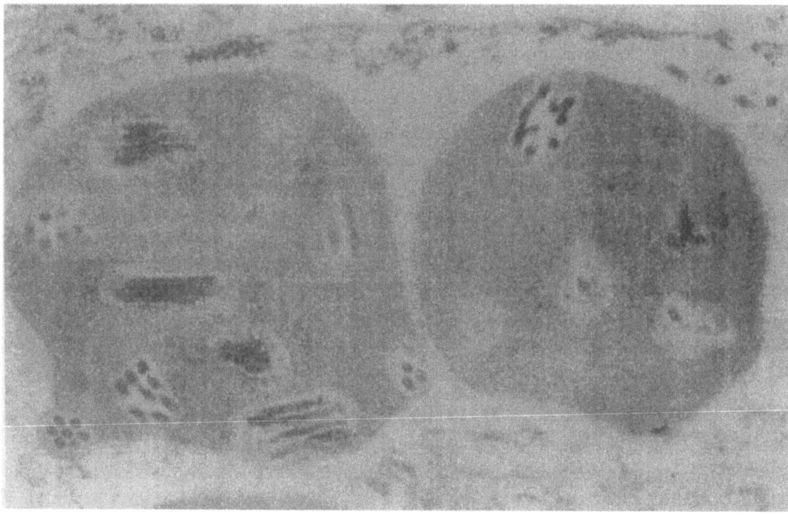

Figure 2.4 Transmission electron micrograph of a thin section of an epidermal cell from a spruce budworm larva, *Choristoneura fumiferana* (Clemens), infected with a nuclear polyhedrosis virus. The polyhedral inclusion bodies are visible in the cell nucleus. Nucleocapsids within virions in these occlusion bodies appear as dark circles or rods, depending on the orientation to the cell section. (Photograph courtesy of J.A. MacDonald, Forestry Canada.)

cides have their greatest potential for use in humid climates or moist environmental strata such as soil. However, new formulation methods such as employing vegetable oils in place of water in spray solutions appear to have potential to widen the range of climates and circumstances under which fungi may be employed successfully (Bateman et al. 1993).

Protozoa. Various protozoa infect insects (Brooks 1988). These protozoa include the microsporidians (Kluge and Caldwell 1992) and the eugregarines (Brooks and Jackson 1990). (See Chapter 4 for an overview of the taxonomy of the protozoa that are pathogens of arthropods.) Some protozoans have been considered for use as microbial insecticides. Species of *Nosema* have been tested as potential biological control agents for grasshoppers (Henry and Onsager 1982) and maize (*Zea mays* Linnaeus) pests (Lublinkhof and Lewis 1980).

Nematodes. Nematodes which have shown the greatest potential for the control of agricultural pests are found in the families Steinernematidae and Heterorhabditidae (Gaugler and Kaya 1990; Kaya 1993), which are mutualistically associated with bacteria that kill the nematode's host through septicemia (Kaya 1985). Nematodes in some other families also attack insects killing their hosts through their growth, much as do parasitoids. These include some mermithids (e.g., *Romanomermis*), phaenopsitylenchids (e.g., *Beddingia* [=*Deladenus*]), iotonchiids (e.g., *Paraiotonchium*), sphaerulariids (e.g., *Tripius*) and tetradonematids (e.g., *Tetradonema plicans* Cobb) (Poinar 1986; Kaya 1993).

Some nematodes have broad host ranges and efforts to use such species for biological control have been through augmentative application of reared nematodes to sites where pests are present. Research for this type of use has been focused on species in the genera *Steinernema* and *Heterorhabditis.* (Literature on nematodes in these genera is confused by many synonyms. See Poinar [1990], Doucet and Doucet [1990], and Kaya [1993] for details on relationships between names.) Nematodes in these families often require high moisture conditions for survival and are sensitive to ultraviolet light, which limits their effectiveness to protected sites such as soil or inside plant tissues.

In addition to augmentative use of nematodes, some species are specific enough for use as introduced agents against immigrant pests. Examples include the phaenopsitylenchid *Beddingia siricidicola* (Bedding), that was introduced into Australia and successfully controlled the European wood wasp *Sirex noctilio* (Fabricius) (Poinar 1986), and the steinernematid *Steinernema scapterisci* Nguyen and Smart that was introduced to Florida to control immigrant mole crickets (Parkman et al. 1993).

Weed-Attacking Herbivores and Pathogens

Herbivores intentionally released for weed control have been primarily insects due to their high degree of host specialization and rapid rates of multiplication (Fig. 2.5) (Andres et al. 1976). Within the Insecta, most releases have been either of beetles (Coleoptera) or moths and butterflies (Lepidoptera) (Chapter 5, see also Julien 1992). The families Chrysomelidae, Curculionidae, Cerambycidae, Pyralidae, Dactylopiidae, and Tephritidae have been used most frequently. In addition to insects, some mites in the Eriophyidae and, more recently, Tetranychidae (Hill et al. 1991) have been employed.

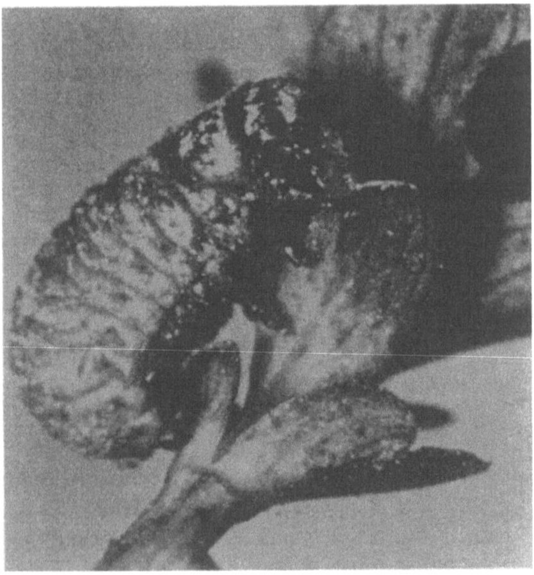

Figure 2.5 Larva of the chrysomelid *Chrysolina sp.*, an invertebrate weed control agent. (Photography courtesy of USAD/ARS.)

In contrast to highly specialized insects targeted at single weed species, some fish have been used as generalist herbivores to control weed complexes in certain habitats, often artificial ones such as canals and water tanks. Among these fish have been several species in the families Cyprinidae, Cichlidae, and Osphronemidae (Julien 1992).

Until recently, few plant pathogens had been used as biological weed control agents (Wilson 1969). Since 1970, however, precedents have been set in both the introduction of new pathogens for control of adventive weeds and the formulation of pathogens for use as biological herbicides. For example, the autoecious rust fungus *Puccinia chondrillina* was introduced in 1971 to Australia from Europe where it controlled narrow-leaf populations of skeletonweed, *Chondrilla juncea* (Hasan and Wapshere 1973; Hasan 1981; Watson 1991). The indigenous fungal pathogen *Colletotrichum gloeosporioides* (Penzig) and Saccardo in Penzig, which attacks the native weed northern jointvetch, *Aeschynomene virginica* (Linnaeus) Brittan, Sterns and Poggenberg, was marketed as a biological herbicide in 1982 (Templeton et al. 1984). TeBeest (1991) reviews the use of pathogens as weed control agents.

Biological Control Agents Suppressing Plant Pathogens

Mechanisms leading to biological control of plant pathogens are complex and may occur by many routes. Consequently, several kinds of agents may be involved (Snyder et al. 1976; Cook and Baker 1983; Campbell 1989; Stirling 1991). Plant pathogens may be suppressed by events that reduce the potential inoculum level of the pathogen in the environment, or by competitive or parasitic interactions among organisms, or by competition for limiting resources. Competition may occur at any point in the infection cycle, from its initiation outside the host, through invasion and growth inside the plant. Some fungi such as species of *Trichoderma* exhibit mycoparasitism, attacking and killing hyphae or other parts of pathogenic fungi (Fig. 2.6). Some bacteria, actinomycetes, and fungi produce chemicals (antibiotics) that actively repress the growth of other species, including pathogens. Some fungi repress the growth of pathogens by outcompeting them for key resources such as minerals, nutrients, oxygen, or water, either at or away from the site of initial infection.

PRINCIPAL BIOLOGICAL CONTROL METHODS

All biological control involves the use, in some manner, of natural enemies to suppress pest population densities to levels lower than they would otherwise be. Three major methods exist for the use of natural enemies: conservation, introduction, and augmentation. Of these, introduction has been the method that has successfully solved the greatest number of insect and weed problems; but has rarely been used to control plant pathogens.

Conservation

Human activities can greatly influence the extent to which natural enemies are able to realize their potential to suppress pests. Conservation as a form of biological control is the study and application of such influences. This approach seeks to identify and rectify negative influences that suppress natural enemies and to enhance agricultural fields (or other sites) as habitats for natural enemies. Fundamental to biological control through conservation is the assumption that species of natural enemies already exist locally that have the potential to effectively suppress the pest if given an opportunity to do so. This assumption is likely to be true for many (but not all) indigenous pests but for adventive pests is usually true only if appropriate natural enemy introductions have already been made.

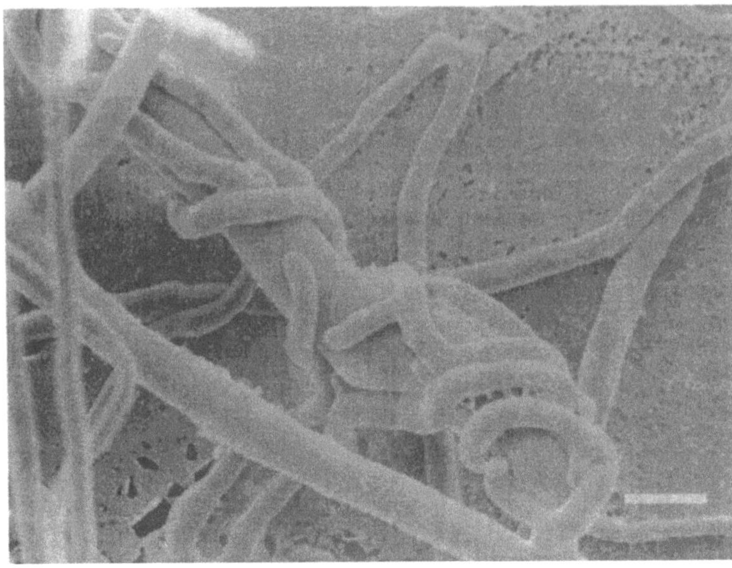

Figure 2.6 Hyphae of the mycoparasitic fungus *Trichoderma* sp. are shown here coiled around the thicker hypha of the fungal root pathogen *Rhizoctonia* sp. Scale bar is 10 μm. (Photograph courtesy of A. Beckett, Department of Botany, University of Bristol.)

Foremost among the negative influences that harm natural enemies are chemical pesticides, especially those with broad spectrum and long residual action (Croft 1990). Other negative forces that affect some kinds of natural enemies are dust on foliage (DeBach 1958; Flaherty and Huffaker 1970) and ants that actively defend such insects as aphids or scales and harvest the honeydew they excrete (DeBach and Huffaker 1971). Other practices that may be negative influences for some natural enemies include date and manner of turning soil, destruction of crop residues, size and placement of crop patches, and removal of natural enemy overwintering sites such as hedgerows.

Positive forms of conservation include efforts to enhance the requisites that natural enemies need to flourish in a system. Such efforts may involve creating or maintaining physical refuges, alternative hosts, or sources of carbohydrates, moderating physical conditions through the use of ground covers, provision of sheltering sites, or use of strip-harvesting methods (van den Bosch et al. 1967; Hance and Gregoire-Wibo 1987; Heidger and Nentwig 1989).

Conservation methods depend primarily on knowing how effective a particular conservation practice is under local conditions. This may require considerable local research and field trials. Once such information is available the method often can be implemented on individual farms independently of the actions of the community as a whole.

Introduction of New Natural Enemy Species

In many areas, adventive ("exotic" or "introduced") species comprise a high proportion of the major pests (Sailer 1978; Van Driesche and Carey 1987). In the United States, for example, immigrant arthropods make up only 2% of the total arthropod fauna, but account for 35% of the most important 700 pest species (Knutson et al. 1990). For such nonnative pests, conservation is likely to be inadequate because sufficiently effective natural enemies will be absent. In such cases, introducing new natural enemy species that are effective against the pest is absolutely essential, and is an approach that historically has been extremely successful. The needed natural enemy species are most often obtained by examining populations of the pest in its native homeland and observing which species of natural enemies attack it there. These natural enemies are then collected, shipped to the country which the pest has invaded and, after being subjected to appropriate quarantine and testing to ensure safety, are released and established. Introduction as a method of biological control has provided complete or partial control of more than 200 pest species (DeBach 1964a; Laing and Hamai 1976; Clausen 1978; Goeden 1978; Julien 1992; Greathead and Greathead 1992). Introduction has a major advantage over other forms of biological control in that it is self-maintaining and, thus, less expensive over the long term. After new natural enemies have been obtained, conservation measures may be required for the new species to be fully effective. To be conducted safely, introduction biological control programs require a high degree of scientific skill and, therefore, typically are conducted by public institutions, using public resources to solve problems for the common good.

Augmentation

When natural enemies are missing (glasshouses, mushroom houses), or late to arrive at new plantings (some row crops), or simply too scarce to provide control, their numbers may be increased through releases (Ridgway and Vinson 1977; King et al. 1985). Such an approach is termed augmentation. Augmentation covers several situations. **Inoculative releases** are those in which small numbers of a natural enemy are introduced early in the crop cycle with the expectation that they will reproduce in the crop and their offspring will continue to provide pest control for an extended period of time. **Inundation**, or **mass-release**, is used when insufficient reproduction of the released natural enemies is likely to occur, and pest control will be achieved exclusively by the released individuals themselves. Inundation with nematodes and pathogens is distinct from mass release of parasitoids and predators only in that pathogens and nematodes resemble pesticides in their packaging, handling, storage, and application methods. However, when pathogen products (such as toxins) are used in lieu of pathogens themselves, it more closely resembles chemical control than biological control, because populations of live organisms are not involved.

Augmentation may be directed against indigenous or adventive pests. The principal limitations on the method have to do with cost, quality, and field effectiveness of the reared organisms. The cost of rearing natural enemies may limit the use of the method either to situations where the natural enemies are inexpensive to rear or to circumstances where crops have relatively high cash value and less expensive alternatives are not available. Only in such circumstances can private companies recoup their production costs and compete economically with alternative methods. Somewhat broader applications are possible when public institutions rear the necessary natural enemies. In both cases, production of high quality natural enemies is essential, as are prior research studies assessing the degree of pest control provided by the reared agent under field conditions.

SECTION II

A Review of the Organisms Employed as Agents of Biological Control

Chapters 3 to 6 present the organisms that have been employed as agents of biological control. These are organized within a taxonomic framework, but the taxonomic literature is not reviewed, except at the summary level. Biologies of these groups are covered in more detail in Section V (Chapters 15–17).

Chapter 3 reviews the parasitic and predacious arthropods, with comments on families of greatest importance and selected examples of important species or genera. In Chapter 4, the pathogens and nematodes attacking arthropods are discussed. Chapter 5 reviews those herbivorous arthropods and plant pathogens employed in biological control of weedy plants. Because of the large number of herbivorous or pathogenic species affecting plants, this treatment is selective, focusing on groups used in past projects. In Chapter 6, organisms useful for suppression of plant diseases are considered. None of these chapters provides in-depth taxonomic treatment of the groups they discuss. Rather, they are intended to provide a broad perspective on the groups to the student of biological control and to point toward the recent summary or review literature. More technical references will be needed by students desiring further study of any particular group.

CHAPTER

3

PARASITOIDS AND PREDATORS OF ARTHROPODS AND MOLLUSCS

INTRODUCTION

The identity and important biological features of the principal orders and families of parasitoids and predators attacking pest arthropods and snails are discussed in this chapter, and additional information may be found in Clausen (1940), Arnett (1968), Askew (1971), Hodek (1973), Foelix (1982), Waage and Greathead (1986), Gauld and Bolton (1988), Borror et al. (1989), Gerson and Smiley (1990), Grissell and Schauff (1990), and Godfray (1994). Information about names and organization of higher taxa of parasitoids and predators has been drawn principally from Askew (1971), Gauld and Bolton (1988), Borror et al. (1989), and Grissell and Schauff (1990). This chapter does not review the taxonomic literature of these groups, but rather introduces students to groups of importance to biological control and cites some projects which employed these groups.

INSECTS PARASITIC ON ARTHROPODS AND SNAILS

Most parasitoids attacking insects are in the orders Diptera and Hymenoptera; some, also, are found in the Strepsiptera and some genera of Coleoptera, Neuroptera, and Lepidoptera. Host-parasitoid relationships have been catalogued from the literature by Thompson and Simmonds (1964–1965) and Fry (1989). Further information is available in regional catalogues such as Krombein et al. (1979).

Terminology

Before discussing biologies of individual groups of parasitoids, some terms commonly used to describe parasitoids and their behaviors need to be introduced. **Parasitoids** differ from parasites (such as Strepsiptera, lice, and tapeworms) in that parasitoids kill their hosts. Different species of parasitoids may attack the various life stages of a pest (Fig. 3.1). Thus, *Trichogramma* spp., which attack the egg stage, are referred to as **egg parasitoids**; Braconidae such as *Cotesia glomerata* which oviposit in caterpillars are termed **larval parasitoids**; and so on for **adult** and **nymphal parasitoids**. Parasitoids that oviposit in one stage, but emerge from a later stage, are named accordingly; for example, the encyrtid *Holcothorax testaceipes* (Ratzeburg) is an egg-larval parasitoid of gracillariid leafmining moths.

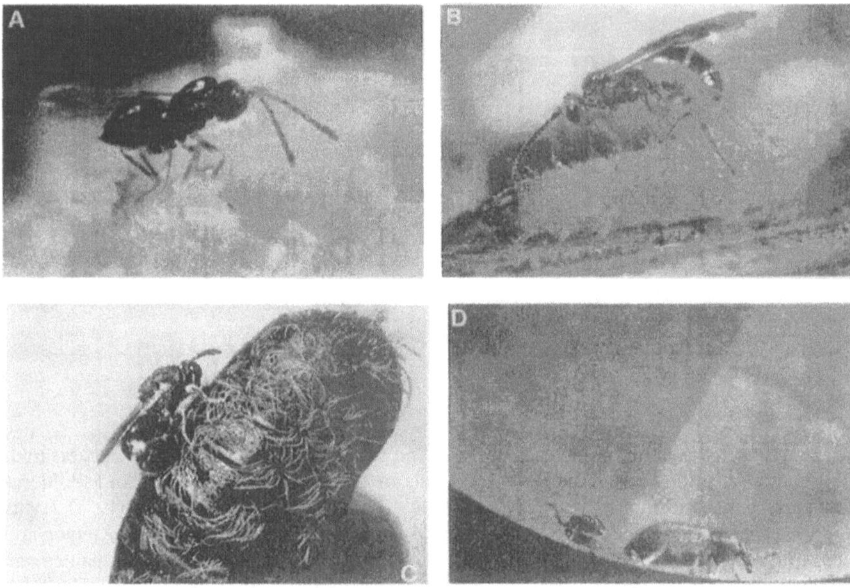

Figure 3.1. All stages of insects may be attacked by parasitoids: (A) mymarids such as *Anaphes iole* Girault, are parasitoids of eggs; (B) the braconid *Bracon hebetor* Say attacks lepidopteran larvae; (C) chalcidids, such as *Brachymeria intermedia* (Nees), are parasitoids of pupae; (D) some braconids attack adult insects as, for example, *Microctonus aethiopoides* Loan which parasitizes the alfalfa weevil, *Hypera postica* (Gyllenhal). (Photographs by M. Badgley [A,B], and courtesy of R. Weseloh [C], and M. Pendrak, USDA, APHIS [D].)

In some parasitoid species, only a single parasitoid can develop to maturity in a single host, while in others several parasitoids (of a single species) develop on or in a single host. The former are termed **solitary** and the latter, **gregarious** (Fig. 3.2). In some parasitoid species, adults deposit a single egg per host which subsequently divides into many cells, each of which develops independently, a process called **polyembryony**. If more eggs are deposited by a single species in a host than can survive, the result is termed **superparasitism**. If two or more species of parasitoids both attack the same host, the resultant condition is described as **multiparasitism**.

Parasitoids that insert their eggs into hosts are called **endoparasitoids**; those that lay their eggs externally and whose larvae develop externally are called **ectoparasitoids** (Fig. 3.3). The adult ectoparasitoid paralyzes the host and then lays its egg on or near the host, which may be exposed in the environment (such as a lepidopteran larvae on a leaf) or concealed in some manner, such as in a leaf mine, leaf roll, or gall. Parasitoids may also be classified by how parasitism affects host physiology. Species (such as braconid endoparasitoids of lepidopteran larvae) which develop inside living, mobile hosts and which benefit from the continued life and feeding of the host are referred to as **koinobionts**. In contrast, those species such as ectoparasitoids, or endoparasitoids of eggs and pupae, which kill their hosts before the parasitoid egg hatches and then develop in or on dead or paralyzed hosts, are termed **idiobionts**.

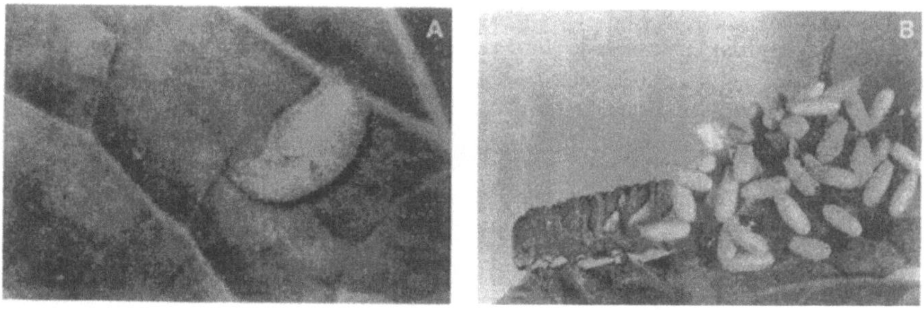

Figure 3.2. (A) Solitary parasitoids are those which produce only one offspring per host as, for example, *Cotesia rubecula* (Marshall). (B) Gregarious parasitoids such as *Cotesia congregata* (Say) produce more than one offspring per host. (Photograph [A] courtesy of D. Biever; and by M. Badgley [B].)

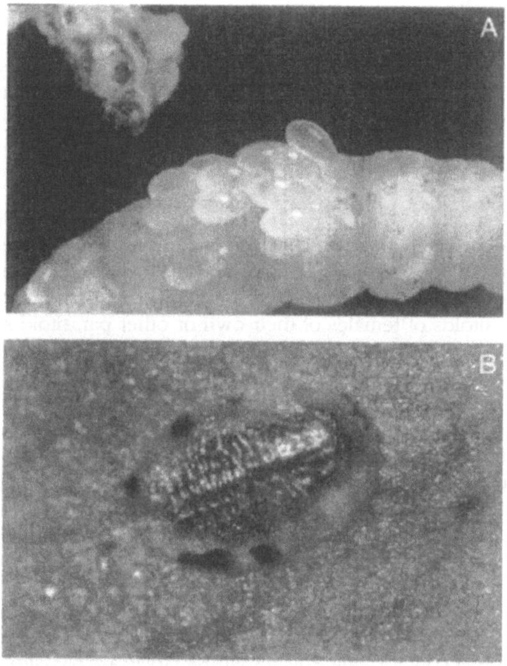

Figure 3.3. Parasitoids may develop inside (endoparasitism) or outside hosts (ectoparasitism). (A) Larvae of the ectoparasitoid *Goniozus* sp. on a larva of the navel orangeworm, *Amyelois transitella* (Walker). (Photograph [A] by M. Badgley.) (B) A pupa of the endoparasitoid *Encarsia strenua* (Silvestri) (Aphelinidae) inside, a nymph of the whitefly *Dialeurodes citri* (Ashmead) (Photograph courtesy of C. Meisenbacher).

Figure 3.4. Parasitoids may themselves be attacked by parasitoids, a process termed hyperparasitism, as seen in this view of a chalcidid wasp (*Conura torvina* [Cresson]) ovipositing in the cocoon of the braconid *Cotesia rubecula* (Marshall). (Photograph courtesy of R. McDonald.)

Those parasitoids that attack other species of parasitoids are termed **hyperparasitoids** or **secondary parasitoids** (Fig. 3.4); parasitoids of nonparasitoid hosts (mostly herbivores) are **primary parasitoids**. In some species, females and males have different hosts; such parasitoids are termed **heteronomous** (Walter 1988). In some heteronomous parasitoids males develop as hyperparasitoids of females of their own or other parasitoid species.

In some species of parasitoids, all eggs are present in a mature state when the adult emerges and eggs can be laid rapidly without need for egg development. This condition is termed **pro-ovigeny**. In contrast, some species emerge with few developed eggs and eggs mature gradually, a process called **synovigeny**. In parasitoid groups in which eggs mature gradually, adult parasitoids need protein in their diets. In some species, parasitoids gain protein by acting as predators, killing some hosts by ovipositor probing, and then consuming hemolymph that exudes from the wound. This process, termed **host feeding**, is found widely among hymenopteran parasitoids (Fig. 3.5) (Bartlett 1964a; Jervis and Kidd 1986). Many adult parasitoids require carbohydrates and longevity is severely reduced in the absence of such food sources.

Diptera

Twelve families of Diptera contain some species whose larvae are parasitoids of arthropods or snails: Acroceridae, Nemestrinidae, Bombyliidae, Phoridae, Pipunculidae, Conopidae, Pyrgotidae[1], Sciomyzidae, Cryptochetidae, Calliphoridae[1], Sarcophagidae, and Tachinidae. Of these,

---

[1]Minor groups that are not discussed further.

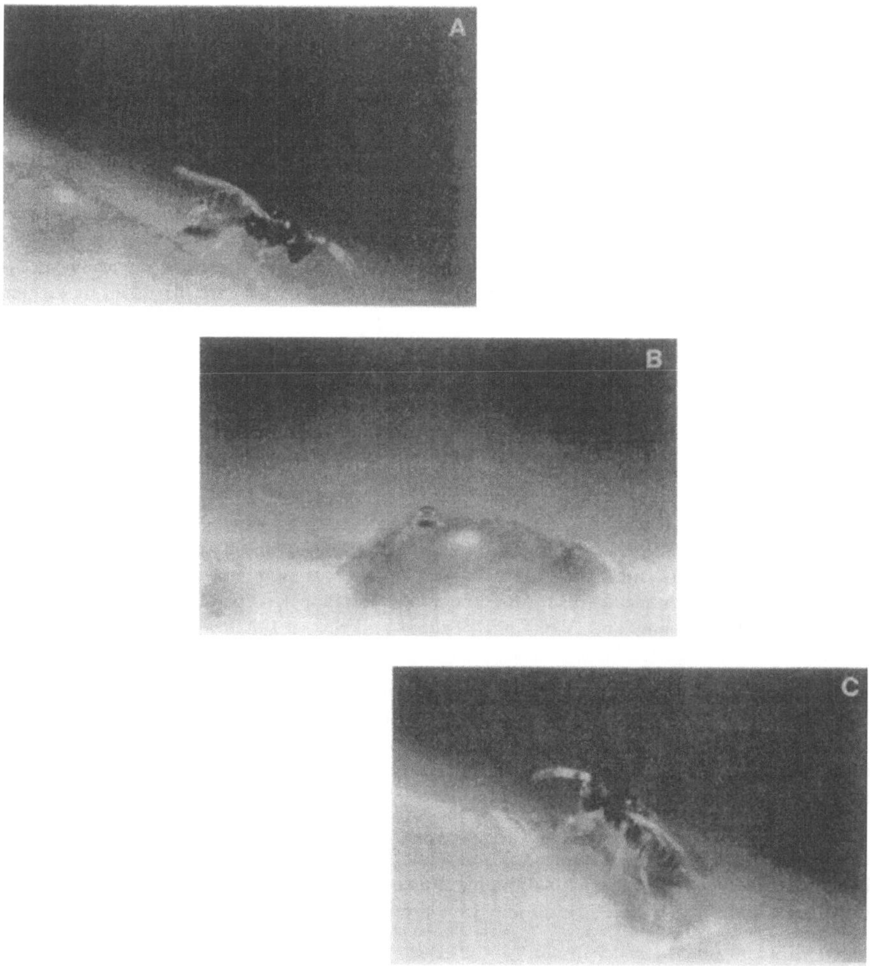

Figure 3.5. In many families, adult parasitoids kill hosts and feed on body fluids for nutrition, a process termed host feeding. This process involves piercing the host with the ovipositor (A), after which host hemolymph exudes from the wound (B), and is fed upon by the parasitoid (C). This series shows *Physcus* sp. feeding on the armored scale *Aonidiella aurantii* (Maskell). (Photographs courtesy of M. Rose. [B] and [C] from Rosen, D. (ed.) 1990. Armored scale insects: their biology, natural enemies, and control. Elsevier Science Publishers. With permission.)

only the Tachinidae have been of major significance in biological control as introduced natural enemies.

Acroceridae. Acrocerid larvae are internal parasitoids of spiders. Although some venomous spiders are considered pests, this parasitoid family has not been used in biological control.

Nemestrinidae. This family is mostly tropical. The six species which occur in North America whose biologies are known are endoparasitoids of either grasshoppers or larvae of scarabaeid beetles. Prescott (1960) records suppression of the grasshopper *Melanoplus bilituratus* (Walker) in Oregon by the nemestrinid *Trichopsidea clausa* (Osten Sacken). While apparently of importance as naturally occurring parasitoids in some cases, no success has been achieved to date with nemestrinids as introduced parasitoids. Palaearctic species are discussed by Richter (1988).

Bombyliidae. This family contains many species parasitic on caterpillars, scarabaeid larvae, hymenopteran larvae, and egg pods of grasshoppers. They have not yet been used to suppress pests through introduction to new areas, but the family is probably important as naturally-occurring mortality agents.

Phoridae. Phorids are thought to enter insects through existing wounds and have been reared from termites, bees, crickets, caterpillars, moth pupae, and fly larvae. The family is noteworthy as parasitoids of ants (Donisthorpe 1927; Wojcik 1989), and its members are of some interest as possible biological control agents for fire ants (*Solenopsis* spp.), some of which are important adventive pests (Williams and Banks 1987; Feener and Brown 1992).

Pipunculidae. Pinpunculids are parasitoids of various Auchenorrhycha (Homoptera), mainly leafhoppers and planthoppers. For example, *Verrallia aucta* (Fallén) is believed important in repressing populations of the meadow spittlebug, *Philaenus spumarius* (Linnaeus), in Europe (Whittaker 1973). Because the latter is an immigrant pest in North America and the parasitoid is not present, introduction of *V. aucta* has been suggested.

Conopidae. Conopids are endoparsitoids of adult bumble bees and wasps. To date there has been no applied use of this group, although in some circumstances some social wasps are pests of concern and further investigation may be warranted (Gambino et al. 1990).

Sciomyzidae. Sciomyzids may be either parasitoids or predators attacking a variety of molluscs, including land snails, water snails, slugs, and clams (Greathead 1980). Interest exists in their use to reduce such human and animal trematode infections as schistosomiasis and fascioliasis (Greathead 1980; Gormally 1988). Some introductions of sciomyzids to new locations have been undertaken; *Sepedomerus macropus* (Walker) and *Sepedon aenescens* (Wiedemann) were introduced to Hawaii against the lymnaeid mollusc *Galba viridis* (Quoy and Gaimard) (Funasaki et al. 1988a).

Cryptochetidae. This small family consists of the single genus *Cryptochetum*, all species of which are believed to be parasitoids of margarodid scales. Cryptochetids are found in Europe,

Asia, Africa, and Australia, but were not originally present in North or South America. One species, *Cryptochetum iceryae*, was introduced successfully into California from Australia and contributed towards the control of the immigrant pest *Icerya purchasi* (Fig. 3.6) (Bartlett 1978). Species of *Cryptochetum* have been reported from other margarodid scales in Japan, the Seychelles Islands, Italy, and Australia.

Sarcophagidae. Some species of sarcophagids are parasitoids of insects, and a few have been introduced to control immigrant pests, for example *Agria affinis* (Fallén) from France and the former Yugoslavia into the United States to control the gypsy moth, *Lymantria dispar* (Linnaeus) (Reardon 1981).

Tachinidae. Most tachinids are solitary endoparasitoids and none are hyperparasitic (Askew 1971). The group is the family of dipteran parasitoids most important to biological control (Fig. 3.7). Many species have been introduced for control of immigrant pest species. For example, *Lydella thompsoni* Herting was introduced to the United States to control the European corn borer, *Ostrinia nubilalis* (Hübner) (Burbutis et al. 1981), and *Cyzenis albicans* (Fallén) was introduced into Canada to control the immigrant species *Operophtera brumata* (Linnaeus) (Embree 1971). Tachinids such as *Lixophaga diatraeae* (Townsend) have been used for augmentative releases (Bennett 1971), while other species have been of interest as indigenous parasitoids, for example, the tachinid *Bessa harveyi* (Townsend), a parasitoid attacking the larch sawfly, *Pristiphora erichsonii* (Hartig) (Thompson et al. 1979). Grenier (1988) reviews the role of the tachinids in applied biological control.

Tachinids have been divided into several groups with respect to their oviposition strategies (O'Hara 1985). Adults of some species deposit their eggs on or in their hosts, whereas others retain their eggs and deposit first instar larvae on or in their hosts. Still others place eggs or

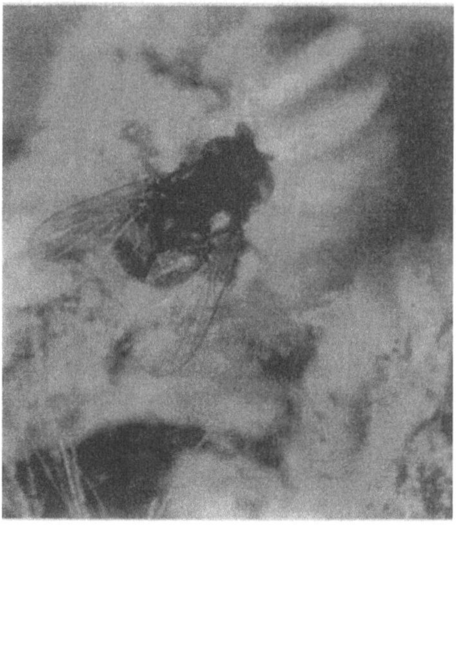

Figure 3.6. *Cryptochetum iceryae* (Williston) (Diptera: Cryptochetidae) is an important parasitoid of the margarodid scale *Icerya purchasi* Maskell. (Photograph by M. Badgley.)

Figure 3.7. *Myiopharus* spp. (Diptera: Tachinidae) attack the chrysomelid *Leptinotarsus decemlineata* (Say). (Photograph courtesy of D. Ferro.)

larvae on foliage or soil. Eggs laid on foliage are placed where they are likely to be consumed later by a host. In such cases, plant volatiles from damaged plant tissue may be attractive to ovipositing flies (Roland et al. 1989). Eggs laid on foliage are often very small (microtype) and deposited in greater numbers than the larger (macrotype) eggs of species which oviposit directly on their hosts (Askew 1971).

Hymenoptera

Hymenoptera are grouped into two suborders, the Symphyta which are sawflies and the Apocrita which include the wasps, ants, bees, and parasitoids. The sawflies, with the exception of the Orussidae, are phytophagous and have played no role in biological control of arthropods. The suborder Apocrita is composed of two divisions, the Parasitica and the Aculeata. The Aculeata are considered a monophyletic lineage and include the wasps, ants, and bees. Bees feed on plant pollen and nectar and are not important in biological control. The wasps and ants are very significant as predators and will be discussed later in this chapter. Some families of Aculeata in the Chrysidoidea are parasitoids (for example, Bethylidae and Dryinidae). The Parasitica are not a monophyletic lineage, but rather an assemblage of all Apocrita that are not within the Aculeata.

The Parasitica contains a rich array of species, of which at least 36 families have members that are parasitoids of arthropods. These families vary greatly, however, in the degree to which they have been used in biological control, due to family size and the types of insects they attack. Most cases of complete or substantial control from parasitoid introductions have involved just two families, the Encrytidae and the Aphelinidae.

A detailed appreciation of the various groups of parasitic Hymenoptera is essential for the biological control practitioner. In the organization of the families of the Apocrita into superfamilies, we follow Gauld and Bolton (1988); within families of the Chalicidoidea we follow Grissell and Schauff (1990). Townes (1988) summarizes the more important sources of taxonomic literature for parasitic Hymenoptera.

Symphyta: Orussidae. This is a small family whose members are parasitoids of larvae of wood-boring beetles in the family Buprestidae. Orussids have not been widely used for biological control introductions, in part because few buprestids have been targets of biological control programs.

Apocrita, Parasitica. The majority of the groups within the Parasitica are parasitoids, but a few are phytophagous (for instance, some of the Cynipidae). Parasitoids occur in eight superfamilies: Trigonalyoidea, Evanioidea, Cynipoidea, Chalcidoidea, Proctotrupoidea, Ceraphronoidea, Stephanoidea and Ichneumonoidea (Gauld and Bolton 1988). Most parasitoids of greatest importance to biological control are in the Chalcidoidea and Ichneumonoidea.

The superfamily **Trigonalyoidea** contains one family, the Trigonalidae, members of which are either parasitoids of social vespids or hyperparasitoids in caterpillars. Eggs are laid on foliage and hatch after they have been eaten by a caterpillar. The trigonalid larva then develops as a hyperparasitoid in parasitoids found in the caterpillar or, if the caterpillar is caught by a vespid, the trigonalid may develop as a parasitoid of the immature vespid.

The superfamily **Evanioidea** may be an artificial grouping. The three families in the superfamily (Evaniidae, Gasteruptiidae, and Aulacidae) are distinct in their biologies:

Evaniidae. These are mainly parasitoids of egg capsules of tropical species of cockroaches. Some species have been investigated for biological control of pest cockroaches (Thoms and Robinson 1987).

Gasteruptiidae. Gasteruptiid eggs are laid in larval cells of solitary wasps and bees. Gasteruptiid larvae first consume the host egg and then feed on the stored food in the cell (Gauld and Bolton 1988). This group has not been important in biological control through introduction.

Aulacidae. Aulacids are thought to be endoparasitoids of wood-boring beetles and Hymenoptera.

The superfamily **Cynipoidea** is comprised of small wasps, some of which are parasitoids while others are phytophagous gall makers. Cynipid galls of many types are well known on oaks (*Quercus* spp.) in temperate regions. Four families contain parasitic species, two of which (Ibaliidae and Eucoilidae) are important in biological control, and two of which are parasitoids of beneficial predators or parasitoids (Figitidae and Charipidae). One family (Cynipidae) contains mostly herbivorous gall makers, or are gall inquilines, species that eat the food resources of primary gall makers.

Ibaliidae. These are parasitoids of horntail sawflies. *Ibalia leucospoides* (Hochenwarth) has been introduced to New Zealand against the wood wasp *Sirex noctilio* to reduce damage in pine plantations (Nuttall 1989).

Figitidae. These are thought to be parasitoids of larvae of such beneficial predatory families as Hemerobiidae, Syrphidae, and Chamaemyiidae and are considered harmful from the point of view of biological control.

Eucoilidae. All eucoilids are believed to be parasitoids of fly pupae. An example is *Conthonaspis rapae* (Westwood) which parasitizes *Delia antiqua* (Meigen), a pest of onions (Sundby and Taksdal 1969).

Cynipidae. Most cynipids are gall formers and are common on oaks (*Quercus* spp.) and roses (*Rosa* spp.), and thus are not biological control agents of arthropods. Some members of the family, however, are inquilines which exploit the food supply of other cynipid gall makers.

Charipidae. Most members of this small family are hyperparasitoids of braconids found in aphids, although some species are primary parasitoids of Psylloidea (psyllids and their relatives).

The superfamily **Chalcidoidea** includes the families that have contributed most to biological control. Sixteen chalcidoid families contain significant numbers of parasitoid species. Of these, the Encyrtidae and Aphelinidae have been used most frequently in biological control. Grissell and Schauff (1990) provide a pictorial key to chalcidoid families for the Nearctic Region as well as notes on family characteristics and biology.

Leucospidae. Leucopsids are parasitoids of bee and wasp larvae. They have not been used for biological control, but might be valuable against pest Aculeata.

Chalcididae. Many chalcidid species attack pupae of Lepidoptera and Diptera. Some are parasitoids of Coleoptera and Hymenoptera. Some species, such as *Brachymeria intermedia* (Nees), have been successfully introduced to new regions in biological control programs (Leonard 1966).

Eurytomidae. Some eurytomids are parasitoids and others are phytophagous. The parasitic species attack hosts feeding within galls, stems, or seeds.

Torymidae. Some torymids are parasitoids of such gall forming insects as Cynipidae and Cecidomyiidae. Others attack mantid eggs or agaonids inside fig receptacles.

Ormyridae. These are parasitoids of gall insects or fig wasps.

Pteromalidae. This poorly defined group has varied behavior. The major subfamilies include: the Spalangiinae, parasitoids of fly pupae, some of which have been introduced to various countries or are reared for augmentative releases against various manure-inhabiting flies (Fig. 3.8) (Patterson et al. 1981); the Cleonyminae, parasitoids of wood-boring beetles or stem- or mud-nesting Hymenoptera; the Microgasterinae, parasitoids of various Diptera, including the Agromyzidae, Cecidomyiidae, Tephritidae, and Anthomyiidae; and the Pteromalinae, a diverse group containing parasitoids of Lepidoptera, Coleoptera, Diptera, and Hymenoptera.

Eucharitidae. All eucharitids are parasitoids of ants. The group has not been used in biological control, but may merit consideration against pest ants, including imported fire ants (*Solenopsis* spp.) in the southern United States.

Perilampidae. This group contains mainly hyperparasitoids of parasitoids of Lepidoptera. As such, members of this family would generally not be considered for use in biological control introductions unless studies indicated a particular species of perilampid was a primary rather than secondary parasitoid.

Tetracampidae. Tetracampids are egg parasitoids of chrysomelid beetles and diprionid sawflies, or of larvae of Agromyzidae.

Figure 3.8. *Sphegigaster* sp. (Diptera: Pteromalidae) is a parasitoid of dipteran puparia. (Photograph by M. Badgley.)

Eupelmidae. There are three subfamilies of eupelmids: the Calosotinae attack insects in stems or wood; the Eupelminae are egg parasitoids of various insects or hyperparasitoids; and the Metapelmatinae are parasitoids of wood-boring insects and cecidomyiids.

Encyrtidae. Encyrtids are of major importance to biological control. Many kinds of arthropods are parasitized by encyrtids, including scales, mealybugs, eggs or larvae of Coleoptera, Diptera, Lepidoptera, larvae of Hymenoptera, eggs or larvae of Neuroptera, eggs of Orthoptera, spiders, and ticks. This family, together with the Aphelinidae, accounts for half of the cases of successful biological control by means of introduction. Important genera in the family include *Comperia, Hunterellus, Ooencyrtus,* and *Epidinocarsis.* One species, *Epidinocarsis lopezi,* is credited with controlling the South American mealybug *Phenacoccus manihoti* and ending devastating crop losses in cassava throughout much of tropical Africa (Fig. 3.9) (Neuenschwander et al. 1989).

Signiphoridae. Some signiphorids are primary parasitoids of diaspine scales and others are hyperparasitoids of scale and whitefly parasitoids.

Eulophidae. This biologically diverse family is of major importance to biological control. Members of this family attack a wide range of hosts, including spider egg cases, scales, thrips, and many species of Coleoptera, Lepidoptera, Diptera, and Hymenoptera. Some species attack leafminers or wood boring insects. *Pediobius* and *Sympiesis* are important genera (Fig. 3.10).

Aphelinidae. This family, together with the Encrytidae, accounts for nearly half of the successful cases of biological control of insects that have been achieved through natural enemy introduction. Aphelinids are parasitoids of armored scales, mealybugs, whiteflies, aphids,

Figure 3.9. The encyrtid *Epidinocarsis lopezi* (De Santis) successfully controlled the mealybug *Phenaccocus manihoti* Matile-Ferrero on cassava (*Manihot exculenta* Crantz) in Africa. (Photograph courtesy of PHMD-IITA.)

psyllids, and eggs of insects of various orders. Genera of major importance include *Aphelinus*, *Aphytis*, and *Encarsia* (Fig. 10.2a) (Rosen and DeBach 1979). Viggiani (1984) reviews the bionomics of the Aphelinidae.

Trichogrammatidae. All members of this family are parasitoids of insect eggs. Species names in much of the older literature are incorrect because of poor taxonomic characterization of species and errors in identification. Species such as *Trichogramma minutum* Riley and *Trichogramma pretiosum* Riley have been widely used in attempts to control various species of Lepidoptera through releases of large numbers of insectary-reared wasps (Fig. 3.11). See Pinto and Stouthamer (1994) for a review of the systematics of the Trichogrammatidae, with special discussion of issues in the genus *Trichogramma*.

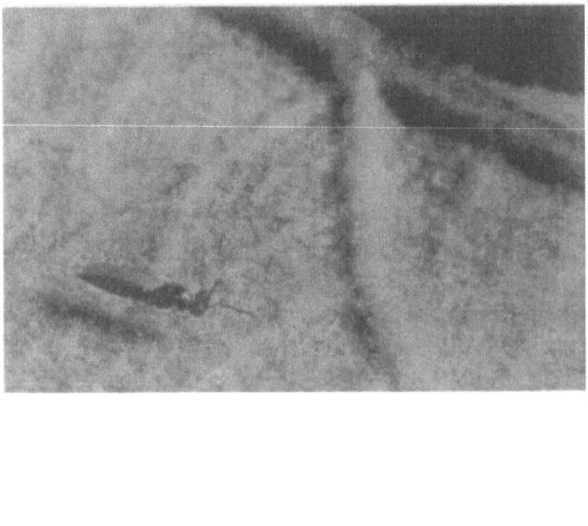

Figure 3.10. The eulophid *Sympiesis marylandensis* Girault is an important parasitoid of leaf-mining moths in the genus *Phyllonorycter*, some of which are pests of apples (*Malus pumila* Miller). (Photograph courtesy of C. Maier.)

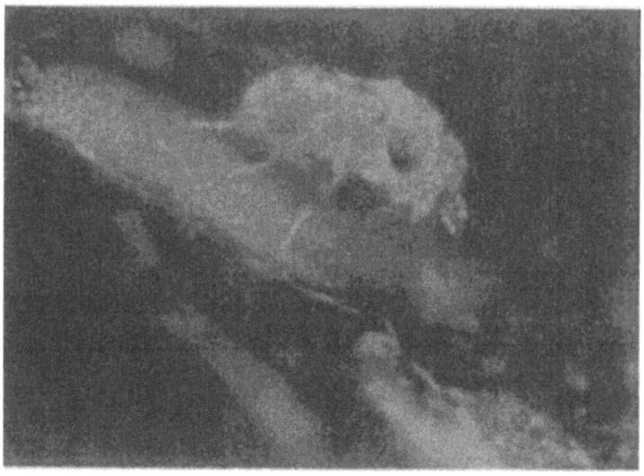

Figure 3.11. Trichogrammatids are parasitoids of lepidoptera and hemipteran eggs. *Trichogramma platneri* Nagarkatti attacks the codling moth, *Cydia pomonella* (Linnaeus). (Photograph by M. Badgley.)

Mymaridae. All mymarids (Fig. 3.1a) are egg parasitoids of insects in various orders, including Hemiptera, Homoptera, Psocoptera, Coleoptera, Diptera, and Orthoptera. Mymarids have been used in biological control projects through introduction, as in the use of *Anaphes flavipes* (Förster) against the cereal leaf beetle, *Oulema melanopus* (Linnaeus) (Maltby et al. 1971).

The superfamily **Proctotrupoidea** appears to be an artificial grouping of parasitoids that do not fit easily elsewhere. Ten families are recognized by Gauld and Bolton (1988), but only four families (Proctotrupidae, Diapriidae, Scelionidae, and Platygasteridae) are of interest in biological control.

Proctotrupidae. Most proctotrupids are larval parasitoids of beetles and gnats which live under bark, in leaf litter, on fungi, and in other similar damp habitats in the temperate latitudes. A few species are of interest as parasitoids of pests, such as *Paracodrus apterogynus* (Haliday) which attacks pest wireworms (Elateridae) in Europe (Gauld and Bolton 1988).

Diapriidae. These are pupal endoparasitoids of various Diptera, including members of the Mycetophilidae, Sciaridae, Chloropidae, Muscidae, and Tephritidae. Some species are hyperparasitoids. Some species attack pest species, such as *Basalys tritoma* (Meigen) which attacks the carrot rust fly, *Psila rosae* (Fabricius), in Europe; *Psilus silvestrii* (Kieffer) which attacks the tephritid *Ceratitis capitata* (Wiedermann); and species of *Trichopria* which are parasitoids of chloropid eye gnats in the genus *Hippelates* (Clausen 1978).

Scelionidae. This large family is poorly understood taxonomically. Its members are all egg parasitoids, and some species have been used in biological control, for example, *Trissolcus basalis* (Wollaston), a parasitoid of the southern green stink bug, *Nezara viridula* (Linnaeus) (Jones 1988). Other important genera are *Telenomus* and *Scelio*.

Platygasteridae. This large but poorly known family has many species in most regions of the world. Most species attack Diptera, especially gall forming cecidomyiids. Some genera, however, attack mealybugs (*Allotropa*) and whiteflies (*Amitus*).

The superfamily **Ceraphronoidea**, previously grouped within the Platygasteroidea but now considered distinct, has two families, the Ceraphronidae and the Megaspilidae.

Ceraphronidae. This small family includes some primary parasitoids of beneficial species such as predacious cecidomyiids, some primary parasitoids of phytophagous insects, and some hyperparasitoids.

Megaspilidae. These are endoparasitoids of a variety of hosts. Some megaspilids develop as primary parasitoids of scales. Others attack beneficial species such as predacious hemerobiids, chrysopids, chamaemyiids, or syrphids. Still others (e.g., *Dendrocerus* spp.) are hyperparasitoids in aphids through aphelinid or braconid primary parasitoids.

The superfamily **Stephanoidea** contains one small family, the Stephanidae, whose members attack larvae of wood-boring beetles. There are six species in North America. It is not a group that has been widely used in biological control.

The superfamily **Ichneumonoidea** is comprised of two families, the Ichneumonidae and the Braconidae. The Aphidiinae are sometimes elevated to the family level but here are kept within the Braconidae.

Ichneumonidae. This large family's (Townes 1969) members parasitize many different kinds of hosts (Fig. 3.12). Most species have long antennae and long ovipositors which are always visible. The most important subfamilies can be grouped by the types of hosts which they attack (after Askew 1971) (these generalities do not hold for every species): ectoparasitoids of larvae or pupae of diverse orders in plant tissue (Ephialtinae, e.g., *Pimpla*); ectoparasitoids of exposed larvae of Lepidoptera and sawflies (Typhoninae, e.g., *Phytodietus*); ectoparasitoids of insects in cocoons—some are hyperparasitoids (Gelinae, e.g., *Gelis*); endoparasitoids of lepidopteran larvae (Banchinae, e.g., *Glypta*; Porizontinae, e.g., *Diadegma*; Ophioninae, e.g., *Ophion*); endoparasitoids of lepidopteran pupae (Ichneumoninae, e.g., *Ichneumon*); endoparasitoids of sawfly larvae (Scolobatinae, e.g., *Perilissus*); and endoparasitoids of syrphid larvae (Diplazontinae, e.g., *Diplazon*).

Braconidae. Braconids have been widely used in biological control, especially against aphids and various Lepidoptera, Coleoptera, and Diptera. Braconids often pupate inside silk cocoons outside the body of their host. Wharton (1993) discusses the bionomics of the Braconidae.

Figure 3.12. Many ichneumonids are parasitoids of larvae or pupae of Lepidoptera. *Campoplex frustranae* Cushman is a parasitoid of the Nantucket pine tip moth, *Rhyacionia frustrana* (Comstock). (Photograph by M. Badgley.)

Twenty one subfamilies have been recognized. The most important of these may be grouped by the types of hosts which they attack (after Askew 1971) (these generalities do not hold for all species): endoparasitoids of aphids (Aphidiinae, e.g., *Aphidius*, *Trioxys*) (for details on the biology of this group, see Starý 1970); endoparasitoids of larvae of Lepidoptera and Coleoptera (Meteorinae, e.g., *Meteorus*; Blacinae, e.g., *Blacus*; Microgasterinae, e.g., *Apanteles*, *Microplitis*; Rogadinae, e.g., *Aleiodes*); endoparasitoids of adult beetles or nymphal Hemiptera (Euphorinae, e.g., *Microctonus*); egg-larval endoparasitoids of Lepidoptera (Cheloninae, e.g., *Chelonus*); endoparasitoids of fly eggs or larvae (Alysiinae, e.g., *Dacnusia*; Opiinae, e.g., *Opius*); and ectoparasitoids of lepidopteran larvae in concealed places (Braconinae, e.g., *Bracon*).

Apocrita, Aculeata. Most species in this portion of the Hymenoptera are either predacious, such as the ants and wasps, or phytophagous, such as the bees. Some families in the Chrysidoidea and Vespoidea, however, have members which are parasitoids.

The superfamily **Chrysidoidea** consists of seven families; we will consider three: Dryinidae, Bethylidae, and Chrysididae.

Dryinidae. Dryinids are parasitoids of leafhoppers and some of their relatives in the homopteran superfamilies Cicadelloidea and Fulgoroidea. Adult dryinids capture host nymphs and, while gripping them tightly, lay an egg between two abdominal sclerites. Some dryinids have been introduced for control of pest leafhoppers, such as the sugarcane leafhopper, *Perkinsiella saccharicida*, and have established, but are not credited with control of the target pest (Clausen 1978).

Bethylidae. Bethylids attack beetle and lepidopteran hosts, mostly larvae, in confined habitats such as leaf rolls and under bark. Some species have been studied for use in biological control as, for example, parasitoids of the coffee berry borer, *Hypothenemus hampei* (Ferrari), (Abraham et al. 1990) and *Goniozus legneri* Gordh, which controls the pyralid moth *Amyelois transitella* (Walker) in almond (*Prunus dulcis* [Miller] D.A. Webb var. dulcis) orchards in California (Legner and Gordh 1992).

Chrysididae. Chrysidids are heavily sclerotized, ruggedly sculptured wasps whose bodies are bright metallic colors. There are three subfamilies: the Cleptinae, parasitoids of cocooned prepupae or pupae of tenthredinid sawflies; and the Elampinae and Chrysidinae, parasitoids (if their young eat the immature stages of the host species) or cleptoparasitoids (if their young consume only the stored foodstuffs of the host species and do not eat the host itself) of various families of wasps and bees. The Elampinae and Chrysidinae are able to roll up into a protective ball when attacked. For a world overview of the family see Kimsey and Bohart (1990).

The superfamily **Vespoidea** differs from the other two distinct superfamilies in the Aculeata (the Apoidea and the Chrysidoidea), but probably does not represent a single lineage within the aculeates. Four vespoid families with parasitic members are Tiphiidae, Mutillidae, Scoliidae, and Sphecidae.

Tiphiidae. Tiphiids are parasitoids of beetle larvae. Species of the subfamily Tiphiinae burrow into soil to attack the larvae of scarabaeid beetles in their earthen cells. *Tiphia popilliavora* Rohwer and *Tiphia vernalis* Rohwer were introduced into the United States to assist in the suppression of the Japanese beetle, *Popillia japonica* Newman. Establishment was achieved and levels of parasitism were high for some years, but ultimately declined; the species are now rare while their host is still common (King 1931; Ladd and McCabe 1966).

Mutillidae. Mutillids are parasitoids of the larvae and pupae of wasps and bees. Since these host groups are rarely encountered as pests, this group of parasitoids has been little used in biological control.

Scoliidae. Scoliids are parasitoids of scarabaeid larvae in soil. One species, *Scolia oryctophaga* Coquille, has been introduced to Mauritius for the control of a scarabaeid sugarcane pest, *Oryctes tarandus* (Olivier), with apparent success (Clausen 1978).

Sphecidae. Most members of this family are predacious, but species in the genus *Larra* are ectoparasitoids on mole crickets.

Lepidoptera

This order is mostly herbivorous. One family, the Epipyropidae, is parasitic. Larval epipyropids attach to Fulgoridae (and other families of Hemiptera and Homoptera) and feed on honey dew and, probably, host body fluids (Borror et al. 1989).

Coleoptera

This order contains many herbivores and predators, but only a few parasitoids. Individual genera in scattered families have developed a parasitic life history by reducing their food requirements to one prey and limiting their mobility. Examples of this pattern are found in the Carabidae (*Lebia* and *Pelecium*), Rhipiceridae (*Sandalus*), Cleridae (*Hydnocera, Trichodes*), Colydiidae (*Deretaphrus, Dastarchus, Sosylus, Bothrideres*), Passandridae (*Catogenus*), and Staphylinidae (*Aleochara*) (Askew 1971). The staphylinids in the genus *Aleochara* are endoparasitoids of dipteran pupae; some species have been used in biological control, for example *Aleochara bilineata* (Gyllenhal) which attacks the vegetable pest *Delia radicum* (Linnaeus) (Read 1962).

Two families (Meloidae and Rhipiphoridae) are more characteristically parasitic. The Meloidae, as larvae, feed on grasshopper and locust eggs or live in cells of solitary bees. Species that attack bees find their hosts by means of triungulin larvae that are phoretic on adult bees. Rhipiphoridae are endoparasitoids of Hymenoptera (such as *Vespula* and various groups of bees) and, in some cases, of cockroaches.

Strepsiptera

This is a small order, nearly all members of which are parasites (not parasitoids) of orthopterans, bees, wasps, ants, cicadellids, cercopids, delphacids, pentatomids, homopterans, and thysanurans (Kathirithamby 1989). The term **stylops** is used to refer to the parasite and **stylopised** to the condition of being parasitized by a stylops. Hosts are encountered by actively dispersing larvae called **triungulins**. These are produced in large numbers by female Strepsiptera, which remain attached to their hosts. Female stylops have reduced body parts and spend all of their adult lives inside their hosts' bodies, with only their fused head and thorax protruding from the host (Askew 1971). Stylopised hosts live out most of their natural life span because the stylops feeds by absorption and causes little or no mechanical damage to tissues. Stylopised hosts often exhibit mosaics of male-female sexual characters, and female hosts fail to produce viable eggs. This order has not played any important role in biological control.

INSECTS PREDACIOUS ON ARTHROPODS

The predacious habit is widespread in the classes Insecta and Arachnida. Juvenile predators use prey for growth, while adults use prey for maintenance and reproduction. This section provides an overview of this diversity and discusses those groups that have been important in biological control. Predacious insects of importance are found in nine orders: Orthoptera, Dermaptera, Thysanoptera, Hemiptera, Neuroptera, Coleoptera, Lepidoptera, Diptera, and Hymenoptera. Of greatest importance are the Hemiptera, Coleoptera, Diptera, and Hymenoptera.

Orthoptera (Grasshoppers, Cockroaches, Mantids)

Most orthopterans are herbivorous (most grasshoppers and locusts) or scavengers (cockroaches). The only group that is consistently predacious is the Mantidae, a family of large predators that are mostly tropical and have raptorial forelegs. Some species have been introduced to new regions, for example the Chinese mantid, *Tenodera aridifolia sinensis* Saussure, and the European mantid, *Mantis religiosa* Linnaeus, both of which were introduced to the United States. While commonly sold by insectaries, there is little reason to believe that these predators provide useful pest control after release.

A few other species of Orthoptera in other families are also predacious, such as the tettigoniid *Conocephalus saltator* Saussure which eats aphids and scales. In tropical regions, some grasshopper and cricket species are predators of rice pests. Overall, the value of orthopterans to biological control is limited.

Dermaptera (Earwigs)

The Dermaptera are easily recognized by their caudal pincers. They are mostly scavengers, but a few species are predacious on aphids and other soft bodied insects. Cans filled with dry grass and tied to trees have been used successfully to enhance earwig predation on aphids in fruit trees (Schonbeck 1988).

Thysanoptera (Thrips)

Most thrips are phytophagous, and some species are pests of cultivated plants. Two families, however, contain predacious species: the Aleolothripidae, as for example *Aleolothrips fasciatus* (Linnaeus) which feeds on thrips, aphids, and mites, and the Phlaeothripidae, such as *Leptothrips mali* (Fitch) (Fig. 3.14g) which feeds on mites, and *Aleurodothrips fasciapennis* Franklin which feeds on whiteflies. The biological significance of this group is probably not yet fully appreciated.

Hemiptera (True Bugs)

This order contains many families with predacious members. Some of the more important of these are as follows.

Predacious Hemipterans in Aquatic Environments. Several families of predacious hemipterans live in aquatic environments: Notonectidae, Pleidae, Naucoridae, Belostomatidae, Nepidae, Gerridae, Veliidae, and others. These are considered generalist predators and, therefore, have not been introduced into new regions to combat specific immigrant pests; consequently they are viewed as of little importance. However, like other generalist predators (e.g., spiders), these groups are probably important for suppression of pest complexes, whose exact species composition changes from place to place and between years. These groups may be important in suppressing pests of aquatic crops such as rice, and for suppressing some medically important pests such as mosquitoes and snails. Conservation of these predators in rice paddies and other habitats where aquatic pests breed is likely to be important (Sjogren and Legner 1989).

Anthocoridae. These small hemipterans (less than 5 mm in length) are important predators of phytophagous thrips and eggs of pests such as *Ostrinia nubilalis* (Fig. 3.13a) (Coll and Bottrell 1991, 1992). *Orius tristicolor* (White) is reared for control of thrips on greenhouse flower crops (Gilkeson 1991), and *Anthocorus gallarum-ulmi* (De Geer) is reared for use against aphids (Ruth and Dwumfour 1989). A few anthocorids have been moved to new locations for control of immigrant pests, such as *Montandoniola moraguesi* (Puton), which was introduced into Hawaii for control of *Gynaikothrips ficorum* (Marchal) (Clausen 1978).

Miridae. This family contains many herbivorous species, including such major pests as *Lygus lineolaris* (Palisot de Beauvois). Some groups, however, are predacious, such as *Deraeocoris* spp. which feed on aphids and other small insects. Some mirids have been imported into new regions for biological control purposes. For example, *Tytthus mundulus* was introduced to Hawaii (U.S.A.) and in conjunction with some introduced parasitoids, successfully controlled the sugarcane leafhopper, *Perkinsiella saccharicida* (Clausen 1978). The mirid *Macrolophus caliginosus* Wagner is used to control whiteflies in glasshouse and outdoor tomato crops (Minkenberg, pers. comm.).

Nabidae. Many nabids (Fig. 3.13c) are predacious and are most common on grass and low, herbaceous plants. Nabids feed on insect eggs, aphids, and other small, slow, or soft bodied insects. *Nabis ferus* Linnaeus is a well-known predator of the potato psyllid, *Paratrioza cockerelli* (Sulc), and the sugar beet leafhopper, *Circulifer tenellus* (Baker).

Reduviidae. Assassin bugs, such as *Arilus cristatus* Linnaeus, attack aphids, leafhoppers, and caterpillars, among other prey.

Phymatidae. Phymatids are ambush predators with raptorial forelegs that wait in flowers to feed on bees, wasps, flies, and other prey. The group has not been of any great significance to biological control.

Lygaeidae. Many members of this family are herbivorous, including such well-known species as the chinch bug, *Blissus leucopterus leucopterus* (Say), and the large milkweed bug, *Oncopeltus fasciatus* (Dallas). Some genera are predacious, including the big-eyed bugs (*Geocoris*

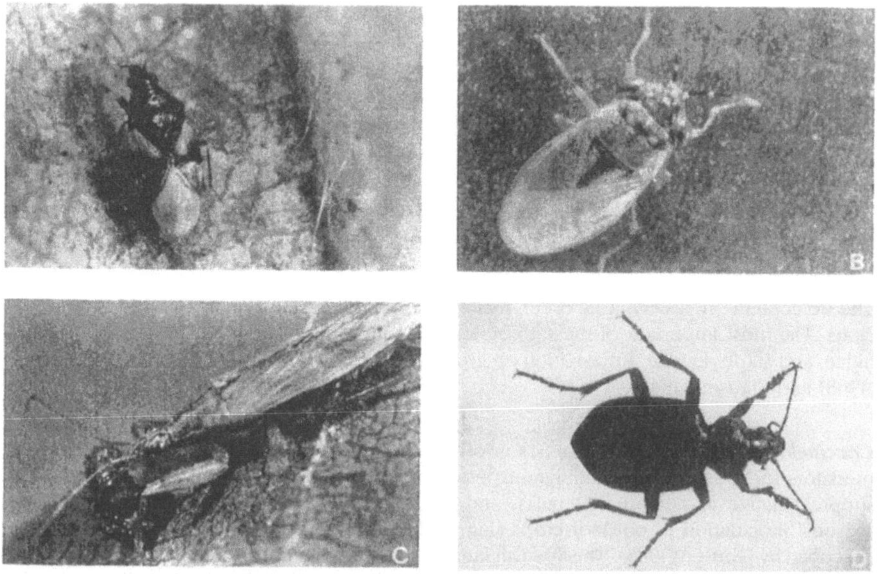

Figure 3.13. Many families of true bugs (Hemiptera) and beetles (Coleoptera) have predacious species of importance. These families include: (A) Anthocoridae, such as *Orius tristicolor* (White); (B) Lygaeidae, such as *Geocoris pallens* Stål; (C) Nabidae, such as *Nabis americoferus* Carayon; (D) Carabidae, such as *Calosoma syncophanta* Linnaeus; (E) Coccinellidae, such as *Hippodamia convergens* Guerin, and (F) Staphylinidae. (Photographs by M. Badgley [A,B,C,F]; and courtesy of R. Weseloh [D] and M. Raupp [E].)

spp.) (Fig. 3.13b), which feed on turf-inhabiting insects and pests of cotton (Gravena and Sterling 1983).

Pentatomidae. This family contains many important herbivorous pest species, such as *Nezara viridula*, and some predacious species. *Podisus maculiventris* (Say) and *Perillus bioculatus* Fabricius are important predators attacking vegetable pests such as *Leptinotarsa decemlineata* (Say).

Neuroptera (Nerve-Winged Insects)

Some families in the Neuroptera are predacious in aquatic habitats (Sialidae, Corydalidae), but because of the types of organisms which they eat, they have not so far been of interest in biological control. The main groups of Neuroptera used in biological control are the Chrysopidae, Hemerobiidae, and the Coniopterygidae. The Chrysopidae are the well-known green lacewings, such as species of *Chrysopa* and *Chrysoperla* (Fig. 3.14ab). Green lacewings are commonly found in vegetation containing grasses and herbs, and are predacious as both adults and larvae on aphids, whiteflies, and eggs of various insects, including species of

Helicoverpa. Several species of green lacewings are reared commercially. Release of lacewings, as eggs or larvae, can be an effective control strategy for aphids in greenhouses (Scopes 1969; Tulisalo and Tuovinen 1975; Hassan 1976/77, 1978) and, potentially, for some outdoor crop pests such as *Aphis pomi* De Geer (Hagley 1989), *Helicoverpa zea* (Boddie), and *Helicoverpa virescens* Fabricius (Ridgway and Jones 1969). The Hemerobiidae, the brown lacewings, are less common than green lacewings, and are found more commonly in wooded areas. The Coniopterygidae, the dusty lacewings, are less than 3 mm in length and feed on mites, small insects, and eggs.

Coleoptera (Beetles)

The Coleoptera are a very large order, including over 110 families, many of which are predacious. The most important of these to biological control have been the Coccinellidae, Carabidae, and Staphylinidae. Information on groups of beetles not treated here appears in Arnett (1968) and Clausen (1940).

Coccinellidae. This group of beetles is of importance to biological control as introduced predators for the control of immigrant pests, as native species that, when conserved, help suppress native and immigrant pests, and for augmentative releases in greenhouses and seasonal inoculation in outdoor crops (Fig. 3.13e). The U.S. and Canadian fauna has been described by Gordon (1985). The Spanish fauna is treated by Infante (1986). Introduced species of coccinellids that have controlled target pests include the vedalia beetle, *Rodolia cardinalis*, which controlled the cottony cushion scale, *Icerya purchasi*, in California and elsewhere (Caltagirone and Doutt 1989); *Cryptognatha nodiceps* Marshall, which controlled the coconut

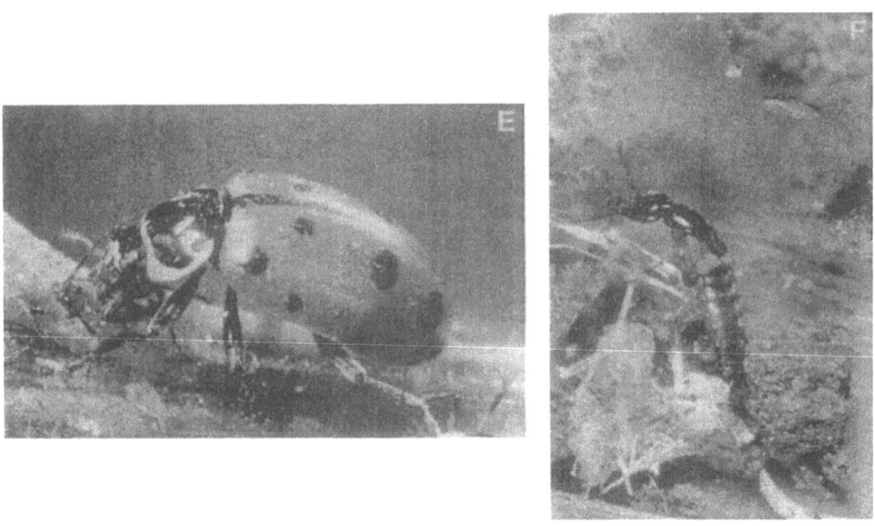

Figure 3.13. (*Continued*)

scale *Aspidiotus destructor* Signoret in Trinidad (Clausen 1978); and *Chilocorus distigma* Klug and *Chilocorus nigritus* (Fabricius), which successfully reduced several other diaspidid scales on coconut in the Seychelles (Clausen 1978). Many introductions of coccinellids, however, especially those made in the years following the success in California against cottony cushion scale, failed because little or no attention was given to gaining a true appreciation for the diet and ecology of the coccinellids being introduced (Caltagirone and Doutt 1989). Native species of coccinellids in many locations are common predators of aphids, scales, eggs of various insects, spider mites, and other pests. Their conservation in crop systems is useful to help suppress pests, as in the case of *Coleomegilla maculata* (De Geer), a predator of eggs of *Leptinotarsa decemlineata* (Hazzard et al. 1991). Some species of coccinellids are reared commercially to control whiteflies, spider mites, scales, mealybugs, or aphids in greenhouses or for seasonal inoculation in outdoor crops. Examples of augmented coccinellids include *Stethorus punctillum* Weise (for spider mites), *Chilocorus nigritus* (for scales), and *Cryptolaemus montrouzieri* Mulsant (for mealybug control) (Hunter 1992). A review of the biology of the Coccinellidae and their role in pest management is given by Hodek (1970, 1973).

Carabidae. Most carabids are generalist predators (Den Boer 1971; Erwin et al. 1979; Thiele 1977; Den Boer et al. 1979) and live on or near the ground, where they feed, mostly at night. Some species leave the soil to climb and feed on plant foliage. Ground beetles are small to large beetles (8–25 mm in length) and are often dark or metallic in color. Many species are important predators in forage, cereal, and row crops (Hance and Gregoire-Wibo 1987). Agricultural practices that have been investigated for conservation of carabids include strip (rather than area wide) pesticide application (Carter 1987), the influence of weeds, and the application of manure (Purvis and Curry 1984). In some instances, carabids with relatively specialized habits have been introduced to new regions for control of immigrant pests as, for example, *Calosoma sycophanta* (Linnaeus) (Fig. 3.13d) which was introduced to North America for the control of the gypsy moth, *Lymantria dispar*. Some carabids, such as *Scaphinotus* spp., feed on snails.

Staphylinidae. Most staphylinids are predacious and some are important predators of eggs and larvae of manure-breeding flies (Axtell 1981) and flies that attack the roots of young onions, cabbage, and broccoli (Fig. 3.13f) (Read 1962).

Aquatic Predacious Beetles. Several families of beetles, such as the Gyrinidae and Dytiscidae, are predacious in aquatic habitats. These families may be significant in crops grown in shallow water, such as rice, or for the control of pests in wet areas (Mogi and Miyagi 1990).

Histeridae. Some species of this family are predators of pests such as *Musca domestica* Linnaeus that breed in manure. *Carcinops pumulio* (Erichson) has been reported to be an important predator of *Musca domestica* Linnaeus eggs and larvae in poultry facilities in North Carolina (Axtell 1981).

Cantharidae. Larvae of some species of this family are predacious. Adults of some groups (*Cantharis*, *Chauliognathus*) feed on aphids and other prey such as eggs of locusts (Clausen 1940).

Cleridae. Most members of this family are predacious as both larvae and adults. *Thanasimus* spp. are important predators of *Ips typographus* (Linnaeus) in central Europe (Mills and Schlup 1989).

Cybocephalidae. Species in this family are predators of scales and whiteflies. *Cybocephalus* nr *nipponicus* Endrody-Younga has been introduced into North America from South Korea and China to control the Asian diaspidid scale *Unaspis euonymi* (Comstock), a pest on various species of euonymus plants (*Euonymus* spp.) used in landscape plantings (Drea and Carlson 1990).

Lepidoptera (Moths and Butterflies)

The vast majority of species in this order are herbivores or scavengers. The predacious habit, however, is found in a few species scattered across a number of families (Balduf 1931). Typical prey are scales, aphids, and other sessile or slow moving insects found on plants. Predacious species are found in at least six families, from which we cite the following examples (after Clausen 1940). Some species of Lycaenidae feed on Pseudococcidae, Cicadellidae, Membracidae, and larvae of Formicidae. *Feniseca tarquinius* Fabricius, for example, eats aphids, and *Spalgis epius* Westwood attacks mealybugs. Some species of *Holcocera* in the Blastobasidae feed on lecaniine scales. In the Heliodinidae, some Indian and Australian species of *Stathmopoda* feed on coccids and *Euclemensia bassettella* Clemens is a common predator of the scale *Kermes galliformis* Riley. Other families with predacious members include the Psychidae, Olethreutidae, Pyralidae, and Noctuidae. The arctiid species *Amata pascus* (Leech) has been used in China via augmentation as a predator of the bamboo scale *Kuwanaspis pseudoleucaspis* (Kuwana) (Li 1989). Some efforts have been made to use predacious Lepidoptera as introduced agents, but the order has been of minor importance compared to the Coleoptera, Diptera, and Hymenoptera.

Diptera (Flies)

The predacious habit is common in many families of Diptera. Many groups, however, are either generalists in regards to the species on which they feed, or are habitat specialists found only in certain environments. Diptera of these two types may be important in natural control (biological control which occurs independent of human activities), and their conservation through favorable management practices may be useful, but they are likely to be of limited value for introduction. Examples of generalists or habitat specialists include the Tipulidae, a few species of which are predacious on insects in wetlands; the Culicidae, a few of which are predacious on other culicids; and the Chironomidae, some of which (in the Tanypodinae) are predacious on larvae of other chironomids. The Rhagionidae are generalist predators both as adults and larvae of a variety of insects, often in decaying organic matter. Asilidae are predacious as both adults and larvae, preying on a wide range of insect species in the habitats in which they occur. Other families that contain predacious members include Empidae, Dolichopodidae, Otitidae, Lonchaeidae, Drosophilidae, Chloropidae, Anthomyiidae, and Calliphoridae. Individual species from these groups can, with respect to a particular biological control project, be important, depending on the biologies involved (Dysart 1991). The dipteran families which have been most significant in biological control are the Cecidomyiidae, Syrphidae, and Chamaemyiidae, which include species that attack aphids and other important herbivorous pests. Some species in these groups have been of value for introduction against immigrant pests.

Cecidomyiidae. Many species in this family are gall makers. However, some species are predacious on aphids, scales, whiteflies, thrips, and mites (Barnes 1929) (Fig. 3.14cd). The best known of these is *Aphidoletes aphidimyza* (Rondani), which is reared and sold for aphid control in greenhouses (Markkula et al. 1979, Meadow et al. 1985). In addition to this augmentative use, predacious cecidomyiids are common predators of aphids in outdoor crops, whose effectiveness could potentially be enhanced through appropriate conservation practices.

Syrphidae. Adults of many syrphid species are brightly-patterned mimics of vespids or bees. Syrphids are thought to be important predators of some aphid species (Fig. 3.14ef) (Hagen and van den Bosch 1968), and some syrphids have been introduced against immigrant species of aphids.

Chamaemyiidae. The larvae of most species in this family are predacious on aphids, scales, and mealybugs and are thought to be important in the natural control of some pest aphids (e.g., *Leucopis* sp. nr. *albipuncta* Zetterstedt which feeds on *Aphis pomi*, Tracewski et al. 1984). Some species have been introduced to new regions for control of immigrant pests as, for example, *Leucopis obscura* Haliday, which was introduced into Canada for control of the pine aphid, *Pineus laevis* (Maskell) (Clausen 1978).

Hymenoptera (Wasps, Ants)

The most important predacious groups in the Hymenoptera are the ants (Formicidae) and two families of aculeate wasps (Vespidae and Sphecidae).

Formicidae. This family contains a large number of species which exhibit biologies ranging from herbivory to scavenging to predation (Hölldobler and Wilson 1990). All ants are social and the number of individuals per colony may be very large. Predacious ants can be an enormous source of nonspecific predation. Ants, including such pest species as the adventive fire ant *Solenopsis invicta* (Buren), have been shown to be important in the suppression of pests in forests and crops (Adlung 1966; Fillman and Sterling 1983; Way et al. 1989; Weseloh 1990; Perfecto 1991). Conservation of native species of ants can be an important source of natural pest control and deliberate manipulation of ants for control of citrus pests was practiced in China over 1900 years ago (Coulson et al. 1982). However, ants should not be introduced to locations outside of their native range because the generalist nature of their predation and the numbers some species can attain per hectare are such that native invertebrates may be endangered via predation or competition (Howarth 1985). Ant species such as the Argentine ant, *Linepithema humile* (Mayr) (formerly *Iridomyrmex humilis*), which tend and defend honeydew-producing Homoptera, can interfere with the effective action of other natural enemies, such as parasitoids of scales and aphids.

Vespidae. Most vespids are brightly patterned social species whose adults capture various insects, including many species of caterpillars, as food for their larvae. (The fauna of the United States and Canada has been described by Akre et al. [1980].) Vespids are believed to be of value in the suppression of some insects, for example, the cassava hornworm, *Erinnyis ello* Linnaeus (Bellotti and Arias 1977). Their action is not, however, directed at any particular pest species. Like other generalist predators, they are valuable as a general restraint on complexes of

Figure 3.14. Important predacious insects are found in Chrysopidae (Neuroptera) (A, adult *Chrysoperla rufilabris* [Burmeister]) (B, larval *Chrysoperla carnea* [Stephens]); Cecidomyiidae (Diptera) (C,D—adult and larva of *Feltiella acarivora* Ruebsaamen); Syrphidae (Diptera) (E,F—adult and larva of *Syrphus opinator* Osten Sacken); and Thysanoptera (G, adult *Leptothrips mali* [Fitch].) (Photographs by M. Badgley.)

insects in the habitats in which they forage. Some attempts have been made to enhance vespid densities in crop fields by providing nesting structures (Gillaspy 1971). As with ants, vespids, by virtue of their large colony sizes and the wide range of prey species they attack, have the potential to threaten nonpest native invertebrates and should not be introduced to areas outside their native ranges. Invasions of immigrant vespids in Hawaii and New Zealand are thought to have endangered indigenous invertebrates and a forest bird (Gambino et al. 1990; Beggs and Wilson 1991). Some immigrant vespids are now targets of biological control programs (Grossman 1990). An additional drawback to this group as biological control agents is that they readily sting people.

Sphecidae. Sphecids are nonsocial wasps that provision their individual nests with a wide range of larger arthropods, including various caterpillars and spiders, as food for their larvae.

ARACHNIDS PREDACIOUS ON ARTHROPODS

Acari (Mites)

Of the 27 or more families of the order Acari that prey on or parasitize other invertebrates, eight appear to be of potential significance to biological control: Phytoseiidae, Stigmaeidae, Any-

Figure 3.14. *(Continued)*

stidae, Bdellidae, Cheyletidae, Hemisarcoptidae, Laelapidae, and Macrochelidae. Of these, the Phytoseiidae are the most important and best known. Other families may become recognized as valuable as our knowledge of their habits increases. Information on these families is summarized by Gerson and Smiley (1990) and Gerson (1992), who provide an excellent introduction to the lesser known groups. Information on phytophagous families of mites mentioned in the section is found in Jeppson et al. (1975). Information on groups of mites important in the biological control of weeds is given in Chapter 5.

Phytoseiidae. This family is used primarily to suppress tetranychid spider mites (Fig. 3.15a) (Hoy 1982a; Gerson and Smiley 1990). Species of phytoseiids have been introduced widely for biological control (McMurtry 1982). Methods for conservation and enhancement of indigenous phytoseiids have been studied extensively in apples (Hoyt and Caltagirone 1971), grapes (*Vitis vinifera* Linnaeus) (Flaherty and Huffaker 1970), strawberries (*Fragaria x ananassa* Duchesne) (Huffaker and Kennett 1956), and other crops. Pesticide-resistant strains of a few species have been isolated and used to inoculate apple orchards and other regularly-sprayed environments (Croft and Barnes 1971). Mass-reared phytoseiids are commercially available and are used against spider mites in greenhouses and such outdoor crops as strawberries (Huffaker and Kennett 1956; Overmeer 1985; De Klerk and Ramakers 1986). Indigenous phytoseiid faunas on crop plants can be rich in species. For example, surveys in South America identified over 40 phytoseiids associated with just one crop (cassava) (Bellotti et al. 1987). Kreiter (1991) records 36 species of phytoseiids in France. Basic understanding of the ecology of phytoseiid species is essential for their successful use, including an understanding of the seasonal ecology and

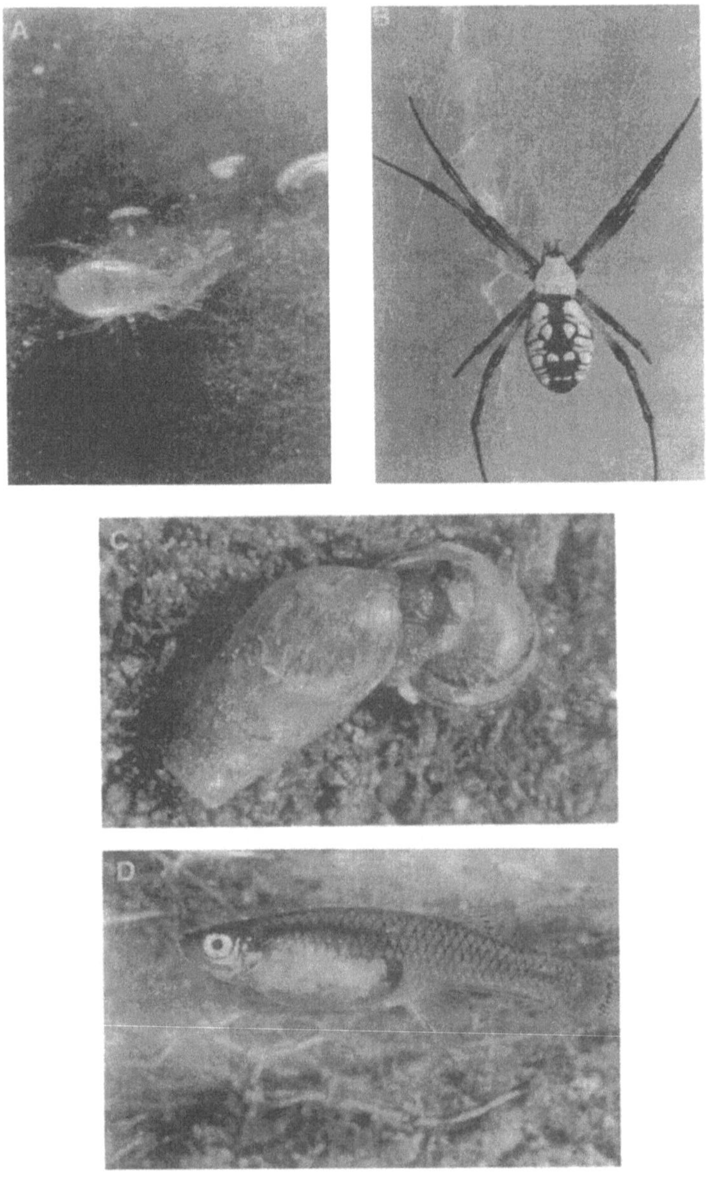

Figure 3.15. Predators in groups other than the Insecta play important roles in biological control, including; (A) predacious mites in the family Phytoseiidae (*Eusieus hibisci* [Chant]); (B) spiders, (C) predacious snails (*Rumina decollata* Risso); and (D) fish (*Gambusia affinis* [Baird and Girard]). (Photographs by M. Badgley [A,B,C] and courtesy of C. Pierson [D].)

movements on and off crops and surrounding vegetation, requirements for refuges to pass unfavorable seasons (Gilstrap 1988), and need for foods other than spider mites.

Stigmaeidae. These are predators of eryiophyid, tetranychid, and tenuipalpid mites. *Zetzellia mali* (Ewing) is an important species for pest mite control in apple orchards in some areas (Woolhouse and Harmsen 1984).

Anystidae. One species in this family has been used successfully in biological control through introduction. *Anystis salicinus* (Linnaeus) was introduced from southern France to Australia where it successfully reduced populations of the red-legged earth mite, *Halotydeus destructor* (Tucker) (Wallace 1981).

Bdellidae. *Bdella longicornis* (Linnaeus) and other species are important early season predators of spider mites in grapes in California (Sorensen et al. 1983). *Bdella lapidaria* Kramer has been shown by the insecticidal-check method (see Chapter 13) to suppress an important pest collembolan in Australia for whose control *B. lapidaria* was originally introduced (Wallace 1954).

Cheyletidae. *Cheyletus eruditus* (Schrank) has been suggested for biological control of pests of stored food products (Zdárková 1986).

Hemisarcoptidae. These are predators of armored scales, and some species are of demonstrated importance, such as *Hemisarcoptes malus* (Shimer), which is an important predator of oystershell scale, *Lepidosaphes ulmi* (Linnaeus), in apple in eastern North America (Pickett 1965) and which was introduced to British Columbia, Canada, where it apparently controlled this pest (LeRoux 1971). *Hemisarcoptes coccophagus* Meyer has been introduced into New Zealand, where it has contributed to the control of the diaspidid scale *Hemiberlesia latania* (Signoret) on kiwi fruit (*Actinidia deleciosa* Planchon) (Hill et al. 1993).

Laelapidae. In maize, populations of *Androlaelaps* sp. and *Stratiolaelaps* sp. caused 63% mortality of eggs of corn rootworms, *Diabrotica* spp., in soils amended with animal manure (Chiang 1970).

Macrochelidae. Species of *Macrocheles* appear to be important predators of fly eggs in manure (Axtell 1963).

Araneae (Spiders)

All spiders (Fig. 3.15b) are predacious, and a few are sufficiently poisonous to be dangerous to people. Spiders are grouped into 60 families, whose higher classification is uncertain. Eighteen different classification systems have been proposed since 1900. An introduction to spider morphology, ecology, and physiology is given by Foelix (1982). The relation of spiders to biological control is discussed by Riechert and Lockley (1984). The role of spiders in biological control is different from that of the hymenopteran parasitoids which have been used extensively through introduction to new locations to control specific immigrant pests. Most spiders lack host specificity, but do show habitat specificity. As such, they are poorly suited for introduction to new regions to control specific pests, but rather can best be employed by use of agriculture practices that conserve native spiders for suppression of groups of pests in crops. Reichert and Bishop (1990) document experimentally the impact of spiders on pests. Features of spider biology that have important influences on their action as biological control agents include the ability of many species to colonize new areas through ballooning as spiderlings, the relatively high numbers of spiders per unit area of land, and the responsiveness of their movements in and out of crops to local conditions of heat and moisture. Five of the more important families are discussed.

Agelenidae. These spiders construct sheet webs on vegetation with a funnel tube at one point (e.g., *Tegenaria* spp.).

Araneidae. These spiders build aerial webs, often circular in design with symmetrical weaving patterns (e.g., *Zygiella* spp.).

Lycosidae. These spiders do not build webs but rather hunt freely over the habitat (e.g., *Arctosa* spp., *Lycosa* spp.). They respond to the vibrations of potential prey. Lycosid spiders have been shown to suppress populations of rice brown planthopper, *Nilaparvata lugens*, and their conservation is critical to rice pest management in Asia (Heinrichs et al. 1982).

Thomisidae. These colorful spiders remain quietly in ambush and do not build webs or move about in search of prey (e.g., *Misumenoides* spp.). They often choose vegetation, especially flowers, as locations in which to await prey.

Salticidae. These spiders respond strongly to visual stimuli and leap at prey (e.g., *Heliophanus* spp.). They do not build webs and are most active on sunny days.

PREDACIOUS OR COMPETITOR SNAILS

Snails have been employed as agents of biological control to displace (through competition) snails that are intermediate hosts of schistosomiasis pathogens and (through predation) to suppress snails that are pests of agricultural crops or other plants (Fig. 3.15c). Aquatic snails, such as species of *Thiara*, *Marisa*, *Tarebia*, and *Melanoides*, that do not serve as intermediate hosts of the liver fluke that causes schistosomiasis in humans, have been introduced to compete with snails, such as species of *Biomphalaria*, that host the schistosome (Pointier and McCullough 1989). Results have been mixed, with introduced species replacing target species in some habitats, but not in others in some countries (as in Venezuela, Pointier et al. 1991), or being more generally successful in others (as in the Dominican Republic, Vargas et al. 1991). The predacious snail *Euglandia rosea* (Ferrusac) has been introduced into various countries (in some cases with negative consequences, see Chapter 8) for the control of the giant African land snail, *Achatina fulica*, a pest herbivore attacking crops (Laing and Hamai 1976). The facultative predator *Rumina decollata* Risso has been used with success against the brown garden snail, *Helix aspersa* Müller (Fisher and Orth 1985).

VERTEBRATES (MAMMALS, BIRDS, FISH)

Birds and predacious small mammals for many years were believed by some to be important forces suppressing populations of pests insects, especially in forests (Bruns 1960), but there are few experimental demonstrations of the effectiveness of terrestrial vertebrate predators for the control of specific pests (Bellows et al. 1982a; Campbell and Torgersen 1983; Torgersen et al. 1984; Atlegrim 1989). Terrestrial predators of interest have included a wide variety of insectivorous birds and small mammals such as mice and shrews. The wide dietary range of such vertebrates and the flexibility of their food-collecting behaviors make the introduction of vertebrate predators to new regions potentially more dangerous than the introduction of other taxa of biological control agents (Legner 1986a; Harris 1990). The principal method through which birds and mammals are used in biological control should, therefore, be conservation and enhancement of existing native species, rather than the introduction of new species. Zhi-Qiang Zhang (1992) summarizes the literature on birds as pest control agents in China.

Some aquatic vertebrates (various species of fish) have been used to control insects that breed in water, such as mosquitoes and chironomids. The species most widely used for control of mosquitoes have been two species of top-feeding minnows in the family Poecilidae, the mosquito fish (*Gambusia affinis* Baird and Girard) and the common guppy (*Poecilia reticulata* Peters) (Fig. 3.15d) (Bay et al. 1976). Use of these species has been successful in many cases (Legner et al. 1974; Bay et al. 1976) but ineffective in others (see Blaustein 1992). Some species of *Tilapia* (Cichlidae) have also been used to suppress mosquitoes by reducing plant biomass, thus rendering the habitat less favorable (Legner 1986a). Introductions of poecilids and other nonnative fish species may affect native fish through competition or hybridization and this possibility should be carefully considered before releasing an adventive fish into a new area (Arthington and Lloyd 1989; Courtenay and Meffe 1989). Use of native fish should be considered as an alternative to such introductions. A rating system describing species characteristics that influence the potential of fish as mosquito and weed control agents exists that can guide the selection of species (Ahmed et al. 1988). Native status should be assessed for regions, not countries, as moving fish between distinct zones within a country may have effects similar to those from introducing species from other countries.

CHAPTER

4

Pathogens and Nematodes of Arthropods and Pathogens of Vertebrates

PATHOGENS OF PEST ARTHROPODS

More than 1500 species of pathogens are known to attack arthropods (Miller et al. 1983). These include bacteria, viruses, fungi, protozoa, and nematodes. Historical developments in the recognition and understanding of pathogens of arthropods are summarized by Steinhaus (1956) and are briefly reviewed in Chapter 1. Detailed treatment of the taxonomy and biology of arthropod pathogens is given by Tanada and Kaya (1993) and, for nematodes, Kaya (1993). Much of the research on these pathogens is directed at development of methods for their culture, mass production, formulation, and use through augmentation as biological pesticides (see Chapter 11).

Bacterial Pathogens of Arthropods

The taxonomy of bacteria is outlined in Bergy's Manual of Determinative Bacteriology and Bergy's Manual of Systematic Bacteriology (Buchanan and Gibbons 1974; Holt 1984, 1986, 1989ab). For a description of the general life cycle of entomopathogenic bacteria and symptoms of bacterial disease in insects, see Chapter 16. While many species of bacteria that cause disease in arthropods are certainly still undiscovered, those studied to date can be divided into two broad groups. One group is the non-spore-forming species in such families as Pseudomonadaceae (*Pseudomonas*) and Enterobacteriaceae (e.g., *Aerobacter*, *Cloaca*, *Serratia*) that are found in soil and also occur in the gut of arthropods. These bacteria can become pathogenic, particularly in conjunction with other pathogens or when the host is physiologically stressed. Some of these organisms can infect mammals and so have not been pursued in most cases as commercial pest control agents against arthropods. The only exception to this is *Serratia entomophila* Grimmont, Jackson, Ageron, and Noonan which is marketed in New Zealand for control of the pasture pest scarab *Costelytra zealandica* (White) under the name Invadet (Jackson 1990). The second group are spore-forming bacteria in the Bacillaceae (*Bacillus*, *Clostridium*), of which most attention has been focused on several species of *Bacillus*, including *B. popilliae*, *B. sphaericus*, and *B. thuringiensis* (Cherwonogrodsky 1980; Lüthy 1986; Tanada and Kaya 1993).

Bacillus popilliae and related species are pathogens of scarab larvae and are of interest as potential agents against turf and pasture pests, such as the Japanese beetle, *Popillia japonica*. Infections by this pathogen are referred to as "milky spore disease" because of the milky color of the hemolymph of diseased hosts (Fig. 4.1). *Bacillus popilliae* does not produce spores (the stage used in pesticide formulations) abundantly *in vitro* (Lüthy 1986). The requirement for living insect hosts for spore production has made production of this pathogen costly and limited its commercial use. *Bacillus popillae* is difficult to separate from closely related species. Species determinations are based on the morphology of the sporulated cell and the host spectrum. A review of the taxonomy of this group of species is given by Splittstoesser and Kawanishi (1981).

Bacillus sphaericus occurs as different strains, some of which are pathogenic to insects. Singer (1990) provides a table of strains and a historical review of studies on *Bacillus sphaericus*, and the properties of this bacterium and its toxins have been reviewed by Baumann et al. (1991). Pathogenic strains have been isolated in a number of countries and are amenable to commercial production because the organism can be produced via fermentation. Insecticidal activity is due to crytalline toxins associated with the cell wall. These are released by digestion after the host insect has consumed bacterial spores. Bacterial reproduction occurs saprophytically after the host has died (Singer 1987). The host range of this bacterium is very narrow and appears to be limited to some genera of mosquitoes (Wraight et al. 1981; Singer 1987; see Osborne et al. 1990 for a review). The molecular biology of genes coding for the toxin has been explored by Baumann et al. (1987, 1988), and the gene has been expressed in *Bacillus subtilis* (Ehrenberg) Cohn (Baumann and Baumann 1989).

Bacillus thuringiensis is the most extensively marketed bacterial pathogen of arthropods (Fig. 4.2). Comprehensive information on this important species has been assembled by Entwistle et al. (1993), and the history of research on this species is summarized by Beegle and Yamamoto (1992). Watkinson (1994) and a series of companion articles synthesize current knowledge on many topics related to *B. thuringiensis*. Many isolates have been collected and classified into 34 serovars based on flagellar antigens by de Barjac and Frachon (1990). Most serovars of this species affect Lepidoptera and have been used to suppress a number of fruit, vegetable, and forest pests. More recently, serovars or isolates have been encountered that affect insects in other orders, expanding the range of targets for *B. thuringiensis* products. The serovar *israelensis* is effective against dipteran larvae, including mosquitoes, blackflies, sewage flies, and fungus gnats, among others (de Barjac 1978; van Essen and Hembree 1980; Mulla et al. 1982). An isolate, *tenebrionis* (serovar *morrisoni*), has been found that is effective against Coleoptera in the family Chrysomelidae and has been developed as a commercial control

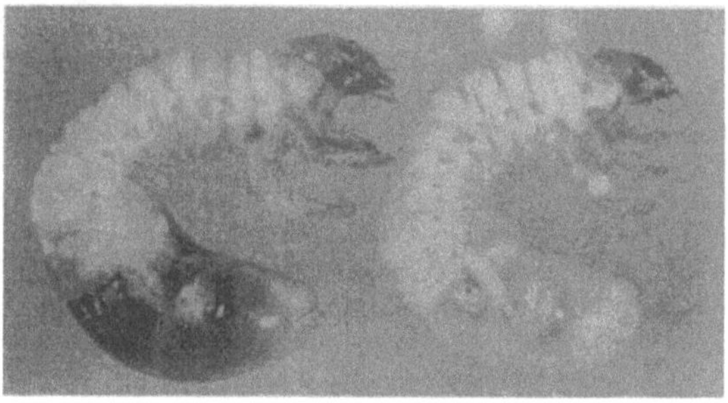

Figure 4.1. *Bacillus popilliae* Dutky infections in scarabaeid larvae are termed "milky spore diseases" because host hemolymph becomes cloudy, as seen here in the infected host on the right, compared to the healthy one on the left. (Photograph courtesy of R. Milner.)

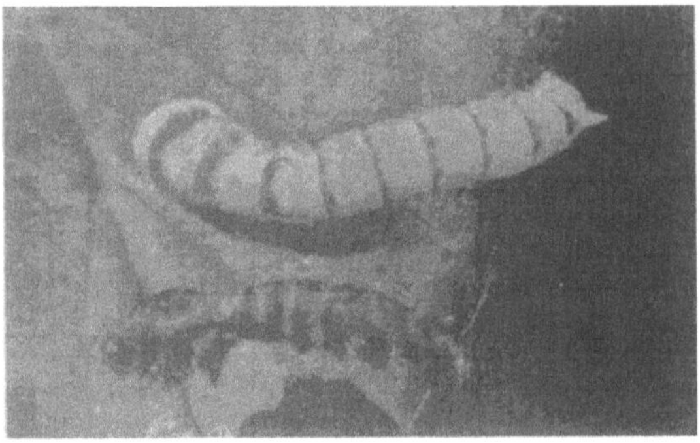

Figure 4.2. Larva (bottom) of the silkworm, *Bombyx mori* (Linnaeus), killed by *Bacillus thuringiensis* Berliner, a spore-forming bacterium that produces toxic proteins. (Photograph courtesy of H.K. Kaya.)

for the vegetable pest *Leptinotarsa decemlineata* and other species (Herrnstadt et al. 1987). Some isolates of *B. thuringiensis* are effective against nonarthropods, including plant-parasitic nematodes and snails (Feitelson et al. 1992; Osman et al. 1992). Several methods are available for detection of new strains of *B. thuringiensis*, including characterization of flagellar antigens, extrachromosomal plasmids, morphology of crystalline inclusion bodies, and gene sequences (Carlton et al. 1990).

Death of hosts which ingest *B. thuringiensis* cells results from the combined effects of poisoning by toxins and bacterial multiplication. Most commercial products of *B. thuringiensis* contain living spores, but some, such as the tablet formulation of the toxin from *B. thuringiensis israelensis*, do not (Becker et al. 1991).

Genes for *B. thuringiensis* toxins have been isolated and incorporated into other bacteria for the commercial production of toxins and also into crop plants, causing the plants to produce toxins at levels adequate to provide pest protection (Vaeck et al. 1987). Populations of pests resistant to the toxins of *B. thuringiensis* have developed in some species of insects subjected to frequent applications of the material (Tabashnik et al. 1990).

Viral Pathogens of Arthropods

More than 1270 insect-virus associations had been recognized by 1981 (Martignoni and Iwai 1981), most (70%) involving Lepidoptera as hosts. See Chapter 16 for a description of a generalized virus infection and its symptoms. Viruses pathogenic in arthropods may be classified based on the nature of their molecular form (single- or double-stranded DNA or RNA) and other characteristics into 16 families: Baculoviridae, Polydnaviridae, Poxviridae, Ascoviridae, Iridoviridae (all five families of which are double-stranded DNA forms), Parvoviridae (a single-stranded DNA form), Reoviridae, Nodaviridae, Picornaviridae, Tetraviridae, Birnaviridae, Rhabdoviridae, Caliciviridae, Togaviridae, Bunyaviridae, and Flaviviridae (all of which are RNA viruses, all single-stranded forms except the Reoviridae and Birnaviridae) (Entwistle 1983; Moore et al. 1987; Tanada and Kaya 1993). Moore et al. (1987) and Tanada and Kaya (1993) provide descriptions of many of these families. Viral nomenclature is different from that of zoological and botanical nomenclature and is set forth in the Fifth Report of the International Committee on Taxonomy of Viruses (Francki et al. 1991). This system has no hierarchical levels above family and does not imply phylogenetic relationships. Phylogenetic relationships between selected baculoviruses have been investigated using DNA sequencing (Zanotto et al. 1993).

The Baculoviridae and Tetraviridae are limited in their host ranges to arthropods. Most other families contain viruses associated with mammals or other nonarthropod groups and thus are of less interest as potential insect control agents because of concerns over possible effects on mammals. All viruses are obligate parasites of living cells. Virus production, therefore, can be achieved only in living hosts or cell cultures.

Baculoviridae have been of greatest interest as potential biological control agents. This family includes the nuclear polyhedrosis viruses (Fig. 4.3ab), the granulosis viruses (Fig. 4.3cd), and, formerly, the nonoccluded viruses (e.g., *Oryctes* virus, now unclassified at the family level) (Entwistle 1983; Tanada and Kaya 1993). Bilimoria (1986) discusses the taxonomy and identification of members of the Baculoviridae. The biology of baculoviruses is discussed by Granados and Federici (1986). Blissard and Rohrmann (1990) present information on the molecular aspects of the baculovirus infection cycle and the organization of the baculovirus genome. Fuxa (1990a) lists 28 cases in which baculoviruses have been used in attempts to suppress insect pests. Of these, 15 cases involve introduction of a virus followed by its permanent establishment and 13 cases were based on augmentative use. Most of these cases have involved nuclear polyhedrosis viruses, four being granulosis viruses and one a nonoccluded virus. Some of the viruses that have been employed successfully for biological control are: a nuclear polyhedrosis virus of *Gilpinia hercyniae* (Hartig), a forest sawfly in Canada which was permanently suppressed after the accidental introduction of the virus (Balch and Bird 1944); the

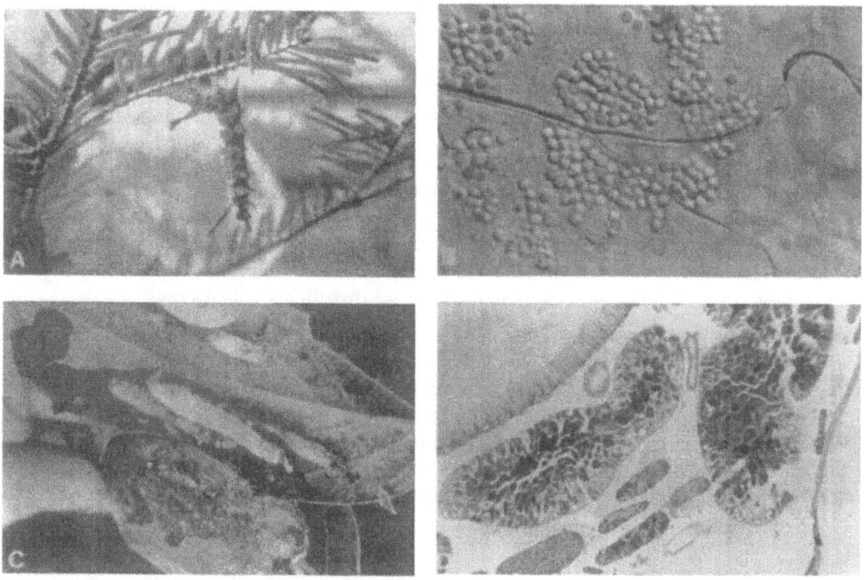

Figure 4.3. Important members of the family Baculoviridae include the nuclear polyhedrosis and the granulosis viruses: (A) larva of Douglas-fir tussock moth, *Orgyia pseudotsugata* (McDunnough), killed by nuclear polyhedrosis virus; (B) micrograph of nuclear polyhedrosis virus in hypodermis tissue of beet armyworm, *Spodoptera exigua* (Hübner); (C) larva of a geometrid (*Sabulodes* sp.) killed by granulosis virus; (D) micrograph of granulosis virus in fat body tissue of fall armyworm, *Spodoptera frugiperda* (J.E. Smith). (Photographs courtesy of C.G. Thompson [A], J.V. Maddox [B], B. Federici [C], and J.J. Hamm [D].)

nonoccluded virus of *Oryctes rhinoceros*, a coleopteran pest of coconut palms which was suppressed for up to 3.5 years following deliberate introduction of the virus (Zelazny et al. 1990; Mohan and Pillai 1993); and the nuclear polyhedrosis virus of *Anticarsia gemmatalis* Hübner, a lepidopteran defoliator of soybeans (*Glycine max* [Linnaeus] Merrill) in Brazil, which is managed by augmentative applications of virus (Moscardi 1983).

Another family of viruses that is important to biological control is the Polydnaviridae, which are mutualistic associates of parasitoids in the Ichneumonoidea (Fleming and Fleming 1992). These viruses replicate only in the ovarial calyx epithelium of their wasp hosts, but cause no noticeable pathology. They are injected into the wasp's lepidopteran hosts during oviposition by the wasp. They are important to successful parasitism because they help suppress the defensive response of the host by depressing its immune system (see Chapter 15; see also Fleming and Fleming 1992; Tanada and Kaya 1993).

Fungal Pathogens of Arthropods

The fungi have generally been regarded as comprising the Division Eumycota within the plant kingdom (Ainsworth 1971; Ainsworth et al. 1973). In Chapter 16 a generalized fungal life cycle and disease symptoms are described. Entomopathogenic species are found in five subdivisions: Mastigomycotina, Zygomycotina, Ascomycotina, Basidiomycotina, and Deuteromycotina. Brady (1981) and McCoy et al. (1988) provide overviews of the taxonomy of fungal groups known to attack arthropods. Fungal biology, pathology, and use for pest control is covered by Steinhaus (1963), Müller-Kögler (1965), Ferron (1978), Burges (1981a), McCoy et al. (1988), and Tanada and Kaya (1993). Burge (1988) covers a broad range of topics related to biological control uses of fungi, both through introduction and augmentation. The role of naturally-occurring entomopathogenic fungi in the population dynamics of insects is reviewed by Carruthers and Hural (1990).

While over 400 species of entomopathogenic fungi have been recognized (Hall and Papierok 1982), most attention with respect to potential as biological control agents has been focused on about 20 species (Zimmermann 1986). Most of these are included in 12 genera: *Lagenidium, Entomophaga, Neozygites, Entomophthora, Erynia, Aschersonia, Verticillium, Nomuraea, Hirsutella, Metarhizium, Beauveria*, and *Paecilomyces* (Roberts and Wraight 1986). Some important entomophagous genera in these subdivisions are as follows:

Mastigomycotina. This subdivision includes two classes with entomopathogenic species. In the Chytridiomycetes are found *Coelomomyces* spp. which attack mosquito larvae. Members of this genus have complicated life cycles with a required alternate copepod or ostracod host. Species in this group have not been used commercially for insect control. In the Oomycetes is the genus *Lagenidium*, whose members are also pathogens of mosquito larvae but which require only one host, the mosquito. *Lagenidium giganteum* Couch (Fig. 4.4) is registered as a pest control product in the United States.

Zygomycota. The family Entomophthoraceae includes important insect pathogens, such as species of *Entomophthora* (Fig. 4.5), *Entomophaga, Erynia*, and *Neozygites*. Information on the taxonomy and biology of this family is found in MacLeod (1963), Waterhouse (1973), Remaudière and Keller (1980), Humber (1981), Ben-Ze'ev et al. (1981), and Wolf (1988). The taxonomic organization of the group has undergone extensive restructuring since the 1960s, which is reviewed by McCoy et al. (1988) and Humber (1990). Hosts of Entomophthoraceae include various Lepidoptera, Coleoptera, aphids, and mites. Spore production for many of the

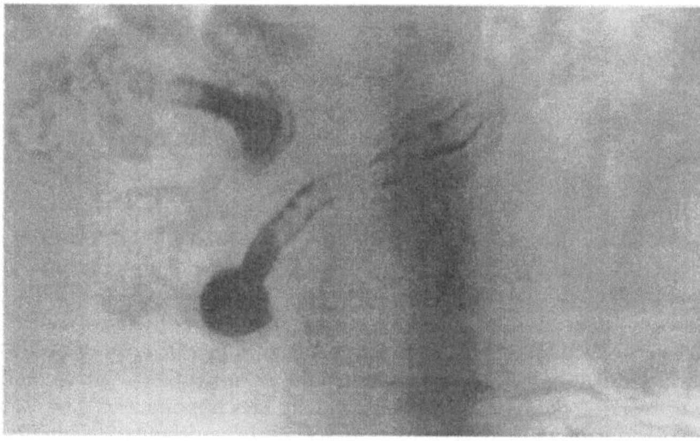

Figure 4.4. Encysted zoospore of *Lagenideum giganteum* Couch with a germ tube penetrating its host, a larva of the culicid *Culex pipiens* Linnaeus. (Photograph courtesy of A.J. Domnas.)

species in these groups has not been obtained independent of living hosts, a factor limiting their mass production and their use as agents of augmentative biological control. Spore production on nonliving media, however, has been achieved for some species such as *Zoophthora radicans*, a species which has been considered for use as a mycoinsecticide (Soper 1985). Recent advances in the use of this group are discussed by Wilding (1990).

Ascomycotina. This subdivision includes *Ascosphaera apis* (Maasen *ex* Claussen) Olive and Spiltoir, the causative agent of chalk brood disease of honey bees (*Apis melifera* Linnaeus). It also includes species of *Cordyceps* that are well-known historically because of the showy nature of the fruiting bodies that grow from infected hosts. No member of this group is of major importance as a pest control agent.

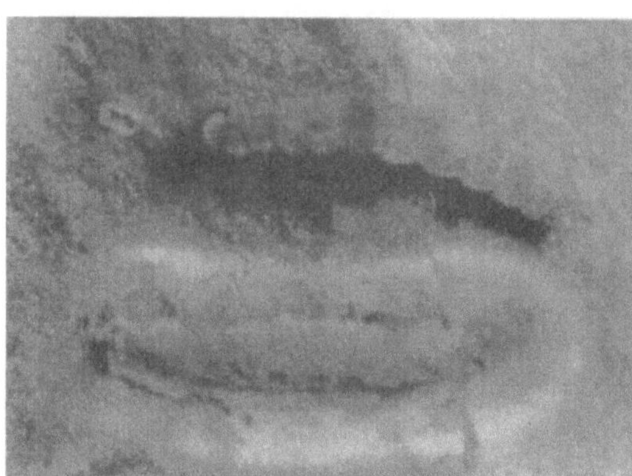

Figure 4.5. Diamond-back moth (*Plutella xylostella* [Linnaeus]) larva killed by *Entomophthora* sp., showing "halo" of conidia. (Photograph courtesy of C.W. McCoy.)

Basidiomycotina. The entomophagous forms within this subdivision are principally members of the genus *Septobasidium*, which are associated with various species of scale insects. Scale-fungus relations are often symbiotic, but verge into parasitism in some species. Entomopathogenic species are also found in the genera *Filobasidiella* and *Uredinella*.

Deuteromycotina. This group (also called the Imperfect Fungi) is comprised of species known only from asexual forms. It is a heterogenous group, the true placement of whose members can only be determined if sexual forms are eventually encountered and correctly associated with the asexual form. The majority of genera of entomopathogenic fungi of importance to pest control are found in this group. These include the genera: *Hirsutella*, in which *H. thompsonii* Fisher is a well-studied pathogen of eriophyid rust mites (McCoy 1981); *Beauveria*, a genus with a wide host range, which includes *B. bassiana* (de Hoog 1972), a species with important biological control potential; *Metarhizium*, a group with a wide host range; *Nomuraea*; *Paecilomyces*, members of which attack important species of soil insects and an important species of whitefly; *Verticillium*, species of which attack whiteflies and aphids; and *Aschersonia*, some of which attack whiteflies and scales (Fig. 4.6). Information on the biologies of many of these fungal pathogens can be found in Ferron (1978), McCoy et al. (1988), and Carruthers and Hural (1990). Information on the commercial development of such fungi as pest control products is given by Roberts and Wraight (1986) and, with special reference to pests in tropical areas, by Maniania (1991).

Protozoan Pathogens of Arthropods

The higher classification of protozoa has changed dramatically in recent years. In 1980, protozoa were raised to a subkingdom, with seven phyla. More recently, Margulis et al. (1990) have combined the protozoa, together with other groups, into the kingdom Protoctista. Entomopathogenic protozoa occur in six phyla: Zoomastigina (flagellates), Rhizopoda (amoebas), Apicomplexa (gregarines, eugregarines, neogregarines, and coccidia), Microspora (micro-

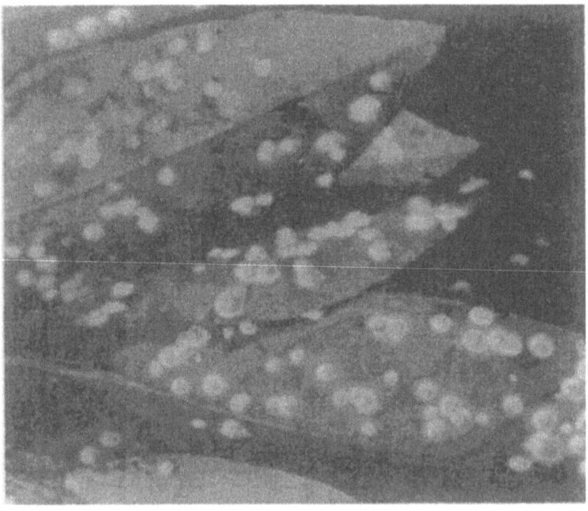

Figure 4.6. Citrus whitefly (*Dialeurodes citri* [Ashmead]) killed by the fungus *Aschersonia aleyrodis* Webber. (Photograph courtesy of C.W. McCoy.)

sporidia), Haplosporidia, and Ciliophora (ciliates). For a general description of the life cycle of entomopathogenic protozoa see Chapter 16. For detailed treatments of the biologies of individual groups see Tanada and Kaya (1993). Of these, most forms of interest as biological control agents are microsporidia.

Zoomastigina (Flagellates). The entomopathogenic trypanosomatids are found principally in the genera *Herpetomonas*, *Crithidia*, and *Blastocrithidia*. They occur most frequently in Diptera and Hemiptera and most often infect the Malpighian tubules or the gut. They have not been employed for biological control purposes. Some typanosomes in insects also cause diseases in vertebrates (*Leishmania* spp., *Trypanosoma* spp.).

Rhizopoda (Amoebas). The best known example in this group is *Malamoeba locustae* (King and Taylor) which infects the Malpighian tubules and midgut epithelium of grasshoppers. Infected hosts have lowered reproductive potentials (Henry 1990).

Apicomplexa (Gregarines, Eugregarines, Neogregarines, and Coccidia). An example of this group is the neogregarine *Mattesia trogodermae* Canning which infects the fat body of various stored-product beetles such as *Trogoderma* spp. Infected hosts are sluggish and survive for shorter periods. Brooks and Jackson (1990) provide a review of the eugregarines.

Microspora (Microsporidia). Microsporidia of potential importance to biological control include species of *Nosema* (Fig. 4.7), *Pleistophora*, and *Vairimorpha*. Larsson (1988) provides a guide to the identification of microsporidian genera. Microsporidia, formulated in baits, have been tested for control of locusts and *Ostrinia nubilalis*, a lepidopteran pest of maize (Henry 1990). When present in species such as coccinellids or herbivorous arthropods being imported for weed control, microsporidia and other protozoans constitute a potentially serious contaminate reducing the fitness of the beneficial insects being introduced (Kluge and Caldwell 1992).

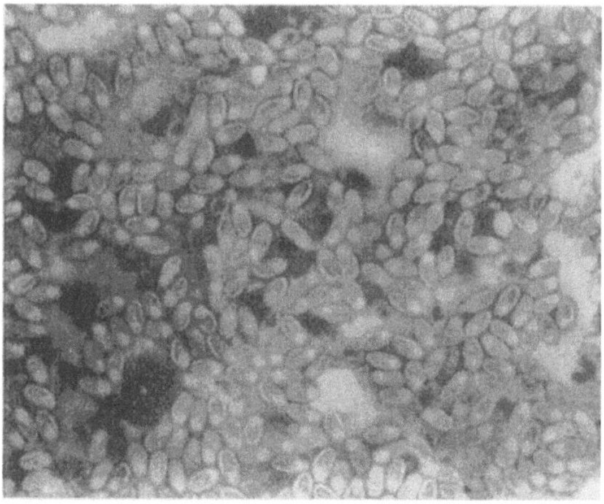

Figure 4.7. Spores of *Nosema* sp. in midgut of the cabbage looper, *Trichoplusia ni* (Hübner). (Photograph courtesy of G.L. Nordin.)

Haplosporidia. Members of this group infect the digestive tract, fat body, oenocytoids, and Malpighian tubules of insects.

Ciliophora (Ciliates). A few species of Ciliophora, such as *Lambornella clarki* Corliss and Coats, are pathogens of mosquito larvae.

Nematodes Attacking Arthropods

The classification of entomopathogenic nematodes at the family level has changed significantly in recent years. Recent treatments are given by Maggenti (1991) and Remillet and Laumond (1991). Reviews of nematodes associated with arthropods are provided by Poinar (1986), Kaya (1993), Kaya and Gaugler (1993), and Tanada and Kaya (1993). Of the 30 or more families of nematodes associated with insects, only nine have members with potential as biological control agents: Tetradonematidae, Mermithidae, Steinernematidae, Heterorhabditidae, Phaenopsitylenchidae, Iotonchiidae, Allantonematidae, Parasitylenchidae, and Sphaerulariidae. Most attention has been focused on two families, the Steinernematidae and Heterorhabditidae. These are associated with pathogenic symbiotic bacteria that enable them to rapidly kill a wide range of hosts. Members of five other families (Tetradonematidae, Mermithidae, Phaenopsitylenchidae, Iotonchiidae, and Sphaerulariidae) also merit discussion. For comments on the life cycles of entomopathogenic nematodes and symptoms of infection, see Chapter 16.

Steinernematidae and Heterorhabditidae. Several species in the genera *Steinernema* and *Heterorhabditis* have been the focus of most efforts to develop commercial uses of nematodes (Fig. 4.8ab) (Gaugler and Kaya 1990; Kaya 1993; Kaya and Gaugler 1993; Tanada and Kaya 1993). Synonomies of species names in these genera complicate the interpretation of literature citations. Ten species are recognized in *Steinernema* (Doucet and Doucet 1990) and three in *Heterorhabditis* (Poinar 1990). Kaya and Gaugler (1993) list these and discuss their nomenclature, and Smith et al. (1992) give a bibliography of research on these families.

These families have been used as commercial pest control agents because they have the following attributes (Poinar 1986):

- a wide host range
- an ability to kill the host within 48 hours
- a capacity for growth on artificial media
- a durable infective stage capable of being stored
- a lack of host resistance
- apparent safety to the environment

These nematodes invade hosts through natural openings (mouth, spiracles, anus) or wounds and penetrate into the hemocoel. Bacteria in the genera *Xenorhabdus* or *Photorhabdus*, symbiotic to the nematode but pathogenic to the host, are released into the hemocoel and quickly kill the host. The nematodes then develop saprophytically on the decomposing host tissues. Gaugler and Kaya (1990) and Kaya and Gaugler (1993) provide information on rearing and using these groups of nematodes for pest control. These nematodes work best in moist environments such as soil. Commercial markets for some strains have been established and large scale production systems developed (Kaya 1985; Gaugler and Kaya 1990).

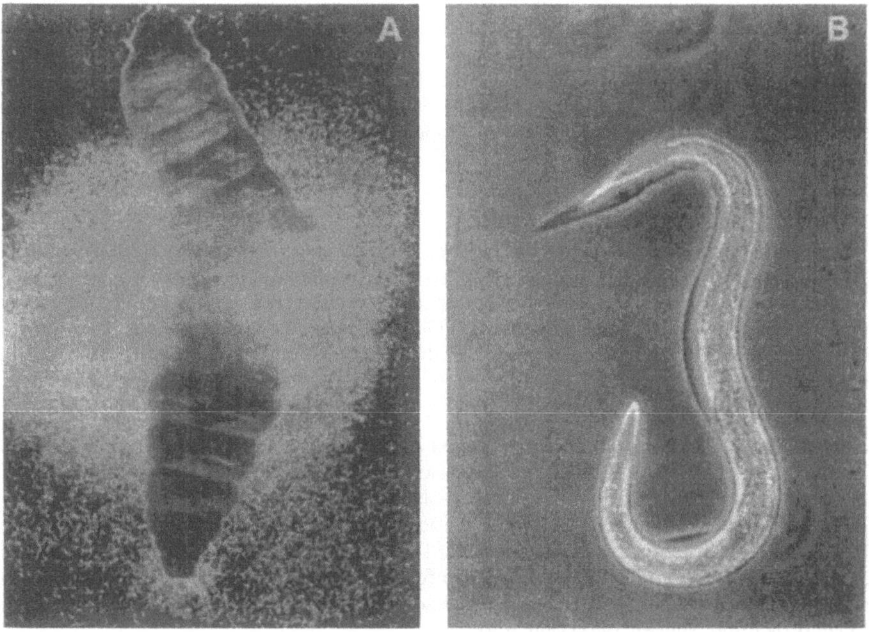

Figure 4.8. (A) Infected host larva (*Galleria mellonella* [Linnaeus]) opened to show juveniles of *Steinernema* sp. nematodes; (B) single infective-stage juvenile nematode. (Photographs courtesy of R. Gaugler.)

In addition to the augmentative use of nematodes in these two families, some species with more narrow host ranges have been imported for the control of immigrant pests; for example, *Steinernema scapterisci* was introduced into Florida in a program to provide biological control of immigrant species of *Scapteriscus* mole crickets (Parkman et al. 1993).

Tetradonematidae. *Tetradonema plicans* is a tetradonematid that infects larvae of *Sciara* spp. (Diptera: Sciaridae), which are pests in glasshouse and mushroom crops. Effective control of these root gnats was obtained by Peloquin and Platzer (1993) with applications of *T. plicans* eggs at a ratio of 10 eggs per host larva. Cultures of the nematode can be maintained on hosts reared on a composted media of sphagnum moss, shredded paper, and commercial rabbit food. Host larvae ingest eggs or, perhaps, young larvae. Nematode larvae then penetrate the gut and enter the hemocoel, where they mature and cause the death of the host. Adult nematodes mate in the host and females either exit and lay eggs outside the host or remain in the host, in which cases eggs are released into the soil as the host and female nematode decay.

Mermithidae. The nematode *Romanomermis culicivorax* Ross and Smith attacks larvae of various mosquitoes. The life cycle is divided between stages that occur within the host and others that occur outside of the host. Infective nematode larvae find hosts, penetrate the integument, and enter the host's hemocoel. Larvae partially mature inside hosts, penetrate through the host integument, and emerge into the environment as fourth stage larvae. They later transform into adults, mate, and lay eggs, which hatch and produce infective juveniles to resume the cycle. This nematode was developed as a commercial product under the name Skeeter Doomt, but the product failed commercially because of storage and transportation problems (Poinar 1979).

Phaenopsitylenchidae. The nematode *Beddingia siricidicola* was introduced from New Zealand into Australia where it contributed substantially to the suppression of a major pest of conifer plantations, the European wood wasp *Sirex noctilio* (Bedding 1984). The nematode invades the ovaries of the adult wood wasp, killing the eggs. The wasp, however, continues to oviposit, with the result that nematodes rather than eggs are deposited in new trees, spreading the nematode.

Iotonchiidae. The iotonchiid *Paraiotonchium autumnale* (Nickle), a native of Europe, was discovered in New York in 1964 attacking the face fly, *Musca autumnalis* De Geer. The nematodes mate in dung and enter fly larvae. Nematodes develop through two generations but do not kill the host, which develops to the adult stage. At this point nematodes are present in the fly's ovaries, reducing egg production. The adult fly, instead of laying eggs, deposits nematodes. A laboratory production method for this species was developed by Stoffolano (1973). A species of iotonchiid attacking *Musca domestica* has recently been recorded from Brazil (Coler and Nguyen 1994) (Fig. 4.9).

Sphaerulariidae. The nematode *Tripius sciarae* (Boiven) attacks sciarid and mycetophilid flies in Europe and North America. The nematode infects its host in the larval stage but does not

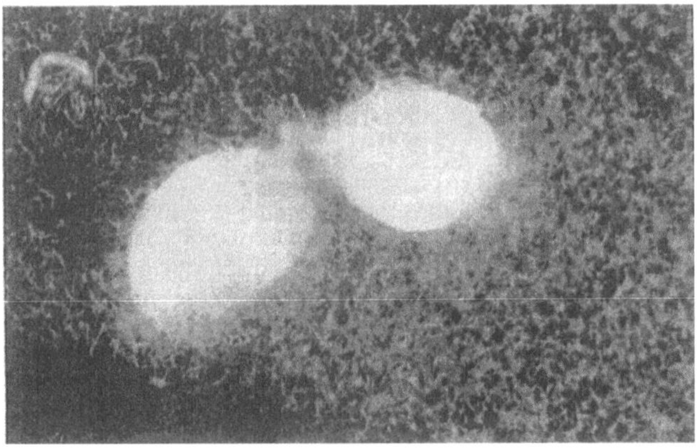

Figure 4.9. Thousands of the iotonchid nematode *Paraiotonchium muscadomesticae* Coler and Nguyen exiting from infected ovaries of its dipteran host, *Musca domestica* Linnaeus. (Photograph courtesy of R. Coler.)

immediately kill it. Some infected hosts develop to the adult stage, but their reproductive systems are destroyed by nematodes; nematodes are spread by oviposition attempts of infected flies. The nematode is easily reared and capable of controlling fungus gnat infestations in glasshouses (Poinar 1965).

PATHOGENS OF PEST VERTEBRATES

Vertebrates have been the targets of biological control efforts in relatively few cases. In those efforts that have been made, pathogens have frequently been chosen. Compared with other taxa, relatively few cases of biological control of vertebrates have been attempted. Several important projects have involved viruses. The European rabbit, *Oryctolagus cuniculus* (Linnaeus), was controlled in Australia through the introduction of a myxoma virus from *Sylvilagus* rabbit species from South America (Fenner and Marshall 1957; Ross and Tittensor 1986). Populations of domestic cats, *Felis cattus* Linnaeus, preying on seabird colonies on oceanic islands have been reduced through the introduction of feline panleucopaenia virus (van Rensburg et al. 1987). A liver nematode, *Capillaria hepatica* (Bancroft), is currently being considered as a means to counter house mice (*Mus domesticus* Linnaeus) in Australia (Singleton and Redhead 1990). A venereal disease of feral domestic goats (*Capra hercus* Linnaeus), caused by the protozoan *Trichomonas foetus* Donné, may offer a means to alleviate the destruction of native vegetation of uninhabited oceanic islands caused by introduced goats (Dobson 1988).

CHAPTER

5

Herbivores and Pathogens Used for Biological Weed Control

INTRODUCTION

Biological weed control has been achieved through two routes: introduction of natural enemies against adventive and native weeds (usually using agents collected from an adventive weed's native range), and augmentation of natural enemies which are released or applied at specific locations where control is needed (Wapshere et al. 1989). Introduction of natural enemies has employed mostly insects, a few other arthropods, some species of fungi, and one nematode. Biological control of weeds through the augmentation of natural enemies has employed almost exclusively various pathogenic fungi and generalist vertebrate herbivores, almost entirely fish. Potential may exist to exploit additional categories of plant pathogens. Fish have been employed for suppression of whole aquatic weed communities in irrigation canals, tanks, ponds, and lakes. This chapter reviews the taxonomic groups of herbivores and pathogens which have been employed in weed biological control and discusses basic elements of their natural history. Chapter 17 discusses key aspects of herbivore biology that are pertinent to weed biological control.

Herbivory is common among invertebrates (Strong et al. 1984; Borror et al. 1989). Rather than discuss all herbivorous groups, the focus in this chapter is on those that have been utilized for biological control. While this focus helps to narrow the discussion, it is also important to bear in mind that a summary of the historical record of weed biological control (Julien 1992) is a description of past events, not the results of an experiment designed to evaluate the efficacy of these various groups of herbivores. Many factors affect the degree of success that has been attained from the use of particular natural enemies. These factors include the biological properties of the organisms, the nature of the weed targets against which they were deployed, and the amount of effort that went into specific projects. Consequently the historical record, while instructive, should be used to compare the effectiveness of agents in different groups only in a very general sense. An open mind should be retained, as agents in groups with little previous success may be important in future projects, especially as plants in new taxonomic groups become of concern as pests. A listing of plant families against which biological control efforts have been made is given in Table 2.1.

Biological control of plants differs in some important ways from biological control of invertebrates. Predation and parasitism of invertebrates are direct causes of mortality of the individuals attacked, and the action of natural enemies is often couched in terms of the amount of this mortality caused to a population of invertebrate pests. Biological control of plants, in contrast, can be achieved by a variety of mechanisms which may or may not include directly-caused mortality of the target plants. Many flowering herbaceous plants, for example, have limited life spans, often a single season or one or two years. Biological control of such plant populations can be achieved if individual plants are subject to sufficient herbivore attack to limit reproduction below the level which permits replacement of the population. In this context, biological control might be successfully achieved by foliage feeders which reduce the productive plant biomass, and thereby render the plant unable to flower or reproduce; by flower feeders which destroy flowers before they can set seeds; and by seed feeders which destroy the seeds themselves. Plants which are prevented from successfully reproducing and then die naturally in the course of a season have been as effectively eliminated as those which are killed outright by herbivore attack. Other plants, such as those which reproduce vegetatively, might require natural enemies which inflict sufficient damage to cause plant death. The control of populations of woody species may take somewhat longer than herbaceous species, if only by virtue of the greater longevity of individuals, but in many cases repeated herbivore attack on woody species can result in premature death of the plant.

Because biological control of plants does not require that death of a plant be the direct result of herbivory, there are various mechanisms by which successful agents affect their target plants. These include direct feeding on foliage, formation of galls, burrowing in plant tissues, and causing diseases of plant tissues. Herbivores found on most plants in natural settings are not usually considered to be at sufficient densities to broadly suppress plant populations; indeed the populations of many herbivore species appear to be limited by natural enemies themselves (Strong et al. 1984), rather than limiting the growth of their host plant populations. When used in biological control, herbivores are often introduced into settings where they have fewer natural enemies than in their natural settings, thus freeing the herbivore population from the controlling influence of these natural enemies to the point where it can suppress the target weed.

INSECTS AND MITES

Julien (1992) provides a world catalogue of the invertebrate herbivore species that have been imported and released in various countries for control of adventive plants. Approximately 259 species of invertebrates have been employed, including 254 insects and 5 mites. Releases of these species have resulted in establishment at one or more locations for 161 (62%) of the species used. Control (at least partial, including cases in which the agent was part of the complex that gave control) of the target weed was achieved in at least one location by 65 species (25% of all species released). The numbers of families and species employed in various insect orders are given in Tables 5.1 and 5.2. These tables only describe what occurred historically, and should not be interpreted as providing rigorous comparisons of the relative probabilities of successful establishment or control among these natural enemy groups. One final caution concerning the historical record is that many of the species whose releases are catalogued in Julien (1992) and which were not listed as having controlled their target pests were released only recently and are still under evaluation. Furthermore, which groups of herbivores become important for biological weed control depends in large part on which plant species become pests. For these reasons, the number of families or species recognized as important may increase with time.

Of 65 species of herbivorous insects that have provided control of their target weeds, 40 are included in six families of insects: three families of Coleoptera, two of Lepidoptera, and one of

Table 5.1 A Summary of the Historical Record of Movement of Groups of Invertebrates between Countries for the Control of Adventive Weeds, Taken from Julien (1992), by Orders

Order	No. Families Employed	No. Species Employed	Establishments (no. species)	Successes (no. species)
Coleoptera	7	109	66 (61%)	33 (30%)
Lepidoptera	21	82	46 (56%)	15 (18%)
Homoptera and Hemiptera	8	19	15 (79%)	8 (42%)
Diptera	6	35	25 (71%)	4 (11%)
Thysanoptera	2	4	2 (50%)	1 (25%)
Hymenoptera	3	4	3 (75%)	2 (50%)
Orthoptera	1	1	1 (100%)	0 (0%)
Acarina	3	5	3 (60%)	2 (40%)
Fungi	4	8	8 (100%)	5 (63%)
Nematodes	1	1	1 (100%)	0 (0%)
Overall	56	268	170 (63%)	70 (26%)

Homoptera. These include Chrysomelidae (12 species), Curculionidae (14), Cerambycidae (4), Pyralidae (3), Arctiidae (3), and Dactylopiidae (4). Relatively few generalities about the biology of these groups are possible at the family level because within most families of insects diverse biologies are exhibited by member species. Broader descriptions of herbivorous insect families are given by Arnett (1968), Borror et al. (1989), and CSIRO (1970), and insect-plant interactions are discussed by Strong et al. (1984).

Coleoptera (Beetles)

Chrysomelidae. Members of this family feed on foliage and flowers as adults. Larvae feed on foliage, are leafminers, or are stem or root borers. Most species pass the winter as adults and many are brightly colored. Thirty-nine species of chrysomelids have been used for biological

Table 5.2 A Summary of the Historical Record of Movement of Groups of Invertebrates between Countries for the Control of Adventive Weeds, Taken from Julien (1992), by Families

Order	No. Species Employed	Establishments (no. species)	Success (no. species)
Coleoptera	109	66 (61%)	33 (30%)
Chrysomelidae	39	21	12
Curculionidae	36	24	14
Cerambycidae	14	9	4
Apionidae	9	3	1
Bruchidae	7	6	1
Buprestidae	3	3	1
Anthribidae	1	0	0
Lepidoptera	82	46 (56%)	15 (18%)
Pyralidae	23	11	3
Noctuidae	10	4	1
Tortricidae	9	5	2
Gelechiidae	5	3	0
Arctiidae	4	3	3
Gracillariidae	4	3	1

Table 5.2 (*Continued*)

Order	No. Species Employed	Establishments (no. species)	Success (no. species)
Lepidoptera (*continued*)			
Pterophoridae	3	3	2
Lyonetidae	3	2	1
Cochylidae	3	2	0
Geometridae	3	2	0
Sesiidae	3	1	0
Aegeriidae	2	0	0
Coleophoridae	2	2	0
Oecophoridae	1	1	0
Pterolonchidae	1	0	0
Carposinidae	1	1	1
Heliodinidae	1	1	1
Sphingidae	1	1	0
Lycaenidae	1	1	0
Dioptidae	1	0	0
Hepialidae	1	0	0
Homoptera and Hemiptera	19	15 (79%)	8 (42%)
Dactylopiidae	6	5	4
Tingidae	5	3	1
Coreidae	3	2	1
Pseudococcidae	1	1	1
Miridae	1	1	0
Aphidae	1	1	0
Psyllidae	1	1	1
Delphacidae	1	1	0
Diptera	35	25 (71%)	4 (11%)
Tephritidae	17	11	2
Cecidomyiidae	6	5	1
Agromyzidae	5	4	1
Anthomyiidae	4	3	0
Ephydridae	2	2	0
Syrphidae	1	0	0
Thysanoptera	4	2 (50%)	1 (25%)
Phlaeothripidae	3	2	1
Thripidae	1	0	0
Hymenoptera	4	3 (75%)	2 (50%)
Tenthredinidae	2	1	0
Eurytomidae	1	1	1
Pteromalidae	1	1	1
Orthoptera	1	1 (100%)	0 (0%)
Pauliniidae	1	1	0
Acarina	5	3 (60%)	2 (40%)
Eriophyidae	3	2	1
Tetranychidae	1	1	1
Galumnidae	1	0	0
Fungi	8	8 (100%)	5 (63%)
Uredinales	5	5	3
Hyphomycetes	1	1	0
Ustilaginales	1	1	1
Coelomycetes	1	1	1
Nematodes	1	1 (100%)	0 (0%)
Anguinidae	1	1	0
Overall 56	268	170 (63%)	70 (26%)

control, and 12 have successfully controlled their target pests in at least one location. *Agasicles hygrophila* Selman and Vogt has been used successfully in Australia and other countries for aquatic forms of alligator weed, *Alternanthera philoxeroides* (Martius) Grisebach (Fig. 5.1, 5.2) (Julien 1981), an adventive species native to South America. Larvae and adults of this species feed on foliage of the weed. The species is multivoltine with as many as five generations per year in Argentina, and adults lay an average of more than 1100 eggs during their 7-week adult lifespan (Clausen 1978). The terrestrial weed *Senecio jacobaea* Linnaeus was controlled by two herbivores, one of which was the chrysomelid *Longitarsus jacobaeae* (Waterhouse) (Fig. 5.3).

Curculionidae. Members of this family are nearly all plant feeders. Larvae feed either externally on foliage or feed internally in fruits or stems. Adults are characterized by an elongated snout with which they drill holes in plant tissues, including fruits and nuts. Thirty-six species of curculionids have been used in biological control projects. Of these, 14 species have provided control of their target species, including *Rhinocyllus conicus* (Frölich), which controlled nodding thistle, *Carduus nutans* Linnaeus, in Canada (Harris 1984); *Neohydronomus affinis* Hustache, which controlled water lettuce (*Pistia stratiotes* Linnaeus) in several locations including Florida (Dray et al. 1990); and *Microlarinus lypriformis* (Wollaston) (Fig. 5.4) which together with *Microlarinus lareynii* (Jacquelin du Val) partially controlled puncture vine (*Tribulus terrestris* Linnaeus) in the southwestern United States and Hawaii (Huffaker et al. 1983).

Cerambycidae and Buprestidae. Larvae of these families typically bore in the stems of woody plants. Adults often feed on flowers. The Cerambycidae usually have long antennae, deeply notched eyes and bright colors. Buprestid adults have short antennae and many species are metallic blue, black, green, or copper-colored. Seventeen species of cerambycids or buprestids have been used in biological control projects and five have provided some control of their target weed species.

Bruchidae. The larvae of this family typically develop inside seeds, especially of legumes. While seven species of bruchids have been employed in biological control projects and some have resulted in the dramatic reduction of seeds and seedlings, only one species suppressed

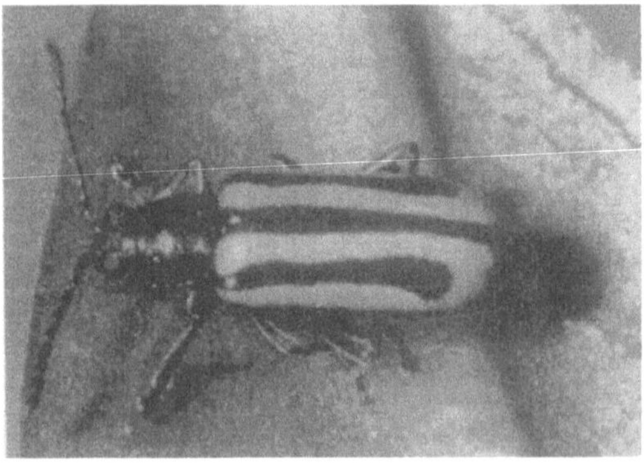

Figure 5.1. *Agasicles hygrophila* Selman and Voyt, a foliar-feeding chrysomelid feeding on alligator weed, *Alternanthera philoxeroides* [Martius] Grisebach. (Photograph courtesy of USDA-ARS.)

Figure 5.2. Populations of alligator weed before (A) and after (B) control of the plant by the chrysomelid beetle *Agasicles hygrophila* Selman and Voyt in the Ortega River in Jacksonville, Florida (U.S.A.) (Photograph courtesy of USDA-ARS.)

Figure 5.3. *Longitarsus jacobaeae* (Waterhouse), a chrysomelid contributing to the control of the weed *Senecio jacobaea* Linnaeus in Oregon (U.S.A.) (Photograph courtesy of P. McEvoy.)

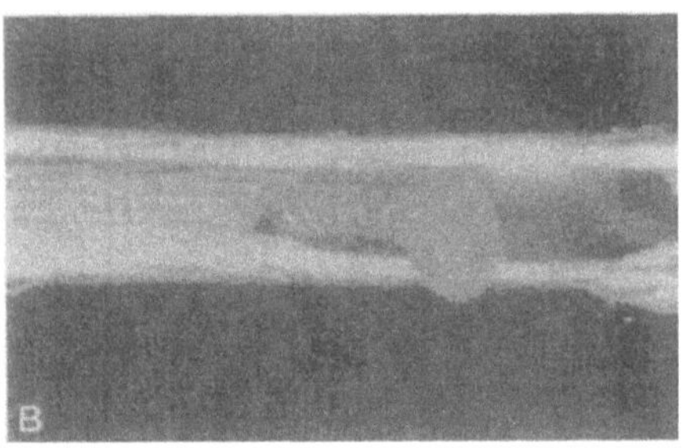

Figure 5.4. Adult (A) and larva (B) of the curculionid *Microlarinus lypriformis* (Wollaston), which contributed to control of puncture vine (*Tribulus terrestris* Linnaeus) in California (U.S.A.) (Photographs by M. Badgley.)

its host weed. In South Africa, two species of seed-feeding bruchids are important in biological control of *Prosopis* spp., a plant which has some desirable characteristics but which has become an invasive weed. By limiting seed production through biological agents, the plant can be preserved but its reproduction severely restricted, reducing its invasiveness and pest status (Zimmermann 1991).

Apionidae. Larvae of this family bore in seeds, stems, and other plant structures. Many members of the genus *Apion* attack legumes. Of nine species of apionids used in biological control, one has suppressed its target weed.

Lepidoptera (Moths and Butterflies)

Pyralidae. This is the third largest family in the Lepidoptera. The adults are small, delicate moths with a scaled proboscis. Larval feeding habits are quite varied and species may be foliage feeders, borers, or feeders on stored products, including beeswax. One of the most famous members of this family in the history of biological control is the Phycitinae species *Cactoblastis cactorum* (Fig. 5.5), which successfully controlled prickly pear cacti (*Opuntia* spp.) in Australia. Of 23 pyralid species employed in biological control, three have suppressed their target weeds.

Noctuidae. This is the largest family in the Lepidoptera. The adults are heavy-bodied, night-flying moths. The larvae are usually smooth bodied and dull in color. Most feed on foliage, but some are stem borers or feed on fruits. One species of the ten employed as biological control agents has controlled its target weed.

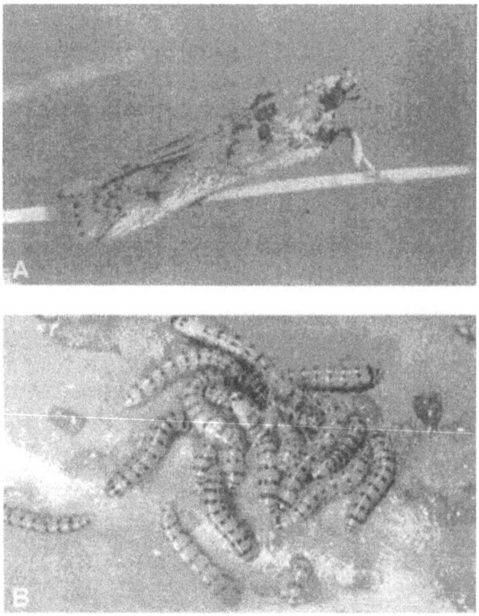

Figure 5.5. Adult (A) and larvae (B) of the pyralid *Cactoblastis cactorum* (Bergroth), which successfully controlled *Opuntia* spp. cacti in Australia. (Photographs by M. Badgley.)

Gelechiidae. Adults of this microlepidopteran family often have the tip of the hind wing pointed and slightly recurved. The feeding habits of the larvae are varied. Some are foliage feeders that roll or tie leaves. Others are leaf miners or gall makers. Of five species that have been introduced, three have become established, but none has controlled its target weed.

Tortricidae. This is one of the largest families of microlepidoptera. The adults are small moths with banded or mottled wings. The larvae are foliage feeders, most often of perennial plants. They may tie or roll leaves and often bore into plant parts. Of the nine species employed in biological control, two have controlled their target weeds.

Arctiidae. This is a moderate-sized family with four rather distinct subfamilies (the Pericopinae, Lithosiinae, Arctiinae and Ctenuchinae), which are given family status by some authors. The Arctiinae are the tiger moths, which comprise most of the species in the family. These are often brightly colored as adults. Larvae are often coated with dense, long, colorful setae. Of the four species that have been used as biological control agents, three have successfully suppressed their target weeds as, for example, *Tyria jacobaeae* (Linnaeus) (Fig. 5.6), which together with the chrysomelid beetle *Longitarsus jacobaeae* successfully controlled tansy ragwort, *Senecio jacobaea*, in Oregon (McEvoy *et al.* 1990).

Gracillariidae. Members of this family are minute moths, with lanceolate wings, whose larvae are leaf miners. Mines are typically blotch in form, and the leaf is often folded. Many mines may occur per leaf. Pupation often takes place in the mine or on the surface of the mined leaf. One of the four species employed for biological control suppressed its host.

Figure 5.6. Adult of the arctiid moth, *Tyria jacobaeae* (Linnaeus), which contributed to the control of *Senecio jacobaea* Linnaeus in Oregon (U.S.A.) (Photograph courtesy of P. McEvoy.)

Homoptera and Hemiptera (Aphids, Scales, True Bugs, Others)

Dactylopiidae. These insects, once important as a source of dyes, resemble mealybugs (Pseudococcidae) in appearance and habits. Guerra and Kosztarab (1992) review the biosytematics of the family. Dactylopiids feed on cacti and while the number of species in the family is not large, the family has played an important role in the successful biological control of several species of cacti (Fig. 5.7, Fig. 5.8). Four of the six species of dactylopiids employed for biological control have provided control of their target cacti.

Tingidae. Adults are small (less than 5 mm in length) and have highly sculpted dorsal surfaces and forewings. Nymphs have spiny bodies. Nymphs and adults feed together on the same host plant by sucking plant fluids from leaves and other tissues, causing foliage to become bronzed; when bug densities are high, plants can die. Some species are relatively specific in their choice of host plants, which may be either herbaceous or woody species. One species, *Teleonemia scrupulosa* Stål, played an important role in the control of lantana, *Lantana camara*, a woody ornamental of great importance as a pasture pest in tropical areas around the world (Julien 1992).

Coreidae. Most coreids are phytophagous. Some species feed on seeds, others on foliage. Some species have enlarged hind tibia or femurs. Many species produce noticeable odors. One of the three species employed in biological control has controlled its host plant.

Diptera (True Flies)

Cecidomyiidae. These are small (usually under 5 mm in length), delicate flies with long legs and long antennae. About two thirds of these form galls, and most of the remaining species

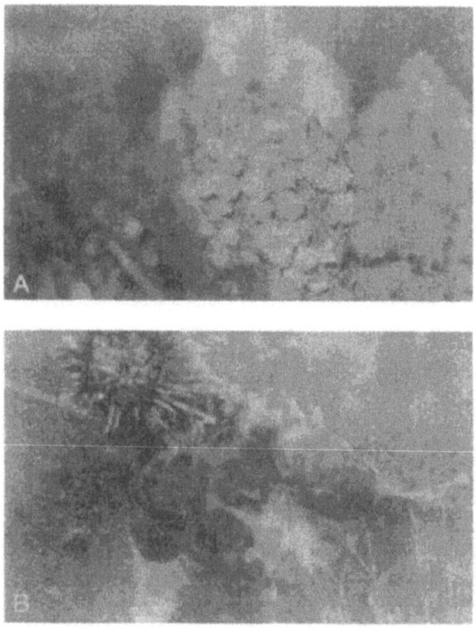

Figure 5.7. (A) Pad of the prickly pear cactus *Opuntia* sp. infested with the cochineal insect *Dactylopius* sp.; (B) close up of Dactylopius sp. with wax removed. (Photographs by M. Badgley.)

feed on decaying vegetation. A few species are predacious. Galls are formed on all parts of plants, including leaves, flowers, and twigs. One of the six species employed for biological control has been successful. The effects of some species have been reduced by local parasitoids.

Tephritidae. The members of this family are small to medium-sized flies, most of which have banded wings. Larvae feed inside plant tissues, tunneling in seed heads, forming galls or feeding in fruits. A few species are leaf miners. Seventeen species have been used in biological control projects, mainly species that feed in seed heads of thistles, knapweeds (Fig. 5.9), and other plants or species that form galls. Establishment has occurred in a substantial proportion of cases, but only two species have successfully controlled their target weeds.

Anthomyiidae. These are small, dark colored flies. The larvae of most species feed in the soil on the roots of plants.

Agromyzidae. These are small flies, with black or yellow bodies. The larvae of most species are leaf miners. Most species make serpentine mines, but some species make blotch mines. Species in the family attack a wide range of plants and are found in most habitats. Of the five species employed for biological control, one has been successful.

Thysanoptera (Thrips)

Phlaeothripidae. These thrips are mostly dark in color, with lighter-colored or mottled wings. Some are predacious; others are spore feeders; and some feed on plants. Of the three species

Figure 5.8. (A) Destruction of *Opuntia oricola* Philbrick by *Dactylopius opuntiae* Cockerell on Santa Cruz Island, California. (A) 1965, (B) 1969, (C) 1975, (D) 1978. (Photographs courtesy of R. Goeden, from Goeden, R.D. & D. W. Ricker [1980], *Proceedings of the Fifth International Symposium on the Diological Control of Weeds, Brisbane, Australia,* 355–65).

employed for biological control, one has suppressed its target weed. *Liothrips urichi* Karmy controlled Koster's curse, *Clidemia hirta* (Linnaeus), in some habitats in Fiji (Rao 1971).

Hymenoptera (Sawflies and Seed Wasps)

Tenthredinidae. This is the most common group of sawflies. Most species are small-to-medium size (rarely more than 20 mm in length). The larvae are caterpillar-like in shape and are external feeders on foliage. Typically there is one generation a year, and the insect passes the winter in a pupal cell or cocoon, in the ground or some protected niche. Most species feed on woody shrubs or trees. A few species are gall makers or leaf miners. Of the two species that have been employed, neither has been successful.

Eurytomidae. These are minute, metallic-colored wasps. They vary in their biology. Some species are parasitic, but others feed in seeds or make galls. The single species of this family that has been employed for biological control suppressed its target weed.

Pteromalidae. The Pteromalidae comprise a highly diverse group of Hymenoptera, which includes some gall formers. One of these, *Trichilogaster acaciaelongifoliae* (Froggatt), has been important in the biological control of *Acacia longifolia* (Andrews) Willdenow in South

Figure 5.9. Adult (A) and larva (B, arrow) of *Urophora affinis* Frauenfeld, a tephritid fly attacking seed heads of diffuse knapweed, *Centaurea diffusa* Lamarck, a range weed in the western United States. (Photographs courtesy of R. D. Richard, USDA-APHIS.)

Africa (Dennill and Donnelly 1991). The wasp is univoltine, and females oviposit in floral buds of the target plant. Extensive attack can occur, and attacked buds do not produce inflorescences or seeds. Attack by the wasp has resulted in 30% mortality of adult trees in some locations. Its final impact is still under evaluation, but it appears promising as a control agent of this species.

Acarina (Gall Mites, Spider Mites)

Eriophyidae. These are extremely small mites (about 0.15 mm in length) that feed on plant tissues. Some species cause galls; others feed externally on plant tissues and, thereby, discolor fruits or other plant parts. The potential biological control uses of the family are reviewed by Gerson and Smiley (1990), who note that eriophyids, while slow acting, are often highly specific as to the hosts on which they feed. One of the three species that have been employed for biological control has suppressed its target weed.

Tetranychidae. Spider mites, which are larger than eriophyids, feed on plant foliage but do not cause galls. Many species produce visible webbing. Gerson and Smiley (1990) summarize past biological control uses of the family, which have been extremely limited. The only target weed against which a spider mite (*Tetranychus lintearius* Dufour) has been deliberately introduced has been gorse (*Ulex europaeus* Linnaeus) in New Zealand (Hill et al. 1991) (Fig. 5.10), and evaluation of this program is continuing.

FUNGAL PATHOGENS

Fungi of importance to biological weed control are found in the subdivisions Basidiomycotina and Deuteromycotina, and potentially in the Ascomycotina. The use of phytopathogenic fungi

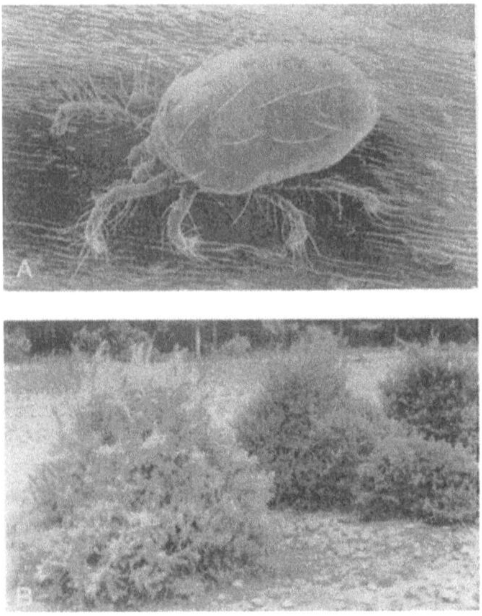

Figure 5.10. *Tetranychus lintearius* Dufour, a tetranychid mite (A) which was introduced to New Zealand for control of gorse, *Ulex europaeus* Linnaeus; (B) Plant in the foreground is lighter in color due to mite damage. (Photographs courtesy of H. Gourlay.)

for introduction to new areas for biological control purposes has been reviewed by Watson (1991) and for development as mycoherbicides by Charudattan (1991).

Ascomycotina. No ascospore forms have been employed as biological weed control agents. However, some species of imperfect fungi in the genera *Fusarium* and *Colletotrichum* have been tried as mycoherbicides. Because other members of these genera have been found to be in the Ascomycotina, it is likely that all members of these genera are also in the Ascomycotina.

Basidiomycotina. This subdivision is divided into three classes: the Holobasidiomycetes (mushrooms and their allies), the Phragmobasidiomycetes, and the Teliomycetes (rusts and smuts). Members of the Holobasidiomycetes are mostly saprophytes; therefore, most species are not important as biological control agents of plants, although some exceptions occur (de Jong 1988). The Phragmobasidiomycetes includes some members (*Septobasidiales* spp.) that are pathogens of scale insects, but the group is not important as weed pathogens. The Teliomycetes is a class that groups together two large but distantly related orders, the rusts and the smuts, each of which are very important for weed biological control through introduction against adventive weeds.

Rusts (order Uredinales) are highly host specific pathogens of vascular plants. Rusts are obligate parasites and, therefore, cannot be cultured in nutrient media. Inoculum must usually be produced on host plants. One of the various spore types produced by the group is termed urediniospores. These are rust-colored and easily distributed by air. Rusts are highly specific in their host relationships and, as such, are likely candidates for use as introduced agents for control of adventive weeds. Three of the five cases of successful introductions of fungi against adventive weeds listed in Julien (1992) are rusts. The most important of these has been the control of skeletonweed (*Chondrilla juncea*) by *Puccinia chondrillina* (Fig. 17.2e) (see Chapter 1 for details). Other *Puccinia* species have been employed in augmentative biological control programs (Phatak et al. 1983).

Smuts (order Ustilaginales), like the rusts, are obligate pathogens of vascular plants. Many smut fungi infect host plants systemically; such infections weaken plants and may disrupt seed production. Spores are dark in color and easily dispersed by air. Like the rusts, smuts show high levels of host specificity and are potentially good candidates for weed biological control. The white smut pathogen *Entyloma ageratinae* Barreto and Evans was introduced into Hawaii, where it successfully controlled its target, hamakua pamakani (mistflower, *Ageratina riparia* [Regel] King and Robinson) (Trujillo 1985) (Fig. 8.5, see Chapter 8 for details).

Deuteromycotina. Because sexual forms of these species are not recognized, these fungi cannot be classified. For convenience they are grouped on the basis of morphology into two form classes, the Hyphomycetes and the Coelomycetes. Most species in these groups are readily grown on culture media and some have been studied as potential mycoherbicides (Charudattan 1991). In addition, some species have been introduced to new locations for biological control in the same manner as rusts. For example, *Colletotrichum gloesporioides* f. sp. *clidemiae* (Coelomycetes) has been introduced into Hawaii to control *Clidemia hirta*, while *Colletotrichum gloeosporioides* f. sp. *aeschynomene* has been developed as a mycoherbicide in Arkansas (U.S.A.) to control northern jointvetch (*Aeschynomene virginica*) in rice and soybean fields (Trujillo et al. 1986; Charudattan 1991; Templeton 1992). Reproduction of Imperfect Fungi is by means of asexual spores called conidia. Conidia of Hyphomycetes are produced freely exposed to air and are usually dispersed by air. Coelomycetes produce conidia in fruiting bodies called acervuli and pycnidia. These conidia are usually sticky and are dispersed by water or insects. While most species require particular environmental conditions for initiation of disease, some require very highly specific conditions (Agrios 1988).

NEMATODES

A number of plant-parasitic nematodes in the Tylenchina, especially the family Anguinidae, induce galls on plant foliage. The taxonomy of this group is reviewed by Siddiqi (1986). Some nematodes in this group are resistant to dehydration, a feature that enhances their survival. A number of species in the group have been considered for use in biological control through augmentation (Parker 1991). One species (*Subanguina picridis* Kirjanova and Ivanova) has been introduced against an adventive plant, *Centaurea diffusa* Lamarck (Julien 1992).

VERTEBRATES

A variety of domestic animals such as goats, sheep, or geese are used to some degree for terrestrial weed control. Fishes have been extensively manipulated for control of aquatic weeds. Julien (1992) lists eleven fish species that have been introduced into various countries to suppress aquatic vegetation. These occur in three families, the Cyprinidae (which include various carp species), the Cichlidae, and the Osphronemidae (Table 5.3).

All of these species are generalist herbivores that suppress whole communities of aquatic macrophytes. Therefore, these species are limited in their application to enclosed water bodies, such as irrigation ditches or ponds, where total or near total weed suppression is desired. The

Table 5.3 Species of Fish Introduced into One or More Countries for the Control of Aquatic Plants

Cyprinidae
1. *Aristichthys nobilis* (Richardson) (big head)
2. *Ctenopharyngodon idella* (Cuvier and Valenciennes) (grass carp, white amur)
3. *Hypophthalmichthys molitrix* (Cuvier and Valenciennes) (silver carp)
4. *Puntius javanicus* (Bleeker)
Cichlidae
1. *Oreochromis aureus* (Steinbachner)
2. *Oreochromis mossambicus* (Peters)
3. *Oreochromis niloticus* (Linnaeus)
4. *Tilapia macrochir* (Boulenger)
5. *Tilapia melanopleura* Dumeril
6. *Tilapia zillii* (Gervais)
Osphronemidae
1. *Osphronemus goramy* Lacepedes (giant gourami)

potential for introductions of such fish to cause damage to natural systems, both the macrophytic and native fish communities, is high. Each introduction must be very carefully considered, taking into account the potential for subsequent spread to other water bodies by flood or casual relocation by people. In some instances, sterile hybrids or sterile triploids (Fig. 17.3) are used to minimize the risk of establishing breeding populations of introduced fish.

CHAPTER

6

BIOLOGICAL CONTROL AGENTS FOR PLANT PATHOGENS

INTRODUCTION

Biological control of plant pathogens is fundamentally a matter of ecological management of a community of organisms, as is all biological control. In the case of plant pathogens, however, there are two distinctions from biological control of organisms such as insects and plants. First, the ecological management occurs at the microbial level, typically in microcosms of the ecosystem such as leaf and root surfaces (Andrews 1992). Second, biological control agents include competitors, as well as parasites. While hyperparasites of plant pathogens and natural enemies of nematodes function in much the same way as do natural enemies (parasitoids) in arthropod systems (by destroying the pest organisms), competitors function by occupying and using resources in a nonpathogenic manner and in so doing exclude pathogenic organisms from colonizing plant tissues. Microbes which negatively affect pathogenic organisms are referred to as antagonists.

Diseases of roots, stems, aerial plant surfaces, flowers, and fruit are caused by a wide variety of pathogens. Because of this diversity, the antagonist species which negatively affect plant pathogens and the mechanisms by which they accomplish their beneficial action are also quite varied (definitions and examples of these mechanisms are given in Chapter 12). Their biology and taxonomic diversity is covered in some detail in several texts and reviews, including Cook and Baker (1983), Fokkema and van den Heuvel (1986), Campbell (1989), Adams (1990), and Stirling (1991). This chapter briefly introduces the antagonists of some important plant pathogens as representative of the broad taxa which are important in this field, beginning with agents affecting microbial pathogens of roots, and proceeding through pathogens of stems, leaves, flowers, and fruit. Natural enemies of plant parasitic nematodes are treated in the last section.

ROOT PATHOGENS

Root diseases are caused by a wide variety of fungi, and by some bacteria, in many crops and plant systems. Biological control agents recognized as significant in suppression of these diseases are largely antagonists which can occupy niches similar to the pathogens and either naturally or through manipulation outcompete the pathogens in these niches. Antibiotic production is also important in a few cases, as are mycoparasitism and induced resistance.

Streptomyces scabies, the causative organism of potato scab, is suppressed by naturally-occurring populations of *Bacillus subtilis* and saprotrophic *Streptomyces* spp. Other microorganisms recognized as suppressing fungal diseases include species of *Pseudomonas* and *Bacillus*. Saprotrophic *Fusarium* fungi are able to suppress populations of pathogenic *Fusarium* spp. through competition for nutrients. There are few well-documented cases of induced resistance for soil-borne pathogens, and these are mostly of wilt diseases. Examples of organisms that induce resistance in plants to pathogens include nonpathogenic strains of *Fusarium* spp., *Verticillium* spp., and *Gaeumannomyces* spp. Mycoparasitic flora such as *Arthrobotrys* spp. (Fig. 6.1), *Coniothyrium minitans* Campbell and *Sporidesmium scerotivorum* Uecker et al. can be added to soil against fungal diseases. *Bacillus* spp. and especially *Pseudomonas* spp. are among bacteria that have properties particularly suited to effective suppression of root-infecting pathogens in soil, such as antibiotic production and competition for Fe^{3+} ions. Mycetophagous soil amoebae have also been noted feeding on pathogenic fungi (Fig. 6.2). These amoebae generally require moist conditions in which to function, and may be important in natural control of some fungi.

STEM PATHOGENS

Stem diseases produce symptoms which include decay and cankers on forest and orchard trees, and such wilts as Dutch elm disease and chestnut blight (caused by the fungus *Cryphonectria parasitica* (Murrill) Barr of Asian origin infecting the American chestnut, *Casta-*

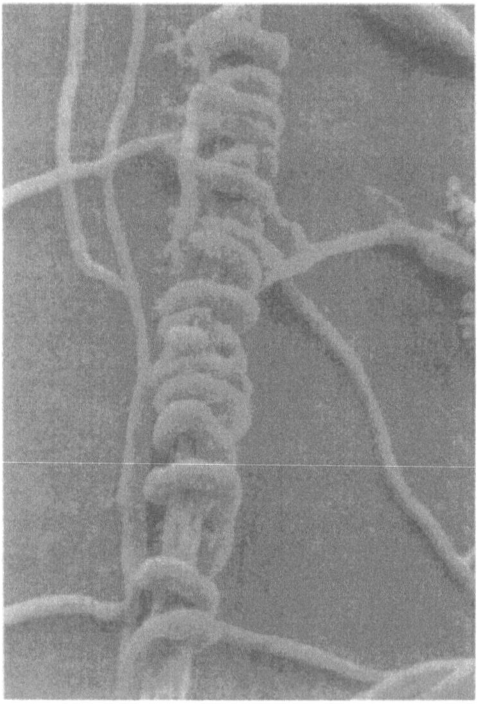

Figure 6.1. Example of a mycoparasitic fungus: hyphae of *Arthrobotrys* sp. coiled around a hypha of *Rhizoctonia* sp. that has died and collapsed. (Photograph courtesy of R. Campbell. *Biological Control of Microbial Plant Pathogens*, Cambridge University Press, with permission.)

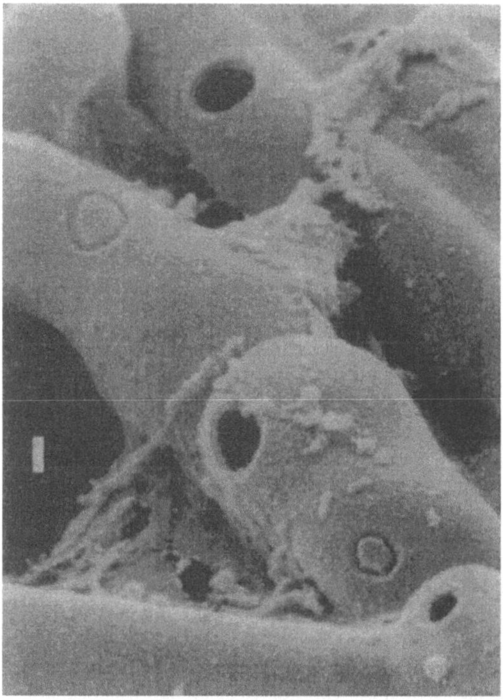

Figure 6.2. Holes in hyphae of the fungal pathogen *Gaeumannomyces graminis* (Saccardo) Arx and Olivier caused by mycetophagous amoebae. Scale bar is 1.0 μm. (Photograph courtesy R. Cook, from Homma et al. 1979. *Phytopathology* 69: 1118–22, with permission.)

nea dentata [Marsham] Borkjauser). Because the etiologies of stem diseases vary, the taxa involved in biological control also vary. In many stem diseases, the pathogen colonizes a part of the host which initially is relatively free of microorganisms, such as a pruning wound. Successful biological control in such circumstances depends on rapidly colonizing this pristine environment with a nonpathogenic antagonistic competitor. Primary among these are competitively antagonistic fungi, including saprotrophic members of the genera *Fusarium, Cladosporium, Trichoderma,* and *Phanerochaete,* and such antibiotic producing bacteria as *Bacillus subtilis* and *Agrobacterium* spp. In the case of chestnut blight, hypovirulent strains of the pathogen itself are crucial in bringing about biological control. In this case, hypovirulence is transmitted cytoplasmically to virulent strains already infecting trees, and disease symptoms decline and disappear.

LEAF PATHOGENS

The growth of microorganisms on leaves is normally severely restricted by environmental factors (Fig. 6.3). Nutrient levels generally are low on leaf surfaces, and microclimate variables, especially leaf surface moisture, temperature, and irradiation, are often unfavorable for microbial development. In temperate climates and in arid tropical regions, water will be intermittent on leaf surfaces, but may be continually present in humid tropical regions. Temperatures on leaf surfaces exposed to direct radiation may rise to several degrees above ambient. The result

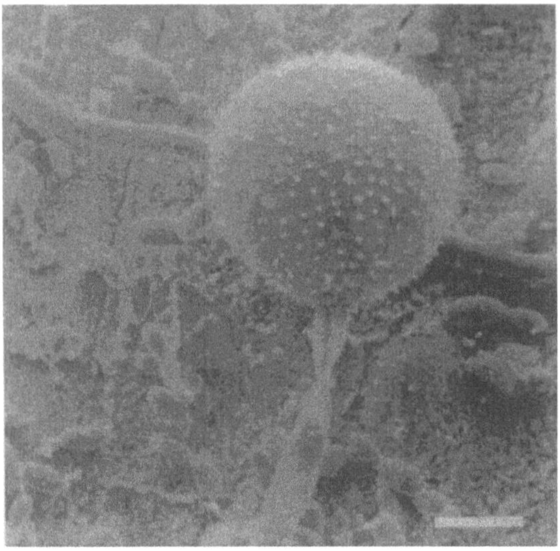

Figure 6.3. A rust spore (*Uromyces vicia-fabae* Persoon: Schröter) together with numerous yeast cells on a leaf surface; this is an unusually dense flora for a temperate leaf surface. (Photographs courtesy of A. Beckett, from Campbell, R. 1989. *Biological Control of Microbial Plant Pathogens*, Cambridge Univeristy Press, with permission.) Scale bar is 10 μm.

of such variation is that microbial floral development on leaf surfaces varies from general scarcity in temperate climates to more extensive microbial films in tropical rain forests (Campbell 1989).

The microbes most frequently recorded as saprotrophs on surfaces of crop plants in temperate conditions and, consequently, the species which are candidates as antagonists of pathogens, include the fungi *Aureobasidium pullulans* (de Bary) Arnaud, *Cladosporium* spp., and such yeasts as *Cryptococcus* spp. and *Sporobolomyces* spp. Beneficial bacteria in the phyllosphere include members of such genera as *Erwinia, Pseudomonas, Xanthomonas, Chromobacterium*, and *Klebsiella*. These lists, based on microbial surveys, usually give no indication of activity of the organisms, but this information can be obtained from experimental studies. For example, early studies on control of botrytis rot in lettuce (Wood 1951) indicated that several organisms were successful in suppressing the disease when sprayed on lettuce (*Lactuca sativa* Linnaeus) plants, among them *Pseudomonas* sp., *Streptomyces* sp., *Trichoderma viride* Persoon: Fries, and *Fusarium* sp. Similar studies show varying degrees of effectiveness in other cropping systems (Peng and Sutton 1991; Sutton and Peng 1993a,b; Zhang et al. 1994). The microbial composition and biological activity of phylloplane microbes can vary with season, position on the top or bottom of the leaf and on location in the plant canopy, depending on the degree of exposure relative to prevailing winds and rain (Campbell 1989).

Biological control of the black-crust pathogen (*Phyllachora huberi* Hennings) on rubber tree (*Hevea brasiliensis* Müller Argoviensis) foliage is accomplished by the hyperparasites *Cylindrosporium concentricum* Greville and *Dicyma pulvinata* (Berkeley and Curtis) Arx (Junqueira and Gasparotto 1991). Botrytis leaf spot in onion (*Allium cepa* Linneaus) was suppressed by *Gliocladium roseum* Link: Bainier (Sutton and Peng 1993a). Other examples include control of powdery mildews, other botrytis rots, and turfgrass diseases (Sutton and Peng 1993a).

Nonpathogenic species of the fungal genus *Colletotrichum* (Kúc 1981; Dean and Kúc 1986) can be used to induce resistance in cucumbers against pathogenic species of the same genus. Inoculation with a nonpathogenic strain of a virus confers protection to plants from pathogenic strains in many diseases. The bacterium *Bdellovibrio bacteriovorus* Stolp and Starr is a parasite of pathogenic bacteria. Finally, there are numerous parasitic fungi which attack pathogenic fungi (Kranz 1981). Among those which have been studied in detail, principally as agents against leaf rusts and mildews, are *Sphaerellopsis filum* (Bivona-Bernardi ex Fries) Sutton, *Verticillium lecanii* (Zimmerman) Viegas, and *Ampelomyces quisqualis* Cesati ex Schlechtendal.

FLOWER AND FRUIT PATHOGENS

Flowers are ephemeral structures and as such have limited opportunity to become infected. One major disease of flowers which has received attention is fire blight of rosaceous plants, caused by the bacterium *Erwinia amylovora* (Burril) Winslow et al. Biological suppression of the disease has been achieved through use of the nonpathogenic species *Erwinia herbicola* (Lohnis) Dye (Beer et al. 1984; Lindow 1985b), sometimes in combination with *Pseudomonas syringae* van Hall (Fig. 6.4). *Erwinia herbicola* was used successful by spraying aqueous suspensions of it onto the flowers just before the time of potential infection (Campbell 1989). The mode of action is primarily competitive exclusion, with the antagonist competing with the pathogen for a growth-limiting resource and possibly other effects such as induced cessation of nectar secretion or accumulation of a host toxin (Wilson and Lindow 1993a).

The fruit diseases addressable through biological control include diseases of fruit on the plant and post-harvest diseases. One of the first systems developed was against *Botrytis cinerea* Persoon: Fries in vineyards, where sprays with spore suspensions of the antagonist *Tricho-*

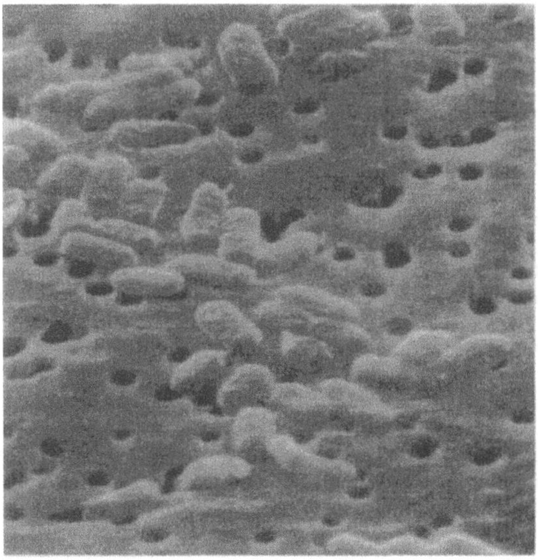

Figure 6.4. The bacterium *Pseudomonas syringae* van Hall, used as a biological control agent of fire blight and other diseases, shown here collected on a millipore filter surface. (Photograph courtesy of S. Lindow.)

derma harzianum Rifai were effective in suppressing disease incidence. Several organisms, including *Gliocladium roseum*, *Penicillium* sp., *Trichoderma viride*, and *Colletotrichum gloeosporioides* were as effective as fungicides in suppressing *B. cinerea* on strawberries (Peng and Sutton 1991). A number of other examples also have been reported (Sutton and Peng 1993a).

Post-harvest diseases, which can be responsible for 10–50% loss of produce (Wilson and Wisniewski 1989; Jeffries and Jeger 1990), have received considerable attention. Numerous reports deal with suppression of post-harvest disease in fruit crops (Campbell 1989; Wilson and Wisniewski 1989; Jeffries and Jeger 1990) by such organisms as species of *Penicillium*, *Bacillus*, *Trichoderma*, *Debaryomyces*, and *Pseudomonas*. The mode of action of many of these is generally antagonism, often through the production of antibiotics which reduce the longevity and germination of spores of pathogens. Others appear to suppress pathogen growth through nutritional competition or induction of host resistance (Wilson and Wisniewski 1989). Post-harvest rots include major diseases caused by *Botrytis cinerea*, *Rhizopus* spp., and other fungi in several crops. Competitive and parasitic fungi, including *Trichoderma* spp., *Cladosporium herbarum* (Persoon: Fries) Link and *Penicillium* spp., give control as good as commercial fungicides. *Enterobacter cloacae* (Jordan) Hormaeche and Eduards reduces rots by *Rhizopus* spp., but there are hesitations about its use on uncooked food products.

PLANT-PARASITIC NEMATODES

Plant-parasitic nematodes inhabit many soils and attack the roots of plants. They are affected by a range of natural enemies, including bacteria, nematophagous fungi, and predacious nematodes and arthropods. There is some limited evidence for virus association with nematodes (Loewenberg et al. 1959), but the etiology of these viruses is not well-known (Stirling 1991). The biologies of natural enemies of nematodes have been reviewed by Sayre and Walter (1991) and Stirling (1991).

Bacteria Affecting Plant-Parasitic Nematodes

Several bacterial diseases of nematodes have been reported (Saxena and Mukerji 1988); other bacteria produce compounds that are detrimental to plant-parasitic nematodes (Stirling 1991). The most widely studied of the bacterial pathogens of nematodes are in the genus *Pasteuria*. Early work was focused on *Pasteuria penetrans* (Thorne) Sayre and Starr *sensu stricto* Starr and Sayre (Fig. 6.5). Recent evidence indicates that this taxon represents an assemblage of numerous pathotypes and morphotypes, and probably represents several taxa (Starr and Sayre 1988). This bacterium has been found infecting a large number of nematode species (more than 200 in about 100 genera, Sayre and Starr 1988; Stirling 1991), does not attack other soil organisms, and is the most specific obligate parasite of nematodes known. Its spores attach to and penetrate the nematode cuticle. Most attention has been centered on populations (*Pasteuria penetrans sensu stricto*, Starr and Sayre 1988) that attack root-knot nematodes (*Meloidogyne* spp.). The spores of *P. penetrans* germinate a few days after a contaminated nematode begins feeding on a root (Sayre and Wergin 1977). The bacterium reproduces throughout the entire female body, and the female may either be killed or may mature but produce no eggs. Bacterial spores (about 2 million from each infected nematode, Mankau 1975) are released when the nematode body decomposes, and they remain free in the soil until contacted by another nematode. They tolerate dry conditions and a wide range of temperatures, and may remain viable in the soil for more than six months. Because it is an obligate parasite, it has not

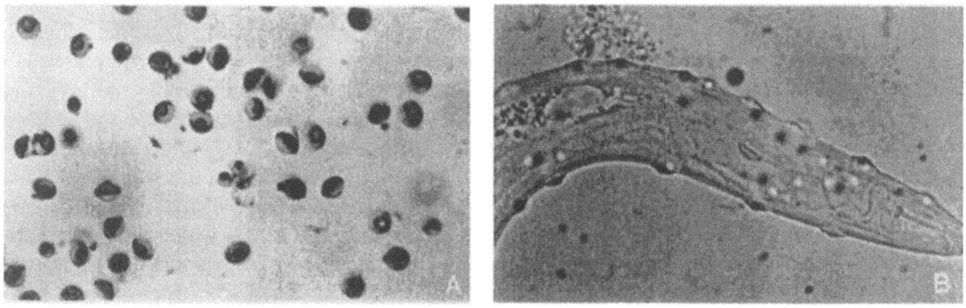

Figure 6.5. The bacterium *Pasteuria penetrans* (Thorne) Sayre and Starr, a pathogen of plant-parasitic nematodes. (A) spores of the bacterium; (B) spores attached to the exterior of a nematode, where they germinate once the nematode enters a plant root. (Photographs courtesy of R. Mankau.)

yet been possible to develop *in vitro* culturing techniques for this bacterium. Different populations of the bacterium show varying degrees of specificity to small numbers of nematode species, but the mechanisms and degree of specificity remain to be elucidated (Stirling 1991). *Pasteuria penetrans* appears responsible for some cases of natural regulation of nematode populations (Sayre and Walter 1991).

A few strains of *Bacillus thuringiensis* are also known to have activity against nematodes, including plant-parasitic species. Zuckerman et al. (1993) report efficacy of a strain against *Meloidogyne incognita* (Kofoid and White) Chitwood, *Rotylenchus reniformis* Linford and Oliveira, and *Pratylenchus penetrans* Cobb in field and glasshouse trials. The body openings of these nematodes are too small to permit the ingestion or other ingress of the bacterium, and Zuckerman et al. (1993) suggest that the mode of action is either a beta exotoxin (Prasad et al. 1972; Ignoffo and Dropkin 1977) or a delta endotoxin released following bacterial cell lysis. A strain of *B. thuringiensis* with a nematotoxic delta endotoxin is the subject of a European Patent Application by Mycogen Corporation of San Diego, California (Zucherman et al. 1993).

Fungi Affecting Plant-Parasitic Nematodes

A large group of fungi attack nematodes in the soil (Barron 1977; Stirling 1991). Numerous species have been reported from all types of soils. The taxonomy of the group has been subject to revision, and we use here the generic names recognized in Stirling (1991).

Some nematophagous fungi are endoparasitic in nematodes. Among these are genera which reproduce through motile zoospores (e.g., *Catenaria anguillulae* Sorokin, *Lagenidium caudatum* Barron, *Aphanomyces* sp.), which generally appear only weakly pathogenic in healthy nematodes (Stirling 1991). Other endoparasitic fungi possess adhesive conidia, and the infection process begins when conidia adhere to a nematode's cuticle (e.g., the genera *Verticillium, Drechmeria, Hirsutella, Nematoctonus*). In *Nematoctonus* spp., the germinating spores secrete a nematotoxic compound which causes rapid immobilization and death of nematodes (Giuma et al. 1973). A few species (*Catenaria auxilaris* [Kuhn] Tribe, *Nematophthora gynophila* Kerry and Crump) parasitize adult females or nematode eggs rather than juveniles.

Other fungi capture nematodes through use of special trapping structures, and have been termed "predatory." Among the more common of these fungi are species in such genera as *Monacrosporium, Arthrobotrys,* and *Nematoctonus.* These fungi consist of a sparse mycelium, modified to form organs capable of capturing nematodes. These organs include adhesive structures, such as adhesive hyphae, branches, knobs, or nets (Stirling 1991). There are also nonadhesive rings, the cells of which expand when touched on their inner surface, constricting the interior of the ring and trapping nematodes. Most of these fungi are not specific and attack a wide range of nematode species. They are widely distributed (Gray 1987, 1988) and most are capable of saprotrophic growth, but often appear limited in this phase in the soil. Many soils suppress the growth of these fungi (a condition called soil fungistasis or mycostasis). This is possibly due to two different causes. Mankau (1962) concluded that a water-diffusible substance was responsible for inhibited germination in tests of soil from southern California (U.S.A.). Other studies have indicated increased activity following soil amendments with nutrients (Olthof and Estey 1966) or organic material (Cooke 1968), which implies fungistasis may be a result of resource limitation. Following saprotrophic growth, formation of trapping structures occurs and is, apparently, stimulated by nematodes (Nordbring-Hertz 1973; Jansson and Nordbring-Hertz 1980). Stirling (1991) suggests that this phase of predacious activity is followed by diversion of resources to reproduction, followed by a relatively dormant phase.

Several species of fungi are facultatively parasitic on nematodes. Of the few of these fungi that are significant pathogens of root-knot and cyst nematodes, *Verticillium* spp. are among the most important. These fungi can parasitize nematode eggs, and *Verticillium chlamydosporium* Goddard plays a major role in limiting multiplication of *Heterodera avenae* Wollenweber in English cereal fields (Kerry et al. 1982a,b). *Paecilomyces lilacinus* (Thom) Samson parasitizes eggs of *Meloidogyne incognita* (Jatala et al. 1979) and *Heterodera zeae* Koshy, Swarup, and Sethi (Dunn 1983; Godoy et al. 1983). *Dactylella oviparasitica* Stirling and Mankau, a parasite of *Meloidogyne* eggs, is thought to be at least partly responsible for natural decline of root-knot nematodes in Californian peach orchards (Stirling et al. 1979).

Predacious Nematodes Affecting Plant-Parasitic Nematodes

Predatory nematodes are found in four main taxonomic groups—Monochilidae, Dorylaimidae, Aphelenchidae and Diplogasteridae—each with a distinct feeding mechanism and food preferences (Stirling 1991). The monochilids have a large buccal cavity that bears a large dorsal tooth; all species are predacious, feeding on protozoa, nematodes, rotifers, and other prey, which may be swallowed whole, or pierced and the body contents removed. The dorylaimids are typically larger than their prey and possess a hollow spear which is used either to pierce the body of the prey or to inject enzymes into the food source and suck out the predigested contents. The group is considered omnivorous, but the feeding habits are known only for a few species (Ferris and Ferris 1989). Almost all the predatory aphelenchids are in the genus *Seinura*. Although small, they can feed on nematodes larger than themselves by injecting the prey with a rapidly-paralyzing toxin through their stylet. The diplogasterids, typically a bacteria-feeding group, have a stoma armed with teeth, and the species with large teeth prey on other nematodes. Species in all these groups are generally omnivorous, feeding on free-living as well as plant-parasitic nematodes. The role of individual species in the population dynamics of plant-parasitic nematodes in the soil has been difficult to quantify, but it is possible that a number of species may act together to produce a significant impact (Stirling 1991).

Insects and Mites

Several microarthropods in the soil, including mites and Collembola, prey on nematodes, and high predation rates have been recorded *in vitro* (Stirling 1991). A few genera are obligate predators of nematodes, while other genera are more general feeders and consume nematodes as well as other foods (Moore et al. 1988; Walter et al. 1988; Sayre and Walter 1991). The information available suggests that as a group, microarthropods are probably significant predators on nematodes in some soils and habitats. Limited information about predation rates in soil is available, however, and more work will be necessary to assess the impact of this group on nematode populations.

SUMMARY

This overview touched briefly on groups of organisms which are antagonistic to plant pathogens and nematodes. These antagonists vary both in their innate ability to suppress plant pathogens and in their ability to thrive and compete in different environments. Consequently the selection of an organism or organisms for any particular biological control program will be a compromise among these parameters and abilities. In addition, the selection of organisms will depend on the approach taken for their use (inoculative augmentation, inundative augmentation, or natural control through conservation). These approaches are discussed in more detail in Chapter 12.

SECTION

III

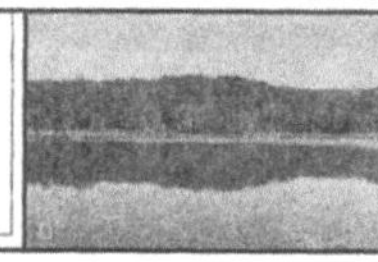

Methods for Biological Control Implementation

In Chapters 7 through 12, the methods by which biological control can be implemented are discussed. This section is the heart of the book. Above all, biological control is an applied discipline intended to provide tools to suppress pests of crops or natural systems in an ecologically sound manner. Chapters in this section discuss how such pest control can be achieved.

Chapter 7 presents methods to conserve existing natural enemies and make them more effective. Chapters 8 and 9 address the introduction of new species of natural enemies to suppress adventive pests. Concepts are treated in Chapter 8, while Chapter 9 provides further practical details on some of the steps in the introduction process. In Chapter 10, the augmentative use of parasitoids and predators is discussed. In both Chapter 10 and Chapter 11 (which address the augmentative use of predators and parasitoids, and microbial pathogens, respectively), the material has been organized around steps in the process of finding, producing, evaluating, and marketing agents, rather than taxonomically by kind of agent. In Chapter 12, the biological suppression of plant diseases is reviewed. Separate discussion of this topic was dictated by the many differences in concepts, terminology, and methods used when plant diseases are the targets of biological control. Biological weed control, while often treated separately in other texts, in this text has been integrated into Chapters 7 through 11.

CHAPTER

7

Natural Enemy Conservation

INTRODUCTION

Agricultural cropping systems consist of patterns of crops and sets of commonly used practices, and these define the conditions under which natural enemies encounter their hosts in these systems. The details of these practices interact with the biologies and needs of natural enemies in many ways. The goal of conservation as a form of biological control is to enhance conditions for natural enemy survival and reproduction relative to pests so that pest population growth rates are lowered and pest densities reduced over time.

For conservation to be an effective form of biological control, appropriate natural enemies must be present. For many native pests, local natural enemies may potentially be effective. Effective natural enemies are less likely to be present, however, if the pest is an adventive species that originated elsewhere. In such cases, new species of natural enemies must first be introduced (Chapters 8 and 9), and then methods to conserve these species can be employed. To control pests, natural enemies must arrive on time, be sufficiently abundant, and attack the pests of interest. Much remains to be discovered as to how these processes can be enhanced within the context of economically viable farming practices.

Farming practices with potential effects on natural enemies may be grouped into five categories: use of agricultural pesticides; management of soil, water, and crop residues; crop patterns; manipulation of non-crop vegetation within or adjacent to crop fields; and direct provision of food or shelter to natural enemies or control of their antagonists.

CONCEPTS

Before considering the various agricultural practices that shape natural enemy environments, the concepts of pest resurgence, pest creation (=secondary pest outbreak or pest upset), and pesticide resistance need to be discussed.

Pest Resurgence

Recognition that agricultural practices can strongly affect the degree to which existing natural enemies exert control of pests developed in part from the observation that routine use of broad spectrum insecticides and acaricides was sometimes followed by a quick return of pests to damaging levels. This phenomenon is termed pest resurgence. The rebound of pests in these situations is due to the more severe and prolonged damage done by the pesticide to the natural

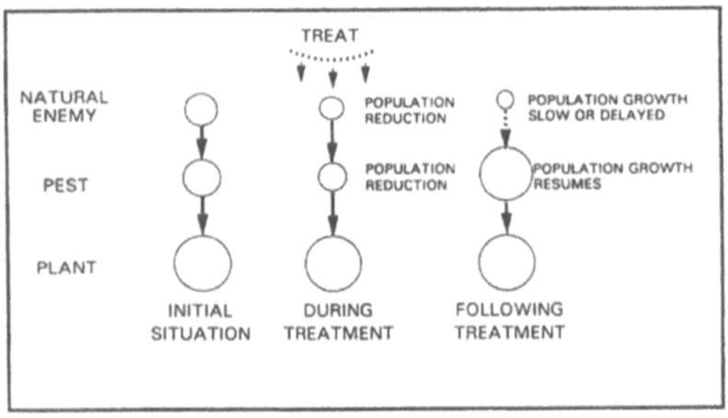

Figure 7.1. Conceptual diagram of pest resurgence following pesticide treatment. Pesticide treatments reduce populations of both the natural enemy and the pest. Pest population growth resumes, while natural enemy population growth is delayed either through longer action of pesticide residues on the natural enemy, or through reduced growth rates because of scarcity of hosts. Subsequently, the pest population exceeds previous levels because of the reduced action of the natural enemy population.

enemies of the pest than to the pest itself (Fig. 7.1). Once the residues of the pesticide have dissipated, those pests that have survived quickly begin to reproduce. Because most natural enemies have been killed or disabled, pests and their offspring experience increased survival rates. Consequently, the rate of growth of the pest population is higher, and the density of the pest quickly returns to or exceeds pre-treatment levels. Pest resurgence has been observed in various crops and many kinds of pests (Gerson and Cohen 1989; Buschman and DePew 1990; Talhouk 1991; Holt et al. 1992). One of the more widespread and dramatic recent examples has been that of *Nilaparvata lugens*, a planthopper found in rice crops in Asia (Fig. 7.2) (Heinrichs et al. 1982). Pest resurgence does not occur with all pests, but rather is most likely for species which are at least partially suppressed by natural enemies, especially if these are more exposed or susceptible to the pesticides than are the pests themselves (Waage 1989).

Secondary Pest Outbreak

Another phenomenon associated with the use of broad spectrum pesticides has been outbreaks of herbivorous species which are not usually pests. These species are termed secondary pests. The mechanism that causes these species to become pests is the destruction of the natural enemies which hold these herbivores at innocuous levels in the absence of pesticide treatments (Fig. 7.3). If these natural enemies are killed by pesticides, the herbivore populations survive better and reach pest densities. Some species of tetranychid mites, scales, and leaf miners are examples of such secondary or pesticide-created pests (Fig. 7.4) (Luck and Dahlsten

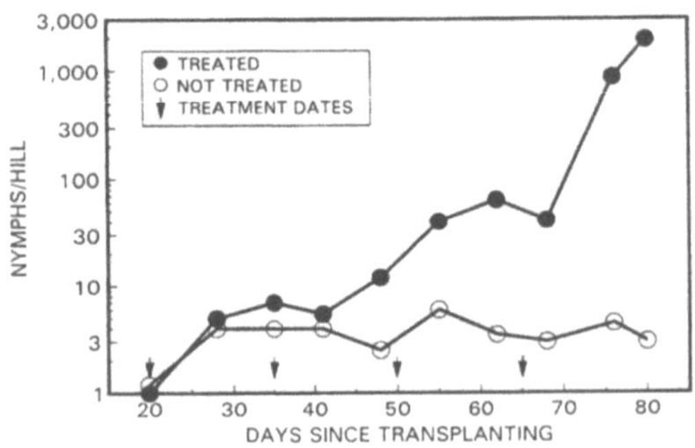

Figure 7.2. Resurgence of rice brown planthopper *Nilaparvata lugens* (Ståhl), in rice fields treated with insecticide versus fields not receiving treatments (after Heinrichs et al. 1982).

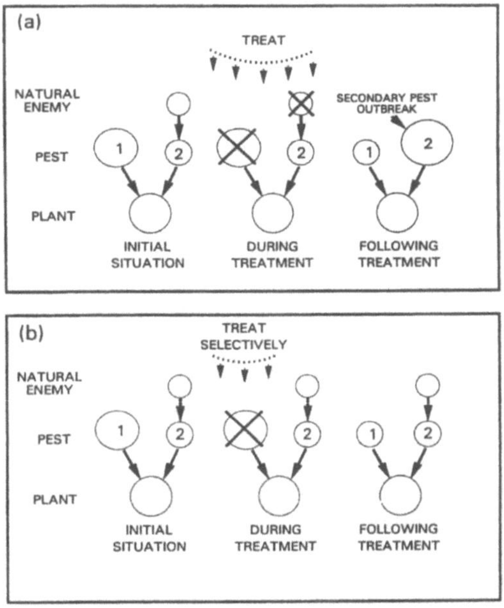

Figure 7.3. Conceptual diagram of secondary pest outbreak. In (a), a general pesticide affects both the target pest, species 1, and also the natural enemy limiting species 2. Following treatment, species 2 undergoes population growth to pest levels in the absence of its natural enemies. In (b), a selective pesticide affects only the target pest, resulting in a situation where natural control of species 2 is not disturbed, and there is no secondary pest outbreak.

1975; Van Driesche and Taub 1983; DeBach and Rosen 1991). Secondary pest outbreaks differ from pest resurgence only in that the pesticide applications are not directly targeted at the secondary pest, but at some other pest species in the crop.

Pesticide Resistance

Pesticide resistance develops in a population when certain individuals possess genes which allow them to better avoid or survive contact with pesticides. Treating such a population with a pesticide confers differentially greater survival or fitness on these tolerant individuals, and the

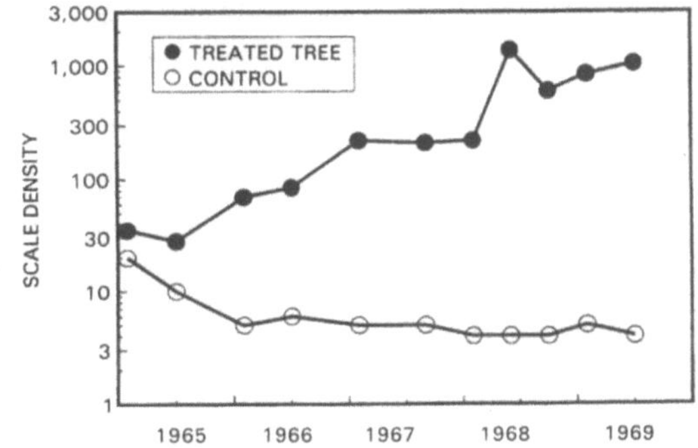

Figure 7.4. Secondary pest outbreaks occur when pesticides are used against one pest and eliminate the natural enemies of another. In this case, treatments of the pesticide DDT on citrus severely reduced populations of the natural enemy *Aphytis melinus* DeBach, so populations of the California red scale, *Aoniidiella aurantii* (Maskell), increased substantially (after DeBach and Rosen 1991).

frequency of the resistant genotype increases when the tolerant individuals reproduce. For species in which these surviving individuals remain together as a new breeding group, undiluted by addition of susceptible individuals from outside the pesticide-treated area, pesticide resistance may develop. An increasing number of pests of several types have become resistant to pesticides (Fig. 7.5). When pests develop resistance, farmers may respond by increasing dosage, changing or alternating pesticides, or combining several pesticides. If resistance is sufficiently severe to prevent control of the pest, chemical control may be abandoned and management systems based on biological control, including the conservation of native natural enemies, may be implemented instead (see Chapter 14). Alternatively, when natural enemies develop resistance to pesticides commonly used on a crop, this resistance may make it possible to conserve such natural enemies as important mortality agents contributing to the control of pests in crops even with continued pesticide use.

USE OF AGRICULTURAL CHEMICALS

Pesticides can reduce natural enemy effectiveness either by directly causing mortality or by influencing the behavior, foraging, or movement of natural enemies, their relative rate of reproduction compared to that of the pest, or by causing imbalances between host and natural enemy populations such as catastrophic host synchronization (Table 7.1) (Jepson 1989; Waage 1989; Croft 1990).

Pesticide-Induced Mortality

Many classes of pesticides are directly toxic, to one degree or another, to some categories of natural enemies. Insecticides and acaricides, for example, are likely to be damaging to most parasitoids and predacious arthropods (Bartlett 1951, 1953, 1963, 1964b, 1966; Bellows and Morse 1988, 1993; Bellows et al. 1985, 1992a, 1993; Morse and Bellows 1986; Morse et al. 1987), while fungicides would generally not affect these organisms but may inhibit fungi pathogenic to pest arthropods (see Yasem de Romero 1986; Saito 1988; Majchrowicz and Poprawski 1993) or fungi antagonistic to plant pathogens (Vyas 1988). Agricultural chemicals such as soil sterilants drastically alter soil microbial, fungal, and invertebrate communities, affecting the influence of such soils on plant pathogens. Other materials may be toxic outside of their intended scope of use. A bird repellent, for example, may also be insecticidal. A fungicide may also kill arthropods (sulfur is damaging to phytoseiid mites) or affect their reproduction or

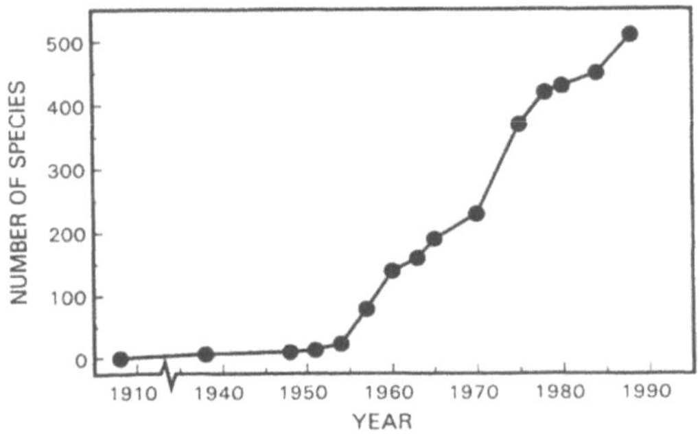

Figure 7.5. Cumulative number of cases of resistance to pesticides in arthropods (after Georghiou and Lagures-Tejeda 1991).

TABLE 7.1 Types of Pesticide Effects on Natural Enemies

Mortality
- from the spray
- from residues on foliage

Other Effects
- reduced longevity
- reduced fertility
- repellency
- lack of hosts
- hormolygosis in host population
- catastrophic synchronization

movement (some dithiocarbamate fungicides reduce phytoseiid reproduction rates). Herbicides may kill beneficial nematodes applied for insect control (e.g., Forschler et al. 1990). Therefore it is important to assume that any pesticide, of whatever type, might affect a natural enemy until data are available to demonstrate that it does not (Hassan 1989a). Even materials often thought of as nontoxic, such as soaps or oils, which may be safe to humans, may be injurious to natural enemies. Oils, for example, when applied to scale species, are likely to reduce emergence of scale parasitoids as well as cause mortality to scales (Meyer and Nalepa 1991).

The degree of effect on a natural enemy population caused by any given pesticide will depend on both physiological and ecological factors. Physiological selectivity consists of the intrinsic relative toxicities of the compound to the pest and the natural enemy. Chemicals vary greatly in their inherent toxicity to given species (Fig. 7.6) (Jones et al. 1983; Smith and Papacek 1991). A few insecticides or acaricides have been discovered that are effective against pests and also relatively harmless, physiologically, to some arthropod natural enemies. Toxicity varies with species of natural enemy, but some examples include pirimicarb, toxins of *Bacillus thuringiensis*, fenbutatin oxide, and diflubenzuron (Hassan 1989a); certain plant alkaloids, mevinphos, and cyrolite (Bellows et al. 1985; Bellows and Morse 1993); avermectin and narrow range oils (Morse et al. 1987); and the systemic materials demeton and aldoxycarb (Bellows et al. 1988). Ecological selectivity results from those aspects of the use of the material that

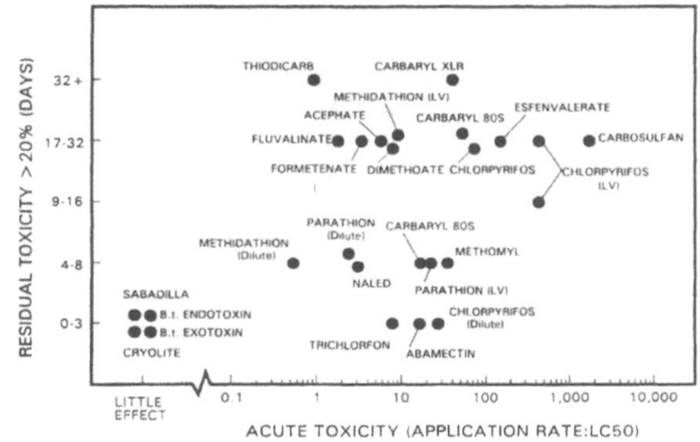

Figure 7.6. Toxicity to *Aphytis melinus* DeBach of pesticides used in citrus agriculture (after Bellows and Morse 1993). The horizontal axis is immediate (acute) toxicity on leaves bearing freshly-deposited residues, and the vertical axis is length of residual toxicity. Note the wide range of both acute toxicities and length of residual action.

determine the degree of contact that actually occurs between the pesticide and the natural enemy. Contact is affected by the formulation and concentration applied, the persistence of the material in the environment (as affected by such abiotic factors as temperature and rainfall), the mode of action of the material (contact versus ingestion), the spatial pattern of application, and the timing of application.

Other Pesticide-Induced Effects

In addition to suffering mortality, natural enemies may become less effective following pesticide use because of other effects of the pesticides on the natural enemies or on the pest (Waage 1989). Exposure to sublethal doses of pesticides may shorten the longevity of natural enemies, decrease developmental rates, reduce natural enemy foraging efficiency, lower reproductive or germination rates, either in absolute terms or relative to that the pest, or alter sex ratios of offspring. At the population level, pesticide use may reduce the effect of a natural enemy by changing host density, distribution pattern, or population age structure (Godfray et al. 1987; Vyas 1988; Croft 1990).

Some indirect effects, such as reduced longevity of adults, altered foraging efficiencies on pesticide treated surfaces (Gu and Waage 1990), sterility, and repellency can be noted in appropriately designed laboratory assays (Croft 1990). Hislop and Prokopy (1981a), for example, noted that benomyl, a fungicide of little or no direct toxicity to phytoseiids, caused sterility to female *Amblyseius fallacis* (Garman) (Fig. 7.7). Similarly, the fungicides thiophanate-methyl and carbendazim have been observed to inhibit oviposition by *Phytoseiulus persimilis* (Dong and Niu 1988). Some materials, not themselves directly toxic to given natural enemies, may make treated surfaces or hosts repellent to natural enemies, which may leave the area. The herbicides diquat and paraquat, for example, make treated soils in vineyards repellent to the predacious mite *Typhlodromus pyri* (Scheuten) (Boller et al. 1984).

In addition, population-level processes can also affect the ability of natural enemies to suppress their target pests. Field tests are required to detect these effects. Natural enemies may become less effective if: pesticides decrease host populations below levels needed to sustain natural enemies (or alter local pest distribution patterns), causing natural enemies either to emigrate, reproduce at reduced levels, or starve; pesticides kill all but one stage of a pest with overlapping generations, causing synchronization of the host's life stages (referred to in the literature as catastrophic synchronization, Godfray et al. 1987; Godfray and Chan 1990); pesticides enhance pest reproduction rates (hormoligosis), or reduce natural enemy reproduction rates, causing pest densities to rise relative to natural enemy densities and allowing pest populations to escape, even if only temporarily, control of natural enemies.

Pesticides that are relatively harmless to individual natural enemies themselves, such as stomach poisons of mineral origin (e.g., cryolite) or biological origin (e.g., some bacterial or plant toxins), can still affect natural enemy populations by destroying their food or host resources. If a very high percentage of the pest population is killed, or if the natural distribution pattern of the pest population is altered in ways that make natural enemy foraging less effective, natural enemy populations may decline through emigration, starvation, or reduced reproduction. Croft (1990) and Flexner et al. (1986) discuss details of effects of *Bacillus thuringiensis* and other microbial pesticides on natural enemies.

Pesticides that kill all but one life stage of a pest with overlapping generations may cause the pest population to become highly synchronized, an effect that may persist for several generations until natural variation in developmental time and reproduction among individuals causes stages to again overlap. For natural enemies that have life cycles shorter than that of the pest,

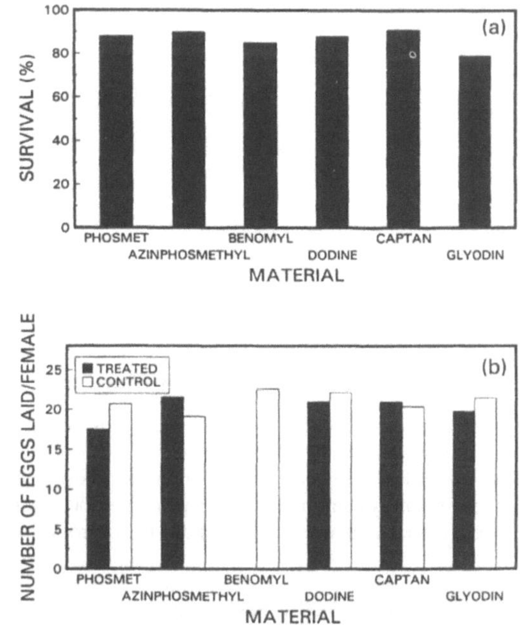

Figure 7.7. Survival (a) and fertility (b) of *Amblyseius fallacis* (Garman) as affected by two insecticides (phosmet and azinphosmethyl) and four fungicides (after Hislop and Prokopy 1981a). Note that benomyl completely inhibited predator reproduction.

such synchronization can cause a temporary absence of suitable hosts. In the absence of suitable hosts, natural enemies may starve, emigrate, or experience very poor rates of reproduction. Examples of such catastrophic synchronization have been reported for pests of coconut (Perera et al. 1988) and perhaps other tropical plantation crops (Godfray et al. 1987). Features such as the relative durations of host and parasitoid life cycles have been suggested as forces that may reinforce catastrophic synchronization once it has arisen (Godfray and Chan 1990).

Factors that alter the relative rate of population increase between the pest and its controlling natural enemies can also decrease the effectiveness of natural enemies. The fecundity of citrus thrips, *Scirtothrips citri* (Moulton), for example, increased significantly when thrips were reared on leaves with 21-d-old dicofol residues and 32-, 41-, and 64-d-old residues of malathion (Morse and Zareh 1991) (Fig. 7.8). Lowery and Sears (1986) found that treatment of green peach aphid (*Myzus persicae* [Sulzer]) adults with sublethal doses of azinphosmethyl increased their fecundity 20–30%. For some kinds of herbivorous arthropods, increased nitrogen levels in foliage from high levels of fertilization can also cause higher survival rates, more rapid growth and increased fecundity. This is especially true for spider mites (van de Vrie and Boersma 1970; Hamai and Huffaker 1978; Wermelinger et al. 1985) and is likely to also be the case for sap-feeding insects such as scales, whiteflies, and aphids. Such increases in pest population growth rates may exceed the ability of the natural enemy population to suppress the pest population. The same effect could also result from pesticides that reduce the reproductive rate of the natural enemy population (Hislop and Prokopy 1981a; Dong and Niu 1988).

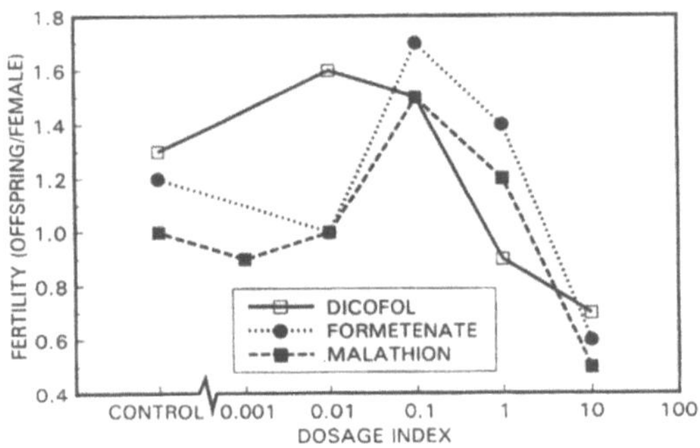

Figure 7.8. Fertility of citrus thrips, *Scirtothrips citri* (Moulton), as affected by dosage of three pesticides (after Morse and Zareh 1991). At low doses, fertility was not different from controls; at high doses fertility was reduced, but at intermediate doses fertility was significantly higher than control.

Solutions to Reduced Natural Enemy Effectiveness Caused by Chemical Use

When both pesticides and natural enemies are employed in a crop, conflicts can be reduced by use of physiologically selective pesticides, use of pesticides in ecologically selective ways, or use of pesticide-resistant natural enemies. These solutions will be needed especially in crop management systems in which key pests are not currently controlled by biological methods but where natural enemies are important in control of secondary pests.

Physiologically Selective Pesticides. These are discovered by systematic testing to identify which pesticides, if any, of those available and effective for the control of the pests of the crop are also relatively harmless to the natural enemies to be conserved. Because populations of natural enemy species collected from different locations may differ in their susceptibility to a pesticide (Rosenheim and Hoy 1986; Rathman et al. 1990; Havron et al. 1991), susceptibilities must be measured for the local populations of natural enemies actually of interest. Also, information about effects of one pesticide is often not useful in predicting the toxicity of other pesticides to a given natural enemy or to other natural enemies (Bellows and Morse 1993). These facts dictate that only comprehensive local testing of pesticide-major natural enemy combinations can fully define which materials may be safely used in a crop (for spiders and brown planthopper on rice in the Philippines, see Thang et al. 1987). In western Europe, all pesticides are tested against eight standard species of natural enemies to partially characterize their likely risk to natural enemies.

Test methods are sensitive to the precise conditions selected for the assay. Careful attention must be given to standardizing the source, age, sex, and rearing history of the natural enemies used in tests, as well as the temperature, relative humidity, and degree of ventilation of the test environment, and the formulation, purity, and dosage of the test material (Croft 1990). The use of standardized assay conditions, such as those developed by the IOBC (International Organization for Biological Control) is critical if studies are to be compared (Hassan 1977, 1980, 1985, 1989a; Hassan et al. 1987; Morse and Bellows 1986). Basic to many such tests is the simultaneous testing of the pest organism under the same conditions as the natural enemies to determine whether differences in susceptibilities exist. As a general rule, pests are less susceptible to pesticides than are their natural enemies.

Methods for such screening range from laboratory tests, through semi-field tests to field studies. Laboratory methods include treatment of natural enemies through ingestion of pesticide or pesticide-treated materials, topical application, and placement of natural enemies on freshly dried pesticide residues on surfaces on which natural enemies are compelled to rest. The slide-dip technique in which organisms are immersed in a pesticide solution is commonly used for tests with mites. Exposure to residues on test surfaces can involve glass, sand, or leaves as the test surface. Foliage may be sprayed in the laboratory or field, and used either immediately after drying, or after aging for various lengths of time under field or standardized laboratory conditions. Semi-field tests involve confining test organisms on parts of plants or whole plants, after treatment of foliage with pesticides. Field tests involve assessing impacts on natural enemy populations when whole fields or plots are treated with pesticide. In field tests, the use of small, replicated plots is often unsatisfactory because natural enemies are mobile and poor separation of treatment effects occurs. The use of large unreplicated plots, with repetition over time, often gives more satisfactory results (Brown 1989; Smart et al. 1989).

Measures used to express degrees of susceptibility to pesticide include the size of the dose that kills half of a sample of the test organisms (LD_{50}). Where organisms are not orally or topically dosed, but rather confined on a treated surface, the measure LC_{50} is used, which is the concentration of solution applied to a treated surface that kills half of the test organisms in a defined period of time (usually 24 or 48 h). Tests which incorporate measurement of effects of pesticide residues of various ages (aged under either natural or defined environmental conditions) are especially helpful in defining the period of risk that particular species of natural enemies experience after a pesticide application (Bellows et al. 1985, 1988, 1992a, 1993; Morse et al. 1987; Bellows and Morse 1988). The ratio of the LC_{50} values of the natural enemy and the pest, or that of the natural enemy to the recommended application rate for a pesticide is a useful comparative measure of the selectivity of a pesticide (Morse and Bellows 1986, Bellows and Morse 1993).

Assessment of natural enemy performance (ability to encounter and subdue prey successfully or, for parasitoids, to locate and oviposit in hosts) is a better indicator of the total effect of pesticide residues than is mortality because it also incorporates the sublethal effects of pesticides on natural enemies.

Ecologically Selective Ways of Using Pesticides. Pesticides can be used in various ways that reduce contact with natural enemies (Hull and Beers 1985), as are discussed below:

Reduced Dosages. Effects of pesticides on natural enemies can be decreased by reducing the dosage applied (Poehling 1989). Use of half or quarter rates of pesticides often provides adequate pest control while reducing natural enemy mortality.

Selective Formulations and Materials. The physical characteristics of pesticide formulations influence their impact on natural enemies. Granular formulations applied to the soil, for example, do not contact natural enemies on foliage or in the air and hence many natural enemies are unaffected by such applications (Heimbach and Abel 1991). Such materials, however, are often designed for the purpose of producing pesticide residues in the topsoil and, in that zone, contact with natural enemies may be prolonged and extensive; such applications would be expected to significantly reduce susceptible natural enemy populations that live in the soil or forage on its surface. Systemic pesticides do little direct damage to natural enemies which do not consume plant sap and thus do not contact the pesticide (Bellows et al. 1988). Pesticides that kill only if ingested, rather than by mere contact with the integument, are less likely to harm natural enemies (Bartlett 1966). Stomach poisons such as some pathogen-derived materials (e.g., *Bacillus thuringiensis*), plant-derived materials (certain alkaloids, Bellows et al. 1985; Bellows and Morse 1993), or mineral compounds (such as, cryolite, Bellows and Morse 1993) are usually not damaging to predators and parasitoids which do not eat plant tissues. However, even stomach poisons can be harmful to natural enemy populations if they cause drastic reductions in host or prey densities.

Limiting Areas Treated. The extent of the area treated with pesticides can be adjusted to reduce exposure of natural enemies. For example, treatment of alternate rows in apple orchards instead of entire blocks controls mobile orchard pests, yet allows greater survival of the coccinellid mite predator *Stethorus punctum* (LeConte) (Hull et al. 1983). DeBach (1958) successfully controlled purple scale, *Lepidosaphes beckii* (Newman), in citrus by applying oil to every third row on a six-month cycle. This approach provided satisfactory control of this species without destroying the natural enemies of other citrus pests. Velu and Kumaraswami (1990) found treatment of alternate rows in cotton to provide effective pest control and, for some of the chemicals tested, to enhance parasitism levels of key pests. In contrast, Carter (1987) found that strip spraying of cereals in the United Kingdom did not provide satisfactory control of aphids when strips were 12 meters wide because natural enemies did not colonize the sprayed strips in time to suppress aphid resurgence.

Limiting Applications in Time. Contact between pesticides and natural enemies can be limited by using either nonpersistent materials, making less frequent applications, or applying materials in periods when natural enemies are not present or are in protected stages. Using nonpersistent pesticides reduces damage to natural enemy populations because natural enemies which emerge after toxic residues have declined (from inside protective structures such as cocoons or mummified hosts) are likely to survive. Also, natural enemies that arrive from untreated areas can recolonize treated fields sooner. Persistence of pesticides varies greatly. Materials such as diazinon or azinphosmethyl leave residues on foliage and other surfaces for one or more weeks at levels that kill natural enemies. Some herbicides, such as the triazines, applied to soil last for months. Other materials, such as the insecticide pyrethrin, degrade in hours or days. Weather conditions strongly affect persistence of pesticide residues. Most important among these are rain, which can wash residues off, and temperature, which can influence both the toxicity of the pesticide and the rates of dissipation and degradation of residues.

Adjusting the timing of pesticide applications to protect natural enemies is a matter either of reducing overall spray frequency so that there are times when the crop foliage is not toxic to natural enemies, or changing the exact timing of particular applications to avoid periods when natural enemies are in especially vulnerable life stages. Gage and Haynes (1975), for example, successfully used temperature-driven models of insect development to time pesticide applications against adult cereal leaf beetle, *Oulema melanopus*, treating after beetles had emerged, but prior to emergence of the parasitoid *Tetrastichus julis* (Walker). This system conserved this important parasitoid, whereas the previous approach of directing pesticide applications at the first generation of cereal leaf beetle larvae (the stage attacked by the parasitoid) did not. Efforts to redirect pesticide applications to periods when natural enemies are less vulnerable may require that natural enemy populations be monitored to determine when susceptible natural enemy stages are present, with the goal of creating pesticide-free times around critical periods. (For more on methods for monitoring natural enemies see Chapter 13 on evaluation methods). Monitoring methods have been employed to detect adults of some parasitoids to aid in their integration into crop management systems as, for example, with parasitoids of California red scale, *Aonidiella aurantii*, on citrus in South Africa (Samways 1986) and parasitoids of San José scale, *Quadraspidiotus perniciosus* (Comstock), in orchards in North Carolina (U.S.A.) (McClain et al. 1990). If many pesticide applications are required, it becomes increasingly difficult to avoid periods when natural enemies are in vulnerable life stages.

System Redesign. Options for the conservation of natural enemies are increased when the need for repeated use of broad spectrum pesticides is eliminated through the development of nontoxic pest control methods (such as use of natural enemies or other methods including traps, mating disruption with pheromones, and cultural methods). Reduced frequency of pesticide use in a crop is likely to greatly increase the survival and population densities of natural enemies, as in pear (*Pyrus communis* Linnaeus) orchards in Oregon, when mating disruption (based on pheromones) was substituted for organophosphate pesticides for control of codling moth, *Cydia pomonella* (Linnaeus). This substitution raised the densities of the predacious hemipteran *Deraeocoris brevis piceatus* Knight and the lacewing *Chrysoperla carnea* (Stephens), resulting in an 84% drop in densities of the pear psylla, *Psylla pyricola* Förster, and a reduction of fruit contamination by honeydew from 9.7% to 1.5% (Westigard and Moffitt 1984).

Creation and Use of Pesticide-Resistant Natural Enemy Populations. Where pesticides are applied to crops and no sufficiently selective material or method of application can be discovered, attempts have been made to release and establish pesticide-resistant strains of key natural enemies. The intent of such releases is to permanently establish the pesticide-resistant form of the natural enemy so that pesticides may continue to be applied for other pests, while not disrupting control of the pest suppressed by the resistant natural enemy.

Pesticide-resistant strains of several species of phytoseiid mites have been developed by laboratory selection or recovered from field populations, including *Metaseiulus occidentalis* (Nesbitt) (Croft 1976; Hoy et al. 1983; Mueller-Beilschmidt and Hoy 1987), *Phytoseiulus persimilis* (Fournier et al. 1988), *Typhlodromus pyri* and *Amblyseius andersoni* (Chant) (Penman et al. 1979, Genini and Baillod 1987), and *Amblyseius fallacis* (Whalon et al. 1982). Resistant strains of parasitic Hymenoptera have also been isolated from field populations and resistance levels to some pesticides further augmented by laboratory selection. Species have included an aphid parasitoid (*Trioxys pallidus* Haliday, Hoy and Cave 1989), a leaf miner parasitoid (*Diglyphus begini* [Ashmead], Rathman et al. 1990), and some scale parasitoids (*Aphytis holoxanthus* DeBach, Havron et al. 1991 and *Aphytis melinus* DeBach, Rosenheim and Hoy 1988).

Studies of these organisms have demonstrated that for many natural enemies genetic variability exists that permits the development of pesticide-resistant populations under field or laboratory selection. In several instances, it has been demonstrated that these strains can establish and survive for one or more years in commercial fields or orchards where pesticide applications are made (Hoy 1982b; Hoy et al. 1983; Caccia et al. 1985). Initial establishment of resistant strains is fostered by prior destruction through pesticide application of any existing susceptible population of the same species (Hoy et al. 1990). Long term persistence of the resistant strain is needed if economic costs of strain development are to be offset by prolonged benefit. In some cases, such as the use of *Phytoseiulus persimilis* for mite control in greenhouse crops, no susceptible strain is present, and it is sufficient merely for the resistance to last for the life of the crop (usually 3–6 months), because new predators will be released in future crops (Fournier et al. 1988). In outdoor crops, maintenance of the resistant strain may require regular pesticide application. Where such applications are employed, introductions of pesticide-resistant natural enemies can lead to their replacement of existing, pesticide-susceptible species (Caccia et al. 1985). In the absence of such ongoing pesticide usage, the introduced strain of resistant natural enemy may be displaced by other, pesticide-susceptible species (Downing and Moilliet 1972). The importance of the level and sustained nature of pesticide selection to the establishment of resistant strains of natural enemies in the field has been pointed out by Caprio et al. (1991). In some cases, the need for continued treatments in the field to retain resistance in natural enemies may be met by pesticide treatments made for other pests in the crop system. Trials in the United Kingdom with an organophosphate-resistant strain of *Typhlodromus pyri* showed survival of the predator in orchards treated with organophosphate insecticides at levels sufficient to control *Panonychus ulmi* (Koch) and *Aculus schlechtendali* (Nalepa). In a pyrethroid-treated orchard this strain of *T. pyri* was scarce and did not suppress pest mites (Solomon et al. 1993).

Conservation Philosophy. Effective conservation of natural enemies through either physiological or ecologically selective pesticides involves changes by growers in outlook as well as technological changes in procedure. Crop production systems based on biological control seek to use pesticides as supplements to natural enemies, not substitutes for them. Emphasis on obtaining a high level of pest control from pesticide application is likely to be detrimental when biological control agents are part of the system. Pesticides can be integrated more effectively with natural enemies when used so as to inflict only moderate levels of mortality (30–60%) on unacceptably high pest populations, when natural enemy action has been insufficient. If pesticides, of whatever degree of physiological or ecological selectivity, are used at rates and frequencies designed to provide the first and basic means of control, natural enemy populations are likely to be too disrupted by loss of their host or prey to provide any significant level of control in the system. How to build pest management systems based on biological control is considered further in Chapter 14.

Various agricultural practices (Table 7.2) other than using agricultural chemicals also affect natural enemies and are considered in the remainder of this chapter.

TABLE 7.2 Practices that Promote Natural Enemy Conservation

1. Limited and selective use of pesticides
2. Refuges adjacent to crop
   a) for hosts and alternate hosts
   b) for supplemental foods (pollen, honeydew)
   c) for overwintering of natural enemy or passing periods between host availability
3. Within-crop habitat improvement
   a) cover crops
   b) intercropping
4. Between-crop natural enemy transfer
   a) crop residue management
   b) strip harvesting
   c) alternate row pruning
   d) landscape crop patterning
5. Direct Provisioning
   a) shelters
   b) food

MANAGEMENT OF SOIL, WATER, AND CROP RESIDUES

Soil Management

Management of soil in the production of crops may involve tillage, addition of manures or chemical fertilizers, treatments to conserve desired pH, or other practices. These can affect the pest problems that subsequently develop in the crop, in some cases by directly affecting the pests themselves and in others by affecting their natural enemies. The direct effects of soil management on pests are a type of cultural pest control and are discussed by Stern et al. (1976). Here we discuss effects of soil management on natural enemies of insect pests. Effects of soil management and other cultural practices on biological control of plant pathogens are discussed in Chapter 12.

The conservation of natural enemies in soil is not different in principle from the conservation of natural enemies in the crop canopy. However, compared to the above-ground zone, the soil has been little studied in terms of the dynamics of organisms living there. Additional information is needed about population interrelationships among various organisms in soil, and between these and pests or natural enemies that pass some part of their life cycles in the soil. Nilsson (1985), for example, noted that ploughing reduced the per m^2 emergence of parasitoids of rape pollen beetles, *Meligethes* sp., in Sweden by 50 and 100% from spring and winter rape crops, respectively. He suggested that direct-drilling of winter wheat crops following rape (*Brassica napus* Linnaeus), rather than ploughing, would act to conserve pollen beetle parasitoids regionally. Ellis et al. (1988) linked an outbreak of cereal leaf beetle, *Oulema melanopus*, in one area of Ontario, Canada, to local cropping patterns in which cereal crops were tilled immediately after harvest instead of the more common pattern of use of cereals as a companion crop with alfalfa with no tillage. Tillage is known to kill 95% of the parasitoid *Tetrastichus julis*, which was absent in the outbreak area, but parasitized 74–90% of the pest's larvae in other parts of Ontario. Beneficial effects of manure as a fertilizer source, in lieu of chemical fertilizers, have been noted for such predacious arthropods as carabid beetles (Purvis and Curry 1984; Hance and Gregoire-Wibo 1987) and the laelapid predacious mite genera *Androlaelaps* and *Stratiolaelaps* which feed on eggs of *Diabrotica* spp. that are pests of maize (Chiang 1970). Also, soil is a major reservoir for inoculum of viral and fungal pathogens and its characteristics affect pathogen persistence. For example, the nuclear polyhedrosis virus of the noctuid cabbage pest *Trichoplusia ni* (Hübner) is more persistent in less acid soils and liming of soils for virus conservation has been recommended (Thomas et al. 1973). In New Zealand, hepialid pasture pests (*Wiseana* spp.) cause greater damage in recently tilled pastures in which cultivation has buried the nuclear polyhedrosis virus of these pests. Virus builds up as pastures age, resulting in greater mortality of the pest (Longworth and Kalmakoff 1977).

Disturbance of dry soils by farm machinery or vehicles in some climates can lead to dusty foliage on orchard crops. Dusty foliage in citrus is sometimes associated with outbreaks of pest insects such as scales (Bartlett 1951; DeBach 1958). The mechanism leading to this condition is believed to be the effect of dusty surfaces on the behavior of parasitic Hymenoptera and predators. When foraging on dusty surfaces, parasitoids spend more time grooming and less time locating and ovipositing in hosts. Dusty surfaces are more quickly abandoned by foraging parasitoids, reducing the likelihood that host encounters will occur with the consequent intensification of local search for additional hosts. Effects on predators are variable. Fleschner (1958) found that the coccinellid mite predator *Stethorus picipes* Casey suffered increased mortality due to desiccation and lowered searching efficiency on dusty foliage. In contrast, the effectiveness of the phytoseiid mite *Metaseiulus occidentalis* as a predator of the Pacific spider mite, *Tetranychus pacificus* McGregor, did not differ in greenhouse and laboratory trials between clean and dusty foliage of two plant species (Oi and Barnes 1989).

In general, where dusty conditions and pest outbreaks occur together, the dust should be viewed as a possible reason for the outbreak. Investigations should then be undertaken to determine whether the natural enemies involved are less effective under dusty conditions. This would be especially likely for scale insects with locally-established, effective parasitoid species. Other potential causes for higher pest populations of some species on dusty edge rows of orchards may include higher temperatures on these trees as they are exposed to greater solar radiation.

Water Management

Irrigation raises humidity in the crop, and this may be important in making the environment more favorable for some kinds of natural enemies. For example, it may be possible to promote epidemics of insect fungal pathogens by manipulation of irrigation or greenhouse watering patterns. Efficacy of *Verticillium lecanii* applications in greenhouses for aphid or whitefly control can be enhanced by manipulating the crop foliage density, watering, and night time temperatures so as to maintain high humidities needed for germination of pathogen spores (Hall 1985). Epizootics of the entomopathogenic fungi *Erynia neoaphidis* Remaudière and Hennebert and *Erynia radicans* (Brefeld) occurred in pea aphids, *Acyrthosiphon pisum* (Harris), on ground covers in Georgia (U.S.A.) pecan (*Carya illinoensis* Koch) orchards which employed overhead, but not drip, irrigation (Pickering et al. 1989). Normal background levels of diseases of pests may also be enhanced. While difficult to quantify, mortality from such diseases is likely more important than generally recognized. Daily loss rates of only a few percent, when sustained over stages with developmental times of weeks or months can, in aggregate, be sources of substantial levels of mortality. Ekbom and Pickering (1990), for example, measured daily losses in the blackmargined aphid, *Monellia caryella* (Fitch), on pecan trees and found that while daily rates of mortality were typically 5–10%, in aggregate over weekly periods, losses were 45–50% and would be even larger when expressed in life table terms over the lifespan of an aphid cohort. Little, however, has yet been done to develop techniques to systematically enhance such processes. Significant potential exists to manipulate crop relative humidities and wetting periods (through crop spacings and irrigation practices), as well as other factors such as phylloplane chemistry (pH, nutrient levels) to enhance arthropod disease levels (Harper 1987). Flooding is used also in some crops to control pests. Flooding was evaluated by Whistlecraft and Lepard (1989) as a means to control the onion pest *Delia antiqua*, but was found to be damaging to the key parasitoid *Aleochara bilineata*.

Crop Residue Management

In many crops, residues left after harvest are disposed of through methods such as burning or tillage. In some instances, these practices may be done to gain an explicit benefit. In other cases, however, crop residue destruction may have no definite function other than being the traditional method to clear the soil surface to facilitate reuse of the land for the subsequent crop.

In some instances, studies have shown that crop residue management can be important in conserving natural enemies of the crop's pests. The phytoseiid *Amblyseius scyphus* Shuster and Pritchard, a predator of the tetranychid mite *Oligonychus pratensis* Banks in Texas (U.S.A.), for example, overwinters inside sorghum (*Sorghum bicolor* [Linnaeus] Moench) straw and is killed if the crop residue is destroyed. Conservation of the straw in a nontillage system has potential

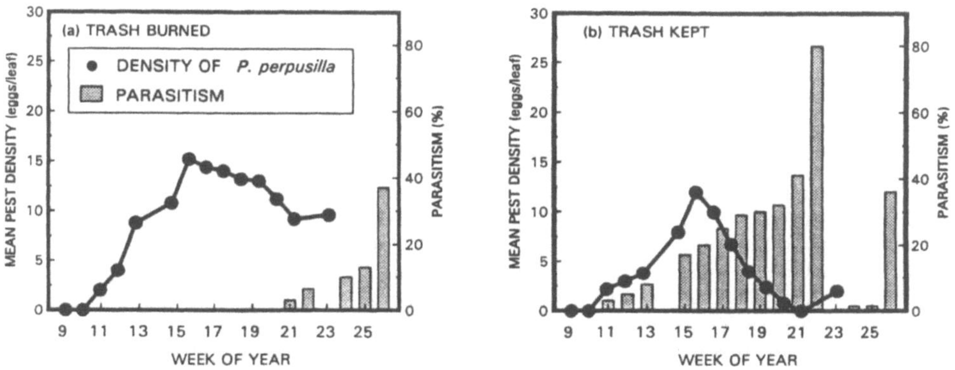

Figure 7.9. Population density of *Pyrilla perpusilla* (Walker) and egg parasitism of it by *Parachrysocharis javensis* (Girault) in sugarcane fields where the trash was burned following harvest (a) or was left in the field following harvest (b) (after Mohyuddin 1991).

to increase predator survival between sorghum plantings, enhancing predator numbers early in subsequent crops (Gilstrap 1988). In India, several parasitoids (*Epiricania melanoleuca* Fletcher, *Ooencyrtus papilionis* Ashmead, *Parachrysocharis javensis* [Girault]) of the sugarcane leafhopper *Pyrilla perpusilla* Walker are eliminated when crop residues are burned. Studies showed that if crop residues were left unburned and spread back on the field after burning, the parasitoids could be conserved at levels able to control the pest (Fig. 7.9) (Joshi and Sharma 1989, Mohyuddin 1991). In rice, preservation of bundles of rice straw in harvested paddies was found to conserve natural enemies of rice pests between plantings (Shepard et al. 1989). In cole crops, retention of live, postharvest plants helps conserve the parasitoid *Cotesia rubecula* (Marshall) which attacks larvae of the imported cabbageworm, *Pieris rapae* (Linnaeus) (Van Driesche, unpubl.). These examples illustrate that crop residue management is a farming practice that, in some cases, significantly affects important natural enemies. It should not be assumed that "clean" farming (weed-free, crop residue-free) is necessarily the best approach as a matter of principle; rather, practice should be guided by local experimentation that compares effects of various crop residue management practices on a variety of factors, including pests, natural enemies, plant diseases, and crop yield.

CROP PATTERNS

Perennial crops such as orchards and vineyards provide relatively permanent host plants for pest populations and their natural enemies; annual crops are potentially less similar from year to year in location. However, just as farmers can plan crop rotations between fields and over a series of years to enhance soil fertility and suppress pests, cropping patterns can be used to either enhance or discourage colonization by insects, including natural enemies. Cropping patterns can be manipulated to promote earlier, more abundant colonization of new fields by natural enemies, or enhance their survival and reproduction once in the crop. While such cropping patterns can improve the performance of natural enemies populations, such patterns are likely to be most valuable in settings where natural enemies contribute significantly to pest suppression. Where natural enemies are lacking, designing cultivation to permit early insect colonization may lead primarily to early pest populations and should be avoided in these cases. Cropping patterns that enhance natural enemy colonization and efficacy may occur with respect to: (1) patches of one crop in an area over time, (2) growing more than one crop in a single field, and (3) effects among distinct crops at a larger (such as an entire field) level.

Single-Crop Patterns to Enhance Natural Enemy Colonization

Perennial crops such as coconuts, apples, and citrus persist in the same physical location for many years. This stability allows local development of natural enemy populations that can consistently be present in the crop, year after year, without a colonization phase in which new patches of the crop must be discovered, first by the pests and then by their natural enemies. This habitat stability may promote biological control because it eliminates the time lag often seen in annual row crops, in which natural enemies arrive too long after pest populations to maintain or suppress pests to acceptable levels during short cropping cycles. To enhance biological control in annual crops, those characteristics of perennial crops that affect natural enemies positively may be recreated by using patterns within and between years that favor natural enemies.

In some crops, such as rice and sugarcane in some areas, habitat stability results because a large part of the landscape is devoted to cultivation, with several plantings each year, resulting in continuous presence of the crop throughout the year (Mogi and Miyagi 1990). Furthermore, in these crops fields are planted at different dates such that all stages of the crop overlap on a local scale. Under these circumstances, those natural enemies adapted to the crop habitat have it available continuously, and new fields can be colonized soon after planting from nearby, more mature plantings.

Opportunities exist to pattern annual crops in ways that would provide opportunities for early colonization of new plantings. Vorley and Wratten (1987) showed that biological control of aphids could be improved if some cereal fields were sown earlier in the preceding fall so that they acquired and retained overwintering parasitized aphids. Parasitoids from these overwintering aphids emerged early and colonized adjacent fields of later-sown cereals when their aphid populations begin to develop the following spring. By acreage, only 4% of cereals needed to be sown early to serve as early season sources of colonizing parasitoids for other fields. Permanent, ungrazed grasslands were also effective sources of early season parasitoids of cereal aphids.

Similar possibilities exist for other crops. Cole crops (*Brassica* spp.), for example, are produced as early and late season crops in some areas, or grown continuously in others. However, only a small portion of the landscape in most areas is devoted to this class of crops. Fields are thus likely to be isolated from one another, and new plantings are often established in distinct locations with little regard for the location of previous plantings of the crop. In addition, destruction of residues of early crops and planting dates of late crops are not linked. These conditions delay colonization of new plantings by natural enemies. Early colonization of cole crop plantings by natural enemies can be promoted by coordinating plantings in time and space to build bridges between crops for natural enemies. The closer new plantings are to old ones, the more quickly they can be located and colonized by insects (pests and natural enemies) emigrating from old plantings. For rapid colonization to occur, plantings must also overlap in time. In some cases, new plantings can be established well before the harvest of old plantings, allowing ample time for transfer between plots. In cases where the old crop is harvested before the new plot is established, the destruction of the old crop's residues (through plowing or other practices) can be delayed long enough to allow natural enemies to emerge and colonize the new planting. In cabbage and broccoli (*Brassica oleracea* Linnaeus) plantings in Massachusetts, for example, early crops (May–July) and late crops (July–September) often do not overlap and destruction of residues from early plantings occurs before or at the same time as planting of late plots even though land used for the spring crop is not replanted the same year. Natural enemies of the lepidopteran pests on these crops can be better conserved by delaying plowing of spring crops until 3–4 weeks after harvest to allow movement to new plantings by natural enemies such as the parasitoid *Cotesia rubecula*, which is found in the cocoon stage on the foliage of the crops. After crops are harvested, regrowth of cole crop plants provides sites for continued development of the insect pests of these crops which in turn serve as the food resource for the natural enemies. In rape crops, management choices of farmers can strongly affect the number of parasitoids that successfully locate and colonized new crop plantings. Hokkanen et al. (1988) noted that spring parasitoid colonization of new rape fields in Finland was enhanced by locating them as closely as possible to fields sown to rape the previous year.

When large fields are completely harvested in a short period, natural enemies active in the crop are totally dispossessed of the means to maintain their populations and must emigrate or die for lack of hosts or prey. Harvest methods that divide fields in smaller units for staggered harvest conserve natural enemies. Strip-harvest of alfalfa, for example, helps retain populations of parasitoids of aphids, alfalfa weevil (*Hypera postica*), and *Lygus* spp. (van den Bosch et al. 1967). Nentwig (1988) found that when German hay meadows were strip-harvested, predacious and parasitic arthropods, especially spiders, became more abundant and herbivores less abundant. Alternate row pruning (which staggers growth of new succulent foliage, attractive as oviposition sites for pests such as whiteflies) can also be used to enhance biological control by parasitoids in some cases (Rose and DeBach 1992).

The best pattern for a crop at the farm level must be worked out for each crop-pest-natural enemy combination, to determine the distance between crop patches and the overlap in time needed to secure effective colonization of new plantings by natural enemies. Other factors that also may bear on the crop pattern selected are movement of natural enemies (or pests) between crops, build up of pests in soil (for example, plant pathogens), and effects of cropping patterns on rates of plant diseases in the crop (Vorley and Wratten 1987).

Intercropping

Whereas the goal of the single-crop sequencing strategy is to promote earlier discovery and colonization of new crop patches by both pests and natural enemies (to achieve a better ratio of the two), crop diversification strategies seek, among other effects, to delay or diminish the number of pests colonizing the crops, or reduce their retention in the crops. Intercropping is the growing of two or more crop species in the same field at the same time. Crops may be either completely mixed or may be segregated into separate rows, which are alternated in some pattern (Marcovitch 1935; Andow 1991b). Intercropping is related to the broader concept of vegetational diversification in agricultural fields (Andow 1991a), which also includes cover cropping and weedy culture.

Two beneficial effects are theorized to result from such vegetational diversification: reduced pest discovery and retention in the crop, and enhanced natural enemy numbers and action (Root 1973). Whether or not these effects are realized in any real cropping pattern is dependent on the biologies of the pests and natural enemies involved, the crops, and the physical environment. Andow (1986, 1988) in reviewing studies of intercropping found that herbivore species were reduced in density in 56% of cases, increased in 16%, and not affected in 28% of cases.

Determining the reasons for observed effects (the relative importance of reduced pest colonization and retention versus increased mortality from natural enemies) is difficult and both mechanisms may operate together. Andow (1990) illustrates a method to separate the relative importance of these factors for Mexican bean beetle, *Epilachna varivestis* Mulsant, in bean-mustard-weed mixtures. Russell (1989) reviewed the effects of intercropping on natural enemy action and found higher levels of mortality from natural enemies in 9 of 13 cases, lowered levels in 2, and no effect in 2 cases. Sheehan (1986) suggested that intercropping may be more beneficial to generalist species of natural enemies than specialists, which may perform better in pure cultures of the crop attacked by their host or prey species. No general characteristics exist that can be used to construct pest-suppressive crop mixtures. Rather, each potential combination of crops must be evaluated in the local environment to determine if it is of value in light of the specific crops, their pests, and natural enemies.

Further, the economic value of pest reduction from vegetation diversification may, in any specific case, be potentially offset by competition among the crop species and by reduction in mechanization of the farming system. In intercrops reviewed by Andow (1991a,b) where herbivores were reduced, yields were not improved for cole crops, whereas in most bean intercrops yields improved, and results were mixed in alfalfa.

Patterns that Enhance Natural Enemies Across Crops

Just as sequential patterns of one crop may be arranged so as to facilitate colonization of new plantings by natural enemies immigrating from older plantings, spatial and temporal patterns of several distinct crops may be arranged to enhance natural enemies. Some natural enemies occur in several crops, feeding on one or several hosts or prey. In such cases, natural enemies may be enhanced in one crop by planting it near or subsequent to another crop which acts as a source of the natural enemy. Gilstrap (1988), for example, noted that in Texas the Banks grass mite (*Oligonychus pratensis*) is found on sorghum, wheat, and grass and that an effective phytoseiid mite moves among these crops feeding on the pest. He proposed methods to facilitate this movement to promote earlier increase of the predator in sorghum. Similarly, in Massachusetts the coccinellid *Coleomegilla maculata* is an important predator of Colorado potato beetle eggs on potato (*Solanum tuberosum* Linnaeus) but also occurs in maize, feeding on aphids and pollen. Both are major crops in the region and proximity of maize to potatoes has the potential to enhance levels of this predator in potatoes. Xu and Wu (1987) report successful use of a rape crop to increase the numbers of a coccinellid, which then moved into adjacent bamboo stands when the rape was harvested and controlled a pest scale on the bamboo. Corbett et al. (1991) reported that alfalfa planted next to cotton served as a reservoir for *Metaseiulus occidentalis*, which (if inoculated into the alfalfa early in the season) increased in number in the alfalfa and migrated into adjacent cotton areas.

MANIPULATION OF NON-CROP VEGETATION

Non-crop vegetation, either within the crop as ground cover or vegetation adjacent to crop fields, can influence the levels of biological control that occur in the crop in the following ways.

Improving Physical Conditions or Food Availability Within Crops

The principal types of within-crop vegetation that can be manipulated to enhance natural enemies are weeds and deliberately planted ground covers. In some instances, additional crop species may also be added to fields in small numbers to conserve key natural enemies. For example, water chestnut (*Eleocharis* sp.) can be planted into rice paddies to maintain the parasitoid *Tetrastichus schoenobii* Ferriere, an important egg parasitoid of the rice pest *Tryporyza incertulas* (Walker) which also attacks *Scirpophaga* sp. on water chestnut (Jiang et al. 1991). For some natural enemies, varieties of the crop itself may serve as a food resource, enabling species such as the phytoseiid *Amblyseius andersoni* (which eats both pest mites and crop pollen) to maintain higher populations during periods when mite prey are scarce (Gambaro 1988).

Cover crops, when compared to areas of bare, tilled soil, can lower soil temperature, raise relative humidity, and may make free water more readily available. In some systems, ground covers may enhance the spaces between crop rows as habitat for ground-living predators such as carabid and staphylinid beetles, as well as coccinellids, syrphids, and other species that may feed on arthropods living on the ground cover vegetation. In the United Kingdom for example, cabbage intersown with clover (*Trifolium* sp.) supported more effective, and in some cases, more abundant populations of soil-dwelling predacious arthropods (principally carabids and staphylinids) (O'Donnell and Croaker 1975; Ryan et al. 1980), which were associated with lower populations of cabbage root maggot, *Delia brassicae* Weidemann, and other pests. Whether or not ground covers enhance natural enemies will vary between crops and locations, being valuable in some cases and having no effect in others. Ground covers are more likely to enhance generalist predators than specialist parasitoids. Where ground covers reduce pest levels, both enhanced natural enemy levels and lowered rates of pest colonization or retention may be involved, and their effects may be difficult to separate. Ground covers also have the potential to compete with crops for moisture or nutrients, reducing crop yield. Ground cover species and sowing densities must be selected and tested locally to determine their value.

In some tree crops, ground covers may be widely used for reasons other than pest control and varying the plant species to enhance their effects on natural enemy conservation may be done with little increase in competition to the crop. In other areas, where bare ground culture is the norm (due, for example, to water limitation) this may not be the case. Ground covers in tree crops have been used most often to enhance natural enemies of spider mites and aphids. Studies of European red mite, *Panonychus ulmi*, in apple orchards in Michigan (U.S.A.) indicated that control by the phytoseiid *Amblyseius fallacis* was linked to the existence of predator populations in early spring on ground cover vegetation (McGroarty and Croft 1975). In citrus groves in south and south central China, control of the citrus red mite, *Panonychus citri* (McGregor), is enhanced by the use of tropical ageratum, *Ageratum conyzoides* Linnaeus, as ground cover within groves (Zhang and Olkowski 1989). The pollen of this ground cover and psocids found on it provide food for predatory mites (*Amblyseius* sp.). In addition, the ground cover lowers the temperature and raises the relative humidity, making the habitat more favorable in dry regions such as Jiangxi and Hunan Provinces for *Amblyseius eharai* Amstai and Swirski. In southeast Queensland, Australia, the use of Rhodes grass, *Chloris gayana* Kunth, ground covers in citrus orchards enhanced food supplies for *Amblyseius victoriensis* (Womersley), an effective predator of the native eriophyid *Tegolophus australis* Keifer, because *A. victoriensis* fed on Rhodes grass pollen (Smith and Papacek 1991). In contrast, levels of two-spotted spider mite, *Tetranychus urticae*, in the early season in peach, *Prunus persica* (Linnaeus) Batsch, orchards in North Carolina increased when orchards floors were sown to narrow leaf vetch, *Vicia angustifolia* Reichard, instead of being kept as bare ground (Meagher and Meyer 1990). Ground covers in apples in British Columbia, Canada, reduced aphid populations up to 4-fold in some years, but also reduced tree growth rates (Halley and Hogue 1990).

Weeds in crops also affect natural enemies (Altieri and Whitcomb 1979). Weeds between crop rows or on orchard floors serve as volunteer ground covers and have the potential to be beneficial, depending on the properties of the weed species, associated natural enemies, and level of competition with the crop. In sugarcane in Louisiana (U.S.A.), subcompetitive stands of broadleaf weeds enhanced predators, especially the red imported fire ant, *Solenopsis invicta*, a major predator of the key insect pest, the sugarcane borer, *Diatraea saccharalis* (Fabricius) (Ali and Reagan 1985). Weeds were killed by crop competition as canopy closure occurred, and yields were enhanced 19% as compared to weed-free plots. In contrast, weeds in maize in New Zealand raised levels of the pest armyworm *Mythimna separata* (Walker) tenfold, reducing yield by 30% (Hill and Allan 1986). Parasitism of this armyworm by *Apanteles ruficrus* (Haliday) was greater in weed-free plots. As with cover crops, effects of weeds are likely to be variable among locations and will require local testing to identify useful crop-weed combinations. Because weed flora is likely to vary between years (in contrast to sown ground covers), year-to-year results also may vary.

Resources Adjacent to Crops

Vegetation adjacent to crops may benefit natural enemies of crop pests by providing sources of plant-derived foods such as nectar and pollen. Such vegetation may also host arthropods such as aphids that provide carbohydrate sources such as honeydew. Arthropod populations on border vegetation may also serve as food for predators or as hosts for parasitoids that also attack the crop's pests (Dennis and Fry 1992). Nectar from wild flowers is eaten by adults of some species of Hymenoptera, and abundance of such flowers may raise parasitism levels in some instances (Leius 1960, 1967). Cereal fields in the United Kingdom which were divided by strips of weeds and wild flowering herbs increased food availability and reproduction for the most abundant carabid, *Poecilus cupreus* Linnaeus (Zangger et al. 1994). *Rubus fruticosus* Linnaeus (blackberry) plants in hedges in Switzerland support large populations of the important mite predator *Typhlodromus pyri*. These hedge populations, when located upwind from vineyards, served as sources of immigrants that repopulated vineyards that had lost predators due to pesticide use (Boller et al. 1988). In Queensland, Australia, windbreaks of *Eucalyptus torelliana* F. Mueller act as reservoirs of *Amblyseius victoriensis* that repopulate citrus orchards (Smith and Papacek 1991). These trees support few prey mites, but provide abundant pollen on which the predatory mites survive. In Californian vineyards, the grape leafhopper, *Erythroneura elegantula* Osborn, is attacked by the egg parasite *Anagrus epos* Girault, which occurs in adequate numbers only in vineyards near riparian areas which support wild blackberry (*Rubus* spp.) plants. These blackberry plants support another leafhopper, *Dikrella californica* (Lawson), that serves as an overwintering host for the parasitoid (Doutt and Nakata 1973). Enhancement of the grape leafhopper parasitoid in vineyards has been achieved more effectively by planting rows of French prunes (*Prunus* sp.) adjacent to vineyards. These trees support a third leafhopper, *Edwardsiana prunicola* (Edwards), that also serves as an overwintering host for the parasitoid (Wilson et al. 1989). The prune leafhopper system responds better to agricultural manipulation than the blackberry leafhopper whose populations do not develop well on blackberries away from riparian habitats (Pickett et al. 1990). A similar system occurs in Turkey, with higher levels of leafhopper parasitism occurring when wild *Rosa* spp. and *Rubus* spp., which harbor alternate leafhopper hosts for egg parasitoids, occur near vineyards (Yigit and Erkilic 1987).

Experimental demonstration that natural enemy individuals from border vegetation attack pest species on crop plants should be sought as some parasitoids appear to exhibit a certain degree of fidelity to their natal host, natal plant species, or both. Starý (1983), for example, noted that the aphids *Aphis urticata* Fabricius and *Microlophium carnosum* (Buckton) on stinging nettles (*Urtica* sp.) are attacked by *Aphidius ervi* Haliday, an important parasitoid of such pest aphids as *Sitobion avenae* (Fabricius) and *Acyrthosiphon pisum*. Cameron et al. (1984) found, however, that when pest aphids on potted crop plants were placed in the wild habitat few were parasitized, suggesting that *Aphidius ervi* populations on various aphids and plants were relatively isolated from one another, with low rates of switching. This conclusion was further confirmed by Powell and Wright (1988), who supported the existence of strains of *Aphidius ervi*, but also provided a note of caution about the use of laboratory parasitoid cultures to evaluate the potential for host switching. They recommended that switching tests be done in the field, as laboratory colonies of parasitoids may lose genetic diversity, and field populations of parasitoids on alternative hosts on wild plants may consist of both individuals with narrow plant foraging preferences and others with wide preferences (Powell and Wright 1992). Proof then that border vegetation provides emigrating natural enemies that enter crops and attack pests there must entail more than the co-occurrence of the same parasitoid species in each habitat. Habitat- and host-switching experiments with trap host exposures in the field are needed, as are verification of dispersal from source to crop. Also, for some natural enemy groups, immigrants may come from more distant sources rather than natural habitats immediately adjacent to crop patches. Approximately half of the spiders, for example, arriving to crop patches in a test in Tennessee (U.S.A.) were species not found in nearby natural vegetation, but rather had entered the area via long distance ballooning (Bishop and Riechert 1990).

Providing Refuges for Natural Enemies in Unfavorable Periods

In many cropping systems, periods occur that are unsuitable for active growth and reproduction of natural enemies, such as winter in high latitudes, dry seasons in some tropical areas, or periods when the crop is not grown. Natural enemy conservation requires that the needs of natural enemies be considered for the entire year, including such periods of inactivity. Some natural enemies pass these seasons in fields on crop residues (see section on crop residue management). In such cases, management of these materials will be important in determining the survival rate of the natural enemies. In other cases, important natural enemies pass unfavorable seasons outside crop patches. For such species, it is important to know where and under what conditions these periods are spent to ensure that favorable sites are available close to crop fields. Research on the overwintering habitat requirements of carabid and staphylinid predators of cereal aphids in the United Kingdom (Thomas et al. 1992; Dennis et al. 1994) and of coccinellids in Belgium (Hemptinne 1988) illustrate the kind of studies needed to define the ecological needs of particular species. Once these needs are known, conditions near crops can be modified to increase the number or quality of such sites. Artificially-created, grass-sown raised earth banks in English cereal fields provided overwintering sites for predators of cereal aphids, enhancing their numbers in adjacent crop areas the following year (Thomas 1990; Thomas et al. 1991). Windbreaks of *Eucalyptus torelliana* around peach (*Prunus persica* orchards in southern New South Wales, Australia, provided overwintering refuges and enhanced colonization of orchards by predatory mites in the spring (James 1989). In general, studies of the overwintering, dry season, or other "off-season" needs of key natural enemies should be routinely investigated as part of comprehensive efforts to maximize populations of such beneficial species.

PROVIDING FOOD OR SHELTER

In some cases, natural enemy survival and reproduction can be enhanced by directly providing foods, hosts, or shelter, or by controlling antagonists or parasitoids of the natural enemies themselves. This concept is related to manipulating ground covers and border vegetation, which also aims to augment such resources, but differs in that the resources are not provided by modifying the vegetation, but are provided directly by the farmer. To be successful, such practices must be relatively inexpensive and produce a large enough increase in pest mortality to merit the cost in materials and labor.

Among the food resources required by many natural enemies are carbohydrates for energy and proteins for growth and reproduction. In nature, carbohydrates may be obtained from prey or host fluids, homopteran honeydew, plant nectar, and other plant materials rich in sugars. Wild flowers, flowering weeds in the crop, extrafloral nectaries in the crop (Rogers 1985), homopterans on the crop, ground covers, or bordering vegetation are all potential sources of such resources. When these sources are inadequate, they may be artificially provided. For example, in Massachusetts, *Edovum puttleri* Grissell, an egg parasitoid of the Colorado potato beetle, *Leptinotarsa decemlineata*, is ineffective in the first half of the cropping period, when potatoes lack aphids (a source of honeydew) (Idoine and Ferro 1990). Under such circumstances artificial application of the missing resource in the form of a sugar or molasses spray might be beneficial.

Protein sources such as protein hydrolysate, yeast, and pollen, required for reproduction by many adult natural enemies, may also be artificially added to crops. Hagen et al. (1970) enhanced reproduction of *Chrysopa* spp. in cotton through application of hydrolyzed proteins, mixed with water and sugar. In contrast, applications of mixtures of sucrose and yeast failed to increase predator numbers in apples (Hagley and Simpson 1981). Pollen increased developmental rates of some phytoseiids (McMurtry and Scriven 1964) and enhanced the proportion reaching the adult stage (Osakabe 1988). Greater numbers of the predator *Amblyseius hibisci* (Chant) on citrus were correlated to increased concentrations of cattail (*Typha latifolia* Linnaeus) pollen from natural sources (Kennett et al. 1979). Artificial application of cattail (*Typha* sp.) or tea (*Camellia sinensis* [Linnaeus] O. Kuntze) pollens appears to be a possible means of increasing the population densities of some phytoseiids and to increase their impact on prey species that are otherwise not adequately controlled.

Shelters and refuge areas for natural enemies, like food and hosts, can either be enhanced through modification of vegetation within crops or border areas, or may be provided artificially. Wooden shelters have been used to increase densities of native *Polistes* wasps in desired locations (Gillaspy 1971; Lawson et al. 1961). Artificial nests made of polyethylene bags have been used to manipulate ant (*Dolichoderus thoracicus* Smith) populations in cocao (*Theobroma cacao* Linnaeus) plantations in Malaysia (Heirbaut and van Damme 1992). Empty cans placed in fruit trees have been used to augment earwig (Dermaptera) numbers in fruit trees (Schonbeck 1988), and straw bundles have been used to enhance numbers of spiders in new plantings of rice (Shepard et al. 1989). Boxes have been used to provide overwintering sites for adults of *Chrysoperla carnea* (Sengonca and Frings 1989). Overwintering of *Metaseiulus occidentalis* in apple orchards in China (introduced from California) was achieved only after overwintering sites around tree trunks were provided. These consisted of either waste cotton held against tree trunks by plastic sheets, or piles of leaf and grass litter piled at the base of trees (Deng et al. 1988). Populations of insectivorous forest birds have been enhanced through the provision of nesting boxes (Bruns 1960). (Note, however, that little proof exists that enhanced bird populations suppressed the target pest insects). Barn owl (*Tyto alba*) densities in Malaysian oil palm plantations have been increased by providing nesting boxes, enhancing rat control (Modh 1990).

Where specific antagonists of important natural enemies exist, their control can be important in preserving the effectiveness of the beneficial agents. For example, some species of ants such as the Argentine ant (*Linepithema humile*), the bigheaded ant (*Pheidole megacephala* [Fabricius]), and *Lasius niger* Linnaeus interfere with the action of natural enemies by physically attacking and removing immature stages of some predators (such as larvae of coccinellids) and interfering with the host searching and oviposition activities of some parasitoids. In many cases, ants are present because they are collecting the honeydew from homopteran colonies of insects such as scales, mealybugs, whiteflies, and aphids. The suppressive influence of ants on natural enemy effectiveness has been demonstrated for scales (DeBach et al. 1951, 1976; Steyn 1958; Samways et al. 1982; Bach 1991) as well as aphids and mealybugs (Banks and Macaulay 1967; DeBach and Huffaker 1971; Cudjoe et al. 1993). Even some pests which do not produce honeydew, such as citrus red mite, *Panonychus citri*, may be affected through ant predation on larvae of such mite-feeding coccinellids as *Stethorus picipes* (Haney et al. 1987).

Restoration of effective biological control in such cases depends on control of the ant species involved, often through the application of pesticides to ant nests or tree trunks, or the application of sticky barriers to tree trunks. Musgrove and Carman (1965), Markin (1970 a,b), and Kobbe et al. (1991) provide information on the biology and control of the Argentine ant, one of the species most frequently interfering with natural enemies. Samways (1990) describes a method of sticky-banding trees to control pest-tending ants that is not phytotoxic to tree bark. Where pest ants are immigrant species, biological control of the ant species may also be a viable course of action.

SAFETY

Biological control through the conservation of natural enemies is safe. The techniques used are largely forms of enhancement of natural or agricultural habitats, or reductions in pesticide usage.

CHAPTER
8

INTRODUCTION OF NEW NATURAL ENEMIES: PRINCIPLES

INTRODUCTION

The intentional introduction of new natural enemies species to suppress populations of adventive pests has long been an important part of biological control. The process has also been applied to native species which are pests. The objective of the tactic is to introduce safe, effective natural enemies to suppress pest populations, and it has been applied in a wide variety of natural, agricultural, and urban settings. Introduced natural enemies have included invertebrates, vertebrates, and microbes, and these have been employed against pest plants, arthropods, and vertebrates. The potential for control of many pest organisms in diverse environments using introduced natural enemies is substantial. The approach has been applied to only a very small percentage of the world's pest species, and broader application of the approach would certainly be beneficial.

Several steps are involved in the development of a program for the introduction of natural enemies (Bartlett and van den Bosch 1964; Boldt and Drea 1980; Klingman and Coulson 1983; Schroeder and Goeden 1986; Waterhouse and Norris 1987; Coulson and Soper 1989; Harley and Forno 1992; Bellows and Legner 1993, Van Driesche and Bellows 1993). These include: identifying candidate or target pest species; selecting favorable search locations; identifying potential candidate natural enemy species; conducting the exploration, collection and shipment; quarantine processing of shipped material, followed by rearing and safety testing; field colonization; monitoring release locations for establishment of natural enemies; and program evaluation.

This overall process of natural enemy discovery, introductions, and evaluations is a series of steps, each of which is based on various ecological principles which help guide the process. This chapter first introduces the method through three case histories; considers the historical record of these activities, and describes ecological and biological bases for how and why these activities are carried out. The steps involved in conducting a program of natural enemy introduction are then discussed. (Chapter 9 provides more detail on the practical aspects of these activities). Finally, the chapter discusses the costs and resources necessary to undertake such programs, the benefits of such programs, and issues regarding the safety of introducing natural enemies.

CASE HISTORY EXAMPLES

There are hundreds of species to which the technique of natural enemy introduction has been applied successfully, including arthropods, aquatic plants, terrestrial plants, and vertebrates. Reviews of such programs are found in DeBach (1964a), DeBach (1974), Clausen (1978), DeBach and Rosen (1991), and Nechols et al. (1995). We present here the histories of three programs which illustrate important features of biological control through natural enemy introductions.

An Aquatic Weed, *Salvinia molesta*

Salvinia molesta, generally referred to as salvinia, is a free-floating aquatic fern indigenous to southeast Brazil. It has received attention as a botanical curiosity and as an aquarium plant, and has been spread widely in the past 50 years throughout the tropics. It has become one of the world's two worst aquatic weeds, and has been the target of several biological control efforts. The successful resolution of this problem illustrates many of the steps necessary for effective biological control by introduction of natural enemies. A more detailed review of the program is given by Thomas and Room (1986).

Salvinia molesta is sterile, but undergoes rapid vegetative reproduction by growth and fragmentation. It can double in size in as little as 2.2 days in favorable conditions and can quickly cover lakes and slow-moving rivers with mats up to 1 m thick. Thick mats of the plant can prevent the passage of even large diesel-powered boats, while a single layer of the plant can impede the passage of canoes. Transport by water, as well as commercial and recreational fishing, can be severely affected or halted. Mats of weed block access to drinking water by humans, domestic stock, and wildlife and can clog irrigation and drainage canals. During floods, moving mats destroy fences and other light structures. The plant is a major weed of rice production, and harbors the snail hosts of schistosomiasis, a major human health threat in the tropics. The covering of the water surface by mats of plants alters the normal aquatic environment, leading to the elimination of most of the benthic flora and fauna. The situation in Papua New Guinea illustrates some of the most severe sociological and economic effects caused by salvinia. In the floodplain of the Sepik River, nearly 80,000 people depend on water-borne traffic (via canoes) for food, fishing, access to markets, schools, and medical care. This floodplain became so infested with the weed that it became a major threat to the existence of these people by reducing or in some cases halting access to villages by canoe.

Early control attempts included chemical treatments, harvesting or removal, and containment with booms placed in the water. None of these was successful on a long-term basis, primarily for economic reasons. The methods were too expensive to permit the repeated treatments required to maintain suppression of the weed population.

Taxonomy played a critical role in the development of a biological control program against salvinia. The weed was initially (1959) thought to be *Salvinia auriculata* Aublet. Searches for natural enemies of this plant in South America revealed three possible agents (a moth, a grasshopper, and a weevil), but establishment of these agents in affected countries was rare and no significant impact on the plant was achieved. By 1970, it had become clear that the weed was not *S. auriculata*. Specimens in a South American herbarium were located which were identical with the weed, which was later described as *Salvinia molesta* (Fig. 8.1a). The native range of *S. molesta* was discovered in southeast Brazil, and collections of insects in this area yielded three promising natural enemies, two of which were identical with those tried previously, together with a third natural enemy, a beetle in the family Curculionidae. This weevil at

Figure 8.1. (A) The aquatic weed *Salvinia molesta* D.S. Mitchell; (B) the phytophagous weevil *Cyrtobagous salviniae* Calder and Sands; *Salvinia*-infested water body before (C) and after (D) suppression by the introduction of *Cyrtobagous salviniae* weevils. (Photographs courtesy of P. M. Room, CSIRO.)

first was thought to be the same species used previously, but was later determined to be a new species, *Cyrtobagous salviniae* (Fig. 8.1b). The initial introduction of this beetle at a lake in Australia resulted, in 14 months, in the complete destruction of a 200-ha salvinia infestation (Fig. 8.1 c,d; Fig. 8.2a). Establishment of the beetle was promoted by manipulating plant host quality through urea applications (Room and Thomas 1985). The weevil has been responsible for cost-effective, environmentally sound, permanent biological control of salvinia in several locations, including Papua New Guinea and Australia. Progress is good at additional locations, including India, East Africa, Zambia, and Namibia.

The successful program against salvinia illustrates the crucial importance of taxonomy, both of the target pest and the natural enemies, in biological control. Correct identification of the pest was vital both to discovering the correct place to search and to obtaining the correct natural enemy. In the absence of this taxonomic work, proper natural enemies might never have been located. The program also emphasizes the significance of finding natural enemies suitable for or adapted to the target pest. Nuances of plant-insect (and other pest-natural enemy) interactions, however subtle, may be critical in permitting a natural enemy to be effective against a particular pest.

Ash Whitefly, *Siphoninus phillyreae*

The whitefly *Siphoninus phillyreae* (Haliday) (Fig. 8.3a) is a multivoltine, sternorrhynchan homopteran, a group which has been the source of many immigrant pest species. Its recent

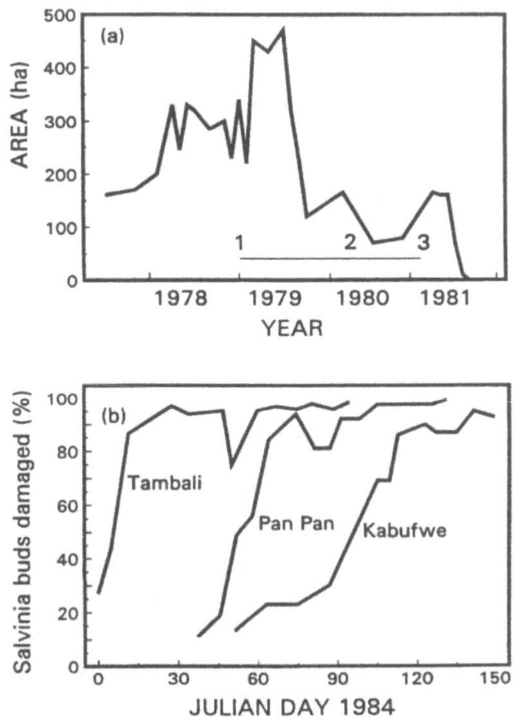

Figure 8.2. (a) Changes in area covered by *Salvinia molesta* D.S. Mitchell population on Lake Moondarra, Australia (after Thomas and Room 1986). (1) period of drought which reduced lake area by 55%; (2) *Cyrtobagous salviniae* Calder and Sands introduced into field cages; (3) beetles released from cages. (b) Increase in damage caused by the weevil *C. salviniae* to salvinia on three lakes in the Sepik floodplain, Papua New Guinea, in early 1984. In each case, damage to buds reached about 50% about seven months after beetles were introduced and most of the salvinia was destroyed in the following five months (after Thomas and Room 1986).

invasion of the Western Hemisphere, and the biological control program mounted against it, are described in Bellows et al. (1990, 1992b) and Gould et al. (1992a,b).

Siphoninus phillyreae is widespread throughout the Eastern Hemisphere from the Iberian peninsula to India, and from Ireland and Poland to sub-Saharan Africa. It feeds on several tree species and, although there have been a few reports of it causing damage to pomegranate, *Punica granatum* Linnaeus, (in Egypt) and pear (in the Balkan countries), these reports appear to be the exception rather than the rule and may indicate areas where the species is a recent immigrant rather than native.

The species was discovered for the first time in the Western Hemisphere in Los Angeles, California in 1988. At that time, the infested area exceeded 700,000 ha; eradication was not attempted. The insect was found infesting more than 50 tree species in its new range, and many of its favored hosts were widely planted as ornamental trees in urban settings. Nymphal populations reached densities which occupied 100% of the available leaf area on many plants. Aerial densities of adults interfered with normal human outdoor activities in areas near infested plants. No natural enemy of aleyrodids native to the region appeared to be providing any significant degree of suppression. The insect spread rapidly throughout most of California and into other portions of the United States, including Arizona and Nevada.

Unlike *Salvinia*, this pest was relatively well known in the literature. Early records included

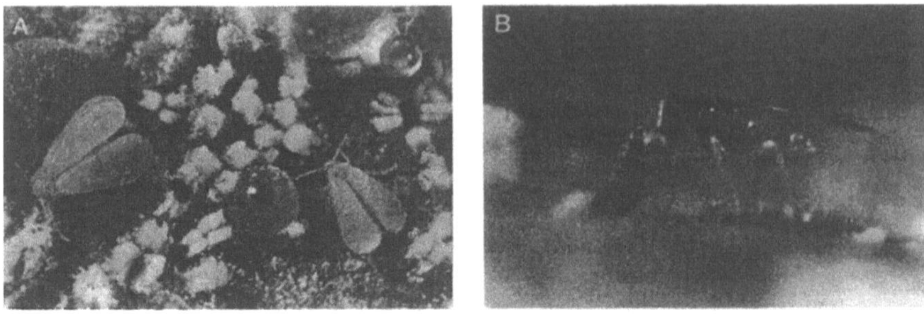

Figure 8.3. (A) The whitefly *Siphoninus phillyreae* (Haliday); (B) the aphelinid parasitoid *Encarsia inaron* (Walker). (Photographs courtesy of J.K. Clark.)

mention of natural enemies and accounts of possible earlier invasions in the Balkans which were followed by appearance of natural enemies and suppression of the pest. Contact with entomologists in countries surrounding the Mediterranean Sea, together with foreign exploration, resulted in introduction into quarantine (in California) of several species of aphelinid wasps and coccinellid beetles.

Of these, the aphelinid *Encarsia inaron* (Walker) from Israel (Fig. 8.3b) was reared in large numbers and released. It established at nearly every release site, and populations grew rapidly, decimating whitefly populations (Fig. 8.4). This parasitoid was distributed to all infested areas in the western United States within two years, established widely, and provided remarkable control in every location. During this time, the coccinellid beetle, *Clitostethus arcuatus* Rossi, was also distributed in a more limited number of areas. It also established widely and continued to reproduce and was an additional source of suppression of the whitefly. The range of the whitefly subsequently expanded to include Hawaii and New Mexico (U.S.A.), but the wasp was also found in these areas, indicating that the population of whiteflies which had been transported to the new locations had included parasitized nymphs. Subsequent infestations on the eastern coast of the United States (North Carolina) were also combated by introducing parasitoids from California.

The program against ash whitefly demonstrates several important phases of an introduction program. Reference to literature was effective in laying the foundation for the exploration and introduction program, and assistance from other entomologists was important in obtaining a

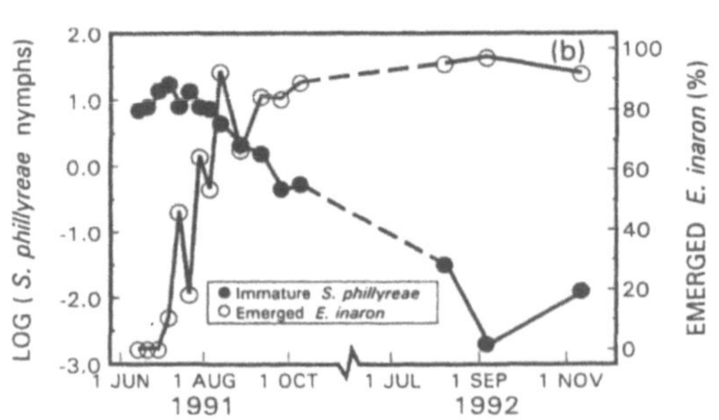

Figure 8.4. Dynamics of populations of the whitefly *Siphoninus phillyreae* (Haliday) in three locations in southern California (U.S.A.) before, during and following the establishment of the natural enemy *Encarsia inaron* (Walker) (after Bellows et al. 1992b).

wide array of candidate natural enemies. Rapid development of an introduction and quarantine program, together with a region-wide effort for redistribution, contributed to effective, rapid suppression of the target pest over a wide geographic range. Finally, the project impressively demonstrates that not all the natural enemies of a particular pest need be present in the new environment to bring about satisfactory control, because control was complete and permanent wherever the single wasp species was introduced.

Mistflower, *Ageratina riparia*

Mistflower (*Ageratina riparia*), also known as Hamakua pamakani, was brought to Hawaii in the 1920s as an ornamental shrub. It subsequently became a major weed of cool, moist forests and upland grazing lands (Trujillo 1985). The white smut pathogen *Entyloma ageratinae* was imported from Jamaica and was demonstrated to be suitably host-specific (Barreto and Evans 1988). It was released at three sites on the island of Oahu (Hawaii) in 1974. The pathogen provided more than 95% control of the weed less than one year after inoculation in areas with optimal temperatures (18–20°C) and high rainfall (Fig. 8.5). Control reached similar levels (95%) in other areas of high rainfall but less optimal temperatures after 3–4 years. Control in low-rainfall areas was less than 80% after 8 years.

The program against mistflower, together with the example of rust fungi used against

Figure 8.5. (A) Region of Hawaiian (U.S.A.) landscape infested with Hamakua pamakani (mistflower), *Ageratina riparia* [Regel] King and Robinson; (B) previously infested region, with reduced populations of mistflower following introduction of a pathogenic rust (*Entyloma ageratinae* Barreto and Evan) and herbivorous insects; (C) close up of mistflower infected by *E. ageratinae*; (D) plants killed by *E. ageratinae*. (Photographs courtesy of E. Yoshioka, State of Hawaii, Department of Agriculture.)

skeleton weed (discussed in Chapter 1), illustrates many aspects of successful programs, including identifying suitable search locations for natural enemy acquisition. They also demonstrate the substantial potential of imported plant pathogens for biological weed control. Work on biological control of weeds with plant pathogens has focused primarily on fungi, especially rusts and smuts, because of their relatively high host specificity.

OUTCOMES OF BIOLOGICAL CONTROL PROGRAMS

Historical Frequency of Success

Although the intentional manipulation and introduction of natural enemies into specific environments may have been practiced for centuries (see Chapter 1), reliable records of such activities begin with the widely known introduction of two natural enemies against cottony cushion scale, *Icerya purchasi*, in California in 1888–1889. Since that time, introductions of natural enemies to control pests have been undertaken approximately 1200 times, in numerous countries around the world (Greathead 1986a). For many of these cases, records exist of the outcome, including whether or not the species of natural enemies became established and, in some cases, what degree of pest suppression resulted. Summaries of projects of biological control based on the introduction of natural enemies include Clausen (1978), Luck (1981), Greathead (1986b), Julien (1992), and Greathead and Greathead (1992). These summaries indicate that the rate of introduction of natural enemies has increased during the 100 years of available records. While the rates of establishment and success have varied during this period, they have been increasing in the last 40 years (Fig. 8.6). The use of pathogens as introduced biological control agents against weedy plants is a more recent phenomenon than is the use of arthropods, dating only from the 1970s, but has been successful in several cases (Watson 1991).

The historical record has occasionally been used to provide comparisons of rates of establishment and control associated with different groups of natural enemies or pests (Hall and Ehler 1979; Hall et al. 1980; Hokkanen and Pimentel 1984; Greathead 1986a; Waage 1990). The value of such comparisons is limited because the historical record is only a record of occurrences, not a collection of data arising from a designed experiment. Significant confounding variation may occur in these data from such sources as the amount of effort expended on various programs and the number and diversity of sources of the natural enemies employed. Information is not available on these important aspects for many biological control projects. Waterhouse and Norris (1987) point out that some projects may have been undertaken with insufficient resources to permit a successful conclusion. Consequently, the historical record contains failures due to insufficient effort or resources, in addition to failures due to biological or ecological factors. Thus, the historical record must be viewed with caution as an information source on establishment rates, project success rates, or success rates for particular natural enemy or target pest groups. It is a record of what has occurred, incomplete in detail, and not an experiment designed to test particular hypotheses regarding biological control success. Rates reported may be minimums for certain activities (such as introductions followed by establishment of a natural enemy), but probably are not comparable across taxa.

Given these limitations, we note that over the 100-year recorded history of biological control through natural enemy introduction, some 1200 projects have been undertaken against target species in many different taxonomic groups (Greathead 1986a; Waage and Mills 1992). Of these projects, approximately 17% provided complete control of their target pests, and another 43% have achieved substantial or partial reductions in pest numbers. In some cases, projects which failed to control their target pest did so because projects were of too limited duration. In other cases, projects have failed even after sustained effort. Such failures may be due to an inability to

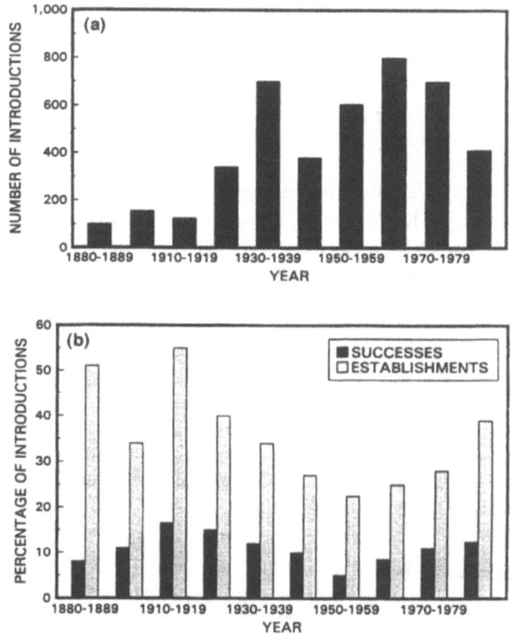

Figure 8.6. (a) Number of new introductions of natural enemies of insects by decade. (b) Proportion of introductions resulting in successful biological control or establishment of a natural enemy (after Greathead and Greathead 1992).

identify ecological parameters critical to the successful functioning of natural enemies in a particular environment, and consequently may yield to further research. Significantly, there appears to be an important relationship between the degree of effort or time expended on a problem and the degree of success (Greathead 1986a; Waterhouse and Norris 1987).

The number of programs conducted against insects varies among insect orders (Waage and Mills 1992); the order most frequently targeted has been the Homoptera. This may reflect the different frequency with which certain taxonomic groups are transported or arrive in new areas.

Much can be learned about the factors which contributed to the success of particular programs (from idiosyncrasies of natural history through degree of effort involved) by reference to the original literature regarding these programs. Reviews of such programs include Bennett and Hughes (1959), Wilson (1960, 1963), Franz (1961a), Turnbull and Chant (1961), DeBach (1964b), Greathead (1971), Rao et al. (1971), Andres and Davis (1973), DeBach (1974), Beirne (1975), Greathead (1976), Clausen (1978), Kelleher and Hulme (1984), Cock (1985), Waterhouse and Norris (1987), Funasaki et al. (1988a,b), Cameron et al. (1989), Julien (1989, 1992), Nafus and Schreiner (1989), Schreiner (1989), DeBach and Rosen (1991), Hoffman (1991), and Nechols et al. (1995). Such works should be referred to for further information on biological control programs in general, or on programs against specific pests.

Magnitude and Speed of Pest Suppression

The degree of suppression achieved in successful biological control programs against arthropods has been quantified in relatively few cases. Early projects often were so successful in reducing pest abundance that a previously devastating pest was relegated to a historical footnote following a biological control program (Dalgarno 1935; Edwards 1936; Byrne et al. 1990) and few quantitative data on population dynamics were published. In such cases, data were more readily available in terms of changes in the yield of the affected crops, as in the case for cottony cushion scale in California where the amount of exportable citrus increased approximately 200% in a single year following introduction of the natural enemies (DeBach 1964a).

More recently, quantifying population dynamics has been an objective in biological control studies. The degree of suppression of pest populations in successful biological control programs has typically been between 2-3 orders of magnitude, or 100- to 1,000-fold reductions in population density (Beddington et al. 1978; Bellows et al. 1992a; Bellows 1993). This level of reduction has been documented for such diverse organisms as Lepidoptera (Embree 1966), aphids (van den Bosch et al. 1970), diaspidid scales (Debach et al. 1971), sawflies (Ives 1976), and whiteflies (Summy et al. 1983; Bellows et al. 1992b) (Figs. 8.4, 8.7). In most of these programs, the declines in pest densities took place over a span of time which occupied between 6-10 generations of the pest (host) population. This length of time appears to be a general phenomenon, and may be related to the time required for a natural enemy population to grow to a level where it can exert its influence on the overall population dynamics of the pest. This time, although perhaps brief for multivoltine arthropods, may be several years for univoltine or bivoltine pests.

Economic Benefits

Economic analyses of the results of biological control programs by introductions are uncommon, in part because the effects of an adventive pest before and after natural enemy introduction are often vastly different, and economic analyses based on equilibrium-point analyses or marginal value analyses do not apply. Tisdell (1990) reports an average benefit-to-cost ratio of 10.6:1 for biological control efforts in Australia, with a maximum exceeding 100:1. Benefits continue to accrue as time passes because of the absence of pest damage in each subsequent production season. Additional benefits also accrue from terminating the use of alternative treatments such as pesticides. These values do not include consideration of the economic benefit of jobs retained which otherwise might have been lost should an agricultural industry fail in a particular region. Perhaps most significant is the value of programs which protect subsistence agriculture in areas where alternatives are unavailable, as in the case of cassava mealybug, *Phenacoccus manihoti*, in Africa (see Chapter 1) (Herren and Neuenschwander 1991), or in the protection of other such vital matters as transport and communication, as in the case of salvinia in New Guinea (Thomas and Room 1986). Further discussion on economic evaluation of programs of biological control is presented in Chapter 13.

Research and development for biological control by introductions of new natural enemies is typically financed from the public sector. This is because such biological control programs do not provide products with repeated marketability, but rather solve problems permanently, at the regional or national level. The economic returns fully justify such investment.

ECOLOGICAL RATIONALE FOR NATURAL ENEMY INTRODUCTION

The ecological basis for biological control through the introduction of new species of natural enemies rests on the principle that many populations are limited in their native habitat by the action of upper trophic level organisms or, in the case of some plant pathogens, by competitors

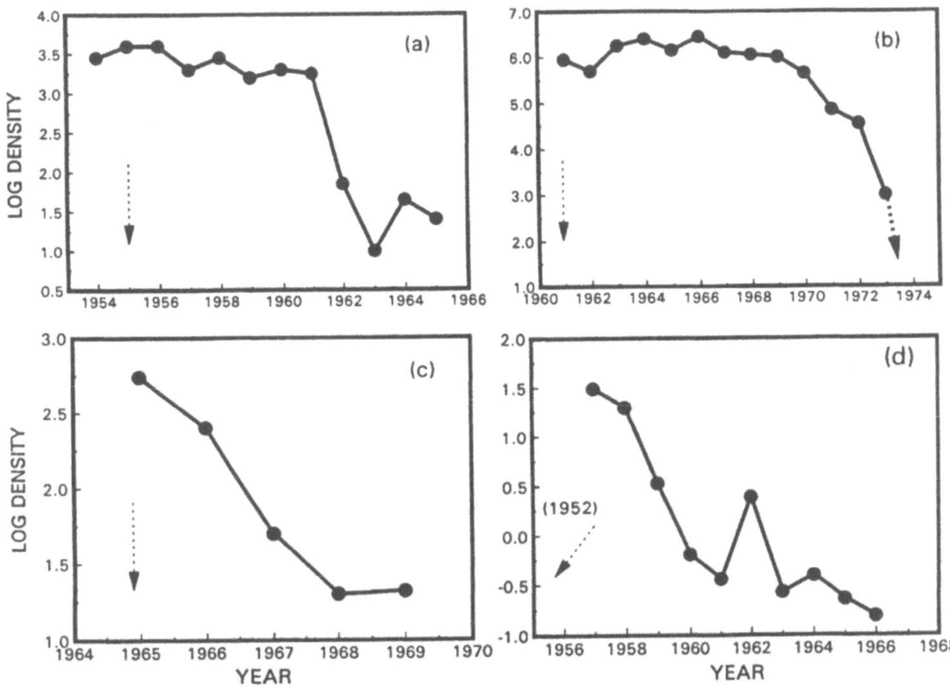

Figure 8.7. Populations densities of targeted species following introductions of natural enemies. (a) Decline in winter moth (*Operophtera brumata* [Linnaeus]) following introductions of *Cyzenis albicans* (Fallén) (after Embree 1966). (b) Decline in larch sawfly (*Pristiphora erichsonii* [Hartig]) following introductions of *Olesicampe benefactor* Hinz (after Ives 1976). (c) Decline in California red scale (*Aonidiella aurantii* [Maskell]) following introduction of *Aphytis melinus* DeBach in California (after DeBach et al. 1971). (d) Decline in olive scale (*Parlatoria oleae* [Colvée]) following introduction of *Aphytis paramaculicornis* Howard and *Coccophagoides utilis* Doutt (after DeBach et al. 1971). Dotted arrows indicate approximate time natural enemy entered each system.

within the same trophic level. An upper trophic level organism is one which feeds directly on an organism occupying a lower position in a food chain; so, in the case of a pest weed an upper trophic level may consist of herbivores and plant pathogens, while in the case of an insect or nematode pest the upper trophic level may consist of parasitoids, predators, and pathogens (Fig. 8.8). Species in these upper trophic levels are often termed beneficial organisms because of their action in suppressing a pest population. Beneficial organisms may be fed upon by additional trophic levels, as in the case of hyperparasitoids attacking primary parasitoids of insects (Price 1972).

Many species achieve pest status in situations where levels of mortality from their natural enemies are reduced. This often occurs when a species is moved (accidentally or intentionally) to a new location without its own natural enemies, and where local beneficial species are infective in suppressing it (Fig. 8.8). Species which have rarely or never been known as pests

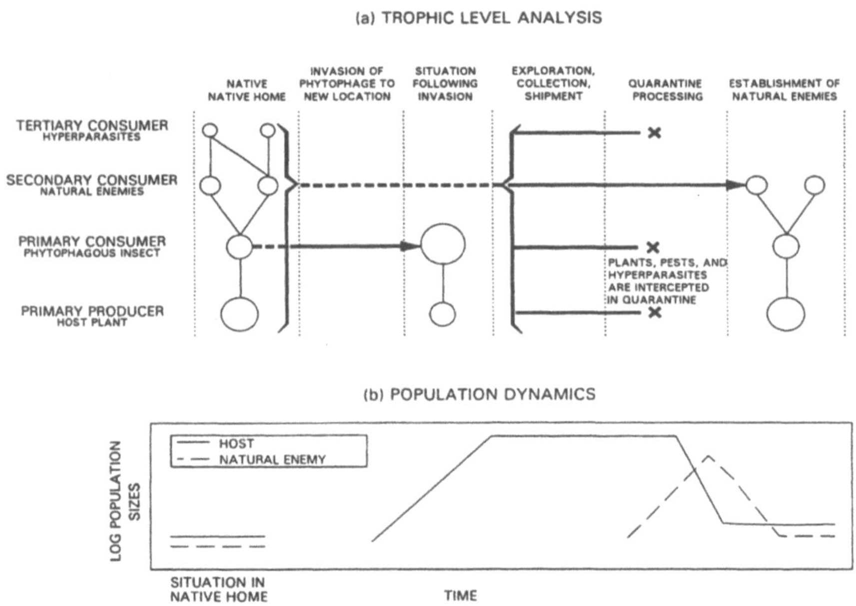

Figure 8.8. Diagram depicting the phases in the invasion of a species to a new area, followed by its successful biological control by introduced natural enemies. (a) Trophic level diagram showing changes in trophic structure during different phases of the program. (b) Population sizes on log scales of pest and natural enemy during different phases of the program.

where they are native may thus become pests of devastating proportions in areas outside their native range. Typical examples include cottony cushion scale in many locations, citrus spiny whitefly (*Aleurodicus spiniferus* [Quaintance]) in Asia, alfalfa weevil (*Hyperica postica*) in the United States, *Carduus* spp. thistles in North America, salvinia in Australia and Papua New Guinea, water hyacinth (*Eichhornia crassipes* [Martinus] Solms-Laubach) in Africa, and the tree *Hakea sericea* Schrader in South Africa.

Biological control can be used to suppress new invading pests through the exploration for and importation of natural enemies from the pest's native location, followed by colonization of the natural enemies in the new habitat (Fig. 8.8). Such introductions, hopefully, bring about reduced population densities of the pest species over a series of generations. The anticipated outcome of the method—the establishment of a new balance with both pest and natural enemy coexisting in the environment at low densities—must be acceptable in view of the goals established by society or industry regarding the particular pest. The method is not appropriate, for example, where eradication of the pest is the primary goal.

While the approach of natural enemy introduction has been used principally against adventive pests, it has also been used against native species which are pests. In these cases, the introduction of new natural enemies is carried out in the hope that one or more of the newly introduced natural enemies will provide more significant suppression of pest populations than do existing ones. Documentation of significant successes in this area are less common than for cases of adventive pests.

PROGRAMS FOR NATURAL ENEMY INTRODUCTION

Synopsis of Method

Introducing new natural enemies against a particular pest has distinct steps, the results of each of which feed directly into the next (see also Waage and Mills 1992). Careful completion of each step is important in providing information needed for subsequent steps (Table 8.1).

Step 1. Target Selection and Assessment

Before developing a program of natural enemy introduction against a pest, the target should be clearly specified and an assessment should be made of its suitability as a target for a biological control program. Such an assessment should consider a number of facets of the problem (Waterhouse and Norris 1987; Harley and Forno 1992; Barbosa and Segarra-Carmona 1993), including biological, economic, social (conflict of interest), and administrative and institutional issues. Some workers have proposed a set of questions defining these issues and assigning numerical ratings for responses; using these questions provides a measure of the suitability of a particular pest species as a biological control target. Often these issues can be as successfully addressed qualitatively as quantitatively.

One list of such questions has been formulated for insects (Barbosa and Segarra-Carmona 1993) (Table 8.2). Similar lists have been proposed for evaluation of biological control of weed

TABLE 8.1 Summary of Steps Normally Part of Programs for Introduction of Natural Enemies

Step	Objectives
1. Target selection and assessment	Identify target pest, define biological, economic, and social attributes which relate to biological control; establish objectives for introduction program; resolve any conflict of interest
2. Preliminary taxonomic and survey work	Determine current state of taxonomic knowledge of pest and natural enemies; conduct literature review on natural enemies of target species and relatives; survey in target area for any existing natural enemies
3. Selecting areas for exploration	Define native home of target pest and other possible areas of search for natural enemies
4. Selecting natural enemies for collection	Choose which of various candidate natural enemies encountered may be appropriate to collect for further study in quarantine
5. Exploration, collection, and shipment of candidate natural enemies	Obtain and introduce into quarantine candidate natural enemies
6. Quarantine and exclusion	Process shipped material to destroy any undesirable organisms
7. Testing and selecting of natural enemies for additional work	Conduct research as necessary in quarantine on natural enemies to define their host associations and biologies
8. Field colonization and evaluation of effectiveness	Release natural enemies in field and monitor for establishment and efficacy
9. Agent efficacy and program evaluation	Evaluate degree of achievement of overall program goals and objectives

TABLE 8.2 Evaluation Index for Setting Priorities among Potential Candidate Pests for Which No Biological Control Program Presently Exists*

Issue	Candidate Score
Biological Control Feasibility	
1. Pest origin and recency of introduction	
a. Introduced	[1]
b. Native	[0]
2. Crop habitat stability	
a. Perennial	[1]
c. Annual	[0]
3. Pest feeding habits	
a. Exposed habit	[1]
b. Concealed habit	[0]
4. Pre-introduction studies: natural enemies present in area of pest origin	
a. Taxonomic identity of natural enemies known	[2]
b. Some knowledge of natural enemies present	[1]
c. No surveys have been conducted	[0]
5. Pre-introduction studies: natural enemy impact	
a. Life table studies available	[2]
b. Some data available on mortality caused by natural enemies	[1]
c. No mortality data available	[0]
6. Status of biological control projects on pest in other areas	
a. Active	[2]
b. Inactive	[1]
c. Nonexistent	[0]
7. Level of accomplishment of biological control programs in other areas	
a. Sustained suppression phase	[4]
b. Colonization and impact assessment	[3]
c. Importation, rearing, and testing	[2]
d. Foreign exploration phase	[1]
e. No biological control program exists	[0]
8. Availability of biological control agents	
a. Local or commercial source	[3]
b. Source in country	[2]
c. Foreign sources	[1]
d. No known source	[0]
Economic Assessment	
A. Crop importance	
1. Current importance—total cash receipts	
a. Among top commodities	[2]
b. Lesser commodity	[0]
B. Pest importance	
1. Consistency of pest problem	
a. Most years	[2]
b. Some years	[1]
c. Rare	[0]
2. Severity of damage per pest individual under current practices	
a. Light	[2]
b. Moderate	[1]
c. Severe	[0]
3. Revenue loss by pest under no control practices	
a. Low	[2]
b. Moderate	[1]
c. High	[0]

(Continued)

TABLE 8.2 (*Continued*)

Issue	Candidate Score
B. Pest importance (*continued*)	
4. Pest status in crop	
a. Sole key pest	[2]
b. No key pests on crop	[1]
c. Other key pests exist	[0]
5a. Type of pest (not applicable for landscape pests)	
a. Indirect pest	[2]
b. Direct/indirect pest	[1]
c. Direct pest	[0]
5b. Type of pest (for landscape pests only)	
a. Low aesthetic damage	[2]
b. Intermediate aesthetic damage	[1]
c. High aesthetic damage	[0]
6. Chemical control is cost effective	
a. No	[1]
b. Yes	[0]
C. Cost/duration/feasibility (increasing cost)	
1. Cost of implementation and complexity	
a. Redistribution and inoculation	[3]
b. Importation/research and development	[2]
c. Augmentative/inundative releases[a]	[1]
d. No information available	[0]
Institutional/Administrative Assessment	
A. Institutional resource base	
1. Resident staff with expertise in pest	
a. Yes	[1]
b. No	[0]
2. Ongoing research on the pest or related pests	
a. Yes	[1]
b. No	[0]
3. Existence of rearing and quarantine facilities	
a. Yes	[1]
b. No	[0]
4. Existence of pest monitoring programs	
a. Yes	[1]
b. No	[0]
5. Existence of collabortion agreements with federal, state, or local agencies	
a. Yes	[1]
b. No	[0]
6. Length of institutional commitment needed for biological control	
a. Short term (1–3 yr)	[2]
b. Medium term (3–5 yr)	[1]
c. Long term (.5 yr)	[0]
B. Desirability of alternative control methods	
1. Is the recommended chemical control agent:	
a. Restricted with recent cancellations	[2]
b. Restricted with no cancellations	[1]
c. Not restricted	[0]
2. Prevalent chemical control material is a potential ground-water contaminant	
a. Yes	[1]
b. No	[0]

(*Continued*)

TABLE 8.2 (*Continued*)

Issue	Candidate Score
B. Desirability of alternative control methods (*continued*)	
3. Other biological control projects are currently active in the crops	
a. Yes	[1]
b. No	[0]
4. An IPM program is currently active in the crop	
a. Yes	[1]
b. No	[0]
5. Pest is currently of public concern	
a. Yes	[1]
b. No	[0]

[a]When biological control agent is commercially available, treat as redistribution/inoculation project and assign score of [3].
•Adapted from Barbosa and Segarra-Carmona 1993

species (Harris 1991). The idea behind such lists is that both biological and social issues play a role in decisions regarding initiation, conduct, and eventual success of biological control programs. In a numerical system such as in Table 8.2, target species with higher scores are expected to have a higher likelihood of successful control.

Biological issues of importance include knowledge about the pest and about its natural enemies (Table 8.2). Initially, the origin of the pest must be ascertained. If the pest is found to be of foreign origin, then new species of natural enemies may be sought for introduction against it. If it is found to be indigenous, then reasons for its pest status should be explored before an introduction program is undertaken, and if human activities have contributed to its pest status, options for ameliorating these effects should be explored. Once a pest's origin has been determined, information should be sought on its pest status in its area of origin. Species which are not pests in their native areas are better candidates for biological control through introductions than are species which are pests throughout their range. For plants, a correlation exists that suggests that biological control has been achieved against asexually-reproducing species more frequently that against sexually-reproducing species (Burdon and Marshal 1981).

Economic issues include: whether the pest has sufficient economic impact to warrant the development of a program against it, and whether a sufficiently low density of pests will result after biological control is in place to alleviate the need for other treatments while allowing economically satisfactory yield. Even in cases where the answer to the second point is not conclusively yes, a program may still be a significant contribution in the overall management of a pest species by reducing pest densities, and thereby reducing the frequency and expense of other treatments. Pests which attack nonharvested parts of plants (indirect pests) are sometimes viewed as more appropriate targets for biological control programs than pests which attack harvestable fruits (direct pests) because the density at which indirect pests cause economic losses is often higher than for direct pests. Even for direct pests, biological control may suppress pest numbers sufficiently to avoid economic damage. In other cases, the degree of suppression from biological control alone might be insufficient, but biological control may be a part of a larger management system for the pest (see Chapter 14).

Social issues, such as conflicts of interest, may arise when a plant or other species is viewed as harmful by some members of society but as valuable by others. Although not specifically addressed by the rating scheme of Barbosa and Segarra-Carmona (1993), such conflicts can give rise to questions about whether or not a particular species should be the target of a biological control program. An example of such a conflict was the different view taken in Australia regarding a toxic, adventive flowering plant, *Echium plantagineum* Linnaeus, by beekeepers (who called the plant Salvation Jane and viewed the plant as desirable) and ranchers (who termed it Patterson's curse and viewed the plant as a toxic weed). Such conflicts should be resolved before initiating a program against any particular species. Other issues may also arise when initially evaluating candidate natural enemies for release against a particular target. If the natural enemy is capable of attacking species other than the target, and these other species are considered valuable, then the potential effect of the natural enemy on these non-target organisms may need to be estimated (a biological issue), and this effect weighed against the potential benefit of release (a social issue).

Finally there are institutional issues, such as the availability of personnel, facilities, and contacts, both domestic and foreign, to conduct the program.

A list of questions such as those suggested by Barbosa and Segarra-Carmona (1993) (Table 8.2) may be used as an aid in establishing priorities among a list of candidate pest species. An additional valuable aspect of using such a list is that it permits an assessment of possible impediments to the implementation of a biological control program and, consequently, can help define the steps necessary to eliminate such obstacles.

Step 2. Preliminary Taxonomic and Survey Work

Once a pest has been selected as a target for natural enemy introduction, the next step is to collate information from the literature and initial field surveys about the species. The taxonomic identification of the species is used to compile the literature records on the distribution, biology, host range, pest status, and natural enemies of the target species and its relatives. A general knowledge of related species may be vital during exploration, because in a native area there may exist a complex of related species, each of which may have specific natural enemies. When a new pest is an undescribed species, its closest relatives must be determined. If the related host species are difficult to distinguish, errors in identification may lead to collecting or propagating a natural enemy from a wrong host species, thus investing in a less effective (or completely ineffective) natural enemy. Taxonomic information, then, forms the initial basis for strategic planning in defining areas suitable for searching for natural enemies and in identifying candidate natural enemy species or groups.

In addition to reviewing the literature for information regarding natural enemies of the target pest, a field survey should be undertaken in the area where introduction of natural enemies is proposed to determine what existing natural enemies, if any, attack the pest in its new home. This information can be useful in defining what natural enemies might have adopted the pest species as host or prey. While such natural enemies may not contribute substantially to control, the taxonomic survey and correct identification of such reference material will help avoid confusion during surveys subsequent to release of new natural enemies. Previous occurrence of a natural enemy in the area where introductions are proposed would not preclude the introduction of the same species from foreign populations from the pest's native home, which may be better adapted to attack the pest species. Discovery of such species in the initial survey, however, will make clear the importance of careful evaluation of later survey results to document establishment of new introductions of the same or morphologically similar species. Such evaluations may require genetic techniques to distinguish among morphologically similar populations.

Step 3. Selecting Areas for Exploration

The objective in identifying potential search locations is to identify geographic regions where effective natural enemy species occur. There are four general paradigms which may be used in searching for natural enemies (Table 8.3). The most frequently adopted approach is to seek natural enemies from the target species. For such searches, several approaches are suitable for defining regions where search might be productive. In some cases this can be accomplished by reference to the literature on natural enemies of the target species. More often, the location of suitable areas for search must be inferred from biogeographical, zoological, and ecological evidence. Search methods include identifying the native range of the pest (a task which itself may take into consideration several different types of information) or identifying effective natural enemies of related species. Once such regions have been identified, search locations within these regions may be selected based on climatic factors (chiefly, similarity to the area into which natural enemies will be released), accessibility, and availability of local support.

The geographic region of origin of a species can be determined by reference to several types of information (Table 8.4), including the center of the current or historical geographic distribution of the species; the area where the preferred or principal host plant originated; the area where the pest is present but kept at low densities by natural enemies; the area where the number of natural enemies or number of host-specific natural enemies is high (Pschorn-Walcher 1963); and the area where many species closely related (congeneric) to the pest species exist. A large complex of natural enemies is often believed to indicate the site of longest residence (native home) of a species, especially if one or more natural enemies are host-specific. For many of these ideas, important information often may be inferred from published literature and from collection and museum records. Proper taxonomic identifications (of both the target organism and of records in the literature) are crucial in this phase of the program.

Searching the Pest's Native Range. One of the first, widely known biological control successes, suppression of the cottony-cushion scale, *Icerya purchasi*, by *Cryptochetum iceryae*

TABLE 8.3 Paradigms for Locating Natural Enemies

Paradigm	Features
1. Natural enemies from target host species	Likely to be an effective search paradigm, with high likelihood of locating successful natural enemies. Often most directly implementable method of search. Applicable to many situations where host-specificity is necessary (as for weed natural enemies) or is expected in the natural enemy group (some parasitoids).
2. Natural enemies from taxonomically-related species	Possibly an effective search paradigm, although will not uncover natural enemies which are host-specific to target pest. Often used as a strategy when target pest cannot be located in search locations.
3. Natural enemies form ecologically homologous species	Some natural enemies are focused on types of host niches rather than taxonomic host ranges. Such natural enemies might include parasites of concealed borers or seed weevils, for example. In these cases, searching on ecologically similar species may uncover suitable candidate natural enemies.
4. Natural enemies from species not related either taxonomically or ecologically to the target pest	Some effective natural enemies come from hosts unrelated to their targets. Some of these associations have been accidental. This paradigm for natural enemy finding may be difficult to conduct because there is little information to associate potential natural enemies with the target host.

TABLE 8.4 Factors Indicating Likely Locations of the Native Home of an Adventive Species

- Center of the current or historical geographic distribution of the species
- Area where the preferred or principal host plant originated
- Region where the pest is present but kept at low densities by natural enemies
- Region where the largest number of natural enemies of the species occurs
- Region where the largest number of host-specific natural enemies occurs
- Area with an abundance of species that are congeneric with the pest
- Center of distribution of all species in the genus (or related genera)

and *Rodolia cardinalis*, followed a pattern typical of successful search in the native range of a pest (Quezada and DeBach 1973). The scale invaded California and natural enemies were subsequently sought in its original range, southern Australia. Some of the more dramatic and repeatable successes in biological control, where the target pest's population density was permanently reduced to below economically damaging levels, have involved the introduction of one or more species of natural enemies from the presumed original native home of the pest (Franz 1961a,b; DeBach 1964a, 1974; van den Bosch 1971; Hagen and Franz 1973; Clausen 1978; Franz and Krieg 1982; Luck 1981; Gould et al. 1981a, 1992b; Bellows 1993). One explanation for these successes is that this approach employs natural enemy-host associations of long standing where the host is native, and that such associations will be characterized by natural enemies suitable for suppression of the target pest population. Successful biological control following this approach has been achieved many times and has been applied in many taxonomic groups. Notable examples include biological control of Klamath weed (*Hypericum perforatum*) in the United States, *Opuntia* spp. cacti in India and Australia, olive scale (*Parlatoria oleae* [Colvée]), walnut aphid (*Chromaphis juglandicola* [Kultenbach]), Comstock mealybug (*Pseudococcus comstocki* [Kuwana]) (Ervin et al. 1983), ash whitefly (*Siphoninus phillyreae*) (Gould et al. 1992a,b), and many more. Reviews of many such cases are presented in Clausen (1978), Hoffmann (1991), DeBach and Rosen (1991), Julien (1992), and Nechols et al. (1995).

Searching on Related or Analogous Species. In a few instances, natural enemies have caused significant declines in population densities of organisms with which they had never had previous contact. Such cases include the devastation of desirable native species by organisms which invaded a region, such as the fungus *Cryphonectria parasitica* of Asian origin which nearly eliminated the American chestnut, *Castanea dentata*, and *Icerya purchasi* from Australia which nearly destroyed citrus in California. There is also a history of successful biological control against adventive pests by organisms secured from areas other than the pest's native home. Examples include regulation of the lepidopteran forest defoliator *Oxydia trychiata* (Guenée) in Colombia by the egg parasitoid *Telenomus alsophilae* Viereck from eastern North America (Bustillo and Drooz 1977; Drooz et al. 1977).

Other examples exist (see Pimentel 1963) which have led to the suggestion that natural enemies suitable for biological control might be found among species that have not experienced close prior relationship with the target organism. Although some examples have been noted of host-suppression by natural enemies that lacked prior contact with the host species (Hokanen and Pimentel 1984), success has been achieved at least as frequently or more frequently by natural enemies with prior association with the target pest, that is, enemies from the native home (Goeden and Kok 1986; Schroeder and Goeden 1986; Waage and Greathead 1988; Waage 1990). In addition, defining regions for search which are likely to produce effective natural enemies is usually possible only for a native home. Undertaking world-wide searches for non-associated natural enemies is not likely to be an efficient strategy for locating and obtaining effective natural enemies. While productive searches can take place both within and outside the presumed native range of the target pest, more successes have come from the former category. However, taxonomically related species, or unrelated species that have life histories and ecologies similar to the target pest and which occur on the target plant in the native home, may be useful sources of natural enemies that merit investigation when the target pest cannot be located or when natural enemies from it have proven ineffective. Examples of successful biological control following this approach include control of European rabbit, *Oryctolagus cuniculus*, in Australia with a virus from a species of rabbit from South America (Fenner and Ratcliffe 1965) and control of the sugarcane borer *Diatraea saccharalis* in Barbados by *Cotesia flavipes* (Cameron), collected in India where it attacks other groups of graminaceous borers (Alam et al. 1971).

Step 4. Selecting Natural Enemies for Collection

Classes of Organisms. In many cases the problem of what type of natural enemies to seek will be defined by what the target pest is. Many pest taxa have only a limited breadth of natural enemies, and candidates will naturally be selected from that group. More broadly, however, some choices may be necessary when planning a biological program.

Natural enemies from a wide variety of taxonomic groups have been employed in biological control programs. These include terrestrial and aquatic vertebrates; molluscs; arthropods (including predacious insects and mites, insect parasitoids, and herbivores); pathogens of arthropods, vertebrates, and weeds; and antagonists of plant pathogens. Different organisms have been targeted against different sorts of pests. Over the last several decades some general guidelines have emerged about the utility and safety of these different taxonomic groups. Most of these issues center around the degree of specificity, and thereby likely safety (i.e., low risk of unforeseen consequences of the introduction) of the natural enemy. The history of past biological control introductions provides some background against which to evaluate the general suitability of these various groups as natural enemies for future programs.

Vertebrates. Terrestrial vertebrates (mammals, birds, reptiles, amphibians) have been introduced into many different locations worldwide, often with undesirable results. Because of the broad range of food-selection and other behaviors exhibited by vertebrates, their use for biological control of terrestrial pests has high potential risk and generally is no longer practiced. Aquatic vertebrates such as fish have been used widely against aquatic weeds or as predators of mosquito and chironomid midge larvae. The minnows *Gambusia* spp. and *Poecilia* spp. are used worldwide against mosquitoes (Legner and Sjogren 1984). However, effects of such introductions on populations of native minnows have caused widespread concern so that use of native fishes is recommended as an alternative (Walters and Legner 1980; Courtenay and Meffe 1989). Because fish can be manipulated readily, the potential to increase the effectiveness of resident species as natural enemies is greater than for resident terrestrial organisms where widespread natural dispersal may have already covered most possibilities.

Molluscs. Molluscs have been introduced as predators of herbivorous snail species and as competitors of snails that serve as vectors of schistosomiasis. Predacious snails can be polyphagous, and their introduction must be considered carefully in relation to possible impacts of the predacious snail on nontarget native snails (Clarke et al. 1984; Murray et al. 1988). Introductions should be made only if the predator has feeding habits that restrict its impact, either ecologically or taxonomically, to the target species, or to that species and other species which are not of concern (Fisher & Orth 1985).

Arthropod Terrestrial Scavengers. Scarab beetles have been introduced to such areas as Australia to remove cattle dung from grazing areas to improve pastures and reduce populations of symbovine flies such as *Haematobia irritans* (Linnaeus) and *Musca* spp. Although dramatic successes have been achieved in removal of dung, widespread concurrent reduction in fly densities has occurred less often (Legner 1986a,b).

Herbivorous Arthropods. Herbivorous arthropods are a group of natural enemies which include both generalist and specialist species, and, consequently, the host ranges of introduced species must be evaluated with special care to safeguard desirable plant species in the areas of introduction. Screening has been so successful that among the numerous importations of beneficial phytophagous arthropods and pathogens around the world (Julien 1992), there has never been any widespread occurrence of harmful results from the organisms imported. Rare reports of beneficial phytophagous arthropods feeding on desirable plant species (Greathead 1973) involved temporary alterations in behavior when a population was undergoing size reduction following the reduction of the target host plant population. Only 20 cases have occurred worldwide in which an intentionally introduced phytophagous species showed an alteration or expansion of its host range, and in none of these cases was there any long-term environmental or economic damage (Dennill et al. 1993, citing a personal communication from C. Moran).

Arthropod and Plant Pathogens. These are natural enemy groups within which there is considerable variation in breadth of host range. Generally, however, the host range is definable with a substantial degree of confidence through laboratory experiments with the pathogens. Present practice focuses on relatively host-specific pathogens (such as rusts and smuts) in introductions for weed biological control, to avoid the risk of introducing a pathogen which may harm desirable plants. Work on insect pathogens intended as species to be introduced to new locations is similarly focused on relatively specific pathogens. Few successes have been associated with pathogens of arthropods, but work on pathogens of weeds indicate substantial potential for success (Bruckart and Dowler 1986; Bruckart and Hasan 1991).

Parasitic and Predacious Arthropods. The opportunity for unforeseen outcomes is perhaps lowest in the group of natural enemies which includes the relatively specific parasitic and predacious arthropods. The impact of the release of such natural enemies is usually largely confined to the target organisms. Generalist predacious arthropods such as social wasps (Vespidae) and ants (Formicidae) are generally not introduced, even though they may be significant predators of certain species, because of the potential impact of these arthropods on human activities and on nontarget organisms. Similarly, widely polyphagous, nonsocial predators or parasitoids are less often suitable for introduction.

Judging Natural Enemy Potential. For several decades biological control scientists and ecologists have debated whether or not there is a need for preliminary selection, from a suite of candidates, of a single species or small group of "best" natural enemies for initial release. This concern is based on the idea that the proper initial choices may be important to the subsequent rate of natural enemy establishment and, ultimately, the degree of pest suppression achieved (Turnbull 1967; May and Hassell 1981, 1988; Ehler 1982, 1990). In spite of considerable discussion in the literature (Turnbull and Chant 1961; van den Bosch 1968; Zwölfer 1971; Ehler 1976; Zwölfer et al. 1976; Pimentel et al. 1984; Legner 1986b), both theoretical treatments (Hassell and May 1973; May and Hassell 1981; see also Chapter 18) and practical experience provide little evidence that any introduction has ever caused the eventual failure of a biological control program, or that introductions made in any particular order led to greater or lesser control than any other order of introduction.

While the full effect of introduced natural enemies which have never been studied cannot be predicted with complete certainty before establishment (DeBach 1964a, 1974; Coppel and Mertins 1977; Ehler 1979; Miller 1983) and, therefore, involves empirical judgment, nonetheless the process of natural enemy introduction is best guided by an educated empiricism (Legner 1986b). Two general approaches have been proposed: *a priori* selection of one (or a few) natural enemy species based on existing information; or the simultaneous introduction, sometimes accompanied by preliminary evaluations, of several potentially effective natural enemies. The debate between the two schools has been lengthy, but both ideas have merit when applied judiciously. However, the manner in which the biological control candidates should be selected is not clearly delineated for most groups of organisms (DeBach 1974; Coppel and Mertins 1977; Pimentel et al. 1984). The best argument for attempting to select for introduction only the most suitable candidates is probably efficiency; proper early selection may lead to a more rapid solution. In cases where only a few candidates are available, however, field trials in the area of release may be undertaken to provide evaluations of several species in place of a more theoretical or laboratory-oriented selection.

In practice, it is likely that selection of candidate agents occurs in nearly every program, even if simply from a pragmatic point of view. When more natural enemies are known that can be processed with available resources, some choices must be made as to which species to consider first. Such choices typically are made based on some sense of value and impact of the natural enemy in its place of discovery, on the degree of association with the target species (Waage 1990), similarity in climate with the target area (van den Bosch et al. 1970), or other criteria such as past effectiveness of similar natural enemies in other projects. In other programs it may be possible to introduce and evaluate more than just a few natural enemies. Historically, in a number of such programs the most valuable or successful species of natural enemy was not recognized as such during the exploration and importation phases (Smith et al. 1964), and its value or contribution became clear only after colonization in the target country.

Parasitic and predacious arthropods and arthropod pathogens are in a category which usually defies accurate prediction of their potential as introduced species for biological control of outdoor pest populations. Guidelines based on laboratory studies or mathematical models are of limited use to judge performance in nature. Laboratory assessments often are poor predictors of the field efficacy of predacious and parasitic arthropods, because their high dispersal capacity, responses to host densities and climate, and dependence on alternate hosts are difficult to model in the laboratory (Messenger 1971; Eikenbary and Rogers 1974; Mohyuddin et al. 1981; Legner 1986b). In addition, release from attack by hyperparasitoids, cleptoparasitoids, or other predators (present in the pest's native range) may make predictions of performance in the new location uncertain. Inability to predict the impacts of natural enemies, however, has not been a major obstacle to their use in biological control. Godfray and Waage (1991) describe how preliminary observations on life-history characteristics may be combined using mathematical models of population interactions. While such population models are not equivalent to field trials of natural enemy potential, they may suggest possible outcomes of different introductions of various species of natural enemies, and in this way help guide the initial process of identifying and selecting candidates for further work.

One area where laboratory evaluation of biological control agents has more potential for direct application is in the selection of natural enemies for use in augmentative programs, especially in protected cultivation such as in glasshouses. The attributes which lead to efficacy in such environments can often be duplicated relatively easily in the laboratory or in small-scale glasshouse trials, and comparisons among candidate natural enemies in these cases may have more relation to their eventual efficacy than agents destined for use in an open environment. These issues are covered in more depth in Chapter 10.

In some cases, particularly when knowledge has been developed concerning the suite of natural enemies available for introduction against a particular target species, selection of natural enemies is either possible (because comparative information is available) or necessary (because resources will permit the initial use of only a certain number of new organisms). Coppel and Mertins (1977) proposed a list of 10 desirable attributes of beneficial organisms to aid in assessing their capabilities. The list reflects ecological and biological attributes notable in species used in successful biological control programs, and considers ecological capability, temporal synchronization, density responsiveness, reproductive potential, searching capacity, dispersal capacity, host specificity and compatibility, food requirements and habits, and hyperparasitism. Additional categories of information of value in obtaining a broad understanding of the niche of a potential biological control agent are systematic relationships and anatomical attributes, physiological attributes (synovigeny or proovigeny), ontogeny, need for alternate hosts, cleptoparasitism, and genetic data (number of founders, strain characteristics, and so on). For parasitoids for use in temperate areas, it has been suggested that it may be advantageous to choose species with lower developmental thresholds (Bernal and González 1993). Even incomplete data can aid in elevating one's understanding of a system toward the "educated empiricism" of Coppel and Mertins (1977).

Finally, studies on natural enemies and their hosts in the native country can provide insight into the nature of interspecies relationships, including differences in species guilds associated with elevation, climate, and seasonality (Legner and Gordh 1992). Thus it may be possible to estimate which species will be best suited for use in particular areas based on knowledge of their ecological requirements, and such information can help guide the development of a program. Examples of programs which have benefited from this approach include: biological control of the chestnut gall wasp, *Dryocosmos kuriphilys* Yasumatsu, in Japan, guided by data acquired about the natural enemies in their place of origin (China) (Murakami et al. 1977, 1980); the program against Comstock mealybug (*Pseudococcus comstocki*) in California (Meyerdirk and Newell 1979; Meyerdirk et al. 1981) guided by basic research on natural enemies in Japan; and the success of the egg parasitoid *Telonomus alsophilae* from North America against the geometrid *Oxydia trychiata* (Guenée) (Bustillo and Drooz 1977) in South America, based on results of studies on the parasitoid in North America (Fedde et al. 1976).

Biotypes and Strains. Natural enemy species can vary widely in attributes which affect their ability to control a particular target, even among different populations of the same species (González 1988). Different populations of a natural enemy species, indistinguishable morphologically or genetically, may differ in some significant, but perhaps cryptic, biological characteristic. Where such populations have stable, heritable biological differences, they are termed biotypes, subspecies, or races. Such populations may vary in habitat choice, host preferences or range, physiology, life cycle parameters, or behavior. Because each of these attributes can affect the suitability of an agent for a particular target species or environment, it is appropriate to introduce different populations, especially those found in areas of different climate or host community composition.

In a number of cases, introducing distinct biotypes of a natural enemy from different locations has proven crucial to both establishing the natural enemy and to suppressing the pest species. One such example is the introduction of two races of *Trioxys pallidus* to control walnut aphid (*Chromaphis juglandicola*) in distinct climates in the United States (van den Bosch et al. 1970). Another example is the repeated, unsuccessful introduction of *Cotesia rubecula* into northeastern North America from Europe against the cabbage pest *Pieris rapae*, which had arrived from Europe in the 1800s. Although none of the early introductions from Europe of *C. rubecula* resulted in establishment, the later introduction of the same parasitoid species from northern China (with a climate and latitude more similar to the target area) did result in establishment (Van Driesche, unpublished). Biotypes can also be important in the case of plant pathogens. In skeletonweed (*Chondrilla juncea*), for example, there are three known biotypes or varieties of the weed, and the rust pathogen which has been used with wide success in Australia attacks only one of the weed varieties. Different strains of the rust are known to attack the other varieties of skeleton weed and are under investigation as additional agents against this weed. The case for introduction of different biotypes is discussed in more detail by González et al. (1979) and Caltagirone (1985). Infraspecific variation in behavior, host range, or climatic tolerances may also be accompanied by identifiable variation in allozymic or genomic frequencies. Such techniques as electrophoresis to examine allozymic polymorphisms, and other techniques which examine differences in the genome directly (such as restriction fragment length polymorphisms and polymerase chain reaction analyses) may be valuable in distinguishing among biotypes (Berlocher 1979; Gargiulo et al. 1988, Izawa et al. 1992; Pinto et al. 1992; Roehrdanz et al. 1993).

Step 5. Exploration, Collection and Shipment of Candidate Natural Enemies

Once the areas to be explored have been identified and the taxa to be sought considered, the exploration trip can be undertaken. This involves several detailed steps, and includes obtaining the necessary authorization and permits to collect and ship natural enemies, traveling and exploring to obtain natural enemies, and shipping natural enemies to the quarantine facility. Details regarding these activities are discussed in Chapter 9.

It is often important to search with both a taxonomic and an ecological view of the target organism, so that searches may include natural enemies of closely related organisms or, in some cases, taxonomically unrelated organisms which occupy similar ecological niches. Searches should include various seasons, elevations, and climates because the natural enemy fauna in each may vary significantly. Introductions of distinct populations of the same species should be made from different areas to include possible cryptic races or biotypes. Searches for natural enemies may include candidates attacking any life stage of the target species. Exploration is a continuing process which may require considerable long-term effort, including multiple exploration trips, to secure suitable natural enemies.

An important purpose of the collection process is to gather as much information about the natural enemies of a particular target pest as possible. In this context, the general approach to collection of natural enemies during exploration is to collect any potentially suitable natural enemy discovered and ship it to the receiving laboratory for cataloging and further evaluation or study. Geographically-broad searches may allow collection of a wider taxonomic range of natural enemies, and collections from a small series of well-separated locations may provide a full compliment of natural enemies for a particular pest species (Waage and Mills 1992). Early season collections may contain fewer parasitoids (of phytophagous natural enemies) or hyperparasitoids (of natural enemies of arthropods) than late-season collections. Notes should be kept on natural enemies found but not shipped which may be of future value.

Step 6. Quarantine and Exclusion

Quarantining materials arriving in an area to exclude unwanted organisms is a concept of long standing. Natural enemies collected in a foreign location must be shipped to a quarantine facility to ensure that no harmful organisms are also introduced during the collection and importation process. To this end, laws have been developed which regulate the shipment and processing of imported goods, especially plants, foods, and animals. Special buildings in which imported materials can be safely inspected are required for this process. Much of the information on excluding unwanted plants and plant pests from commerce through the use of quarantine has been reviewed by Kahn (1989). The facilities and procedures employed in the United States for these activities are reviewed by Coulson et al. (1991). Many countries have enacted legislation regarding quarantining incoming materials, and the Food and Agriculture Organization of the United Nations has published guidelines for quarantine procedures during introductions of biological control agents (Anon. 1992).

Quarantine laboratories for introduced natural enemies serve two purposes for biological control programs. First, they provide a place where all organisms entering a country can be contained in a primary quarantine facility until unwanted organisms can be screened out and desirable organisms cultured. Second, they provide a location for safe testing of potentially beneficial organisms. Although a primary concern of quarantine is the prevention of unintentional release of foreign organisms, it is also the place where the desired beneficial organisms must be bred and initially cultured. Consequently, the process of quarantining materials must provide an environment conducive to successfully breeding natural enemies (and consequently their hosts) while at the same time preventing their escape.

Protection against liberation of known undesirable organisms is generally achieved by a conservative approach to handling incoming material. Material received into quarantine is initially opened inside an observation chamber (such as a cage with a glass top and closed sides). This allows inspection, separation of contents, and containment of any undesirable organisms. Potentially beneficial organisms are then separated by species or collection locality and set up in the quarantine facility for initial propagation, as discussed in Chapter 9.

For candidate natural enemies to be demonstrated as beneficial, they must be able to complete their life cycle with only primary host material (the target pest) available; tests which confirm this are generally taken as evidence that the candidate species are primary parasitoids, predators, or disease agents, and not hyperparasitic or saprophagous in habit. Where the target pest (such as a weed) is related to organisms (such as other plants) of importance in the environment, tests on host specificity may also be warranted. For phytophagous arthropods or plant pathogens introduced as natural enemies, a number of additional tests are usually conducted. These tests are aimed at determining the potential host range of an introduced natural enemy and as an aid to preventing the release of a natural enemy into an environment where desirable plants might be affected (see Safety Testing following). The mechanisms for conducting these tests are discussed in Chapters 9 and 17.

Establishing colonies free from contamination is a major objective of quarantine operations.

Small populations of natural enemies may be at substantial risk of extinction if their culture is invaded by another organism. Desirable parasitoids, predators, and phytophagous species may be lost or eliminated from cultures by the appearance of some other natural enemy which competes for the same host. Natural enemies of phytophagous species may also be lost when their host culture is contaminated by another phytophagous species. Phytophagous natural enemies may be subject to attack by parasitoids or diseases. In all cases, protecting a small, unique culture in quarantine is a paramount objective in processing imported material (see Chapter 9).

The issues regarding safe handling of pathogens are similar to those for other groups of organisms: security against unauthorized release from the containment facility and security of the imported material from contamination. The practical matters of implementing procedures designed to meet these needs are different, however, and focus more nearly around the functioning of a microbiological laboratory, including special filtering of incoming and exhaust air supplies, containment and treatment of all waste water, and special culturing conditions to minimize contamination of microbial cultures (Melching et al. 1983).

Step 7. Safety Testing and Selection of Natural Enemies for Further Study

For some groups of natural enemies, particularly for phytophagous arthropods and plant pathogens introduced against weed pests, specific information about their host ranges must be developed to determine their suitability for the new environment before their release (Zwölfer and Harris 1971; Frick 1974; Wapshere 1974a, 1989; Woodburn 1993) (see Chapter 17 for detailed discussions of such tests). The objective of these tests is usually to determine the potential for infection or feeding, survival, and reproduction on a suite of plant species, usually related to the target weed, and other plant species that may be of particular concern regarding potential susceptibility to the natural enemy. Such other plants might include crop plants unrelated to the weed but grown in the area in which the release is planned, if these crops do not occur in the area from which the herbivore was collected.

These tests are conducted in the quarantine facility while the candidate natural enemies are still confined. Such tests on the performance of biological organisms in laboratory settings clearly have limited applicability to behavior in field settings, but provide information about the potential physiological host range of a particular natural enemy. Field studies performed in the country of origin of the agents whose specificity is being studied are valuable if the test plant species can safely be introduced into the desired location. Such "garden-plot" tests have been used, for example, to measure host ranges of capitulum-feeding insects attacking yellow starthistle, *Centaurea solstitialis* Linnaeus, in Greece (Clement and Sobhian 1991). Results of laboratory and field tests such as these must usually be interpreted together with other biological and ecological information to assess the potential host range under field conditions (the ecological host range) (Cullen 1990).

Release from quarantine is authorized only for natural enemies whose anticipated ecological host range in the new environment is acceptable. The acceptable degree of host range breadth is not fixed, but may vary from one program or area to another, even for the same target or natural enemy species depending on the circumstances in the release area. Generally, natural enemies are cleared for release if the release will not put at risk desirable or valuable plant (or animal) species either in the area of release or in adjacent areas within the likely natural spread of the agent. Small amounts of feeding on nontarget plants are not usually considered prohibitive impediments to release, and natural enemy species with a narrow range of polyphagy (oligophagy) are often acceptable for release.

Parasitic and predacious arthropods are less often subjected to host-range testing, and their are several reasons for this. Many such natural enemies are introduced against adventive pests which have few or no close taxonomic relatives in the new target region. In such cases, newly introduced natural enemies may be sufficiently specific in the new environment to avoid any significant impact on other nontarget taxa. The host ranges of many parasitic insects, and to a lesser degree of some predacious arthropods, are limited to a few related species or genera within a family. These host species are typically not at risk of extinction, and the addition of new natural enemies only rarely causes substantial mortality to nontarget hosts. Exceptions may occur, particularly when the addition of natural enemies occurs together with other factors, such as habitat destruction, especially on islands (Tothill et al. 1930), but definitive data and examples are rare (Howarth 1991). In general, laboratory evaluations of host ranges of parasitic or predacious insects may be difficult to interpret and are subject to the same limitations as for phytophagous natural enemies; consequently behaviors in the laboratory are unlikely to fully define either the behavior or impact of a natural enemy under field conditions. While in general, agents with higher host specificity are preferred for reasons of greater effectiveness and safety to nontarget organisms, in some cases such as phytoseiid mites, use of species with broader feeding habitats (generalists), but close association with the crop plant on which the pest occurs, may be an effective introduction strategy (McMurtry 1992). When assessing the suitability of a candidate natural enemy, the risk of environmental impact from potential natural enemy attack on related host species must be compared to the continued damage to the environment caused by the uncontrolled population growth of the target pest.

Introduction of vertebrates, although less practiced, involves additional evaluation of their feeding habits and other behaviors. In the case of fish, tests conducted in experimental ponds or artificial streams permit study in isolated conditions. There are still possibilities that undesirable traits, such as feeding on the spawn of other desirable species, may appear once populations are allowed to establish widely (Legner and Sjogren 1984). Extensive efforts should be made to recognize such potential problems when considering vertebrate species for release.

Step 8. Field Colonization and Evaluation of Effectiveness

A critical step in the introduction of new natural enemies, following the establishment of pure cultures and the authorization for release, is colonization of the new organisms in the field. The historical record shows that 34% of attempts to colonize natural enemies have succeeded (Hall and Ehler 1979); this rate is a minimum, subject to the cautions discussed previously. Ability of agents to establish in the field can be related to several factors, including ecological, human, and technical or financial issues (Beirne 1984; Van Driesche 1993).

Many failures of new natural enemies to become established are attributable to the selection of agents unsuited to the intended purpose, either because they are unable to survive the local climate, or because they have no preference for attacking the target pest. When such limitations are known, researchers should seek other natural enemy species or races which may be more suitable. Some natural enemies may be found attacking the target host, but may be doing so in a casual or accidental way. In such cases, the natural enemy may not search well for or reproduce well on the target host if its primary association is with some other host species. For natural enemies of phytophagous arthropods, the host plant on which the natural enemy and its host are found may be an important factor. Plant characteristics such as chemical composition, leaf texture and pubescence, and plant architecture all can affect the ability of natural enemies to attack otherwise suitable hosts (Elsey 1974; Keller 1987). If the agent has been collected from the target pest on the principal crop affected by the pest, then host plant suitability for the agent is likely assured. If, however, the target pest is found on a wide range of plant species or on a species with many varieties or local populations which differ in their physical or chemical characteristics, different natural enemies may have different efficiencies on the host on various plant species.

An additional consideration in the establishment of candidate natural enemies is climatic similarity between the originating location and the release location. Agents have, in some instances been transferred successfully between dissimilar climates (Bustillo and Drooz 1977). Such cases are rare, however. More typically, natural enemies have been more successful when selected from areas with climates at least broadly similar to that of the target area (Messenger et al. 1976). Climatic factors likely to be of significance include extremes of temperature and humidity, effects of seasonal rainfall patterns on host and host plant availability, and photoperiod. Climatic maps or computerized meteorological data can be used to map similarities between regions to help direct foreign collecting to areas with climates similar to the intended release areas (Yaninek and Bellotti 1987). The limits of adaptability of new agents, however, may eventually be discovered only by trial in the field under various conditions.

Finally, some agents must enter diapause successfully to complete their seasonal cycle if success in colonization is to be achieved. If the agent uses its target host as the diapause host, problems of host availability are unlikely to arise. If, however, the natural enemy employs another species not present in the target area as a diapause host, the new agent may not establish. Correct interpretation of climatic signals for induction of diapause may also be important and may vary among populations of natural enemies.

The likelihood of establishing a natural enemy can be enhanced by maximizing several factors during a release program. These factors include genetic factors, health of individuals in the colony, numbers and life stage released, employing optimal release methods, and protecting release locations. Generally, releases consisting of large numbers of healthy individuals in robust stages, released in secure environments with suitable host populations, have better opportunities to establish. Techniques for optimizing these factors are discussed in Chapter 9. Hopper et al. (1993) review management of genetics of natural enemies during biological control projects.

Sites for release should reflect the diversity of climates where the pest occurs to maximize the possibilities that a climate suitable for reproduction of the natural enemy will be discovered. If possible, separate sites should be used for each natural enemy species or biotype introduced. Sites should not be so isolated that natural enemy populations would have difficulty dispersing to additional infested habitats.

Following release of a natural enemy, populations must be monitored to determine whether or not the natural enemy becomes established. Establishment is usually defined as the presence of a breeding population of the natural enemy after 12 or more months have passed since the last release. Establishment of the natural enemy is usually determined by sampling the host population for the presence of the new species. Where closely-related populations of natural enemies are present, such sampling may require genetic identification to discriminate among biotypes. Monitoring must be conducted in each of the release sites because the sites may differ in microclimatic suitability for natural enemy reproduction. Records should be maintained on the climatic conditions at each site to determine if conditions and other edaphic factors have changed during the colonization program. Records of establishment should be maintained and published to provide public records of the presence of the new species in the region where introductions were made.

The colonization of an agent may require repeated attempts before success is achieved. Colonization is often a long-term program, rather than a single event. Once an effective method for establishing a particular species has been discovered, establishment in other locations may become more efficient. One of the most common errors in technique which leads to failure in colonization is that efforts are too few, with no attempt at reassessment and further action. Thorough programs, with suitable levels of resources to follow through as necessary, will produce more frequent establishments than less complete programs. However, after it has become clear from repeated trials that an agent does not establish in a particular environment, further efforts at establishing that natural enemy in that environment are no longer appropriate.

Step 9. Agent Efficacy and Program Evaluation

Once establishment of natural enemies is documented, the effect of specific natural enemies on the host population can be evaluated. This step is crucial to the completion of the introduction process because it quantifies the effect the natural enemy has on its target pest in the new environment. (Various approaches to agent evaluation are discussed in Chapter 13.) This information contributes directly to program planning. If the natural enemy is having a beneficial effect in reducing population survival or reproduction, its further distribution in the infested region would be warranted. If it is not having a beneficial effect, resources can be directed at other candidate agents.

Beyond evaluating the impact of the various natural enemies, longer-term evaluation should also be made of the overall biological control program and its effect on the target pest. Typically, populations require several generations for a newly-established host-natural enemy system to reach a new balance. Once such a timeframe has passed (or earlier if quantitative evaluations of natural enemies permit), an assessment can be made of the degree of suppression provided by the introduced natural enemies. If the suppression is sufficient, the program can turn its resources to region-wide distribution of the natural enemies. If the suppression is insufficient, the program objectives should be reconsidered in light of what has been discovered. In particular, the need for different natural enemies should be addressed, especially if natural enemies were discovered during the exploration phase of the program which have not yet been introduced or colonized. If it appears that additional natural enemy species or biotypes can be successfully introduced, then programs to evaluate them once they are established should be conducted. If no further natural enemies are known, additional exploration may be needed.

COSTS, SOCIAL BENEFITS, AND SAFETY

Resources Required for Introduction Programs

Biological control programs through introductions have higher costs during the research and development phases than once the natural enemy is established in the environment. When effective natural enemies are established, pest suppression continues without additional cost, and benefits continue to increase as the degree of control and geographic range over which a pest is controlled grow (see Chapter 13).

Biological control programs which begin with little information about either the target or its natural enemies may demand substantial resources. For insect pests, new programs may require an average of 6 to 7 or more scientist-years plus supporting staff, depending on the origin and type of pest and the amount of information available about its natural enemies. More effort is required for plants. In Canada, successful weed control has generally required 4 or 5 natural enemies per weed, with investment of 4 to 5 scientist-years per agent (Harris 1979), a range of 11 to 24 scientist-years per pest (Andres 1976, Harris 1979), and program costs in the range of U.S.$1–2 million dollars (Tisdell et al. 1984).

When a biological control program is being copied after one already developed elsewhere on the same target pest and employs the same natural enemies, much of the research and development cost has already been borne. The cost of repeating the program can be comparatively small, particularly if extensive retesting of the organisms involved is not required. Nevertheless, in these cases the costs of initially obtaining adequate numbers of the natural enemies, ensuring that they are free from undesirable organisms, and of additional mass rearing are not negligible. Unless adequate stocks of the organisms can be obtained from an agency able to supply them free from parasitoids and diseases, it is desirable to rear them in quarantine through at least one generation to ensure that they are free from such organisms. Field assessment of the degree of damage occurring prior to the introduction is also important, so that comparisons can be made following establishment of the natural enemies. This range of activities would require at least 0.5 scientist-year per natural enemy, with appropriate support by assistants (Waterhouse and Norris 1987), and a commitment should be present for several years to prevent waste of investment caused by early termination of suitable projects.

Biological control programs also require the physical resources of quarantine facilities, glasshouses, and laboratories to support the continuing work. While these facilities are frequently part of existing agricultural ministries or universities, the cost of establishing and maintaining them is also part of the cost of biological control programs.

Social Benefits

Biological control is, in its larger context, one of many ways in which people manage and alter the environment. Plants and animals are moved to or invade new parts of the world. These movements sometimes have unforeseen negative consequences, and organisms become pests. Biological control is a response to such events that involves movement of additional, beneficial species to correct damage from adventive species that have become pests. Benefits associated with biological control through introduced natural enemies have been very great (Clausen 1978). Controling pests through biological control introductions has been crucial in the continued existence of many cultivation systems, in the use of many ornamental plants, and in the preservation of waterways and other natural environments in countries around the world. Beneficial effects include reduction of pest numbers and elimination of harmful effects of pest damage, reductions in pesticide use and environmental contamination, increased productivity of agricultural or recreational areas, and recovery or protection of natural environments. In addition, for biological weed control programs, there are increases in numbers and diversity of desirable species of plants or animals following a successful program (see Chapter 21), because dense weed populations often suppress other plant species, which recover once the pest species is brought under control. Thus there are social, economic, and ecological benefits to the reduction of pest impact through biological control.

Safety

Risks of biological control through the introduction of new species of biological control agents are related primarily to potential effects on nontarget organisms. Some evidence exists, in one case, of a target species extinction (the coconut moth *Levuana iridescens* Bethune-Baker on the island of Fiji) following the intentional introduction of an insect parasitoid (Tothill et al. 1930; Howarth 1991), and for possible contributory effects in a few other cases. One other program resulted in extinctions of some precinctive species of *Partula* land snails in the Pacific island of Moorea following the introduction of the predacious snail *Euglandia rosea* to control the land snail *Achatina fulica* (Murray et al. 1988). These land snails were not target species in the program, and neither their extinction nor control was an objective in the program. These latter extinctions are examples of unintentional environmental effects caused by the release of an introduced biological control agent.

Most of the instances of environmental harm from biological control reflect early practices, such as the introduction of vertebrates, that are no longer followed. There is also substantial evidence that strategies to avoid risk have been extremely successful. For example, there is no evidence that any phytophagous organism, tested prior to release and then used in a weed biological control program, has ever caused any permanent harm to a food crop (Waterhouse and Norris 1987). Indeed, biological control through introduction of new species of natural enemies is a form of direct ecological management over which we have substantial control. Issues of safety, in the sense of preventing or minimizing possible negative impact of natural enemy introductions, arise in three areas during the introduction process: contaminants in imported materials, unintentional release of quarantined materials, and unforeseen negative impacts of agents chosen for release. Safety in the first two of these issues, which requires minimizing the potential impact of contaminants in shipments and the unintentional release of quarantined materials, is achieved by effective and watchful handling of material in quarantine. Procedures for proper sorting of material when first imported, destruction of undesirable organisms, and strict quarantine handling of all material until released from quarantine, where properly implemented, have reduced to effectively zero the risk of harm from either of these first two issues (see Chapter 9).

Many potential negative effects of candidate natural enemies can be tested for, thus maximizing the safety of an introduction with regard to the third issue. Unforeseen effects of natural enemies might be grouped in three categories. First, an intended release might negatively affect the biological control program itself. For example, if a culture of natural enemies is released which carries a pathogen or hyperparasitoid of the natural enemy, it may introduce a biological factor which acts against the natural enemy. Careful biological study of introduced cultures can limit the risk of such biological factors affecting an introduction program. Second, there may be damage to crops or other economically desirable plants or organisms following the release of a natural enemy. Careful host-range testing conducted prior to release, however, minimizes the likelihood of such occurrences. The effectiveness of established protocols for such testing is underscored by the lengthy record of introductions of phytophagous natural enemies without any permanent harm to any nontarget system. In only one case has there been temporary feeding on a food crop by a natural enemy population which had increased on its target weed and, following collapse of the weed population, temporarily had little else to feed upon. The situation was temporary, and the herbivore population soon declined and was no longer recorded damaging the food crop (Waterhouse and Norris 1987). Third, there may be unanticipated effects on nontarget organisms in the environment. The potential for such effects is minimized by criteria employed in selecting agents for release, including current practices which avoid release of vertebrates (except in certain aquatic systems), social invertebrates, and broadly polyphagous invertebrate herbivores, parasitoids, or predators. Additionally, we may consider the composition of the ecological community into which the natural enemy is to be released. When the community includes species or groups of special value or at special risk (perhaps of extinction in the case of precinctive populations), evaluations of the program's objectives and of the biology of the candidate natural enemies can include specific consideration of their potential impact on these species. Natural enemies may be sought which do not or cannot attack these high-value groups. Unforeseen negative impacts on nontarget organisms through the release of natural enemies appear to be rare and are usually outweighed by substantial environmental improvement brought about by the reduction in density of the target pest.

CHAPTER
9

INTRODUCTION OF NEW NATURAL ENEMIES: METHODS

INTRODUCTION

Carrying out biological control programs based on the introduction of new species of natural enemies involves several steps, each of which contributes to the overall effectiveness of such programs. Many of the technical and biological considerations related to acquiring and shipping entomophagous and phytophagous arthropods have been described by Bartlett and van den Bosch (1964), Boldt and Drea (1980), Klingman and Coulson (1982), Schroeder and Goeden (1986), and Coulson and Soper (1989). This chapter discusses the following steps: (1) conducting the exploration and collection; (2) shipping natural enemies; (3) quarantine processing of foreign material; (4) ensuring the safety of natural enemy introductions; and (5) field colonization of natural enemies and initial program evaluation.

THE COLLECTING TRIP

Planning and Preparation

Permits, Regulations, Quarantine Facilities. The first steps in planning and organizing an exploration trip are to obtain the permits necessary to collect, export, and import the collected material and to coordinate with an authorized quarantine facility for the handling of the shipped material. Permits may be necessary from both national agencies and local (state or provincial) agencies. Permits are required not only for importation into the home country, but may be required by the country of origin, where regulations may govern the collection and the export of living material or dead (museum) specimens. While there may not be national laws governing specifically the shipment of natural enemies, such shipments may be regulated by interpretation of other more general laws governing the importation or receipt of nonnative organisms (Coulson and Soper 1989). Similarly, states or provinces may also have regulations governing the movement and release of nonnative organisms, and these may apply to introduced natural enemies. A valuable source for determining whether collection or export permits are required is usually a colleague in the source country. Inquiries at the provincial- and national-level equivalents of departments of agriculture can also provide the necessary information about required permits.

Separate permits may be required for importation into quarantine and for release from quarantine. Coordination with a quarantine facility should include copies of these permits, expected dates of shipments, arrangements for customs clearance of shipped material, and agreements on expectations for handling, sorting, and subsequent shipping of natural enemies to the researcher. Arrangements with the receiving quarantine laboratory must also assure the timely availability of any necessary host material.

Funding. Programs of natural enemy introductions usually require stable, long-term funding. Searches for natural enemies may require 12 to 18 months of preparation, several months of searching, and months or years of subsequent work to evaluate the efficacy of the introduced natural enemies. Budgets should include funds for support staff in the home institution for processing material received while the explorer is out of the country, support for host culture development and maintenance for a period of several months before and after the exploration, as well as funds for subsequent maintenance of laboratory cultures of newly imported natural enemies.

Funds for the actual exploration trip should include travel and subsistence costs as well as support for local guides, unanticipated travel costs, air transport for shipped material, and local purchases of supplies. In some countries, local travel to and from field sites must be provided by official government drivers, and funds for this service must also be incorporated into the planning process. Additional funding is often desirable to establish cooperative projects of extended collection and shipment with colleagues in foreign locations.

The Explorer/Collector. The explorer must have a broad knowledge of the target pest and its natural enemies, including their geographic and host ranges, host plants, biology, and taxonomy. Foreign exploration duties have often been assigned to full time foreign explorers or research scientists on long term assignment in a foreign country. They are also undertaken by academic or research staff from state or federal agriculture experiment stations or other agencies. These staff are usually scientists already working on the target pest or its close relatives.

The explorer must be prepared to travel under often adverse circumstances and will need knowledge of the languages and customs of the areas where explorations are to be conducted. The services of a local bilingual guide are often invaluable. Cooperation of colleagues or contacts in the area to be searched are important for effective use of the time available for the search.

Collections of natural enemies for a particular program may also be obtained through the services of intermediary organizations (USDA-ARS, CAB International Institute for Biological Control, and others), or through cooperation with colleagues in other areas.

Planning the Trip. Before departing on the trip, a detailed itinerary should be developed for coordinating with the quarantine facility and other staff involved in the program. The places selected for searching should be in accord with the procedures for identifying favorable search locations as discussed in Chapter 8. Passports and visas must be obtained, often significantly in advance, and for some countries this may require letters of cooperation or requests from national or provincial officials. Contact with foreign collaborators can provide valuable information on suitable seasons and locations to be included in the exploration. Where guides are needed, foreign colleagues can often advise on the availability and costs of these services.

The use of the itinerary in planning with other staff of the program should not preclude alterations in the itinerary once the trip is under way. However, the itinerary should include dates on which the quarantine laboratory should expect shipments and, if possible, dates and times for planned contact by telephone, facsimile transmission (FAX), or telegram. These should be used frequently to inform the staff of the quarantine facility and home institution of any new developments, changes in the expected dates of shipments, or changes in travel plans.

Equipment. Arrangements should be made before departure for all needed equipment. Usually the explorer must carry this equipment, but in cases where work is being conducted in conjunction with a laboratory in the foreign country, some items such as microscopes may be available there.

Equipment may be grouped under three headings: collection equipment, identification and handling equipment, and packaging and shipping supplies (Table 9.1). The specific needs for each of these categories will vary depending on the target organisms. Collecting equipment may include plant-harvesting or soil-digging tools, sample storage vials or bags, maps and notebooks to record collection locations, and ice chests or other insulated storage containers. Cameras are also often useful for recording the condition of collection locations. Identification and handling equipment may include forceps, scalpels, scissors, probes, microscopes, hand lenses, lights, vials or other containers to house collected natural enemies, petri dishes with prepared media for inoculation with pathogens, and literature pertaining to identification of host plants, hosts insects, and natural enemies. Packaging and shipping supplies will include primary containers, external containers, wrapping materials, cushioning material, shipping boxes, tape, shipping labels, and permits. Expedition supplies can require substantial storage space, and these needs must be considered when planning luggage allowances and transport.

Conducting the Exploration

Documents. The initial stages of setting up a foreign exploration trip typically involve a literature search, taxonomic research and study of museum material or possibly voucher specimens and explorers' notes from earlier trips dealing with the same pest or search areas, and correspondence with collaborators abroad to determine suitable seasons and locations to search. Important correspondence may include recommendations from colleagues, local agricultural extension people, botanists, and personnel from botanical gardens and nature pre-

TABLE 9.1 List of Equipment for Exploration for Natural Enemies

(a) *Collection equipment*	(b) *Identification and handling equipment*	(c) *Packaging and shipping supplies*
maps	microscope and light	Styrofoam or insulated outer containers
spades, shovels, trowels	magnifying glass and lenses	cold or hot packs
pruning shears, saws	reference texts for local floras, etc.	shipping boxes (cardboard or wood)
gloves	small vials for isolating specimens	address labels
camera	labels	quarantine labels
collecting bags (paper, plastic)	rubber bands	copies of shipping permits
ice chest	cards for labeling groups of vials	tape, string
	record notebook or forms	
	scissors, ruler	
	honey	
	insect pins	
	forceps (fine and large)	
	alcohol	

serves, especially in the search area. A letter of introduction from the host country's consulate to institutions requesting their cooperation and possibly temporary use of their facilities may also be desirable.

The explorer should have a valid passport, necessary visas, immunization and medical records, and a supply of personal business cards. Letters of authorization from institutions or officials in the countries to be visited which show the names of the cooperating institutions and individual collaborators may also be necessary.

Travel. Natural enemy collectors must travel relatively "self-contained." Necessary supplies include drinking water, water purification tablets, high-energy foods, and any necessary medications. Such medications may include those for intestinal disorders which, although unlikely to be life-threatening, can lead to restrictions from travel and productive field work. Materials for shipping packages may be difficult to find in some locations, so it is prudent to carry containers, string, tape, and labels. The trip's itinerary should be well-planned, and all tickets should be obtained before leaving the home country, because tickets can be difficult to obtain in some areas. Nonetheless, the explorer can be flexible and ready to alter the itinerary should circumstances (such as particularly productive or unproductive collecting in a particular location) warrant. In some locations, it may be appropriate for the explorer to notify the embassy or consulate of his presence and local address in a foreign country to expedite communication from embassy officials should the need arise. Collectors may find extended periods of field work taxing. Suitable precautions, such as carrying water in the field, wearing hats in warm climates, and other protective measures, should be followed.

Collection. The explorer should collect as much material as possible. A local colleague can be helpful in locating habitats or agricultural plantings suitable for collecting. Local graduate students often are excellent in this role. Searches are usually for native populations of natural enemies and their hosts, and these may occur at substantially lower (often very low) densities than in areas where the target species is a pest. Field notes should be kept which include dates of all collections; names of contacts, villages, and farms; notes on habitat types, plant communities, host plant species, and host species located, as well as natural enemies found. Notes should also be kept on other pests found and possible sources of natural enemies for other crops. Maps (obtained either in the home country or abroad) and photographs can be a valuable addition to the field journal and very helpful in defining search locations precisely. These records are especially valuable in planning future trips. Provision should be made to maintain collected material in optimal condition while in the field, including a small, insulated ice chest or other similar container to limit variations in temperature, together with suitable collecting containers or media.

Processing. Adequate time should be allowed each day to sort and process collected material for storage and shipment. Depending on the amount of material found in the field, time for processing collections may exceed the time spent collecting. Access to a local laboratory facility can provide working space, lights, microscopes, and other facilities, but in general the explorer should take all the materials and equipment necessary to process collections in a hotel room or similar accommodation.

Shipping. Collected material should be shipped as frequently as necessary to ensure optimal condition of the material on arrival at the quarantine facility. Where possible, natural enemy stages least susceptible to the rigors of transport should be selected for shipment. Such stages would include pupae of holometabolous insects, larvae in diapause (and, therefore, not requiring food), or eggs. For pathogens, suitable stages might include cadavers, spores, and fungal hyphal colonies on agar. Special provisions (such as ice packs or heat packs, and humidity-moderating materials) should be available to package material if variations in temperature or humidity must be restricted during shipment. Additional details regarding packaging and shipping natural enemies are considered later.

Details regarding shipping natural enemies are covered in the section on shipping procedures, but some issues should be mentioned here which may affect planning and conducting exploration. Shipment of collected material should be by air freight or, if this is impossible, airmail. The collector may have to hold material for several days before reaching a city with air freight service. Once in the city with such service, traveling to an airport facility and making arrangements for shipping can require several hours; substantial time should be planned for this in the itinerary. Arrangements at the airport can be complicated by language barriers, so the assistance of a local colleague may be helpful. If airmail is the method used for shipment, the shipper should witness application of stamps to the package and (most important) their hand cancellation.

When possible, air freight and airmail shipments should be dispatched so that the material will arrive in the home country early in the working week. This will facilitate rapid handling of the package and minimize delays, especially avoiding the delay characteristic of weekend arrivals. Personnel at the quarantine facility should be notified of the shipment by telephone, telegram, or facsimile transmission (FAX), including the name of the carrier, routing information, time of arrival, and airway bill number. The external packaging material should bear the necessary permits and address labels to facilitate recognition and handling by customs and inspection personnel. Labeling procedures in particular must be meticulously performed to avoid delay at the port-of-entry customs and agriculture inspection.

Local Assistance. The assistance of a local colleague is helpful to maximize the collection effort, especially if a relatively short stay in the area is mandated by travel itineraries, unexpected delays from local holidays or strikes, inclement weather, or other circumstances. Such a person can serve as guide, driver of a rental vehicle, mentor regarding local customs, expedite access to private or public properties, and assist in organizing shipment of collections. A local host will be able to allay the suspicions or perhaps outright hostility of the local people toward the presence of a strangely-acting foreigner. Obviously, the collector must be able to pay promptly for expenses incurred. Fluctuating exchange rates can be a problem difficult to anticipate when making long range plans. Surges in the inflation rate can also create unexpected expenditures. Because of such contingencies, the collector is advised to carry 20% more than the estimated expenses in the form of travelers' checks in small denominations. Internationally-accepted credit cards should also be carried.

Assistance may be obtained in some circumstances at embassies or consulates. Inquiries at regional, national, or international laboratories can often yield very useful information and assistance.

SHIPPING PROCEDURES

Shipment of live natural enemies is a critical part of the introduction of natural enemies (and the later distribution of natural enemies once suitable species have been designated for release). Shipment of natural enemies involves three fundamental steps: being aware of and complying with the necessary regulations governing the collection, shipment, and importation of natural enemies (as previously discussed); selecting stages, packaging, and shipping the natural enemies; and, receiving natural enemies and clearing them through official channels until they are received by the quarantine facility. Several papers have treated portions of each of these three steps (Bartlett and van den Bosch 1964; Bolt and Drea 1980; Bellows and Legner 1993). They are treated here, in a general context together with some detail for illustration, acknowledging that each of the steps may apply somewhat differently to particular projects or portions of projects, or to work conducted in different countries.

Packaging and Shipping Natural Enemies

The permits necessary for export from the country of origin and importation into the target country should be obtained before shipping and, if possible, before exploration. Shipment labels referencing such permits are often provided by regulatory agencies and should be attached to the outside of any shipments. Any foreign shipments received under permit should be handled according to the permit restrictions, which usually require processing in a primary quarantine facility designed for this purpose.

During packaging and shipping, steps should be taken to avoid including undesirable material, such as known pest species not involved directly in the exploration's mission, unnecessary plant material (to reduce the risk of unintended shipment of plant pathogens), and soil and manure (to reduce the risk of unintended shipment of plant seeds and pathogens).

Different stages of arthropod natural enemies selected for shipping vary in their requirements, both physical and biotic, for optimal survival during shipping. Where possible, stages selected for shipping should be nonfeeding stages, or those with the least requirement for food or other sustenance. Less active stages of an organism often survive the rigors of transport better than active stages. Such stages may include, for arthropods, natural enemy eggs and pupae, or in the case of egg, larval, or pupal idiobiont parasitoids, parasitized hosts. Natural enemy adults in some groups can also be shipped successfully. Arthropod pathogens may be shipped in living or dead hosts or on artificial media. Plant pathogens may be shipped with pieces of infected plant material or, in the case of some fungi, as cultures on artificial media.

Shipping Eggs. When eggs are shipped, they should be kept cool to reduce hatching, so long as this does not jeopardize their survival. Where eggs are attached to plant material, excess plant material should be removed. The plant substrate should be fastened securely in small containers, or packed in layers separated by soft tissue to avoid movement that may cause the eggs to become dislodged from the substrate. Organisms should be packaged in several small containers (vials, petri dishes) rather than one large container to minimize loss should one container break, get wet, or otherwise be damaged. Such containers as vials and bottles should be closed with nonabsorbent cotton, or if they are sealed, should contain some absorbent material to prevent condensation of moisture caused by changes in temperature during shipment.

Shipping Feeding Stages. Stages of immature natural enemies which need to eat (larvae or nymphs), and feeding stages of immature hosts bearing natural enemies, pose special challenges for shipping and should be avoided if possible. When species to be collected are encountered as larvae, it may be possible to arrange to remain in the area or to return at a time when such stages are nearing pupation, or they may be collected by the explorer and fed until they mature. Pupae or adults are often shipped more successfully than larvae or nymphs, since pupae do not require food or moisture during shipping and adults are often very tolerant of the stresses of shipping.

When shipping immature arthropods, they should be individually isolated in containers to prevent cannibalism, which can occur in many taxa, particularly when food is scarce. Larvae or pupae present inside insect hosts or in plant stems, roots, or other plant parts should be left *in situ*, and shipped inside their hosts.

It may be necessary to provide actively feeding stages with a small amount of plant material on which to feed during transit. Individuals should be isolated in these cases, providing a small amount of food material for each individual in separate small vials or petri dishes, taking care in shipping plant parts to maintain the integrity of the plant material and prevent secondary bacterial and fungal growth on the plant tissue. After removing excess soil, roots bearing natural enemies should be packaged in lightly moistened sphagnum moss, wrapped in burlap and a cloth bag, and placed in an unwaxed cardboard or wooden container. The material should not be enclosed in plastic, as this may lead to excessive condensation of moisture. Plant leaf, stem, or flower galls containing actively feeding or developing stages can be packed with sphagnum moss or excelsior. Ample space should be provided for air circulation to prevent overheating and decomposition of materials. It may be advantageous to seal the ends of cut plant stems with melted paraffin wax. Some Hemiptera may be shipped as immatures or adults; predacious forms should be fed before shipping if possible.

Shipping Pupae. Pupae should be divided into a number of small lots in separate containers. Lepidopteran pupae collected from the soil can be packed in slightly moistened sphagnum moss; dipterous pupae extracted from soil can be repackaged in moist sterilized soil in a series of small containers. Pupae on twigs or foliage, or in cocoons on other substrates, are best left attached to the substrate and packaged in shredded newsprint or sphagnum moss to prevent movement during shipping. Dried flower heads containing diapausing larvae or pupae can be shipped in bulk in cloth bags enclosed in a cardboard or wooden container. Natural enemies of sternorrhynchan homopterans (whiteflies, mealybugs, aphids, and relatives) are usually best shipped as pupae inside their hosts. Isolation of individual hosts is critical in these groups to prevent loss of pupae to hyperparasitism by early emerging females, because many species of natural enemies of these groups produce male progeny as hyperparasitoids on female pupae. Bits of leaves or stems bearing these pupae can be placed in small (¼ dram) vials, which can then be placed in groups in a styrofoam container. Humidity can be somewhat controlled, and desiccation of the material probably lessened, by also enclosing a smaller container (petri dish) containing a saturated salt solution, sealed with semipermeable membrane (such as Opsite Woundt, available from medical suppliers) which is adhesive and seals such containers well (Hendrickson et al. 1987).

Shipping Adults. Adults of some natural enemies are sufficiently robust to be shipped, particularly when time in transit is minimal (48 h or less). Adults should be provided with access to a moist, absorbent material during shipment, although care must be taken to prevent excessive moisture. A small amount of moistened shredded wood (avoid chemically-treated wood and such odoriferous woods as cedar), pipe cleaners, or other material can be provided as resting substrate for the adults.

Access to food is important for survival in some groups. Small spots of honey placed on the inside of the container often provide suitable food for most adult Hymenoptera. Honey can also be added to vials containing pupae, so that any adults emerging during transit will have access to food. Access to sugar water or honey, either on the inside of the glass containers or on wicks of cotton or sponge, also should be provided for adult Lepidoptera and Diptera. Moistened raisins have also been used (Bartlett and van den Bosch 1964). Pollen may be provided to adult predacious mites. Adult Coleoptera can survive several days without food, particularly if well fed before shipment.

Shipping Pathogens. Shipment of pathogens requires special care to limit the growth of saprotrophic organisms on the diseased arthropod or plant. Pathogens in some cases can be isolated on artificial media prior to shipping. When shipping pathogens on host material, such material should be divided between several small containers to limit the potential for spread of undesired infection arising from any particular specimen.

Additional Guidelines. Other practices that help ensure successful shipping of natural enemies include the following:

- Minimize extreme temperatures and the length of time organisms must remain in the package, as these are the principal causes of loss during shipment. Cool packages before shipment and include in the final package a frozen artificial ice pack (or dry ice in cases where exceptional cold is warranted) or, in cold periods, a heat pack, to protect against excessive changes in temperature.
- Limit the amount of fresh plant material; such material breaks down rapidly when packages are inadvertently exposed to sunlight or warmth, leading to loss of pathogen cultures or death of other natural enemies.
- Control the amount of free moisture in packages.
- Package natural enemies separately as individuals whenever possible to limit loss from breakdown of substrate and to prevent cannibalism of arthropod natural enemies or contamination among pathogen sources.
- Avoid fresh plastic materials in collecting and shipping containers; these may be toxic to natural enemies.
- Avoid the use of gelatin capsules; they extract moisture from the environment, contributing to desiccation of collected specimens.
- Enclose groups of vials (grouped together by natural enemy species or stage or collection location) or other containers in closed Styrofoam containers, cloth bags, or other additional wrapping. Enclose these secondary packages in an insulated container such as a Styrofoam chest and then in a more robust shipping container, either cardboard, wood, or other firm material (Fig. 1.8). This final container must bear shipping and permit labels, routing instructions, and instructions for handling on arrival (telephone numbers for the primary quarantine facility).
- Ship by the most direct secure routing, which usually means air freight or other express services such as international couriers.
- Delay sealing the package before shipment as long as possible.
- Contact the primary quarantine facility as soon as details of the routing are known. If necessary, delay shipment in the country of origin until a suitable arrival time can be arranged to facilitate processing the shipment through regulatory agencies at the port of entry.

Receiving Shipments From Abroad

Receiving and handling shipments from abroad pose no serious problems if suitable arrangements are made in advance. The principal concern is rapid clearance of the material through the official interception agency at the port of entry. Two general approaches are used: collecting the shipment directly by personnel from the receiving quarantine facility, or using a customs broker.

If a point of entry is available near the receiving quarantine facility, every effort should be made to ship directly to this port of entry. Delays at points of entry far from the receiving quarantine, where experience in handling and clearing material may be limited, can lead to substantial loss of live material. When material arrives at the port of entry, quarantine personnel can be contacted directly by both the air cargo company and by personnel from the official intercepting agency, ensuring the material is collected from the port of entry immediately upon arrival and taken directly to quarantine, minimizing delays. In addition, a long-term relationship between personnel at the interception agency and the receiving institution will lead to efficient and rapid handling of shipments.

If there is no port of entry which can be reached by direct shipment from abroad, shipments must be collected by a customs broker. The efficiency of this procedure depends largely on communicating the need for rapid handling to the broker. An experienced broker can can gain permission to obtain the shipment and to forward it immediately (by courier or air cargo) to the receiving quarantine facility. Personnel at the receiving quarantine can facilitate this transfer by providing correspondence to the broker on official letterhead authorizing the collection of the shipment, and by contacting personnel at the interception agency at the port of entry in advance to explain the circumstances and particular details of the shipment. In all cases, developing a working relationship with a broker and agency personnel will improve efficiency in processing shipments.

Shipping Within a Country

Some states or provinces govern the movement and release of arthropods in addition to the regulations imposed at a national level. When such shipments are made, permits should be obtained authorizing these shipments and packages containing biological control agents should be appropriately labeled.

PROCESSING MATERIAL IN QUARANTINE

Facilities and Equipment

A quarantine facility requires, at a minimum, a secure room in which to open, examine, and process incoming shipments of beneficial organisms prior to further study. Quarantine facilities may be as small as a two-room facility housed in a larger building, or may be separate buildings with a suite of rooms for processing and handling several shipments simultaneously (Fig. 9.1).

Facility Structure. The structure of a quarantine facility will be determined by both its intended uses and local construction codes. Functionally, a quarantine facility may be viewed as a series of boxes fitting inside each other, with the most secure room equivalent to the most interior box. Leppla and Ashley (1978) give floor diagrams for five biological control quarantine facilities. Interior structural details will also depend on the intended use of the facility.

Rooms for handling beneficial arthropods require controlled environments (temperature, humidity, lighting). The air exchange systems in such rooms must provide for such control, and additionally must not permit escape of organisms. Escape is usually prevented through the use of extremely fine screen filters on the air exchange ducts, absence of opening windows (windows may be present, but should be nonopening and well-sealed), and sealed doors opening only to interior, sealed hallways. Rooms used for initial opening of packages of

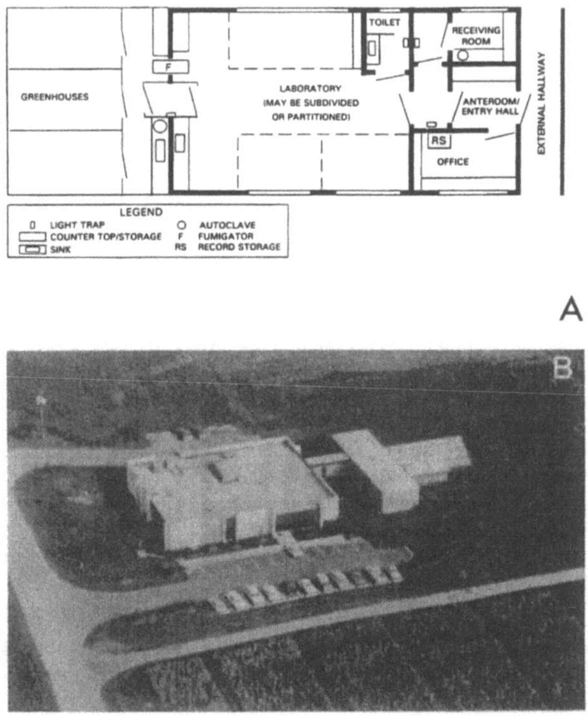

Figure 9.1. (A) Layout of a quarantine facility includes an anteroom, secure doors between the quarantine hallway and the external hallway, and an airlock separating the hallway from a primary receiving room where packages are first opened upon receipt. Additional rearing rooms may be included in the quarantine suite (after Fisher 1978). (B) Photograph of a biological control research laboratory; quarantine facility is inside the left side of the largest building. (Photograph [B] courtesy of R.M. Hendrickson, USDA-BIRL.)

arthropods are often situated with a north-facing (in the northern hemisphere) window to attract any escaped organisms and facilitate their recapture. When several rooms are available, they should have independent environmental controls to permit the maintenance of a variety of conditions in the facility. Personnel should wear laboratory coats while in the facility and remove them before leaving, leaving the coats in the facility.

Rooms designed for use with pathogens of arthropods or weeds require additional precautions because of the size of the organisms under study. Design criteria for a pathology laboratory are discussed by Melching et al. (1983) and Watson and Sackston (1985). The building (or a portion of one) must be sealed and recirculating air conditioning must include a double set of filters designed to remove particles larger than 0.5 μm, capable of removing airborne bacterial and fungal spores. Exhaust air must pass through a third, deep-bed filter before being passed to the outside. Air pressure within the facility must be negative to the outside atmosphere to prevent unanticipated air exchange from inside the facility to the outside. The laboratory and glasshouse facilities are typically divided into smaller cubicles to limit contamination among study areas. A culture of the target or rearing host must be maintained in another facility to prevent its contamination by the pathogens being reared in quarantine.

Equipment. Equipment needs in quarantine facilities vary with the class of organisms being handled. Space in quarantine is often limited, so there is a need to avoid overstocking with equipment. When equipment in quarantine is no longer needed, it must be thoroughly decontaminated before removal. Decontamination in a facility processing arthropod parasitoids, predators, and herbivores may be accomplished by wiping down the equipment with ethyl alcohol. Where this is not practical, or for facilities handling pathogens, fumigation will be necessary. Equipment and notes are decontaminated at a USDA-ARS facility for importation of pathogens in Frederick, Maryland, by processing them through a cold gas fumigation chamber utilizing 10% ethylene and 90% carbon dioxide for a minimum of two hours (Melching et al. 1983).

Identification of imported organisms typically requires a microscope. A binocular dissecting microscope (10–120×) and high quality, fiber optic illuminator are generally required for identification of arthropods. The microscope should be mounted on a pedestal with an adjustable arm for versatility in positioning. A compound microscope will be necessary to screen cultures of beneficial arthropods for infection by entomopathogens, and to identify imported pathogens. This microscope can be located outside the facility, so long as specimen mounting materials are available in the facility and only mounted, sterilized slides are removed from the facility.

A reference file of critical literature should be available in quarantine (or nearby for use with preserved organisms) because correct identification of introduced species is crucial to initial handling, establishment of cultures, and rearing in quarantine. Available reference works should include keys to families and genera of imported organisms (both adults and immatures), works on host associations and life histories, and on rearing procedures for pathogens. In addition, access to identified voucher specimens can greatly increase the speed with which an identification can be made.

Other equipment needs include:

- Rearing cages or containers to maintain the integrity of separate natural enemy species or collections. These cages may be located together in a single quarantine room, or located in individual rooms or cubicles.
- Glass vials with stoppers or cotton plugs to isolate arthropods.
- Facilities to prepare media for culturing microorganisms.
- Autoclaves for sterilization in facilities where pathogens are handled and for disposal of unwanted materials received in foreign shipments of natural enemies.
- Dry heat ovens for sterilization where autoclaves are not used.
- Fluorescent lights suitable for the maintenance of plant growth.
- Refrigerators or cold rooms to maintain organisms in a diapause state. Refrigerators can also permit shipments to be cooled before initial processing, which will facilitate handling of arthropods.
- Carbon dioxide to induce short-term anesthesia of arthropods, and, in some cases, to stimulate mating in Hymenoptera.
- Various hand tools, ranging from hammers and screwdrivers to repair and adjust cages, through microforceps and probes for handling minute arthropods.

Personnel and General Procedures

Each quarantine facility should have a designated quarantine officer responsible for its operation. This consolidation of responsibility increases security, ensuring that each organism is properly handled and tested and that adequate records of all organisms received, shipped, or otherwise processed are kept. An assistant should also be designated who can assume responsibility in the absence of the supervisor. The quarantine officer should be familiar with the regulations and laws governing the conduct of the facility, develop and maintain contact with the regulatory personnel at the local ports of entry through which shipments arrive, process all incoming shipments, maintain the necessary records, and oversee the functioning of the facility.

Maintaining contact with the network of explorers, foreign colleagues, commercial shipping and custom clearance agencies, and federal and provincial regulatory officials is the joint responsibility of the quarantine officer and project director. Maintaining communication with the various agencies and personnel involved can be important in facilitating the arrival and handling of shipments at ports of entry. Ability to recognize and resolve problems is a critical skill in maintaining efficient movement of shipments through regulatory channels into the hands of quarantine personnel.

Only when a facility is large and has a number of varied programs being conducted concurrently should additional personnel have access to the security area and become involved with the quarantine handling of introduced material. In these cases, the quarantine officer should ensure that each person is familiar with quarantine operating procedures and that all persons who work regularly in the security area are certified regarding familiarity with federal, state, and local regulations for the importation and release of biological control agents and quarantine procedures.

Quarantine personnel should have knowledge and skill in several key areas pertaining to the conduct of biological control related to quarantine operations. Workers should be familiar with the concepts and practices of biological control and how they apply to the specific projects and organisms with which they are involved. The quarantine officer and his assistant should have a broad knowledge of all projects conducted in the containment area. Project directors should have knowledge of the taxonomy and life histories of targeted pests and possible natural enemies. Such information can be critical in planning host specificity studies and in verifying the nature of the relationship between a presumed primary natural enemy and its host. Literature references on candidate species or their relatives can be vital sources of information and should be readily available. The project director should provide the quarantine staff with appropriate information on specific natural enemy groups anticipated in shipments.

All workers should be familiar with the physical operation of the containment facility and its equipment, particularly ventilation, power and utilities, and function of the autoclave or other sterilizer. Names and telephone numbers of maintenance and repair personnel to be contacted in cases of needed repairs should be available for normal workdays as well as weekends and holidays. Janitorial needs within the quarantine facility should be handled by quarantine personnel. Equipment and facility service or repair personnel must be apprised of the importance of quarantine security and should be accompanied by quarantine personnel when working in the security area. Visiting scientists and regulatory personnel should also be accompanied in the facility. Casual visits by individuals or groups should not be permitted in the containment area. Fire, earthquakes, vandalism, and illness may interrupt quarantine operations. Instructions should be posted on entryways advising emergency personnel which entry and exit procedures will be least likely to breach quarantine security. Telephone numbers for contacting the quarantine officer and assistant, during both business and off-duty hours, should be posted at the entrance to the facility. If shipments are expected outside of normal working hours, it may be advantageous for the quarantine officer to carry a pager keyed to the quarantine telephone.

SAFETY

The principal issues regarding the safety of programs for the introduction of new natural enemies may be considered in three groups: protecting the environment from unwanted organisms by recognizing and destroying in quarantine undesirable organisms which may be accidentally grouped with beneficial organisms in the initial shipment; establishing laboratory cultures of potentially beneficial organisms and protecting them from contamination and from parasitoids, hyperparasitoids, or diseases; and evaluating the suitability and safety of the candidate beneficial organisms for release into the target environment. If initial findings are favorable, candidate organisms may be then released to a laboratory for further study or for additional propagation and release against the target pest.

Quarantine Protocols and Initial Screening

Shipments that may contain live organisms received from abroad initially are opened inside an observation cage with a glass top and closed sides and fitted with one or two cloth sleeves through which the worker manipulates the material. This design permits initial separation and containment of undesirable organisms. Live organisms are collected into glass vials; packaging materials are heat treated in a dry oven or autoclave and discarded. The live organisms are then screened taxonomically; those known to be undesirable (hyperparasitoids or unwanted phytophagous arthropods) are killed and preserved for the voucher collection. Potentially beneficial organisms are separated by species or collection locality, observed for mating, and placed in isolation cages with the appropriate host for propagation.

If shipments consist of plant pathogens, precautions against unauthorized release are those of a microbiological laboratory, and include special air-filtering requirements as discussed earlier, and sterilization of water and soil supplies (both entering and exiting the quarantine laboratory), typically by autoclave, before discharging them from the area. Personnel must shower before leaving the laboratory, leaving their work garments inside the facility.

Initial Culture Establishment

The establishment of laboratory cultures free from contamination is a major objective of quarantine operations. Some quarantine cultures, particularly those maintained on plants, may be subject to infestation by other phytophagous organisms (aphids, whiteflies, and mites are among groups often found on plants maintained in culture). The control of these organisms must be undertaken with care, and a particular objective should be avoiding the use of pesticides where possible.

Quarantine cultures can also be subject to infestation by parasitoids (of phytophagous natural enemies) or hyperparasitoids, and infection by pathogens. Vigilance by the quarantine personnel is critical in identifying such infestations early and isolating the affected cultures. Isolation is perhaps the only mechanism to eliminate parasitoids and hyperparasitoids from laboratory cultures. Additional measures to eliminate arthropod pathogens include use of sterile rearing containers and frequent changes of containers (Etzel et al. 1981). Surface sterilization of eggs, for example by briefly dipping in 10% solution of bleach (sodium hypochlorite solution) in water, may also be successful (Briese and Milner 1986).

Protecting the imported pathogen lines from contamination requires provision for isolation of microbial cultures. Where microbes are grown in artificial media, isolation may be accomplished by maintaining cultures in sealed containers, in different growth chambers, or in different rooms. Frequent evaluation of the state of cultures and early isolation of cultures which show contamination are crucial in maintaining imported lines. Special equipment, such as laminar flow hoods and apparatus to filter air supplies, may be vital to the integrity of the cultures. Where cultures are maintained on live hosts, provision must be made for the maintenance of uninfected host material (either plant or arthropod) outside the quarantine facility. Where particular biotypes of certain pathogens are being studied, phenotypic (by allozyme analysis or related procedure) or genotypic (by DNA characterization) identification may be advisable. Type specimens of such cultures may be kept in cryogenic storage for comparison with later introductions or recoveries, or to evaluate isolated lines for genetic drift or contamination.

Host Range and Safety Testing

For phytophagous arthropods or pathogens introduced against weed pests, tests defining the host-range of candidate natural enemies are essential. A frequently sought characteristic of a natural enemy of plants is that it be relatively specific to the target plant in the target environment. While such testing must be completed before natural enemy release from quarantine, it need not necessarily be conducted in a quarantine laboratory (Harley and Forno 1992). In some cases, such testing can be conducted in the country of origin of the natural enemy, if the test plants may be grown in that country. In cases involving a natural enemy used in previous programs, such testing may have already been conducted and the feeding or infectivity range may already be known. In the absence of this information, tests are conducted in quarantine following introduction and initial culture establishment.

For herbivorous arthropods, tests are typically conducted as either "no-choice" or "choice" tests. "No-choice" tests are conducted by isolating the agent in a cage with a nontarget plant species to determine whether any feeding or reproduction can occur. In "choice" tests, the agent is given a choice, in a single cage, of two or more plant species, usually including the target species, and feeding and reproduction on each is recorded. "Choice" tests may be conducted in cages in quarantine or, in the country of origin, in open fields where plantings of the test plant species are maintained and records of feeding and reproduction from natural populations are kept. For plant pathogens, only no-choice tests are applicable.

Plant species selected for testing are typically taken from a group of plants taxonomically related to the target weed. Plant species may first be selected which are most closely related to the target species (same genus), and then additional tests may be conducted on other plants more distantly related (in the same tribe, family, or order). Tests are sometimes conducted on important crop or wild plants even though they may not be closely related to the target plant. Agents are only released for further testing or propagation when the results of initial tests indicate a favorable degree of host specificity. Acceptable natural enemies need not be entirely species-specific, but must be sufficiently narrow in their host ranges that they do not feed on desirable nontarget plants in the area planned for release, or in other areas which might be reached by natural movement of the natural enemy over time.

The steps typically followed in studying and approving weed biological control agents for release have been discussed by Frick (1974). Zwölfer and Harris (1971) and Wapshere (1974a,b) discuss the importance of host-specificity tests and host range determination. Several authors have questioned the applicability of results obtained under artificial or laboratory conditions to the behavior of a natural enemy in field settings, and the validity of such procedures for determining host plant acceptance by a phytophage is somewhat tenuous. Nonetheless, such tests have normally been the basis for deciding whether or not a natural enemy may be released and have probably been valuable as a conservative approach to choosing agents for biological control of weeds. This approach has contributed to the exceptional record of safety and absence of negative unanticipated consequences which have characterized biological weed control.

Host-range evaluations have been applied less frequently to natural enemies of arthropod pests. The relevancy of such tests to predators or parasitoids is subject to the same criticism as in the case of herbivores; the range of behaviors in a laboratory setting is often not reflected in an ecological setting, and species may be much more selective or host-limited in the environment than in a laboratory.

FIELD COLONIZATION PROCEDURES

Initial Selection of Agents

As discussed in Chapter 8, several initial factors may prevent a natural enemy from establishing, including a poorly-developed relationship with the host or host plant, dissimilarities in climate between the areas of origin and the target areas, and the absence of such ecological requirements as diapause cues or alternate hosts. When such restrictions are known or strongly suspected in a particular candidate natural enemy, it should be given only limited resources for propagation and release, and efforts should be made to obtain more suitable natural enemies. Some of these restrictions may not be evident until after releases have taken place. In some cases, laboratory or glasshouse studies may help predict these problems, but results from laboratory studies must be viewed carefully, as they rarely match field performance directly.

Selection and Protection of Release Sites

Release sites must be large enough to support populations of the target pest and natural enemy indefinitely and not be heavily influenced by pest or natural enemy movement into or out of the area. The size may vary for different natural enemies. Release sites should contain enough hosts to permit reproduction of the natural enemy. Augmentation of host populations, if needed, may be accomplished by release of host arthropods from laboratory cultures, or for weed biological control agents by overseeding or fertilization of target plants. Release sites should be protected from physical and chemical disruption. This typically implies that a crop must not be sprayed with pesticides and must be left unharvested; in the case of pest weed populations, they must be left unmanaged by customary practices. Security of the release site is maximized when the site is located on property owned by the institution conducting the research. In other situations, written agreements with the owners or managers of the property describing site management may be advisable.

Some habitats, such as forests and orchards, have a longer crop cycle than such habitats as annual cropping systems. The temporal duration of annual systems can be enhanced, however, by multiple plantings which overlap in time, by delayed harvesting, or by sequential cutting and regrowth. This temporal duration can provide newly introduced natural enemies the time necessary to complete development in a habitat which might otherwise be removed or destroyed during harvest.

Quality of Released Agents

Release Numbers. In many programs, the number of individuals available for release may be limited, and researchers may be faced with a choice between several sites with few individuals released in each or fewer sites with more individuals released in each site. The latter approach has generally been followed, although the distinction between "few" and "many" individuals probably varies with the taxa involved. Some species or taxa apparently establish more easily than others. The best guide is past experience with the taxa involved.

Agent Quality. Several factors can contribute to the quality of natural enemies when they are released. These include genetic, health, and behavioral aspects. In general, we seek to optimize these factors for the agents being released.

Usually, little is known about the genetics of particular populations of natural enemies being held in a quarantine laboratory. Most genetic principles indicate that early release from quarantine into larger propagation cultures and into the field will not negatively effect the population genetics, while lengthy confinement in conditions of limited population size or artificial environments in quarantine could possibly give rise to selection or drift. Therefore, it is advantageous to release populations from the constraints of quarantine and move them into field settings rapidly.

Released individuals should be as healthy as possible. Cultures should be maintained with optimal numbers of hosts, and natural enemies should be reared on the best stages of their host. Before release, individuals should be cared for in ways that enhance their longevity, fertility, and natural host preferences in the field. Adults should be fed and allowed to mate before release. Adults should be offered water and a carbohydrate source such as honey and, in most cases, should be held until they have reached reproductive maturity (which in many species occurs very shortly after emergence). Large cages and natural light may be necessary to obtain mating in some natural enemies. Some natural enemies may benefit from exposure to the target host prior to release. For many organisms this will happen naturally in the culture. For organisms reared on alternative hosts, exposure to the target host can be arranged in the laboratory prior to release. Pathogens in artificial media which lose pathogenicity away from the target host should be cultured through at least one generation in the target host immediately prior to release.

Natural enemies of all stages must be protected from adverse conditions during transport to the release site. Use of insulated containers to prevent extreme temperatures is essential. If transporting or shipping requires more than a few hours, food and possibly moisture must also be provided. Releases should take place, if possible, in the early morning or evening. Natural enemies should be placed or released onto plants in sheltered positions. Releases should not be planned immediately after rainfall (when foliage is wet) or when rainfall is expected.

Life Stage Released. Releases of arthropod natural enemies may be made of several different life stages. Adults are often preferred because they require less protection in the environment than other stages, they are mobile, and they can begin reproduction immediately upon release. Pupae are sometimes easier to transport, and eggs and other immature stages may be transported directly on foliage where necessary. If adults are particularly prone to emigration upon release from confinement, immature stages can be advantageous in establishing a field colony.

Parasitized hosts may be used for releases of parasitoids. In many cases these hosts will come from rearing programs at the institution directing the program. Although many species of parasitoids may be released as immatures in their hosts, this approach is particularly valuable for groups with delicate adults such as egg parasitoids; Moorehead and Maltby (1970) describe the placement of host eggs parasitized by the mymarid *Anaphes flavipes* in the field in special containers. In other cases, it may be possible to collect parasitized hosts in numbers sufficient to redistribute them to new locations, as was the case for cereal leaf beetle, *Oulema melanopus*, larvae parasitized by *Tetrastichus julis* (Dysart et al. 1973). When such a program is undertaken, care should be taken to evaluate the condition of each shipment or individual to certify that diseased individuals and hyperparasitoids are not also being redistributed.

Plants bearing parasitized hosts can also be used for colonizing natural enemies. This approach has the advantage that natural enemies will emerge over a period of time providing a continual inoculation of adults into the environment. This strategy may be particularly crucial for some heteronomous parasitoids where males are produced as hyperparasitoids of conspecific females. In these cases, the females emerging on a plant bearing parasitized hosts can produce male offspring on female pupae and, also, disperse to parasitize (with female offspring) the yet unattacked field population of the target species (Fig. 9.2).

Pathogens should be released in those life stages most likely to infect hosts. For plant pathogens, this may entail the collection and direct distribution of the infective stages themselves. For insect pathogens, infective stages of the pathogen may be similarly distributed, or infected hosts may be released in the environment, as in the case of the coconut rhinoceros beetle (Waterhouse and Norris 1987).

In cases where it is not clear what stage will lead most readily to establishment, the release in the same plot of several stages may provide the best opportunity for a species to become established.

Optimizing Release Methods. Agents may be released into cages or liberated freely in the larger environment (Fig. 9.3). Cages should be large enough to enclose a suitable number of hosts, and robust enough to withstand wind, rain, or other conditions likely to be present at the release site. Cages usually should be removed a few days after the individuals have been

Figure 9.2. Colonies of hosts on potted plants may be used to make releases of heteronomous parasitoids, which require previously parasitized hosts for production of males. (Photograph courtesy of M. Moratorio.)

Figure 9.3. Natural enemy releases may be made into sleeve cages (A) to confine natural enemies to specific areas; or may be made freely into the pest's habitat (B,C). (Photographs [B,C] courtesy of USDA/APHIS.)

released to prevent overexploitation of the resource and to free the population from any potentially harmful environmental effects of cages (see Chapter 8).

Releases of agents should be made so that the active (adult or larval) or infective stages appear in the environment coincident with susceptible plant or host stages. For some natural enemies such as predators released into populations of prey which contain many susceptible stages, timing of releases to coincide with a particular prey stage is less important. For species which attack only short-lived stages of their hosts, such as stage-specific parasitoids, timing is more critical. In general, synchrony of releases can be best assured by directly sampling the host population to detect suitable stages.

Releases for initial colony establishment may be accomplished by hand-placing natural enemies individually or in groups into favorable locations in the environment. Mechanical release systems have been used for widespread establishment of initial colonies, especially in situations where the host locations were not reachable by surface shipment or transport (Fig. 9.4). *Epidinocarsis lopezi*, for example, was released successfully against cassava mealybug in tropical Africa from airplanes by dropping vials containing adult wasps, which were able to escape after vials reached the ground (Herren et al. 1987).

Confirming Establishment

Monitoring for establishment of released natural enemies is the final step in initial colony establishment. Repeated assessment of populations in the released area may be necessary to determine if a released agent is reproducing in the environment (see Chapter 13). Monitoring

Figure 9.4. When large, or remote, areas require natural enemy releases, these may be made by use of aircraft (A), which release vials of natural enemies (B), via special machinery (C), as was done for natural enemies of cassava pests in Africa. (Photographs courtesy of PHMD-IITA.)

may take one of several forms. If no other similar natural enemy is present in the system, as may be the case for the first natural enemy introduced into a pest population, simple visual inspection in the field (or examination of specimens reared in the laboratory from samples collected at the release site) may be suitable to determine if a new natural enemy is established. Voucher specimens of adult natural enemies should be taken to confirm identification. In systems in which similar natural enemies are already established, larger samples may be necessary to permit rearing of many natural enemies. These natural enemies must then be identified to determine if the newly introduced species are present. Where different populations of the same species of natural enemies exist together, identification may require analysis of allozymes or genetic markers. Samples must be taken over a period of at least one year to conclude that a natural enemy is established in a new environment. In some cases, establishment may occur but be undetectable for several years.

For diseases of arthropods, hosts may be collected and reared to detect the development of disease; some disease organisms may require further rearing on selective media for proper identification, while others can be identified from direct mounts of pathogens. In some cases, antigen-antibody techniques have been developed to permit detection and identification of disease organisms at early stages of infection. Identification of plant pathogens follows similar steps, with samples of pathogens reared from plants in the target population collected for identification through microscopic examination or antigen-antibody techniques.

Colonization of a new agent may require repeated efforts before establishment can be confirmed. Such repeated efforts are warranted after the investment required to secure the species and clear it through quarantine for release. Only after concerted efforts at establishment in all available environments have failed should the conclusion be drawn that a species is not likely to establish in a particular region.

Evaluation of Agent Efficacy and Program Success

Following the establishment of natural enemies, it is important to quantitatively evaluate their efficacy against the target pest in the new environment. Effective natural enemies will warrant further distribution. If the initially introduced natural enemies are not effective, introducing additional species may be appropriate. Quantitative evaluations of natural enemy performance in the field may be conducted in a variety of ways, as described in Chapter 13. Based on the outcome of such quantitative evaluations, decisions can then be made on directions for the program, such as further distribution of certain organisms or recommendation for additional introductions. Evaluating biological control programs is discussed briefly in Chapter 8 and more fully in Chapter 13.

CHAPTER

10

AUGMENTATION OF PARASITOIDS, PREDATORS, AND BENEFICIAL HERBIVORES

INTRODUCTION

When natural enemies are absent, occur too late, or in numbers too small to provide effective pest control, releases may be made of reared natural enemies. This process is called augmentation and may be conducted either as a private commercial business or a government-provided service. Augmentation of natural enemies is practical in circumstances where permanent colonization of suitably adapted natural enemies is not feasible, for example, in protected culture or where existing natural enemies are not sufficiently effective. In the latter case, it may be appropriate to secure new species or strains of natural enemies from the pest's homeland and permanently colonize these species. However, where such efforts are not productive in resolving the pest problem, augmentative liberations of natural enemies may be undertaken to supplement the action of existing natural enemies. A spectrum exists from releases which are meant only to inoculate the crop with the natural enemy (with most control being provided later by offspring of released organisms), termed **inoculative augmentation**, to those releases in which all control is expected to be exerted by the released organisms themselves, with little or no contribution by their offspring, a procedure termed **inundative augmentation**. Both large organisms such as hymenopteran and dipteran parasitoids, predacious arthropods, and herbivorous arthropods and small organisms such as plant or arthropod pathogens, plant disease antagonists, and arthropod-attacking nematodes can be augmented. In this chapter, we consider the augmentation of large organisms. In Chapter 11 we discuss the augmentation of small organisms. Several major issues affect the augmentative use of parasitoids, predators, and herbivores for pest control: agent selection, quality, cost, shipping and storage, application methods, product evaluation, market development and continuity, and assurance of safety.

AGENT SELECTION

The first requirement for biological control through augmentation is to find a species of natural enemy capable of suppressing the target pest under the intended conditions of use which also can be mass reared. Natural enemies used in augmentative programs must be able to recognize, locate, and successfully attack the target pest. The value of moderately effective agents may be enhanced in several ways: by increasing the number of individuals released, by selecting plant cultivars that are resistant to the pest (Boethel and Eikenbary 1986), and by selecting plants on which the searching efficiency of the natural enemy is improved (Boethel and Eikenbary 1986). Where these approaches are insufficient, it may be necessary to find a better agent or use a mixture of complementary agents, such as predators released together with parasitoids. Worldwide, there are many hundreds of species of crop pests. The natural enemies of only a tiny fraction of these have been investigated. To expand the use of augmentative biological control, knowledge of the natural enemies of more pest species will be needed.

Natural populations of beneficial organisms vary in many important characteristics that strongly influence their potential for successful use as biological control agents. Some species may be highly effective in controlling a given pest, while other, closely-related species are partially or completely ineffective. Disparities may be due to differences in host preferences, rates of increase, climatic tolerances, or other causes. Within species, populations or strains may also exhibit important differences in features such as critical photoperiod for diapause induction (Havelka and Zemek 1988) and parasitism level (Pak and van Heiningen 1985; Antolin 1989), as well as features such as degree of resistance to particular pesticides (Rosenheim and Hoy 1986; Inoue et al. 1987). To help choose useful agents, laboratory tests (such as those to measure attack rates on the target host or the ability to forage on the crop plant) can be applied to colonies of potential candidate organisms (Bigler et al. 1988; Hassan 1994).

Highly host specific agents are likely to perform better, per released individual, than generalist agents. Host-specific natural enemies are preferred by both the natural enemy producer and the crop grower, even if they must produce or use several different natural enemies for different pests in the crop (van Lenteren 1995). For example, effective control of many Lepidoptera is best achieved when particular species (and perhaps races) of *Trichogramma* are produced and marketed for control of a specific pest. *Trichogramma nubilale* Ertle and Davis is an appropriate species for control of the noctuid *Ostrinia nubilalis* (Curl and Burbutis 1978), whereas some other species of *Trichogramma*, while useful for different pests, are ineffective against *O. nubilalis.*

Efforts have been made to identify attributes that might be measured in the laboratory that would predict which of a set of natural enemies might be most suitable for augmentative use against a given pest (van Lenteren and Woets 1988; Hassan 1989b; Janssen et al. 1990). Van Lenteren and Woets (1988) present a scheme for screening natural enemies for seasonal inoculative programs (in which small numbers of the agent are released and the resulting population is expected to provide season-long control of the pest). The testing scheme consists of a set of steps, with failure at any step resulting in the elimination of the species from further consideration. Step one assesses whether the species has any obvious negative aspects that would lower the species' effectiveness. Step two requires that the agent develop to the adult stage on the target pest. Step three requires that the agent attack the pest on the crop and develop successfully under the climatic conditions at the site. Step four suggests that the rates of population increase of selected agents should be greater than the growth rate of the target pest population when it is reproducing in the presence of the natural enemy. Van Lenteren and Woets (1988) also note that a candidate natural enemy should not attack other beneficial organisms in the same environment (which may occur with some predators) and should possess a well-developed ability to locate the target pest; host location ability, however, is difficult to measure realistically in the laboratory. When natural enemies are to be used in an inundative manner (nematodes for control of pest larvae in soil), with little or no dependence on progeny for continued suppression, some of these criteria become less pertinent. Inundative agents need not be intrinsically well-synchronized with their hosts because the time of release can be chosen through monitoring the host's phenology. Similarly, the number of agents released can be adjusted in accordance with host density. In contrast, under seasonal inoculative use (for example, *Encarsia formosa* used for control of *Trialeurodes vaporariorum* in glasshouse vegetable crops), the abilities to find the host efficiently and reproduce are critical. Selection or testing schemes are more likely to be effective at spotting species that are clearly not suited for an intended use, than at picking the best species from a set of plausibly good species.

Ultimately, demonstration that any particular candidate natural enemy species is a good agent for a particular use must be shown by testing the agent under the conditions of use in the field. The effectiveness of any agent may differ between crops, cropping systems, or geographic regions; therefore, the agent must be tested locally on the target crop as discussed in the section on testing the efficacy of mass-reared agents in this chapter.

QUALITY

Natural enemies intended for augmentative use must be of acceptable quality if they are to perform satisfactorily. While many attributes affect quality, management of processes affecting quality may be grouped into three categories: securing agents with acceptable quality to initiate rearing for commercial production; use of rearing procedures that retain agent quality under commercial production conditions; and applying genetic selection methods to improve the quality of natural enemies.

Securing Quality Agents to Initiate Rearing

Natural enemies used in augmentative programs are effective (have acceptable quality) to the extent that they perform well against the target pest on the target crop under typical climatic and agronomic conditions. The physical scale of the crop is also important as some agents perform differently in small scale tests than in full-sized fields or glasshouses. Therefore, in defining quality, the intended use of the agent must be known. Given an appropriately chosen agent for the intended purpose, a large, healthy founder population must be obtained that will embody the desired genetic diversity, be free of pathogens, hyperparasitoids, and other contaminants, be well nourished and highly fecund. Having secured healthy founding stock, the quality of the culture must be maintained over the period of commercial production. For reviews of concepts of quality in mass-reared organisms see Boller (1972), Chambers (1977), van Lenteren (1986), Ochieng-odero (1990), Bigler (1991), and Nicoli et al. (1994).

Retaining Quality Under Commercial Production Conditions

Trade-offs typically exist between conditions that favor rearing efficiency and those that produce agents which retain the highest degree of natural behavior (Boller 1972). Therefore, over time, a laboratory culture of an agent may become different from the original strain. To ensure production of organisms with acceptable quality from commercial cultures, several factors must be considered: genetics, nutrition, prevention of contamination, and provision of opportunities for exposure to appropriate host or prey semiochemicals and learning opportunities.

Genetics. Three genetic processes can occur in reared populations that may reduce quality: inbreeding depression, drift, and selection (Mackauer 1972; Roush 1990). The first two occur only in very small populations; the third may occur in both small and large populations and is the factor of greatest concern in mass rearing of natural enemies.

Inbreeding depression occurs when close relatives mate, leading to an increase of homozygosity in the offspring with the result that deleterious recessive traits become expressed (see Fabritius 1984). In populations with random mating, this usually occurs only in small (fewer than 100 individuals) populations. Drift, also a small-population process, involves the change of allele frequency between generations through the loss of some alleles which by chance are not transmitted to the next generation because the few individuals carrying these alleles fail to breed sufficiently to pass them on. Because most commercial colonies of organisms are large and are expanded rapidly from any small initial founding groups, neither inbreeding depression nor drift are likely to be major factors in commercial rearing facilities (Roush 1990).

In contrast to drift and inbreeding depression, selection for conformity to laboratory conditions, with concurrent reduction in fitness for the wild environment, is more likely to occur in natural enemy rearing facilities. The types of adaptations that may occur in commercial colonies are varied and will depend on both the biology of the specific organism being reared, and the physical and chemical conditions employed. Typically, large commercial rearing operations involve elevated densities of the reared organism, use of unnatural foods, prey or hosts which are often presented at high densities, unnatural conditions of light, and absence of a wide range of physical and chemical clues which normally are part of the organism's environment. Thus, for example, because hosts are extremely easy to find in commercial colonies, parasitoids may be selected for reduced flight, reduced walking speed, and acceptance of hosts lacking semiochemicals normally needed to elicit oviposition. Similarly, pathogens reared on fermentation media are often selected for characters which lead to a reduced ability to attack living hosts (Hajek et al. 1990, see Chapter 11).

Not all losses of quality result from genetic changes. Factors such as diseases or inadequate nutrition may be involved. To determine whether genetic selection is responsible for any observed reduction in quality, outcrossing to wild individuals may be employed. If problems are genetic in origin, offspring should show improved characteristics. Large insectaries can monitor genetic changes in reared populations by periodically assessing the frequencies of selected allozymes (or use other methods to chemically or genetically type populations) to detect shifts away from patterns present in the founder population.

To remedy effects of selection, periodic replacement of colonies with wild individuals or introduction of a significant proportion of wild individuals into the commercial colony can be done (Roush 1990). Such individuals may be collected from nature, or may be drawn from a pool of individuals held in long term storage, provided that some stage, such as eggs or pupae, can be stored successfully. This remedy is most practical for research colonies but becomes increasingly difficult as the size of the rearing colony increases because relatively few wild individuals are likely to be available for collection. Consequently, considerable time is needed to increase new collections to levels needed for commercial production, causing long unproductive periods. In addition, wild organisms pose risks of introducing individuals of a wrong species (as in rearing of *Trichogramma*) or disease to a colony and, initially, may perform poorly under laboratory conditions due to lack of adaptation. If, however, the commercial culture can be periodically renewed in some manner, the organisms actually sold would be subjected to fewer generations of laboratory selection. Viability of organisms held in storage may decline with time, however, and the length of time over which storage may be employed for any given natural enemy must be determined on a case-by-case basis.

Nutrition. The foods, prey, or hosts used to rear biological control agents in commercial cultures can influence the size, vigor, fecundity, sex ratio, and host recognition ability of the agents produced, significantly influencing their effectiveness for their intended purposes. The natural diet of an agent can be assumed, in general, to be the best diet for the production of a vigorous agent with natural behaviors. The target pest, however, may not be the best or complete diet for some predators, such as generalist phytoseiids which may benefit from access to such additional foods as pollens, molds, or alternative prey mite species (James 1993). In addition, strict use of the natural diet, prey, or host may be prohibitively expensive or intractable for use under commercial conditions. To solve this problem, alternative rearing procedures may be developed based on the use of more easily reared alternative species of hosts or artificial diets. Agents produced using alternative diets or hosts should be compared to individuals reared on the natural diet to see if detectable changes occur. Such changes may require a number of rearing cycles to become apparent; therefore, comparisons should be made periodically between lines reared on the alternative diet and the natural diet or hosts. Colonies of *Lixophaga diatraeae*, for example, when reared continuously on the greater wax moth, *Galleria mellonella* (Linnaeus), declined in female fertility and mating ability of males such that the colony failed after six generations (Etienne 1973). Dicke et al. (1989) found that the spider mite *Amblyseius potentillae* (Garman) when reared on a vegetable diet (pollen of the bean *Vicia faba* Linnaeus) showed greatly reduced rates of predation on apple rust mite, *Aculus schlechtendali*, compared to a colony of the same predator reared on two-spotted spider mite, *Tetranychus urticae*. Chabora and Chabora (1971) noted that two parasitoids of house fly pupae (*Musca domestica*) exhibited improvement (more offspring and lower developmental mortality) when two host lines were hybridized. In contrast, *Geocoris punctipes* (Say), reared for six years on artificial diet showed no change in host preferences compared to wild individuals (Hagler and Cohen 1991). *Podisus* spp. reared 15 generations on a meat-based artificial diet showed no differences from those reared on live prey (De Clercq and DeGheele 1993). Comparisons of *Trichogramma brassicae* Bezdenko reared on two different alternative hosts, *Anagasta kuehniella* and *Sitotroga cerealella* (Olivier), showed both species to be equally suitable hosts (Bigler et al. 1987).

In general, it should be expected that diets may progressively affect a number of important natural enemy characteristics in colonies reared for many generations. Loss of quality can be detected by periodic comparison of performance to that of individuals collected from the wild or, if available, to stored members of the original founding population. Nutritional, as opposed to genetic, basis should be suspected if altering diet or host species causes rapid improvement. With appropriate testing before the adoption of an alternative rearing host or artificial diet, and with continued monitoring of the quality of the strain as produced over time in the commercial colony, the potentially damaging effects of inappropriate rearing diets or hosts can be avoided or corrected should they arise.

Preventing Contamination. Cultures of reared organisms are vulnerable to contamination of several types. Contagious pathogens may invade cultures because high densities promote extensive contact among individuals and between individuals and feces and other products that may harbor pathogenic microbes. Adding wild individuals to a commercial colony may introduce pathogens to a colony; for example, granulosis virus of the lepidopteran *Pieris rapae* is easily introduced to colonies by adding either wild butterfly larvae or wild adults of its parasitoid, *Cotesia rubecula*. In this system, transmission can occur among host larvae through fecal contamination of food if larvae are not reared individually, or between adult parasitoids and host larvae through mechanical contamination of the parasitoids' ovipositors. Microsporidia pose special problems in that they may be transmitted both horizontally and vertically (for explanation of terms see Chapter 16) and reduce fertility and longevity of infected organisms without causing immediate death (Kluge and Caldwell 1992). Special efforts are needed to rid a colony of microsporidia, employing separately-reared family lines and destruction of all infected lines (Kluge and Caldwell 1992). Disease transmission in rearing colonies may also be enhanced by superparasitism, as each oviposition by a new female parasitoid carries a possibility of infecting the entire set of immature parasitoids within a host with the infectious agent (Geden et al. 1992). It should be noted, however, that not all microbial agents associated with natural enemies are damaging. Efforts to rid colonies of microbes are justified only when proof is available that the organism is reducing the survival or fecundity of either the natural enemy or the host.

Hyperparasitoids of various species may invade cultures of such beneficial species as aphid parasitoids and predacious cecidomyiids (Gilkeson et al. 1993). Rearing procedures and cage construction must be adjusted to exclude such contaminants should invasion occur. In addition, cross contamination between cultures of two or more beneficial species can be a problem. *Amblyseius cucumeris* (Oudemans) and *Amblyseius mckenziei* Schuster and Pritchard, for example, are difficult to rear in the same facility because of such cross-contamination problems.

Semiochemical Exposure and Learning Opportunities. Chemicals associated with hosts or prey have been shown to be important sources of information for natural enemies (Vet and Dicke 1992). (See Chapter 15 on natural enemy interactions with plants and semiochemicals for more on this point.) If natural enemies are reared on artificial diets or alternative hosts, rearing conditions may present no opportunities for contact with critical host chemical clues (Noldus 1989). *Trichogramma brassicae* reared on *Anagasta kuehniella* was less effective in parasitizing its intended host, *Ostrinia nubilalis*, in cage trials than wasps reared on *O. nubilalis* (van Bergeijk et al. 1989). Rearing on such alternate hosts may lead to poorer performance against the target host under field conditions. Where understood, prerelease conditioning of natural enemies with key host or host-plant compounds may be feasible. The simplest, most direct form of such conditioning would be to allow natural enemies to contact the target weed or pest (on the crop plant), including opportunities to reproduce, before release. For parasitoids, such experiences are believed to promote intensified local host search, improve host recognition and attack, and reduce dispersal (Noldus et al. 1990).

Monitoring Cultures for Quality. Regular assessment of the quality of the organisms produced during rearing is essential in producing natural enemies. Cultures should be checked periodically against their original characteristics by reference to data on the organism's performance in specific bioassays or allozyme frequencies (Morgan et al. 1988). Reared-organisms can also be compared to living specimens which have either been stored in an inactive state or recently collected from field populations. Comparisons may be based on overall performance of the agent in glasshouse tests, or on measurement of specific behaviors such as walking speed (Bigler et al. 1988) or parasitism rates. For example, the quality of various species of *Trichogramma* reared for use against *Ostrinia nubilalis* can be checked by releasing wasps from the production colony into glasshouses in which host eggs masses have been placed on plants. The quality of the culture can then be assessed by measuring the ability of the wasps to fly to the maize plants, find the host eggs, oviposit, and successfully develop (Bigler 1994). Various measures of success in performing these tasks (numbers of wasps reaching the plant in a fixed time, numbers of egg masses discovered, percentage of eggs attacked, percentage of wasps emerging from parasitized eggs) can be compared to the performance of the wasps originally used to establish the colony. Quality control tests have been developed for a number of natural enemies used to control pests of glasshouse crops, and these tests are used on a regular basis to monitor quality (Nicoli et al. 1994). Quality can be further preserved by rearing wasps at lower host densities, followed by storage in diapause for later use (allowing accumulation of production), and by limiting the number of generations an agent is reared on an alternate host before the production colony is reinitiated with material from a parent colony reared on the target host (Bigler 1994).

Improving Natural Enemies Through Genetic Selection

Genetic selection methods may be applied to natural enemy cultures to enhance or alter specific characteristics in desired ways, without impairing other essential aspects of the organism's biology. Features that have been selected for include enhanced movement and host-finding ability in nematodes (Gaugler et al. 1989), altered diapause characteristics (Gilkeson and Hill 1986), and more rapid developmental rates in parasitoids (Weseloh 1986). In addition, various natural enemies have been selected for resistance to one or more pesticides (Roush and Hoy 1981; Hoy and Standow 1982; Markwick 1986; Grafton-Cardwell and Hoy 1986; Hoy and Cave 1988). Whereas genetically-selected organisms intended to form permanent free-living populations in the field after release may be displaced by other species of natural enemies, or hybridize with natural populations, agents used augmentatively will be less affected by these problems to the degree that the pest control is achieved by the released individuals or their immediate offspring. In such cases, issues of long-term fitness for the wild environment are reduced in importance. Phytoseiid mites may be selected for pesticide resistance, for example, and used for control of spider mites in glasshouses concurrent with pesticide applications. Selection of non-diapausing strains of agents such as the aphid predator *Aphidoletes aphidimyza*, the thrips predator *Amblyseius cucumeris*, and the predacious mite *Metaseiulus occidentalis* for use during winter in glasshouses in temperate areas can be a method to improve an agent for a specific task (Field and Hoy 1985; Gilkeson and Hill 1986; Morewood and Gilkeson 1991). Use of pesticide-resistant natural enemies for the purpose of including pesticide treatments during crop production can discourage the use of biological control for other pests. Whether or not genetically modified strains or organisms retain sufficient overall fitness to maintain themselves under field conditions as distinct, reproducing populations is a more complex matter, involving matters of fitness and competition (Xiong et al. 1988).

COST

To be economically successful, natural enemies offered for sale must be priced competitively with other control options. Several factors affect the cost-competitiveness of reared biological control organisms, including: cost of rearing the agent, cost of competing controls such as chemicals, and value of the crop and ability of purchased natural enemies to enhance the profitability of the crop.

Cost of Rearing the Agent

Commercially reared biological control agents will be competitive with chemicals only if productivity of rearing systems, in terms of saleable value of reared organisms, is high relative to costs of capital, labor, and raw materials needed to produce the beneficial species. In any given country, production costs can normally be lowered only by reducing costs of raw materials, largely by finding cheaper alternatives and by automating rearing processes, often in conjunction with increasing the volume of total production so as to make most efficient use of investments in machinery or materials.

Natural Rearing Systems. Natural rearing systems are those in which the agent is produced using the natural host or prey, which itself has been reared on one of its normal food plants. Mass rearing of the predatory mite *Phytoseiulus persimilis* on the two-spotted spider mite, *Tetranychus urticae*, reared on bean plants, for example, is such a natural rearing system, and is economically successful (Fig. 10.1) (Fournier et al. 1985; see Gilkeson 1992 for a review of mass rearing methods for phytoseiids). Similarly, other phytoseiids may also be reared efficiently using natural systems (Friese et al. 1987), as can *Encarsia formosa*, a parasitoid of the greenhouse whitefly, *Trialeurodes vaporariorum*, (Fig. 10.2) (Popov et al. 1987). Other systems in which use of natural hosts is likely be economically feasible include parasitoids and predators of some leafminers (e.g., *Liriomyza* spp.), thrips, scales, aphids, and mealybugs (Fig. 10.3).

However, for many natural systems, labor costs for rearing plants and herbivores will make natural enemies uncompetitive with chemical controls or too expensive relative to the value of the crop yield protected by release of the agent. This is especially likely for systems in which plants needed in the rearing system are slow-growing or expensive, or the required herbivores are cannibalistic or especially susceptible to disease when crowded, or the natural enemy itself is cannibalistic. Such rearing systems produce relatively expensive natural enemies which consequently are either not economically competitive or are so only for use on high-value crops, or for uses requiring small numbers of beneficial organisms per unit of crop. Agents that

Figure 10.1. *Phytoseiulus persimilis* Athias-Henriot (A), a predacious mite used to control pest tetranychids in outdoor strawberries and in glasshouse crops, can be commercially reared on bean plants infested with spider mites (B). (Photograph [A] by M. Badgley and [B] courtesy of L. Gilkeson.)

Figure 10.2. *Encarsia formosa* Gahan (A), a parasitoid employed for control of greenhouse whitefly on glasshouse vegetable crops can be produced commercially on whitefly-infested tobacco plants (B). Parasitized whiteflies turn a distinctive black color (C), facilitating collection. (Photographs by M. Badgley [A] and courtesy of L. Gilkeson [B,C].)

are difficult to rear inexpensively in natural systems include tachinid parasitoids of Colorado potato beetle (Tamaki et al. 1982), staphylinid predators of onion maggot (*Delia antiqua*) (Whistlecraft et al. 1985), and eulophid parasitoids of Mexican bean beetle (*Epilachna varivestis*) (Stevens et al. 1975). In some instances, producing biological control agents in natural hosts may be economical if required host species are being mass reared for other reasons. For example, some species of fruit flies (Tephritidae) are produced for sterile male release programs; therefore, using some of these already available hosts for parasitoid rearing makes the production of parasitoids in natural hosts feasible (Ramadan et al. 1989).

Alternative Rearing Systems. Costs may be reduced by finding cheaper substitutes at either the "plant" or "herbivore" trophic level in a rearing system. For example, winter squash of various kinds may be used in lieu of plant foliage to rear some species of diaspidid scales and their associated natural enemies (Rose 1990). For parasitoids, alternative hosts are commonly used. *Trichogramma* spp., for example, are widely employed in augmentative programs because they are readily produced in eggs of cheaply-reared, stored product moth pests such as *Anagasta kuehniella* or *Sitotroga cerealella.* These insects may be reared in large containers on inexpensive grains, requiring little labor in comparison to systems in which potted plants must be produced and manipulated. Laing and Eden (1990) describe such a rearing system for *Trichogramma minutum*, a parasitoid considered for use in Canada against the spruce budworm, *Choristoneura fumiferana* (Clemens). Converting to alternative rearing systems, in

Figure 10.3. *Aphidoletes aphidimyza* (Rondani) (A), a cecidomyiid used for control of aphids in glasshouses is reared commercially on aphid-infested pepper plants (*Capsicum sativum* Linnaeus) (B). Mature *A. aphidimyza* larvae drop from plants into water-filled trays from which they are collected (C). (Photographs courtesy of L. Gilkeson.)

addition to reducing labor costs, may also permit greater mechanization if rearing modules or materials can be manipulated by machinery.

When parasitoids are reared in their original pest hosts, concerns may exist that unparasitized hosts in released material might contribute new pests to the field population. This concern may be overcome by rearing parasitoids in hosts that have been killed by ultraviolet light, radiation, heat, or cold (Barton and Stehr 1970; Brower 1982; Goldstein et al. 1983) or through use of mechanical separation procedures (flotation, brushing, sieving) that allow the desired organism to be separated from hosts.

Foods or hosts used to rear natural enemies can diverge to varying degrees from those used by field populations of the agent. A parasitoid, for example, might be reared on a host which is produced on an artificial diet. Such hosts might fail to produce important kairomones used by parasitoids for host recognition. Kairomonal activity in such cases might be restored to host frass by adding dried plant material to the diet. For example, adding dried maize foliage to artificial diets used to rear *Ostrinia furnacalis* (Guenée) restored kairomonal activity to larval frass for the parasitoid *Macrocentrus linearis* Hadley (Qiu et al. 1988).

Some species of tachinids may be reared by artificially extracting maggots from gravid females by shredding females in a mechanical food blender. Maggots are then placed in agar-filled cups that contain host larvae (King et al. 1975; Gross and Johnson 1985). Rearing of some tachinids can be enhanced by using host kairomones to stimulate larviposition, as for *Bonnetia comta* (Fallén) (Allan and Hill 1984). In other systems, parasitoid rearing may be made more efficient by using mechanical separators to isolate individuals and by using artificially killed hosts (Syed 1985). Rearing predators may be made more efficient by substituting nonliving foods for living prey. The lygaeid bug *Geocoris punctipes* has been successfully reared on a meat diet composed of liver and ground beef at costs as low as U.S. $0.63 per 1000 insects (Cohen 1985). The phytoseiid mite *Amblyseius teke* Pritchard and Baker has been reared on a diet composed of honey, egg yolk, Wesson's salt, and water (Ochieng et al. 1987). Abou-Awad et al. (1992) found that two species of phytoseiids reared on artificial diets laid 25 to 50% fewer eggs per day, but the offspring obtained developed normally on the diet and became normal sized adult mites. Alternating live prey with pollen is an effective method of rearing the phytoseiid *Amblyseius fallacis* at lower cost than using only live prey (Zhang and Li 1989).

Efforts to rear arthropod parasitoids on nonliving media have been made with several species, including *Trichogramma pretiosum*, the chalcidid *Brachymeria lasus* (Walker), and the tachinid *Eucelatoria* sp., among others (Hoffman et al. 1975; Nettles et al. 1980; Thompson 1981). *Bracon mellitor* Say and *Catolaccus grandis* Burks have been reared *in vitro* on diets containing only defined biochemicals, minerals, and chicken egg yolk (Guerra et al. 1993). Steps in successful artificial rearing of egg parasitoids included the development of a non-host medium suitable for all phases of development (Hoffman et al. 1975), development of a membrane and associated chemical stimuli capable of eliciting oviposition into artificial eggs (Morrison et al. 1983; Nettles et al. 1985), and field trials of effectiveness of artificially-reared parasitoids (Liu et al. 1985; Dai et al. 1988). Initially, development of artificial hosts focused on physiological requirements for parasitoid growth, pupation, and emergence. For some species, systems now exist that meet these requirements (Bratti 1991; Masutti et al. 1993). Next, artificial rearing systems will need to incorporate host kairomones and other stimuli needed by the parasitoid to elicit host recognition and attack behaviors. The utility of such artificial rearing systems, apart from research applications, is based on assumed lowered production costs permitted by greater mechanization when live hosts are not employed.

Competing Controls Such as Chemicals

The cost of reared natural enemies must be judged in terms of value of the crop production protected by using the agent and in comparison to the cost of competing pest control options such as chemicals. For example, use of the egg parasitoid *Trichogrammatoidea cryptophlebiae* Nagaraja against citrus pests in South Africa is economically viable, even though the level of pest control is only 54 to 60%, because the method performs as well as the competing chemical option and is only one-third as expensive (Newton and Odendaal 1990). In glasshouse tomatoes in western Europe, biological control was less expensive than chemical control for all major pests (whiteflies, spider mites, thrips, and leaf miners) (van Lenteren 1989) (Fig. 10.4). In some instances, social policies or attitudes can also be factors in the relative competitiveness of biological and chemical pest control approaches. Prohibitions in Germany, for example, against use of pesticides on farmlands near urban areas created a market for the use of *Trichogramma* spp. for control of the maize borer *Ostrinia nubilalis*. Similarly, problems that arise from the use of pesticides may limit their application in certain situations as, for example, on farm lands over public groundwater recharge areas or near municipal wells. In some

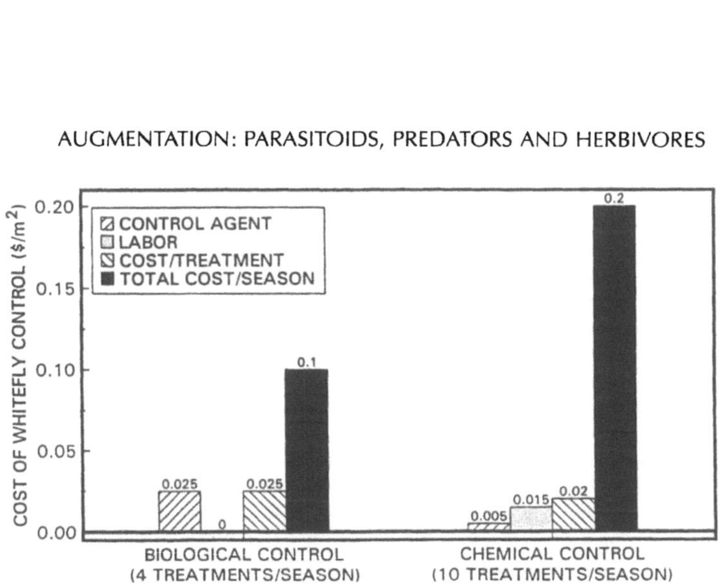

Figure 10.4. Relative costs of chemical and biological pest management in glasshouse tomatoes (*Lycopersicon lycopersicum* [Linnaeus] Karsten ex Farwell) in Europe (after van Lenteren 1989).

circumstances, pesticide use may conflict with grower objectives as, for example, pollination in seed crops or sale of crops as pesticide-free produce. On the other hand, subsidies of pesticide purchases by governments or other agricultural policies can artificially increase the competitiveness of pesticides. The effect of various governmental policies on the competitive balance between biological and chemical pest controls is discussed in greater detail in Chapter 20.

Crop Value and Cost of Controls

If purchased natural enemies are expensive, their use will be limited to high value crops with significant pest losses. In such crops, farmers will have sufficiently large profit margins to be able to afford expensive pest controls. In Massachusetts (U.S.A.), for example, cranberries are a high value crop, with potential profits of $11,000/ha/an. (1992 dollars). Replanting costs exceed $80,000/ha, with 4–5 years required after replanting to achieve full production. Given these high costs, root weevils (which kill sections of vines) are serious pests which merit substantial pest control costs. Consequently, growers are willing to use nematodes at $600/ha to control this pest on a spot treatment basis. In contrast, potatoes in Massachusetts have a profit margin of only $400–600/ha (1992 dollars); producers of this crop cannot afford to use biological control agents as expensive as these nematodes. Trumble and Morse (1993) calculated the net economic benefit of control of *Tetranychus urticae* on strawberries in California by either acaricides alone, releases of the predator mite *Phytoseiulus persimilis* alone, or combinations of both acaricides and predator mite releases. They found the highest economic return to the grower was the combination strategy, but only for acaricides that were compatible with the predator mite.

The value of a biological control agent may be increased, however, if it allows growers currently using biological control for other pests to address a new pest in a manner that does not jeopardize their existing pest control system. (See Chapter 14 for more on the channeling effects of pest control strategies).

STORAGE AND SHIPPING

Storage plays an important role in commercial rearing schemes for producing natural enemies by allowing facilities to operate throughout the entire year producing natural enemies that may be in demand for only a few months of the year. In the absence of effective methods to store natural enemies, rearing colonies must be increased during the production season to meet demand, then reduced in size, shut down, or converted to rearing another agent if one exists whose period of demand differs from that of the first. Seasonal fluctuations in production increase costs through underuse of facilities which are idle in the off season and create scheduling problems if complicated schemes of seasonal switching between rearing systems are used. Year-long, uniform usage of facilities is important because the capital costs of the business (costs of leasing or purchasing space) are likely to be fixed and cannot be avoided when an agent is not being produced. Similarly, year-round production simplifies training and retention of full time staff. The period of demand for products of a commercial insectary can be broadened by producing several kinds of agents, with different agents targeted toward crops or regions that have seasonally different needs; by marketing one agent for several pests with different seasons (Laing and Eden 1990); or by having customers in both the northern and southern hemispheres, thus gaining markets virtually the year round. (See, however, the section in this chapter on quarantine issues in international marketing of natural enemies).

Large organisms for biological control (predators, parasitoids, herbivores) have relatively poor storage qualities compared to nematodes and microbial species, some of which can be kept for many months or years. At best, storage of large organisms is a matter of a few months, usually only weeks. Cold temperatures are used to prolong the period over which agents may be held and remain viable. Each species is likely to have its own special characteristics such that tests will be required to find the best combination of cold temperatures, relative humidity, and lighting for a given agent. Effective storage times vary. The coccinellid *Synonycha grandis* Mulsant remained 50 to 100% viable for up to 5 months at 18–20°C (Deng et al. 1987). The predacious midge *Aphidoletes aphidimyza* could be stored at 1°C for up to 2 months with less than 10% mortality, but required a conditioning period of 10 days at 5°C (Gilkeson 1990). Diapausing individuals may be stored longer than nondiapausing individuals. Diapausing adults of *Chrysoperla carnea* could be stored with little mortality at 5°C for 31 weeks (Tauber et al. 1993), and diapausing *Trichogramma evanescens* Westwood could be stored up to a year with little or no loss of quality (J. van Lenteren, pers. commun.). Nondiapausing pupae of *Trichogramma pretiosum* could be stored only 12 days at the most favorable temperature range of 15 to 17°C (Stinner et al. 1974). *Trichogramma minutum*, reared in Canada for control of spruce budworm (*Choristoneura fumiferana*), were shipped immediately following production and could be stored (at 10°C) only for 1 to 2 days (Laing and Eden 1990). In contrast, the predacious mite *Amblyseius cucumeris* could be stored for 10 weeks at 9°C with 63% survival (Gillespie and Ramey 1988). Morewood (1992) presents information on storage and shipment of the widely used predacious mite *Phytoseiulus persimilis*, noting that addition of food, even at low temperatures, is useful, but that addition of bran or vermiculite promoted mold and reduced survival in storage.

In addition to understanding and employing proper storage conditions for natural enemies, effective methods must exist for shipping natural enemies so that they arrive in good condition (Fig. 10.5). Besides using sturdy packaging materials to prevent crushing, shipping factors that must be controlled are temperature, humidity, time in transit, and preventing escape of what are often very small organisms. To prevent agents from desiccating during shipment, humidi-

Figure 10.5. Once collected, natural enemies are packaged for safe commercial shipment. For example (A), *Encarsia formosa* Gahan pupae are shipped glued to small cards in a protective box. Some natural enemies may be stored at cool temperatures for various periods prior to shipment (B), facilitating processing of orders. (Photographs courtesy of M. Hoddle [A] and L. Gilkeson [B].)

trons made of wet salt inside plastic bags with membranes permeable to water vapor may be employed, providing humidity for up to 11 days (Hendrickson et al. 1987). For shorter periods, moist sponges may be used. To prevent overheating in summer, cold packs may be included in packages. To prevent freezing in winter, vials of hot water or chemical packs such as those used as handwarmers by skiers may be included in packages for overnight protection. Wherever possible, rapid delivery services should be employed to keep time in transit to no more than 1 to 3 days. Including foods (such as honey) in shipping containers so that natural enemies can feed immediately upon emergence can increase longevity and fecundity.

APPLICATION METHODS

Once appropriate natural enemies, of acceptable quality, have been produced and delivered to the user, application systems are needed that maximize the pest control potential of the agents by efficiently placing them when and where they are needed (see Smith 1994 for a review). Methods to time the release of natural enemies and concentrate their activity in periods when pests are present include the use of various types of traps which collect one of the life stages of the target pest. For example, in avocado (*Persea americana* [*P. gratissima* C.F. Gaertner]) and citrus in California, pheromone-trap catches of the key lepidopteran pests may be used to time the release of *Trichogramma platneri* Nagarkatti (Anon. 1987). Because host eggs, the target stage, will occur shortly after increases in catches of pest moths, traps indicate the best time to release egg parasitoids. Also, in citrus in California, pheromone and visual traps are used together to time the release of *Aphytis melinus* against California red scale (*Aonidiella aurantii*) (Phillips 1987). In sweet corn (maize) in Nova Scotia, blacklight catches of *Helicoverpa zea* adults are used to time releases of *Trichogramma pretiosum* (Neil and Specht 1990). Pheromone traps for cabbage moth (*Barrathra brassicae* Linnaeus) are used in Germany to time releases of *Trichogramma evanescens* (Terytze and Adam 1981). Release of *Encarsia formosa* in glasshouse crops can be timed using number of degree-days accumulated since first whitefly oviposition, so as to coordinate parasitoid releases with the occurrences of the older nymphal instars (Osborne 1981). Proper timing is likely to be important with releases of many biological control agents because they may attack only certain stages of the pest and may themselves be short-lived. Consequently, if released agents are to be effective, appropriate hosts must be present at the time of release. Information on the timing and duration of the pest in the target stage can also be used to determine the total number of natural enemy releases needed over the course of the season. If host oviposition, for example, is distributed fairly evenly over 3 to 4 weeks and egg parasitoids survive only a few days in the field, then a series of weekly releases are likely to be needed to expose most hosts to attack.

The physical methods used to disperse or release natural enemies must be nondamaging, give good dispersion, be quick, and be low in cost. Tests to verify that a given release approach has these characteristics should be carried out under conditions of field use, as was done by Messing et al. (1993) for release systems used for the fruit fly parasitoid *Diachasmimorpha longicaudatus* (Ashmead) in Hawaii. Depending on the circumstances, releases may be made by hand, as when cards bearing *Encarsia formosa* pupae are placed in glasshouses for whitefly control, or by machinery from either the ground or air. Phytoseiid mites may be released by dispersing bran containing the mites by hand, with a granular-pesticide dispenser (Ables 1979; Fournier et al. 1985), or by air through special release systems for use in remote areas (Drukker et al. 1993). *Trichogramma* sp. releases in forests have been made by placing cartons with parasitized hosts on branches in release areas (Houseweart et al. 1984; Smith et al. 1987) or parasitized eggs may be broadcast from aircraft using a sling-type small grain seeder (Hope et al. 1990). Voblyl et al. (1988) describe a tractor-drawn device for mechanical distribution of *Trichogramma* spp. in crops. In France, a capsule (Pyratypt) has been developed for use in applying eggs parasitized by *Trichogramma* spp. The capsule protects immature wasps from predators before adults emerge, yet allows adults to disperse through small holes in the capsule (Anon. 1988a).

Other considerations related to methods of release may include the need to expose agents being released to hosts or host stimuli before release, or to release agents early or late in the day. Both of these activities could be helpful in reducing rapid emigration away from the release area. Gross et al. (1981) increased by sixfold the retention of *Trichogramma pretiosum* in a laboratory release area by exposing the wasps to host eggs. If the host stage to be attacked has a prolonged period of occurrence in the field and repeated releases of the natural enemy are not possible, it may be beneficial to adjust release methods so that natural enemies from a single release emerge to the adult stage over a more prolonged period. This may be achieved either by releasing natural enemy immature stages of varying ages, or by using release containers that contain host stages that permit natural enemy breeding within the release container for a period of time (Shi et al. 1988). Alternatively, natural enemy releases may be enhanced by the concurrent (or previous) release of hosts, either killed target hosts or live alternative hosts, to increase the host population available for natural enemy attack. This latter approach has been used to increase reproduction of parasitoids of manure-breeding flies, using freeze-killed fly pupae (Petersen 1986), and of *Trichogramma semblidis* (Aurivillius), a parasitoid of grape vine pests such as *Eupoecilia ambiguella* Hübner in Germany (Sengonca and Leisse 1989). Finally, plants bearing populations of alternative hosts which are not a threat to the production crop may be placed in glasshouses. These plants (called banker plants) provide habitat and hosts to maintain natural enemy populations. This method has been used for aphids and leafminers (Bennison 1992; van Lenteren 1995).

PRODUCT EVALUATION

In-depth evaluation of the effects that augmentative release programs have on their target pests are basic to successful use of the method. Approaches and issues for evaluating the effects of natural enemies are discussed at length in Chapter 13. Many factors can potentially affect the level of pest control produced by the release of a natural enemy. Field tests are essential to determine the level of control produced by a particular pattern of release of an agent against a pest in a particular area. Factors that must be considered in such tests include the species and strain of agent released, the number released, the number and sequence of releases in relation to the phenology of the presence of the pest, and the manner of release. For natural enemies that have not been subjected to such trials, or only in other, dissimilar locations, there is no firm basis for recommendation to growers as to the effectiveness of the agent. Lack of such knowledge can lead to employment of ineffective agents. *Trichogramma fasciatum* (Perkins), for example, was released in Barbados for control of *Diatraea saccharalis*, with annual releases from 1930 to 1958 of up to 300 million individuals on the island. However, when releases were stopped in 1958, the level of damage caused by the pest did not change, indicating that the releases were not appreciably decreasing damage (Alam et al. 1971).

Field trials of natural enemies provide necessary information on the effectiveness of particular species and rates of release. Sale of biological control agents in Holland is allowed only after demonstration of their efficacy. Producers must submit test results obtained under practical conditions (often done in commercial growers' holdings) which demonstrate the agent can reduce the pest sufficiently. Tests for effectiveness have been conducted for such systems as *Trichogramma* spp. to control some maize, apple, citrus, sugarcane, and forest pests (Fig. 10.6)

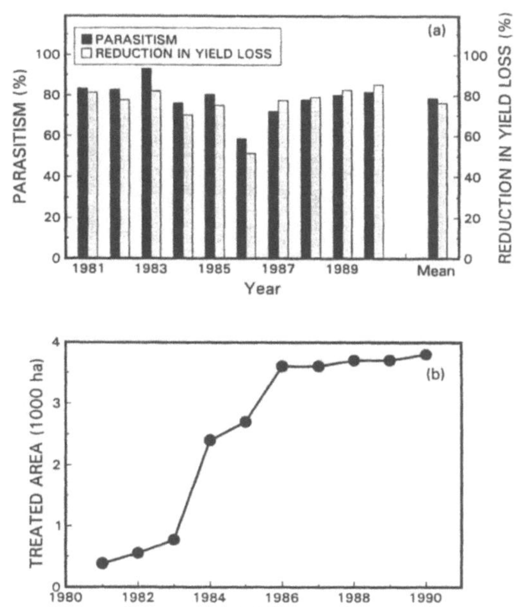

Figure 10.6. (a) Effect of *Trichogramma brassicae* Bezdenko releases on parasitism and yield loss to *Ostrinia nubilalis* (Hübner) in Europe and (b) area receiving *T. brassicae* treatments (after Bigler et al. 1989).

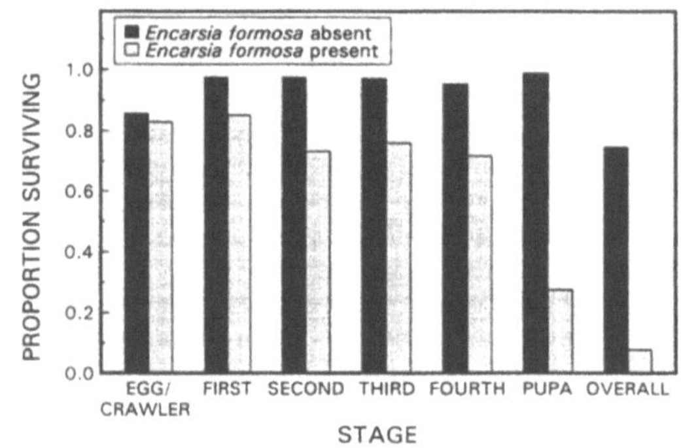

Figure 10.7. Survival of *Bemisia argentifolii* Bellows and Perring on poinsettia (*Euphorbia pulcherrima* Willdenow ex Klotzch) as affected by weekly releases of three *Encarsia formosa* Gahan adults per plant (M. Hoddle, unpublished).

(Bigler 1986; Misra et al. 1986; Hassan et al. 1986, 1988; Smith et al. 1987; Neil and Specht 1990; Newton and Odendaal 1990; Andow and Prokrym 1991); for *Encarsia formosa* to control whiteflies in glasshouses (Fig. 10.7) (Helgesen and Tauber 1974; Woets 1978); for *Aphytis melinus* to control *Aonidiella aurantii* in lemons (*Citrus limon* [Linnaeus] Burman) in California (Moreno and Luck 1992); for the eulophid *Edovum puttleri*, a parasitoid of the potato pest *Leptinotarsa decemlineata* (Van Driesche et al. 1990); and for phytoseiid mites to control spider mites in strawberries and grapes (Fig. 10.8, 10.9) (Osman and Zohdy 1976; Oatman et al. 1977;

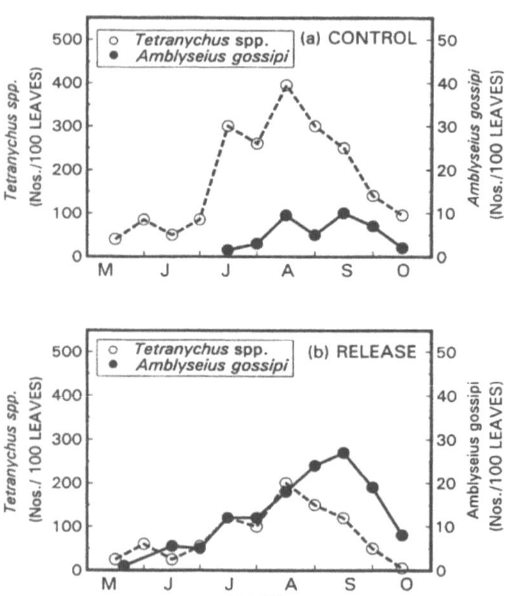

Figure 10.8. Populations of *Tetranychus* spp. (pest mites) and the predatory mite *Amblyseius gossipi* (Elbadry) on cotton in non-release and release plots (after Osman and Zohdy 1976).

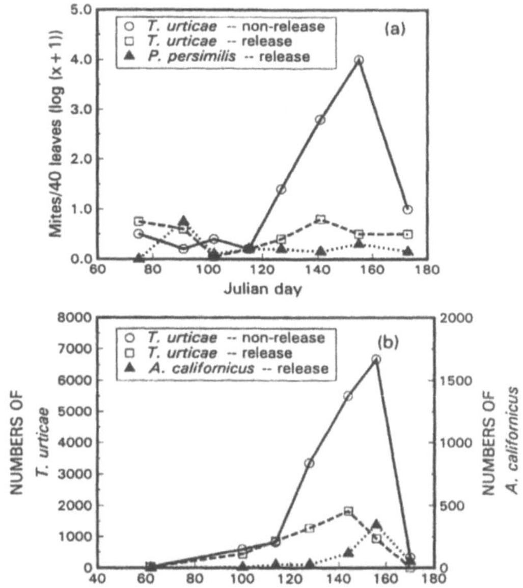

Figure 10.9. (a) Numbers of *Tetranychus urticae* Koch and *Phytoseiulus persimilis* Athias-Henriot in release and non-release glasshouse strawberries (after Port and Scopes 1981); (b) numbers of *Tetranychus urticae* in field strawberries with and without releases of *Amblyseius californicus* McGregor (after Oatman et al. 1977).

Duso 1989). Early season release of the mealybug parasitoid *Leptomastix dactylopi* Howard in citrus in Queensland, Australia, led to parasitoid populations that reproduced earlier in orchards and supplemented the action of the naturally occurring population of this parasitoid. Releases caused parasitism levels to rise earlier and suppress pest infestation levels on fruit to less than 5% (Smith et al. 1988).

For parasitoids, efficacy tests generally consist of establishing release and nonrelease plots and assessing parasitism levels either in naturally occurring hosts or artificially deployed ones. Additional useful measures include levels of parasitism in the test plots in the year before the releases and the year after to assess background levels of parasitism and carryover from released parasitoids through reproduction in the field. Where natural populations of the released species occur, methods are needed to separate the observed levels of parasitism into amounts caused by the released parasitoids and those from the field population. This was done by Feng et al. (1988) for *Trichogramma dendrolimi* Matsumura using radiolabeling of released wasps so that their offspring were also labeled in the field. Studies in nonrelease areas may be helpful in determining whether a parasitoid to be released occurs in the area naturally, and what levels and patterns of parasitism are characteristic of nonrelease plots (Bai and Smith 1994). To evaluate the releases of predators, densities of both predators and prey should be measured in both release and nonrelease plots before and after the releases.

In addition to measuring the level of kill inflicted (percentage parasitism or equivalent measure) on the host, densities of the damaging stage of the pest should also be measured. With lepidopteran pests of maize subjected to *Trichogramma* sp. releases, it is important not only to measure the percentage parasitism of host eggs in release and nonrelease plots, but also to measure the densities of pest larvae in these plots. These measurements are important because if a density dependent mortality factor exists which acts on a stage after that attacked by the released natural enemy, the effect of the release may be partially or wholly compensated for by reduced mortality in the subsequent stage. In sugarcane, for example, failure of releases of *Trichogramma* spp. to reduce larval densities (and hence damage) of sugarcane borers is thought, at least in some cases, to have resulted from the presence of density dependent mortality acting on the larvae, the effect of which was lowered if high egg parasitism rates reduced larval densities (Hamburg and Hassell 1984). In contrast, use of *Trichogramma nubilale* to parasitize eggs of *Ostrinia nubilalis* on maize in Minnesota (U.S.A.) was effective in suppressing larvae (the damaging stage), since larval mortality was not density dependent (Prokrym et al. 1992).

All evaluations should include an economic assessment of the costs of the natural enemies relative to the benefits from crop protection and the cost of competing forms of control, so the overall financial value of the natural enemy can be estimated. One factor that has a strong bearing on the financial impact of the augmentative use of a biological control agent is the number of years of control that may occur following the year of release. In German swine facilities, for example, release of the predator *Ophyra aenescens* Robineau-Desvoidy suppressed populations of *Musca domestica* effectively for up to nine years (Betke et al. 1989). One application of the copepod *Macrocyclops albidus* (Jurine) to piles of automobile tires in Louisiana (U.S.A.) suppressed larvae of *Aedes albopictus* (Skuse) for over a year (Marten 1990). Another factor is that replacement of broad spectrum pesticides by biological control agents may allow other minor pests (formerly suppressed by the broad spectrum pesticides) to become occasional pests which may need to be controlled. These control costs must be included in calculating the cost of the biological control program.

MARKET DEVELOPMENT AND CONTINUITY

Business, legal, and management factors affect augmentative biological control directly. Among these are: the need to develop markets for new products; the choice between public and private rearing facilities; issues related to seasonal and geographic need for reared natural enemies; legal constraints on shipment of natural enemies between countries; and integration of augmentative biological control agents into pest management systems.

Market development is the process by which new agents are brought into commercial production and availability of the agent is advertised to growers or other customers who then adopt their usage (Fig. 10.10). Large companies can initiate new product lines that are not

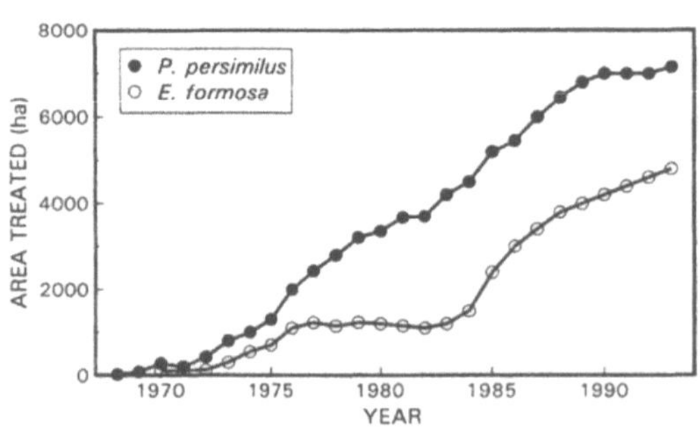

Figure 10.10. Area of glasshouse culture treated with releases of the natural enemies *Encarsia formosa* Gahan and *Phytosieulus persimilus* Athias-Henriot (after van Lenteren 1995).

expected to show a profit for some years while products gain customer acceptance. Smaller companies must initiate new product lines in small steps, establishing a market concurrently with investment in production. Government subsidies for product development are another approach that may be used to develop markets. Public recognition of the value of such subsidies and the mechanisms to enact them vary greatly between countries.

Both private and public sector rearing facilities provide methods for making natural enemies available for a range of pest problems. In some countries (for instance China and India), government or community insectaries have produced natural enemies for sale or use by local farmers (Balasubramanian and Pawar 1990). Other marketing models include grower cooperative insectaries, such as those of the Filmore District citrus growers in California, and corporate businesses that run insectaries to service the pest control needs of their own agricultural properties. Each of these marketing systems is subject, in differing ways, to economic forces and public policies that affect the kinds, numbers, costs, and availabilities of the natural enemies reared (Chapter 20).

Seasonal and geographic availability of and demand for natural enemies affects the development of market systems for natural enemies. Natural enemies need to be reliably available from year to year to inspire grower confidence and stimulate adoption of the approach. Successful natural enemies for augmentation will be those that have markets that are sufficiently large geographically and seasonally so that producers can obtain reliable profits from sales.

In some countries, international movement of reared natural enemies is subject to governmental requirements for importation permits. Producers and distributors may be required to obtain permits before shipping new products to customers. Issues that must be addressed to gain access to some countries include the need to demonstrate that shipments are correctly identified, consistent in their content, and free from contaminants of all types. (For more on quarantine issues see the section in this chapter on safety issues.) In some countries (such as the United States) permits are also needed for movement between states within the country.

An additional factor that can affect the adoption of augmentative biological control methods is the need to control pest complexes in crops. Problems may arise if natural enemies are available for only some pests. Growers may abandon biological controls because chemicals used for one pest may destroy the biological control agents released for other species, or may adequately control the entire pest complex. In some glasshouse crops, for example, biological control of whiteflies by natural enemies may be disrupted by chemical controls applied for thrips. In other circumstances, chemical and biological control methods may be compatible. Biological control of whiteflies, for example, can be combined successfully with the use of *Bacillus thuringiensis* to control lepidopteran larvae. In some cases, the advantages of chemical control might be offset by problems such as pest resistance, residues, or worker exposure to pesticides. The attractiveness of biological control may prompt grower support for pursuit of biological control agents for all members of a pest complex. Grower education about the technical methods to employ the biological control agents and about the advantages of biological control in improving worker safety and increasing convenience of scheduling of other tasks in the crop will be critical in promoting adoption.

NATURAL ENEMIES IN CURRENT USE

The number of species of natural enemies offered for sale, and the number of firms engaged in sales, is increasing. Some species offered for sale are reared, but others may be collected from field populations. Some firms rear and distribute natural enemies, while others act only as distributors for natural enemies produced by other facilities. Guides to suppliers include

Hunter (1992), which covers North American suppliers only, and Anon. (1994a) which is international in coverage. Purchasers of natural enemies should seek to verify that the species bought is appropriate for control of the intended target and that the material received is indeed the species advertised.

SAFETY

To date there have been no documented cases of harm resulting from the routine sale and release of beneficial arthropods (parasitoids, predators, or herbivores) for augmentative biological control. These organisms are particularly safe to human health compared to pesticides which they replace. Their use in enclosed environments such as glasshouses, for example, can make a significant contribution to improved human health by reducing worker exposure to pesticides.

Augmentative use of vertebrates consists mainly of releases of fish for mosquito or weed control. These uses do have the potential for undesired effects on nontarget species or the environment in some circumstances. Use of grass carp (*Ctenopharyngodon idella*), for example, to reduce macrophytic vegetation (for mosquito habitat reduction) can result in increased nutrient and algal densities and reduced water clarity (Maceina et al. 1992). Such releases may or may not be intended to result in the permanent establishment of the released fish in the target habitat, and as such may be considered either as augmentative biological control or control through natural enemy introduction.

Safety Issues

Several safety issues related to augmentative biological control can be identified for discussion. Understanding these can help ensure that beneficial arthropods (parasitoids, predators, weed control agents) and vertebrates will continue to be produced and used safely. Among these issues are: allergic reactions of workers to arthropod parts in natural enemy rearing facilities; permanent establishment of released natural enemies; nuisance problems from released organisms; harm to economic plants, invertebrates, or other organisms; and harm to noneconomic native species of plants, invertebrates, or other organisms.

Allergies. Natural enemy rearing facilities, of necessity, handle large quantities of insects or other organisms, whose body parts (scales, setae, etc.) are handled or which become airborne and can be inhaled by workers. Repeated inhalation exposure to such materials can cause allergic reactions. Dermal exposure may cause skin rashes. These are common problems, for example, in facilities devoted to rearing spider mites and their predators. Worker health in natural enemy production facilities can be protected by providing adequate ventilation, filtering particles from contaminated air, and introducing fresh air. Workers, also, should reduce inhalation of allergenic material by using a face mask respirator and reduce dermal contact by using protective clothing such as long sleeved shirts and caps.

Establishment. Because organisms used in programs of augmentative biological control are released freely into the environment, they may become established in the country of release. In the United States, for example, the European mantid, *Mantis religiosa*, became established following sale of egg cases. Similarly, the predator mite *Phytoseiulus persimilis*, released on field strawberry crops, became established in California (McMurtry et al. 1978). Establishment may or may not be of concern for a given species, but before allowing the routine sale and subsequent release of an organism in an area to which it is not indigenous, the potential consequences of establishment should be considered. The sale of nonnative species of lacewings, for example, in most locations would not be of concern, but may be undesirable in certain areas with precinctive species of lacewings which have limited distributions and unknown susceptibilities to competition, as is the case in the Hawaiian Islands (U.S.A.) (Tauber et al. 1992). In contrast, the sale of colonies of nonindigenous species of ants, for pest control or as pets, is likely to pose risks to native invertebrates in most locations, particularly if ants species sold are aggressive generalist predators or vigorous competitors that reach high densities.

Nuisance Problems. Most natural enemies in commerce are physically innocuous. This is important, because released organisms should not bite, sting, contaminate food, or cause allergic reactions. Vespids, for example, would be a poor choice for sale as a biological control agent because they sting people. Contamination potential may depend on local rules. Until recent revision of governmental rules on food contaminants in the United States, sale of biological control agents for release into grain storage facilities was prohibited, because the small quantity of insect parts resulting from such use was officially viewed as food contamination (Cox 1990).

Economic Species. Generally, augmentative releases of organisms are very unlikely to attack commercially important species. However, this might occur in special circumstances. For example, *Trichogramma* spp. may, for example, be capable of attacking eggs of silkworm moths (*Bombyx mori*) and thus would be unwelcome in silk-producing areas. Similarly, village production and sale of birdwing butterflies (Papilionidae) in Papua New Guinea might suffer from the release of some species of parasitoids. While augmentative release of herbivorous insects for control of weeds is not currently practiced, should it be in the future, such releases would require prior screening of released agents for their ability to feed on local flora prior to sale, using the same criteria that are employed to assure safety of herbivores intended for establishment.

Non-Target Invertebrates. Augmentative releases of parasitoids and invertebrate predators have not been shown to cause harm to non-target invertebrate populations. However, just as *Trichogramma* spp. releases might affect cottage farming of birdwing butterflies, the same releases might affect densities of wild populations of these species. Allegations of such effects have been made relative to native moths and other insects in Hawaii (Howarth 1983, 1985, 1991), but confirming evidence is lacking.

Mechanisms to Ensure Safety

Threats to worker health that might occur in natural enemy rearing facilities can be minimized under workplace safety laws, such as ventilation standards to ensure that airborne contaminants from organisms are kept below levels likely to induce allergies together with protective clothing. Potential problems associated with augmentative releases of nonnative natural enemies can be avoided by employing a governmental review process to assess the safety of any natural enemy proposed for sale as a prior condition to extending permission for its sale. If national governments have no set policy on the importation of natural enemies, the FAO guidelines (Anon. 1992) may be followed. Governmental reviews should seek to assure that each proposal for natural enemy importation and release is compatible with the public interest. Mechanisms should exist to allow for groups with varying concerns or interests to participate in the review process. Laws governing this process may explicitly address biological control as a regulated activity (in Australia, see Cullen and Delfosse 1985; Delfosse 1992), or in many cases such laws may be those governing plant protection and quarantine that are applied to natural enemy introductions as well (as in the United States, Coulson and Soper 1989; Coulson et al. 1991).

CHAPTER

11

Augmentation of Pathogens and Nematodes

INTRODUCTION

Numerous species of nematodes and microorganisms have potential for use as commercial products (Starnes et al. 1993). Successful commercial development of pathogens for augmentative biological control involves: agent selection (to obtain the best species and strains for the target pest); development of cost-effective methods for mass rearing; effective methods for storage and shipping of the agent; creation of formulations to protect and deliver the agent to the target pest's location; field testing of the product's efficacy and methods for its application; economic factors affecting product development and markets; and demonstration of safety of products to man and the environment.

AGENT SELECTION

Selection of a pathogen for the control of a target pest involves choices at two levels. First, a choice may be possible among different agent groups (viruses, bacteria, fungi, protozoa, and nematodes). Second, within a given group of agents, the particular species, strain, or isolate must be identified that has the best properties for the desired use. Patent and registration requirements may differ among agents, also affecting choice of agent.

Choosing Among Groups of Pathogens

Choice among agent groups, given that candidates exist for a given target pest, can be guided by three factors: ease and cost of production; degree of pathogenicity and host specificity; and environmental or habitat features influencing effectiveness.

Ease and Cost of Rearing. The most basic factor affecting differences in ease and cost of rearing of different organisms is whether or not living hosts are required for pathogen production. The microbial pesticide of greatest commercial application in the United States, *Bacillus thuringiensis*, for example, can be grown on fermentation media. In contrast, *Bacillus popilliae* which requires living hosts for effective production, has not been as successful commercially. Some bacteria, fungi, and nematodes can be grown in nonliving media, a factor promoting their use. Other species within these groups, for example, many species of Entomophthoraceae fungi, require living hosts. However, all viruses and the protozoa (principally, microsporidia) of biological control interest must be grown in living hosts or host cell cultures, increasing the cost of producing these agents and limiting their use. Other aspects of production, such as use of liquid media in place of solid media, or development of simple systems of on-farm pathogen production by farmers, can also affect the cost of labor and machinery needed for production. Ease of production is a function of the technology available for the task, which is subject to improvement. Development of higher yielding cell lines for virus production, for example, might in the future reduce the cost of virus production enough to make commercial *in vitro* production feasible. Similarly, development of rearing media that employ cheaper ingredients, such as locally produced cereals, in place of chemically-defined but more costly media, can reduce the cost of the production of fungi (Hoti and Balaraman 1990).

Degree of Pathogenicity and Host Specificity. In choosing a pathogen for development as a commercial product, the degree of pathogenicity and level of host specificity of any particular agent are important considerations (Charudattan 1989). The degree of pathogenicity directly affects the cost of pathogen-based pesticides by determining the quantity that must be applied to achieve control. Because production costs of many pathogens are relatively high, selecting a highly pathogenic strain, which is effective in smaller doses, is essential to increasing the cost competitiveness of the pathogen. High levels of pathogenicity may also be important for controlling a range of instars of a pest, as some strains of a pathogen may be more effective than others in killing less susceptible stages. Such characteristics can be important in making the use of a product commercially successful.

The host specificity of a pathogen is important in that it determines the size of the potential market for the product. For highly specific agents to have commercial value, they must attack pests affecting widely grown or high-value crops to support sufficient sales. Many viruses, for example, are relatively host specific. Many are limited to hosts in just one or a few genera, for example, the virus of the brown tailed moth, *Euproctis chrysorrhoea* (Linnaeus), which is limited to a single host (Kelly et al. 1988). Such viruses currently have no commercial potential (unless their host is a major pest in a high-value crop) because their markets are typically too small to permit economies of large-scale production. Viruses with broader host ranges do exist, such as the *Autographa californica* nuclear polyhedrosis virus which attacks at least six species (rigorously confirmed) and perhaps up to 43 species in 11 families of insects (Payne 1986). Genetic engineering can be used to broaden the host spectrum of some types of pathogens. This has been done for the bacterium *Bacillus thuringiensis*. Strains that are specific for certain types of hosts (subsp. *kurstaki* for Lepidoptera, subsp. *israelensis* for Diptera, subsp. *tenebrionis* for Coleoptera) can be genetically manipulated so that the host ranges of several strains are combined in the newly created form (Crickmore et al. 1990; Gelernter 1992).

Environmental or Habitat Features Affecting Effectiveness. The choice between major groups of agents may be dictated in some cases by similarities in environmental conditions in the pest's microhabitat to those favoring pathogenicity or reproduction of the agent. Nematodes, for example, have enjoyed greatest success in moist habitats such as soil and, to a lesser degree, inside plant tissues for control of leafminers or stem borers. While means may be found in the future to make nematodes work in drier environments such as on leaf surfaces, current circumstances dictate that the ecological requirements of the agent be met by developing products targeted at pests in favorable microhabitats. In other cases, formulation methods may be developed that overcome some of the environmental limitations of agents. Viruses, for example, are destroyed by exposure to ultraviolet light and may be protected by adding chemicals to the product's formulation that absorb radiation of damaging wavelengths. Similarly, fungal requirements for moisture may be overcome by formulation in oil or oil-water invert emulsions. The role of formulations in extending the effectiveness of pathogens is covered in more detail in the section on formulation methods.

Choosing the Best Species or Strain

Discoveries of new microbial agents may be the result of chance discoveries, laboratory screening efforts, or field surveys. Chance discoveries are those in which an agent with important new properties, not specifically being sought, is encountered and its value recognized. The discovery, for example, of a *Bacillus thuringiensis* isolate (later termed *israelensis*) that is pathogenic to dipteran larvae was such an unforeseen event. This discovery demonstrated the possibility of finding isolates of this pathogen effective against important new types of pests and stimulated screening programs to search for other useful isolates. Other chance encounters of useful new organisms include the finding in Texas of a new nematode, *Steinernema riobravis* Cabanillas et al., that is highly effective against pupae of the maize pest *Helicoverpa zea* (Raulston et al. 1992; Cabanillas and Raulston 1994).

Screening programs may also be used to find pathogens effective against a specific pest. Screening may be done by examining the activity of existing laboratory collections of pathogen isolates for activity against pests of concern. Kawakami (1987), for example, screened 61 isolates of *Beauveria brongniartii* (Saccardo) Petch for pathogenicity against the mulberry pest *Psacothea hilaris* (Pascoe).

Field surveys, however, are the basic source of new pathogen isolates. New isolates effective against a specific target pest may be encountered by collecting large numbers of the pest in the field, searching for dead or moribund specimens, and examining them by microbial culturing techniques. Koch's postulates must then be followed to determine which of these isolates are pathogenic in the target organism. Isolation of a pathogen from the actual target pest can be (but isn't always) important because pathogenicity may vary for the same species of pathogen depending on which host the isolate comes from. *Verticillium lecanii* strains, for example, originating from whiteflies, aphids, or thrips, are most virulent to the host from which the isolate was collected (Hirte et al. 1989). New pathogens may also be encountered in broad field surveys of pathogen groups. Efforts to find new nematodes, for example, have been made in Hawaii, Northern Ireland, Italy, and Sweden, among other locations, using wax moth larvae as baits to collect nematodes from randomly selected soils (Burman et al. 1986; Blackshaw 1988; Deseo et al. 1988; Hara et al. 1991). Surveys conducted to detect new natural enemies of adventive pests in their countries of origin may also be a source of new pathogens that are used either as agents of introduction or for augmentative use as commercial products.

Patents and Legal Issues

Sale of microbial pesticides, unlike the sale of arthropod parasitoids, predators, or nematodes, is regulated in the United States and some other countries under laws that govern the use of chemical pesticides. Such regulation affects strain selection because pathogens sold for pest control must be registered as pesticides and, consequently, with the exception of fungi, may gain patent protection. Patent protection gives finders of new pathogen isolates proprietary interests in their discoveries. Availability of patent protection stimulates the search for or laboratory creation of new strains.

PRODUCTION

Pathogens may be reared either in intact living hosts (*in vivo*), or in fermentation media or live host cell lines in media (both approaches termed *in vitro*). From the earliest days, pathologists have recognized that dependency on living hosts limits large scale production. Some groups of pathogens, however, are difficult to rear apart from living hosts. These pathogens include all viruses, microsporidia, many of the Entomophthoraceae fungi, a few bacteria such as *Bacillus popilliae* (which can be reared on fermentation media but does not produce spores efficiently apart from living hosts), and some families of nematodes. Pathogens that must be reared in living hosts require more labor for their production because rearing systems based on live hosts are difficult to automate and often lack economies of scale. Production costs for *in vivo* systems, therefore, are relatively high in countries where labor costs are high (Huber and Miltenburger 1986). *In vitro* production systems can usually take advantage of automation of rearing and economies of scale. Taborsky (1992) provides technical information on small-scale pathogen production for use in tropical areas.

In vivo Rearing Systems

Examples of agents that have been produced commercially in the United States in living hosts include the microsporidium *Nosema locustae* Canning, the bacteria *Bacillus popilliae*, *Bacillus lentimorbus* Dutky, and five nuclear polyhedrosis viruses (Huber and Miltenburger 1986). The process of *in vivo* rearing may be divided into four steps (Huber and Miltenburger 1986): mass-rearing the host insect; propagating the pathogen in the host; processing the pathogen; and controlling pathogen quality.

Mass-Rearing the Host. The insect hosts used for production are normally taken from a laboratory colony. In some cases, the cost of maintaining such a colony can be avoided if the host is easily obtained in the field in sufficient numbers. For example, the bacterium *Bacillus popilliae* is reared in field-collected larvae of *Popillia japonica*, since these are more easily collected than reared (Ignoffo and Hink 1971). Field collections have the limitation that hosts may only be available in certain seasons and may be contaminated by parasitoids or other pathogens. If the virus or other pathogen to be reared will grow in more than one host, it may be possible to rear the pathogen in an alternative host, choosing the most easily reared species from the host range.

Propagating the Pathogen. Production of viruses and microsporidia is typically initiated by contaminating the host's food with pathogen inoculum. If oral inoculation is ineffective, other methods may be used, as in the case of *Bacillus popilliae*, in which each host larva must be individually inoculated by injection. The methods used to rear the host during the growth of the pathogen must be such that dead, infected hosts at the end of the rearing period are easily harvested. Hosts cannot be allowed to burrow deeply into masses of diet. For hosts with such habits, alternative methods of diet presentation must be developed.

Processing the Pathogen. Harvesting and purification of the pathogen must be done inexpensively and must protect the viability of the pathogen. Hosts should be collected after death but before putrefaction (to minimize contamination by other microbes). Hosts can be collected manually or by vacuum suction (in the case of fragile cadavers). If necessary, cadavers may first be frozen to make them more durable for collection, but caution should be used as some pathogens (certain viruses) experience reduced viability if frozen. For small-scale rearing systems (as for research), pathogens such as viruses may be separated from host remains via centrifugation, but this approach is too expensive for commercial production. In commercial systems, infected host cadavers are dried and ground and the resultant mixture of viruses and host tissue particles is the final product. This has the advantage that presence of host tissue increases the shelf-life of the virus (Huber and Miltenburger 1986). For nematodes and microsporidia, other methods of collecting the reared pathogens from host cadavers are required. Nematodes, for example, may be collected by allowing them to move out of host cadavers into water.

Quality Control. Tests must be performed regularly to ensure that the culture has not become contaminated, especially by microbes pathogenic to humans (Podgwaite and Bruen 1978). In addition, the quantity and virulence of the pathogen being produced must be determined frequently. The quantity of pathogens being produced can be assessed by counting the number of nematodes, bacterial spores, or viral inclusion bodies per host or unit of medium. Virulence can be measured by conducting a bioassay on the target host, with comparison to an appropriate standard such as a properly stored sample of the original isolate of the pathogen. The activity of different batches of the *Anticarsia gemmatalis* baculovirus in Brazil, for example, was monitored and results indicated that 85% of batches had activity equal to the original strain. Partial loss of activity in 15% of the batches was related to excessive drying times, and the production process was modified to correct the problem (Moscardi et al. 1988). Only such tests, conducted regularly, can prevent the unknowing production of an ineffective product should the strain change in virulence or be overtaken by a competing contaminant organism superficially similar to the one being reared. The quality of agents reared by means of several different methods can be measured to determine if differences exist between rearing methods (Gaugler and Georgis 1991). For *Bacillus thuringiensis,* potency of the produced agent can also be indirectly measured by determining the size of the crystal mass of the bacterial toxin (Muratov et al. 1990).

In Vivo *Rearing of Viruses* (Fig. 11.1). Bell (1991) describes an *in vivo* system for the nuclear polyhedrosis virus of *Helicoverpa armigera* (Hübner), reared in larvae of *Helicoverpa virescens.* Host larvae were reared in cups of artificial diet and were exposed to virus through its application as a spray to the diet one week after host eggs were added. At the end of the second week most larvae were dead, at which time they were collected, homogenized, strained through cheese cloth, and the virus particles harvested via centrifugation. Optimal viral inoculation rates were determined by comparing yields from a series of different viral doses per cup. Low doses failed to infect all larvae. High doses killed larvae while still small, reducing viral yield per larva. The cost of rearing was calculated at about $0.02 (U.S.) per host, 80% of which cost was for labor. Shapiro (1986) gives a detailed discussion of procedures for *in vivo* rearing of baculoviruses.

In Vivo *Rearing of Microsporidia.* Microsporidia are reared in living hosts, using procedures similar to those for viruses (Brooks 1980, 1988; Kurtti and Munderloh 1987). Productivity of *in vivo* systems for microsporidia depends strongly on the agent being produced and the host used. Much of the rearing cost is that of rearing the insect host. The next biggest determinant of

Figure 11.1 (A) *In vivo* rearing of baculovirus using live larvae of spruce budworm, *Choristoneura fumiferana* (Clemens); (B) harvest of virus from host cadavers; (C) centrifugal mill used for grinding lyophilized, virus-infected cadavers. (Photographs courtesy of J. C. Cunningham.)

rearing cost is the quantity of microsporidia produced per infected host, which can vary greatly depending on a variety of factors. For example, some microsporidia produce general systemic infections, whereas others are limited in their attack to specific tissues. Yield per host will be reduced in the latter case because only a portion of the host is actually infected. Henry et al. (1978) describe an *in vivo* production system for *Nosema locustae* for grasshopper control in which sufficient spores to treat one ha could be produced at a cost of less than U.S. $0.25 (1978 dollars) (costing $1.86 once formulated with bait). This was an economically acceptable price for application on range lands producing at least $12.00 per year of forage.

In Vivo *Rearing of Nematodes.* All nematodes can be reared in living hosts. For example, heterorhabditid and steinernematid nematodes, the groups of greatest commercial interest, may be reared in larvae of the greater wax moth, *Galleria mellonella*. Methods for rearing the insect host on a laboratory medium, initiating nematode infection, harvesting, and storing juvenile nematodes of the Heterorhabditidae and Steinernematidae have been described (Dutky et al. 1964; Woodring and Kaya 1988; Lindegren et al. 1993). Harvesting nematodes is achieved by allowing nematodes to migrate towards a water trap away from host cadavers (Fig. 11.2). This system is relatively expensive with costs of about U.S. $1.00 (1990 dollars) per million infective juveniles. Economies of scale can not be achieved for commercial production levels because procedures cannot be automated. Yields of *Steinernema glaseri* (Steiner) can be increased 3 to 4 times by rearing in dead, rather than live, wax moth larvae (Leite et al. 1990). *In vitro* methods are available for large scale, automated rearing (Friedman 1990). Nematodes in some families such as the Mermithidae (for example, *Romanomermis culicivorax*, a parasitoid of mosquito larvae) must be reared in living hosts (Poinar 1979).

In Vivo *Rearing of Fungi.* Most entomopathogenic fungi are facultative parasites which also exist as saprotrophs and therefore can be grown apart from living hosts. Only a few groups are obligate parasites which must be reared in living hosts. Ignoffo and Hink (1971) recognized eight species of entomopathogenic fungi that had been cultured successfully in living hosts. Among these were several species of *Entomophthora*, including a species which attacks the spotted alfalfa aphid, *Therioaphis maculata* (Buckton), in California (U.S.A.) and of *Coelomomyces*, which is grown either in mosquito larvae or copepods (McCoy et al. 1988). Fungi are not, however, generally reared commercially *in vivo* due to the higher cost of such rearing compared to *in vitro* systems.

In Vivo *Rearing of Bacteria.* *Bacillus popilliae* production involves both an *in vitro* and an *in vivo* step. Sporulation cannot be effectively obtained on artificial media (Stahly and Klein 1992). Therefore, the final step in obtaining large quantities of *B. popilliae* spores involves inoculation of scarab larvae with vegetative cells of the pathogen (Fig. 11.3). Dulmage and Rhodes (1971) describe the process of *in vivo* spore production. However, because only one spore is obtained

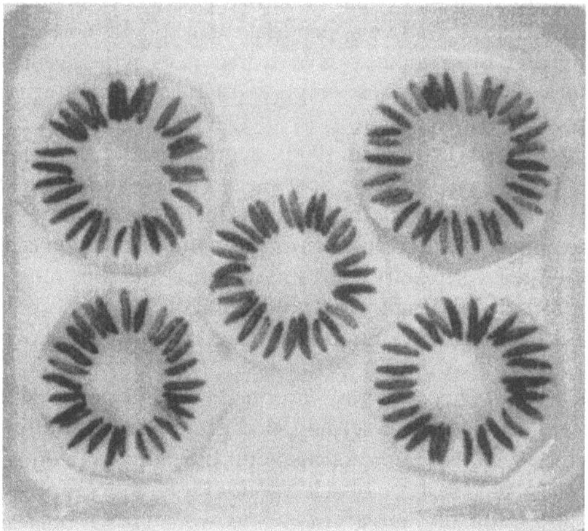

Figure 11.2. Heterorhabditid and steinernematid nematodes may be reared *in vivo* and collected by use of water traps from host cadavers. (Photograph courtesy of R. Gaugler.)

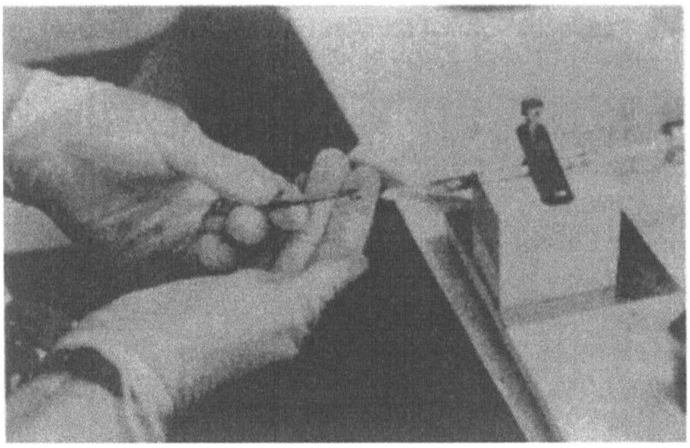

Figure 11.3. Hand inoculation of scarab grubs with vegetative cells of *Bacillus popillae* Dutky to obtain spores through *in vivo* rearing. (Photograph courtesy of S. Roy.)

from each infective vegetative cell, it is necessary to first rear large quantities of vegetative cells on fermentation media. Methods for this are also described by Dulmage and Rhodes (1971).

In Vitro Rearing Systems

In vitro rearing employs either fermentation media or insect cell cultures. Many fungi, bacteria, and nematodes can be grown in fermentation media, which permits large-scale production using relatively inexpensive materials and automated systems. Virus and microsporidia can be produced in insect cell cultures, but commercial *in vitro* production is not yet economical.

In Vitro *Rearing of Viruses.* Insect cell cultures may be used to produce insect viruses. Granados et al. (1987) noted that at least four cytoplasmic polyhedrosis viruses, two entomopoxviruses, four nuclear polyhedrosis baculoviruses, one granulosis virus, and two nonoccluded viruses could be grown in insect cell cultures. Virus production depends on having cell cultures from an appropriate host species and host tissue. The first insect cell cultures were initiated from hemocytes or ovarian cells, whereas most viruses in nature replicate in midgut or fat body cells. Initial attempts to establish cell cultures from these tissues from Lepidoptera were unsuccessful (Granados et al. 1987), but subsequently, cell cultures from such tissues were developed (Lynn et al. 1990). Rearing productivity of a virus can differ greatly in different cell cultures from the same host species and tissue (Lenz et al. 1991). Screening programs are often required to identify the most productive cell cultures in which to rear any particular virus. For example, to find cells of *Cydia pomonella* in which its granulosis virus would reproduce, 81 cell cultures were established and screened and only nine were susceptible to the virus (Naser et al. 1984). Concentrations of polyhedral occlusion bodies produced by two different *Helicoverpa zea* cell cultures differed 85-fold and the number of occlusion bodies produced per cell differed 10-fold (Lenz et al. 1991).

Reducing cost of virus production will require: developing systems suitable for large batch size production, increasing viral productivity per unit of media, and lowering the cost of cell culture media. Some of these problems have been partially solved in recent years.

The potential use of cell cultures for large scale production is discussed by Weiss and Vaughn (1986). Initially, most insect cells were grown only as monolayer cultures on surfaces. This culture form is unsuitable for commercial use because it lacks economy of scale that would allow for large production volumes. More recently, however, a number of insect cell suspension cultures have been developed in which cells exist free in liquid media without physical support (Huber and Miltenburger 1986). Such suspensions, however, must be oxygenated. Bubbling oxygen through the media was found to be mechanically damaging to the insect cells, but diffusion across the wall of a silicon tube placed inside the fermenter was both safe and effective (Huber and Miltenburger 1986).

The yield of virus particles per unit volume of cell suspension in view of the cost of the rearing media is an economic issue of great importance. Field studies suggest that effective application rates for various viruses are in the range of 10^{11}–10^{12} polyhedral inclusion bodies per ha. Many *in vivo* systems are able to produce this quantity of virus in about 1000 diseased caterpillars, each costing about U.S. $0.02 for a cost of $20 per ha per application. In contrast, the cost of producing the same quantity of virus in insect cell culture in 1986 was prohibitive, about $900. Virus yields differ between cell cultures. Media costs are a major factor in determining the cost of viruses reared in cultures of insect cells. Initial media required the use of fetal bovine serum, and this factor alone accounted for 90% of the medium's cost (Huber and Miltenburger 1986). Alternative media using less expensive ingredients are actively being developed and several defined media exist which do not require bovine serum. Continued improvements in methods for virus production in cell cultures are likely. King et al. (1988) reported a method of culturing the *Autographa californica* virus in cells of *Spodoptera frugiperda* J. E. Smith in multiple-membrane alginate-polylysine capsules. This method increased virus yield tenfold compared to conventional cell suspension cultures. Similarly, production of the nuclear polyhedrosis virus of *Lymantria dispar* is possible in cultures of fat body cells in batches of up to 40 liters (Lynn et al. 1990). These and other improvements may eventually reduce *in vitro* virus production costs sufficiently to be competitive with chemical pesticides.

In Vitro *Rearing of Microsporidia.* At least six species of *Nosema* and two of *Vairimorpha* have been grown successfully in insect cell cultures (Kurtti and Munderloh 1987). However, the use of insect cell cultures to mass produce microsporidia is not yet economically feasible because of limitations on the mass culture of insect cells themselves, and because the yield of microsporidia spores per infected cell is too low. Field application rates for microsporidia have been in the range of 10^9–10^{13} spores/ha, while yields of microsporidia from cell cultures have been 10^6 and 10^7 spores per ml (Kurtti and Munderloh 1987).

In Vitro *Rearing of Nematodes* (Fig. 11.4). Heterorhabditid and steinernematid nematodes may be grown in media that do not contain any live hosts. Glaser et al. (1940) were the first to establish a successful, large-scale *in vitro* production process for a steinernematid nematode. Three problems must be solved for *in vitro* rearing of nematodes: suppression of microbial contaminates, maintenance of the symbiotic bacterium (*Xenorhabdus* spp., *Photorhabdus* spp.) which actually kills the host insect, and provision in the medium of all key nutrients for nematode growth. The microbial contamination problem can be addressed by developing axenic nematode cultures (ones containing no other live organisms but the nematodes) either by using antibiotics or rigorous sterilization of the media and surface sterilization of the nematodes eggs used to initiate the colony (Lunau et al. 1993). Such axenic cultures can then be inoculated with the microbial symbiont needed by the nematode to kill hosts, resulting in mixed cultures of the desired nematode and symbiotic bacterium (Lunau et al. 1993).

To achieve larger scale nematode production at commercially competitive prices, three factors were important historically: identifying inexpensive nutrients, identifying culturing conditions that promoted high yields, and using liquid rather than solid culture media

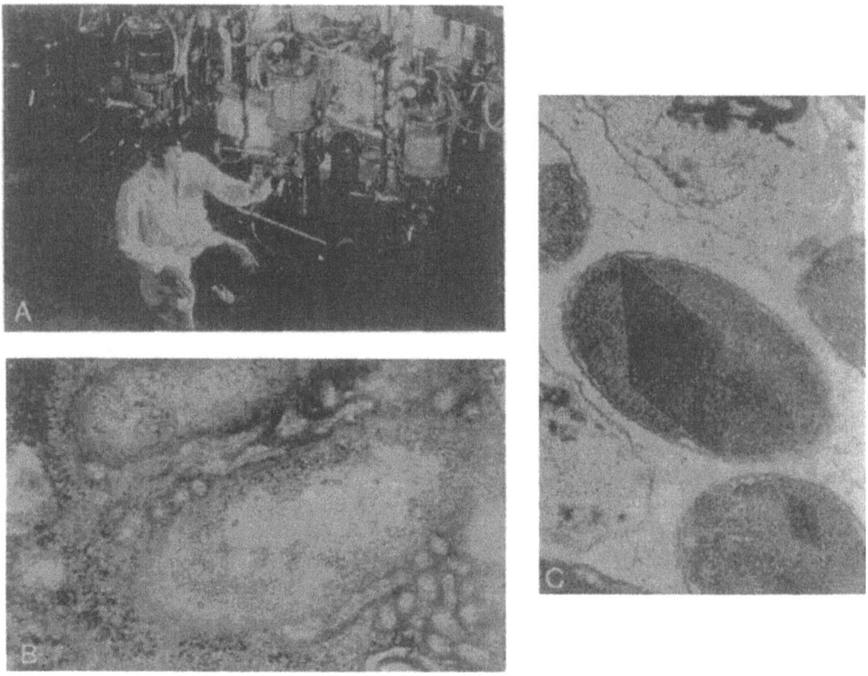

Figure 11.4. Heterorhabditid and steinernematid nematodes can be reared in nutrient broth (A); effectiveness requires retention of symbiotic bacteria (*Xenorhabdus* spp. [B], *Photorhabdus* spp.). Toxin crystal visible in cell (C). (Photographs courtesy of R. Georgis [A]; and R.J. Akhurst, CSIRO [B,C].)

(Friedman 1990). The search for better nutrients (cheaper, yet high yielding) led to the consideration of many materials. Ultimately, however, it was recognized that incorporating the symbiotic bacteria (*Xenorhabdus* spp., *Photorhabdus* spp.) in the rearing media was beneficial because these bacteria produced enzymes that broke down proteins into materials usable by the developing nematodes. With these bacteria in the media, dog food was found to be an acceptable medium. Culture of nematodes together with their symbiotic bacteria is termed monoxenic culture.

The search for optimal rearing conditions for high yield also considered many physical types of rearing systems. Both solid and liquid media can be successfully employed, but liquid culture has better economies of scale and is more amenable to commercial use. In liquid media, however, as the volume of the reaction vessel increases, oxygen may become limiting in some portions of the media. Methods to add oxygen need to take into account susceptibility of nematodes to mechanical damage from shearing caused by stirring the medium (for aeration) or bubbling oxygen into it. Acceptable limits for these procedures have been identified and Biosyst, one of the major commercial producers of nematodes, now uses large-scale liquid monoxenic culture systems employing vats of 15,000 liters or greater (Friedman 1990).

In Vitro *Rearing of Fungi.* Small scale *in vitro* production of entomopathogenic fungi is typically done on solid culture media, with the fungus growing as a mat on the surface of the medium and then producing spores on aerial hyphae (Fig. 11.5A). Media may be either defined agar-based media or natural substances such as rice or bran. McCoy et al. (1988) list media for the production of a number of fungi that are candidate biological control agents. Conidial spores (the stage typically used as the pest control propagule) are harvested from solid-media cultures by washing fungal mats with distilled water.

Surface cultures on solid media, while suitable for small scale production, lack the economies of scale and potential for automated handling that are desirable for large scale commercial production. For large scale production of microbial agents, liquid (submerged) culture systems are desirable (Fig. 11.5B). Entomopathogenic fungi of several types can be grown in liquid cultures. However, only a few (*Beauveria bassiana, Hirsutella thompsonii*) will sporulate in submerged culture (Dulmage and Rhodes 1971; van Winkelhoff and McCoy 1984). This problem can be resolved by a two-step culturing process in which submerged cultures are first used to produce large quantities of mycelia, which then are placed onto solid culture media for the production of conidial spores (McCoy et al. 1988). Commercial two step culture systems (termed diphasic) have been developed for a number of fungi, including *Beauveria bassiana* (Miao et al. 1993), *Hirsutella thompsonii, Nomuraea rileyi* (Farlow), and *Verticillium lecanii.*

An alternative method for the commercial production and use of entomopathogenic fungi is to develop application methods to use mycelial fragments, rather than conidial spores, as the infective propagule to be applied. This approach has been explored with *Hirsutella thompsonii,* and a patented process has been developed in which mycelia can be produced in submerged culture, dried, and stored under refrigeration until applied (McCoy et al. 1975; McCabe and Soper 1985). Bartlett and Jaronski (1988) give further details on commercial production of entomopathogenic fungi. Methods for rearing *Beauveria bassiana* that require

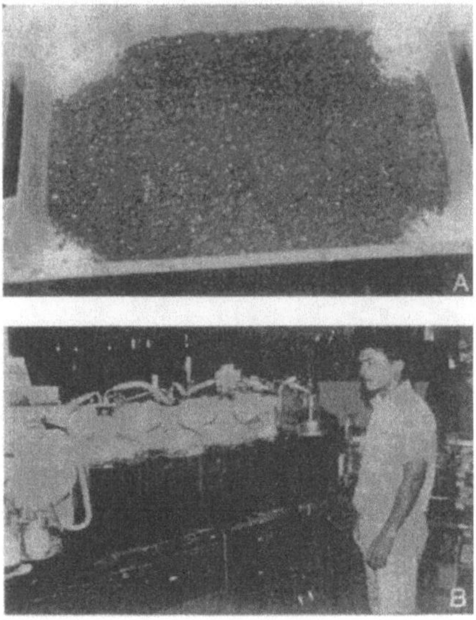

Figure 11.5. Entomopathogenic fungi may be reared *in vitro* using either solid (A) or liquid media (B). (Photographs courtesy of R.M. Pereira [A] and D.W. Roberts [B].)

minimal training and equipment have been developed in Colombia for fungus production on individual farms to control coffee pests (Antía-Londoño et al. 1992). Feng et al. (1994) review the production, formulation, and application of *B. bassiana*.

Methods and issues of concern in the *in vitro* production of plant pathogenic fungi to be used as mycoherbicides are essentially the same as for entomopathogenic fungi. Boyette et al. (1991) review culture techniques for plant pathogenic fungi, and Stowell (1991) discusses large-scale industrial production systems.

In Vitro *Rearing of Bacteria*. The principal bacterium reared *in vitro* for pest control is *Bacillus thuringiensis*. Dulmage and Rhodes (1971) present detailed descriptions of methods for production of *B. thuringiensis*. This species can be reared either on semisolid media or liquid media (Fig. 11.6). Semisolid systems utilize bran for the growth of the organism, and the rearing material is simply dried and ground at the end of the rearing process. Bran, however, swells when wetted and so is unsuitable for the production of wettable powder formulations. Semisolid systems are used primarily to produce material for dust formulations. Material for wettable powder formulations can be produced by liquid culture, based on such ingredients as molasses, corn-steep liquor solids, and cottonseed flour. The reared bacteria and associated toxins may be recovered by a variety of methods including filtration, centrifugation, and precipitation (Dulmage and Rhodes 1971). Methods for *in vitro* production of the vegetative cells of *Bacillus popilliae* are also given in Dulmage and Rhodes (1971). A method for rearing *Bacillus*

Figure 11.6. Small scale liquid fermenter used for production of *Bacillus thuringiensis* Berliner. (Photograph courtesy of D. Cooper.)

thuringiensis subsp. *israelensis* in coconuts has been developed for use in remote areas of Peru that allows villagers to rear this bacterium to treat ponds for mosquito control (Metcalfe 1991).

Maintaining and Improving Genetic Quality

Preventing Deterioration of Strain Quality. Microbial agents reared on fermentation media can lose infectivity upon repeated *in vitro* reproduction. *Entomophaga maimaiga* Humber, Shimazu and Soper, for example, declined in infectivity to *Lymantria dispar* after 15 to 50 cycles of reproduction on fermentation media (Hajek et al. 1990). Repeated rearing of *Nomuraea rileyi* by conidial transfer led to the loss of virulence to larvae of *Anticarsia gemmatalis* in 16 generations. Loss of virulence was associated with the conidial stage as no loss of virulence in this species was seen in up to 80 passages based on mycellia transfers (Morrow et al. 1989). Attenuation has also been observed in at least seven other fungi species (Hajek et al. 1990 and references therein). Similarly, baculoviruses produced in alternative hosts may have reduced infectivity in the original host. Such a reduction occurred with the nuclear polyhedrosis virus of the silkworm *Bombyx mori* when reared continuously in the Asiatic rice borer, *Chilo suppressalis* (Walker), for 18 or more generations (Aizawa 1987). In other species of pathogens, frequent passage *in vitro* has not been associated with decline in infectivity (Hajek et al. 1990). For cases in which loss of infectivity from continuous rearing in fermentation media does occur, it can be prevented by maintenance of a parent culture of the agent on living hosts. Material from the parent colony can be used to inoculate production batches, and thus the product that is sold is always only one generation away from pathogens reared *in vivo*. If methods are available to store the stock culture, for example, under refrigeration or in liquid nitrogen, then the frequency with which the stock culture must be recycled in living hosts can be reduced, decreasing cost and increasing convenience. Quality of stock cultures may decline with time, however, and the length of time over which storage may be employed for any given natural enemy must be determined on a case-by-case basis.

Genetic Improvement of Nematodes and Pathogens. Nematodes and microbes can potentially be improved in a variety of characteristics, such as infectivity rate to a given host, host range, and pesticide resistance. Gelernter (1992) discusses several examples of possible improvements in *Bacillus thuringiensis* strains through a variety of techniques. Improvements are also possible for characteristics affecting production, such as yield of spores or rate of growth under production conditions. For nematodes (but generally not for microbes), host finding behaviors may also be subject to improvement (Gaugler et al. 1989).

Aizawa (1987) expanded the host range of silkworm nuclear polyhedrosis virus to include the Asiatic rice borer, *Chilo suppressalis*, through repeated rearing in the new host. Using genetic engineering, Crickmore et al. (1990) expanded the host range of an isolate of *Bacillus thuringiensis* by combining genes for toxins allowing for attack of Diptera and Coleoptera into a single strain of the bacterium. Gaugler et al. (1989) reported a 20 to 27 fold improvement in host-finding ability by *Steinernema carpocapsae* (Weiser) through laboratory selection. Thirteen rounds of selection increased the host finding distance of *S. carpocapsae* from 3.5 cm to 20 cm/hr (Gaugler et al. 1991). However, the selected strain did not provide increased field efficacy (Gaugler unpublished, in Kaya and Gaugler 1993). Little else has been done in the area of selection for improvement of nematodes (Hominick and Reid 1990), but the concept appears sound and awaits future development. Plasmids have been used to genetically transform the fungus *Metarhizium anisopliae* to be resistant to the fungicide benomyl (Goettel et al. 1990).

Such fungicide-resistance in entomopathogenic fungi would allow their use to control insects in crops also subject to plant diseases for which fungicides must be applied.

STORAGE AND SHIPPING

Some pathogens can be stored for months or years at room temperature (Fig. 11.7). This is an important advantage for such products because it allows for year-round production, with storage until the season of use. This improved storage ability also allows agents to be shipped by slower (lower cost) methods without concern for deterioration in route. Such microbial products may approximate chemicals in regards limits on their shipping, storage, and shelf life. Other pathogens such as nematodes and some fungi are more delicate and can be stored only for several months and may require refrigeration. Improvement in the storage characteristics of many pathogens is an essential step for large-scale commercial use of these organisms.

Storage of Nematodes. Heterorhabditid and steinernematid nematodes survive well for a number of months if refrigerated and stored in thin, moist, well-aerated layers. Optimal temperatures for nematodes storage vary between species but, in general, steinernematids survive best when stored at 5–10°C and heterorhabditids at 10–15°C (Georgis 1990). Refrigeration, however, is not a strict requirement. The Biosyst product Biosafet, for example, is a formulation of nematodes on a thin layer of gel supported by a mesh screen, and in this form, nematodes could be successfully stored for up to one month at 25°C (Freidman 1990). *Steinernama carpocapsae* can be stored up to five months at room temperature, and 12 months under refrigeration (Georgis 1990; Georgis and Hague 1991).

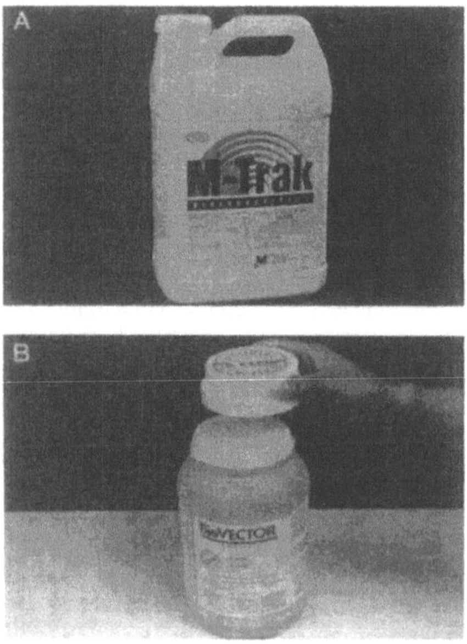

Figure 11.7. Formulation and packaging are essential steps to make pathogens and nematodes available for use as bio-pesticides. (A) *Bacillus thuringiensis* Berliner and (B) heterorhabditid and steinernematid nematodes. (Photographs courtesy of Mycogen [A] and Biosys [B].)

Storage of Fungi. Storage properties of fungi used for insect or weed control vary depending on the species and what form the infective unit takes. Fungi may be marketed as conidial spores, chlamydospores, blastospores, or mycelial granules or fragments, among other possibilities. The mycoherbicide DeVine™, formerly marketed for control of strangler vine (*Morrenia odorata* Lindle) in citrus in Florida contained chlamydospores of the fungus *Phytophthora palmivora* (Butler) Butler formulated as a liquid concentrate. These had to be held under refrigeration until applied and had a shelf life of only about 6 weeks (Boyette et al. 1991). Commercial use was possible because the product was marketed in a small region to a small set of users (Kenney 1986). McCabe and Soper (1985) describe a process for the production and use of fungal mycelia of those entomophthorans amenable to culture on artificial media. The mycelial product, however, has to be stored at or below 4°C to maintain its viability. The fungal mosquito pathogen *Lagenidium giganteum* produces oospores which can be harvested and stored in dry form for many months, producing infective zoospores when re-wetted (Latgé et al. 1986). Spores of Deuteromycete fungi are often easier to store than those of Entomophthoraceae. Blastospores of the Deuteromycete *Verticillium lecanii*, marketed as Vertalec™ and Mycotal™, must be stored under refrigeration and are viable only for a few months (Bartlett and Jaronski 1988).

Storage of Microsporidia. Survival of microsporidian spores is highly variable depending on the species, temperature, and nature of the preparation stored. Various species survive from 2 to 18 months at temperatures from 0 to 6°C (Brooks 1988). Spores of some species (*Vairimorpha necatrix* Kramer), when lyophilized (dried slowly and then frozen) in 20–50% sucrose or in host tissues (cadavers), have retained rates of infectivity of 80% for up to two years (Brooks 1988). This method appears to have potential for enhancing the storage properties of microsporidian spores.

Storage of Viruses. Unformulated entomopoxvirus of the grasshopper *Melanoplus sanguinipes* (Fabricius) retained high activity for up to 9 months when stored at 4°C. The same virus, if formulated as a bait in bran, lost activity over time, possibly due to bacterial antagonism. This did not occur if the virus was encapsulated in starch granules in place of bran (McGuire et al. 1991). In general, polyhedra of most nuclear polyhedrosis viruses are stable when frozen or refrigerated, but nonoccluded virus are not (Young and Yearian 1986).

Storage of Bacteria. *Bacillus thuringiensis* spores and toxins are stable at room temperature and do not require refrigeration for storage or distribution (Dulmage and Rhodes 1971).

FORMULATION AND APPLICATION

Formulation and application methods for nematodes and microbial pathogens seek to place the agent in the correct location to maximize contact with the target pest. In addition, they help protect the agent from adverse physical conditions, enhancing the survival of the infective propagule and extending the period over which an application of the agent is effective. Angus and Lüthy (1971) present a general discussion of the various formulations used for microbial pesticides and the functions of the various ingredients.

Formulation of Nematodes. Nematodes have been formulated in many different ways, including being combined with alginate, clay, activated charcoal, gel-forming polyacrylamides, vermiculite, peat, evaporetardants, ultraviolet protectants, placed on sponges and in baits, and stored in anhydrobiotic form (Georgis 1990). Some of these formulations are aimed at extending nematode survival in storage, enhancing ease of handling, or improving performance after application is made. Development of a flowable concentrate formulation, for example, eliminates the need to dissolve a carrier matrix and suspend nematodes prior to application (Georgis, personal communication in Kaya and Gaugler 1993).

Some nematode formulation practices seek to raise nematode survival under unfavorable physical conditions at the application site, increasing effectiveness. The principal factor limiting nematode performance in many habitats is desiccation. Gel-forming polyacrylamides have been used to retain water at the application site, which improved control of the citrus pest *Diaprepes abbreviatus* (Linnaeus) (Georgis 1990). Glazer et al. (1992) report that addition of the antidesiccant Folicotet enhanced nematode control of the foliar pest *Earias insulana* (Boisduval) on cotton foliage in glasshouse tests in Israel. Ultraviolet light also kills nematodes; *Heterorhabditis bacteriophora* Poinar is more sensitive than *Steinernema carpocapsae* (Gaugler et al. 1992). Such sensitivity to ultraviolet light suggests that nematodes are going to be of greatest value in soil and other protected environments. Some protection against damage from ultraviolet light can be obtained by adding protective materials to the formulation (Gaugler and Boush 1978). Successful use of nematodes for control of foliage pests may be achieved through further development of materials to protect nematodes from desiccation and ultraviolet light. An alternative approach that is also under investigation is the use of desiccated (anhydrobiotic) nematodes. Research has shown that some nematodes can survive slow desiccation, and when rehydrated inside an insect's gut, revive and successfully attack the host (Georgis 1990). Desiccated nematodes are less sensitive to dry conditions at the application site. The approach, however, is not developed well enough for commercial use.

Certain formulations seek to deliver the nematode to a particular target pest or habitat that might not be reached by more common methods such as a broadcast water spray. Baits, for example, attempt to target groups of pests such as cutworms (Noctuidae) and crickets (Gryllidae) that might not be contacted by a simple broadcast spray but which are likely to locate and eat baits. The application method used can also be part of the process of increasing the likelihood of contact with the target pest. Borers in stems of cane berries, for example, can be targets for nematodes because nematodes, applied as a spray to the canes, enter tunnels in canes where pest larvae feed (Miller and Bedding 1982). Nematodes may be directed against insects that attack the roots of seedlings of such plants as cabbage (*Brassica oleracea*) by applying nematodes to the roots of seedlings prior to planting, so that nematodes are in position to protect the plants immediately. In turf, penetration of nematodes through the thatch into the plant root zone is critical for effective control. Nematode movement downward can be enhanced on small acreages such as golf courses by irrigating after the application is made (Shetlar et al. 1988). At a larger scale, such as pastures, irrigation may not be possible because of the cost of applying large quantities of water. Berg et al. (1987) describe a mechanical device that uses drill action to introduce nematodes into the root zone, reducing the water needed from 20,000 to 1520 liters/ha.

Formulation of Fungi. Fungi used for plant or arthropod control can be applied in a variety of life stages, and these stages may differ in their environmental sensitivities and physiological requirements for initial survival and initiation of infection. The effectiveness of fungi is likely to be affected by the degree of spread and adhesion of the applied material to the target substrate. Fungal spores need close contact with the plant surface or arthropod integument to initiate infection. Stickers are, therefore, likely to be important components of many fungal biopesticides. Surfactants (wetting agents) may, however, reduce spore attachment to hosts and spore viability; each surfactant-fungus combination should be checked for compatibility (Connick et al. 1990).

For mycoherbicides that are intended for application to the soil, alginate granules provide a uniform medium that supports fungal growth well, allowing spore production for an extended period after application, up to 6 weeks in some cases (Connick et al. 1990). Spores produced by such granules on the soil are not likely to redistribute themselves to aerial plant parts if the spores are large. This approach, therefore, works best where spores are intended to contact their target in or on the soil, or when target weeds are still short (less than 2–4 cm). Granular formulations of vegetative cells of entomopathogenic fungi such a *Metarhizium anisopliae* have also been developed (Storey et al. 1990) and appear promising. Nongranular formulations must be used for products intended to deliver fungal spores to aerial parts of tall weeds or to insects feeding on plants.

Some fungal spores germinate rapidly in water, so liquid formulations are not usable because spores germinate prematurely. In such cases, dust or wettable-powder formulations may be used. Other fungi are applied as conidia that germinate and initiate infection after exposure to either very high relative humidities or free water such as dew for a critical period of time. Because natural free water on treated surfaces may be present for an insufficient amount of time, formulations have been examined that retain water around the applied fungal spores or other structures. For fungi which require a specific period of exposure to free water or a carbon source for spore germination, invert emulsions of water in oil can be used (Connick et al. 1991). Use of invert emulsions, under conditions lacking a natural dew, increased control of the weed *Cassia obtusifolia* Linnaeus (sicklepod) by a conidial suspension of *Alternaria cassiae* Jurair and Khan from 0% to 88% (Connick et al. 1990). Commercial use of invert emulsions is limited by the large quantities of liquid that must be applied to obtain the droplet sizes required for adequate water retention (greater than 600 microns) (Egley et al. 1993). Other formulations that have similar water-retention action include the use of vegetable oils, and the ultralow-volume application of such oils. Bateman et al. (1993) found that formulation of *Metarhizium flavoviride* Gams and Rozsypal in cottonseed oil reduced the LD_{50} of the pathogen to the desert locust, *Schistocerca gregaria* Forskal, over 100-fold. Performance of oil formulations was especially enhanced with respect to water formulations in arid environments (relative humidities less than 35%) and preliminary trials in Niger gave satisfactory results under arid conditions (Bateman 1992). Formulation of fungal spores in oils also provides partial protection against destruction caused by ultraviolet light (Moore et al. 1993). In some cases, efficacy of mycoherbicides can be enhanced by the inclusion in the formulation of nutritional materials, such as sucrose or soy flour to support growth of the fungus prior to host invasion (Walker 1981; Weidmann and Templeton 1988).

Entomopathogenic fungi for control of outdoor arthropod pests may be manipulated by farmers by moving infected hosts between sites. Branches bearing cadavers of the whitefly *Dialeurodes citri* (Ashmead) killed by *Aschersonia aleyrodis* Webber were used successfully in China to create fungal epidemics on new trees (Gao et al. 1985). In Japan, polyurethane sheets impregnated with conidia of *Beauveria brongniartii* were used to control the whitespotted longicorn beetle, *Anoplophora malasiaca* (Thomson), by wrapping the trunks of citrus trees (Hashimoto et al. 1989). Electrostatic application devices have been used to increase deposition of the blastospores of *Verticillium lecanii* on the undersides of plant foliage in greenhouses for aphid control (Sopp et al. 1989). Many of the same issues affecting formulations for fungal pathogens of arthropods also are important for fungal pathogens of plants, especially the need to contact the target host, by virtue of correct placement and good adhesion, and the need for adequate water or high relative humidity for spore germination.

Formulation of Microsporidia. Spores of microsporidia have been applied both as water suspensions and in baits. Control with baits has been better than with water sprays (Brooks 1988). Microsporidia are sensitive to ultraviolet light and ultraviolet light protectants have been added to some water formulations. Bait formulations of *Nosema locustae*, used for control of grasshoppers, have consisted of bran sprayed with spores and thickened with 0.2% hydroxymethyl cellulose as a sticker (Brooks 1988). The neogregarine *Mattesia trogodermae* has been formulated as spores applied to paper disks treated with pheromone of the target pest. *Trogoderma glabrum* (Herbst) becomes inoculated with spores in the process of walking on the pheromone-treated surfaces (Shapas et al. 1977).

Formulation of Viruses. Formulations of viruses seek to create products that have stable physical properties (no caking or clogging) suitable for application with conventional pesticide application machinery. In addition, formulations often contain spreaders, ultraviolet light protectants, and food items intended to stimulate consumption by the pest (Young and Yearian 1986).

Baculovirus formulations consisting of filtrates of crushed host cadavers mixed with water, if stored under refrigeration or frozen, usually perform as well or better than more complicated formulations. However, such a simple approach is not useful for production of a product for commercial use, which must be stored for up to six months and which must have physical characteristics that allow the material to be applied with various types of application machinery. Several methods have been used to formulate commercial products. The first of these is lyophilization of the virus. Clumping may be prevented by mixing the host cadavers with lactose prior to lyophilization. A second approach is mixing attapulgite clay with the virus in a water solution, which is then sprayed and allowed to dry. This process yields a stable wettable powder in which the virus is microencapsulated by a coating of clay. A third approach is to microencapsulate virus occlusion bodies with materials such as methyl cellulose or gelatin (Young and Yearian 1986).

Materials that act as ultraviolet light protectants for viruses include a variety of dyes, especially Congo red (Shapiro and Robertson 1990), starch encapsulation (Ignoffo et al. 1991), and optical brighteners (Shapiro and Robertson 1992). Adding such optical brighteners as Leucophor BSt and Phorwite ARt reduced the LC_{50} concentration for the virus of *Lymantria dispar* 400 to 1800-fold, depending on the material. Such an increase in efficiency, if realized under field conditions, has potential to drastically lower the amount of material needed for control, reducing the cost.

Application methods for viruses other than broadcast foliar treatments have been less often considered, but other methods exist to deliver virus to target pests. Ignoffo et al. (1980) found that if cabbage seedlings were dipped at planting in a solution of *Trichoplusia ni* virus, pathogen activity remained high for up to 84 days. This approach reduces the quantity of virus needed for treatment, and minimizes labor and machinery costs. Jackson et al. (1992) experimented with the placement of *Autographa californica* virus in *Helicoverpa virescens* pheromone traps to determine if male moths would effectively disperse the virus within the population. Although autodissemination did occur, transmission rates and subsequent larval mortality rates were low. Honey bees have been used to disseminate *Heliothis* nuclear polyhedrosis virus, using special pathogen-applicator devices attached to hives (Gross et al. 1994).

Formulation of Bacteria. The majority of bacterial pest-control products that have been marketed have been based on *Bacillus thuringiensis*. Angus and Lüthy (1971) provide summaries of kinds of additives that have been components in formulations of this bacterium. Most products contain both live spores and toxins. Spores are relatively stable and are marketed as both wettable powders and liquids. Most products of *B. thuringiensis* are formulated to be mixed with water and applied as foliar sprays. In the case of *B. thuringiensis* subsp. *israelensis*, used for control of mosquito, blackfly, and other dipterous larvae, formulations exist that are intended to be applied as liquids to aquatic habitats (Mulla et al. 1990). Other formulations exist, such as briquettes, which are intended to dissolve over an extended period and which can be tossed into mosquito breeding areas by hand. Superabsorbent-polymer, controlled-release systems have been developed for both *B. thuringiensis* subsp. *israelensis* and *Bacillus sphaericus* for use in aquatic habitats that extend the effective control period from an application from a few days to over a month (Levy et al. 1990).

Other formulations of *B. thuringiensis* use starch granules to encapsulate spores, together with such additives as stickers, ultraviolet light protectants, or feeding stimulants. McGuire et al. (1990) report that starch encapsulation together with the phagostimulant Coaxt enhanced control of the maize borer *Ostrinia nubilalis*.

Formulations of *B. thuringiensis* must be ingested to be effective, and most products are directed against larval stages. A bait formulation combining *B. thuringiensis* spores with liquid sugars was found to control adult moths of some species (Potter et al. 1982). More recently, genes for the toxic proteins of *B. thuringiensis* have been introduced into other organisms, using these as, in effect, novel formulation systems to deliver the *B. thuringiensis* toxins in new locations or ways. A sporefree formulation of *Bacillus thuringiensis* toxins, encapsulated in dead cells of *Pseudomonas* bacteria, has been created by transferring *Bacillus thuringiensis* genes into a *Pseudomonas* bacterium, which produces the toxins and is then killed at the end of the production cycle (Gelernter 1990). Genes from *B. thuringiensis* coding for various toxins have been introduced into a maize root bacterium (*Pseudomonas fluorescens* [Trevisan] Migula) to suppress damage from larvae of root-feeding beetles. Genes of *B. thuringiensis tenebrionis* have also been introduced into potatoes and tobacco (*Nicotiana tabacum* Linnaeus), causing toxins to be produced in plant foliage and protecting plants from foliage-feeding pests (Vaeck et al. 1987; Peferoen et al. 1989). The evolutionary implications of such manipulations on the development of resistance to *Bacillus thuringiensis* need to be considered before widespread use of such cultivars occurs (Gould 1988).

EVALUATION

Because pest control products based on microbial pathogens or nematodes are applied to specific locations at particular times, measuring their effects is like evaluating the efficacy of chemical pesticides in many regards (Fig. 11.8 through 11.12). Differences, however, include the ability of some pathogens to reproduce in the field for additional generations after the initial application and greater variability in efficacy due to sensitivity to abiotic and biotic conditions. Evaluations of microbial pathogens and nematodes address: differences among agents; differences of formulations and application methods; effects of environmental factors on product effectiveness; and persistence of an agent's effect, resulting from agent reproduction. Longer term, use of microbial pathogens or nematodes may change the dynamics of the pest management system in the crop, increasing the potential for additional biological control in the system. Product evaluation should incorporate assessment of such changes as well as the immediate consequences of the product's use.

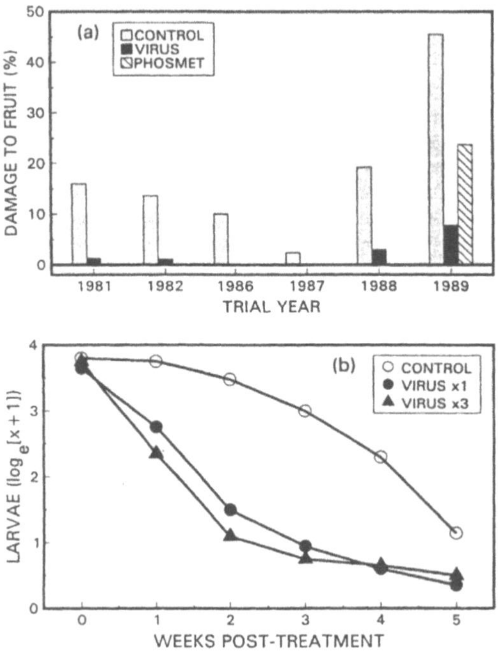

Figure 11.8. (a) Effect of granulosis virus on apple infestation by *Cydia pomonella* (Linnaeus) (after Jaques 1990); (b) effect of granulosis virus on populations of the cabbage pest *Pieris rapae* (Linnaeus) (after Tatchell and Payne 1984).

Comparisons among Agents

When a number of biological control agents are available that will attack a target pest of interest in the laboratory, tests will be needed to identify which might be best suited to control the pest under field conditions. Many heterorhabditid and steinernematid nematodes, for example, are broad enough in their host ranges that for pests in favorable habitats, such as soil, it would be reasonable to include several different nematode species or strains in field trials. Capinera et al. (1988) compared three species of nematodes for control of *Agrotis ipsilon* (Hufnagel) and

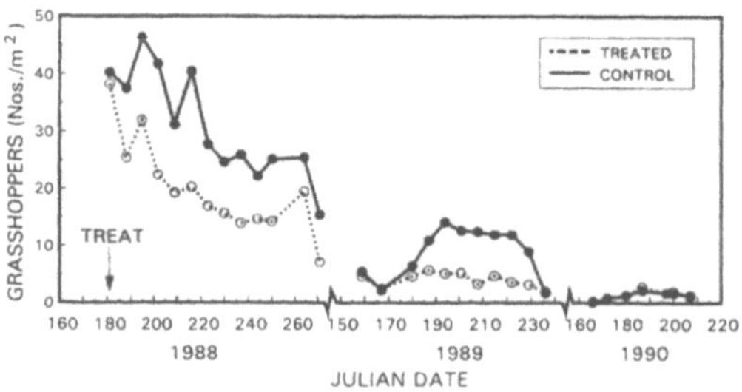

Figure 11.9. Population densities of several grasshopper species in shortgrass prarie following single treatment with bran bait containing *Nosema locustae* Canning (after Bomar et al. 1993).

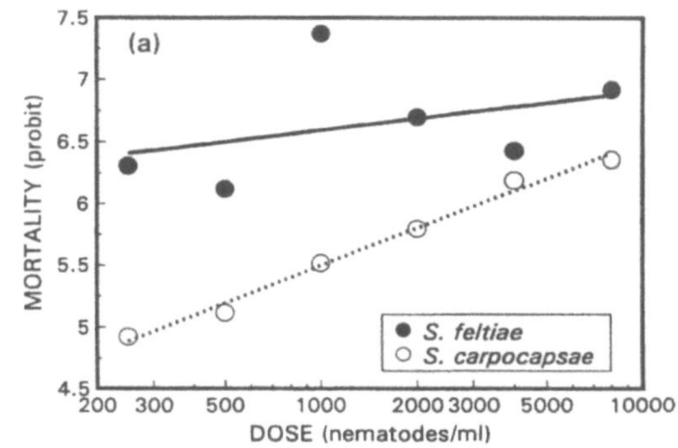

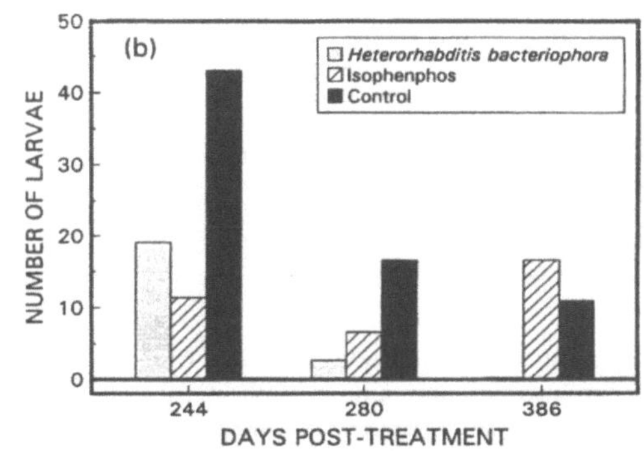

Figure 11.10. (a) Relationship between dose and mortality for the nematodes *Steinernema feltiae* (Filipjev) (= *S. bibionis*) and *Steinernema carpocapsae* (Weiser) when used in treatments against the currant borer *Synanthedon tipulioformis* (Clerck) (after Bedding and Miller 1981); (b) Number of live Japanese beetle (*Popilia japonica* Newman) larvae in control plots and plots treated with isophenphos or the nematode *Heterorhabditis bacteriophora* Poinar (after Klein and Georgis 1992).

Wright et al. (1988) compared four species for control of the turf pests *Popillia japonica* (Japanese beetle) and *Rhizotrogus majalis* (Razoumowsky) (European chaffer).

Comparisons of Variations in Formulation and Application Methods

To take best advantage of whatever biological possibilities a species of nematode or pathogen may have, field research is needed to identify the most effective application rates, formulation, manner and timing of application, and seasonal pattern of applications. Because pathogens are relatively expensive, optimizing these variables is crucial in reducing the amount of inoculum required and thus reducing cost. Such trials are also valuable in determining what formulations and application methods perform consistently so that growers can use them with confidence, thus encouraging their adoption. Wright et al. (1988), for example, in their tests of nematode species considered rates of nematodes spanning an eightfold range. Capinera et al. (1988) compared three methods of delivery of nematodes for cutworm (Lepidoptera: Noctuidae) control: calcium alginate capsules, wheat-bran baits, and aqueous suspensions. While baits might be supposed to be a more efficient method of delivery for nematodes (in view of the common use of baits as means to deliver chemical pesticides to this group of pests), no improvement was noted compared to water sprays. Kard et al. (1988) compared two methods

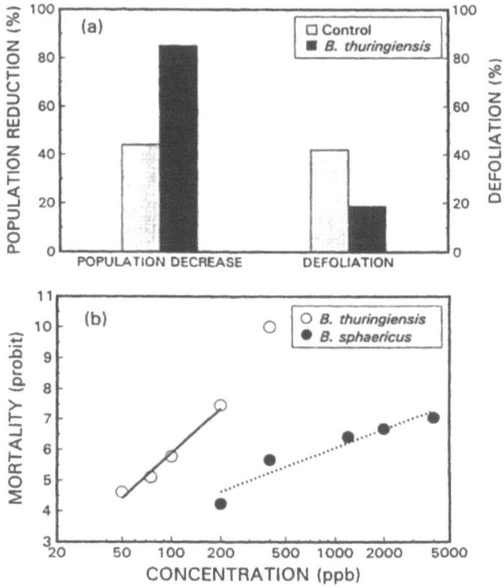

Figure 11.11. (a) Effect of *Bacillus thuringiensis* Berliner on population reduction of the spruce budworm, *Choristoneura fumiferana* (Clemens), and subsequent tree defoliation (after Cadogan and Scharbach 1993); (b) mortality of *Aedes stimulans* (Walker) larvae caused by *Bacillus thuringiensis* Berliner subsp. *israelensis* and *Bacillus sphaericus* Neide in a field test (after Wraight et al. 1982).

for application of nematodes to soil in Christmas tree plantations, but found no differences between injection and surface application.

A microbial pathogen or nematode must also be evaluated to determine the host stage most susceptible to the agent, methods to time applications, and to assess the number of applications needed to protect the crop for its entire period of risk from the pest. Bari and Kaya (1984), for example, found that older larvae of the artichoke (*Cynara scolymus* [Linnaeus]) pest *Platyptilia carduidactyla* (Riley) (artichoke plume moth) were more susceptible to *Steinernema carpocapsae* than were first and second instar larvae. Webb and Shelton (1990) found that late second and early third instar larvae of the cabbage pest *Pieris rapae* were more susceptible to granulosis virus infection than were either first, fourth, or fifth instar larvae. Microbial products,

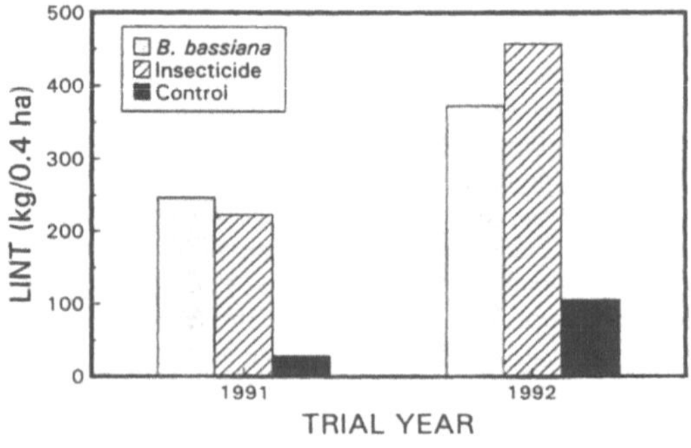

Figure 11.12. Cotton lint yield in trials of the fungus *Beauveria bassiana* (Balsamo) Vuillemin against boll weevil, *Anthonomus grandis* Boheman (after Wright 1993).

because they are often effective only against specific life stages and because they may have short residual times in the field, must be carefully timed. *Pieris rapae*, in tightly synchronized populations, was well controlled with a single granulosis virus application applied against first instar larvae, and under these conditions one treatment was as effective as three weekly applications (Tatchell and Payne 1984). However, if the field population was not tightly synchronized, so that oviposition occurred over an extended period, then multiple applications gave better control than a single application. Monitoring becomes increasingly important when microbial pesticides are used. Pheromone traps, for example, were used in Kenya to time applications of *B. thuringiensis* to control neonate larvae of *Spodoptera exempta* (Walker) (Broza et al. 1991).

Effects of Environmental Factors

Environmental factors can strongly influence microbial agents and nematodes by affecting the degree of contact between pest and agent achieved by the initial application (coverage), by reducing the life span of the microbe or nematode at the application site (survival), or by reducing the rate at which pathogens that contact hosts successfully initiate infections (infectivity).

Coverage. Degree of contact between nematodes and the target pest can be influenced by the nature of the soil and plant cover at the application site. Thatch, for example, reduces penetration of nematodes applied as water applications onto turf (Georgis 1990). Because nematodes must reach the root zone to contact hosts in the soil, thicker thatch, which reduces the numbers of nematodes reaching the root zone, reduces effectiveness. Penetration of thatch may be improved by applying greater volumes of water during or after the application. Nematodes are also strongly affected by the physical characteristics of the soil. They perform best in sandy loam soils that retain water and which have adequate spaces between soil particles to permit movement by nematodes (Kaya 1990). For nematodes to be used reliably, effects of these factors must be tested and product labeling and use recommendations written to list appropriate application sites, rates, amounts of irrigation, or other factors as needed.

Foliar applications of microbial products directed against such pests as whiteflies, aphids, and some lepidopteran larvae such as *Plutella xylostella* (Linnaeus) must adequately cover the undersides of leaves to provide effective control. Dense canopies or hairy leaf surfaces may reduce deposition rates on undersurfaces of leaves and thus reduce effectiveness in some crops.

Survival. Many microbial agents are killed by exposure to ultraviolet light or other environmental conditions such as excessive dryness. The granulosis virus of *Pieris rapae* was more than 67% deactivated in one day in trials on cabbage in the United Kingdom (Tatchell and Payne 1984). Nematode survival is affected by moisture and temperature, among other factors (Kaya 1990). Efficacy trials must attempt to assess the sensitivity of any given species of pathogen and its various formulations to the normal range of abiotic factors likely to be encountered at typical application sites, given the uses recommended by the product's label.

Infectivity. Nematodes that survive and contact a host are typically able to attack the host without being strongly influenced by the environment. Similarly, microbes such as bacteria, viruses, and microsporidia that infect the host after being ingested by it are relatively unaffected by the external environment. Some fungi, however, depend on exposure to free water in the form of dew, or very high humidities, for critical periods after contacting the host in order for spores to germinate and penetrate the host's integument (Connick et al. 1990). The effect of this environmental limitation can be reduced by modifications of product formulation, such as the use of invert emulsions or oils as diluents (see section on Formulation and Application). Product evaluation, however, would require that such formulations be fully tested under field conditions under the full range of normal moisture levels in the crops and countries for which the product is being sold.

Persistence of Agent Impact Due to Agent Reproduction

Discussion of factors affecting the evaluation of nematodes and microbial pathogens has been based on the view that control will result solely from the nematodes or microbes actually applied. In some cases, however, pathogens or nematodes may successfully reproduce at the field application site causing subsequent cycles of attack over some period of time (termed "recycling of the pathogen"). For example, Allard et al. (1990) found that levels of infection by the fungus *Metarhizium anisopliae* in the sugarcane froghopper, *Aeneolamia varia* Fabricius var. *saccharina*, remained higher in treated plots than in control plots for up to six months after a single application. In sugarcane in Australia, a single application of the same fungus provided commercial levels of control of the pest *Antitrogus* sp. for more than 30 months (Samuels et al. 1990). Young and Yearian (1989) showed that following death of larvae of *Anticarsia gemmatalis* from nuclear polyhedrosis virus, viral particles persisted on soybean plants for several weeks at levels effective in initiating epizootics. The increase in persistence of virus in this test, as compared to short residuals often observed from water applications of viral particles to tops of plants, was attributed to protection from ultraviolet light provided by the host cadaver and later to position on the plant, often lower and in cracks or crevices protected from direct light. Jackson and Wouts (1987) found that the degree of control of the grass grub *Costelytra zealandica* provided by applications of the nematode *Heterorhabditis* sp. in New Zealand increased from 9 to 56% over an eighteen-month period, indicating an increase in the numbers of nematodes at the site over time through reproduction. Kaya (1990) discusses nematode recycling and suggests that in cases in which hosts are present throughout the year, are moderately susceptible to nematodes, occur on a crop which can tolerate a significant pest load, and occur at sites where the soil characteristics are favorable to nematodes, nematodes could be used in an inoculative rather than an inundative manner.

While manufacturers of nematode and microbial pathogen products of necessity must focus on measurement of the immediate impact of use of their materials, evaluations should also include study of longer term effects. Persistence, where it occurs at high levels, will affect both the economics of the use of the product (by reducing frequency of application) and the product's role in the pest management system, because long-term conservation of the agent may be possible in the crop. An economic analysis of the potential to control damage to pastures in Tasmania caused by *Adoryphorus couloni* (Burmeister) showed recycling (with control from single treatments persisting 5–10 years) to be a critical factor in making the use of the fungus economical in view of the annual costs of renovation of damaged pastures, and in view of the competing cost of chemical control (Rath et al. 1990). Persistence of pathogens and nematodes may also have implications for protection of nontarget invertebrates, especially if the pathogens are not native to the country where they are applied (see Safety in this chapter).

Changes to the Pest Management Dynamics of the Crop System

Using arthropod pathogens or nematodes in place of chemical insecticides for key pests in a crop can significantly increase the possibility of conserving native parasitoids and predators of other pests in a crop and the ability to import and establish new beneficial species. Evaluation of augmentative uses of nematodes and pathogens should, therefore, include an assessment of how the overall dynamics of the pest management system for a crop will be changed.

In potatoes in Massachusetts, for example, substituting *B. thuringienis* subsp. *tenebrionis* to control *Leptinotarsa decemlineata* in place of broad spectrum contact insecticides greatly expanded the possibility of conserving natural enemies of pest arthropods in the crop. The use of this microbial product, which affects no other insects in the crop except the pest, conserved the beneficial coccinellid *Coleomegilla maculata*, a predator of aphids and of the eggs of *L. decemlineata*, and several tachinid parasitoids of the pest in the genus *Myiopharus*. In addition, it made possible the conservation of existing braconid parasitoids of the aphids in the crop and the opened the possibility of importation of new species of aphid parasitoids. In Europe, use of granulosis virus to control the apple pest *Cydia pomonella* reduced outbreaks of European red mite, *Panonychus ulmi*, by better conserving predacious mites (Huber 1986). The effect of such substitutions on the dynamics of a pest management system, however, depends on the structure of the entire pest and natural enemy complex in the crop. The substitution of steinernematid nematodes for chemical pesticides to control the black vine weevil, *Otiorhynchus sulcatus* (Fabricius), in Massachusetts cranberries had little effect on the control of other pests in the system, because applications for the weevil were not suppressing important natural enemies of other pests. Therefore, control of most other pests in the system remained based on the use of chemicals (Shanks and Agudelo-Silva 1990).

PRODUCT DEVELOPMENT

A variety of factors affect whether or not an agent that is biologically successful can also be commercially successful. Market potential (potential for a producer to make a profit) determines the amount of research on the commercial development of a given microbial agent that is likely to occur (Falcon 1985). Many factors affect the market potential of agents over and above the degree to which they are biologically effective. These include the volume of potential sales of the product and the likelihood that growers will choose the product instead of chemical pesticides that may already be in use. In addition, legal factors affecting use, such as requirements for product registration and availability of patent protection for new products, also affect the economic viability of developing an agent for commercial use. The influence of such forces on product development is illustrated by Huber (1990) who recounts the twists and turns between the 1963 discovery of the granulosis virus of the codling moth, *Cydia pomonella*, in Mexico, and the marketing of it decades later in Germany as Granupomt. Cunningham (1988) lists 19 species of target pests against which baculoviruses have been produced for use as microbial insecticides.

In some locations, local production of microbial pesticides is likely to increase their use, compared to imported products, by reducing costs and need for foreign currency (Bhumiratana 1990). Implementing a program to rear and use the *Anticarsia gemmatalis* virus in Brazil to control soybean pests has increased the area treated with this virus from 2000 ha in 1982–1983 to over 1,000,000 ha in 1989–1990 (Moscardi 1990).

Size of the Market

Perhaps the biggest factor influencing investment in development of microbial pathogens or nematodes as pest control products is the potential volume of sales, given that an effective product can be produced and that it is widely adopted. Products with narrow activity ranges must be directed at important pests and be relatively inexpensive to produce. The manufacturer of an effective mycoherbicide (DeVine®) ceased production because the market was too specialized to be profitable (Heiny and Templeton 1993). (For more on the economics of mycoherbicides, see Heiny and Templeton [1993].) In contrast, *Bacillus thuringiensis* subsp. *tenebrionis*, directed against the potato pest *Leptinotarsa decemlineata*, has been commercially successful because the pest was of considerable importance on a widely grown crop and because competing control options (chemical pesticides) were failing rapidly in many areas due to resistance. An opening into the market existed because growers were ready to try something new when existing methods failed. In contrast, *Bacillus popilliae*, marketed for control of the turf pest *Popillia japonica*, was commercially unsuccessful because it was relatively expensive (having to be reared in live insects), worked well only under certain, poorly defined soil and temperature conditions, and was in competition with a variety of pesticides that still provided effective control of the pest. Currently in North America, there is commercial interest in pathogens for control of domestic cockroaches because the economic value of this market is very large.

For highly specific agents, such as viruses, commercial development is likely only for key pests of crops grown in large acreage such as cotton, maize, and soybeans (Huber 1986). Development of viruses for smaller markets, such as codling moth (*Cydia pomonella*) on apples, will require a reduction in the cost of producing and registering viruses. Products for specialty crops grown on limited numbers of ha have little or no chance for commercial development unless a product exists which is already being produced for another, larger market. The use of *Bacillus thuringiensis* subsp. *israelensis* for control of flies in mushroom houses and sewage plants, for example, is feasible only because this agent is already being produced for mosquito control, a much larger market. Products for public sector uses, such as for the control of defoliators of public forests or grasshoppers on public grazing lands, may be feasible if public funds can be used to support the development, registration, and perhaps production of the product as, for example, the viral product Gyp-Check® developed by the U.S. Forest Service to control the gypsy moth, *Lymantria dispar* (Podgwaite and Mazzone 1981) and baits containing *Nosema locustae* to control grasshoppers in the North American prairie region. Development of broad spectrum pathogens is feasible to some degree by genetic modification of pathogens (Gelernter 1992), but if carried too far would eliminate one of the major ecological advantages offered by use of microbial pesticides (their relative selectivity) and could raise important concerns about safety to nontarget invertebrates. Some nematodes are relatively broad in their host range and one or a few species may be produced for use against a variety of pests.

Competition with Pesticides

If chemical pesticides did not exist, development of microbial pesticides would be given a higher priority and would occur more rapidly. However, except for markets that are legally closed to pesticide competition (for example public forest lands in many provinces of Canada), microbial products must compete with existing chemical pesticides, with which growers may be more familiar. Opportunities for microbial pesticides to take over markets from competing chemical products exist if: chemical use is prohibited by government; chemicals in existence fail to provide control due to resistance; a microbial pesticide is highly effective and cheaper than existing chemical pesticides; or pesticide-caused problems, such as secondary pest outbreaks, become severe and are correctly linked to pesticide use in the minds of growers, who then are directed to use microbial products that do not cause secondary pest outbreaks. The out-of-crop environmental harm sometimes caused by chemical pesticides (bird kills, fish kills, water contamination) is likely to slowly create societal interest in alternative pest control methods. This interest, combined with government restrictions on pesticide use that make pesticides more expensive (increased registration costs, taxes on pesticides, applicator insurance) or less convenient to use (regulations requiring posting of treated areas, reentry times for treated areas, applicator training, pesticide application record keeping), in the long term may create demand for efficacious, nonchemical products. Conversion of such generalized interest in alternatives into actual new microbial products, however, is an uncertain process.

Issues that strongly affect growers' views of the relative merits of chemical versus microbial pesticides are level of pest control achieved, speed of control, and predictability of control after application. Chemicals are believed to be, and often are, able to act rapidly, kill a high percentage of the pests, and do so reliably every time they are used. In practice, of course, chemicals vary in these characteristics and often are not as effective or reliable as commonly believed. Microbial pesticides are believed to act slowly, be less effective, be somewhat erratic in performance, and be strongly influenced by circumstances that vary from application to application. Development of microbial pesticides should attempt to address these problems in several ways. First, the variability of microbial pesticide performance should be reduced by developing, through research, a detailed understanding of the factors that affect efficacy and then adjusting either the formulation or the directions for use to eliminate variation in results as completely as possible. Second, extension agents must educate growers to understand that neither extremely high levels of kill nor rapid kill are truly necessary for effective pest control. Extension agents should stress, rather, the importance of rapid cessation of pest feeding, long-term reduction in pest reproduction rates, sustained mortality at more moderate levels, and leaving some pests to be killed by parasitoids and predators so as to sustain the populations of these beneficial organisms in the crop.

Another issue affecting the adoption of formulated pathogens as pesticides is the occurrence of multiple pests in a crop in cases where only some of the pests may be controlled by a pathogen but many of the pests are susceptible to a single pesticide. While pathogens and chemical pesticides are often compatible, the use of two products (a pathogen and a chemical) when one (the chemical) will satisfactorily accomplish the task is unlikely to be adopted.

Legal Factors

Legal factors can also affect the feasibility of commercial success of a microbial pesticide or nematode product. These factors include the requirements and costs of registering pest control products with governments, the availability of patent protection, and taxes on or subsidies for pesticides.

Chemical pesticides are subject, in many countries, to government requirements for registration as a condition of sale. Such registration typically includes the obligation of the company selling the product to submit a body of scientific evidence on the identity, efficacy, safety, and environmental fate of the product. In the United States and many other countries, microbial pesticides, but not nematode products (Hominick and Reid 1990), are also subject to product registration. The exact body of information that must be submitted is, however, different and in general less costly to produce (Environmental Protection Agency 1983; Aizawa 1990; Betz et al. 1990; Kandybin and Smirnov 1990; Quinlan 1990). Podgwaite and Mazzone (1981) discuss the process required for the registration of the virus of *Lymantria dispar* as a pest control product in the United States.

Chemicals, when registered for use as pesticides, are able to obtain patent protection in many countries, which prohibits other companies from selling, for a set number of years, the same product in competition with the company that did the developmental work. Microbial pathogens when developed as pesticides are also able to seek patent protection. Patent protection has generally not been available in the case of multicellular pathogens such as fungi and nematodes. However, patents have been granted covering technology for rearing, formulating, or applying such organisms. In addition, in a few cases patents have been granted for novel use patterns, such as the use of the nematode *Steinernema scapterisci* for control of mole crickets.

Taxes and subsidies, by affecting directly the costs of pest control products, can significantly affect the balance between microbial and chemical pesticides. In some countries, for example, pesticides may be taxed. In such countries, if microbial pesticides are exempt from such taxes, their competitive standing against chemical pesticides improves. Similarly, in some countries, governments subsidize pesticides to make pesticides more affordable to farmers. If such subsidies were to be allowed for microbial pesticides, but denied for chemical pesticides, this would encourage the use of microbial pesticides. For more on such issues, see Chapter 20 on the effects of government policies on biological control.

SAFETY

Safety of microbes used as insecticides has been covered in depth by Laird et al. (1990), including registration requirements in various parts of the world, safety of bacteria, viruses, nematodes, and fungi to man and other vertebrates and to nontarget invertebrates. The environmental risks of genetically engineered agents are considered as well. Most microbes and nematodes used as biological control agents occur naturally in many environments, often in large quantities when epizootics occur. The general absence in the medical literature of cases of these agents infecting man is important evidence that these agents do not pose a significant health risk. Further evidence on safety is gathered on each specific microbial agent in the course of its commercial development. For safety testing procedures for microbial agents see the following reviews for various taxa: plants (Campbell and Sands 1992), fish and crustaceans (Spacie 1992), birds (Kerwin 1992), mammals (Siegel and Shadduck 1992), and nontarget insects and acari (Fisher and Briggs 1992).

Safety of Nematodes

Commercial production and application of nematodes for pest control is viewed by government regulators in most countries (except France and Japan) as being equivalent to the augmentative use of parasitoids or predacious arthropods. Nematodes are considered safe to man and other vertebrates. Rats exposed by mouth or intraperitoneal injection to *Steinernema carpocapsae* showed no signs of pathogenicity, toxicity, or infection (Gaugler and Boush 1979). Effects of nematodes on nontarget invertebrates have been reviewed by Akhurst (1990). Based on data from laboratory tests, species of nematodes in the families Steinernematidae and Heterorhabditidae appear to have broad host ranges within the Insecta. However, in such tests high dosages of nematodes are presented with candidate hosts under unnatural physical conditions that are moist and otherwise favorable to the nematodes, which is likely to increase the host range beyond that typically seen in nature. Field data suggest that risks to nontarget species from nematode applications are low because nematodes have limited motility and are restricted to specific environments due to intolerance of dryness and other unfavorable physical conditions (Georgis et al. 1991). *Steinernema carpocapsae*, for example, has been shown to have no effect on intact earthworms (*Aporrectodea* sp.) (Capinera et al. 1982). Georgis et al. (1991) did not observe any harm to nontarget soil arthropods in golf course turf, maize or cabbage fields, or cranberry (*Vaccinium macrocarpon* Aiton) bogs from applications of steinernematid or heterorhabditid nematodes. Nematodes do kill immature stages of parasitic wasps if parasitized hosts ingest nematodes (Kaya and Hotchkin 1981; Akhurst 1990). Jansson (1993) provides a broad discussion of potential effects of non-native nematode species used for augmentative biological control.

Safety of Microbial Pathogens

In the United States, Europe, Russia, and Japan, pest control preparations that are sold commercially and which contain pathogens or other living microbes are treated as pesticides. These products must be registered with the appropriate government agency that regulates chemical pesticides and must demonstrate their safety to that agency before being offered for sale. This process generates information to demonstrate that a microbial product, as actually manufactured and offered for sale, is safe for use as recommended on the label. The information required for registration of microbial products, due to the nature of the products, differs in some regards from the information required for the registration of chemical pesticides. At a minimum, data are needed to: define the active agent taxonomically; define the methods of culture of the active agent; demonstrate that the commercial product is free from contamination by other, potentially dangerous, microbes; and demonstrate that the pathogen is not infectious in man or domestic animals. In addition, studies on the fate of the pathogen in the environment or of its effect on nontarget organisms may be needed as, for example, the assessment of the effect of *Bacillus thuringiensis* subsp. *israelenis* on non-target aquatic organisms (Merritt et al. 1989; Welton and Ladle 1993). Countries with industries based on culture of invertebrates, such as Japan's silkworm industry, may require that preparations such as *Bacillus thuringiensis* do not contain live spores, but only pathogen-derived toxins (Aizawa 1990).

In countries which do not treat microbial pesticides as products requiring governmental registration, local systems for pathogen production may be developed (Metcalfe 1991; Antía-Londoño et al. 1992). Local pathogen production systems, at the village or farm level, or by national in-country producers, should be monitored by government health agencies to ensure that systems, as operated, produce high quality preparations of the intended pathogen, free of other microbial agents.

Requirements for registration of microbial pesticides have been summarized for the United States (Environmental Protection Agency 1983; Betz et al. 1990), western Europe (Quinlan 1990), eastern Europe and Russia (Kandybin and Smirnov 1990), and Japan (Aizawa 1990). While each country's requirements differ somewhat, the broad theme is to treat microbial pesticides under the same laws as pertain to chemical pesticides and to vary the data requirements to allow for differences between chemicals and infectious agents. The United Kingdom and Germany have more extensive regulations; other European countries less so. Only Denmark exempts infectious agents from regulation as pesticides (defining them as biological control agents).

The rationale used to develop regulations in the United States is discussed by Rogoff (1982). Currently some 45 products, representing 26 species of microbes, have been registered for pest control in the United States. The majority of these (18) are targeted against insects, and two are targeted against weeds (Table 11.1; others are targeted against plant pathogens, Table 12.4). The system employed by the U.S. Environmental Protection Agency is to organize testing require-

TABLE 11.1 Microbial Pesticides Registered in the United States Against Insects and Weeds (as of 1993)

Species of Microbe	Pest Controlled
Bacteria	
1. *Bacillus popilliae* Dutky 1 *Bacillus lentimorbus* Dutky	Japanese beetle larvae
2. *Bacillus thuringiensis* Berliner subsp. *kurstaki*	Lepidopteran larvae
3. *Bacillus thuringiensis* subsp. *israelensis*	Dipteran larvae
4. *B. thuringiensis* var. *san diego*	Coleopteran larvae
5. *B. thuringiensis* subsp. *tenebrionis*	Coleopteran larvae
6. *B. thuringiensis* subsp. *kurstaki* strain EG2348	Lepidopteran larvae
7. *B. thuringiensis* subsp. *kurstaki* strain EG2424	Lepidopteran/Coleopteran larvae
8. *B. thuringiensis* subsp. *kurstaki* strain EG2371	Lepidopteran larvae
9. *Bacillus sphaericus* Neide	Dipteran larvae
10. *B. thuringiensis* subsp. *aizawai* strain GC-91	Lepidopteran larvae
11. *B. thuringiensis* subsp. *aizawai*	Lepidopteran larvae
Fungi	
12. *Phytophthora citrophthora* (Smith and Smith) Leonian, Strangler vine race	Citrus strangler vine
13. *Colletotrichum gloeosporioides* (Penzig) and Saccardo in Penzig f. sp. *aeschynomene* ATCC 20358	Northern joint vetch
14. *Lagenidium giganteum* Couch	Mosquito larvae
15. *Metarhizium anisopliae* (Metchnikoff) Sorokin strain ESF1	Cockroach and fly control
Protozoa	
16. *Nosema locustae* Canning	Grasshoppers
Viruses	
17. Polyhedral inclusion bodies of *Heliothis* nucleopolyhedrosis virus (NPV)	Cotton bollworm, budworm
18. Polyhedral inclusion bodies of Douglas-fir tussock moth NPV	Douglas-fir tussock moth larvae
19. Polyhedral inclusion bodies of gypsy moth NPV	Gypsy moth larvae
20. Polyhedral inclusion bodies of pine sawfly NPV	Pine sawfly larvae

ments into tiers, with all agents being to subjected to Tier I tests, and only those agents which give some indications of survival, replication, infectivity, toxicity, or persistence in Tier I tests being required to undergo Tier II or III tests. Tier I tests include a series of acute tests in various laboratory animals by oral, dermal, and inhalation routes. In addition, intravenous infectivity tests are required for bacteria and viruses and an intracerebral test for viruses and protozoa. Intraperitoneal infectivity tests are required for fungi and protozoa. Primary dermal and eye tests are required for all agents. Also, agents must be tested for ability to cause hypersensitivity reactions and immune responses. Tissue-culture tests for infectivity to mammalian cells are required for viruses. Nontarget tests are done for birds, wild mammals, plants, insects, and honey bees. All Tier I tests must be completed before permits are issued for outdoor tests of the agent (Podgwaite 1986). In general, microbial agents studied so far pose no threat to man or vertebrates, with the exception of occasional allergenic responses and, for some fungi, infections. Invertebrates may be killed by some microbial products under some circumstances. More detailed comments are summarized below.

Safety of Bacteria. *Bacillus thuringiensis* strains that produce beta-endotoxin are variably pathogenic in mice and chickens, but all strains currently used for pest control are deficient in ability to produce this particular toxin (Podgwaite 1986). Strains in commercial use do not infect man or other vertebrates under normal circumstances. Two instances have been reported of apparent *B. thuringiensis* infection in man, one finger infection (caused by a needle injury to a laboratory worker manipulating *B. thuringiensis* cultures) and one eye infection. In both cases, infections were not serious and responded to antibiotics. Some possibility exists that these cases, particularly that of the eye infection, represented persistence of the bacterium, but not infection and that another unidentified infectious agent was present and was the actual cause of the infection. This is possible because the diagnosis was based only on occurrence of the *B. thuringiensis* bacterium in the eye, not demonstration of infection. Persistence of *B. thuringiensis* cells for a few weeks in exposed tissues is known to occur (Siegel and Shadduck 1990a).

Laboratory tests on rabbits and other vertebrates for *Bacillus sphaericus* and *Bacillus thuringiensis* subsp. *israelensis* (Shadduck et al. 1980; Seigel and Shadduck 1990a) and for *Clostridium bifermentans* Weinberg and Séguin serovar *malaysia* (Thiery et al. 1992) all indicated that, while these bacteria may persist for varying lengths of time in exposed vertebrate tissues, they do not cause any pathogenic effects and are not harmful. The literature on *B. sphaericus* and *B. thuringiensis* are reviewed in detail by Siegel and Shadduck (1990b,c). Based on these and other tests, these bacteria have been recognized as safe for use as pest control agents in circumstances involving human exposure. However, some concern has been raised about potential *Bacillus thuringiensis* contamination of drinking water supplies in Germany because *B. thuringiensis* is closely related to *Bacillus cereus* Frankland and Frankland, a bacterium implicated in food poisoning in humans (Helmuth 1988). Epidemiological studies of human populations exposed to pathogens used in pest control operations can be conducted to help assess whether or not human populations are at risk, as was done, for example in Oregon (U.S.A.) for rural populations exposed to *B. thuringiensis* subsp. *kurstaki* used in attempts to eradicate *Lymantria dispar* from forested areas (Green et al. 1990).

Nontarget invertebrates may or may not be at risk from these bacteria. Neither *B. sphaericus* nor *B. thuringiensis* affects honey bees under field conditions (Vandenberg 1990). Other invertebrates, such as the silkworm, *Bombyx mori*, nontarget forest Lepidoptera, and immature parasitoids inside target pests, are likely to be killed if exposed to *B. thuringiensis* (Podgwaite 1986). Data on such effects are summarized by Flexner et al. (1986). Concern over potential harm to beneficial insects such as silkworms has led some countries, such as India, to prohibit the use of *B. thuringiensis* products, a position that might now logically be revised, given the diversity of available *B. thuringiensis* strains, some of which do not affect Lepidoptera (Padidam 1991). Miller (1990) assessed the effect of *B. thuringiensis* subsp. *kurstaki* applications on nontarget forest Lepidoptera in Oregon; some species found in control areas were absent from treated areas, but the degree of impact was viewed as lower than that which would have occurred with chemical pesticide applications.

B. thuringiensis subsp. *israelensis*, when applied to aquatic systems, shows little effect on nontarget invertebrates, with the exceptions of those in the families Chironomidae, Dixidae, and Certopogonidae, which may be moderately to severely affected (Flexner et al. 1986). Merritt et al. (1989) provide an example of an evaluation of application of *B. thuringiensis* subsp. *israelensis* to river systems for control of simuliid larvae. Zgomba et al. (1986) recorded some effect of this bacterium on Ephemeroptera and Odonata, but much less than that caused by applications of chemical pesticides in the same circumstances.

Safety of Viruses. The majority of viruses which have been considered for use in pest control are either nuclear polyhedrosis or granulosis viruses, both of which are in the family Baculoviridae, whose members are only known to infect arthropods. Several nuclear polyhedrosis viruses have been extensively tested in over 24 vertebrate species, including a range of mammalian, avian, and fish species to determine if they have the ability to infect vertebrate hosts. None has shown any ability to infect vertebrates (Burges et al. 1980a; Podgwaite 1986). Granulosis viruses have been less extensively tested, but available data suggest that this group, which is only known to infect Lepidoptera, is unable to infect vertebrates. Other families of viruses such as the Parvoviridae, which are not currently being considered for use as pest control agents, are capable of infecting vertebrates and potentially pose risks to man or domestic animals. Saik et al. (1990) summarizes information on non-Baculoviridae effects on domestic animals.

Guidelines for safety tests for baculoviruses of Lepidoptera and sawflies (Hymenoptera) are given by Burges et al. (1980b). Required information includes:

(1) identity and information on the formulated product (name, formulation, composition statement)
(2) identity of agent (name of virus, diagnostic tests for agent, list of known impurities)
(3) biological properties of the active agent (host spectrum, natural geographic occurrence, natural infectivity, stability)
(4) manufacture, formulation, and quality control (production methods, assays to ensure standardization, tests to show freedom from specific human pathogens, assays for general microbial contaminant level)
(5) application (pests controlled, crops to be treated, rates, numbers and timing of applications, method of application, use pattern if nonagricultural)
(6) experimental data on efficacy (laboratory and field tests on efficacy and resistance)
(7) residues (identification method for virus residues, quantification of residues on crop at harvest, residues in the environment)
(8) infectivity and toxicity in mammals (single dose, nonocular by various routes; ocular, single dose; repeated dermal and respiratory exposure to test allergenicity; cell culture studies to test for infectivity to mammalian cells; carcinogenicity and teratogenicity tests)
(9) effects on humans (in manufacturing facilities, applicators)
(10) information on environmental and wildlife hazards (honey bees, parasitoids and predators of target pest, earthworms, two fish, two birds).

Viruses tested so far do not directly infect or harm adult parasitoids of pest species because viruses are infective only if ingested. Immature parasitoids in infected hosts may die, but not as the result of virus infection, but rather of premature loss of the host (through its death from virus) or alteration of the quality of the host (Flexner et al. 1986).

Viruses of arthropods are typically limited in their host ranges such that only species in one genus or in related genera in one family are infected. Therefore, distantly related types of invertebrates are not at risk from virus applications (Podgwaite 1986). The nuclear polyhedrosis virus with the widest known host range is the *Autographa californica* virus, which infects up to 43 species of Lepidoptera (variation in count reflects different levels of proof as to the virus's identification). Cytoplasmic viruses, one of which is being considered for commercial development in Japan, have broader host ranges than most nuclear polyhedrosis viruses. The virus from the caterpillar of the lasiocampid moth *Dendrolimus spectabilis* Butler, for example, also infects caterpillars in several other genera of Lepidoptera (Podgwaite 1986).

Safety of Fungi. Of the various fungi that have been developed for commercial use as pest control agents, most have shown no infectivity to man or other vertebrates (Podgwaite 1986).

Negative findings were made for mice fed or exposed to *Nomuraea rileyi* (Ignoffo et al. 1979), for rats, rabbits, and guinea pigs fed or exposed to *Hirsutella thompsonii* (McCoy and Heimpel 1980), for mice injected with *Verticillium lecanii* (Podgwaite 1986), and for mice injected with *Lagenidium giganteum* (Kerwin et al. 1990).

Health concerns for man have been recognized for a few fungi of potential economic use. *Beauveria bassiana*, for example, has been reported to cause allergies in humans (York 1958) and is at least an opportunistic pathogen in man and other mammals (Burges 1981b). Two species of *Conidiobolus* in the Entomophthorales have been reported to be pathogenic in man (Wolf 1988).

Fungi infect and kill invertebrates which contact or ingest fungal spores. However, mortality of nontarget invertebrates from spore contact is typically less than 10% (Flexner et al. 1986). Higher mortality appears to occur if fungal spores are ingested. Larvae of *Cryptolaemus montrouzieri* suffered 50% mortality when fed Boverint, a commercial preparation of spores of *Beauveria bassiana*. Adult ladybird beetles, however, were not affected (Flexner et al. 1986). Honey bee workers experienced 29% mortality when fed spores of *Hirsutella thompsonii* (Cantwell and Lehnert 1979). Both *Beauveria bassiana* and *Metarhizium anisopliae* infect *Bombyx mori* and, also, have been associated with honey bee kills following field applications (Podgwaite 1986). Some evidence also exists suggesting that parasitized hosts of some species experience increased susceptibility to fungi and that populations of some overwintering carabids and other invertebrates, normally subject to mortality from fungi, may be at increased risk if large amounts of fungal inoculum are added to soils as a consequence of agricultural pest control (Flexner et al. 1986). Granular mycelial formulations of fungi appear relatively safe to nontarget organisms. Furthermore, what effects may occur are likely to be small compared to the relatively large reductions in populations of invertebrates currently caused in agricultural fields by chemical pesticides.

Safety of Microsporidia. Relatively few microsporidia have been studied for commercial development as pest control products. Consequently less information is available on this group of organisms. Studies on *Nosema locustae* showed that this pathogen did not irritate, replicate, or accumulate in rats, guinea pigs, rabbits, rainbow trout, or bluegill sunfish following administration by appropriate routes (Brooks 1988). Existing evidence suggests that *Nosema* and related entomopathogenic protozoa pose no risk to man (Siegel and Shadduck 1992) and little or no risk to other vertebrates (Saik et al. 1990a). Microsporidia are known to infect various invertebrates, reducing longevity and fecundity. Host ranges of individual species of microsporidia may span several orders of insects (Brooks 1988). When parasitoids oviposit in hosts with microsporidial infections, the resulting adult parasitoids reared from infected hosts carry microsporidial infections (Vinson 1990a). Such infections reduce longevity and lower fecundity. Microsporidia have been shown to adversely affect laboratory cultures of various insects, including parasitoids (Flexner et al. 1986). *Nosema locustae* has been shown not to infect honeybees (Podgwaite 1986). Studies of *Nosema furnacalis* Wenn showed that none of nine species of predacious arthropods fed infected prey developed active *N. furnacalis* infections (Oien and Ragsdale 1993). The parasitoid *Macrocentrus grandii* Goidanich, however, reared in infected hosts did become infected.

Genetically Altered Microbes. Genetic engineering and cell fusion methods are being used to alter microbial pathogens for improved characteristics for biological control. In the forefront of such activities are efforts to alter the virulence or host ranges of baculoviruses (Betz 1986; Wood and Granados 1991) and of such bacteria as *Bacillus thuringiensis* (Gelernter 1992). More rapid cessation of feeding by baculovirus-infected hosts has been achieved by incorporating genes in the *Autographa californica* nuclear polyhedrosis virus that code for production of an insect-specific neurotoxin derived from a scorpion (Stewart et al. 1991).

Products arising from new genetic combinations will require registration as microbial pesticides and new regulations may be needed to assess whether these new organisms pose risks to man or the environment. Because these modified organisms may differ very little from their unmodified sources, information available on spread and persistence of natural viruses or other microbes in the environment is useful to determine the likely persistence of such organisms. Fuxa (1989) has summarized existing information on microbial agents from this point of view. Persistence varies greatly depending on the microhabitat, ranging from mere days on surfaces, such as foliage, which are exposed to ultraviolet light, to years in soil.

General concepts stressed in evaluating the risk of genetically modified organisms are persistence, dispersal, host range, and recombination with other microbes (Maramorosch 1987; Cory and Entwistle 1990; Fuxa 1990b). One proposed principle is that genetically modified organisms should be judged based on what they are (the characteristics they have been altered to possess) rather than on the technology used to make the modification (e.g., genetic or classical breeding). Speculations about potential risks have included: that new gene combinations from different species or genera might result in totally unforeseen properties in the modified organism; that modified organisms might become pests by virtue of excessive abilities to survive and spread; and that genetically modified organisms might interact with other species in ways leading to the transfer of genetic material to these other species, with unknown consequences (Fuxa 1990b). Some authors have considered that infectious agents with overly broad invertebrate host ranges might put native invertebrate populations at some risk. Williamson (1991), for example, estimates that 5–10% of Britain's Lepidoptera would be susceptible to a strain of the *Autographa californica* virus which has been modified to increase its host range. He recommends further genetic modifications, such as removal of the polyhedral gene, to render the virus incapable of sustained persistence in the wild. Field trials of modified *Autographa californica* viruses indicate that removing genes for production of polyhedron protein production is successful in making this virus nonpersistent (Possee et al. 1990). Because such nonoccluded viruses are rapidly inactivated, efficacy under field conditions is reduced. Co-occlusion (in which modified viruses and wild type viruses are used to simultaenously infect hosts to produce viruses of both strains in shared occlusion bodies) has been proposed as a strategy for formulating such nonoccluded viruses to permit their effective use (Wood et al. 1994). Wood and Granados (1991) give an overview of the potential uses of genetically modified baculoviruses. Charudattan (1990) summarizes information concerning use and release of genetically modified fungal pathogens of plants and concludes that modified fungi tend to decline after mycoherbicidal use and appear safe.

CHAPTER 12

METHODS FOR BIOLOGICAL CONTROL OF PLANT PATHOGENS

INTRODUCTION

Organisms for biological control of plant disease can be used in various ways, but most attention has been given to their conservation and augmentation in a particular environment, rather than to the importation and addition of new species as is often done for insect or weed control. The choice of these approaches is in part because there is usually a diverse set of microbes already associated with plants. These microbes provide substantial opportunity for development of resident species as competitors or antagonists to pathogenic organisms.

Both conservation and augmentation have some application in each of the main groups of plant diseases. The use of microbes for control of plant pathogens is covered in more detail in several texts, including Cook and Baker (1983), Parker et al. (1983), Fokkema and van den Heuvel (1986), Lynch (1987), Campbell (1989), and Stirling (1991) and in other review articles (Wilson and Wisniewski 1989; Adams 1990; Jeffries and Jeger 1990; Sayre and Walter 1991; Andrews 1992; Cook 1993; Sutton and Peng 1993a).

The first section of this chapter briefly reviews how the paradigms of conservation and augmentation apply to biological control of plant pathogens. The next section describes the general environment in which biological control of plant pathogens takes place. The third section discusses the principal mechanisms by which antagonists of plant pathogens achieve biological control. The two subsequent sections deal with conservation and augmentation of antagonists, respectively. Both of these sections examine the uses of biological control against root, stem, leaf, flower and fruit diseases, and also discuss biological control of plant-parasitic nematodes. The final section deals with some technological issues facing the development of antagonists in agricultural and social systems.

PARADIGMS OF BIOLOGICAL CONTROL OF PLANT PATHOGENS

Biological control of plant pathogens through conservation is accomplished either by preserving existing microbes which attack or compete with pathogens or by enhancing conditions for their survival and reproduction at the expense of pathogenic organisms. Conservation is applicable in situations where microorganisms important in limiting disease-causing organisms already occur, primarily in the soil and plant residues but in some cases also on leaf surfaces. They may be conserved by avoiding practices which negatively affect them (such as soil treatments with fungicides). The soil environment may be enhanced for some beneficial organisms through adding organic matter (soil amendments).

Biological control of plant pathogens through augmentation is based on mass-culturing antagonistic species and adding them to the cropping system. In the context of the paradigms discussed in this text, this is augmentation of natural enemy populations, because the organisms used are usually present in the system, but at lower numbers or in locations different than desired. The purpose of augmentation is to increase the numbers or modify the distribution of the antagonists in the system. In some cases, such organisms are taken from one habitat (for example the soil) and augmented in another (for example the phyllosphere). The activity of augmenting microbial agents is sometimes termed "introduction" in the plant pathology literature, in the sense of "adding" them to the system (Andrews 1992; Cook 1993). The organisms introduced, however, are usually found in a local ecosystem and are not introduced from another region of the world (in the sense of Chapter 8).

Augmentation of antagonists falls naturally into two approaches. The first is direct augmentation, at potential infection sites or zones, with organisms antagonistic or parasitic to the pathogens themselves. In this approach, the antagonist population is directly responsible for disease suppression. A second approach is to inoculate plants with nonpathogenic organisms that prompt general plant defenses against infection by pathogens (induced resistance). Disease control is then achieved through greater plant resistance to infection.

Substantial work has been done in characterizing the role of microorganisms in biological control of plant diseases. The biological mechanisms underlying the success of these antagonists in such settings may include initial competition for occupancy of inoculation sites, competition for limiting nutrients or minerals, antibiotic production, and parasitism.

HABITAT CHARACTERISTICS

To understand the principles that apply to biological control of plant pathogens, we must first consider the ecology of the system at the level of the pathogens and the agents used to control them. Aerial plant surfaces, by and large, present hostile environments to colonizing microbes, in many cases consisting of surfaces protected by cuticular waxes, with very small amounts of nutrients available on these surfaces. Further, surfaces of the above-ground portions of plants may be dry. Consequently, pathogenic microbes attempting to colonize these surfaces may face a number of difficulties, including competition with other, nonpathogenic, microbes.

The rhizosphere (the roots and the region immediately adjacent) is a somewhat richer environment than the phyllosphere because of simple sugars, amino acids, and other materials exuded by the roots, but in the remainder of the soil the growth of microbes is often carbon-limited (Campbell 1989). Moisture in the rhizosphere may be more continuous in time and space than on the above-ground surfaces of plants (the phylloplane), but the rhizosphere may be subject to periodic drying.

Various forms of competition in these environments are important in the ability of any particular organism to increase in numbers and consequently reduce the numbers or activity of other organisms, including plant pathogens (Campbell 1989; Andrews 1992). Microbial competition can be important at two main stages of growth of pathogen populations. First, there may be competition during initial establishment on a fresh resource that was not previously colonized by microorganisms. Second, after initial establishment, there is further competition to secure enough of the limited resources present to permit survival and eventual reproduction. Microorganisms show many traits which may characterize them as particularly adept at either the colonization phase or subsequent phases of competition. Species referred to as *r*-strategists (ruderal species) have a high reproductive capacity. These species produce so many spores or reproductive bodies that there is a high likelihood that some will be found near any newly available resource. These species are effectively dispersed and establish readily in disturbed habitats or in the presence of noncolonized resources. They are found in disturbed settings where easily decomposable organic matter or root exudates are found, and where initial resource capture is crucial for survival. In contrast to these *r*-strategists, species found in more stable situations face competition for space and limited resources (Begon et al. 1986). These organisms, termed *K*-strategists, become more dominant as a community matures and becomes more crowded. These concepts form the endpoints of a continuum, and there are varying degrees of *r*- and *K*-related characteristics in different microbes in various habitats (see Andrews and Harris [1985] for further discussion on these concepts in microbial ecology).

Plant pathogens are spread across this *r-K* range of characteristics and vary in other important biological characteristics. There are opportunistic pathogens that are able to attack young, weakened, or predisposed plants, but may be poor competitors (*Botrytis, Pythium, Rhizoctonia*). There are pathogens that tolerate environmental stresses. These organisms often live in situations with few competitors, because few species are able to exist in such environments. Some pathogens, such as the *Penicillium* species that cause postharvest rots, produce antibiotics that inhibit competitors. Other species (such as *Fusarium culmorum* [Smith] Saccardo) have a very high competitive ability. It is important to understand the ecology of a target pathogen before one can effectively consider what biological control strategy might be most effective. Stress-tolerant and competitive species, for example, would require different biological control strategies and agents than would ruderal ones.

In the same way that antagonists of plant pathogens vary in *r-K* and other characteristics, the properties of an effective biological control agent will depend on the setting in which it is intended to function. In many agricultural settings, disturbance makes new resources available to microbes through crop residue burial, cultivation, or planting. A frequent need, therefore, is a control agent that has the characteristics of an *r*-strategist (Campbell 1989), which can grow quickly and colonize new resources rapidly, with minimal nutrient and environmental restrictions. It should function well in disturbed environments and have some means (such as spores) of surviving in the soil or on the plant near to the pathogen inoculum or the source or site of infection. Biological control agents that are *r*-strategists are an approximate equivalent of a protectant fungicide, being in place before the pathogen infection cycle can begin. In other programs, such as those directed against a pathogen which has already invaded the plant host, a more competitive species will be required. Finally, a biological control agent may have to be tolerant of abiotic stresses, particularly for use in dry climates or on leaves.

Before discussing some of the microbes and other agents used in biological control of plant diseases, we will discuss briefly some of the important ecological properties of plants and their environments. These properties affect the ways in which biological control is implemented, as well as what kind of properties might be important in an antagonist of a plant disease.

Soil and the Rhizosphere

Although there is much variation in soil types in different locations, soils are typically rich in microflora, with propagules numbering in the hundreds of thousands per gram of soil (Campbell 1989). In most soils, growth of microorganisms is carbon-limited, either because what carbon is available is not physically accessible or because the microbes do not possess the enzymes necessary to degrade the carbon-containing molecules that are present. An exception to this general limitation is the region immediately surrounding plant roots. This region, the rhizosphere, contains easily metabolized carbon and nitrogen sources such as amino acids, simple sugars, and other compounds exuded by the roots. Consequently, this region is more favorable than surrounding soil for the support of microflora. Root pathogens and plant-parasitic nematodes may be found growing on or in roots, but many microbes in the soil will be dormant because of resource limitations. Because there are many dormant organisms in the soil prepared to take advantage of any favorable period or opportunity, competition for resources in the soil may be significant and may limit the ability to augment beneficial organisms and have them flourish, unless soils are first sterilized to eliminate potential competitors. Therefore, much research surrounding biological control of root diseases and nematodes has centered around identifying soils which are naturally suppressive to particular disease organisms and investigating the microbial components of the soil responsible for the suppression. Management of such antagonistic organisms for biological control can range from treatment of soil to favor the desirable organisms (conservation) through inoculation of soils or plants with specific beneficial microorganisms (augmentation).

Phyllosphere

The phyllosphere is significantly different from the rhizosphere in its structure, ecology, nutrient availability, and exposure to climatic factors (Andrews 1992). Leaves are relatively hostile to microorganisms. They are generally hydrophobic and covered with cutin and wax, which limits the amount of exudate (and hence nutrients) that reaches the leaf surface. These and other factors impose severe environmental restrictions to microbial growth on leaf surfaces. Fungal pathogens of leaves often enter the leaf tissue very shortly after germination of the pathogen and, consequently, are protected inside the plant for much of their growth. Bacterial pathogens may multiply on the leaf surface before invading leaf tissues. Biological control of disease can take place either through general inhibition and competition on the leaf surface prior to invasion of leaf tissues or through suppression of the disease after the pathogen has invaded. Biological control within leaf tissues can occur through one of several mechanisms, including induced resistance in the plant and hyperparasitism of the pathogen.

Woody stems are habitats low in nutrients and often difficult for pathogens to penetrate. Because the wood itself supports very few saprotrophic microorganisms, pathogens colonizing the wood through wounds, dead branches, or roots find very few competitors. Because there are few organisms present to conserve, protection of the wood from these decay organisms can be achieved by protecting the relatively small, well-defined wound or branch stub through inoculation (augmentation) with specific microorganisms. These wounds are initially very low in sugars or other nonstructural carbohydrates, and antagonists such as *Trichoderma* spp. can successfully compete for these limited resources. Many of the organisms used in the biological control of stem diseases are employed by applying them directly to stem wounds, where they colonize resources and subsequently exclude pathogenic forms. This initial occupancy by antagonists subsequently limits infection by decay-causing organisms, and hence controls the succession of microorganisms in the wood. Of the successful, commercially-available biological control products for plant diseases, several are for diseases of woody stems (Campbell 1989).

MECHANISMS OF BIOLOGICAL CONTROL OF PLANT PATHOGENS

There are several different ways in which a microbial biological control agent can operate against a targeted plant pathogen (Elad 1986). Among these are competition, induction of plant defenses, and parasitism.

Some agents act through competition for limited resources, and through this competition the growth of the pathogen population is suppressed, reducing the incidence or severity of disease. One important component of competition can be competition for Fe^{31} ions. These ions are sequestered by chemicals called siderophores, which are produced by many species of plants and microbes. Highly efficient siderophores from nonpathogenic microbes can remove Fe^{31} ions from the soil, outcompeting siderophores from pathogens and thereby limiting the growth of pathogen populations. Some biological control agents compete through the production of antimicrobial substances such as antibiotics which inhibit the growth of pathogens directly, rather than by preemptive consumption of limiting resources.

Another important mechanism limiting infection is the induction of plant defenses against pathogens by nonpathogenic organisms. Cross-protection and induced resistance are mechanisms in which plants are intentionally exposed to certain (nonpathogenic or mildly pathogenic) microbes, thereby conferring in the treated plants some resistance to infection by pathogens. Induced plant defenses may include lignification of cell walls through the addition of chemical cross-linkages in cell wall peptides which makes the establishment of infection through lysis more difficult, suberification of tissues (where plant cell walls are infiltrated with the fatty substance suberin, making them more corklike), and other general defenses, including production of chitinases and β 1,3-glucanases. These plant defenses then limit later infection by pathogens. The biological control agent employed may be an avirulent strain of the pathogen, a different forma specialis, or even a different species of microorganism.

Parasitism is a third mechanism by which beneficial microorganisms suppress plant pathogens. Some species of *Trichoderma*, for example, attack pathogenic fungi, leading to the lysis of the pathogen (Fig. 2.6). Natural enemies of plant-parasitic nematodes include bacterial diseases and nematophagous and nematopathogenic fungi.

CONSERVATION

Root Diseases

Just as in the case of conservation of natural enemies of pest arthropods and weedy plants, conservation activities for the suppression of plant pathogens consist of either avoiding practices which reduce desirable antagonists or actively modifying the environment to favor or selectively enhance the growth of such species. In the case of soil microflora, species employed for biological control of plant pathogens are often competitive antagonists. Adding amendments to soil is one way in which soil microorganisms may be managed to enhance populations of these beneficial organisms. Addition of organic matter to soils for control of *Streptomyces scabies*, the causative organism of potato scab, is one example. Addition of carbon sources to soil increases general microbial activity which leads to reductions in *S. scabies*. Specifically, *Bacillus subtilis* and saprotrophic species of *Streptomyces* were encouraged by barley, alfalfa, or soy meal (Campbell 1989). Soy meal was also a substrate for antibiotic production against *S. scabies*. A general rise in soil organic matter also gave control of *Phytophthora cinnamomi* Rands in avocado in Australia (Manajczuk 1979). The addition of more than ten tons of organic matter per hectare per year led to general increases in numbers of bacteria. Lysis of the hyphae and sporangia of the pathogen were attributed to species of *Pseudomonas*, *Bacillus*, and *Streptomyces*.

Some soils appear to suppress disease naturally and may contain antagonistic or antibiotic flora which flourish without the need for amendments. One example of such suppressive soils is the *Fusarium*-suppressive soil in the Chateaurenard District of the Rhone Valley in France

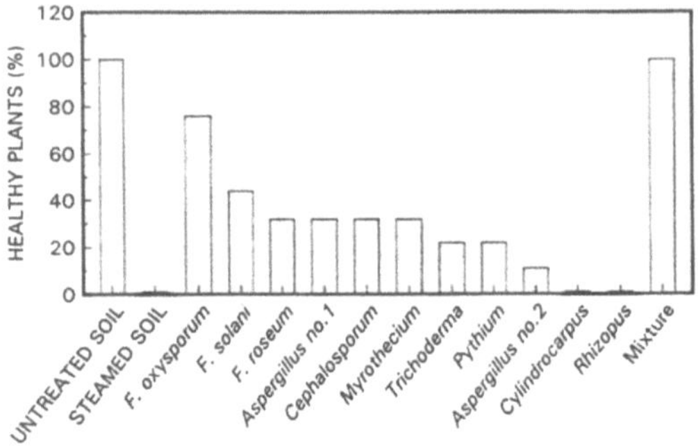

Figure 12.1. The suppressive nature of soil from the Chateaurenard District (Rhone Valley, France) against *Fusarium oxysporum* f.sp. *melonis* Snyder and Hansen is caused by the presence of other non-pathogenic fungi in the soil. Untreated soil and steamed soil with a mixture of the fungi present completely suppress the disease. The species of fungi present vary in their individual ability to suppress the disease, with a non-pathogenic form of *Fusarium oxysporum* Schlechtendal playing an important role (after Alabouvette et al. 1979).

(Fig. 12.1). Here, *Fusarium oxysporum* f.sp. *melonis* Snyder and Hansen is present, but no disease develops when susceptible melon varieties are grown. These soils are suppressive for several other types of *F. oxysporum*, but not to other species or genera of pathogens. The suppressive nature of the soils is clearly biotic, because the soils lose their suppressive ability when steam-sterilized, and the suppressive ability can be transfered to other soils. The antagonists principally responsible for this suppression are nonpathogenic strains of *F. oxysporum* and *Fusarium solani* (Martius) Saccardo. The suppression appears to be due to fungistasis induced by nutrient limitation. The competing fungi appear to have nearly the same ecological niche as the pathogenic forms, and the saprotrophic forms outcompete the pathogens for limiting resources so that dormant chlamydospores of the pathogen do not germinate in the presence of host root exudates. It may be possible to develop systems for other areas using the antagonists from the Chateaurenard area (Campbell 1989), although additional research may be necessary to permit their effective operation in different soils. Other soils suppressive to *Fusarium* wilts are also known. There are numerous other examples of suppressive soils, although some soils or combinations appear to give somewhat variable results.

Leaf Diseases

Conservation of existing flora may be important in limiting the extent of a number of leaf diseases (Campbell 1989). These effects are often revealed through the use of fungicides which deplete extant fungi, permitting the development of previously unimportant diseases. Fokkema and de Nooij (1981), for example, evaluated the effects of various fungicides on leaf surface saprotrophs that have been used in biological control. Wide-spectrum fungicides permitted almost no growth of saprotrophs, while more selective agents permitted some growth of several genera of saprotrophs. In cases where these saprotroph populations play an important role in limiting disease organisms, the application of fungicides would eliminate their contribution to pathogen suppression. One such case is illustrated by Fokkema and de Nooij (1981). Plants treated with benomyl (a systemic fungicide) had fewer saprotrophs and developed more necrotic leaf area when inoculated with *Cochliobolus sativus* (Ito and Kuribayashi) Drechsler ex Dastur than nontreated plants (*C. sativus* is insensitive to benomyl). Another example (Mulinge and Griffiths 1974) is leaf rust of coffee (*Coffea arabica* Linnaeus), caused by *Hemileia vastatrix* Berkeley and Broome. The disease can be controlled by proper application of fungicide. However, if fungicides are applied in one year and not in the next, the disease is

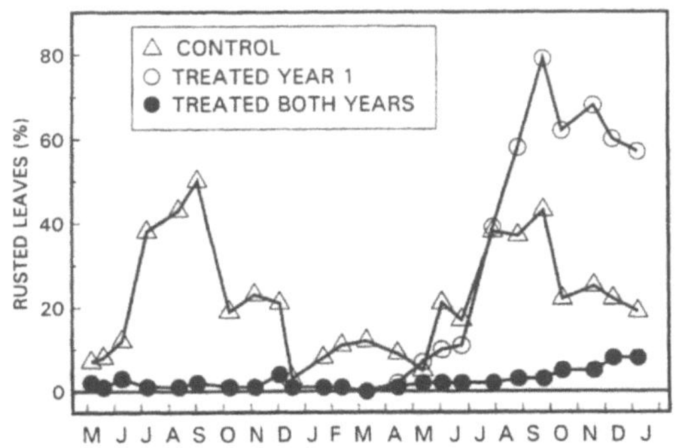

Figure 12.2. The importance of conserving fungi as biological control agents can be demonstrated by the use of fungicides, as in the case of fungicide-treatment effects on incidence of rust (*Hemileia vastatrix* Berkeley and Broome) on coffee (*Coffea arabica* Linnaeus) leaves. The untreated control has seasonally variable incidence with maxima of approximately 50%, while plots treated in both years have very low incidence. Plots treated only in the first year have low incidence in that year, but subsequently have incidences much higher than the untreated control, indicating the presence of microorganisms important in limiting the disease (after Mulinge and Griffiths 1974).

worse on the treated plants than on those which did not receive treatments either year (Fig. 12.2). The elimination of the saprotrophic flora by the fungicide removes their natural suppressing influence on the disease organisms, permitting the disease to be worse. Here, careful use of selective fungicides will be crucial to conserving the important antagonistic flora and permitting their beneficial action.

Plant-Parasitic Nematodes

There are several reports of substantial natural control (control by natural enemies without intentional manipulation) of plant-parasitic nematodes. Stirling (1991) and Sayre and Walter (1991) review several of these; one example is that of the natural suppression of the cereal cyst nematode *Heterodera avenae* in cereal cultivation in the United Kingdom (Gair et al. 1969). In this case, populations of the nematode initially increased for the first 2–3 years of cultivations, and then declined continually during 13 years of continuous cultivation of both oats and barley (a more susceptible crop) (Fig. 12.3). Four species of nematophagous fungi were present in the soil. The two species principally responsible for nematode suppression were *Nematophthora gynophila* and *Verticillium chlamydosporium*. Both fungi attacked female nematodes, either destroying them or reducing their fecundity. The activity of both fungi was greatest in wet soils during laboratory trials (Kerry et al. 1980). Although natural suppression of the nematode population takes some time to develop in these soils, once established it maintains the population below the economic threshold (Stirling 1991).

Conserving nematode antagonists in soils (as opposed to directly enhancing their numbers), is a matter that has received relatively little attention. The application of toxins (insecticides, fungicides) to aerial portions of crops or directly to soils often leads to pesticide activity in the soil. All nematicides are nonselective in their action and, hence, will kill predatory nematodes (Stirling 1991). In addition, herbicides have well-documented effects on soil microorganisms (Anderson 1978) and may well exert some influence on microbial antagonists of nematodes,

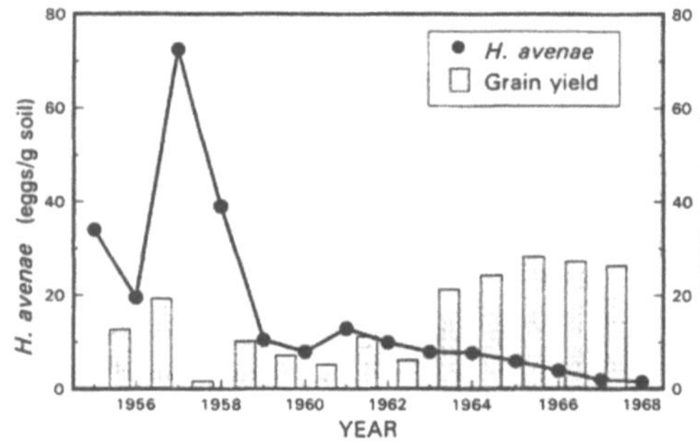

Figure 12.3. Post-cropping levels of *Heterodera avenae* Wollenweber and crop yields under continuous grain (oats [*Avena sativa* Linnaeus] 1955–1962, barley [*Hordeum vulgare* Linnaeus] 1963–1968) culture in England (United Kingdom) (after Gair et al. 1969).

and insecticides may negatively affect soil microarthropods. Many fungicides are known to be detrimental to nematophagous fungi (Mankau 1968; Canto-Saenz and Kaltenbach 1984; Jaffee and McInnis 1990), but at levels higher than would be expected under normal field practice. Among the fumigant nematicides, ethylene dibromide (EDB) and dibromo-chloro-propene (DBCP) appear nontoxic to the nematode-trapping fungi (Mankau 1968), and several herbicides were shown to be not harmful to *Arthrobotrys* sp. (Cayrol 1983). Despite these potentially significant effects on beneficial microflora and fauna and the possibility of conserving these organisms by appropriate choice of material, little has emerged to integrate these ideas into normal farming practice. Perhaps because there has been no serious emergence of nematode problems associated with the use of these materials, this *status quo* is justified. Nonetheless, the opportunities for conserving biologically important agents should be considered in the development of future integrated management programs for plant-parasitic nematodes. Similarly, cultivation practices may be selected to favor natural enemies of nematodes. Among these are minimum or conservation tillage, which reduced the number of cysts of *Heterodera avenae* on roots and the amount of damage caused by the nematode on wheat in Australia (Roget and Rovira 1987). Other practices which may affect populations of natural enemies include normal tillage (which adds crop residue to the soil and thus may favor certain beneficial organisms) and crop rotation sequences (Stirling 1991).

The knowledge that some soils are naturally suppressive to nematodes prompts the question of whether or not the features of these soils can be used to improve biological control. In all documented instances where they have been studied, the suppressive properties of these soils appear to result primarily from the action of one or two specific biological control agents (Stirling 1991). The suppressiveness requires substantial time to develop, and considerable crop loss might be incurred during such an initial phase. Some risk is involved also, because the suppressive nature of the soil may not develop to suitable levels. Careful management of crop varieties, particularly using varieties resistant or tolerant to nematode damage during the initial phases of land use for cropping, is an important part of taking advantage of the potential of these resident natural enemies. Farmers have large amounts of capital invested in land, equipment, and cropping costs, and consequently require a certain degree of reliability in pest control measures. Because of the variable nature of natural suppressiveness of nematodes, any natural control of nematodes in the foreseeable future is most likely to arise fortuitously rather than result from any deliberate actions by scientists or farmers (Stirling 1991).

Where soils are not naturally suppressive to nematode populations, they may be manipulated to enhance what natural control agents are present. Most attention in this arena has been given to the addition of organic matter to the soils. Much of the information regarding the effects of these amendments is circumstantial, but the beneficial effects appear widespread. Many different soil amendments have been considered and evaluated, and the reduction of plant damage from nematodes following such amendments may occur through a variety of mechanisms (Stirling 1991).

One such mechanism is through the general improvement of soil structure and fertility. Addition of crop residue or animal manures increases ion exchange capacity of the soil, chelates micronutrients to make them accessible by the plant, and adds available nitrogen. Grown under such improved conditions, plants are better able to tolerate damage from nematodes. Certain amendments may directly improve plant resistance to nematodes (Sitaramaiah and Singh 1974). Others may contain or release compounds which adversely affect nematodes. Among amendments containing such compounds are those of neem (*Azidirachta indica* A. Jussien) seeds or leaves and of castorbean (*Ricinus communis* Linnaeus) (Stirling 1991 and references therein). Other amendments release nematicidal compounds during decomposition. The most widely studied of these compounds is ammonia. Because nitrogen is a constituent of nearly all soil amendments, ammonia is usually produced during decomposition. A careful balance must be maintained in the carbon:nitrogen ratio, together with sufficient concentrations of ammonia, to provide optimal effect without phytotoxicity (Stirling 1991).

Finally, there is the direct stimulation of nematophagous or antagonistic organisms. Spores of many nematophagous fungi fail to germinate in otherwise suitable but nonamended soils (Dobbs and Hinson 1953), and this soil mycostasis can affect both spores and mycelia (Duddington et al. 1956ab, 1961; Cooke and Satchuthananthavale 1968). Before predation of nematodes can take place, mycelial growth and trap formation must occur. The addition of organic matter provides a substrate which may stimulate spore germination. Organic amendments stimulate a broad range of soil microorganisms, so the effects of amendments on populations of these organisms is complex. Microbial population growth generally increases immediately following the addition of organic matter and, subsequently, as part of the community succession, there is an increase in populations of nematode-trapping fungi. The general hypotheses regarding the beneficial effects of organic amendments center around the stimulation of the saprotrophic growth phase of nematophagous fungi, and stimulation of other general microorganisms which may be detrimental to nematodes, such as antibiotic-producing bacteria. A general rise in enzymatic levels also occurs following soil amendment, and the enzymes may attack the structural proteins in nematode cuticle or egg shell. Chitin amendments in particular have received attention, and addition of chitin to soil is followed by a relatively long-term (4–10 weeks) rise in chitinase activity in the soil. Chitin is the principal structural component of nematode egg shells, and the increase in chitinase activity may be accompanied by decreased survival of nematode eggs. However, the decomposition of chitin also releases ammonia, which may contribute to its beneficial effects. Speigel et al. (1988, 1989) concluded that the beneficial effects of chitin amendments resulted from the action of specialized microorganisms.

A present limitation of the implementation of amendments for nematode control is that such amendments must be applied in large amounts, between 1–10 t/ha to be effective. The use of local resources for such amendments will keep transport costs minimal. One product, the chitin-based Clandosant (derived from crab shells), has been marketed commercially. There is some evidence that the effectiveness of certain amendments may be enhanced by inoculating them with degradative microorganisms (Galper et al. 1991), and Stirling (1991) suggests consideration of systems in which amendments can be inoculated with a specific microorganism as they are applied to the soil.

AUGMENTATION

Augmentation of antagonists of plant disease organisms can generally be of two types, inoculation and inundation (also see Chapter 10). Inoculative releases consist of small amounts of inoculum, with the intention that the organisms in this inoculum will establish populations of the antagonist which will then increase and limit the pathogen population. In inundative releases, a large amount of inoculum is applied, with the expectation that control will result directly from this large initial population with limited reliance on subsequent population growth. Biological control of plant pathogens may also rely on a hybrid of these two concepts. A large amount of inoculum may be applied, both to increase the population of the antagonist and to improve its distribution to favor biological control, and antagonism can result from both these applied organisms and the increased population of antagonists resulting from their reproduction. Biological control of blackcrust (*Phyllachora huberi*) on rubber tree foliage by the hyperparasites *Cylindrosporium concentricum* and *Dicyma pulvinata* (Junqueira and Gasparotto 1991) is one example of long-term control of a plant pathogen by a single augmentation in an agricultural system (Cook 1993). In this case, rubber trees were treated with spore suspensions of the antagonists (inundatively), which resulted in control over more than one season. More generally, beneficial microorganisms are added seasonally or more frequently.

Where the beneficial organisms involved are being placed into a habitat or environment other than where they originated, the organisms are often referred to as "introduced" in plant pathology (Andrews 1992; Cook 1993). There are several examples where such organisms, when moved to a new habitat (for instance, from the soil to the above-ground part of a plant) colonize and serve as successful agents of biological control (Andrews 1992; Cook 1993).

Competitive Antagonists and Antibiotic Production

Root Diseases. One way in which flora may be manipulated to protect against disease is to intentionally inoculate soils or seeds with microbial antagonists. Such antagonists, to be successful in their task, must be able to colonize plant surfaces and survive in the competitive environment of the soil. Flora with demonstrated ability to achieve this under field conditions include fungi, principally *Trichoderma* spp., and, among the bacteria, *Bacillus* spp. and *Pseudomonas* spp.

Among the bacteria, species of *Bacillus* are regularly used for biological control of root diseases. Members of the genus have advantages, particularly that they form spores which permit simple storage and long shelf life, and they are relatively easy to inoculate into the soil. The consequence of this biology is, however, that although the inoculant may be present in the soil, it may be in dormant or resting stages. Nonetheless, species of *Bacillus* have provided good control on some occasions. Capper and Campbell (1986) showed a doubling of wheat yield over wheat plants naturally infected with take-all by those also inoculated with *Bacillus pumilus* Meyer and Gottheil. *Bacillus pumilus* and *B. subtilis* were also used to protect wheat from diseases caused by species of *Rhizoctonia* (Merriman et al. 1974). A major difficulty with the use of *Bacillus* spp. is that the control provided is often variable, with different results in different locations, or even in different parts of a season in the same location (Campbell 1989). *Bacillus subtilis* is used as a seed inoculant on cotton and peanut (*Arachis hypogaea* Linnaeus), with nearly 2 million ha treated in 1994 (Blackman et al. 1994). Treatment promotes increased root mass, nodulation, and early emergence, and suppresses diseases caused by species of *Rhizoctonia* and *Fusarium*.

Of substantially more promise as antagonists of root diseases are species of *Pseudomonas*, particularly the *Pseudomonas fluorescens* and *Pseudomonas putida* (Trevisan) Migula groups (Campbell 1989). These bacteria are easy to grow in the laboratory, are normal inhabitants of the soil, and colonize and grow well when inoculated artificially. They produce a number of antibiotics as well as siderophores. Several have received patents and are marketed commercially for control of root rot in cotton (Campbell 1989). An isolate of another species of *Pseudomonas* has been used as an antagonist of take-all disease of wheat (Weller 1983). Isolates of *Ps. fluorescens* from soils showing some control of take-all can be applied as seed coats and inoculated into fields suffering from the disease. Such treatments give 10–27% yield increases compared with untreated, infected control groups. Evidence points to both siderophore and antibiotic production as important.

Species of the fungal genus *Trichoderma* can be saprotrophic and mycoparasitic and have been used against wilt diseases of tomato, melon, cotton, wheat, and chrysanthemums. The antagonists were applied to seeds or through a bran mixture incorporated into the planting mix at transplanting. Although disease did develop, it did so much more slowly than in untreated soils, resulting in a 60–83% reduction in disease (Siven and Chet 1986). The mode of action against *Verticillium albo-atrum* Reinke and Berthier wilt of tomatoes appeared to be antibiosis.

Stem Diseases. The control of *Heterobasidion annosum*, the causative agent of butt rot in conifer stumps, by *Phanerochaete gigantea* was one of the first commercially available agents for biological control of a plant pathogen (Campbell 1989). The disease caused by *H. annosum* is primarily a disease of managed plantations. The fungus colonizes freshly cut stumps, invades the dying root system, and can then infect nearby trees through natural root grafts, causing death of the trees. *Heterobasidion annosum*, however, is a poor competitor, and when a stump is intentionally inoculated with *Ph. gigantea* (and usually with chemical nitrogen sources which encourage growth of the antagonist) the antagonist rapidly colonizes the resource, excluding future attack by the pathogen and even eliminating existing pathogen infection (Table 12.1). Very little inoculum is needed on a freshly-cut stump, and the shelf life of the pellet formulation is about two months at 22°C. The antagonist is able to outcompete *H. annosum* even when the initial inoculum favors the pathogen by as much as 15:1 (Rishbeth 1963).

The ascomycete fungi *Eutypa armeniaceae* Hansford and Carter and *Nectria galligena*

TABLE 12.1 Colonization of Scots Pine Stumps (*Pinus sylvestris* Linnaeus) After Inoculation with Various Fungi as Antagonists to *Heterobasidion annosum* (Fries) Brefeld (Which Was Inoculated into All Stumps) (After Rishbeth 1963)

	% of Mean Area of Stump Section Colonized After:					
	10 Weeks			6 Months		
Inoculated Antagonist	Species Inoculated	Pg[a]	Ha[b]	Species Inoculated	Pg	Ha
None	—	28	38	—	80	7
Botrytis cinerea Persoon: Fries	5	5	55	0	0	25
Trichoderma viride Persoon: Fries	0	10	65	0	43	40
Leptographium lundbergii Lagerberg and Melin	95	9	5	37	47	0
Phanerochaete gigantea (Fries: Fries) Rattan et al.	80	80	Trace	75	75	0

[a]*Phanerochaete gigantea* (Fries: Fries) Rattan et al.
[b]*Heterobasidion annosum* (Fries) Brefield

Bresadola in Strass infect apricots and apples, respectively, and cause stem cankers and eventual death of the trees. Pruning wounds in apricots are treated with *Fusarium laterium* Nees: Fries through specially adapted pruning cutters. *Fusarium laterium* produces an antibiotic which inhibits germination and growth of *E. armeniaceae*. When applied, the concentration of the antagonist must be greater than 10^6 conidia/ml. Integrated application which includes a benzimidazole fungicide gives better control than either fungicide or antagonist alone. *Nectria galligena* infection can be reduced through sprays of suspensions of *Bacillus subtilis* or of *Cladosporium cladosporioides* (Fresenius) de Vries. These antagonists are not in commercial use because apples are treated for *Venturia inaequalis* (Cooke) G. Winter (apple scab) so frequently that *N. galligena* is controlled by those sprays.

Crown gall is a stem disease caused by the bacterium *Agrobacterium tumefaciens* (Smith and Townsen) Conn. It affects both woody and herbaceous plants in 93 families. Infection is typically from the soil, rhizosphere, or pruning tools. Control can be effected by treating plants with a suspension of a related saprotrophic bacterium *Agrobacterium radiobacter* (Beijerink and van Delden) Conn strain K-84. This strain of the bacterium produces an antibiotic which is taken up by a specific transport system in the pathogen bacterium, which is then killed. The commercially-available formulations of this agent are effective primarily against pathogen strains which attack stone fruits, but other bacteria are under investigation for use against strains pathogenic in other crops. This agent has been altered by gene-modifying technology to produce a new strain (strain 1024) which lacks the ability to transfer antibiotic resistance to the target bacterium.

The fungus *Chondrostereum purpureum* (Persoon: Fries) Pouzar infects stems of fruit trees and produces a toxin which leads to a condition known as silverleaf disease. Stems can be inoculated with a species of *Trichoderma* grown on wooden dowels or prepared as pellets which are inserted into holes bored in the affected stem. Treated stems recover from the disease more rapidly than untreated stems. The *Trichoderma* sp. can be applied to pruning wounds to prevent initial establishment of *C. purpureum*.

Leaf Diseases. Control of leaf diseases at the time of pathogen germination has been shown in the laboratory. This control occurs in the presence of competitive organisms, which may include fungi, yeast, or bacteria. The mode of action in some cases is competition for nutrients which, together with water, are necessary for successful germination and invasion of many pathogens. The germination of *Botrytis* sp., for example, is inhibited by certain bacteria and yeasts (Blakeman and Brodie 1977). This inhibition is less pronounced if additional nutrients are supplied, indicating that the mechanism is, at least in part, resource competition. Studies on control of *Botrytis* rot in lettuce (Wood 1951) indicated that several organisms were successful in suppressing the disease when sprayed on lettuce plants, among them species of *Pseudomonas*, *Streptomyces*, *Trichoderma viride*, and *Fusarium*. Peng and Sutton (1991) evaluated 230 isolates of mycelial fungi, yeasts, and bacteria and tested them as antagonists of *B. cinerea* in strawberry in both laboratory and field trials. Several organisms (including members of each taxonomic group tested) were effective, some as effective as captan (a commercial fungicide) (Table 12.2). Sutton and Peng (1993b) further evaluated *Gliocladium roseum* and determined that the suppression of *B. cinera* by this antagonist was probably a result of competition for leaf substrate. The fungi *Gliocladium roseum* and *Myrothecium verrucaria* (Albertini and Schweinitz) Ditmar were also effective in suppressing *B. cinerea* in black spruce (*Picea mariana* [Miller] Britton Stearns Poggenburg) seedlings (Zhang et al. 1994).

Bacteria may also be used to limit frost damage to leaves and blossoms of plants. Certain

TABLE 12.2 Effects of Various Microorganisms and Captan on Incidence of *Botrytis cinerea* Persoon: Fries in Strawberry. All Plots Were Treated with *B. cinerea* Conidia. Several Antagonists Were as Effective as Captan in Reducing Disease Levels. (After Peng and Sutton 1991)

	Incidence of *B. cinerea* (%)[a]			
	Cambridge Plots		Arkell Plots	
Treatment	Stamen	Fruits	Stamens	Fruits
Water check	4 b	35 c	2 b	59 b
Botrytis cinerea Fries: Persoon check	19 a	59 a	16 a	71 a
Captan	5 b	40 b	3 b	56 b
Bacillus sp.	15 a	56 a	12 a	74 a
Cryptococcus laurentii (Kufferath) Skinner	17 a	43 b	11 a	72 a
Rhodotorula glutinis (Fresenius) Harrison	6 b	33 c	12 a	48 c
Alternaria alternata (Fries) Kessler	9 b	51 a	3 b	52 c
Myrothecium verrucaria (Albertini and Schweinitz) Ditmar	15 a	48 b	13 a	60 b
Fusarium graminearum Schwabe	18 a	47 b	16 a	65 a
Fusarium sp.	17 a	46 b	16 a	64 a
Drechslera sp.	16 a	26 c	9 a	56 b
Trichoderma roseum (Persoon) Link	12 a	27 c	13 a	41 c
Epicoccum purpurascens Ehrenberg ex Schlechtendal	6 b	43 b	4 b	51 c
Colletotrichum gloeosporioides Penzig	4 b	44 b	6 b	41 c
Trichoderma viride Persoon: Fries	5 b	38 b	2 b	37 c
Penicillium sp.	4 b	14 d	1 b	38 c
Gliocladium roseum Link: Bainer	4 b	25 c	3 b	43 c

[a]Numbers in a column with same letters were not significantly different (P = 0.05)

bacterial species such as *Pseudomonas syringae* and *Erwinia herbicola* serve as nucleation sites on leaves for the formation of ice, and, in their presence, ice forms soon after temperatures fall below freezing. If these ice-nucleating bacteria are replaced by competitive antagonists (such as certain strains of *Ps. syringae*) that lack the protein that causes ice-nucleation, frost is prevented even at temperatures from −2 to −5°C (Lindow 1985b). The protective bacteria, after being applied to the leaves, colonize them for up to two months, an interval suitable to protect from frost during the limited season that low temperatures are likely. A naturally-occurring, non-ice nucleating strain of *Ps. fluorescens* is registered in the United States as a commercial product (Frostban B®) for suppression of frost damage (Wilson and Lindow 1994).

Spraying suspensions of propagules, generally at high concentrations, is the principal method for applying biological control agents to foliage (and to flowers), and dusts (such as lyophilized bacterial preparations) are also used. Spray methodology has yet to be refined in terms of sprayer characteristics, droplet size, and pressures, and other methods of application with greater efficiency may be necessary to effectively target certain plant parts (Sutton and Peng 1993a).

Flower Diseases. One of the principal diseases of flowers which has received attention is fire blight of rosaceous plants, which is particularly severe on pear (Campbell 1989). The causal bacterium, *Erwinia amylovora*, also occurs on leaves and may cause stem cankers. The bacterium is transferred by insects to flowers in the spring from overwintering sites on stem cankers, and subsequently from flower to flower. Infection enters the pedicel and from there the stem. Infected flowers and small stems die, and cankers form on other stems. Chemical control is difficult and expensive, and sometimes is ineffective because of resistance to copper compounds and streptomycin. Biological control has been effective using *Erwinia herbicola*, sometimes in combination with *Pseudomonas syringae* (Wilson and Lindow 1993). Suspensions of *E. herbicola* are sprayed onto the flowers just before the period of potential infection. The antagonist occupies the same niche as the pathogen, reducing the numbers of *E. amylovora* by competition, and there is also evidence for the production of bacteriocins (chemicals which suppress population growth of related bacteria) by some strains. Control can be good, comparable to that achieved by commercial bactericides, though repeated application of the bacterium was necessary (Isenbeck and Schultz 1986). Another approach to control is to reduce secondary infections on leaves, which leads to reductions in the overwintering population of the pathogen. This control is achieved by treatment with the antagonists *Ps. syringae* and other bacteria (Lindow 1985b). A novel approach to dissemination of the antagonistic bacteria has been evaluated by Thomson et al. (1992). These workers mixed *E. herbicola* and *Ps. fluorescens* with pollen in a special apparatus at the entrance to honey bee (*Apis mellifera*) hives. Bees emerging from these hives through the mixtures transmitted the antagonists to the flowers efficiently, although disease control was not evaluated because of absence of disease in the test orchards.

Fruit Diseases. Fruits are subject to attack both by general pathogens (*Botrytis*, *Rhizopus*, *Penicillium*) and by a few specialist pathogens such as the coffee berry disease fungus *Colletotrichum coffeanum* Noack and *Monilinia* spp., which cause brown rots of rosaceous fruits. While many of these are controlled by fungicides, *Trichoderma viride* has been shown to limit disease from *Monilinia* spp. Various *Bacillus* spp. also are antagonistic to these fungi through production of antibiotics and by reducing the longevity and germination of spores. Both the bacteria and culture filtrates have been used with some success against these pathogenic fungi, but there has been no commercial development, probably because fungicides used routinely in orchards for control of other diseases give some control of brown rot (Campbell 1989).

Among the most serious diseases of soft fruits are postharvest rots (Dennis 1983), especially that caused by *Botrytis cinerea*. Potential for biological control of postharvest diseases has been reviewed by Wilson and Wisniewski (1989) and Jeffries and Jeger (1990) (also see Wilson and Wisniewski 1994). In strawberries, *B. cinerea* grows saprotrophically on crop debris and from there infects flowers or fruit. Various species of *Trichoderma* have been evaluated and gave control as good as standard fungicides (Tronsmo and Dennis 1977). The antagonists *Cladosporium herbarum* and *Penicillium* sp. gave excellent results in controlling *Botrytis* rot on tomato (Newhook 1957). Honey bees have been used to distribute *Gliocladium roseum* to strawberry flowers (Peng et al. 1992) and raspberry flowers (Sutton and Peng 1993a) to suppress *Botrytis* rot.

Induced Resistance

Root Diseases. Induced resistance is a form of biological control in which the natural defense responses of the plant, which may include production of phytoalexins, additional lignification of cells, and other mechanisms (Horsfall and Cowling 1980; Bailey 1985), are promoted in the plant prior to exposure to the pathogen. These resistance mechanisms are induced by challenging the plant with a nonpathogenic organism. The induced plant defenses then limit later infection by the pathogen. The organism employed may be an avirulent strain of the pathogen, or a different forma specialis, or even a different species. There are few well-documented cases of induced resistance for soil-borne pathogens, and these are mostly of wilt diseases. Dipping tomato roots in a suspension of *Fusarium oxysporum* f.sp. *dianthi* a few days before likely exposure to the pathogen *F. oxysporum* f.sp. *lycopersici* (Saccardo) Snyder and Hansen conferred protection that lasted a few weeks. Cotton may be protected for three months or longer by spraying the roots at transplanting with a mildly pathogenic strain of the disease-causing pathogen *Verticillium albo-atrum* (Fig. 12.4). The role of some fungi against take-all of wheat includes some elements of induced resistance. *Gaeumannomyces graminis* var. *graminis* grows on grass roots and also has been found on wheat, where it occupies a niche similar to that of the pathogen *G. graminis* var. *tritici* Walker. The antagonist invades the root cortex but not the stele, and is halted by the lignification and suberization of the cortex and stele. Root cells with these chemically-changed walls are less susceptible to invasion by the pathogen. Although this interaction produced yield increases in Europe, the strains or species present in the United States did not appear to confer resistance, and in Australia there were only slight yield increases (Campbell 1989). These variable results, while somewhat common for biological control of soil-borne pathogens, do not reduce the value of the antagonists where they do work, but rather indicates some potential challenges in defining the taxonomy, biology, and host-plant relationships important to biological control in this group of organisms.

Leaf and Stem Diseases. Induced resistance can control anthracnose diseases caused by *Colletotrichum* spp. (Kúc 1981; Dean and Kúc 1986). *Colletotrichum lindemuthianum* (Saccardo and Magnus) Lamson-Scribner causes anthracnose of beans, *Colletotrichum lagenarium* (Passerine) Saccardo causes cucumber anthracnose, and *Cladosporium cucumerinum* Ellis and Arthur causes scab in cucumbers. Inoculation of cucumbers with *Colletotrichum lindemuthianum* (which does not cause disease in cucumbers) made plants resistant to both *Colletotrichum lagenarium* and *Cladosporium cucumerinum*. Treatment applied to an early leaf resulted in protection of later leaves, even when the initially inoculated leaf was removed. The factor causing resistance travels systemically through the plant. Variations on this approach include inoculating an early leaf with a pathogen, inducing resistance throughout the plant, and then removing the infected leaf. Induced resistance also occurs in some virus diseases (Thomson 1958) and may last for years, as in the case of healthy citrus seedlings being inoculated with an avirulent strain of citrus tristeza virus.

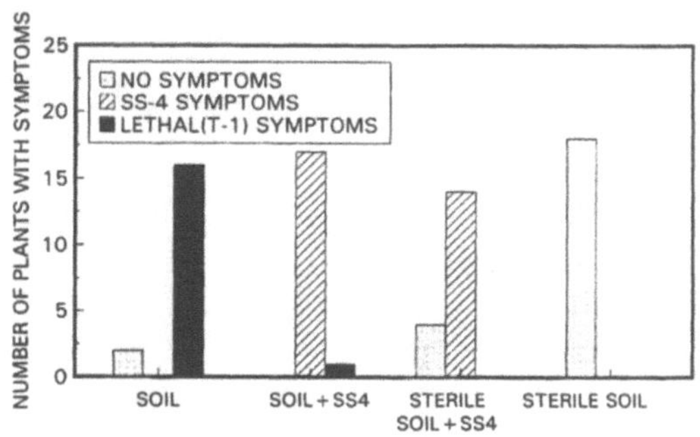

Figure 12.4. Induced resistance with a reduced virulence strain (SS-4) of *Verticilium albo-atrum* Reinke and Berthier in a field of cotton (*Gossipium hirsutum* Linnaeus) infested with a virulent strain (T-1) of the pathogen (after Schnathorst and Mathre 1966).

Stem rot in carnations, caused by *Fusarium roseum* Link: Fries '*Avenaceum*,' can be prevented by inoculating wounds inflicted during propagation with the nonpathogenic *F. roseum* '*Gibbosum*.' This inoculation produced a germination inhibitor and also reduced the time needed for the stems to develop resistance to the pathogen. This hastening of resistance was caused by activation of the host's defense mechanisms, and is another example of induced resistance.

Hypovirulence

Chestnut blight, caused by *Cryphonectria parasitica*, is controlled by employing hypovirulent strains of the disease pathogen. A number of hypovirulent strains are known, and inoculating infected trees with a hypovirulent strain leads to reduced canker size and greater stem survival (Fig. 12.5). In the field, hypovirulent strains are inoculated into infected trees at the rate of 10 inoculated trees/ha. The hypovirulent strain spreads from these locations and, on contacting more virulent strains, fuses with these strains and exchanges a viral element infecting the pathogen. The hypovirus, which causes hypovirulence, is transferred to the virulent strains, attenuating their effects. Active cankers are eliminated in ten years (van Alfen 1982).

Parasitism of Pathogens and Nematodes

Root Diseases. The mycoparasites *Trichoderma* spp. have been used successfully against diseases caused by *Rhizoctonia* and *Sclerotium* pathogens. One example is the pathogen *Sclerotium rolfsii* Saccardo, which attacks many crop plants and survives unfavorable periods by forming sclerotia in the soil. Strains of *T. harzanium* that have β 1–3 glucanases, chitinases, and proteases have been isolated. These enzymes permit *T. harzanium* to parasitize the hyphae and sclerotia of the pathogen, invading and causing lysis of the cells. *Trichoderma harzanium* is grown on autoclaved bran or seed, and this material is then mixed with the surface soil (Chet and Henis 1985). Two other fungi known to parasitize sclerotia are *Coniothyrium minitans* and *Sporidesmium sclerotivorum* (Ayers and Adams 1981).

Sporidesmium sclerotivorum is a hyphomycete that in nature behaves as an obligate parasite of sclerotia of *Botrytis cinerea* and several species of *Sclerotium* (Adams 1990). It has been studied as an agent against botrytis rot in lettuce, where it shows considerable potential. It

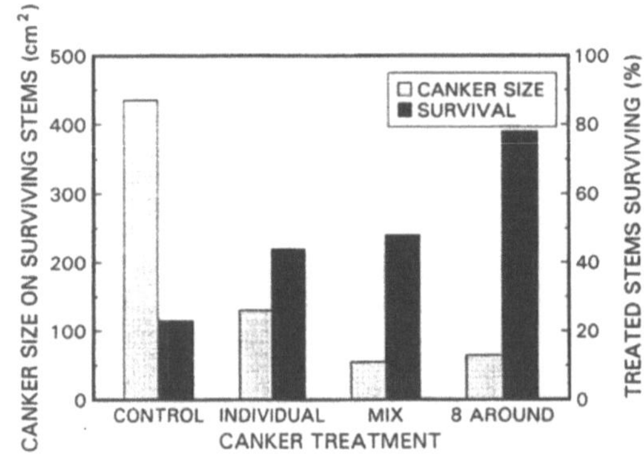

Figure 12.5. Example of hypovirulence in American chestnut (*Castanea dentata* [Marsham] Borkjauser) inoculated with normal (control) and hypovirulent strains of *Cryphonectria parasitica* (Murril) Barr. "8 around" is 8 individual strains placed around a canker (after Jaynes and Elliston 1980).

can be grown *in vitro* on various carbon sources and is efficient in converting glucose into mycelium. Spores produced in mass culture are collected, processed, and applied to infected soil, and field tests are promising (Adams 1990).

Leaf Diseases. Some plant pathogens, including fungi and some bacteria, are known to be attacked by other pathogens. *Bdellovibrio bacteriovorus* is a bacterium that can attack other bacteria by penetrating the cell wall and lysing the host bacterium, subsequently reproducing inside its host. Different strains of *Bd. bacteriovorus* have been examined for virulence against *Pseudomonas syringae* pv. *glycinae* (Coeper) Young, Dye and Wilkie, the cause of soybean blight. By applying *Bd. bacteriovorus* at sufficiently high rates, disease symptoms were reduced more then 95% (Scherff 1973) (Fig. 12.6). Parasites of fungi pathogenic on leaves are numerous (Kranz 1981), but only a few have been studied in much detail, such as *Sphaerellopsis filum*, *Verticillium lecanii*, and *Ampelomyces quisqualis*. The mycoparasite typically penetrates the host hypha or spore and kills it. Some of the control may be from the pathogen overgrowing the sporulating pustules of the pathogen and preventing spore release and thus reducing inoculum in the environment, even if the spores are not killed. A typical problem with implementation of these mycoparasitic fungi is that they often do not affect a large proportion of the pathogens unless humidity and temperature are high. Consequently, although much reduction of spore production may take place, there is still sufficient inoculum of the pathogen remaining to cause disease. These mycoparasites often are seen only at high incidences of disease, which is unsuitable for general control of the target pathogens. They may have some use in particular systems, either in the tropics or in greenhouses, where environmental conditions are more favorable.

Plant-Parasitic Nematodes. The bacterial pathogen of nematodes most studied is *Pasteuria penetrans sensu stricto* Starr and Sayre (Starr and Sayre 1988), which is an obligate parasite of root-knot nematodes (*Meloidogyne* spp.) and has not been successfully cultured in vitro. This restriction in mass culturing has limited attempts to test the bacterium's effectiveness (Stirling 1991). In experimental trials, it has shown potential for controlling root-knot nematodes (*Meloidogyne* spp.) (Mankau 1972; Stirling et al. 1990) (Fig. 12.7), infesting a high proportion of nematodes in soil to which bacterial spores had been added, and in other trials (U.S. Depart-

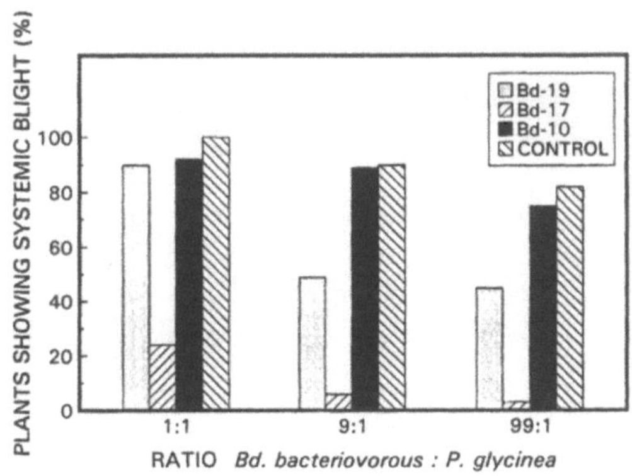

Figure 12.6. Percentage of soybean (*Glycine max* [Linnaeus]) plants showing symptoms of systemic blight after leaves were inoculated with different strains of *Bdellovibrio bacteriovorus* Stolp and Starr mixed in varying ratios with the pathogen *Pseudomonas syringae* pv. *glycinea* (Coeper) Young, Dye and Wilkie (after Scherff 1973).

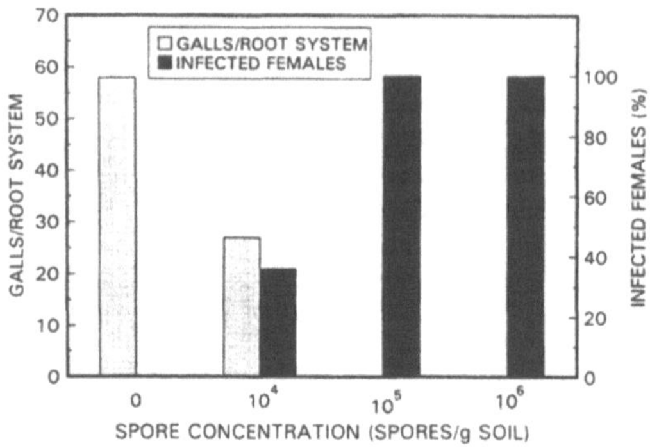

Figure 12.7. Number of galls caused by *Meloidogyne javanica* (Treub) Chitwood (and percentage of females infected with *Pasteuria penetrans* (Thorne) Sayre and Starr *sensu stricto* Starr and Sayre) after movement as juveniles through 4 cm of soil infested at various densities with *P. penetrans* spores (after Stirling et al. 1990).

ment of Agriculture 1978) reducing damage to plants in plots containing the bacterium. Observations by Mankau (1975) indicated that populations of the bacterium did not increase rapidly in field soil. The development of a mass production method in which roots containing large numbers of infected *Meloidogyne* spp. females were air-dried and finely ground to produce an easily handled powder enabled more extensive testing (Stirling and Watchel 1980). When dried root preparations laden with bacterial spores were incorporated into field soil at rates of 212–600 mg/kg of soil, the number of juvenile *Meloidogyne javanica* (Treub) Chitwood in the soil and the degree of galling was substantially reduced (Stirling 1984) (Fig. 12.8); other authors have reported similar results (Stirling 1991). Effective use of this bacterium through such inundative release would require concentrations on the order of 10^5 spores/g soil (Stirling et al. 1990). Such quantities could only be produced on a large scale with an efficient *in vitro* culturing method, a problem which has received attention but has not yet yielded a solution (Stirling 1991). Use in inoculative releases, where smaller numbers of spores are applied and a crop tolerant of nematode damage is grown to permit the increase of both nematode and bacterial populations, has been suggested (Stirling 1991). Conserving the bacterium in the presence of nematicides appears possible. Of seven tested nematicides, only one showed slight toxicity to the bacterium (U.S. Department of Agriculture 1978).

The use of *Bacillus thuringiensis* strains with activity against nematodes is also possible. As

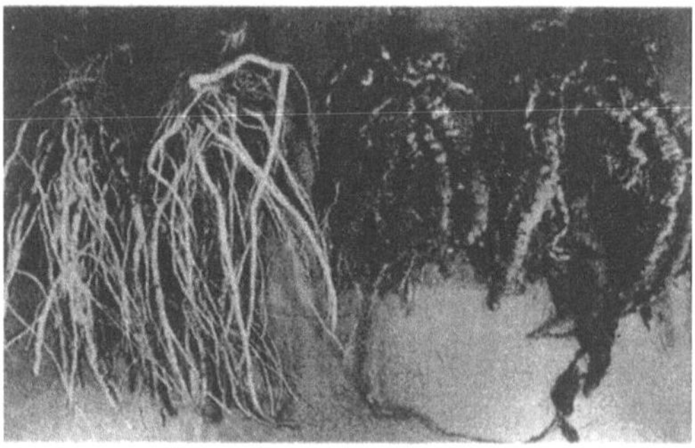

Figure 12.8. Roots of tomato grown in field soil infested with *Meloidogyne javanica* (Treub) Chitwood and treated with 600 mg of a dried root preparation of *Pasteuria penetrans* (Thorne) Sayre and Starr per kilogram of dry soil (left) and grown in untreated soil (right). (Photograph courtesy of G. Stirling, from Stirling, G. 1984, *Phytopathology* 74:55–60, with permission.)

these bacteria may be cultured in fermentation media, their mass culture is simpler than for *P. penetrans*. Suppression of nematodes was possible through drench applications and through incorporating the bacterium into a methyl cellulose seed coat (Zuckerman et al. 1993).

Considerable attention has been given to the nematode-trapping fungi as possible augmentative agents. Mass culture on nutrient media is possible for these fungi. Two cultures of nematophagous *Arthrobotrys* fungi have been developed and tested for addition to soil for specific target environments. Cayrol et al. (1978) reported the successful use of *Arthrobotrys robusta* Cooke and Ellis var. *antipolis*, commercially formulated as Royal 300® against the mycetophagous nematode *Ditylenchus myceliophagus* Goodey in commercial production of the mushroom *Agaricus bisporus* (Lange) Singer. The nematophagous fungus was seeded simultaneously with *A. bisporus* into mushroom compost, which led to 28% increases in harvest and reduced nematode populations by 40%. The results justified the commercial use of the fungus for nematode control in mushroom culture. Cayrol and Frankowski (1979) reported the use of *Arthrobotrys superba* Corda (Royal 350®) in tomato fields, applied to the soil at a rate of 140 g/m^2, resulting in protection of the tomatoes and colonization of the soil by the fungus. Other reports have indicated little efficacy of fungal preparations when added alone to soil (Barron 1977; Sayre 1980; Rhoades 1985). In general, there has been limited success in the use of these agents (see Stirling [1991] for a summary). The fungistatic nature of soil (Mankau 1962; Cooke and Satchuthananthavale 1968) may limit the ability of these fungi to grow even when added in substantial numbers to soil. Additional work is needed, perhaps in the areas of colonization and soil amendments together (for example, Table 12.3), for the use of nematophagous fungi to become suitably reliable for general use as a control method.

Many predacious fungi may be unsuited for control of root-knot nematodes, *Meloidogyne* spp. Stirling (1991) suggested that *Monacrosporium lysipagum* (Drechsler) Subramanian and *Monacrosporium ellipsosporum* (Grove) Cooke and Dickinson, which can invade egg masses, may warrant further investigation. The nematode-trapping fungi are likely to be more effective against ectoparasitic nematodes and such species as *Tylenchus semipenetrans* Cobb, where juvenile stages migrate through the rhizosphere. Little attention has been given to testing predacious fungi against such nematodes.

Fungi which are internal parasites of nematodes have proven difficult to culture on nutrient media, and consequently there have been few attempts to use them for augmentative control of nematodes. Alternative mass-culturing techniques may hold some promise (Stirling 1991). In the few experiments reported, the fungistatic effects of soil often limited fungal growth and the effectiveness of the antagonists. Lackey et al. (1993) report the production and formulation of

TABLE 12.3 Results of a Field Microplot Experiment with *Heterodera schachtii* Schmidt on Sugarbeet Showing the Effects of Soil Amendments and Nematode-Trapping Fungi on Nematode Populations and Yield (After Duddington et al. 1956a)

Treatment	Final nematode population/100 g soil		Yield of roots + tops (t/ha)
	Cysts	Eggs	
1. Untreated	467	114	45.4
2. Bran (20 t/ha)	383	103	56.3
3. *Monacrosporium thaumasia* (Drechsler) de Hoog & van Oorschot mycelium (6.8 t/ha) at planting	488	144	56.3
4. Treatment 2 + treatment 3	333	94	59.2
5. Treatment 4 + *M. thaumasia* mycelium (6.8 t/ha) in mid season	363	128	60.1

Hirsutella rhossiliensis Minter et Brady on alginate pellets (see also Fravel et al. 1985) which, when added to soil, led to transmission of the fungus to the nematode *Heterodera schachtii* Schmidt and suppressed nematode invasion of roots.

Among the facultatively parasitic fungi which attack nematodes, *Paecilomyces lilacinus* and *Verticillium chlamydosporium* have received the most attention as possible augmentative agents. The results of studies on *P. lilacinus* have been variable, with some studies showing some positive effect of the fungus, while others show little or no effect (Stirling 1991). The mechanisms leading to the beneficial effect have not been clearly elucidated, but may be from metabolic products or effects other than direct parasitism of eggs. Studies have generally involved the addition of fungal preparations to the soil at the rate of 1-20 t/ha, which is likely too great for widespread commercial use. Additions at lower rates (0.4 t/ha) in a variety of carriers (alginate pellets, diatomaceous earth, wheat granules) have also shown limited beneficial effects (Cabanillas et al. 1989; Stirling 1991). Tribe (1980) suggested the direct addition of *V. chlamydosporium* to the soil. Kerry (1988) added hyphae and conidia, formulated in sodium alginate pellets or in wheat bran, to soil, and the fungus proliferated in the soil only from granules containing bran. When chlamydospores were used as inoculum, the fungus was able to establish without a food base (De Leij and Kerry 1991). Of three isolates studied, only one successfully colonized tomato root surfaces. This species apparently has considerable promise, but screening programs will be necessary to identify isolates with characteristics suitable for biological control (Stirling 1991).

Predacious microarthropods and nematodes have evoked considerable interest. Most work has been done in simple microcosms, and there have been no attempts to evaluate augmentative release of these organisms in a field setting. In one experiment, Sharma (1971) found nematode numbers reduced by 50% or more in glass jars inoculated with mites and springtails compared with similar jars containing no predators, but the author pointed out possible causes for the reduction other than simple predation. Experiments with predacious nematodes have in general failed to demonstrate a measurable impact of the predator (Stirling 1991). One exception was the reduction of galling by *Meloidogyne incognita* on tomato by predacious nematodes (Small 1979). The general suitability of these groups of organisms for inundative release is questionable, because of the potential difficulties in developing technologies for their rearing, packaging, transport, and delivery beneath the soil in a viable state (Stirling 1991).

Mycorrhizae

Mycorrhizae are nonpathogenic fungi associated with roots in some temperate forest trees. Ectomycorrhizae are mostly basidiomycetes which form a sheath over the root, and hyphae spread out into the soil. These fungi have been studied in relation to nutrient uptake, but they also affect root disease. Because they completely enclose the root, they change the quantity and quality of exudates reaching the soil; consequently, roots with mycorrhizae have a different rhizosphere flora than uninfected roots (Campbell 1985). In at least one case, the mycorrhizal fungus *Pisolithus tinctorius* (Persoon) Coker and Couch, the thick symbiont sheath forms a barrier to infection by such pathogens as *Phytophthora cinnamomi* attacking eucalyptus trees. Other mycorrhizal fungi produce antibiotics effective against *P. cinnamomi* in plate tests. The intentional manipulation of mycorrhizal fungi for disease control has not been widely implemented, but opportunities for selected uses may be possible (Campbell 1989).

Another group of fungi are the vesicular arbuscular mycorrhizae (VAM), which are phycomycete fungi associated with the roots of many plant species including many crops. These fungi do not form a sheath surrounding the root, and their effects on disease are complicated, but are in general beneficial (Campbell 1989). Some of these effects may involve changes in host plant physiology in the presence of the symbiont, as there is no direct evidence of pathogen inhibition by these fungi.

DEVELOPING AND USING BENEFICIAL SPECIES

Growth of our knowledge about biological control of plant diseases has been extensive since the first experimental reports (Hartley 1921; Sanford 1926; Millard and Taylor 1927; Henry 1931), and substantial potential for microbial control of pathogens has been demonstrated. A number of products or programs have reached the stage of commercial development or availability (Table 12.4). Products in current use include both those aimed at specialty markets for control of certain stem or flower diseases (for which chemical control is either unavailable or expensive) and those aimed at larger scale markets such as seed treatments for widely planted crops.

The cycle for research, development, and implementation of antagonists of plant pathogens

Table 12.4 Some Commercially Available Antagonists or Products for Plant Pathogens and Plant-Parasitic Nematodes

Organism	Trade Name	Target
(a) Targeted against plant pathogens		
Agrobacterium radiobacter (Beijerink and van Delden) Conn strain K-84	Agtrol; Galltrol; Norbac 84-C	*Agrobacterium tumefaciens* (Smith and Townsend) Conn (crown gall)
Bacillis subtilis (Ehrenberg) Cohn	Kodiak, Epic	Seed treatment against *Rhizoctonia* spp., *Pythium* spp. and *Fusarium* spp. root diseases
Pseudomonas cepacia (Burkholder) Palleroni and Holmes	Blue Circle; Intercept	*Rhizoctonia* spp. *Pythium* spp., and *Fusarium* spp. diseases of seedlings
Pseudomonas fluorescens (Trevisan) Migula	Dagger G	*Rhizoctonia* spp. and *Phythium* spp. diseases of seedlings
Coniothyrium minitans Campbell	Coniothyrin	*Sclerotinia sclerotiorum* (Libert) de Bary in sunflower
Fusarium oxysporum Schlechtendal (nonpathogenic strains)	Fusaclean; Biofox C	Diseases from pathogenic strains of *Fusarium oxysporum*
Gliocladium virens Millers, Giddens and Foster	GlioGard	*Rhizoctonia* spp. and *Phythium* spp. diseases of seedings and bedding plants
Mycorrhyzae	Vaminoc	*Botrytis* spp. and *Phythium* sp. diseases
Phanerochaete gigantea (Fries: Fries) Rattan et al.	—	*Heterobasidion annosum* (Fries) Brefeld (butt rot)
Pythium oligandrum Drechsler	Polygandron	*Pythium ultimum* Trow in sugar beet
Streptomyces griseovirides Anderson et al.	Mycostop	*Alternaria* sp. and *Fusarium* spp. diseases
Trichoderma harzanium/polysporum (Link) Rifai	BINAB	Wood-rot fungi; *Verticillium malthousei* Ware in mushrooms
Trichoderma harzianum Rifai	F-Stop; Trichodex; Supravit; T-35; TY	*Heterobasidion annosum*; diseases caused by *Rhizoctonia* spp.,, *Pythium* spp., *Fusarium* spp. *Botrytis cinerea* Persoon: Fries, and *Sclerotium rolfsii* Saccardo
Trichoderma lignorum (Tode) Harz	Trichodermin-3	*Rhizoctonia* spp. and *Fusarium* spp. diseases
(b) Targeted against plant-parasitic nematodes		
Arthrobotrys robusta Cooke and Ellis var *antipolis*	Royal 300®	*Ditylenchus myceliophagus* Goodey
Arthrobotrys superba Corda	Royal 350®	*Meloidogyne* spp.
Chitin-based amendment	Clandosan®	Plant-parasitic nematodes

is composed of several steps. These include initial discovery of candidate agents, refinement of knowledge of their biology, ecology, and mode of action, microcosm and field trials of their efficacy, and large-scale development for commercial production.

The first challenge in the development of a biological control program is the discovery process. Many microorganisms show potential as antagonists of particular pathogens. Protocols have been proposed to make the process of screening these candidates more efficient (Andrews 1992; Cook 1993). The principal difficulties are screening out candidates that are effective only during *in vitro* (agar plate) trials but are not effective in natural settings, and in selecting candidates that can be successfully cultured in large quantities. Following discovery of suitable candidates, research focuses on their mode of action and on factors which may enhance or limit their efficacy in targeted settings (glasshouses, field plots). In addition, experimental fermentation and formulations must be developed for production of materials suitable for use in agricultural settings. Finally, issues of large-scale production and delivery must be addressed. Products for use must be effective on an economical basis, and economies of scale may play an important role in the eventual availability of any organism or product. Each must have a satisfactory shelf life, and safe and effective methods for application must be discovered or developed (Cook 1993; Sutton and Peng 1993a). Such application methods might include sprays of suspensions or dusts, contact application, bee vectoring, and production of antagonists in a crop environment (Sutton and Peng 1993a).

The adoption of any biological control agent in commercial agriculture is dependent on its reliability and its availability. Limitations to the process of eventual adoption, therefore, include cost of development and size of potential market. Many pesticides for control of plant diseases have a broad spectrum of activity, are applicable in a variety of crops and settings, and may act either prophylactically, therapeutically, or both. Biological controls, in contrast, often have narrow ranges of activity and may work in only a few crops or soil types, and while they can often act both prophylactically and therapeutically, their action may take some time to develop. Therefore, they may have a narrower market than a chemical pesticide and be unattractive for development by major corporations (Andrews 1992). In this context, it may be appropriate for public institutions such as government experiment stations to undertake the development of such biological controls, in the same way that they take the responsibility for development of new plant varieties (Cook 1993).

Microorganisms intended for use as biological control agents must be viewed in a biological rather than a chemical paradigm (Cook 1993). Where an effective pesticide may work in many places, each place may have unique soil, edaphic, and biological features which limit or enhance the effectiveness of microbial antagonists of pathogens. Consequently, each microbial biological control system may have to make use of locally adapted strains, taking advantage of resident antagonistic flora and fauna and augmenting their effectiveness with additional species or strains, or enhancing resident populations through soil amendments. Although the different strains may use common mechanisms to achieve biological control (such as production of antibiotics), competitive abilities adapted to local conditions may be vital to permit the organisms to compete for resources and effectively control pathogens.

SECTION IV

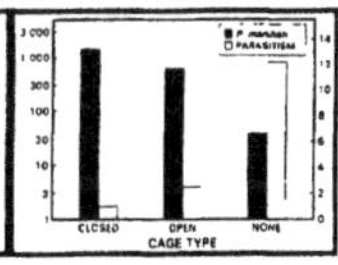

EVALUATION AND INTEGRATION

In this section methods are presented for evaluating the effects of natural enemies on pest populations, and methods by which biological controls may be integrated into larger pest or crop management systems. In Chapter 13, evaluation of natural enemies is considered. Evaluation is a crucial phase of biological control implementation. Evaluations provide guidance to scientists as to what is truly effective, and data to politicians on whether money invested in biological control has been socially productive. Some aspects of evaluation, such as the role of natural enemies in stabilizing densities of target populations, can benefit from theories based on insect population models, which are developed in Chapter 18.

Chapter 14 considers how biological controls fit into pest management systems. Conflicts between types of control are considered, as well as approaches to reduce conflicts or to convert systems from one type of control (for example, chemicals) to another type, such as biological control agents.

CHAPTER
13

Natural Enemy Monitoring and Evaluation

INTRODUCTION

Evaluation is a basic part of all biological control work. Evaluation serves two broad purposes. First, it is the tool through which scientists obtain the biological information needed to separate effective biological control agents or methods from ineffective ones, so that further efforts can be focused on the most effective options. Evaluation methodology is especially critical when biological control is only partially effective, or when the biological effects of several partially effective methods or agents need to be compared. In such cases, the effect of the natural enemy may not be as dramatic as agents (or practices) that provide complete control of the pest. Second, economic evaluations of biological control projects are needed to provide economic planners with data on value returned for funds invested so that further investments can be made in activities with greatest likelihood of productive results.

Evaluation, or sampling, of natural enemies may be undertaken for several specific reasons: to assess what natural enemies are associated with a target pest (in the area of intended release, or in the presumed native range); to monitor the progress of actions taken as part of a biological control project; to determine the current status of one or more natural enemies in a crop for pest management decision making; to evaluate the impact that a natural enemy species, or conservation or augmentation practice, has on a pest population; and to evaluate the economic consequences of a biological control project.

SURVEYS TO DOCUMENT NATURAL ENEMY-PEST ASSOCIATIONS

In many biological control projects, efforts are required to document what natural enemies are present attacking the target pest. Typically such surveys are needed in introduction programs in both the proposed area of introduction and in the area from which natural enemies are to be collected. Surveys in the proposed area of introduction are important in establishing what natural enemies are already present. This information is needed so that subsequent efforts to recover the released natural enemies and to distinguish them from previously existing ones will not be confounded by the previous occurrence of similar species. Surveys in the presumed native home (the area from which natural enemies are being sought for introduction) are needed to develop the list of potential candidates from which the species of natural enemies actually introduced are chosen. In addition, surveys of nontarget hosts in the area of release are needed after the introductions have been made to determine the width of the host range of the new species after it has been released. Surveys of natural enemies are also needed in programs of augmentative biological control, in the area where augmentation is to occur, to determine if the species to be liberated already occurs at the site and if so, at what level of abundance. This information is crucial in quantifying the effect of the natural enemy liberation, separate from the effects of previously existing individuals of the same natural enemy species. Biological control programs based on conservation employ surveys to develop an understanding of the species composition of the natural enemy fauna. This information is used to determine which species are of greatest importance, and which sampling methods will be needed to quantify their abundance.

Methods used to collect natural enemies in surveys vary according to the kind of natural enemy which is the object of the work. For parasitoids of arthropods, the most widely used approach is to collect various stages of the target pest and rear these to obtain adult parasitoids which can be identified to species. This information may sometimes be linked to other samples from the same dates and locations which are dissected to obtain further information on the levels of parasitoid attack, an approach of value when difficulties in rearing prevent accurate determinations of the proportion of hosts parasitized.

For predators of arthropods, direct observation in the field is an important way to discover which species of predators attack a specific pest. In addition, live predators collected in the field from plants inhabited by the target pest can be presented with the pest in the laboratory. Predators which eat the target pest under these conditions can be considered potential predators of the pest species, with confirmation coming later from direct observations made in the field. Bellotti et al. (1987) surveyed the phytoseiids associated with cassava green mite, *Mononychellus tanajoa* (Bondar, *sensu lato, M. caribbeanae* [McGregor]), on cassava, *Manihot esculenta,* in Colombia using this approach. In addition, antigen-antibody methods can be used to identify which predators are consuming a target prey species of interest (Sunderland 1988; also see Assigning Mortality to Particular Species, later in this chapter).

Surveys of arthropod weed control agents are similar to those for predators and are based largely on direct observation, searching the target plant in various locations in the field. Feeding tests in the laboratory (or informally during the foreign exploration trip) can be conducted to confirm each species as a herbivore capable of feeding on the target plant.

Microbial agents (pathogens of arthropods and plants, fungal antagonists) can be isolated either by culturing diseased hosts or infected plant tissue, or by collecting soil samples and incubating them on appropriate media. Nematodes can be recovered either from bodies of infected hosts collected in the field, or by collecting soil samples and incubating them with a substitute host such as larvae of the greater wax moth, *Galleria mellonella.*

MONITORING PROGRESS OF INTRODUCTION PROJECTS

During the beginning phase of a program of natural enemy introduction, evaluation often takes the simple form of determining establishment at the release site and subsequent spread to new areas. Evidence for establishment is obtained by observing the agent in samples taken at the release site and surrounding areas over a several year period. The most commonly used approach, for parasitoids, is to collect samples of the target host, which are then reared or dissected. For predators, direct observation may be employed or collection methods such as sweep netting or trapping. Sticky traps were used in Florida to monitor the distribution and abundance of *Encarsia opulenta* (Silvestri) and *Amitus hesperidum* Silvestri, parasitoids of the citrus blackfly, *Aleurocanthus woglumi* Ashby, following their introduction (Nguyen et al. 1983).

Relative samples are unlikely to provide a quantitative assessment of the level of mortality the new agent has brought to the pest's life system. Rather, such sampling will indicate if the new agent is present and if it is common or rare, either on its own or in relation to other species attacking the same target pest. Positive data indicate establishment has occurred. Negative data, however, do not indicate that colonization has failed. Initial populations may be too low to detect, and several years may be required to determine whether or not colonization has occurred.

Typically, samples need to be taken in the first year of a project at release sites to determine if the agent survives and reproduces in the same season. In the following years, further samples will be needed to determine whether the agent has survived one, two, and more years. In climates with harsh seasons (winters, or prolonged dry periods in tropical areas), establishment is likely to have occurred once the agent has survived for two years.

Following establishment, spread of new agents needs to be documented, as does the range of other species attacked in the field. Spread of a new agent can be assessed by taking host samples, or any other technique that directly detects the agent at a site. For natural enemies of crop pests, the obvious approach is to take samples in crop patches at increasing distances from the release point. In addition, for many natural enemies it is important to determine whether the target pest is also being attacked by the new agent on wild plants outside of crop fields. Parasitoids of the imported cabbageworm, *Pieris rapae*, for example, exist outside crops, attacking host larvae on wild species of crucifers in meadows and along the edges of wooded areas. Finally, it is also important to search for newly established natural enemies on species other than the target pest to determine the field host range of the new agent. It is important, however, to remember that, as with the target pest, levels of attack by the agent on nontarget species in surveys are not a quantitative assessment of the agent's effect on these species. Estimating the quantitative impact of a newly introduced natural enemy (or the use of a new conservation technique or augmentative practice) on the population dynamics of a species is a more complicated matter, and is discussed later.

SAMPLING NATURAL ENEMIES FOR PEST MANAGEMENT

Biological control programs can occur either as efforts that are quite separate from other pest control concerns, or in contexts where many pests must be managed successfully all at the same time. In the latter case, biological control is some portion of a larger pest management program for a crop (see Chapter 14). In pest management systems, evaluation of natural enemies can be undertaken for two basic reasons. First, information may be needed on the value of one released agent versus another (or one natural enemy conservation practice or release rate versus others). Evaluations of this sort seek to quantify the population-level effect of an agent on its host (or prey) under one or more sets of circumstances. This type of evaluation is discussed in a later section in this chapter.

Another type of information about natural enemies is needed when biological control agents are components of pest management systems. Estimates of natural enemy abundance or intensity of natural enemy action are obtained by monitoring. These estimates are then used by farmers (or pest management consultants) to decide whether other forms of pest control are likely to be needed to suppress the pest in the near future. Examples include monitoring ratios of predator-to-prey mites in crops such as apples, grapes, and strawberries (Pasqualini and Malavolta 1985, Nyrop 1988) and the ratio of parasitized lepidopteran eggs to unparasitized eggs on field tomatoes (Hoffmann et al. 1991). There are three key components in natural enemy monitoring: knowing which natural enemy species are important; having a method to accurately assess both natural enemy and pest densities or their ratio; and understanding the relation between natural enemy counts and pest population growth (relative to pest counts and crop stage) well enough to predict the trend in pest numbers in the near term.

Developing Natural Enemy Lists

To institute natural enemy monitoring, the identities of the important natural enemies must be known. This step is vital, but often neglected. Cranberries, for example, are one of the top three cash crops in Massachusetts, with an annual harvest value in excess of $100 million (U.S.). A research station dedicated solely to this crop has been in place since 1910, yet only the barest outlines of the natural enemy fauna affecting the arthropod pests of the crop are known (Van Driesche and Carey 1987). Such neglect is not rare. However, in many crops much progress has been made in the documentation of natural enemy faunas (cassava in South America, Bellotti et al. 1987; apples in the northeastern United States, Maier 1994; maize in east Africa, van den Berg 1993). Existing knowledge is proportional to past effort and more information is typically available in countries with long traditions of agricultural research than in more recently developing countries. Where not available, development of a sound knowledge of the local crop natural enemy fauna and flora is a vital first step. The process begins with surveys to establish which species of pests and natural enemies are present, and, therefore, which species are likely to be most important based on their commonness in the crop, in absolute terms or relative to other species of similar nature (parasitoids of a certain pest life stage, or among generalist predators of a certain size). Once this level of information has been obtained, laboratory observations are needed on the prey preference and biology of the more common natural enemies.

Measuring Natural Enemy Numbers

For each crop, methods must be devised to monitor the density of key natural enemies. With respect to parasitoids and predators of arthropod pests, methods include traps of various types, such as odor and visual traps, pitfall traps, and suction traps. Direct sampling of the density of parasitized versus healthy hosts, or counts of predators per sample unit of crop or as a ratio to prey can also be used. *Aphytis* spp. parasitoids of red scale, *Aonidiella aurantii*, in South Africa, for example, can be monitored either with odor traps baited with its host's pheromone or with visual (yellow) traps (Samways 1988; Grout and Richards 1991). *Pholetesor ornigis* (Weed), a parasitoid of the spotted tentiform leafminer, *Phyllonorycter blancardella* (Fabricius), can be monitored in apple orchards using yellow sticky traps. Trap placement must be considered carefully; traps placed away from the orchard periphery and in tree interiors are more effective than traps in other locations (Trimble 1988). Parasitoids of fruit flies have been monitored by placing fruit in wire cages coated with sticky material (Nishida and Napompeth 1974). In general, many semiochemicals have potential uses as lures to trap natural enemies. Also, aggregation and sex pheromones of the natural enemies themselves have potential as lures (Lewis et al. 1971), for example, the aggregation pheromone of the aphid parasitoids *Lysiphlebus* spp. and the sex pheromone of the soldier bug *Podisus maculiventris*. Coccinellids, an important predator group attacking aphids, can be monitored by means of sweep net counts and timed visual counts (Elliot et al. 1991). Hunger, temperature, and time of day can, however, affect relative counts of coccinellids, making them ineffective measures of coccinellid density in some instances (Frazer and Raworth 1985). Relative sampling methods must be calibrated by reference to absolute sampling methods so that the reliability of relative methods (which are often used for speed and ease of application) is known for specific crop systems (Michels and Behle 1992). Direct counts of predators or parasitized pests are the most common method used to assess the level of natural enemies in pest management systems. Such counts are often made together with counts of the pest to give predator/prey ratios or percentage parasitism values.

Forecasting Pest and Natural Enemy Populations

Information about the abundance of natural enemies is used to calculate the number of damaging pests per sample (revised economic threshold) either at the moment of the sample, or to predict what change in pest numbers is likely to be over a period of time (usually a few weeks at most) for which the need for active pest suppression is being forecast (Ostlie and Pedigo 1987).

For pests affected by pathogens and parasitoids, modified pest thresholds may be calculated by reducing the observed count per sample of the pest by the fraction that are diseased or parasitized, because these are not damaging individuals. This modification is justified if the sampled stage is not the stage that causes damage. For example, the tomato hornworm, *Helicoverpa zea*, is counted in the egg stage, but the damaging stage is the larva; therefore, any parasitized eggs should not be included in estimates of pest numbers. The same will also be true if the pest management decision is based on counts of the damaging stage, if counts are made in the generation prior to damage. For example, parasitism of the apple blotch leafminer, *Phyllonorycter crataegella*, can be measured in the first generation (a nondamaging generation due to low numbers and phenology of crop development), and then used to forecast the need for control prior to the occurrence of the damaging stage in the second generation (Van Driesche et al. 1994). Finally, information on levels of parasitism can be used even if parasitism occurs in the damaging stage and generation, if attack occurs early in the stage's development, and if parasitism reduces the feeding rate of the damaging stage. Parasitism of the variegated cutworm, *Peridroma saucia* (Hübner), for example, can be used to modify the spray threshold for this species in peppermint (*Mentha piperita* Linnaeus) in Oregon (U.S.A.) because parasitism occurs in early instars and decreases feeding of older larvae (Coop and Berry 1986).

Information on densities of key predators is typically used to forecast what change in pest density is likely to occur over a short period in the near future. Ratios of predator and prey mites in apples and other crops are used to determine if circumstances are favorable for biological control over a near-term horizon of a few weeks (Nyrop 1988). For example, in vineyards in the Crimea (Ukraine) ratios of 1 predator mite (*Metaseiulus occidentalis*) per 25 phytophagous mites (*Eotetranychus pruni* [Oudemans]) were associated with mite populations that did not increase to economically damaging levels, and this ratio was used as a forecasting tool (Gaponyuk and Asriev 1986).

Trap catches of adult pests and parasitoids can also be used to forecast the short-term trend in pest populations. Sticky trap catches of adults of the leafminers *Liriomyza trifolii* Burgess and *Liriomyza sativae* Blanchard and their parasitoids in watermelon (*Citrullus vulgaris* Schrader) in Hawaii (U.S.A.) enabled a 3-week forecast of future mine numbers (Robin and Mitchell 1987). The frequency of presence of nematodes (*Beddingia siricidicola*) in wood chips cut by axe from pines infested with *Sirex noctilio* was used with a regression model to predict the proportion of pests infected with nematodes in Australia. This information was then used to plan releases of nematodes in areas with low infection levels (Haugen and Underdown 1991). Counts of parasitized hosts can also be used to forecast the need for pest control measures. In processing tomatoes in California (U.S.A.), ratios of black (parasitized) and white (presumed healthy) eggs of *Helicoverpa zea* are used, together with counts of white eggs per leaf, in a sequential sampling process to reach a decision about the need to apply pesticides (Hoffmann et al. 1991) (Fig. 13.1).

Critical to the use of natural enemy monitoring information in pest management is the accuracy of pest economic threshold information. Such thresholds must be based on factual data, and not mere impressions or past practice. (Past practice may be an effective guide to avoid pest damage, but may be wastefully conservative if injury is assumed at pest levels that are much lower than true economic injury levels.) Furthermore, it is important to understand whether or not natural enemy effects have already been included when the threshold was determined. In some cases, no specific thought may have been given to this issue in the development of the threshold, and so natural enemy action may or may not have been in force in the plots originally used to develop the threshold, and, therefore, may or may not have been included in the threshold. In some crops, crop sale price may be subject to future market fluctuations such that economic thresholds for pest damage cannot be easily applied during the cropping cycle.

EVALUATION OF POPULATION EFFECTS OF NATURAL ENEMIES

Pesticides are intended to cause high levels of pest mortality in relatively short times and must be reapplied periodically to sustain control. Biological control, in contrast, is intended to have a prolonged, often permanent, effect on pest populations, lowering survivorship and pest density and, in some cases, enhancing stability. Regardless of the exact type of biological control action taken, be it introduction of a new species of natural enemy, implementation of a better natural enemy enhancement practice, or release of reared natural enemies, methods are needed to accurately measure the effect achieved by the action taken. Quantitative evaluation

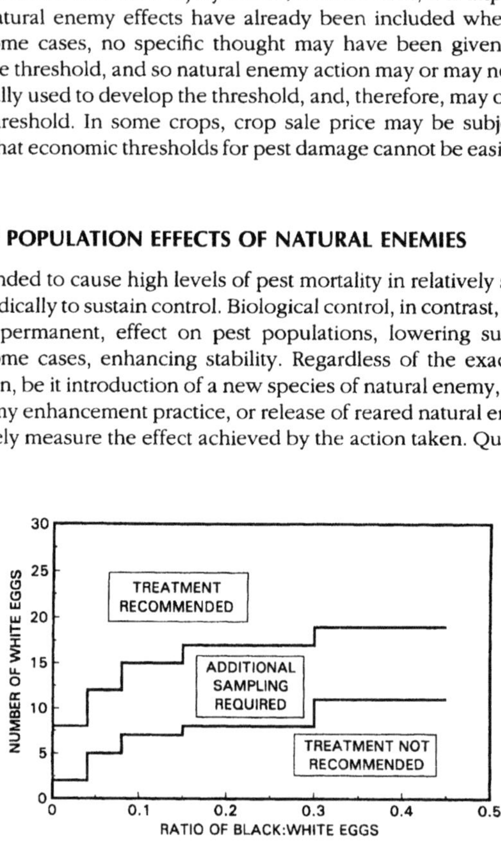

Figure 13.1. A sequential sampling plan for anticipating economic damage by *Helicoverpa zea* (Boddie) in tomatoes, *Lycopersicon lycopersicum* (Linnaeus) Karsten ex Farwell (after Hoffmann et al. 1991). A sample of 30 leaves is collected, and both the total number of white (not showing parasitism) eggs and black (parasitized) eggs are counted. If the number of white eggs exceeds the upper threshold, treatment is appropriate. If the number of white eggs is below a separate threshold, treatment may be withheld. Between these two thresholds is a region where additional sampling is recommended to improve precision in the decision-making process. The thresholds increase as the proportion of the population which is parasitized increases.

allows comparisons to be made among natural enemies or conservation methods so that the most effective can be selected for further application.

Evaluations of natural enemy effects on pest populations seek answers to three questions: (1) are natural enemies reducing the average density of the pest, (2) what are the mechanisms behind this density reduction, and (3) which species of natural enemies are responsible for the observed impacts.

The first of these questions, effect on average pest density, is best addressed through the manipulative (or experimental) approach pioneered by Paul DeBach and others (see Luck et al. 1988 for a review), which integrates all factors bearing on the effectiveness of the natural enemies for the crop and pest under study. The manipulative approach seeks to create two or more populations of the pest, some of which are subjected to the natural enemy (or conservation or augmentation practice) to be evaluated and some of which are not subject to the agent (or practice). Differences in pest mortality in these populations are then correlated to the natural enemy's density after one or more generations and interpreted as the effect of the agent or practice to be evaluated. The manipulative approach is powerful because the contrast is tangible to the observer and, therefore, convincing to growers. All factors affecting the population interaction between the natural enemy and its host are present, whether they are recognized by the researcher or not. In natural enemy introduction programs, the "experimental" approach of DeBach is limited to agents being evaluated at the time of initial introduction (such that time or location can be the basis of plots having and lacking the agent) and to systems where caging or some other form of exclusion can be applied effectively to produce paired plots, some with the natural enemy and some without. The method cannot be used to evaluate the effects of existing natural enemies that cannot be excluded from plots (which is necessary to provide the "without natural enemy" treatment for the test). The "experimental" approach is usually the basis for evaluations of effects of practices intended to conserve natural enemies and of augmentative microbial pesticides.

The second question, determining the mechanisms that underlie the observed effect on average density, requires use of the life table (or analytical) approach proposed by Varley and Gradwell (1970, 1971) (see Manly 1977, 1989; Bellows et al. 1992c for a review). This approach allows mortality from one natural enemy to be compared to other sources of mortality acting on the pest, and it allows the contribution of a given natural enemy to population regulation to be assessed. The analytical approach seeks to estimate numbers entering each stage in the pest's life cycle, the fertility of the adult stage, and the numbers dying in each stage from specific sources of mortality, including the agent, to be evaluated. These data allow the agent's effect on pest survival to be judged relative to other sources of mortality and its effect on the population growth rate of the pest can be calculated. Information from such studies is often organized in life tables. This analytical approach provides a well-organized framework for assembling information on the set of mortalities that shape the life system of the target pest and for defining the relative importance of mortality from a specific agent of interest and its contribution toward stabilizing the population. This approach does not fully answer the question of how much the average pest density would change if the natural enemy were not present (or if a conservation or augmentation practice were not employed). However, if an estimate of the pest's fertility is also available, this allows the pest's population growth rate to be calculated. This estimate includes the action of the natural enemy. If the mortality from the natural enemy of interest is removed from the life table and densities of subsequent stages calculated, the population growth rate for the pest in the absence of the natural enemy can be estimated. The difference between these two population growth rates measures the effect of the natural enemy on the pest. The estimate of the pest's population growth rate in the absence of the natural enemy, however, is not data, but a hypothesis. This hypothesis needs to be confirmed by construction of life tables for pest populations in the field in plots lacking the natural enemy being evaluated (Bellows et al. 1992c).

Once levels of mortality from specific causes have been measured and associated with the density of the pest population present at the time the mortality factor exerted its action, further analyses can be made to define the role the source of mortality plays in regulation of the target pest. A powerful approach to form and test such hypotheses, which capitalizes on the best features of both the manipulative and analytical approaches, is to construct life tables for analysis from sets of populations having and lacking the agent or practice to be evaluated. This strategy allows both the effect on mean density and the mechanisms responsible for the effect to be assessed.

The third question, separate quantification of the contribution of members of a natural enemy complex to the total observed mortality, draws on both experimental and analytical techniques. Manipulative methods, such as serological methods to link predators with specific prey, labeling techniques, and other approaches help define who is killing whom. Analytical techniques, such as the calculation of marginal attack rates, allow contemporaneous sources of mortality from two or more agents to be separately quantified (Elkinton et al. 1992).

In the following discussion, we consider first the experimental evaluation techniques for assessing affects on average density, then analytical techniques for determining mechanisms, and, finally, approaches to segregating the action of a natural enemy complex into the contributions of specific species or species groups.

Measuring Reductions in Pest Density Caused by Natural Enemies

The manipulative or experimental approach provides the clearest assessment of the degree a natural enemy affects average pest density. The essence of the manipulative or experimental approach is the creation of pest populations some of which have the natural enemy or practice to be evaluated and others which do not. The "with" and "without" conditions are the treatments in the experiment. Populations to be compared are presumed to be completely alike in all ways other than the presence or absence of the agent or practice being evaluated.

For nonmobile natural enemies (such as soil applications of plant pathogen antagonists) and for tests of the effects of agricultural practices intended to enhance natural enemies (such as the use of ground covers, reduced pesticide use, etc.), treatments are easily established in whatever location is desired, and the test of the agent or practice is rather simple. In tests of agricultural practices intended to conserve mobile natural enemies, while the practices can be located in specific plots, natural enemy populations cannot. Rather, successful enhancement in one plot will create populations of mobile natural enemies that will spread into adjacent plots not having the same experimental features. Consequently, small plots are often ineffective in discovering the value of practices intended to augment mobile natural enemies because results are blurred across plots by natural enemy movement. A better approach is to use fewer, larger plots, with plot size being based on the foraging or dispersal distances of the key natural enemies. Since large plots may be difficult to replicate, due to expense, biological realism will be obtained at the cost of reduced statistical power. The analytical approach can also be applied in such plots (creating life tables for pest populations having and lacking an agent or practice). While this requires a larger sampling effort, such an approach increases the power of the design.

When the goal is to evaluate the effect of a mobile natural enemy, methods must be employed to create plots where the natural enemy will not occur. For nonnative natural enemies that are just being introduced, time and geography can be used as the basis for the "with" and "without" contrast. For native natural enemies, or species previously introduced, these methods are not feasible and some form of chemical, biological, or mechanical exclusion must be employed to create plots where pest populations are not affected by natural enemies to be evaluated (DeBach and Huffaker 1971). In contrast, for conservation and augmentation studies, plots having and lacking the new conservation technique, or the natural enemy releases, can be created by the experimenter.

Before-and-After Studies. When the introduction of a new natural enemy to a region is contemplated, plots can be established prior to the introduction in which the pest's density and survivorship are assessed for several generations, before the agent's release. These data may then be compared to additional observations made in the same locations at a later date after the new natural enemy has become established at the sites. This approach was used, for example, by Gould et al. (1992a,b) to evaluate the effect of the parasitoid *Encarsia inaron* on the ash whitefly, *Siphoninus phillyreae*, in California. The effect of the natural enemy is determined by comparing the average density of the pest after the natural enemy has become established to the pest density prior to introduction. When the density declines significantly and when this drop is correlated in time with a sharp rise in mortality, with much of the increased mortality being due to the new agent, the new agent is considered to have been the cause of the change of pest density (see Fig. 8.4). This approach is even more powerful when combined with quantitative life tables constructed during this period. This approach is more effective for some kinds of pests and natural enemies than others. The method works well, for example, with pests that show high site fidelity, such as scales and some whiteflies, but does not work for pests, such as imported cabbageworm (*Pieris rapae*) and spruce budworm (*Choristoneura fumiferana*), whose densities at sites are strongly influenced from year to year by movement of adults (or other stages) over long distances. Such movement can lead to large fluctuations in pest numbers between years at particular sites, which may mask the effects of any changes in survivorship induced by a new natural enemy. Demonstrating that increased mortality is due to a particular natural enemy is also easier with some kinds of natural enemies, such as parasitoids, than others, such as predators, whose action, unless directly observed, is seen only as a change in pest density. (See more below on methods to assign mortality to agents.) Before and after studies may also be criticized in that pest density changes across two or more years may be caused by unique, year-specific weather events, since no two years are ever exactly the same. To partially compensate for this problem, before and after studies should be conducted at multiple sites in distinct areas, and studies should include control sites, studied both before and after the year of release but at which no releases of the agent are made (see Geographic Studies).

Geographic Studies. Another approach that can be used with newly introduced biological control agents is to initiate sampling of the target pest in a series of sites in the year that the natural enemy is introduced, but to release the new agent at only some sites and reserve the rest as controls (Van Driesche and Gyrisco 1979). Because sites will inevitably vary in various physical characteristics, several control and release sites will be needed. Also, it is common for control sites to be invaded by the new natural enemy, especially in highly successful projects. To compensate for this, control sites should be located well away from release sites. Unfortunately, when working with new natural enemies whose dispersal powers are unknown, how far is far enough is a matter of guess work. Five to fifteen km is a reasonable figure, but longer distances may be desirable in some cases. It is, however, important that control sites not be placed at such a distance that they are located in an ecological or climatic zone that is different from that of the release sites.

Exclusion Studies. Exclusion of natural enemies from research plots can be achieved, potentially, by five general methods: hand-removal, selective chemicals, caging or other mechanical barriers, dust, and ants. The last two are very limited in application, being useful mainly for parasitoids of homopterans such as some scales and aphids.

Hand-Removal. In a few instances, the action of natural enemies has been demonstrated by mounting a continuous observation of a small segment of a pest population and removing natural enemies as they arrive at the site. This approach is impractical for any extensive use, due to the high labor cost involved, and is limited to homopterans, mites, and other pests that are relatively sessile, show high site fidelity, develop rapidly through a number of generations, and have key natural enemies large enough to be seen and removed by an observer. With hand-removal, there are no cage effects and no effects from chemicals, thus hand-removal is useful to validate results obtained from use of other exclusion methods. Hand-removal was used by Huffaker and Kennett (1956) to demonstrate the impact of phytoseiid mites on the cyclamen mite, *Phytonemus pallidus* (Banks), on strawberries, for comparison to other experiments employing selective pesticides to exclude predators. Fleschner et al. (1955) used hand-removal to demonstrate the effect of natural enemies on several pests of avocado.

Selective Chemicals. If pesticides can be found that are toxic to the natural enemy and yet relatively harmless to the pest, they can be used to exclude the natural enemy from some plots for comparison to other nearby plots not treated with the pesticide. Chemical exclusion in most instances reduces, but does not totally eliminate, the natural enemies in a crop. Consequently, chemical exclusion experiments measure part but not all of the effect of the natural enemies on the pest. Furthermore, whole natural enemy complexes are likely to be reduced by the chemicals applied, and, therefore, the results may not be assignable to any one species, but rather the whole complex.

The validity of exclusion experiments based on the use of selective pesticides depends on three conditions being met: the chemical must be sufficiently toxic to the key natural enemies of the pest that they are greatly reduced; the chemical must cause little or no mortality to the pest; and the chemical must not raise or lower the pest's fecundity, either directly or indirectly by inducing chemical changes in the pest's host plant.

Selective chemicals have several advantages compared to the use of cages to exclude natural enemies. First, the method allows relatively large areas to be used as exclusion plots, allowing the method to be used with species, such as some Lepidoptera, that do not complete their life cycles well in small cages (Brown and Goyer 1982; Stam and Elmosa 1990). Second, the method does not confine the target pest, allowing dispersal of winged adults, scale crawlers, or other life stages to occur in a normal manner, which cannot happen if test populations are confined inside cages.

Selective chemicals have been used to conduct natural enemy exclusion experiments for a variety of arthropod pests, including many species of scales (DeBach and Huffaker 1971), mites (Braun et al. 1989), aphids (Milne and Bishop 1987), thrips (Tanigoshi et al. 1985), and mealybugs (Neuenschwander et al. 1986) among others (Fig. 13.2). The method can also be used to demonstrate the effect of herbivorous arthropods in the control of weed populations

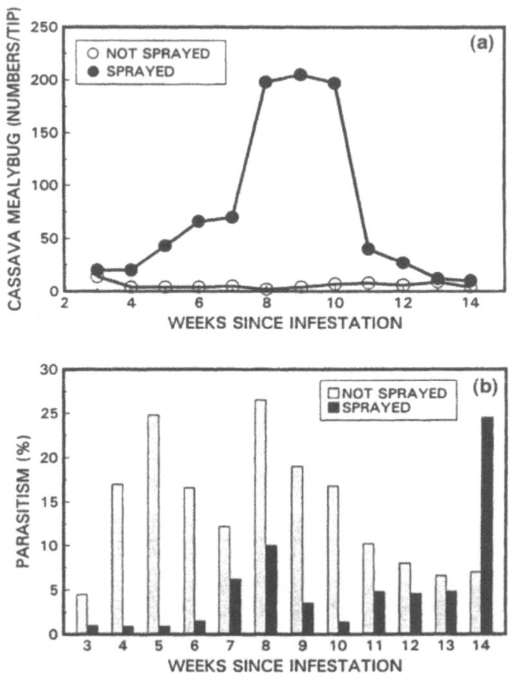

Figure 13.2. Selective chemicals may be used to demonstrate the effect of natural enemies. Densities (a) of cassava mealybug, *Phenacoccus manihoti* Matille-Ferrero, are higher and parasitism by *Epidinocarsis lopezi* (De Santis) (b) is lower, in sprayed versus not sprayed plantings of cassava (*Manihot esculenta* Crantz) (after Neuenschwander et al. 1986).

(Kirkland and Goeden 1978). Balciunas and Burrows (1993) assessed the effect of native insects by growing 60 saplings of *Melaleuca quinquenervia* (Cavanilles) Blake, a target for biological weed control in Florida, in pots in the plant's natural habitat on Australia. Insecticides were used to protect half of the saplings. Treated saplings showed greater height and biomass within six months. Most damage was caused by insects that exhibited only low levels of herbivory, but which collectively significantly and rapidly reduced plant growth. Indeed, the method is even easier to apply in the evaluation of weed biological control agents than insect biological control agents. Selectivity between the natural enemies (arthropods) and the pest (plants) is easily obtained because most insecticides are not herbicidal. The method is also applicable to evaluation of the impacts of other taxa of natural enemies such as fungal pathogens (using fungicides).

When conducting natural enemy exclusion experiments based on the use of selective pesticides, laboratory tests must be conducted to validate each of the key prerequisites on which the validity of the method depends. First, the relative toxicity of a series of potentially selective pesticides must be measured for both the pest and each of the key natural enemies which are to be evaluated. From these results, a chemical must be identified that is of very low toxicity to the pest (in all of its life stages) and of high toxicity to the key natural enemies (in at least one of their life stages) (Braun et al. 1987a). Materials with these properties must then be tested to determine their effect on the pest's fecundity, either directly on the pest or indirectly through pesticide-induced changes in the host plant. Tests should include both direct application of the pesticide to the pest and rearing of the pest on live foliage bearing pesticide residues of various ages. Pesticide stimulation of pest fecundity has been noted for one or more chemicals for aphids (Lowery and Sears 1986), thrips (Morse and Zareh 1991), spider mites (Boykin and Campbell 1982), and planthoppers (Chelliah et al. 1980). Braun et al.'s (1987b) study of permethrin for use in cassava to suppress phytoseiids attacking the phytophagous mite *Mononychellus progresivus* Doreste in Colombia is a good example of how to organize a program of laboratory tests to identify selective pesticides for use in natural enemy exclusion experiments. A final point of validation occurs when field data are collected on natural enemy abundance in the treated and untreated plots, in that only at this stage will it be learned if the field applications suppressed the natural enemy to the expected and desired degree.

Cages and Mechanical Barriers. The use of cages to evaluate the role of resident natural enemies by excluding them from plots, plants, or plant parts infested with the pest has been employed extensively (DeBach et al. 1976). For pests such as scales, whiteflies, and aphids, which are able to pass all their lifestages successfully inside small cages, populations may be caged for one or more generations. For species that are unable to pass some lifestages—for example a highly dispersive adult stage—in small cages, studies may focus on patterns of mortality of cohorts of immatures within a single generation (for example, borers in trees, Mendel et al. 1984).

The extent of the plant area enclosed in cages can vary from parts of leaves or stems enclosed by 1–2 cm dia leaf cages (Chandler et al. 1988), to sleeve cages enclosing tree branches (Prasad 1989), or bucket cages enclosing clumps of cereal crop plants (Rice and Wilde 1988), to small field cages up to a meter on edge (O'Neil and Stimac 1988a), to large field cages (up to 6 or more meters on an edge) to enclose patches of crops such as alfalfa (Frazer et al. 1981) or whole trees (Faeth and Simberloff 1981; Campbell and Torgersen 1983). A range of cage sizes, from leaf cages to field cages may be employed in a single study to test various aspects of pest-natural enemy interactions (Kring et al. 1985).

Cage exclusion experiments often consist of three treatments: a closed cage, an open cage, and a no-cage treatment derived from sampling the unmanipulated population (Knutson and Gilstrap 1989). The open cage is intended to be as physically similar as possible to the closed cage (while allowing natural enemies access to the pests), so that any cage effects on microclimate will also be present in the open cage treatment. Similarity between the pest density or survival in the open cage and in the no-cage samples is taken as evidence that differences between the closed cage and the other treatments (in pest density or survival) are due to natural enemy exclusion and not cage effects. DeBach and Huffaker (1971), for example, used open and closed leaf cages to measure the effect of parasitoids (*Aphytis* spp.) on California red scale, *Aonidiella aurantii*, on leaves of English ivy, *Hedera helix* Linnaeus. Cages differed only in that open cages had minute access slots in sections of the cage/leaf joint. These spaces allowed parasitoids to enter the caged area. In these cages, parasitism reached 80–90% and kept scale densities low compared to closed cages. A similar design of open and closed cages and uncaged hosts was used by Neuenschwander et al. (1986) to evaluate the effect of *Epidinocarsis lopezi* on the cassava mealybug, *Phenacoccus manihoti* (Fig. 13.3).

The principal confounding cage effects of concern are elevated temperatures inside cages (leading to faster pest developmental rates), restriction of pest migration (leading to changes in densities), and higher humidities (potentially leading to increased rates of fungal disease). The first problem, elevated temperatures, can be guarded against by using both open and closed cages of similar design. Effects of cages on temperature can be assessed by placing probes inside test cages (open and closed) to measure temperatures, for comparison with each other

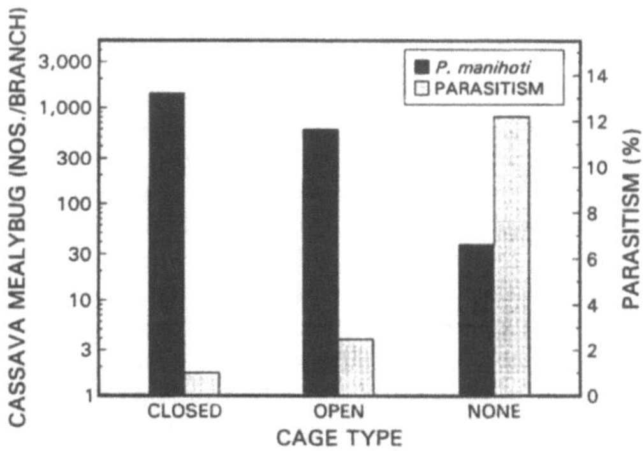

Figure 13.3. Numbers of cassava mealybug, *Phenacoccus manihoti* Matille-Ferrero, and percentage parasitism by *Epidinocarsus lopezi* (De Santis) on branches either covered with a closed sleeve cage, an open sleeve cage, or not covered with a cage (after Neuenschwander et al. 1986).

and to nearby sites on uncaged foliage. The second problem, restricted migration, may not be of concern for single generation studies of sedentary forms. Emigration and immigration may be of concern in other cases, such as studies of mobile stages or in intergenerational studies of sedentary insects as armored scales. The significance of migration should be quantified for any species as part of caging experiments. The third problem, enhanced fungal disease levels, can be evaluated by comparing levels of diseased pests in closed cage and no-cage treatments.

Another important consideration in cage studies is the thorough removal of all natural enemies from the closed cage treatment. How this is achieved depends directly on the biology and seasonal histories of the pest and natural enemies under study. In some instances, cages may be installed over patches of pests when they are free of natural enemies. When this is not possible, natural enemies may be removed from cages through trapping or the application of a short residual insecticide. Following pesticide application, the target pest may have to be reintroduced. A third approach is to first cage plants free of pests and, then, infest these with the pest species, taken from a natural enemy-free laboratory colony. When this latter approach is followed, the resultant study is termed a cohort survival study and is not directly comparable to the events on uncaged plants which are likely to have a different pest population density and instar structure than artificial cohorts.

Cage exclusion experiments measure effects of entire natural enemy complexes. Separating total impact of the complex into portions assignable to specific natural enemy species or groups requires the collection of additional information. Effect of parasitoids and predators is indicated by the level of parasitism and other death in open cage or no-cage treatments. In some instances, the closed cage treatment can be divided into several types of cages with varying size openings in the screening. Rice and Wilde (1988) used this approach to separate the effect of large predators from that of parasitoids and small predators in a study of natural enemy impacts on greenbug, *Schizaphis graminum* (Rondani), on wheat and sorghum. Dennis and Wratten (1991), working with cereal aphids on wheat in the United Kingdom, devised a cage capable of excluding aphid predators and parasitoids, which could be stocked selectively with a species from the natural enemy complex occurring at the site. Observations on the degree of reduction in aphid population growth rates that occurred in these cages, when compared to the field and to aphid populations inside cages which excluded the entire natural enemy complex, allowed the separate contribution of each species to the overall effect to be assessed.

Another approach to mechanical creation of natural enemy-free pest populations is the use of barriers. For ground-dwelling predators that do not readily disperse through flight, barriers such as strips of plastic or metal can isolate patches of a crop, such as wheat or alfalfa, from the rest of a larger field. These patches can then be cleared of the ground-dwelling natural enemies (such as carabids) by using a short residual pesticide or through intensive trapping. This method has been used successfully to measure the effect of nonflying, ground-dwelling predators on various cereal aphids, including *Rhopalosiphum padi* Linnaeus in spring barley in Sweden (Chiverton 1987) and *Sitobion avenae* in wheat in the United Kingdom (Winder 1990).

Caging experiments can also be used to assess the effect of natural enemies on the fecundity of pest populations. Van Driesche and Gyrisco (1979) used caged populations of alfalfa weevil, *Hyperica postica*, adults to measure the impact of the braconid parasitoid *Microctonus aethiopoides* on fecundity. This was done by contrasting the fecundity of a population protected from the parasitoid, but held in an outdoor insectary in an alfalfa field, to the fecundity of the field population exposed to the sterilizing action of the parasitoid.

Relatively little use has been made of cages in assessing the effect of herbivores on pest plants. Opportunities exist, especially for short-term studies such as measurement of reduction in seed production from herbivores, as reflected by caged and uncaged inflorescences (caged after pollination, if relevant). Gilreath and Smith (1988) used small cages on cactus pads to exclude parasitoids and predators of the beneficial herbivore *Dactylopius confusus* (Cockerell). This created high versus low populations of this species and a tenfold difference in the level of cactus pad death from herbivory. The results of the experiment reveal the potential of the herbivore to provide pest control of the affected cactus, if imported into a new region without its parasitoids and predators. In this form, caging experiments have the potential to assist in the detection of new candidate weed biological control agents in the pest's country of origin. Long-term caging studies pose problems of cage effects, such as reduced lighting and increased humidity, that could affect plant health.

Ants. The interference to some natural enemies by ants which tend honeydew-producing homopterans is a form of biological exclusion that has been used successfully to evaluate the effect of natural enemies of such insects as scales, mealybugs, aphids, and whiteflies (DeBach et al. 1951, 1976). Such tending behavior is found in species such as the Argentine ant, *Linepithema humile*, bigheaded ant, *Pheidole megacephala*; *Lasius niger*, among others. Plots having normal and reduced levels of natural enemies of some species can be created by either poisoning ants in selected plots, or mechanically excluding them from individual trees (Musgrove and Carman 1965; Markin 1970a,b; Samways 1990; Kobbe et al. 1991). It should not, however, be assumed that all ant tending is harmful to parasitoids. In some cases, ant tending has been shown to increase parasitism, by protecting parasitized aphids from attack by predators or hyperparasitoids (Völkl 1994).

Dust. As with ants, dust can interfere with certain natural enemies, mainly parasitoids, attacking pests that are not themselves much affected by dust (Bartlett 1951, DeBach 1958). Dust seems to interfere with parasitoid behavior, reducing foraging and oviposition, increasing grooming, and reducing lengths of visits on dusty foliage. The observation that dusty conditions can lead to outbreaks of some kinds of scales and other pests gave rise to the idea of using dust to evaluate the impact of natural enemies of such pests. Diatomaceous earth may be applied to crop foliage to suppress aphelinid parasitoids of whiteflies, an approach useful for experiments assessing effects of such parasitoids.

Mechanisms of Natural Enemy Influence on Pest Density and Stability

Section Overview. Quantifying the mortality in a pest population due to a natural enemy expresses the total number in one or more generations of the pest that, having entered one or more particular life stages, die in those stages because of the action of the natural enemies being evaluated. Such information may be organized by constructing life tables for the pest. (For pests without distinct generations, life tables summarize events over a series of time steps, in place of pest generations). With the use of life tables, mortality from the agent being evaluated can be compared to other sources of mortality operating on the pest's population. When information on the pest's fecundity is available, the effect of the natural enemy under study can be expressed in terms of its effect on the pest's population growth rate. When a series of life tables is available (constructed for populations separated either spatially or temporally), the contribution of the mortality caused by the natural enemy under study to the regulation of the pest's population can be assessed.

This discussion of the analytical evaluation of the effect of natural enemies follows Bellows et al. (1992). First, basic concepts and terms are presented and discussed. Second, four data collection schemes suitable for use in life table construction are reviewed. Third, a discussion is presented of the inferences that may be drawn from one or more life tables, such as the determination of whether or not, and to what degree, natural enemies contribute to the regulation of the pest population.

Concepts and Parameters. Four concepts are important in understanding the collection and compilation of data for the construction of life tables, and the subsequent analysis of the data for the evaluation of the effect of natural enemies. These are: (1) the difference between total number entering a stage for a generation and the density on any single sample date, (2) the concept of rates of recruitment to and loss from a stage, (3) marginal attack rates and their relation to apparent mortality rates and *k*-values, and (4) the population growth rate. A fifth concept, key-factor analysis, is also discussed.

TOTAL NUMBER ENTERING A STAGE VS. DENSITY. The most common type of population data collected from arthropod populations is stage density. Frequently, a series of samples is collected spanning the period from when the stage of the pest under study first appears until it is no longer present. Typically, such data form bell shaped curves as seen in Fig. 13.4. A potential error in population analysis is to equate the peak value in such a stage-density curve to the total number of the pest that entered the stage over the course of the generation. Density of a stage on any date is, rather, the sum of all the numbers of organisms that have entered the stage minus all that have left the stage by death or development to a different life stage up to the sample date. The relation between these values can best be understood by reference to an example, given in Table 13.1, in which data on the egg stage of the Colorado potato beetle are presented. The first column lists dates of a series of intervals covering one generation; the second column lists the number of eggs laid by the population in each interval; the third column lists numbers of eggs that die or hatch in each time interval; and the fourth column lists egg densities from samples at the beginning of each interval. Values in Table 13.1 are shown graphically in Fig. 13.4.

From these numbers several points of interest arise. First, the total number of insects entering the egg stage for the entire generation can be found directly by summing the numbers gained by the population in each interval (column two). In our example, this is 988 eggs. Second,

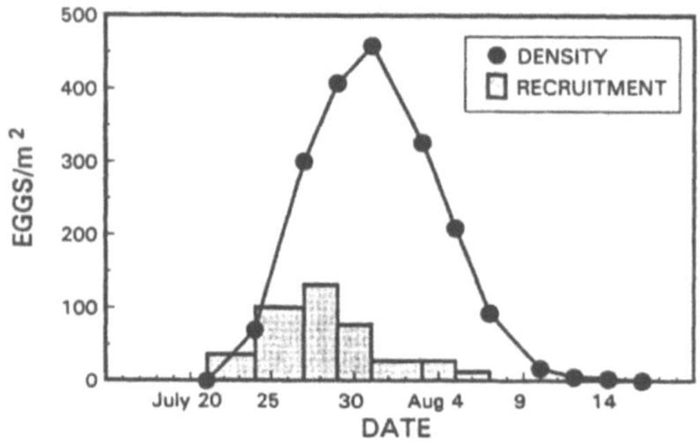

Figure 13.4. Number of eggs recruited and total number of eggs present for the Colorado potato beetle, *Leptinotarsa decemlineata* (Say) (after Van Driesche et al. 1989a).

the highest egg density (459 eggs, on 3 August) was only 47% of the actual number that entered the stage over the course of the generation. The reason is that some individuals have died or hatched before all the eggs have been laid, and, therefore, at no time are all the eggs present to count. The exact relationship of density to total numbers recruited to a stage will depend on how the patterns of gains to and losses from the stage are distributed over time and on whether or not these two processes overlap (see Van Driesche 1983).

Rates of Recruitment and Loss. Stage densities and their related recruitment and loss rates may appear similar, but they differ in a basic way that is signaled by the units associated with each of these parameters. Density values have units that denote a number of animals per habitat unit or per sample, for example, number of eggs/m^2 or eggs/leaf. Density values are determined on specific dates and measure the state of the system at that time. In contrast, both gains and losses to a stage are rates, with units that denote numbers of animals per sample unit per time period, for example, number of eggs laid/leaf/week. Rates of gain to a stage are termed **recruitment**, and can occur from oviposition, live births, or as molts from earlier

TABLE 13.1 Numbers of Colorado Potato Beetle (*Leptinotarsa decemlineata* [Say]) Eggs Recruited to, Lost From, and Total Present in a Population on Potatoes (After Van Driesche et al. 1989a)

Sampling Interval	Eggs Recruited (laid) in Interval (nos./m^2)	Number Lost in Interval (hatch or death) (nos./m^2)	Total Eggs Present at Start of Interval (nos./m^2)
July 21–July 24	108.9	38.9	0
July 24–July 27	301.1	70.1	70.0
July 27–July 29	262.1	155.2	301.0
July 29–July 31	154.6	103.4	407.8
July 31–August 3	83.2	215.6	459.1
August 3–August 5	54.4	171.3	326.6
August 5–August 7	23.9	140.7	209.8
August 7–August 10	0	76.2	92.9
August 10–August 12	0	11.4	16.7
August 12–August 14	0	2.5	5.3
August 14–August 16	0	2.8	2.8
August 16			
Total Recruitment	988.2		

stages. Note that this parameter refers to the rate of gain to the stage for the population as a whole and is not the same as the daily per capita rate of oviposition for individuals.

Losses from a stage are of two sorts, death and advancement to the next life stage. An egg, for example, that is eaten by a predator will no longer be present to contribute to the egg density count on the next sample date. Note that interval-specific death rates are calculated only from deaths during the time period between samples. Samples of larvae, for example, might be collected twice a week and held to determine the rates of parasitism. Some larvae may die from parasitism within the 3–4 day period between samples; others may have been more recently parasitized and die later (day 5 and beyond). While the sample percentage parasitism, as commonly calculated, would count all deaths from parasitism, regardless of when the deaths occurred, if a loss rate from parasitism per 3–4 day period is being calculated, only deaths occurring within that time interval would be counted. Deaths that occurred later would be reflected in the level of parasitism measured in the next sample, for the next interval.

The other manner in which losses to a stage can occur is through growth of some individuals to a life stage other than the sampled stage. If, for example, an egg parasitoid is being studied, eggs that hatch into larvae are gone from the egg population. In studies of natural enemies that attack one life stage but emerge from a later one, it may be useful to define the stage used for counting purposes to be some combination of stages, as for example egg-larva or larva-pupa.

Apparent Mortality, k-Values, and Marginal Attack Rates. Life tables seek to express the total number of individuals that, having entered a life stage, die in that stage from particular sources of mortality. The number dying in a stage as a percentage of the number entering the stage is termed **apparent mortality**. Apparent mortality may be used to calculate the ***k*-value** for mortality, $k = -\log(1 - \text{apparent mortality})$. These ***k*-values express the same information as apparent mortality, but in a form which is additive over stages, so that the** *K*-value for total mortality is given by $K = k_1 + k_2 + \ldots k_n$. This additive property is useful in some types of life table analyses.

When only one source of mortality affects a stage, apparent mortality is an unambiguous measure of this mortality. However, if two or more sources of mortality act together within a stage, apparent mortality does not accurately reflect the force of each agent because some pest individuals will be subject to attack by two or more agents, but will die of only one. In such cases, the mortality rate may be estimated as the marginal attack rate (Royama 1981; Carey 1989, 1993; Elkinton et al. 1992). **Marginal attack rates** are defined as the level of mortality from an agent that would have occurred if the agent had acted alone (Fig. 13.5). For some kinds of agents, such as parasitoids, this is equivalent to the number of hosts attacked (stung), even though some of these hosts will later be killed by other agents (such as predators) in place of being killed by parasitism.

A general equation for calculating the marginal attack rate, *m*, for contemporaneous factors is given by Elkinton et al. (1992):

$$m_i = 1 - (1 - d)^{d_i/d},$$

where m_i is the marginal rate for factor *i*, d_i is the observed death rate from factor *i*, and *d* is the death rate from all causes combined. These calculations apply to a wide variety of cases (e.g., multiple parasitoids, or parasitoids and predators) and any number of contemporaneous factors. Slightly modified calculations provide marginal rates for other special cases (Elkinton et al. 1992). The usual *k*-values may be calculated from marginal rates ($-\log[1 - m_i]$). Marginal rate calculations are required to obtain correct estimates of the mortality caused by specific agents when several sources of mortality act contemporaneously.

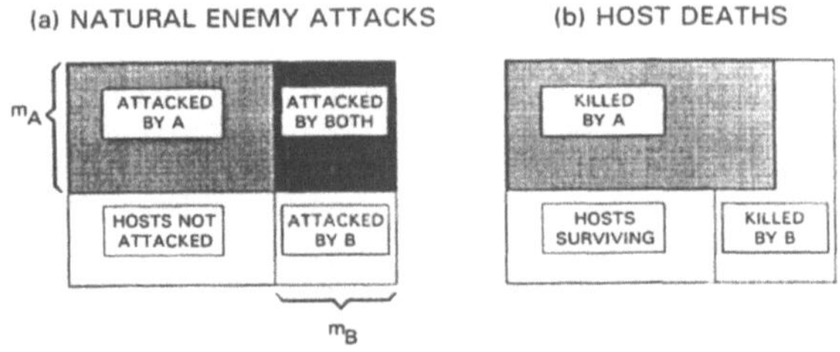

Figure 13.5. In populations subject to two contemporaneous mortality factors (factor A and B), the number attacked by each factor (a) will exceed the number killed by each factor (b) because some individuals will be attacked by both factors, but must necessarily die from only one. Quantifying the effect of each factor requires estimating the marginal attack rate (m) for each factor (after Elkinton et al. 1992).

Population Growth Rate. The population growth rate for a species is the rate at which the population is growing or declining, in view of its reproduction and all the mortality affecting it. Growth of populations may either be expressed as the inter-generational **net rate of increase (R_0)** or the instantaneous **intrinsic rate of natural increase (r_m)** (Southwood 1978). Net growth rates (R_0) above 1 and instantaneous rates r_m above 0 denote expanding populations; net rates below 1 and instantaneous rates below 0 denote declining populations. When growth rates can be calculated for populations both having and lacking mortality from a specific natural enemy, the difference in growth rate reflects the effect of that source of mortality (Van Driesche et al. 1994).

Key Factor Analysis. Key factor analysis is a procedure to identify which mortality factor contributes most to variation in total mortality among generations (Morris 1959; Varley and Gradwell 1960). The form of key factor analysis most widely employed (Varley and Gradwell 1960) is graphical in nature. For each of a set of life tables spanning a series of generations, mortality from each source is expressed as a *k*-value (negative logarithm of [1− the marginal attack rate] associated with each factor, used for computational convenience) and each *k*-value is plotted against time, as is that of total mortality (denoted as the *K*-value). The mortality whose temporal pattern looks most like the pattern of total mortality is termed the **key factor**. Variations on this procedure have been developed that regress individual mortalities against total mortality, with the factor with the greatest slope selected as the key factor (Podoler and Rogers 1975; Manly 1977).

A key factor is not necessarily the factor most important in determining the average density of the pest population, nor is it necessarily important in the regulation of the pest population. Rather, it is simply the factor which contributes most to temporal changes in mortality between generations in the system as observed. Successful biological control agents need not be key factors. This concept can be easily grasped through an example. Consider an insect population subject to two sources of mortality (Table 13.2), a common egg parasitoid that consistently kills 97–99% of all eggs in each generation and a fungal larval pathogen that kills from 0 to 80% of larvae in various years. The fungal disease of larvae is the key factor (Fig. 13.6). Note that the key factor is the factor that acts as the greatest source of the variation in total year-to-year mortality, while the largest amount of mortality comes from the egg parasitoid. The removal (or

TABLE 13.2 Hypothetical Mortalities from Two Agents Acting in Separate Life Stages (k_1 Acting on Eggs, k_2 Acting on Larvae) Over a Series of Generations (Subjected to Key-Factor Analysis in Fig. 13.6)

Generation	Number of Eggs	Proportion Mortality	Number of Larvae	Proportion Mortality	Number of Adults	Fecundity (eggs/female)	k_1	k_2	Total K
1	1000.00	0.985	15.00	0.400	9.00	100	1.82	0.22	2.05
2	900.00	0.985	13.50	0.800	2.70	100	1.82	0.70	2.52
3	270.00	0.975	6.75	0.140	5.81	100	1.60	0.07	1.67
4	580.50	0.980	11.61	0.060	10.91	100	1.70	0.03	1.73
5	1091.34	0.975	27.28	0.800	5.46	100	1.60	0.70	2.30
6	545.67	0.985	8.19	0.260	6.06	100	1.82	0.13	1.95
7	605.69	0.980	12.11	0.000	12.11	100	1.70	0.00	1.70
8	1211.39	0.980	24.23	0.320	16.47	100	1.70	0.17	1.87
9	1647.49	0.985	24.71	0.400	14.83	100	1.82	0.22	2.05
10	1482.74	0.980	29.65	0.600	11.86	100	1.70	0.40	2.10

addition) of the parasitoid would do more to change the average insect density than the removal or addition of the fungus, while removal of the fungus would reduce the amount of year-to-year change in mortality.

Parts of Life Tables. Observations on pest numbers and mortalities can be organized for analysis into tables called life tables (Table 13.3) (Southwood 1978; Bellows et al. 1992). In life tables, rows represent developmental stages of the pest, or time periods in its development.

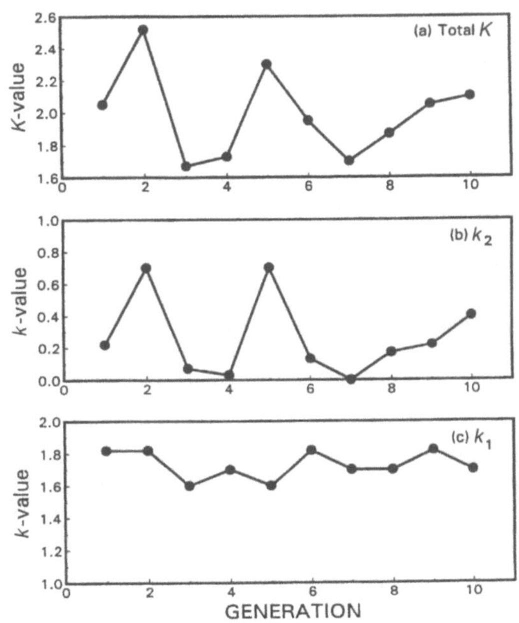

Figure 13.6. Graphical key-factor analysis for the hypothetical population represented in Table 13.4. Note that the key factor is factor 2, the temporal pattern of which most closely resembles that of total K, although factor 1 (k_1) has a much greater average impact on the population (has a greater mean value for mortality).

TABLE 13.3 A Sample Life Table, Using Data for *Pieris rapae* (Linnaeus) (After Van Driesche and Bellows 1988)

		Stage				Apparent Mortality		Real Mortality		
Stage	Factor	l_x	d_x	Factor d_x	Marginal Attack Rate[a]	Stage q_x	Factor q_x	Stage d_x/l_0	Factor d_x/l_0	*k*-value Factor
Egg		10.6690	0.1280			0.0120		0.0120		
	Infertility			0.1280	0.0120		0.0120		0.0120	0.0052
Larvae		10.5410	10.5139			0.9974		0.9855		
	C. glomerata			4.6607	0.8675		0.4422		0.4368	0.8777
	Predation			5.8532	0.9806		0.5553		0.5486	1.7122
Pupae		0.0271	0.0084			0.3100		0.0008		
	Predation			0.0084	0.3100		0.3100		0.0008	0.1612
Adult		0.0187								
Fertility		356.0								
Sex ratio		0.5								
F_1 progeny		3.3285								
R_0 (F_1/l_0)		0.3120								

[a]Marginal rates are calculated by using equations in Elkinton et al. (1992) for the case of a parasitoid and a predator where the predator is always credited for the death of a host attacked by both agents, i.e., where $c = 1$.
[b]Value from Norris (1935).

There are several columns in life tables. These provide data on the numbers of the pest that remain alive long enough to enter a given life stage (l_x), the number that die in each stage (d_x), and the separation of these deaths into the numbers caused by each recognizable kind of mortality. Additional columns may be included that reformulate these data as mortality rates for ease of comparison. Mortality rates may include apparent mortality ($q_x = d_{xi}/l_{xi}$), real mortality (d_{xi}/l_{x0}), marginal attack rates, and *k*-values.

Data Collection Methods for Life Table Construction. Four approaches have been used to estimate parameter values needed for life tables: (1) stage frequency analysis, (2) direct measurement of recruitment, (3) death rate analysis, and (4) growth rate analysis. To evaluate the effects of natural enemies, these methods must estimate the number of pests that enter particular life stages and the numbers that die in those stages from particular natural enemy species or groups.

Stage Frequency Analysis. Density counts of the population of an organism in a specific life stage can be made repeatedly over the time period when the target stage is present. Data of this type are easily obtained and frequently collected in the study of animal populations. The desire to use such observations to develop life tables for the studied species has stimulated mathematical developments for application to such data. Given the validity of the various assumptions made by such approaches, these methods provide estimates of the numbers entering the stage over the course of the generation (Richards and Waloff 1954; Dempster 1956; Richards et al. 1960; Southwood and Jepson 1962; Kiritani and Nakasuji 1967; Manly 1974, 1976, 1977, 1989; Ruesink 1975; Bellows and Birley 1981; Bellows et al. 1982b; see Southwood 1978, McDonald et al. 1989 for reviews).

The construction of life tables based only on numbers entering each of the stages in the life history of the organism permits the apparent mortality for each stage to be calculated. It does not, however, usually allow the effects of specific mortality factors to be separately estimated, unless only one source of mortality acts on a given stage. When two or more sources of mortality act contemporaneously in a stage, additional information (the magnitude and order of attack by each agent, the outcome of attack by two or more agents, and the degree to which one agent avoids attacking hosts previously attacked by another agent) is needed.

Two of the stage frequency analysis methods mentioned have been specifically modified to allow the calculation of both the number of the pest entering a life stage and the number attacked by a specific natural enemy. Methods modified in this manner have included the graphical method of Southwood and Jepson (1962) (Bellows et al. 1989a) and the second method of Richards and Waloff (Van Driesche et al. 1989a). In general, these approaches are applicable to some but not all situations. Potentially, future modifications of additional stage frequency analysis techniques may expand the usefulness of this class of methods for natural enemy evaluation.

Recruitment Estimation. An alternative to estimating total numbers entering a stage from stage frequency data (density estimates) is direct measurement of recruitment to the stage. Rather than measure, for example, the number of eggs present on each of a series of sample dates, the number of new eggs laid per time period can be measured over the entire period during which oviposition occurs. Such values, summed across the whole period during which new members of the stage arise, yield a direct estimate of total numbers entering the stage, as required for construction of life tables. Simultaneous measurement of numbers dying in each time interval from specific causes allows estimates to be made for losses assignable to particular factors. The ratio of these two values (losses:recruitment) then measures the total proportional losses in a stage from the factor (Van Driesche and Bellows 1988). The relationships between recruitment and population densities for healthy and parasitized hosts is illustrated in Fig. 13.7.

Where several mortality factors act together in a stage, loss rates assignable to each factor may be estimated if they are recorded separately at recruitment (for example, for parasitoids, if measured as parasitoid ovipositions). Other calculations are necessary to separate contemporaneous factors if losses are measured at the time of death (Elkinton et al. 1992). Sampling methods to measure recruitment exist for a number of life stages and species, and the method can be applied wherever the biologies involved allow effective sampling approaches to be devised (Lopez and Van Driesche 1989; Van Driesche et al. 1990). For systems in which recruitment into a particular stage cannot be measured easily, it may be possible to obtain an estimate of recruitment from densities of the stage of interest plus estimates of recruitment to the previous stage (Bellows and Birley 1981; Bellows et al. 1982a).

Death Rate Analysis. When several factors act together in one or more stages and their initial attacks cannot be easily observed, a process termed "death rate analysis" can be applied to the observed rates of mortality suffered by the pest from each factor (Gould et al. 1990; Elkinton et al. 1992). This approach calculates the marginal attack rates for each factor (from the observed death rates in individual samples from the population) in each of a series of time intervals (from one sample date to the next). Total mortality associated with each factor over the entire sampling period is found from the product of the rates of survival (= 1 − attack rate) over all the sampling intervals:

$$\text{total mortality for factor} = 1 - \prod_{i=1}^{t} (1 - \text{marginal attack rate for interval i}).$$

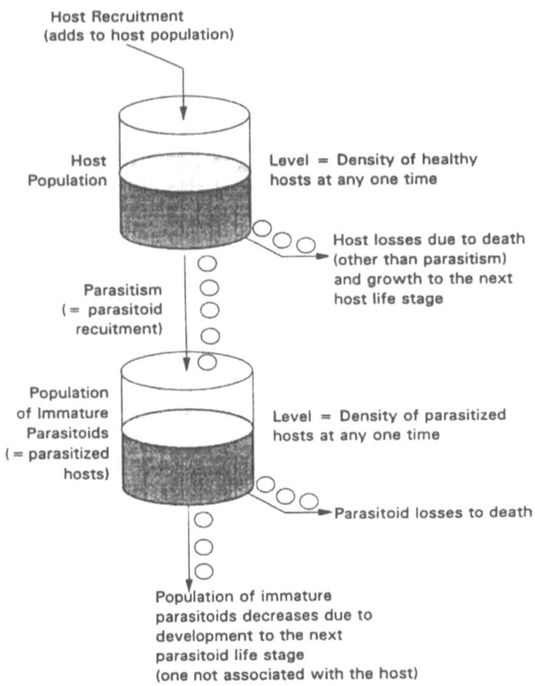

Figure 13.7. A conceptual model of recruitment to a host population, together with losses, resulting in the moment-to-moment host density, with linkage to the population of parasitized hosts (whose density is similarly determined by recruitments, as parasitoid ovipositions in hosts, and by losses).

To use this method, the numbers of deaths from each factor that occur in defined time periods must be determined. Typically, periods are from one sample date to the next, with the proviso that the samples must be reared at field temperatures so that rates of development of natural enemies (parasitoids, pathogens) inside insects in the sample remain the same as for the field population from which the sample taken. A further and final condition is that this method, in the form presented by Gould et al. (1990) and Elkinton et al. (1992), can be applied only to systems in which all recruitment of the host to the earliest sampled stage is completed before sampling and before attack of the mortality agents under study begins. As formulated, the method is not applicable to cases in which host recruitment is widely distributed over time. When independent information on the density of the stages of the pest under study are available, the same analyses can be made for predation.

Growth Rate Analysis. Attempts to evaluate the effect of mortality sources on continuously breeding organisms such as some species of aphids, mealybugs, or other groups, have been based on projections of estimated population growth rates, followed by comparison of the projection to the observed future density. If, for example, on one date, a population of a given density is observed and its growth rate estimated, then a new density can be forecasted for a second, later date, under the assumption that the population is growing at a constant rate. This estimate is then compared to the density of the population actually observed on the second

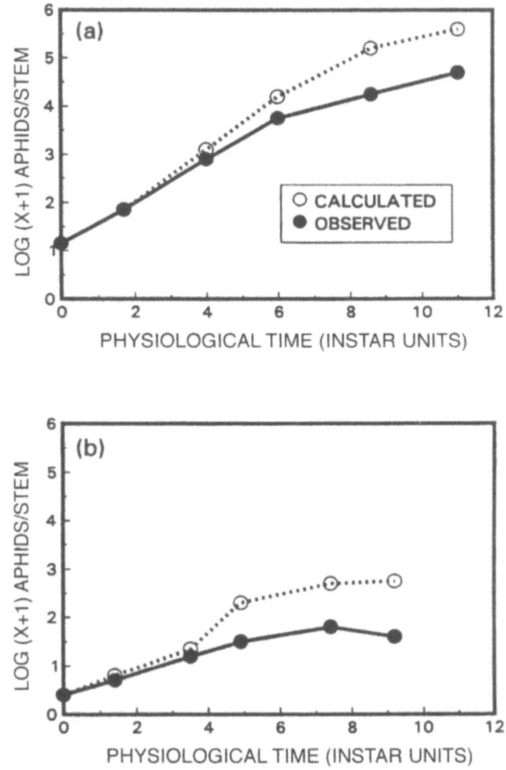

Figure 13.8. The influence of natural enemies can be quantified by calculating the growth rate of a population in the absence of natural enemies and using this growth rate to estimate population increases over time. Differences between the estimated and observed densities may be attributed to the action of natural enemies. In these examples for two populations of pea aphid, *Acyrthosiphon pisum* (Harris), natural enemies are responsible for approximately a tenfold reduction in aphid density over the period of the crop cycle in both (a) 1980 and (b) 1982 (after Hutchinson and Hogg 1985).

date and the difference attributed to mortality (Fig. 13.8). This method depends fundamentally on being able to develop an accurate estimate of the population growth rate in the absence of mortality from natural enemies. Several approaches to do so have been developed (Hughes 1962, 1963; Hutchison and Hogg 1984, 1985). Typically, these methods make the assumption that a continuously breeding population will reach a stable age structure of its life stages and that the population's growth rate can be inferred from the ratio of the life stages. Data on population age structure may be collected either through observing the field population of interest, or under laboratory conditions by establishing cohorts of the pest on the same plant species as the field population and allowing them to reproduce in the absence of mortality. Both approaches assume that a stable age distribution will be reached. The validity of this assumption has been questioned (Carter et al. 1978). Growth rates of small groups of some insects such as aphids may be measured under field conditions using cages to exclude natural enemies from some groups but not others (Morris 1992). Such experiments can also include manipulation of additional factors such as competing herbivores or water stress to separate their effects from those of natural enemies as determinants of population growth rates.

Regardless of how the population growth rate is estimated, it is used to estimate the aggregate effect of all mortality acting on the study population over the time interval, together with the effects of reductions in fertility caused by disease or parasitism. To divide this total effect into components attributable to specific sources of mortality or to lost fertility, further information is needed. This information may be obtained by noting the levels of disease and parasitism present in the samples taken at the beginning of each time interval over which population projections are made. Other methods of segregating mortality into agent-specific components follow in this chapter.

Making Inferences About Effects of Mortality Agents from Life Tables. Once one or a series of life tables has been constructed properly they may be used to make inferences about the importance of a natural enemy or the effect of some agricultural practice intended to enhance a natural enemy's effect. While various inferences are possible, principal points of interest are likely to include: (1) how large one source of mortality is relative to other factors that act either in the same or in a different stage, or the same factor at different sites or under different management practices; (2) whether a newly added factor's effect is offset by the lessening of other contemporaneous or subsequent factors; (3) how close a population is to the point of no net growth ($R_0 = 1$) and how much a given factor contributes to reducing R_0; and (4) whether the presence of a particular factor affects the stability of the pest population.

Comparing Sizes of Mortality Factors. Direct inspection of life tables allows comparisons between the amounts of mortality caused by different agents, or the same agent in different life tables. When more than one mortality factor affects a stage, factors must be compared in terms of marginal attack rates, not apparent mortality rates. For example, consider the case of an internal parasitoid that causes 80% mortality of a pest in each of two plots, but is followed by a predator that kills both parasitized and healthy hosts. If the predator kills 10% of the pests in one location and 70% in the other, the apparent mortality rates from parasitism would be 72% at the low predation site and 24% at the high predation site, because all predator-killed pests would be ascribed to predation, including hosts killed by predators that were also parasitized. If apparent mortality rates were used as the measure of the parasitoid's activity level, the parasitoid would appear to be more active in the site with less predation. This interpretation might lead to attempts to identify factors that differed between locations influencing the parasitoid's abundance or effectiveness, when none existed. Use of marginal attack rates as the measure of mortality would show there to be no difference between the parasitoid populations in the two plots.

Is a Newly Added Mortality Factor's Impact Offset by the Lessening of Another Contemporaneous or Subsequent Factor? When life tables are available for populations both having and lacking a particular mortality agent (as for example where a new parasitoid of a pest has recently been introduced), comparison of such life tables can test to what degree, if at all, the mortality from the new agent is offset by reductions in mortality rates by other factors. Of special concern would be any density dependent mortality that affects a stage in the life history that comes after the stage attacked by the new agent. For example, larvae of some pests may be affected by parasitoids (or other factors) whose intensity is positively related to larval density. In such cases, mortality to the egg stage caused by augmentative release of egg parasitoids such as *Trichogramma* is likely to be partially offset by reduced mortality from subsequent larval mortality factors. Inspection of life tables can be used to form hypotheses about such interactions, based on the density-relatedness of factors as seen in sets of life tables. These are then tested by obtaining life tables from populations which are otherwise comparable, but which differ in the presence or absence of the agent whose effect on the life table is to be evaluated.

How Close is a Population to the Point of No Net Growth (R_0 = 1) and How Much Does a Given Factor Contribute to Reducing R_0? To achieve biological control of a pest population, the pest's power of increase must be reduced enough that the pest is unable to reach harmful densities. In areas where pests breed continuously throughout the year, this implies that the long term average of the net population growth rate must be near one (or if measured as r_m, $r_m \leq 0$). In temperate areas where pests suffer large population reductions during winter, population growth rates above unity may still be low enough that damaging densities do not occur in the seasonally available time prior to the population reduction associated with the winter season. Complete life tables include estimates of fertility, thus allowing population growth rates to be calculated and indicating how much additional mortality would be needed to suppress the pest adequately. The contribution of any agent to the reduction of R_0 could be calculated from such a life table. In general, unless a subsequently-acting factor acts in a density dependent manner (positively or negatively), the marginal rate of any additional source of mortality will result in a corresponding reduction in R_0. For example, if a new parasitoid is added that attacks 50% of any stage, then there will be 50% fewer animals in the next generation and R_0 will be reduced by 50%. To determine whether compensatory interactions between existing and newly added mortality factors are likely, mortalities affecting life stages subsequent in the life history to the point at which the new source of mortality acts should be studied to determine whether any are density dependent.

Does the Presence of a Given Mortality Factor Affect the Stability of the Pest Population? Many characteristics of pests, natural enemies, and their interactions have been proposed as forces acting to regulate pest populations. Foremost among these have been positive density dependent mortality rates. Other possible stabilizing factors include interference among parasitoids, parasitoid aggregation, inverse density dependence, host refuges, specific types of natural enemy searching behaviors such as a sigmoid functional response, invulnerable life stages or invulnerable fractions of host populations, and spatial patchiness (see Bellows et al. 1992c and references therein).

While some of these factors have been associated with increased stability of models (see Chapter 18), demonstration of their action in field populations has been much more difficult. Failure to detect stabilizing influences associated with particular pests or natural enemies can be due to several causes. In some instances, negative results may be obtained precisely because the factor does not contribute toward population stability of the pest. In other cases, data may have been collected improperly or structured inappropriately. For example, in looking for density relatedness of a parasitoid such as *Cotesia rubecula* that attacks first and second instar larvae of its host, *Pieris rapae*, it is the level of attack on these stages and the density of these stages that must be noted to determine whether or not the attack is density dependent. If levels of parasitism are noted, instead, in fourth instar larvae (the stage that dies of parasitism) and associated with the density of that stage, then the actual nature of any density relatedness that might exist cannot be detected because the data were collected incorrectly in view of the biology. Similarly, attempts to detect density dependence in space may be dependent on identifying a spatial scale that is relevant to the organisms involved, given their powers of movement and behavior (Heads and Lawton 1983). Additional difficulties arise in examination of populations for spatially-related density dependence, which is limited to detecting responses based on aggregative rather than intergenerational numerical responses by the natural enemy population. Analysis of temporal series of populations is limited by the problem of autocorrelation in the variables used for analysis.

Several tests have been developed which attempt to detect overall regulation from population census data, and these fall generally into two categories: parametric tests (those relying on such parameters as regression slopes arising directly from data) (Varley and Gradwell 1968; Bulmer 1975; Ricker 1973, 1975; Slade 1977; Pollard et al. 1987; Reddingius and den Boer 1989), and tests based on Monte Carlo or bootstrap techniques. Several studies based on simulated data sets (Slade 1977; Vickery and Nudds 1984; Gaston and Lawton 1987; Holyoak 1993) have generally concluded that no parametric technique is sufficiently robust or powerful to detect regulation from population census data (Bellows et al. 1992c; Holyoak 1993). Some of the Monte Carlo and bootstrap techniques are more powerful and robust in detecting regulation (Pollard et al. 1987; Reddingius and Den Boer 1989; Crowley 1992; Holyoak 1993; Dennis and Taper 1994). Wolda and Dennis (1993) have raised the point that analyzing bounded time series can indicate regulated behavior in the absence of a density dependent mechanism and suggest that this approach may pose problems in interpreting positive findings. Other authors have commented that this problem will be of little consequence in ecological settings, because the factors which could lead to regulated populations are all likely biological (and density dependent) in nature (Hanski et al. 1993; Holyoak and Lawton 1993; Wolda et al. 1994). Combining some of these tests into a multiple-test framework may provide greater generality in the detection of density dependence (Holyoak and Crowley 1993).

Techniques to test the density relatedness of specific mortality factors or agents have been based on various methods of regressing pest mortality from the factor against pest density at the time of attack (Hassell and Varley 1969; Varley and Gradwell 1970; Bellows 1981; Murdoch et al. 1984; Elkinton et al. 1992). Several of these regression techniques have been employed successfully in some cases, but unsuccessfully in others. No statistical method has emerged as superior to all others. Consequently, at this time, failure to detect density relatedness for a factor's action should not be viewed as definitive. Summaries of rates of detection of density relatedness in past studies (Lessells 1985; Stiling 1987; Walde and Murdoch 1988) should also be treated cautiously, in view of the limitations of methods available for such detection.

Further Uses of Life Tables. In addition to the points raised in the preceding sections, life tables have important applications in three additional contexts: in "with and without" experiments on the effects of natural enemies; in studies of the effect of plants on host-natural enemy interactions; and in host addition experiments.

Paired Life Tables. Comparing plots having and lacking a natural enemy, as discussed earlier in this chapter, forms the core of the experimental method of natural enemy evaluation. The strength of the experimental approach is its integration of the total set of factors influencing the population, and the tangible nature of the proof it provides. Its weakness is that it provides little or no insight into the mechanisms involved. When paired life tables are constructed from plots having and plots lacking a natural enemy, the strongest features of both approaches are combined. Life-table data quantify mechanisms responsible for the overall effects demonstrated by the broad experimental design. This approach has been employed to assess the effect of introduced parasitoids of the citrus blackfly, *Aleurocanthus woglumi*, in Florida

(Dowell et al. 1979) and of native parasitoids of the apple blotch leafminer, *Phyllonorycter crataegella*, in Massachusetts (Van Driesche and Taub 1983).

In the *P. crataegella* case, life tables were constructed both for untreated plots, where parasitoids were present (Table 13.4) and for plots treated with pesticides, where parasitoids were nearly eliminated (Table 13.6). Life tables from the untreated plots indicate parasitism was a substantial source of mortality, and yielded a net rate of increase (R_0) of 1.78. This life table may be used to recalculate a hypothetical net rate of increase in the absence of parasitism by setting values for parasitism to zero (Table 13.5). In this table we find a value for R_0 of 7.70 when only parasitism is removed; the value would be higher if some of the residual mortality (possibly due to host feeding) was also removed. This recalculation provides a hypothesis regarding the significance of parasitism in this system, and the hypothesis may be examined in light of data from populations which lack parasitism (Table 13.6). Here we have nearly nonexistent parasitism, and a net rate of increase of 9.41, similar to the hypothesized value. This comparison provides direct, quantitative evidence of the amount of impact parasitism is contributing to the *P. crataegella* system, and illustrates the value of employing paired life tables in evaluating natural enemies.

Paired studies can also be essential to understanding interactions between different sources of mortality, particularly when new natural enemies are being added. Roland (1990, 1994) in his study of winter moth (*Operophtera brumata*), an immigrant forest pest in Canada, found that the magnitude of the effect of introduced parasitoids in British Columbia, Canada, was affected strongly by generalist predators, which fed preferentially on nonparasitized hosts.

Studies of Plant Effects. Plant characteristics have been shown to influence natural enemies significantly (Bergman and Tingey 1979; Price et al. 1980; Shepard and Dahlman 1988).

TABLE 13.4 Life Table of the First Generation of *Phyllonorycter crataegella* (Clemens) in an Unsprayed Apple Orchard at Buckland, Massachusetts (U.S.A.) (From Van Driesche and Taub 1983)

	Stage			Marginal	Apparent Mortality		Real Mortality		k-value	
Stage Factor	l_x	d_x	Factor d_x	Attack Rate[a]	Stage q_x	Factor q_x	Stage d_x/l_0	Factor d_x/l_0	Stage	Factor
Egg	283	6			0.021		0.021		0.009	
Infertility[b]			6	0.021		0.021		0.021		0.009
Sap larvae	277	63			0.227		0.223		0.112	
Parasitism			35	0.134		0.126		0.123		0.062
Residual			28	0.108		0.101		0.098		0.050
Tissue larvae	214	168			0.785		0.594		0.668	
Parasitism			140	0.729		0.654		0.494		0.567
Residual			28	0.206		0.130		0.098		0.100
Pupae[b]	46	0	0	0.000	0.000		0.000		0.000	
Adults	46									
Sex ratio[b]	0.50									
Fertility[b]	22									
F_1 progeny	505									
R_0	1.78									

[a]Marginal rates are calculated by using equations in Elkinton et al. (1992) for the case of a parasitoid and a predator where the predator is always credited for the death of a host attacked by both agents.
[b]Values for *P. blancardella* from Pottinger and LeRoux (1971).

TABLE 13.5 Life Table of the First Generation of *Phyllonorycter crataegella* (Clemens) in an Unsprayed Apple Orchard at Buckland, Massachusetts (U.S.A.), Modified by Deleting Mortality from Parasitism (From Van Driesche and Taub 1983)

	Stage			Marginal	Apparent Mortality		Real Mortality		k-value	
Stage Factor	l_x	d_x	Factor d_x	Attack Rate[a]	Stage q_x	Factor q_x	Stage d_x/l_0	Factor d_x/l_0	Stage	Factor
Egg	283	6			0.021		0.021		0.009	
Infertility[b]			6	0.021		0.021		0.021		0.009
Sap larvae	277	28			0.101		0.009		0.046	
Parasitism			0	0.000		0.000		0.000		0.000
Residual			28	0.101		0.101		0.099		0.046
Tissue larvae	249	51			0.206		0.180		0.100	
Parasitism			0	0.000		0.000		0.000		0.000
Residual			51	0.206		0.206		0.180		0.100
Pupae[b]	198	0	0	0.000	0.000		0.000		0.000	
Adults	198									
Sex ratio[b]	0.50									
Fertility[b]	22									
F_1 progeny	2178									
R_0	7.70									

[a]Marginal rates are calculated by using equations in Elkinton et al. (1992) foi the case of a parasitoid and a predator where the predator is always credited for the death of a host attacked by both agents.
[b]Values for *P. blancardella* from Pottinger and LeRoux (1971).

TABLE 13.6 Life Table of the First Generation of *Phyllonorycter crataegella* (Clemens) in a Sprayed Apple Orchard at Buckland, Massachusetts (U.S.A.) (From Van Driesche and Taub 1983)

	Stage			Marginal	Apparent Mortality		Real Mortality		k-value	
Stage Factor	l_x	d_x	Factor d_x	Attack Rate[a]	Stage q_x	Factor q_x	Stage d_x/l_0	Factor d_x/l_0	Stage	Factor
Egg	433	9			0.021		0.021		0.009	
Infertility[b]			9	0.021		0.020		0.020		0.009
Sap larvae	424	19			0.045		0.044		0.020	
Parasitism			1	0.002		0.002		0.002		0.001
Residual			18	0.043		0.042		0.041		0.019
Tissue larvae	405	34			0.084		0.079		0.038	
Parasitism			17	0.043		0.041		0.039		0.019
Residual			17	0.043		0.041		0.039		0.019
Pupae[b]	371	0	0	0.000	0.000		0.000		0.000	
Adults	371									
Sex ratio[b]	0.50									
Fertility[b]	22									
F_1 progeny	4072									
R_0	9.41									

[a]Marginal rates are calculated by using equations in Elkinton et al. (1992) for the case of a parasitoid and a predator where the predator is always credited for the death of a host attacked by both agents.
[b]Values for *P. blancardella* from Pottinger and LeRoux (1971).

Plant effects can operate at several levels. At the plant species level, a pest found on many plant species may be subject to attack by different natural enemies on different plants. Conversely, parasitoids derived from hosts on alternative plant species may not effectively recognize and attack the same pest on the target crop. At the level of one crop plant, varieties may differ in characteristics, such as abundance of nectaries, hirsuteness, or presence of glandular trichomes, that may affect the efficiency of a natural enemy species foraging on the plant (see Chapter 15 for more on plant effects on natural enemies). Manipulation of the effects of plants on natural enemies is an important method of natural enemy conservation (see Chapter 7). Evaluation of such plant effects is of broad interest (Price 1990) and can be achieved by building life tables for plots with the desired contrasts of plant features and using exclusion cages, with subplots having and lacking the key natural enemies of interest.

Host Addition Experiments. Another purpose for which life tables can be used is to compare mortalities in plots where host densities have been raised by artificial addition of hosts to the field population to mortalities in plots without such host additions. Elkinton et al. (1990) and Gould et al. (1990) increased numbers of gypsy moth larvae (*Lymantria dispar*) by adding egg masses to plots and were able to detect and quantify increased parasitism due to spatial aggregation of the tachinid *Compsilura concinnata* (Meigen) at locations with high host densities.

Assigning Mortality to Particular Species

A third step in the evaluation of the effects of natural enemies is to divide the total losses measured in each pest life stage or in various time periods into the separate contributions made by particular species of natural enemies, or by groups of natural enemies that act in similar ways (ground-dwelling predators, flying predators). Methods for this differ between groups of natural enemies because of the nature of the physical evidence left following their attack on the pest.

Parasitoids of Arthropods. The division of total mortality from parasitism into the separate effects of individual species of parasitoids is usually possible because hosts attacked by parasitoids bear immature parasitoids. When hosts drawn from field populations are reared, these immature parasitoids mature and produce adult parasitoids which can be identified to species and the number of deaths they cause counted. This process then directly divides parasitoid-caused mortality into its species-specific components.

Three potential difficulties exist that affect the accuracy of this process. First, the numbers of hosts from dying from parasitism must be analyzed appropriately to calculate the underlying marginal attack rates for each parasitoid species (Elkinton et al. 1992) because these rates, not apparent mortality rates, correctly measure the significance of each separate parasitoid in the system.

Second, deaths caused by parasitoids through host-feeding, host-mutilation, or stinging without oviposition cannot be detected by rearing live hosts or determining the proportion that bears immature parasitoids. To measure this component of parasitoid-caused mortality, methods similar to those used to measure predation must be used. If the bodies of hosts that have been attacked in these ways remain detectable in samples (for example, for hosts inside plant tissue such as leaf miners and gall makers or hosts attached to plant tissues, such as scales), then these cadavers can be examined. If evidence exists that allow parasitoid attacks to be distinguished from other kinds of mortality, then these hosts can be credited as mortality caused by either a specific parasitoid (if only one parasitoid attacking the host exhibits host feeding), or credited to a group of parasitoids (if several species host feed). If cadavers of attacked hosts are lost, then they will not be detectable in samples except as reduced density. These host deaths will be indistinguishable from deaths from predation and can be inferred only from laboratory studies in which the frequency of host feeding relative to oviposition is observed for particular parasitoids species (Van Driesche et al. 1987). Laboratory experiments may also be needed to demonstrate that parasitoids are responsible for other types of mortality observed in hosts, such as that caused by wounding from repeated ovipositions that may occur when parasitoid-host ratios are high (Ryan 1988).

Third, reductions in host fertility due to sterilization of adult hosts prior to actual death from parasitism must be accounted for by supplemental experiments which directly measure these losses (Beard 1940; Young 1987). Alfalfa weevil (*Hyperica postica*) populations, for example, suffer an approximate 50% decline in total oviposition from attack by the euphorine braconid *Microctonus aethiopoides* (Van Driesche and Gyrisco 1979). Aphids of many species suffer partial declines in adult fertility as a consequence of parasitism by aphidiid parasitoids, with the number of progeny produced by parasitized adults being a function of aphid age at the time of parasitoid attack (Campbell and Mackauer 1975; Liu and Hughes 1984). Parasitism of the pine engraver *Ips pini* (Say) by the pteromalid *Tomicobia tibialis* Ashmead lowered beetle reproduction by 50% (Senger and Roitberg 1992).

Pathogens of Arthropods. Division of total mortality from diseases into portions resulting from attack by specific pathogens of arthropods is very similar to the example of parasitoids, in that samples of the pests can be collected, some of which will be infected at the time of collection. These organisms can be reared and pathogens responsible for each dead host can be obtained for identification. If collections of pests are dissected, well-known pathogens may be recognizable based on presence of features such as viral inclusion bodies or other structures, although a risk of misidentifications exists should a new pathogen, not previously a common mortality factor, be encountered. Other techniques for detecting infections of specific pathogens in hosts in early stages of disease include electrophoresis, antigen-antibody methods (for example, ELISA and related techniques), and DNA detection methods (Keating et al. 1989; McGuire and Henry 1989; Hegedus and Khachatourians 1993; Shamin et al. 1994) (Fig. 13.9). These techniques offer advantages of speed and approach the direct measurement of marginal attack rates for pathogens. However, one potential difficulty with their use is that sublethal infections may be detected and incorrectly counted as contributing to mortality. Conversely, some pathogens may not reach the detection threshold until shortly before death.

Vertebrate Predators of Arthropods. Vertebrates, because of their larger size, longer life span, more varied diet, and greater ability to switch between foods as abundances change, are usually not strongly linked in their population dynamics to single species of pest invertebrates. Buckner (1966), Frank (1967), and Hanski (1990) illustrate some of the approaches that can be used to quantify the effects of such vertebrate predators as small mammals and birds on forest insect populations.

Arthropod Predators of Arthropods. Unlike parasitism and disease, acts of predation often leave no enduring physical evidence, except in cases where the prey is found in some structure

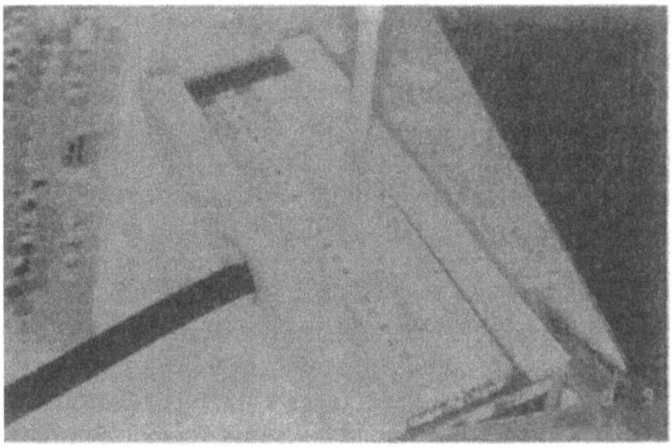

Figure 13.9. Antigen-antibody methods such as ELISA may be used to detect prey antigen in guts of predators or pathogens inside hosts, providing valuable information on identities of natural enemies attacking target species. (Photograph courtesy of J. Burand.)

(plant parts, galls, leaf mines) that is durable and which continues to bear marks of the predator's attack (Sturm et al. 1990). Consequently, indirect methods must often be used to quantify levels of mortality inflicted on pest populations by predators. Two general approaches to this task have been developed. One approach ("top-down method") consists of measuring total losses from predation suffered by the pest population, and then by a variety of methods assigning portions of total losses to specific predators or groups of predators. The other approach ("bottom-up method") starts with observations on the numbers of various types of predators in a system and uses information on the foraging abilities and feeding capacity of specific predator species to develop an estimate of the degree of effect a given predator population is likely to be having on the pest population (O'Neil and Stimac 1988b). The following discussion describes the rationale and techniques for each of these approaches.

Top-Down Method. The first step in this approach is determining the total losses in the prey population that are due to predation. In life table analyses, predation losses are usually estimated as the difference between total losses and all losses that can be attributed to specific sources such as parasitism, disease, or other known causes. Measured in this way, estimates of losses attributed to predation are likely to be inflated to some degree by losses from factors such as rain and other abiotic factors that are not separated from predation. If actual losses from such factors are measured, these can be used to further refine predation estimates (Nuessly et al. 1991 presented methods for estimating effects of rain and wind on *Heliothis* eggs).

An alternative approach to determining the total losses due to predation is to use some form of exclusion to create prey populations with and without exposure to the predator complex present. Differences in survivorship of the prey between these two subpopulations provides a measure of the total mortality from predation, provided the barrier or cage excludes only predators. Chiverton (1986), for example, used barriers to exclude ground-dwelling predators attacking cereal aphids in the United Kingdom. By varying the types of barriers used, or dimensions of mesh in wire or nets, exclusion can be limited to specific predator groups, allowing their effects to be quantified separately. Campbell and Torgersen (1983), for example, were able to use combinations of bird netting and sticky barriers to separately quantify the effects of predation by birds versus ants on western spruce budworm, *Choristoneura occidentalis* Freeman, larvae and pupae. Since this approach often does not lend itself to distinguishing the effects of particular predator species, it must be used in conjunction with other information (such as the relative abundance of predators of various species within a given predator group) to make the final estimate about the importance of individual species.

A third way to measure predation is to follow the fate of either a natural or an artificially established cohort of prey, making the assumption that losses experienced by the experimental group will be the same as those of the field population under study. One way to attempt to meet this assumption is to match features of the exposed cohort to those of the target population as a whole, including such characteristics as density, age, size, naturalness of placement of prey on the host plant, and other factors such as presence of kairomones that might affect predation rates. A relatively direct way in which to achieve this matching is to use naturally occurring individuals of the prey species, selected at random from the field population, as the exposed group for observation. If laboratory-reared individuals are used, they should be chosen so that numbers, ages, dietary histories, and placement in the field mimic the characteristics of the study population at the time of exposure (Hazzard et al. 1991). Alternatively, several cohorts may be exposed, each matching only a portion of the field prey population. Predation rates on these cohorts then are applied only to that fraction of the total field prey population that is in each life stage, as determined by sampling (Elven et al. 1983). Exposed individuals can be marked in various ways, with dye, paint, tags, radioactive isotopes, heavy metals, or other procedures, to help identify them as cohort members upon recapture at the end of the exposure period.

Having measured total losses to predators (with predators either viewed as a single guild or divided into several smaller sets based on size, taxon, or other characteristics), it would be of interest if this estimate could be further separated into components assignable to individual species of predators. There is no rigorous way to do this unless the numbers of predators are such that one or more of the subgroups whose actions can be experimentally measured consist of only one species. Where this is not possible, further separation must be based on estimates of the level of predation potentially associated with each predator species in time periods of interest, using information on each predator's density, consumption capacity, foraging effectiveness, and other characteristics. The description of what each predator's contribution ought to be, based on its potential, is the basis of the "bottom-up method."

Bottom-Up Method. The bottom-up method seeks to construct a picture of the suite of predator species present in the habitat of the prey species and to estimate their relative importance using several types of information (see Whitcomb 1981 for a discussion of this process).

The first step is to develop a list of predator species in the habitat (Bechinski and Pedigo 1981). Such lists may be quite long, running to tens or hundreds of species. In developing lists, it is important to remember that no single sampling method or time of sampling will catch all species of predators in the habitat. Furthermore, numbers caught may be influenced by how the biology of a given species interacts with the sampling method and will not solely be a reflection of a species' density in the habitat. Therefore, it is important to use various sampling approaches in the early phase of the predator survey and to consider the results of several methods in rating the abundance of any single species relative to the other members of the predator guild.

The second step is to determine which predators have potential to consume the target pest species. One approach is simply to combine hungry predators with the target prey in the laboratory and see whether predation occurs. If it does not, the species may safely be discarded as a potential predator of the target species. It is important to remember that under laboratory conditions prey may be accepted that are rejected or never encountered in the field.

The third step is to obtain information on which predators are actually consuming the target prey in the field. Methods to do so are direct observation and detection in predators of some label indicative of the prey species. Direct observation is simple but time consuming, and consists of stationing an observer near either a naturally occurring patch of the prey species or an artificially deployed group of prey and waiting for predation to occur (Kiritani et al. 1972; Godfrey et al. 1989). One important caveat about such observations is that some species of predators feed at night when they may be more difficult to observe.

Detection in predators of labels derived from feeding on the target prey is an important method used to identify which predators eat which prey. One class of labels is those that can be introduced (usually in rearing diet) into laboratory-reared prey which are then exposed to predators in the field. Marks include some fat soluble dyes such as Calco Oil Redt (Elvin et al. 1983), radioactive materials such as isotopes of phosphorus P^{32} (McDaniel and Sterling 1982; Gravena and Sterling 1983), or distinctive elements such as rubidium (Cohen and Jackson 1989). In each case, these materials are ones that can be fed to prey reared in the laboratory and which will pass into predators that eat marked prey in quantities sufficient for later detection. Dyes are detected during dissections of predators by their color. Radioactive materials may be detected with a Geiger counter, and materials like rubidium require use of atomic absorption spectrophotometry.

Another class of labels is the proteins of the tissue of the prey species itself. Antibodies against prey antigens (proteins) can be used to determine if a given predator has recently ingested proteins from the prey. Many techniques have been developed for this type of analysis, including electrophoresis, single radial immunodiffusion, rocket immunoelectrophoresis, enzyme-linked immunosorbent assay (ELISA), tube precipitin test, ring test, Oakley-Fulthorpe test, crossed immunoelectophoresis, crossover immunoelectrophoresis, fluorescence immunoassays, chromatography, double diffusion, immunofixation, standard immunoelectrophoresis, latex agglutination, passive haemagglutination inhibition assay, and complement fixation test (see Sunderland 1988 for a detailed discussion of these tests). Test features of importance are ease of use (speed and cost), sensitivity (minimum detectable quantity of antigen), and specificity of the reaction (freedom from false positives). The last issue is especially important. If antisera are prepared against blends of prey proteins, the probability is very great that other potential prey will cross react to the antiserum, misleading the investigation. Use of monoclonal antibody technology provides a means to prepare antisera to a single protein of the prey species (Greenstone and Hunt 1993; Hagler et al. 1994). Once available, such an antiserum must be tested for cross reactivity against protein mixes from the other potential prey in the habitat to estimate the potential rate of false positives.

Of the tests mentioned, the first four can be used quantitatively, the next six have potential to be used quantitatively given further development, and the last seven are qualitative only. Qualitative tests do not indicate how much antigen was ingested and, therefore, do not give any indication of the number of prey eaten by the predator. Development of quantitative assays (to score the number of prey eaten by a sampled predator) is complicated by many factors, including meal size, time since ingestion, temperature, species differences, and sensitivity of the test. Approaches to quantitative assays are discussed by Greenstone and Hunt (1993).

Antigen-antibody tests do not by themselves evaluate the effect that any one species or set of species of predators has on the target prey population. Rather, they establish the list of predators that are actually eating a particular prey. Such information, however, can be used with additional data, such as predator density, to calculate the daily predation rate associated with each predator species. Formulae exist to calculate minimum and maximum daily rates of predation per unit area of habitat per predator species. The minimum rate formula (Dempster 1967) calculates the rate as predation rate per day = predator density × proportion of predators giving a positive reaction to prey antigen ÷ the number of days antigen remains detectable. The maximum rate formula (Rothschild 1966) calculates the rate as predation rate per day = predator density × proportion of predators giving a positive reaction to prey antigen × the mean number of prey eaten per day in the laboratory by the predator when prey are abundant. Number of prey eaten per day under laboratory conditions can be expressed in relation to prey density (O'Neil 1988). Leathwick and Winterbourn (1984) used these formulae to calculate the effects of a series of predators on lucerne aphids (*Acyrthosiphon* spp.) in New Zealand. Sopp (1987) combined these formulae in modified form, using ELISA methods which allowed quantity of prey antigen to be calculated, as well as prey antigen disappearance rates (from digestion) based on temperature and time. A further improvement in quantifying invertebrate predation rates from ELISA data has been developed by Sopp et al. (1992). Calculations of daily predation rates for individual species of predators over extended portions of time would add a further dimension to such evaluations. Wratten (1987) provides an overview of principles for evaluating of predation through these methods.

Plant Pathogens. The occurrence of pathogens affecting plants can be measured by collecting samples of diseased tissue and culturing the pathogen for identification. For this process, as with identification of arthropod pathogens, the most reliable species identifications are based on rearing and culturing. For well-known species, dissections and microscopic examination may suffice, or the electrophoretic, antigen-antibody, or DNA detection methods mentioned above may be employed. However, unlike attacks by pathogens of arthropods, infections may influence plants in ways that are detrimental, yet fall short of outright mortality. Pathogens may destroy roots, hamper the functioning of conductive tissues, or blight reproductive tissues, with resulting loss in growth, competitive vigor, and reproduction. Analysis of the effects of specific pathogens can be assessed by experiments comparing rates of these processes, plus rates of plant survival, under conditions of exposure and nonexposure to the pathogen, under defined environmental conditions. Results of such experiments, coupled with survey data on infection rates of target plants in field populations, allows the relative importance of pathogens affecting the plant to be rated.

Herbivorous Arthropods. Separation of the total impact of all herbivores affecting a plant into the separate effects of each herbivore species involves the same set of issues as discussed above for plant pathogens. Evaluations start with field surveys to develop a comprehensive list of species in the herbivore complex (Sheppard et al. 1991). Such data must be based on a suitably wide variety of sampling techniques and must be quantified in terms of numbers per unit area so that comparisons can be made between organisms affecting various parts of the plant. These data, when combined with estimates of consumption rates, allow some comparisons to be made between herbivores. These data can further be compared to results of single-species experiments in which effects on plant growth, survival, competitiveness, and reproduction are compared between plots having and plots lacking the herbivore of interest. Development of paired plant life tables for plots having and lacking specific herbivores (such as recently introduced weed control agents) is an effective way to quantify natural enemy effects on plants. Chapter 17 presents a discussion of herbivore impacts on plants.

SIMULATION MODELS AND THE EVALUATION OF NATURAL ENEMIES

Simulation models have been closely linked to the evaluation of natural enemies, although both model construction and its relationship to natural enemy evaluation is somewhat more complicated than the analytical approaches discussed above.

Simulation models are mathematical tools, or constructs, designed to mimic the interactions of different populations. Models are typically framed as a series of mathematical equations containing parameters which define such attributes of population growth as developmental time, adult longevity, fertility, and search rates. Models may be designed to incorporate only a single population, or may be more complex, and include both a natural enemy population and its host, and even the host plant on which the host insect feeds (Crute and Day 1990; Godfray and Waage 1991; Boot et al. 1992; Gillman et al. 1993; Gutierrez et al. 1993; Trichilo and Wilson 1993). Berry et al. (1991), for example, used a simulation model to describe growth of populations of the Banks grass mite, *Oligonychus pratensis*, and its phytoseid predator *Amblyseius fallacis* on maize. Smith and Trout (1994) use a population growth rate model for rabbits (*Oryctolagus cuniculus*) in the United Kingdom to form hypotheses about effects of several possible control options. When models are executed, or run, the values given to each parameter combine in the ways determined by the mathematical equations and produce an outcome or result of the interaction among the populations in the model. In this context the model is an abstraction, or a simplification, of the populations represented, and the modeled outcome is a hypothesis about how the natural system performs. Each model outcome is defined by the structure and form given to the model and the particular parameter values used in its execution.

Simulation (and other models) are hypothesis generators, in that they can synthesize the consequences of a series of assumptions (concerning search rates or behaviors, age-specific effects, and other factors) into a single result—the population trajectory of the pest and its natural enemy. In this context, simulation models can serve to generate hypotheses about very complex situations (such as what the consequences might be of altering the density response of predator searching) in ways that are not possible for many of the simpler analytical methods discussed earlier in this chapter.

These hypotheses, of course, do not evaluate the performance of a natural enemy, but rather propose quantitative and testable ideas about how the system functions. If a model of a pest-natural enemy system is constructed, it might be run, for example, both with and without the natural enemy to produce hypotheses about what impact the natural enemy is having on the pest population. These hypotheses, like any other hypothesis, must be tested against experimental observations and quantitative data. Hence, the link from the model to the process of evaluation is through field experiments to determine if the system behaves in the way the model suggests. This process is often termed "model validation." Model validation can be refined further by examining details about the model's predicted outcomes, such as daily rates of mortality in the pest population, and comparing these results to those actually observed in the field. Such further evaluation of the model permits examining the applicability of the internal structure of the model, in addition to examining the suitability of its overall outcome.

When a model has been validated and found to be in accord with field results for a number of different situations (such as when a model successfully describes the growth of the pest population in the absence of the natural enemy and also describes the system's behavior in the presence of the natural enemy), researchers gain confidence that the model indeed captures essential elements of the biological interactions among the populations. Such "validated" models are then sometimes used to explore the behavior of the system under alternative structures or hypotheses, such as examining how the system might behave following the addition or substitution of another natural enemy (Godfray and Waage 1991) or in the presence of treatments of chemical pesticides (Trichilo and Wilson 1993). Each of these exercises generates new alternative hypotheses about how the system might function under those circumstances. This demonstrates one of the important strengths modeling technology can contribute to biological control: models can emulate the behavior of complex systems and permit complex questions to be asked and hypotheses to be formed. While hypotheses must always be tested by careful experimentation and collection of data, using models to define potential outcomes and to refine hypotheses can be an aid in planning field experiments to examine relationships and dynamics in biological control systems.

DOCUMENTATION, ECONOMIC EVALUATION, AND PLANNING

Documentation

Biological control programs need to be thoroughly documented so that the facts of what was done, when, where, and by whom are clear. Records concerning the importation, release, and recovery of new species of natural enemies are critical in this regard. These records should be listed, if possible, with a national authority charged with maintaining information on biological control and publishing periodic reports of such actions (as is the case for USDA-ARS in the United States, Coulson 1992). Voucher specimens of natural enemies and, in some cases, the pests they attack, should be deposited for future reference in appropriate collections in institutions of sufficient stability and resources that their future curation can be guaranteed.

Economic Evaluation

Biological control projects must be evaluated in economic terms convincing to economists and political administrators who must assess the productivity of the resources invested in the work. The form economic evaluation takes will vary for augmentative, conservation, and introduction projects.

Augmentative projects are successful if pests are controlled, if farmers can afford (and actually adopt) the use of the natural enemies being sold, and if profits from sales of reared natural enemies exceed production costs. Sales, of course, are frequent only if reared natural enemies are effective and costs are competitive with other forms of control. Many factors bear on the costs of reared natural enemies relative to chemicals, some of which (for example, ease of application or desire to reduce worker exposure to chemical pesticides) are difficult to express in economic terms. In part, cost of reared natural enemies is determined by the size of the rearing facility and the size of the market for the facility's products (e.g., Rothenburger and Sautter 1987). Projections of economic feasibility of use of reared natural enemies can assist decisions about investments in facilities for rearing natural enemies. The U.S. Department of Agriculture, for example, made such an assessment for the use of the wasp *Pediobius foveolatus* (Crawford) for control of *Epilachna varivestis*, a pest of soybeans (Reichelderfer 1979). (For more on factors affecting cost of reared natural enemies see Chapters 10 and 11). Costs associated with the development, registration, and marketing of pathogens as pesticides have been estimated, and are 8–16 times smaller than current costs for development of chemical pesticides (Woodhead et al. 1990).

Conservation methods of biological control can readily be evaluated economically by comparing production costs and crop yields under a conservation management system and some other approach. Concurrent efforts are needed to measure to what degree the new practice enhances natural enemies of the pest and to relate this enhancement to changes in pest numbers.

Economic assessments of the benefits of introduced biological control agents have been made for a number of weed and arthropod pests (Andres 1976; Harris 1979; Ervin et al. 1983; Norgaard 1988; Voegele 1989; Tisdell 1990). Estimates of benefit-to-cost ratios for biological control projects have exceeded 100:1, and for projects conducted in Australia they averaged 10.6:1 for biological control work, compared to 2.5:1 for chemical control projects (Tisdell 1990).

Measures used to express benefits from introduced natural enemies include costs savings of producers (saved crop production plus lowered pest control costs), profit increases, increased values of land (pasture improved by control of a weed), and lowered cost of a product to the consumer (Tisdell 1990). Habeck et al. (1993) present an economic model for biological control through introduction that examines how large average expected economic benefits of such projects need to be for benefits to exceed costs.

PLANNING

National or institutional biological control programs should be evaluated periodically to determine if various components of the program are functioning as intended, if target selection methods are identifying targets amenable to actual control, as evidenced by the success record of the national program in achieving its pest control goals, and if the funds invested are justified in view of the data on the economic results of individual projects.

The value of the method to a society should be judged at the level of the overall national program, not the individual project. The case of biological control through introductions may be considered, for example. About one project in six is likely to be completely successful, and an additional two projects in five are likely to be partially successful. The economic return on projects is likely to be on the order of 10:1 or greater. The economic return for the national program as a whole is, therefore, likely to be 3:1 or better, allowing for programs that are not successful (Bellows 1993). This perspective is important because some projects may fail. These, however, should not be a source of disillusionment with the method because benefits from successful projects are large enough to offset the costs of unsuccessful projects.

CHAPTER
14

Integration of Biological Control into Pest Management Systems

INTRODUCTION

In some instances, biological control programs occur in areas where little or no pest control is attempted, such as forests, rangeland, or aquatic areas. In other instances, there may be only one important pest as is the case for many insect pests of woody landscape plants. Biological control programs conducted against pests in such circumstances can stand alone (Huffaker 1985); because no broader system of pest management exists, there is little or no need to harmonize other pest control activities with the biological control program.

Other biological control programs address pests which are elements of pest complexes, all parts of which must be managed successfully for crop production. In such instances, the functioning of biological control agents (and hence the success of programs) will be strongly affected by other pest management actions. Tactical methods to conserve biological control agents in pest control systems are discussed in Chapter 7. The focus of this chapter is on the strategic relationships between biological control and other pest control activities that determine the role biological control plays in crop management systems. Historical perspective on approaches to integrated pest management and the relative role of biological control is provided by Cate and Hinkle (1993).

PEST MANAGEMENT SYSTEMS: FOUNDATIONS AND TRANSITIONS

Foundations

Many crops have key pests, the control of which is essential for production of the crop. Controls employed for such key pests form the foundation for the crop's pest management system around which all other control decisions must be coordinated. Pest management systems may rest on foundations either of chemical control, cultural control, biological control, or control through plant resistance, depending on what is required for control of their key pests (Fig. 14.1).

Chemical Foundations. In apple production in the northeastern United States the apple maggot fly (*Rhagoletis pomonella* [Walsh]), the plum curculio (*Conotrachelus nenuphar* [Herbst]), and the apple scab pathogen (*Venturia inaequalis*) are key pests which directly attack the fruit and which must be controlled to produce a high quality crop. Until the development of

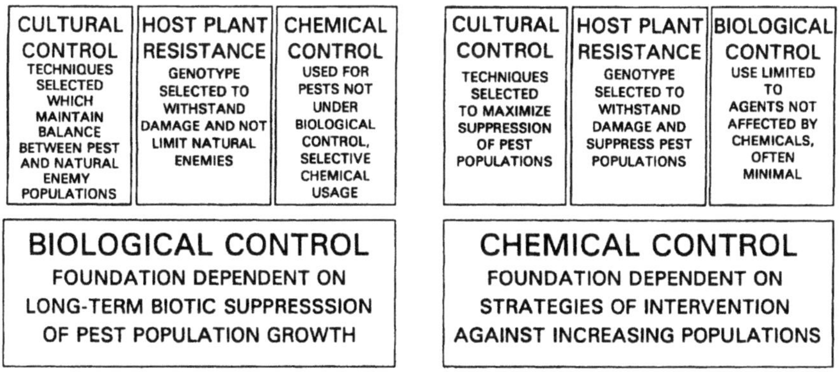

(a) Biologically-based pest control system, with foundation on natural enemy-pest relationships.

(b) Chemically-based pest control system, with foundation on strategies of intervention with pesticides.

Figure 14.1. Pest control systems may rest on biological control (a) or chemical control (b) foundations.

effective trapping methods for suppression of apple maggot flies, control of these key pests rested on a foundation of chemical control through the use of foliar applications of contact insecticides and fungicides repeated at intervals throughout the risk period. Controls for other pests were adjusted around the chemical control of these key pests. In general, the major secondary pests (mites, leaf miners, and aphids) are subject to control by natural enemies. In practice, much of the potential for biological control of these pests is not realized because pesticides applied for control of key pests harm important natural enemies of secondary pests, or enhance reproduction rates of secondary pests such as spider mites. As a consequence, additional chemical controls are often applied for mite or leaf miner control because chemical control of secondary pests is more compatible with chemical controls used for the key pests than are biological controls.

Biological Control Foundations. In crops in which the key pest is controlled by natural enemies, biological control forms the foundation for crop pest management. Biological control-based systems are most likely to develop if a significant key pest is the successful target of biological control early in the history of the production of the crop. In citrus in California, for example, control of the cottony cushion scale, *Icerya purchasi*, by the coccinellid *Rodolia cardinalis* established the need for the conservation of this coccinellid in the crop. The requirement to conserve this key natural enemy acted subsequently to influence choices about methods that could be used to control other pests in the crop. Chemical controls could not be employed except in ways that were not disruptive to the biological control foundation for the pest control system. Other examples of the influence of a biological control foundation include the mealybug *Pseudococcus longispinus* (Targioni-Tozzetti) on avocado in California (Bennett et al. 1976); the mango mealybug, *Rastrococcus invadens* Williams, in West Africa (Agricola et al. 1989); the spotted alfalfa aphid, *Therioaphis maculata*, in Australia (Hughes et al. 1987); the diaspid scales *Aonidiella aurantii* and *Chrysomphalus aonidum* (Linnaeus) (Bruwer and Villiers 1988; Bedford 1989) in South Africa on citrus; and the alfalfa weevil, *Hypera postica*, in the eastern United States (Day 1981). In each of these projects, successful biological control of a key pest of the crop was achieved, creating the need to avoid the use of pesticides in the future in order to preserve natural enemies providing control of these key pests.

Mixed Foundation Systems. Not every crop has a clearly defined control strategy as a pest management foundation. Crops may lack well-defined key pests, or pests may be amenable to either chemical or biological control. In such systems, consequences of choices made about control are likely to accumulate, leading the system towards the development of a foundation based on some dominant control method such as pesticides, biological control, host plant resistance, or cultural practices. Strawberries, for example, in many parts of the United States are affected by spider mites, mirid bugs in the genus *Lygus*, and grey mold (*Botrytis cinerea*). Mites can be suppressed by either acaricides or augmentative releases of predacious mites. *Lygus* bugs cannot be easily suppressed by biological control methods within single fields within a crop cycle, but do have potential to be suppressed regionally by introduced parasitoids from Europe (Day et al. 1990). Finally, even grey mold, traditionally suppressed with fungicides, has potential to be controlled by biological control methods, using antagonistic fungi distributed by pollinators (Peng et al. 1992). In such systems, development of the pest control foundation will depend on choices made by researchers, pest management consultants, and farmers. Practices employed will affect decisions regarding the development of controls for remaining pests. Choosing chemical controls may make use of biological controls less likely. Use of biological controls for one pest may stimulate use of biological control for additional pests. Chemical controls for *Lygus* bugs, for example, may make releases of predacious mites ineffective, resulting in the use of acaricides. Use of some fungicides may reduce the reproductive rates of predacious mites (Dong and Niu 1988), making biological mite control less effective.

Types of Transitions and Openings that Permit Strategic Change

Many pest control systems are based primarily on the use of pesticides. Initial efforts at improving pest management in such cases is often largely pesticide management, in that the quantity of pesticide used is reduced by pest monitoring, use of damage thresholds, and calibration of machinery, but most pest control is still achieved by the use of pesticides. Transforming such systems to ones based on biological control or controls other than the use of chemical pesticides is a subject of interest in many countries. In principle, several routes might lead to such transitions: incremental modifications, new control methods for key pests, development of pesticide resistance by key pests, legal restrictions on essential pesticides, or development of biological controls for key pests.

Incremental Modifications. A commonly held belief is that chemically-based control systems can be modified gradually into ones in which few or no pesticides are used by a series of small changes, requiring only small adjustments by growers in any given year. In practice, this goal is often difficult to achieve because chemical practices are in many ways antagonistic to biological control. Opportunities to modify such practices through use of physiologically-selective pesticides will be limited by the finite number of pesticides registered for use on the crop. Efforts to employ ecologically-selective methods of pesticide application will be limited by constraints on retaining effectiveness of the pesticides against the target pests. Dosages reduced beyond certain limits may be ineffective; alternative row application methods may control some pests, but not others. While incremental methods are most acceptable to growers, their effectiveness is limited. More far-reaching changes may be needed in the form of partial or total redesigning of the controls employed.

New Control Methods for Key Pest. One approach that is often involved in converting a chemical-based control system to a nonchemical system is the development of some alternative method to control the key pest in the crop. In apples in Massachusetts, for example, control of apple maggot (*Rhagoletis pomonella*) with traps (Prokopy et al. 1990), or in peaches in Australia, control of oriental fruit moth (*Grapholita molesta* [Busck]) through pheromone-based mating disruption (Vickers et al. 1985), are such key transformations. Similarly, in cotton a successful eradication program in the southeastern United States against *Anthonomus grandis* Boheman, the boll weevil, removed the need for treatment of cotton with chemicals for this key pest (Lambert et al. in press). In Brazil, use of the *Anticarsia* nuclear polyhedrosis virus for control of the soybean pest *Anticarsia gemmatalis* removed the need for use of chemical pesticides for this key pest in the crop (Moscardi 1990).

In general, any nonchemical control method that replaces regular use of a broad spectrum pesticide in a crop will increase the opportunity to achieve biological control of the remaining pests in the crop. Control methods to replace chemical pesticides include nematodes, pathogens, traps, mating disruption with pheromones, cultural practices such as crop rotation, host-plant resistance, manipulation of cropping patterns, or area-wide pest elimination by eradication programs such as the use of sterile insect releases, or combinations of methods as were employed against the boll weevil, using pesticides and cropping restrictions. Elimination of formerly-required pesticide applications is particularly valuable if these are early season applications. This allows time for natural enemy populations to colonize and reproduce in the crop without immediately being suppressed by pesticide treatments.

Development of Pesticide Resistance by a Key Pest. When key pests become highly resistant to most available pesticides, the ability of growers to effectively suppress the pest with chemically-based pest management systems may be lost. At this point, the system is without definite shape and growers are amenable to changing the basis for pest control in the crop. For example, in the 1980s *Leptinotarsa decemlineata*, the Colorado potato beetle, in parts of the northeastern United States became resistant to virtually all available pesticides. Dosages and numbers of sprays required to suppress the pest were so high that control costs made the potato crop unprofitable. In the absence of workable chemical pest control, interest developed in the *tenebrionis* strain of *Bacillus thuringiensis* which was found to be effective against smaller larvae of the pest (Ferro and Lyon 1991). Elimination of synthetic chemical pesticides for control of this key pest permitted better survival of natural enemies affecting the beetle and opened up the possibility of achieving biological control of such secondary pests as the foliage-feeding aphids in the crop (such as *Myzus persicae*).

In other crops, similar situations have developed with pesticide-resistant Lepidoptera, whiteflies, and mites, among others. The development of pesticide-resistant cotton pests in the Cañete Valley of Peru in the 1950s, following the widespread use of synthetic pesticides, is perhaps the best known example of this process. In Peru, failure of pest control based purely on pesticide application quickly occurred and was replaced by a system incorporating cultural practices such as fixed planting dates and clean fallow periods, and the active conservation of natural enemies (van den Bosch et al. 1976).

Restrictions on Pesticides. The ability to continue to employ chemical pest control systems may also be lost if key pesticides are banned or prohibited in selected areas. In forests in Canada, for example, control of *Choristoneura fumiferana*, the spruce budworm, for many years was based on aerial applications of various chemical pesticides. In recent years, several provinces have made such forest pesticide applications illegal, except for *Bacillus thuringiensis*. This prohibition created the need for a management system based on biological control. Ultimately, both applications of *Bacillus thuringiensis* and releases of the egg parasitoid *Trichogramma minutum* were considered as substitutes. Similarly, prohibitions against the application of pesticides in certain areas (such as near cities in Germany, or on lands in public groundwater resource areas in Massachusetts) may also create the need to develop a pest management system based on nonchemical methods such as biological control. Changes in public preferences concerning pest control may also influence growers' preferences in pesticide usage. Where growers perceive that the public wants produce grown with little or no use of pesticides, growers may initiate the transition to pest management systems with nonchemical foundations. Finally, loss of pesticides with selective properties may make it more difficult to conserve natural enemies in some pest management systems. For example, termination in the United States of the registration of the aphicide pirimicarb reduced the opportunities to integrate chemical and biological control of aphids in some crops.

Development of Biological Controls for Key Pests. Biological control research may proceed parallel to but separate from existing pest management programs which address current pest control needs. Biological control research can seek more basic solutions that must be studied and initially implemented outside of growers' fields. For example, efforts to introduce new natural enemies of a key pest may be needed. These may proceed for a number of years in biological control research groves or plots. When effective natural enemies have been discovered, imported, colonized, and have begun to spread through the region, they offer the opportunity to restructure the crop pest management system so that the new biological control agent for the key pest becomes the foundation for the system.

COMBINATIONS OF FACTORS AND THEIR INTEGRATION

A detailed treatment of how various agricultural practices influence the conservation of natural enemies in crop systems is given in Chapter 7. Here we review concepts concerning factors which might need to be integrated and some difficulties in doing so. In general, four combinations of factors are possible: biological and chemical control, biological and cultural control, biological control and pest-resistant plants, and two or more different types of biological control.

Biological and Chemical Control

To varying degrees, pesticides adversely affect natural enemies (Croft 1990) and the use of some pesticides may prevent the use of biological controls in the same crop. Some conflicts between chemical pesticides and biological control agents may be reduced by either changing the pesticide in ways that reduce the damage it causes or using natural enemies that are resistant to the pesticide (Hoy 1982; Hoy et al. 1990). Chapter 7 contains additional material on pesticide-resistant natural enemies.

The damage done to natural enemies by pesticides can be reduced by either lowering the amount applied, by seeking selective pesticides that are physiologically safer to natural enemies, or altering the way in which pesticides are applied to reduce contact with natural enemies. Methods to reduce the amounts of pesticides applied are varied and are reviewed in Chapter 7. These include: making applications in strips, or alternate rows; using reduced rates; monitoring pests to omit unnecessary sprays; and timing sprays to avoid periods of greatest sensitivity of key natural enemies. Use of physiological selectivity of pesticides may be increased by screening key natural enemies for susceptibility to pesticides registered for use on the crop, and using those least damaging to important natural enemies. Substituting stomach poisons for contact pesticides may protect some types of natural enemies. Substituting inherently less toxic materials such as insect growth regulators, pheromones, or microbial pesticides can also reduce conflicts between natural enemies and pesticides. Pesticide damage to natural enemies is likely to be most significant early in the crop cycle when natural enemies are less numerous. Reducing pesticide use at this point in time is likely to be more valuable than equal reductions later in the cropping cycle.

Biological and Cultural Control

In general, cultural practices are less likely to be in conflict with natural enemies than pesticides. However, some conflicts can exist. Traps, for example, catch natural enemies as well as pests and if deployed in too great a density may depress natural enemy numbers. In olive (*Olea europaea* Linnaeus) groves in Crete, for example, yellow traps used in a mass-trapping effort against the olive fly, *Dacus oleae* (Gmelin), at a level of two traps per tree captured a large proportion of the parasitoids emerging from the diaspidid pest *Aspidiotus nerii* Bouché. Neuenschwander (1982) recommended that trap numbers be restricted to preserve parasitoids of this scale. Soil management (tilling, mulching, manuring), often aimed at suppressing pest weed populations, can affect (positively or negatively) many types of natural enemies, including spiders, ground beetles, nematodes, some parasitoids, and some phytoseiid mites (Purvis and Curry 1984; Nilsson 1985; Riechert and Bishop 1990; Brust 1991). Crop residue management, which may be part of the weed control program or may be intended to suppress crop diseases or arthropod pests, can affect the ability of important natural enemies to pass from one crop to the next. Key parasitoids of some sugarcane pests in India, for example, are reduced if all the cane trash is burned after harvest (Joshi and Sharma 1989) (Fig. 7.9).

Biological Control and Pest-Resistant Plants

The use of pest-resistant plants is usually considered compatible with the use of natural enemies. Indeed, it often is. The suppression of *Therioaphis maculata*, the spotted alfalfa aphid, in California, for example, depended on the combined effects of native natural enemies, three introduced parasitoids, and the use of resistant varieties of alfalfa (Hagen et al. 1976b).

In some instances, however, characters conferring pest resistance on plants may reduce the effectiveness of biological control agents. Some suppressive traits of resistant plants may be internal to the plant (for example, secondary plant compounds) and have little direct contact with natural enemies, but may have indirect effects due to changes in host chemistry or quality. For example, adults of the parasitoid *Diaeretiella rapae* (McIntosh) were smaller (and hence likely to be less fecund) when reared on *Diuraphis noxia* (Mordvilko) (Russian wheat aphid) fed resistant wheat varieties rather than susceptible ones (Reed et al. 1992). Similarly, parasitism of the dipteran wheat pest *Mayetiola destructor* (Say) was lowest on wheat cultivars with the highest levels of antibiosis to the pest (Chen et al. 1991).

Plant traits that have external manifestations may directly affect natural enemies (Bergman and Tingey 1979; Price et al. 1980; Shepard and Dahlman 1988). Tomatoes (*Lycopersicon lycopersicum*, for example, with genes for high trichome density and high levels of methyl ketone production reduced parasitism from tachinid parasitoids of larvae of *Helicoverpa zea* (Farrar et al. 1992) and from *Trichogramma* sp. and *Telenomus* sp. egg parasitoids of the same host (Farrar and Kennedy 1991; Kashyap et al. 1991). These effects resulted from a variety of causes including mechanical entanglement by trichomes, poisoning of tachinid planidial larvae by trichome exudates, reduction of adult parasitoid search time, and direct toxicity to immature parasitoids within hosts from secondary plant compounds. Nectariless cottons, which exhibited lowered nutritional suitability for several species of pest leafhoppers and plant bugs, also reduced *Trichogramma* sp. egg parasitism of *Helicoverpa zea* (Treacy et al. 1987). High trichome densities on stems of tomatoes in glasshouses in the Netherlands acted as barriers to dispersal of the predacious mite *Phytoseiulus persimilis*, a predator commonly released on the crop for mite control (van Haren et al. 1987). Most recently, transgenic plants have been developed which are capable of expressing *Bacillus thuringiensis* toxins in their tissues (Vaeck et al. 1987), increasing the potential for pest-resistant plants to affect natural enemy populations by reducing the host supply.

Van Emden (1991) proposed that very high levels of pest resistance in crop varieties is not desirable because of harmful effects on natural enemies. He suggested that levels of pest resistance in plants should be no higher than needed to supplement the effects of existing natural enemies. If higher, rather than supplementing natural enemy action, the plant resistance is being substituted for natural enemy effects, reducing a multifactor control system to a single factor system (resistant plants). This is likely to be less stable in the long run and hasten the development of resistance to the variety by the pest.

Two or More Different Types of Biological Control

In pest management systems based primarily on biological control, several types of natural enemies may be required for control of various members of the pest complex. Predatory mites, whitefly parasitoids, and predators for thrips may all be required, for example, in a glasshouse vegetable production system. In crops such as cabbage, the conservation of native natural enemies may need to be combined with the release of other parasitoid species and the use of a microbial pesticide such as *Bacillus thuringiensis*. Potential conflicts that may arise in such complex systems include predators that eat one of the other natural enemies and pathogens that kill the hosts of one of the parasitoids. Cloutier and Johnson (1993) found that the thrips predator *Orius tristicolor* also fed on the predatory mite *Phytoseiulus persimilis* when both were used for control of glasshouse pests, for example. In such cases, avoiding conflict is often a matter of timing releases so that predators do not encounter the other natural enemy immediately upon release.

Conflicts between pathogens and parasitoids caused by elimination of the host population for the parasitoids can be managed by adjusting the rate and degree of coverage of the microbial pesticide to reduce the rate of kill, leaving some hosts to be parasitized. Trials may be needed to determine if the actions of two or more biological control agents are compatible. Application of the fungus *Aschersonia aleyrodis* was found by Ramakers and Samson (1984) to be compatible with the use of *Encarsia formosa* in glasshouse crops for control of the whitefly *Trialuerodes vaporariorum* because its use did not alter the ratio of parasitized and nonparasitized whiteflies. Nealis et al. (1992), in studies of biological control of *Choristoneura fumiferana* in Ontario, Canada, found that larvae parasitized by *Apanteles fumiferanae* Viereck could be protected from being killed by *Bacillus thuringiensis* treatments by delaying the application date because parasitized larvae, in the later part of the developmental period of the immature parasitoid, ate relatively little foliage and thus did not consume a toxic dose of the pathogen.

INTEGRATING NATURAL ENEMIES INTO PEST MANAGEMENT SYSTEMS

Several techniques are available to integrate natural enemies into pest management systems, with the goal that biological control then provides the foundation for control in the system. These include natural enemy monitoring; using natural enemy numbers in models to predict the outcomes of natural enemy-pest interactions at the population level; compatibility of natural enemies with pesticides used in the crop, and with new crop varieties; effects of cultural practices on natural enemies; and techniques for direct natural enemy management.

Natural Enemy Monitoring and Thresholds

If control of pests is to be based on the action of natural enemies, methods must be available to measure the abundance of key species or groups of natural enemies at various times in order to determine if they are sufficiently abundant to maintain the pest under control. First, methods must be available to recognize the important natural enemies in their various stages, including adults, immatures, parasitized hosts, and such structures as cocoons, mummified hosts, and hosts bearing parasitoid emergence holes (Zhumanov 1989). Second, thresholds must be developed through local studies that characterize the levels of natural enemy abundance that are associated with effective pest control. These thresholds may take various forms including predator-to-prey ratios (Gumovskaya 1985; Gravena and Da Cunha 1991) or proportions of hosts parasitized (Martin and Dale 1989). This type of information is then used to modify pest thresholds (which by themselves help conserve natural enemies in many instances by reducing pesticide use [Ali and Karim 1990]), substituting a higher threshold, raised according to the level of natural enemy abundance actually observed (Ooi and Heong 1988; Hoffmann et al. 1990; Van Driesche et al. 1994) (Fig. 14.2). For more on methods to monitor natural enemies in pest management, see Chapter 13.

Using Natural Enemy Numbers in Pest Models

Given monitoring methods for both pests and natural enemies, information on natural enemy abundance can be incorporated into models used to make management decisions about pest control. Fleming (1988), for example, incorporated information on the level of larval parasitism of the alfalfa pest, *Hypera postica*, in his management model. Decision charts developed by Martin and Dale (1989) related proportions of leaves in glasshouse crops infested with the greenhouse whitefly (*Trialeurodes vaporariorum*) and levels of parasitism in ways that allowed determinations to be made as to whether or not chemical controls were needed.

Models in their most sophisticated form can also incorporate factors that affect crop profitability. This allows pest management decisions to be based on information that accurately reflects the likely economic benefit of specific pest control options. Such models should

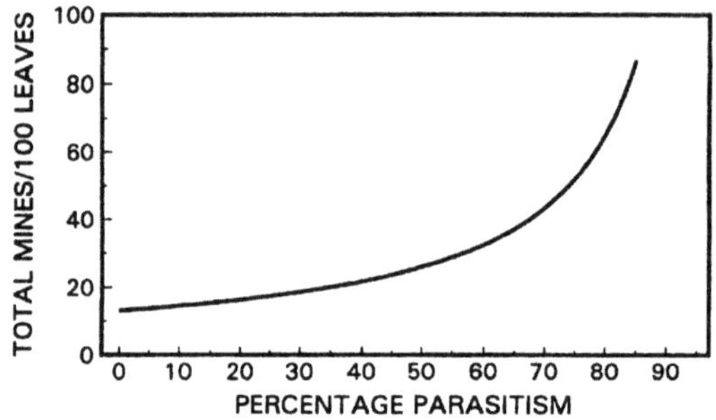

Figure 14.2. Modified action threshold (numbers of tentiform mines per 100 leaves in the first generation) for the gracillariid leafminer *Phyllonorycter crataegella* (Clemens) in relation to the level of parasitism of tissue feeders (instars 4 and 5) in the first generation (after Van Driesche et al. 1994). For mine densities and parasitism combinations above the line, treatment is recommended, but not for those below the line.

include information on: the level of the pest (suggesting possible losses); the levels of key natural enemies (suggesting possibility for biological control to reduce these losses); costs of pest control options; current crop load of marketable produce; and current crop price. Collectively these define the value of the produce at risk which determines the amount of pest control that is economical to implement. A model has been developed for cotton production in Texas that estimates the potential value of suppressing losses from each of the important cotton pests, especially the mirid *Pseudatomoscelis seriatus* (Reuter) (Breene et al. 1989), and incorporates information on levels of the predator *Solenopsis invicta* (Brinkley et al. 1991). Models analyzing the effects of crop prices in celery (*Apium graveolens* Linnaeus var. *dulce* [Miller] Persoon) and tomatoes are given by Trumble and Alvarado-Rodriguez (1990, 1993) that allow the cost effectiveness of various combinations of *Trichogramma* release, mating disruption, *Bacillus thuringiensis* applications, and chemical pesticide applications to be compared as the market price of the vegetables varies seasonally, taking into account the different levels of effectiveness of the various approaches.

Compatability of Natural Enemies and Pesticides

When chemical pesticides are required to supplement the control of one or more members of a pest complex, care must be taken not to employ pesticides that will destroy the biological control systems operating against other pests in the crop. To choose pesticides with appropriate properties, pesticides must be screened to determine their effects on the important natural enemies in the system. Methods for such screening have been reviewed in detail in Chapter 7. These studies must be comprehensive and local. Testing the whole range of pesticides that are registered for use on the crop and which are effective against the pest is needed to identify any differences that may exist between materials. Tests need to be conducted in the region, with local species of natural enemies, rather than relying on data from the literature from related species of natural enemies, because species, and even local populations of the same species, can vary in their sensitivity to particular pesticides. Croft (1990) summarizes much of the literature on natural enemy-pesticide interactions. Hassan (1977, 1980, 1985, 1989a) and Hassan et al. (1987) provide standardized protocols for testing effects of pesticides on various specific types of natural enemies. James (1994) discusses the successful integration of releases of the egg parasitoid *Trissolcus oenone* (Dodd) with selectively-timed, low rates of endosulfan for the suppression of the pentatomid citrus pest *Biprorulus bibax* Braddin. The integrated program reduced bug densities from 10,000–35,000/ha in untreated orchards to <500/ha and lowered fruit damage from 70–90% to less than 5%.

Natural Enemies and Crop Varieties

Less widely recognized is the need to screen crop varieties for their compatibility with key natural enemies. Rice and Wilde (1989), for example, screened sorghum varieties developed for resistance to the aphid *Schizaphis graminum* for effects on the key aphid predator *Hippodamia convergens* Guerin. Gowling (1988) assessed effects of parasitism on aphid population growth rates on resistant and susceptible wheat varieties. In general, plant breeding programs should screen collections of existing varieties for compatibility with a range of important natural enemies and should take such information into consideration when developing new varieties. Natural enemy considerations are likely to influence both the kinds of characters incorporated into new varieties and the optimal level of intensity of their action (van Emden 1991). Models suggest that natural enemies have the potential to affect the rate of adaptation of an herbivore to a resistant crop cultivar, with this process being slowed if a significant part of the pest's mortality is due to natural enemies rather than solely from the resistance of the plant (Gould et al. 1991).

Natural Enemies and Cultural Practices

Just as crop variety and pesticide use affect the environment in which natural enemies must function, so do various cropping practices, such as soil and water management, weed control, planting and harvesting schedules, and crop residue handling practices. Bieri et al. (1989), for example, compared the effects of two different irrigation systems used in glasshouse cucumber (*Cucumis sativus*) production in Switzerland on biological control of mites and thrips. Hokkanen et al. (1988) examined the effects of various soil tilling and seeding systems used in rape production in Finland on the parasitoid of the rape coleopteran pest *Meligethes aeneus* Fabricius. The effects of such factors are covered in depth in Chapter 7.

The role of such factors in shaping the degree to which natural enemies are effective against pests is large and modifications to cultural practices should be explored actively in the process of developing pest management systems based on biological control. Research should be conducted on how various practices affect important natural enemies. These results can then be used to determine whether alternative practices (ones capable of better conserving important natural enemies) might not be equally acceptable to growers. It is important to recognize that crop production is subject to many constraints other than losses from pests in any one group. Weed control, for example, cannot be sacrificed for marginal gains in arthropod pest control, nor can the labor-saving benefits of mechanization easily be foregone. Nevertheless, cultural practices are not all precisely fixed and may be subject to enlightened modification. In sugarcane in India, for example, postharvest burning of crop residues was a standard cultural practice, yet once the pest control benefits of retaining some residues (about 25% of total) were understood, this practice was feasible to adopt (Joshi and Sharma 1989). Similarly, once the benefits of planting French prunes as reservoirs for leafhopper parasitoids in Californian vineyards were recognized, this practice could also be implemented (Wilson et al. 1989). It is important that pesticide-free plots be available for such research and that researchers using these facilities be free to manipulate cropping practices broadly. Such trials would be difficult to conduct in growers' fields because of the risk of crop loss, and, therefore, public supported research stations should be expected to play a major role in such developmental research.

Natural Enemy Management

Efforts to develop pest management systems based on biological control agents can include direct manipulation of the natural enemy populations themselves. Natural enemy behavior may be manipulated, for example, through the application of kairomones to increase parasitoid attraction or arrestment. Alternatively, the genetic composition of existing natural enemy populations may be changed through the release of genetically superior forms. These introductions may consist either of new species, or new strains of existing species with different characteristics as, for example, pesticide resistance.

Behavioral Management. A wide variety of plant- and host-derived chemicals influence decisions made by natural enemies. Kairomones can assist natural enemies in locating prey from a distance, increase retention on host patches, evoke localized intensive search, or in other ways influence the effect of natural enemies on hosts, as reviewed by Tumlinson et al. (1992) and discussed in Chapter 15. Gross et al. (1984), for example, increased parasitism of *Helicoverpa zea* eggs by *Trichogramma pretiosum* by augmenting the density of host eggs in cotton and by applying hexane extracts of host moth scales. Chatelain and Schenk (1984) succeeded in attracting predators of bark beetles to infested trees by treating tree trunks with frontalin or exo-brevicomin, synthetic bark beetle pheromones. Aggregation of predators of lepidopteran pests of maize was increased by applying homogenized prey bodies to whorl stage maize (Gross et al. 1985). Aggregation of adults of *Chrysoperla carnea* was achieved in olive orchards by applying artificial honeydew (L-tryptophan) (McEwen et al. 1994). Oviposition of *Cyzenis albicans*, a tachinid parasitoid of *Operophtera brumata*, was increased on apple trees by applying extracts of oak leaves, the usual oviposition site (Roland et al. 1989).

Responses of natural enemies to such chemicals sometimes are based on learning. Opportunities may exist to use these learning abilities, and the natural enemy response to semiochemicials in general, to manipulate the behaviors of natural enemies for pest management purposes (Dicke et al. 1990a; Lewis and Martin 1990; Vet and Groenewold 1990), but this has not yet been achieved.

New Natural Enemy Species or Genotypes. A basic tool in developing management systems based on biological control is the improvement of natural enemy populations operating in the system. In many instances, introductions of new species of natural enemies may be an essential step in developing effective biological control. Existing natural enemies, particularly for adventive pests, may be inadequate to provide control. Adding new species, or strains of existing species, is a technique that forms one of the foundations of biological control (Chapters 8 and 9).

In instances in which pesticide tolerance is an attribute essential for natural enemy survival, because key pests are currently controlled by chemical means, strains of natural enemies may be introduced that possess pesticide resistance. Such populations may in some instances be found in nature in regularly-sprayed environments (such as apple orchards) or may be developed through laboratory selection (Chapter 7).

SECTION
V

Natural Enemy Biology

Chapters 15, 16, and 17 discuss the biologies of key groups of natural enemies. Detailed knowledge of the biologies of natural enemies is essential for their successful manipulation. Chapter 15 treats parasitoids and predators of arthropods, with special attention to host discovery, foraging, overcoming host defenses, and effects of plants on natural enemies. Chapter 16 discusses the biologies and dynamics of pathogens of arthropods including the basic steps in the pathogen life cycle (host contact, penetration, reproduction, exit, spread, and persistence) and routes of transmission. Chapter 17 covers biological aspects of weed control agents, especially determinants of the host range and factors influencing the effects agents have on plant performance and plant population dynamics. The population biology of interacting natural enemies and their hosts (or targets) is discussed in Chapter 18, specifically with reference to relatively simple mathematical models of these interactions.

CHAPTER

15

BIOLOGY OF ARTHROPOD PARASITOIDS AND PREDATORS

INTRODUCTION

Basic knowledge of natural enemy biology and ecology is the foundation upon which the applications of biological control rest. In this chapter, aspects of the biology and ecology of arthropod parasitoids and predators are considered. The biology of vertebrates used in biological control is not covered. Following chapters cover the biologies of microbial pathogens and nematodes (Chapter 16) and weed control agents (Chapter 17).

Beneficial arthropods possess an array of biological and behavioral attributes. The full range of these factors cannot be covered in this chapter, and the reader is referred to Hagen et al. (1976a), Hodek (1986), New (1992), and Sabelis (1992) for additional information on predator biology, and Clausen (1940), Askew (1971), Doutt et al. (1976), Godfray and Hassell (1988), Waage and Greathead (1986), Wharton (1993), and Godfray (1994) for information on parasitoid biology. Aspects of natural enemy biology considered in this chapter include: (1) finding habitats and hosts, (2) moving between host patches, (3) host recognition and assessment, (4) overcoming host defenses, (5) regulating host development, (6) learning by natural enemies, (7) effects of plant features on natural enemies, and (8) seasonal synchrony of life cycles and local races of natural enemies.

FINDING HABITATS AND HOSTS

Natural enemies may emerge as adults into the habitat in which their host or prey occurs. In such cases, the parasitoid or predator need not locate the habitat, only the host. Should the parasitoid not be reproductively mature on emergence, however, adults may need to disperse out of the host's habitat to seek other resources, creating a later need to rediscover the host's habitat (Vinson 1981). In other instances, adult natural enemies may emerge in a location different from the habitat of their host, or in an area no longer suitable as habitat for the host due to plant senescence or destruction or previous suppression of the host population by natural enemies or other causes. In such cases, the natural enemy must discover habitats suitable for the host or prey it requires, and, thereafter, locate hosts or prey. In some instances, natural enemies may be able to orient directly to hosts from long distances without the need to first locate the habitat, for example, parasitoids that orient to volatile chemicals emitted by their hosts.

Host Signals and Natural Enemy Responses

The ability of parasitoids and predators to detect cues in their environment and use these successfully to locate the resources they require is based on a complex set of long- and short-distance visual, olfactory, gustatory, mechanoreceptor, and auditory signals that operate at several spatial scales (Table 15.1, Fig. 15.1) (Lewis et al. 1976; Vinson 1981, 1984a,b; van Alphen and Vet 1986; Bell 1990; Lewis and Martin 1990; Vet and Dicke 1992). Olfactory, visual, and auditory signals are possible to perceive from a distance and can be used for long distance orientation to either the host's habitat or, in some cases, the host itself. Gustatory and mechanoreceptor cues must be perceived following physical contact with the source and usually are involved only in short distance responses, such as host finding once the host habitat has been entered.

Perception of long distance odor cues can evoke upwind movement or movement up an increasing odor gradient. Perception of visual or auditory cues can result in movement toward the source of the signal. Perception of contact-chemical signals can result in following odor trails on surfaces, arrestment on a patch, or intensified search of a patch. Perception of mechanical vibrations can result in oviposition, probing, drilling, or other attempts to reach hosts within structures or tissues.

A special set of terms describes responses mediated by volatile chemicals (Nordlund 1981). Pheromones function within single species as, for example, the sex pheromones emitted by one sex to attract the other in many species of insects. In contrast, allelochemicals convey information between members of two different species. Several terms are employed to describe types of allelochemicals, according to who benefits from the message. Allomones, for example, are substances that benefit the emitter, but not the receiver, such as venoms and defensive secretions. Kairomones, in contrast, favor the receiver, not the emitter. Volatiles, for example, arising from a caterpillar's frass that are used by a parasitoid to help locate the caterpillar function as kairomones. Synomones are chemicals that favor both the emitter and the receiver, as is the case of plants under herbivore attack that emit substances attractive to upper trophic level organisms that attack the herbivore.

Long-Distance Orientation

Long-distance orientation may be directed either at discovering habitats in which the organism to be attacked by the natural enemy lives, or to discovering the organism directly. Natural enemies using stimuli of various types to locate suitable hosts are faced with what Vet and Dicke (1992) have termed the "reliability/detectability problem." They suggest that signals originating directly from hosts are highly reliable, since they come from the object being sought, but are produced in small quantities making them less detectable from a distance. In contrast, signals originating from the host's habitat, such as its food plant, may be present in larger quantities, making them easier to detect from a distance, but will have lower reliability because some host plants bear no hosts.

TABLE 15.1 Cues for Location of Hosts and Host Habitats

Long or Short Range Cues	Short Range Cues
• airborne chemicals	• contact chemicals
• vision	• vibration
• sound	

Figure 15.1. Levels of spatial scale in finding host habitats and hosts: (A) the host's habitat in the broad sense, here a cabbage field; (B) the target pest's host plant, one cabbage; (C) the local patch on which the host must be located, here one leaf potentially containing a host; (D) the host encounter, in which the braconid *Cotesia rubecula* (Marshall) locates its pierid host *Pieris rapae* (Linnaeus). (Photograph [D] courtesy of D. Biever.)

Long-Distance Habitat Finding. Long-distance orientation to the habitat itself has not been sufficiently studied for upper trophic level organisms. Evidence that parasitoids and predators respond to host habitat cues (separately from host cues or plant/host cue complexes) is circumstantial. One line of evidence is that some species of parasitoids attack polyphagous hosts only on one host plant suggesting a previous attraction by the parasitoid to the plant (Vinson 1981). For example, in South Africa the noctuid *Helicoverpa armigera* is found on many species of plants, but one of its parasitoids, the braconid *Microbracon brevicornis* Westmead, attacks this host on only one group of plants, *Antirrhinum* spp. (Taylor 1932). A second line of evidence is that some parasitoids attack a range of host species or genera if the hosts feed in a certain manner, on a particular type of plant. For example, the eulophid parasitoid *Sympiesis marylandensis* Girault attacks about forty species in three genera of gracillariid moths, all of which are lower-surface leaf miners of rosaceous trees in eastern North America (Maier 1994).

A third line of evidence for long distance habitat attraction by parasitoids is that some parasitoid species respond to plant volatiles in olfactometers in the absence of any hosts, host damage, or host-derived materials (Elzen et al. 1986; Martin et al. 1990; Wickremasinghe and van Emden 1992). Six species of aphid parasitoids responded positively to the odors of uninfested leaves of the plant species from which the mummies of the parasitoids used in the test had been collected (Wickremasinghe and van Emden 1992). *Leptopilina heterotoma* (Thompson), a parasitoid of drosophilid larvae in rotting fruits, responds to volatile compounds produced by yeast (Dicke et al. 1984).

Only limited evidence is available for long-distance response by predators to habitat clues. Aquatic predators in the Dytiscidae and Notonectidae are believed to be attracted to "shiny" surfaces (being perceived as water bodies) (Vinson 1981). A species of predacious carabid beetle in the genus *Bembidion* is attracted to volatiles from cyanobacterial mats of *Oscillatoria* spp. that are the habitat of its prey along beaches of saline lakes (Evans 1984).

Long-Distance Host Finding. In addition to cues originating from the host plant of the prey or host, cues useful for long-distance host finding may also arise from: plants under herbivore attack; herbivore frass or other components of the plant/host complex; associated organisms (such as fungi or microbes in rotting plant tissue); or the host itself (such as host pheromones, visual or auditory signals) (Vinson 1981).

Plants under attack by herbivores, for example, may emit larger volumes and different types of chemicals than healthy plants (Turlings et al. 1990, 1993). Such volatiles may be used by parasitoids and predators to solve the reliability/detectability problem because they are produced in large quantities and yet are uniquely associated with the presence of the host (Fig. 15.2). The predacious mite *Phytoseiulus persimilis* is attracted by volatiles released by lima bean plants (*Phaseolus lunatus* Linnaeus) under attack by *Tetranychus urticae* (Dicke et al. 1990b). These volatiles are not necessarily emitted only by the damaged tissue, but may emanate from the entire plant and thus be produced in large quantity. The mechanisms of long-distance host finding by phytoseiid mites are considered in detail by Sabelis and Dicke (1985). Material regurgitated by the beet armyworm, *Spodoptera exigua* Hübner, when applied to damaged sites on maize, caused the plants to release large quantities of terpenes that were attractive to several parasitoid species (Turlings et al. 1993). Study of the effects of volatile materials on behaviors of natural enemies may be conducted in olfactometers or wind tunnels.

Hosts produce many materials which can be perceived by parasitoids or predators. Some of these can be perceived from a distance; others affect parasitoids only when touched. The reponse of parasitoids to host plants is often stronger if hosts, host damage, or host-related materials are also present (Turlings et al. 1991; Steinberg et al. 1993). In some cases, the herbivore-plant complex is attractive while the plant and host separately are unattractive, as is the case of the encyrtid parasitoid *Epidinocarsis lopezi*, cassava, and the mealybug *Phenacoccus manihoti* (Nadel and van Alphen 1987). Both hymenopteran and dipteran parasitoids respond to herbivore/plant complex materials (Hassell 1968; Navasero and Elzen 1989; Roland et al. 1989).

Other organisms associated with hosts, such as fungi or bacteria in decaying fruits, mushrooms, or sap flows, produce compounds that are used as cues by parasitoids to locate hosts

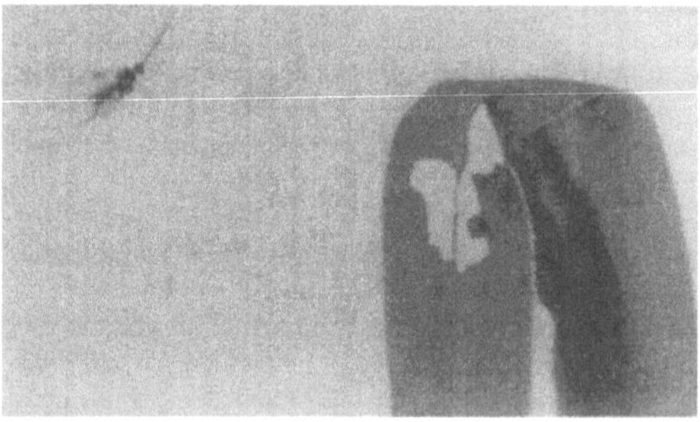

Figure 15.2. Plants damaged by herbivores emit volatile chemicals attractive to parasitic Hymenoptera, which aid parasitoids in locating hosts. (Photograph courtesy of T. Turlings.)

from a distance (Dicke 1988). For example, a fungus often associated with larvae of tephritid flies produces acetaldehyde which attracts *Diachasmimorpha longicaudatus* Ashmead, a parasitoid of the fruit fly larvae (Greany et al. 1977). The parasitoid *Ibalia leucospoides* responds to odors of the wood digesting fungus *Amylostereum* sp., which is a symbiotic fungus associated with the parasitoid's host, the wood wasp *Sirex noctilio* (Madden 1968).

Direct long distance attraction to hosts may also result from auditory or visual cues. Sounds may be used to locate hosts in groups that characteristically communicate by auditory signals, such as crickets and cicadas. Cade (1975) induced the tachinid parasitoid *Ormia ochracea* (Bigot) to larviposit onto dead crickets (*Gryllus integer* Scudder), when cadavers were placed on speakers emitting the song of this species of cricket. Dead crickets associated with various other noises used as controls were not attacked. Soper et al. (1976) reported that the sarcophagid parasitoid *Colcondamyia auditrix* Shewell locates its cicada host (*Okanagana rimosa* [Say]) by sound. Visual cues may be used by various predators (such as Diptera in the family Asilidae, Odonata in various families, and sphecid hunting wasps) to detect and capture prey. Long-distance detection of hosts by parasitoids through visual cues has been suggested (Henriquez and Spence 1993), and vision certainly plays a role in short-range orientation to hosts.

Volatile chemicals such as host sex or aggregation pheromones can be used by parasitoids to locate hosts. *Trichogramma pretiosum* responded in olfactometers to the sex pheromone of its host *Helicoverpa zea* (Lewis et al. 1982; Noldus 1988; Noldus et al. 1990). Large numbers of aphelinid parasitoids in the genus *Aphytis* were collected on sticky traps baited with the sex pheromone of California red scale, *Aonidiella aurantii*, in citrus (Sternlicht 1973). Attractiveness of the sex pheromone of this scale to *Aphytis melinus* and *Aphytis coheni* DeBach was confirmed using olfactometers in the laboratory. Similarly, attraction to aggregation pheromones has been demonstrated for clerid predators of scolytid bark beetles (Payne et al. 1984), for tachinid parasitoids of adult southern green stink bug (*Nezara viridula*) (Harris and Todd 1980), and for a scelionid egg parasitoid of the predacious bug *Podisus maculiventris* (Aldrich et al. 1984). The bug's parasitoid finds adult male bugs via its host's aggregation pheromone and is then phoretic on females to arrive, ultimately, at sites where eggs are deposited. The parasitoid *Leptopilina heterotoma*, which attacks drosophilid larvae in rotting fruits and mushrooms, is attracted from a distance to habitat patches by the aggregation pheromone of the adult flies (Wiskerke et al. 1993). The scelionid egg parasitoid *Trissolcus basalis* is attracted to adults of its host, *Nezara viridula*, through long distance attraction to its defense compounds (Mattiacci et al. 1993). Phoresy is used by a species of *Telenomus* as a means of arriving at the location of its host's eggs, with adult parasitoids being present in the anal tuft of the host lymantriid moth (Arakaki 1990).

Finding Hosts Over Short Distances

Once parasitoids or predators have arrived in the host's habitat, hosts must be located. A variety of stimuli and search strategies exist which aid in host location.

Kinds of Cues. Cues that natural enemies use for short range orientation to hosts include contact kairomones, host vibrations, visual and acoustical signals, and mechanical aspects of host-caused damage. Of these, contact kairomones (chemicals perceived primarily by touch or taste rather than olfaction) play an important role in many parasitoid-host systems. A wide range of materials associated with hosts or host-damaged plants may contain chemicals which

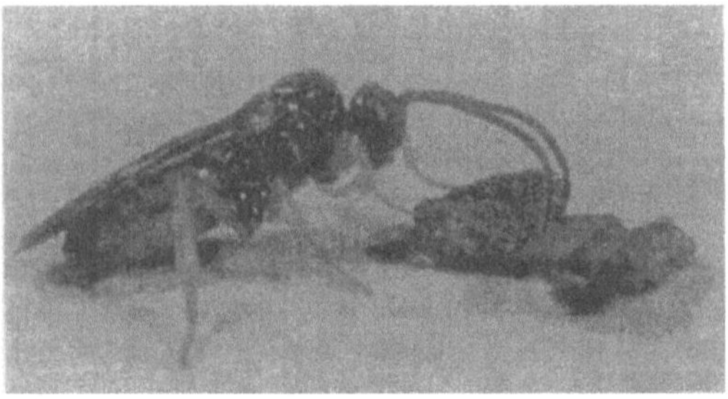

Figure 15.3. Host frass of some herbivorous arthropods contains chemicals used by parasitoids in local host finding. (Photograph courtesy of J. Lewis.)

assist parasitoids in host detection, including: host frass (Fig. 15.3); chemicals associated with host-damaged plant tissue or host-infested media (often including host mandibular or salivary gland secretions); host scales, setae, silk, or other body parts or products; host-marking pheromones; and honeydew.

Host frass (feces) may contain compounds that affect search behaviors of many species of predators and parasitoids, either alone or in combination with other host-derived materials or plant-produced chemicals (Jones et al. 1971). The phytoseiid *Amblyseius fallacis* responds to contact with feces and silk of *Tetranychus urticae* and other prey species by exhibiting more active, local search patterns (Hislop and Prokopy 1981b; Kong and Zhang 1986). The parasitoid *Diadromus pulchellus* Westmead reacts to various plant-produced but host-modified volatile disulfide compounds in the frass of its host *Plutella xylostella* by moving toward the odor source (Auger et al. 1989). Similarly, host-seeking of *Cotesia flavipes* is stimulated by water extracts of frass of its host *Diatraea saccharalis* (van Leerdam et al. 1985). For *Venturia canescens* Gravenhorst, an ichneumonid parasitoid of larvae of *Plodia interpunctella* (Hübner), 2-palmitoyl- and 2-stearoylcyclohexane-1,3-dione have been identified as components of the kairomone assisting this parasitoid in locating its host (Nemoto et al. 1987).

Chemicals in host frass that function as kairomones may be either synthesized by the hosts directly (Nemoto et al. 1987) or may be plant-produced materials that are sequestered by hosts and passed, with or without modification, in feces (Auger et al. 1989). In cases in which plants are sources of chemicals (or precursors of chemicals) that act as kairomones, the species of plant on which the insect feeds can determine whether or not host frass is attractive to parasitoids (Nordlund and Sauls 1981; Inayatullah 1983).

Host-damaged plants or host-infested media can also present nonvolatile compounds that are used by natural enemies for local host finding. Such materials may originate from the plants themselves in response to host feeding, may come directly from hosts in the form of various secretions, or may derive from other organisms in the plant tissue or media whose activities are stimulated by the presence of the host. Nealis (1986), for example, noted that feeding damage by *Pieris rapae* larvae on cabbage (*Brassica oleracea*) caused the parasitoid *Cotesia rubecula* to be attracted to and arrest at feeding sites. Similarly, Loke and Ashley (1984) found that contact with maize leaves damaged by larval feeding of the noctuid *Spodoptera frugiperda* altered thesearching behaviour of *Cotesia marginiventris* (Cresson). The eucoilid parasitoid *Leptopilina clavipes* (Hartig) searched longer on food patches treated with filtrates of host-infested mushrooms than on untreated patches (Vet 1985).

Host body parts can also affect parasitoid searching behavior (Jones et al. 1973). *Trichogramma* egg parasitoids, for example, more actively search areas contaminated with scales of adult moths of the host species (Lewis et al. 1972; Kainoh et al. 1990). For *Trichogramma nubilale*, experiments with extracts of scales of the host *Ostrinia nubilalis* have shown that 13,17-dimethylnonatriacontane is the most important chemical inducing klinokinesis and retention of the parasitoid on treated patches (Shu et al. 1990). Host-marking pheromones are used by some herbivores to reduce the frequency of subsequent ovipositions in discrete resources, such as fruits, by conspecific females. Some parasitoids, such as *Opius lectus* Gahan which oviposits into the eggs of its host *Rhagoletis pomonella*, are influenced by host-marking pheromones, remaining on and antennating longer on marked fruits, provided fruit volatiles are also present (Prokopy and Webster 1978).

Honeydew is a host-produced substance commonly associated with such homopterans as aphids, whiteflies, and scales. Honeydew-contaminated, but host-free, plants arrested larger numbers of the aphid parasitoid *Ephedrus cerasicola* Starý than clean plants in a glasshouse test, suggesting that honeydew acts as a kairomone for this species, causing increased retention time on contaminated plants (Hågvar and Hofsvang 1989).

Host movement or vibrations can be used by parasitoids to either locate concealed hosts or can function to stimulate probing and oviposition (Vinson 1976; Vet and Bakker 1985). Vet and Bakker (1985) found that *Ganapsis* spp. of drosophilid parasitoids, which seek hosts located inside rotting fruits, do so by remaining stationary and sensing vibrations from larval movements. Use of acoustical signals for host detection has also been noted. The braconid parasitoid *Dapsilarthra rufiventris* (Nees) used sounds produced by larvae of *Phytomyza ranunculi* Schrank feeding inside their leaf mines, in combination with kairomones, to locate hosts over short distances (Sugimoto et al. 1988). There was also evidence that this parasitoid could use the color pattern of the mine as one of the cues for short-range host detection. Some parasitoids respond to mechanical aspects of herbivory, apart from contact or volatile chemical cues that herbivory may cause. Faeth (1990) showed that rates of parasitism were higher for leaf miners in leaves paired with herbivore-damaged leaves than with control leaves, even though damaged leaves were sealed in a fluorocarbon telomer that prevented the emission of volatile chemicals or contact with chemicals on the leaves' surfaces. This finding implies that the mechanical aspects of damage alone (cut edges, irregular outlines) influenced parasitoid behavior.

Responses to Local Cues. Cues suggesting the nearby presence of hosts modify natural enemy behaviors, and these changes increase probability of host encounter (Beevers et al. 1981). Several patterns of behavioral changes lead to increased rates of host encounter: behaviors that localize search to a restricted area; behaviors that intensify the level of search activity; and behaviors that lead natural enemies to hosts by trails or plant morphology.

Search can be locally intensified if the parasitoid decreases its walking speed (Waage 1978), changes from straight line walking paths to ones that frequently bend back on themselves (Waage 1979; Loke and Ashley 1984; Kainoh et al. 1990), or perceives patch boundaries and responds by reversing direction (Waage 1978). A consequence of such behaviors would be to increase the number of parasitoids arrested on a host patch and to increase the average time spend by parasitoids in a local patch (Prokopy and Webster 1978; Vet 1985; Nealis 1986). Parasitoids that hunt for concealed hosts in media may be arrested over patches of media containing hosts by contact with kairomones on the surface of the media and may respond by increased time spent motionless waiting to detect host vibrations (Vet and Bakker 1985). Behaviors that act to intensify search activity include increases in the rate of "sampling" movements made by parasitoids. These movements include antennation of surfaces (Prokopy and Webster 1978) and probing of media with the ovipositor (Vet and Bakker 1985). Kairomones that are deposited in linear fashion can evoke trail-following behavior which leads a parasitoid directly to its host's location. The bethylid parasitoid *Cephalonomia waterstoni* Gahan follows kairomone trails deposited by mature larvae of the rusty grain beetle, *Cryptolestes ferrugineus* (Stephens), moving to pupation sites (Howard and Flinn 1990). Other examples are given by Strand et al. (1989) and Cederberg (1983). Behaviors that induce parasitoids to follow morphological features of plants, such as veins, leaf edges, or stems may increase the probabilities of host encounter if hosts are clustered adjacent to such features (Ayal 1987).

HOST PATCH EXPLOITATION AND ABANDONMENT PATTERNS

Many species of herbivorous arthropods are not distributed randomly over space, but occur in clumps. Foraging for hosts includes locating, exploiting, and abandoning such host patches (Emlen 1966; MacArthur and Pianka 1966; van Alphen and Vet 1986; Vet et al. 1991). Initial approaches to the study of foraging activities of arthropods consisted largely of examining *a priori* postulates of mechanisms that would, in theory, optimize foraging activities. Efforts to validate these models were made by studying the foraging activities of individual parasitioids or predators in laboratory settings. More recently, *a priori* modeling has been supplanted by *post hoc* reconstructions based on detailed studies of foraging, most often under laboratory conditions. Because of the small size of most parasitoids and predatory arthropods, field studies of foraging have been rare.

Optimization Models

The *a priori* construction of optimization models for the description of arthropod foraging was based on the assertion that a patch should be exploited until the encounter rate within the patch had decreased to the value of the encounter rates averaged across all patches (Charnov 1976). Various optimization models have been developed to describe insect foraging (Cook and Hubbard 1977; Hubbard and Cook 1978). Three principal rules of exploitation have been advanced to describe how foragers might decide when to abandon a patch (van Alphen and Vet 1986): number expectation (Krebs 1973), time expectation (Gibb 1962), and giving-up time (Hassell and May 1974; Murdoch and Oaten 1975). Foragers that hunt with the expectation of encountering a fixed number of hosts would leave a patch after that number has been encountered, whether or not additional hosts were still available. Strand and Vinson (1982a), for example, found that *Cardiochiles nigriceps* Viereck always abandons tobacco (*Nicotiana tabacum*) foliage after one host larva is attacked. This strategy provides no mechanism for abandoning patches that contain no hosts. It would be most suitable for solitarily-distributed hosts because each patch would contain at most only one host and thus the forager should immediately begin to search for a new patch after each successful host encounter. Foragers that hunt with a fixed time expectation would leave patches after that time has elapsed whether or not hosts had been encountered or additional hosts remained undiscovered. Such a strategy would explain the inversely density-dependent patterns of parasitism often seen in nature. Alternatively, foragers hunting by application of a fixed giving-up time would abandon a patch once a preset time had elapsed without encountering a suitable host. If hosts were encountered, the clock could be reset, and the patch would be abandoned only when no new hosts could be found in this interval. Little evidence exists that either of these two methods of foraging (fixed time, fixed giving-up time) are actually employed by any parasitoid species (van Alphen and Vet 1986).

Laboratory *Post hoc* Study of Foraging Behavior

Development of a *post hoc* understanding of foraging strategies was a response to the inability of the *a priori* modeling efforts discussed above to adequately describe observed foraging patterns. Optimal foraging models suggested what foragers ought to do, given full knowledge of nearby host patches and host densities on these patches. The problem is, of course, that foragers do not have this knowledge. The *post hoc* approach has been to observe foraging and its component behaviors in detail and to vary patch characteristics experimentally to determine what effects these have on foraging behaviors. From the results of such studies a view has emerged that synthesizes the influences of a variety of factors on foraging activities without any need for foragers to have information about other patches, or to be able to compute or keep time.

Patch Factors Affecting Foraging Behavior. Seven factors have been studied and found to have some effect on foraging activities of individual foragers (van Alphen and Vet 1986): patch structure; presence of host kairomones; encounters with unparasitized hosts; encounters with parasitized hosts; encounters with conspecific foragers; previous experiences on other patches; and travel time between patches, as discussed below.

In the laboratory, *Leptopilina heterotoma* searches longer on larger host-habitat arenas (patches of nutrient yeast devoid of hosts) than on smaller arenas (van Lenteren and Bakker 1978). Patches with rough surfaces were searched longer than similar patches with smooth surfaces (van Alphen and Vet 1986).

Waage (1978, 1979) found that the parasitoid *Venturia canescens* increased its patch-time allocation in response to increasing amounts of kairomone left in the media by host larvae (*Plodia interpunctella*). Dicke et al. (1985) presented similar findings for response of the parasitoid *Leptopilina heterotoma* to its host's kairomone in experiments in which no hosts were present. Honeydew produced by scales or aphids has been found to increase the searching time on patches of both parasitoids (*Microterys flavus* [Howard]) (Vinson et al. 1978) and predators (*Coccinella septempunctata* Linnaeus) (Carter and Dixon 1984a). Encounters with unparasitized hosts increase patch search time for some parasitoids whose hosts are clumped (*Venturia canescens*, Waage 1979; *Asobara tabida* Nees, van Alphen and Galis 1983), but may cause patches to be abandoned by parasitoids whose hosts are not clumped (*Cardiochilis nigriceps*, Strand and Vinson 1982a). For some predators, such as *Syrphus ribesii* (Linnaeus), hunger levels act in an analogous manner to host encounter rates for parasitoids. Satiated predators tend to remain where they are. Predators starved for a short period (24 h) engage in intensified but local search, and predators starved for longer periods (48 h) disperse, looking over larger areas for new host patches (Rotheray and Martinat 1984). Intensified local search may be induced in predators by unsuccessful prey encounters in which the prey escapes (Carter and Dixon 1984b).

Encounters with parasitized hosts decreased foraging time on a patch for several parasitoids (*Venturia canescens*, Waage 1979; *Leptopilina heterotoma*, van Lenteren 1991). Removal of parasitized hosts following ovipositions showed that this effect is the result of a negative stimulus from contact with a parasitized host rather than merely a decreasing rate of contact with healthy hosts (van Alphen and Vet 1986). In some parasitoid species, contact with parasitized hosts has no detrimental effect on search time on patches (*Asobara tabida*, van Alphen and Galis 1983), or may increase search time compared to patches containing no hosts at all.

Encounters on patches with conspecifics may reduce foraging time on the patch (Hassell

1971; Beddington 1975). Some parasitoids mark exploited host patches with marking pheromones that reduce search time of subsequent visits made by themselves or by other females encountering the patch within a short period (*Pleolophus basizonus* [Gravenhorst], Price 1970; *Microplitis croceipes* [Cresson], Sheehan et al. 1993). The coccinellid *Adalia bipunctata* (Linnaeus) lays fewer eggs on patches of aphids on which conspecific larvae are encountered (Hemptinne et al. 1992).

Little information is available on the effects of previous experiences on other patches and of travel time between patches. Van Alphen and van Harsel (1982) showed that foraging time of *Asobara tabida* increased when presented with a host species to which it had been conditioned 24 hours previously. Travel time between patches, according to optimal foraging theory, is likely to be associated with risk of mortality and to the extent that this is so, foraging times should increase as the travel time to discover new patches increases. However, few data are available on this subject.

Mechanisms Producing Behavioral Responses to Patch Stimuli. To understand how patterns of foraging exhibited in response to patch characteristics are produced, one must understand how relatively simple behaviors of foragers are affected by stimuli from patches (Lewis et al. 1976). Broadly, to forage effectively, organisms must employ mechanisms that restrict search to profitable patches, that induce foragers to abandon patches before the point of diminishing returns, and to search efficiently (Bell 1990).

Four behavioral mechanisms exist by which search can be restricted to a patch. First, the organism can exhibit a looping or spiraling behavior (with a consistent right or left bias) or a zigzag pattern (alternating right and left turns), in place of more straight line motions. Such turning patterns will tend to prevent large displacements away from the local area. Second, a forager can move less often or move for shorter distances per movement. Third, the forager can depart from each resource item in a random direction, different from the direction of arrival to the resource item; this can be achieved by turning completely around several times on the resource item during its exploitation. Fourth, the forager can reverse direction at patch edges whenever contact is lost with patch stimuli.

Two mechanisms enable a forager to leave a patch before the point of diminishing returns. First, if the organism tends (as many species do) to move in relatively straight lines, except in response to encounters with resource items (which can induce looping movements as noted above), then normal decay of such resource-induced looping patterns with time will permit resumption of straight line movement unless further hosts are encountered. Straight-line movement will be resumed when encounters with prey or hosts become less frequent, and will result in the forager moving out of the patch. Second, for foragers that remain on patches because loss of contact with host kairomones at patch edges induces reversal of direction, habituation to kairomones may eventually cause this direction-reversal mechanism to fail, causing the organism to leave the patch.

Foragers that are able to benefit from past experience can modify their foraging activities to specialize on resource types that are more abundant in a local area or during a given time period (a process termed conditioning) and can become more efficient in finding, recognizing, or attacking a given type of prey based on past experience with the same prey (a process called learning) (see the section on learning later in this chapter).

Field Studies of Natural Enemy Foraging. Few studies exist of parasitoid foraging under field conditions (Waage 1983; Thompson 1986; Casas 1989; Sheehan and Shelton 1989). What studies exist, however, have special importance because they test the degree to which ideas developed and tested in laboratory settings actually describe events which take place in nature. Among the few field studies is that of Waage (1983) in which aggregation of arrested parasitoids (*Diadegma eucerophaga* Horstmann and *Diadegma fenestralis* [Holmgren]) on high-density host patches (larvae of *Plutella xylostella*) was demonstrated. Casas (1989), in a study of the apple leafminer parasitoid *Sympiesis sericeicornis* Nees, observed that the presence of mines on individual leaves could be detected while the parasitoid was in flight adjacent to the leaf, but whether or not mines contained suitable hosts could only be determined after landing. The decision to abandon a leaf appeared to be motivated by encounters with already parasitized (or otherwise unsuitable) hosts rather than the passage of a fixed amount of time. Sheehan and Shelton (1989) found that the braconid wasp *Diaeretiella rapae* did not discover large patches of host plants (collards, *Brassica oleraceea*) faster than small patches, but that it was slower to leave large patches. The number of arrested parasitoids on a patch, therefore, was determined by decisions to leave patches, not factors affecting patch discovery.

Synthesis of Foraging Behaviors and Mechanisms. Synthesis of parasitoid foraging behaviors was initiated by Waage's (1978, 1979) study of *Venturia canescens* foraging for larvae of *Plodia interpunctella* (Fig. 15.4). Foraging time on host patches was highly variable and Waage postulated a behavior-driven mechanism in which contact with host kairomone produced a certain level of search motivation in foraging wasps. This level, if high enough, caused a reversal in walking direction if contact with kairomone was lost at the patch edge. The motivation level, however, decreased with time due to habituation to the kairomone stimulus, but could be renewed to high levels by contact with and oviposition in unparasitized hosts. Conversely, contact with already parasitized hosts caused a decrease in motivational level.

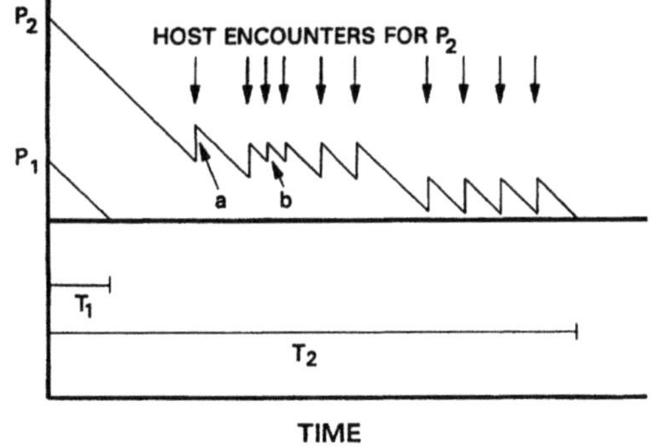

Figure 15.4. One model advanced to describe searching time of a parasitoid on a host habitat patch is that of Waage (1979) in which motivational state, or responsiveness, of the parasitoid affects response to host kairomones. When the motivational state is low, as for P_1, the responsiveness decays to a null level before the parasitoid encounters a host, and so when the parasitoid reaches the edge of the kairomone patch (at time T_1), it does not recognize the edge and leaves. For parasitoid P_2, a host encounter augments the responsiveness, and the parasitoid continues to respond to the patch edge and remains on the patch. The degree of responsiveness increases with time since last oviposition (to a maximum), so that small increases (b) follow recent ovipositions and larger increases (a) result from longer periods with no ovipositions. Repeated host encounters continue to enhance the responsiveness until no additional hosts are encountered before the responsiveness falls to zero, and the parasitoid moves off the patch (at time T_2).

If the patch edge was contacted while the parasitoid was at a sufficiently low motivational level, loss of contact with the kairomone was not perceived and the direction-reversal mechanism was not activated, with the result that the parasitoid left the patch (Fig. 15.4). The essential points of this model (stimulation by contact with healthy hosts, negative stimulation by contact with parasitized hosts) have been confirmed in other systems as previously noted. Recent application of this empirical approach to foraging analysis are the studies of Haccou et al. (1991) on *Leptopilina heterotoma* and Hemerik et al. (1993) on *Leptopilina clavipes*. It is now recognized that the foraging decisions of an individual are influenced by genetic factors (fixed differences among individuals), phenotypic plasticity (variable differences among individuals that reflect past learning and other experiences), and physiological status of the individual at the moment (relative to needs for food, mating, or hosts in which to oviposit) (Lewis and Martin 1990; Lewis et al. 1990).

Related Processes: Egg Load and Superparasitism. Egg load (the number of mature eggs) of parasitoids is considered an important influence on the number of eggs laid per host (clutch size) in gregarious species (Minkenberg et al. 1992). It also affects the decision to oviposit versus host feed on a host (Collier et al. 1994). In the aphelinid *Aphytis lingnanensis* Compere parasitizing *Aonidiella aurantii*, parasitoids with fewer eggs deposited smaller clutches and parasitoids with larger egg loads showed a higher rate of foraging activities (Rosenheim and Rosen 1991).

Superparasitism is the result of multiple ovipositions in hosts that are already parasitized by the same species. When groups of foraging females occur on a patch, encounters are likely with previously parasitized hosts. Whether or not oviposition in such hosts is in the second female's best interests depends (for gregarious species) on the rate at which larval survival declines as total eggs per host increases, the number of eggs already present in the host, and (in cases of contest competition) the probability that the offspring of the second female can defeat the offspring of the first female and take possession of the host. Studies show that it may be in the interests of a second female to lay a small clutch in a previously parasitized host, particularly if transit times to new patches are long or hosts are scarce (Waage 1986; van Dijken and Waage 1987; van Alphen 1988).

HOST RECOGNITION AND ASSESSMENT

Once potential hosts or prey have been located, they must be recognized and evaluated. For both predators and parasitoids, this process entails final determination that the prey (or host) is appropriate (in terms of species, growth stage, and, for parasitoids, previous parasitism). For parasitoids, the host also must be evaluated to determine how many eggs, and of which sex, should be laid.

Host Recognition

Once a parasitoid has physically contacted a potential host, it must decide if it is a correct species and life stage to use as a host. Parasitoids use chemical cues, both on the outside and inside of hosts, as well as physical features of the host such as size, shape, and texture (Fig. 15.5). Recognition of *Helicoverpa virescens* eggs by the scelionid *Telenomus heliothidis* Ashmead illustrates this process (Strand and Vinson 1982b, 1983a,b,c). Host examination and attack in this egg parasitoid can be divided into seven steps: host encounter, drumming, adoption of

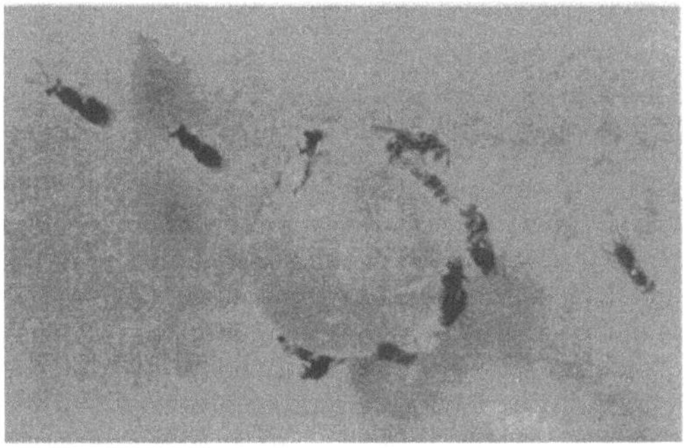

Figure 15.5. Host recognition can be achieved through sensitivity to a variety of physical or chemical cues. Here, females of *Aprostocetus hagenowii* (Ratzeburg) attempt oviposition in a glass bead treated with calcium oxalate and other materials from host glands which serve (along with a curved surface) to elicit host acceptance (after Vinson and Piper 1986). (Photograph courtesy of B. Vinson.)

drilling posture, probing, drilling, oviposition, and marking (Strand and Vinson 1983a). Both physical and chemical factors were necessary to elicit this chain of behaviors. Size and shape were important; the parasitoid only accepted eggs between 0.50 and 0.63 mm in diameter and spherical hosts were preferred to spheroidal ones. Color was not important (Strand and Vinson 1983b). Two proteins from the host accessory gland were found on eggs which acted as recognition kairomones (Strand and Vinson 1983c). The degree of acceptance of glass models was proportional to the fraction of the model's surface that was coated with the kairomone (Strand and Vinson 1983b). Application of this kairomone to eggs of *Spodoptera frugiperda* and *Phthorimaea operculella* Zeller, both nonhosts for this parasitoid, resulted in host acceptance and successful parasitism (Strand and Vinson 1982b), confirming the key role of this kairomone in stimulating host acceptance and oviposition. Similarly, the glue used by females of the brown-banded cockroach, *Supella longipalpa* (Fabricius), to glue ooethecae to surfaces is the host recognition cue for the host-specific egg parasitoid *Comperia merceti* Compere (Van Driesche and Hulbert 1984). In studies in which artificial eggs were offered as hosts to the egg-larval parasitoid *Ascogaster reticulatus* Watanabe, an external kairomone was found to be essential, but not sufficient, to elicit oviposition (Kainoh and Tatsuki 1988). A variety of host structures offer chemical or physical cues to parasitoids and are used in host recognition. Aphelinid parasitoids of some diaspid scales use components of cuticular waxes in scale coverings for host recognition (Luck and Uygun 1986; Takahashi et al. 1990). In *Lymantria dispar*, host recognition by *Cotesia melanoscela* (Ratzeburg) depends on a physical attribute (density of long setae) plus unidentified chemical components in the larval exuviae (Weseloh 1974). In *Trichogramma minutum*, egg size is an important character affecting host acceptance. Egg size is estimated from curvature which is believed to be determined by parasitoids by sensing the scapal-to-head angle while walking on host eggs (Schmidt and Smith 1986, 1987).

Endoparasitoids use sensory structures on the ovipositor to detect chemical cues inside hosts once the ovipositor has been inserted. These cues consist of amino acids, salts, and trehalose (Vinson 1991) and may stimulate oviposition. For some parasitoids, these internal cues are less specific than external ones, and may also occur in nonhost life stages or species (Kainoh et al. 1989).

Host recognition mechanisms do not invariably lead to attack on only physiologically suitable hosts. *Cardiochiles nigriceps*, for example, is stimulated by the salivary secretions of *Helicoverpa zea* (among other species) and oviposits in this species. Parasitism is not, however, successful, as immature parasitoids are consistently encapsulated in all larval instars (Lewis and Vinson 1971).

Host Quality Assessment

Once a host has been recognized as an appropriate species and stage in which to oviposit, the parasitoid must assess the quality of each individual host to determine the number and sex of eggs it should receive. The two most common attributes to be judged are whether or not the host has already been parasitized and the size of the host.

In some host-parasitoid systems, hosts provide only enough resources to support the growth of one parasitoid. In such cases, it is usually advantageous for subsequent female parasitoids of the same species to detect that a host has already been parasitized and to avoid ovipositing in it (as is the case for the braconid *Orgilus lepidus* Muesebeck and its host the potato tuberworm, *Phthorimaea operculella* [Greany and Oatman 1972]). In some host-parasitoid systems, oviposition into an already parasitized host may be advantageous to the second female if her offspring can outcompete the immature stage of the first parasitoid (true in some cases in which the second parasitoid is a different species than the first) or the host can support the growth of more than one parasitoid (with perhaps, some decrease in survivorship of parasitoids or reduction in size of resulting adults).

Larvae of internal parasitoids can eliminate conspecific or other competitors either by physical attack, using mandibles (Hymenoptera) or mouth hooks (Diptera), or by physiological suppression, particularly in later instars, by anoxia, changes in availability of nutrients, poisons, or cytolytic enzymes (Vinson and Iwantsch 1980a). For example, larvae of *Pieris rapae* that have recently been parasitized by *Cotesia glomerata* are still suitable hosts for oviposition by *Cotesia rubecula*, for a period of a few days, because its mandibulate first instar larvae can kill the nonmandibulate larvae and eggs of *C. glomerata* (Laing and Corrigan 1987). In the same system, larvae parasitized by *C. glomerata* are also suitable hosts for oviposition by a second conspecific female (for a short time following the first oviposition), because host larvae can support widely varying numbers of this gregarious species (Le Masurier 1991). Dissections from field populations of *Pieris rapae* in Massachusetts showed that second clutches of this parasitoid occurred in 8% of parasitized larvae (Van Driesche 1988).

The ability of parasitoids to detect previous parasitism varies. The ability to detect previous ovipositions of conspecifics has been demonstrated frequently. The ability to detect previous ovipositions of other species of parasitoids has been documented less often. Interspecific discrimination (in the sense of bias against oviposition by the second species) occurs in some species combinations (Bai and Mackauer 1991; Scholz and Höller 1992) but not in others (Ables 1981). In some systems, a female of a second species may discriminate only if the progeny of the first species are older (larvae versus eggs) (Tillman and Powell 1992). Whether on not a species discriminates against hosts parasitized by another species may depend on the intrinsic competitiveness of immatures of the second versus first species for possession of the host. A highly competitive species may have little reason to exhibit discrimination (Scholz and Höller 1992).

Cues used to detect previous parasitism in a host may either be external marks on hosts, or internal changes in host haemolymph or tissues. The former are typically detected by contact of the antennae with the host's surface. The latter are detected through sensory structures on the tip of the ovipositor after it has been inserted in a host. The scelionid egg parasitoids *Telenomus podisi* Ashmead and *Trissolcus euschisti* (Ashmead), for example, both mark egg masses of their host *Nezara viridula* with a water soluble chemical (Okuda and Yeargan 1988). The braconids *Cardiochiles nigriceps* and *Micropletis croceipes* mark hosts with secretions from their alkaline (or Dufour's) glands (Vinson and Guillot 1972). Such external marks typically are short-lived (1–2 days). Some parasitoids, such as *Ooencyrtus nezarae* Ishii, use egg stalks as a more enduring cue by which to recognize parasitized hosts (for up to 8 days in this case) (Takasu and Hirose 1988). The aphid parasitoid *Ephedrus cerasicola* uses its antennae to detect external cues to avoid conspecific parasitism immediately (less than 6 hours) after the first parasitization, but later (6–24 hours) relies more on detecting internal cues by short probes with its ovipositor (Hofsvang 1988).

The decision as to the number of eggs to lay in a host will depend both on the previous parasitism status of the host and other factors such as host size. Superparasitism may be beneficial if supernumeraries have some chance of survival, if hosts rather than eggs are limiting to the female's opportunities to oviposit, and if the discovery of new hosts is either time consuming or exposes the female to increased risks from predation or other mortality factors (Waage 1986). For idiobiont parasitoids that suspend development of their hosts at the time of oviposition, host size at the moment of oviposition is the size of the resource. For gregarious parasitoids which can attack the host in several instars, hosts represent varying quantities of resources and the number of eggs is often adjusted accordingly. *Anagyrus indicus* Shafee et al. attacks the spherical mealybug *Nipaecoccus vastator* (Maskell) and oviposits in all three nymphal stages and the adult female, but varied the number of eggs per host from an average of 1.2 for first instar nymphs to 2.8 for adults, when hosts were presented in a no-choice design (Nechols and Kikuchi 1985).

Hymenopteran parasitoids which have arrhenotokous (haplo-diploid) reproduction can control the sex of each offspring by fertilizing or not fertilizing eggs as they are laid. Fertilized diploid eggs give rise to females, while unfertilized haploid eggs give rise to males (Fig. 15.6).

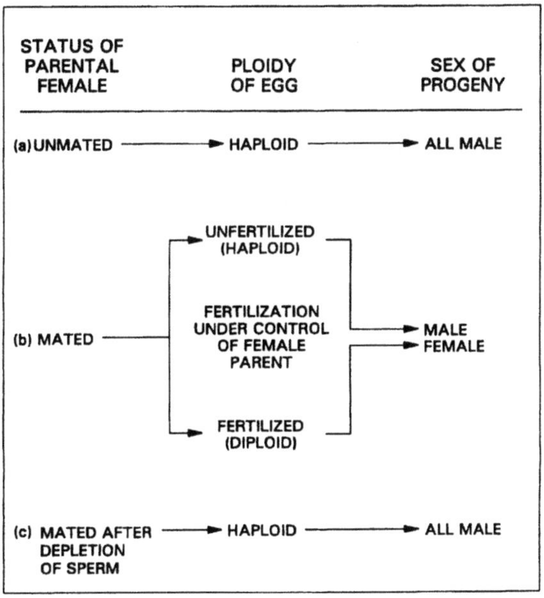

Figure 15.6. Sex ratio of offspring can be an important component of the ovipositional behavior of a female hymenopteran parasitoid. While virgin (a) and mated females with no sperm reserves (c) lay unfertilized (male) eggs, mated females (b) have control over the sex of the offspring. The decision regarding what sex of egg to lay in a particular host can involve the female's perception of host quality, size, or previous state of parasitism.

Consequently, in addition to determining the number of eggs to assign to a given host, females must decide which of these eggs should be female. The proportion of eggs that should be male depends on both host quality (especially size and, for gregarious species, previous parasitism) and, in theory, the degree that females will be mated by siblings (Waage 1986 and references therein).

In gregarious idiobiont species that attack hosts in various instars, smaller hosts often receive single, male eggs; mid-sized hosts single, female eggs; and large hosts may receive several eggs. For example, *Aphytis lingnanensis*, a parasitoid of California red scale (*Aonidiella aurantii*), laid male eggs in small scales and female eggs in large scales (Fig. 15.7) (Opp and Luck 1986). Ovipositions by second females of gregarious species into previously parasitized hosts may be biased toward males because these hosts provide relatively fewer resources than do unparasitized hosts (Werren 1984a, Waage and Lane 1984).

For solitary idiobionts, smaller hosts generally receive male eggs, while larger hosts may receive either male or female. The more frequent allocation of male eggs to smaller life stages of a host species has been observed in parasitoids of aphids (Cloutier et al. 1991), agromyzid fly larvae (Heinz and Parrella 1990), and tephritid fly larvae (Avilla and Albajes 1984). A possible explanation for this pattern is that fitness of female wasps is believed to increase faster with increase in adult size than does fitness of male wasps, so depositing female eggs in larger hosts is advantageous (Charnov 1979).

In addition to the size and quality of a host, the likely mating structure of the offspring population can influence the optimal sex ratio. Fisher (1930) pointed out that in populations with panmictic (random) mating, males and females should be produced in equal proportions because if they are not, a fitness advantage will be conferred on parents producing the rarer sex, which over time causes the sex ratio to tend toward equality. Hamilton (1967, 1979) described a model of local mate competition (nonrandom mating) that predicts that as the number of females ovipositing in a host patch decreases, the proportion of males should decrease as well. In patches in which only one female oviposits, only as many males should be produced as are physically needed to mate all their sisters. In practice, parasitoids may use particular sequences of sex allocation in order to ensure the presence of males in each host patch. For example, some solitary scelionid egg parasitoids that oviposit in host egg masses often begin by laying a male egg in the first or second host, followed by a series of female eggs, and then repeating the process (Waage 1986). This allocation pattern ensures that a minimum number of males will be present on the patch, without regard to when the oviposition sequence ends for lack of hosts or due to disturbance.

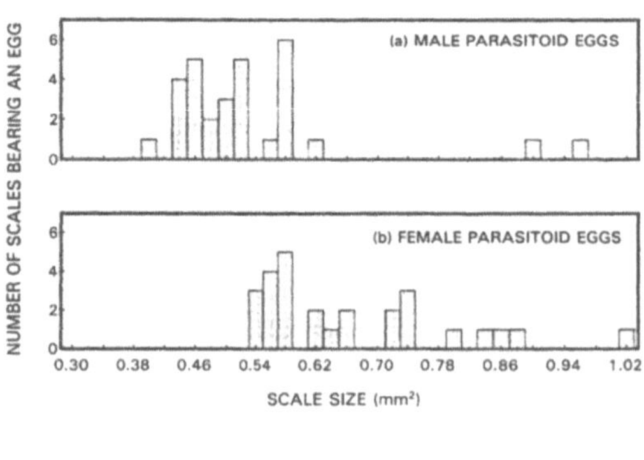

Figure 15.7. *Aphytis lingnanensis* Compere allocated male and female eggs selectively among individuals of its host, California red scale (*Aonidiella aurantii* [Maskell]), with more smaller scales receiving male eggs and more larger hosts receiving female eggs (Opp and Luck 1986).

Factors affecting sex-ratio allocation have been reviewed by Waage (1986) and Wrensch and Ebbert (1993). Werren (1984b) presents a sex-ratio allocation model that combines the predicted effects of local mate competition and host quality variation. While variation in sex-ratio allocation in response to host quality (such as size) has been demonstrated in many species (see above for some examples), experimental confirmations of local mate competition are less convincing. King (1989), for example, did observe an increase in male bias in broods from patches with two versus one female of *Spalangia cameroni* Perkins, but did not observe any differences between two and any higher numbers of females per patch as predicted by local mate competition theory. One of the key tenets of the theory, that males produced in broods with highly female-biased sex ratios (from patches exploited by one female) should restrict their mating to their natal patches, was found not to be true for *Pachycrepoideus vindemiae* (Rondani) (Nadel and Luck 1992). Further evaluation of the applicability of local mate competition theory is warranted (see Hardy 1994).

One of the practical problems to which sex-ratio allocation theories apply is the problem of superparasitism and extreme male bias which can arise in some laboratory colonies when parasitoid adults become excessively numerous relative to their hosts. Such conditions are detrimental, leading to low productivity and poor quality wasps, a very large fraction of which are males.

In some parasitic Hymenoptera, microbial infections appear to be responsible for the existence of all-female (thelytokous or deuterotokous) lines (Stouthamer 1990).

OVERCOMING HOST OR PREY DEFENSES

Once a host has been located, recognized, and its value assessed, it must be successfully attacked. Arthropods that are prey or hosts for predacious arthropods or parasitoids employ several types of defenses (Fig. 15.8). First are behaviors or morphological characters that reduce the likelihood of being discovered. Second are features that reduce the probability, once found, of being successfully attacked. Third, for parasitoids only, are mechanisms used to kill parasitoid eggs or larvae after attack has occurred (Gross 1993).

Avoiding Discovery

The first sort of defense consists, broadly, of residing in refuges in time or space where natural enemies are absent or scarce (Sheehan 1986; Russell 1989; Hawkins 1990; Hochberg and Hawkins 1992). Occurrence in less accessible structures, or on host plants or in plant species assemblages not searched by important parasitoids are examples. Another strategy includes behaviors that tend to disassociate a host from cues (such as kairomones) that might assist a predator or parasitoid in locating prey or hosts. The tendency of some caterpillars to frequently change positions while feeding on a leaf, or to flick frass away from themselves so it falls in a spot removed from a larva's true location, are examples of such behaviors.

Preventing Attack

Features that reduce the rate of successful attack of hosts or prey once they are discovered can be chemical, behavioral, or morphological in nature. Mechanisms used by arthropods to avoid predation include defensive chemicals, fighting, crypsis, escape, mimicry, aposematism, posture or size, dilution, mutualism, armor, acoustic effects, and feigning (Witz 1990).

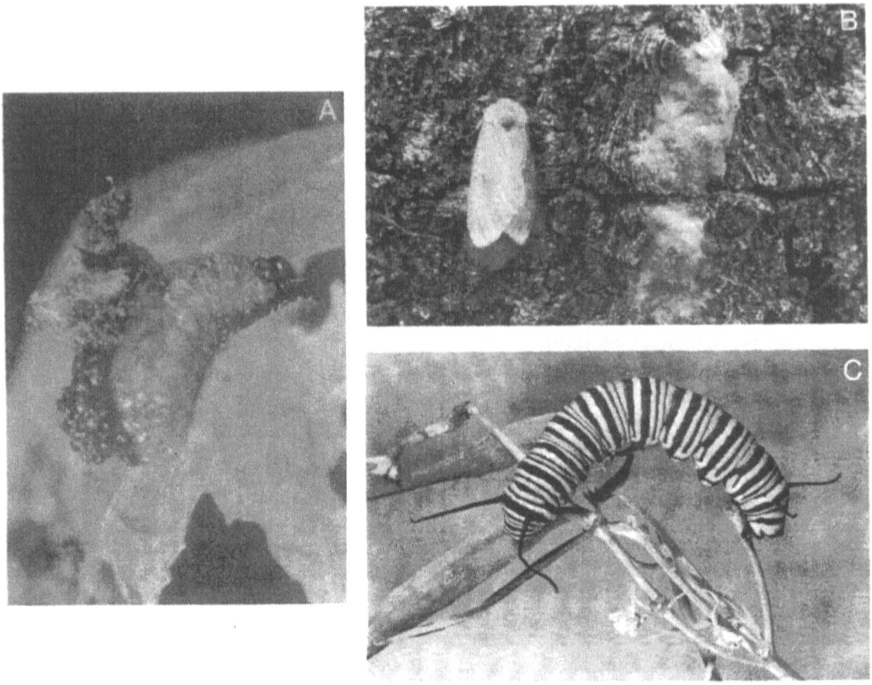

Figure 15.8. Mechanisms used to deter predation or parasitism include: (A) behavioral defense—the creation of trash packets by chrysomelid beetle larvae; (B) architectural defense—deposition of eggs in masses such that inner eggs are inaccessible to parasitism, as in *Lymantria dispar* [Linnaeus]; and (C) chemical defenses—sequestering of plant substances to render herbivore immature stages unpalatable to predators; here milkweed toxins make monarch butterfly (*Danaus plexippus* [Linnaeus]) larvae unpalatable to birds. (Photograph by M. Badgley [A,C] and courtesy of J. Elkinton [B].)

Chemical Defenses. Many chemical substances are used by herbivorous arthropods to reduce attacks by natural enemies (Pasteels et al. 1983), including compounds that are forcefully ejected at attackers and substances that are present in the cuticle or internally which are distasteful. Plant compounds, for example, may be sequestered by herbivores and incorporated into tissues to make herbivores less palatable to predators (Duffey 1980). The large milkweed bug, *Oncopeltus fasciatus*, when reared on seeds of the milkweed *Asclepias syriaca* Linnaeus sequesters dietary cardenolides and stores them in bilateral thoracic and abdominal spaces. These cardenolides are effective in inducing aversive learning in invertebrate predators which prey on protected milkweed bugs, conferring protection on the bug at the population level (Berenbaum and Miliczky 1984). Larvae of the pyralid *Uresiphita reversalis* (Guenée) are protected against attack by predacious ants and vespids by alkaloids (present in larval integument) which are sequestered from the leguminous hosts plants on which the larvae feed (Montllor et al. 1991).

Behavioral or Structural Defenses. Morphological and behavioral factors which hosts use for defense against parasitoids have been discussed by Gross (1993) who recognized five categories: morphological features, evasive behaviors, aggressive behaviors, protection by attending ants, and parental care.

Morphological Features. Features of eggs which affect their vulnerability to attack by parasitoids include the thickness of the egg chorion, the presence of coverings such as setae, silk, or scales, and the architecture of the egg mass. Egg mass architecture, for example, functions by burying some eggs beneath layers of other eggs, rendering them inaccessible to most types of parasitoids. Friedlander (1985), for example, calculated that for the nymphalid *Asterocampa clyton* (Boisduval and LeConte) a greater percentage of the eggs in small egg masses are exposed to potential parasitism than in large eggs masses. Larvae and nymphs may gain protection against parasitism by inhabiting protective bags, cases, or cocoons, by having thick cuticles, armor, or (for colonial species) webbing. The nymphalid *Euphydryas phaeton* Drury forms colonial webs that provide protection from parasitism to early instars and molting larvae (Stamp 1981, 1982, 1984). Protection from predation can also be obtained from similar features. The larva of chrysomelid *Cassida rubiginosa* Müller carries a trash packet of cast skins and feces (a fecal shield) over its back. This packet forms a maneuverable armored shield that is highly effective in blocking the bites of ants (Eisner et al. 1967). Pupae may gain protection from parasitiods from thick cuticles, suspension from threads, or occurrence in large or difficult to penetrate cocoons. Adult insects obtain protection from parasitism from thick cuticles. The general freedom of adult insects from parasitism (relative to other stages) is emphasized by specialized attack behaviors that are employed by euphorine braconids, one of the few groups that attack adult insects. These species have developed abilities to oviposit in regions not protected by thick cuticle, such as the anus, the mouth, and intersegmental areas (Gross 1993).

Evasive or Aggressive Behaviors. The ability to kick, jerk, wriggle, or roll can provide some protection against parasitism. Older instars of aphid nymphs, for example, are sometimes able to deter parasitism by kicking (Gerling et al. 1988). Similarly, head jerking by larvae of *Euphydryas phaeton* (Nymphalidae) is sometimes effective in knocking aside the ichneumonid parasitoid *Benjaminia euphydryadis* Viereck (Stamp 1982). Even some pupae can execute effective evasive movements. Pupae of *Lymantria dispar* use leverage provided by attachment to a silk sheath to arch and move in a spinning fashion. This tactic causes adults of the chalcidid *Brachymeria intermedia* to become entangled in the sheath and break off its attack (Rotheray and Barbosa 1984). The ichneumonid *Bathypectes anurus* (Thompson) reduces its own rate of hyperparasitism by *Dibrachys cavus* (Walker) by exhibiting a jumping motion caused by movements of the larva inside the cocoon (Day 1970). More aggressive forms of defense against parasitoids, such as emission of defensive excretions, are also employed. *Helicoverpa virescens* foul the bodies of the braconid *Cardiochiles nigriceps* with an oral exudate by lunging and vomiting at the parasitoids. Parasitoids hit with the liquid cease oviposition attempts and engage in grooming (Hays and Vinson 1971). Detection of adult parasitoids in the nests of the social wasp *Polistes exclamans* Viereck and *Polistes instabilis* de Saussere causes wasps to search out and destroy adults and immatures of the eulophid parasitoid *Elasmus polistis* Burks (Lutz et al. 1984). Small workers of the ant *Atta cephalotes* Linnaeus ride on pieces of leaves being transported back to the nest by larger workers and actively defend the larger workers against attacks of parasitic phorid flies (Eibl-Eibesfeldt 1967).

More examples of the kinds of aggressive behaviors used by arthropods for defense against parasitoids are given by Gross (1993).

Protection by Attending Ants. Some species of myrmecophilous caterpillars such as *Jalmenus evagoras* Schmett, which feeds on Australian acacia trees, have been shown to have reduced rates of parasitism when trees are colonized by ants (Pierce et al. 1987). Some species of honeydew-producing Coccidae, Pseudococcidae, and Aphididae also show lower rates of parasitism when honeydew collecting ants are present (Gross 1993 and references therein). Not all species of ants provide effective protection. Some ants are more aggressive than others, and their abilities to recognize parasitoids of different species may vary (Völkl 1992). Indeed, in some instances ant-tending may lead to enhanced rates of parasitism (Völkl 1994).

Parental Care. Maternal care is found in many Hemiptera, Membracidae, and Coleoptera (Gross 1993). Females of such species defend their offspring (egg masses, groups of nymphs) against parasitoids (Maeto and Kudo 1992), but removal experiments so far have not shown that these behaviors actually confer any protection (Gross 1993).

Killing Immature Parasitoids after Attack

Idiobiont parasitoids, which permanently paralyze their hosts during oviposition, generally are not subject to host cellular or other defensive responses (Askew and Shaw 1986; Godfray 1994). Eggs and larvae of koinobionts, the hosts of which remain alive and active for a substantial period following parasitism, exist inside still-living hosts, which may mobilize physiological mechanisms against them (Vinson 1990a). Insect immune systems differ from those of mammals in two important ways. First, they lack specificity and do not produce antibodies capable of recognizing and binding to specific foreign antigens. Second, most species lack increased responsiveness to foreign materials as a result of prior exposure. Insect immune systems do, however, recognize and attack foreign materials. Both serum and cellular responses are involved. Serum responses consist of the production of relatively nonspecific substances such as phenol oxidase and agglutinins that have antimicrobial or antibacterial effects. Host responses to parasitoid eggs or larvae may involve a combination of cellular and serum responses. Cells called hemocytes isolate parasitoid stages in a process called encapsulation (Fig. 15.9), which is a coordinated response involving the aggregation, adhesion, and flattening of hemocytes over foreign bodies too big to be engulfed by single cells, resulting in the isolation of the parasitoid inside a cellular capsule (Nappi 1973). Encapsulation is sometimes accompanied by deposition of dark pigment called melanin, the precursors of which are thought to be toxic to parasitoids. Ratcliffe (1982) describes six common types of hemocytes. Several factors affect the strength and rapidity of the encapsulation and melanization responses (Vinson 1990a; Pathak 1993; Ratcliffe 1993). Encapsulation is influenced by host age (older hosts generally are more effective at encapsulation, Blumberg and DeBach 1981; Debolt 1991; Van Driesche 1988), parasitoid strain (Blumberg and Luck 1990), superparasitism (excess eggs tending to exhaust the host's encapsulation capacity, Blumberg and Luck 1990), and temperature (encapsulation being more effective at higher temperatures, Blumberg 1991). Parasitoids subject to potential encapsulation employ mechanisms to evade or reduce rates of encapsulation (Vinson 1990a). One mechanism is selecting young host stages (stages least effective at encapsulation) for oviposition (Debolt 1991). Another mechanism employed by a few specialized parasitoids is to place eggs or larvae in nervous tissue or other organs where they are inaccessible to

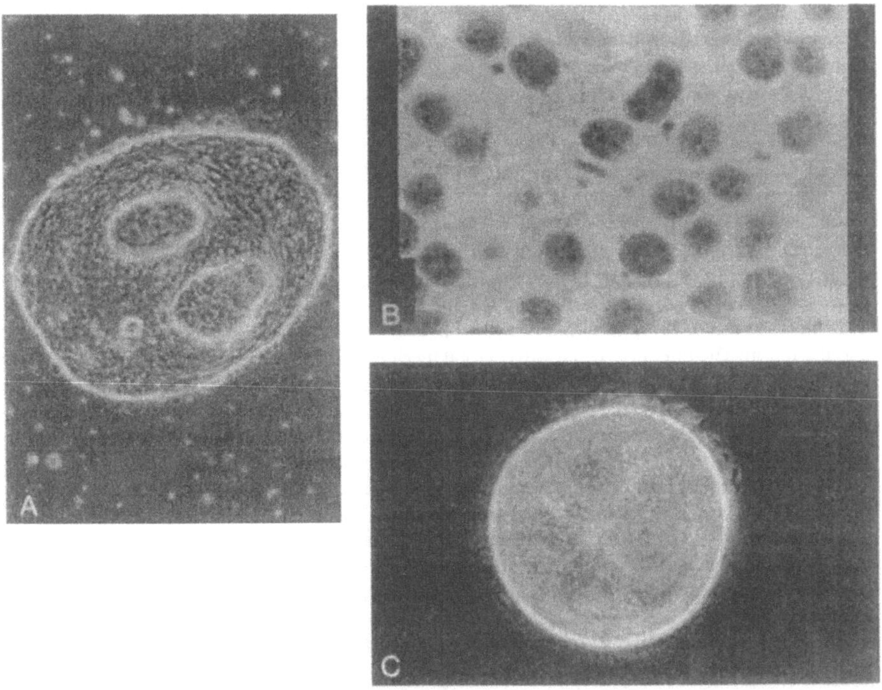

Figure 15.9. Some hosts defend against parasitoid attack by walling off (encapsulating) eggs or larvae of parasitoids, as seen in (A) in which two eggs of *Cotesia congregata* (Say) appear within such a capsule. Parasitoid suppression of host defenses may involve various factors including venom, polydnaviruses (B), and teratocytes (C). (Photographs courtesy of N. Beckage [A] and I. de Buron [B,C].)

hemocytes (Hinks 1971; Godfray 1994). A third mechanism is the employment by various species of Braconidae and Ichneumonidae of viruses (Polydnaviridae) which are transmitted to hosts in calyx fluid injected into hosts at the time of oviposition (Stoltz and Vinson 1979; Edson et al. 1981; Dover and Vinson 1989; Vinson 1990b,c; Whitfield 1990; Fleming and Flemming 1992; Stoltz 1993). These viruses prevent the encapsulation process and protect the parasitoid eggs or larvae, in some cases by directly destroying lamellocytes, one of the hemocytes important in encapsulation (Rizke and Rizki 1990; Davies and Siva-Jothy 1991). In addition, these viruses appear to also play a part in regulating the host's physiology and subsequent development in ways that create circumstances favorable for parasitoid development (Whitfield 1990) (see also the following section). Federici (1991) suggests that these virus-like structures may be parts of the parasitoids' genomes, of viral origin, but no longer actually viruses.

HOST REGULATION AND PHYSIOLOGICAL INTERACTIONS

In addition to successfully evading host attempts to kill their eggs or larvae, koinobiont endoparasitoids in many cases must regulate host physiology to successfully complete their development (Vinson 1975; Vinson and Iwantsch 1980b; Lawrence and Lanzrein 1993). Endoparasitoids especially benefit from the ability to control host moulting, feeding and movement. Consequences that hosts may experience as a result of parasitism include lengthening of the feeding stage; induction of extra larval stages; precocious metamorphosis, blockage of moulting, or induction of developmental arrest (Jones 1985; Lawrence and Lanzrein 1993); and induction or breaking of host diapause (Moore 1989). Even predators may regulate the physiology of their prey as does the mantispid *Mantispa ubleri* Banks which prevents hatching of spider eggs in egg masses which it inhabits, effectively preserving them for later consumption (Redborg 1983).

Exactly how parasitoids regulate their hosts has implications for their use as biological control agents (Jones 1986). For example, if parasitism induces an extra larval instar, the duration of the larval feeding period may be increased within the host generation (Jones 1986). This increase might be detrimental, unless offset by depression in host density by the parasitoid's action. The effect of parasitism on total feeding by pests varies, with solitary parasitoids often killing hosts earlier in larval development than gregarious parasitoids, whose offspring need more resources for their development (Senthamizhselvan and Muthukrishnan 1989). The degree of reduction in feeding by a pest population caused by parasitism can be important in determining the success of a parasitoid as a biological control agent (Hill 1988).

A detailed understanding of the mechanisms behind endocrine regulation of hosts by parasitoids is not yet available in most cases. Multiple components, including viruses, venoms, and parasitoid-derived serosal cells called teratocytes, appear to be involved. The ovarian calyx fluid containing polydnavirus is not sufficient to suppress moulting in *Helicoverpa virescens* attacked by the braconid *Cardiochiles nigriceps*. Moulting suppression requires both calyx fluid and venom. Furthermore, venoms of other species of *H. virescens* parasitoids were not effective as substitutes when combined with *C. nigriceps* calyx fluid (Tanaka and Vinson 1991). A third component, teratocytes, has been implicated in causing arrested development in *H. virescens* larvae parasitized by the braconid *Microplitis croceipes* (Zhang and Dahlman 1989). Teratocytes play several important roles in host-parasitoid interactions, including immunosuppressive and secretory functions (Dahlman 1991). The relative importance of polydnaviruses, venom, and teratocytes vary in different groups of parasitoids. In general, in braconids both polydnavirus and venom are required to suppress moulting, but in ichneumonids, virus alone is sufficient.

The endocrine interactions between endoparastic insects and their hosts have been reviewed by Beckage (1985) and Lawrence and Lanzrein (1993). Parasitoid regulation of host feeding, moulting, and movement patterns provides a variety of potential benefits to the parasitoid. These include: seasonal linkage of life histories; timing of parasitoid development; arrestment or advancement of the host to the stage needed to support parasitoid growth; diversion of nutrients for use by immature parasitoids (through suppression of host egg production); and use of host movements to reduce risk of predation or hyperparasitism.

Some parasitoids use endocrine and physiological cues from the diapause condition of hosts to regulate their diapause (Schoonhoven 1962), with the result that emergence of adult parasitoids occurs at times appropriate to encounter the susceptible stage of the host for oviposition. When the tachinid *Carcelia* sp. develops in a univoltine species, it enters diapause, but when the same parasitoid develops in a bivoltine species, the parasitoid does not enter diapause at the end of the first generation. Rather, it continues to develop and enters diapause when its host diapauses at the end of the second generation (Klomp 1958).

Parasitoids may attack hosts in stages earlier than the stage ultimately consumed by the immature parasitoid. Development of these parasitoids is usually delayed until the host has reached the stage appropriate to support parasitoid growth. Parasitoids may use endocrine cues associated with host development to time their initiation of moulting or rapid growth (Pennacchio et al. 1993). For example, *Diachasma* sp., a parasitoid of *Drosophila* larvae, moults from first to second instar within eight hours after puparium formation of its host, regardless of whether oviposition in the host larva occurred two, four, or eight days earlier (Pemberton and Willard 1918; also see Beckage 1985 and references therein).

Unlike the above two situations, in which parasitoids react to host endocrine events in order to time their own development, in other cases parasitoids act directly to control the developmental stage or activities of their hosts. This effect may be achieved either by preventing the host from moulting out of a particular stage, or by forcing the host to prematurely moult into a particular stage (Lawrence and Lanzrein 1993). The gregarious parasitoid *Copidosoma truncatellum* (Dalman), for example, causes its host *Trichoplusia ni* to undergo an extra larval moult (Jones et al. 1982). This lengthens the feeding period and increases the resources the host provides to the brood of developing parasitoids. Another parasitoid of *T. ni*, *Chelonus* sp., causes its host to prematurely initiate metamorphosis; hosts spin cocoons, but do not actually pupate (Jones 1985). This ensures that the protective structure of the cocoon is provided to the developing parasitoid before the host's death. These effects appear to be mediated by polydnaviruses injected into hosts by parasitoids, as pseudoparasitized hosts (injected with virus but lacking a parasitoid egg) exhibit behaviors similar to those of parasitized hosts.

Antigonadotropic effects of parasitoids on their hosts include the partial or complete suppression of egg maturation in some species, such as parasitism of *Anasa tristis* (De Geer) by *Trichopoda pennipes* Fabricius (Beard 1940; Beckage 1985). While not explicitly demonstrated, this effect is believed to benefit the parasitoid by making nutrients available that would otherwise be sequestered in developing oocytes (Hurd 1993). Nutritional benefits also appear to accrue to parasitoids from castration of male hosts (Reed-Larson and Brown 1990).

Selective movement of parasitized hosts away from feeding sites at the end of the parasitoid's development has been demonstrated. Brodeur and McNeil (1989) showed that parasitized individuals of the aphid *Macrosiphum euphorbiae* (Thomas) moved from the foliage of the host plant to concealed microhabitats, a movement pattern not exhibited by healthy hosts. Movement away from feeding sites reduced the rate of hyperparasitism (Brodeur and McNeil 1992). Similar influence of host movement was found in diapause-bound braconids (*Microplitis mediator* [Haliday]) in bertha armyworm larvae (*Mamestra configurata* [Walker]) (Pivnick 1993). The mechanisms by which such control of host movement is achieved are not known.

LEARNING IN NATURAL ENEMIES

The ability of arthropods to learn has been demonstrated in both parasitoids and predators (Arthur 1966; Murdoch 1969). The ability to learn has important implications for biological control in areas such as the quality of reared organisms used in augmentation and the effectiveness of parasitoids emerging from hosts on noncrop vegetation.

Many studies of arthropod learning have been conducted using parasitoids. Among the effects of learning are strengthening of a response to a single stimulus (either improving speed or efficiency of the response, or strengthening preference for the stimulus over other possible sitmuli [conditioning]), or linking two or more different stimuli that occur together (associative learning).

The strengthening of a response to a single stimulus occurs, for example, when previous contact by a parasitoid adult with a host increases the frequency, speed, or accuracy of its orientation, handling, and ovipositional responses toward the same host species. *Brachymeria intermedia* in olfactometer tests walked upwind more often, more rapidly, and probed more often in air ladened with host kairomone if parasitoids had had prior experience with host pupae (Cardé and Lee 1989). Such prior host contact can also influence host preference. Parasitoids, either in mass-rearing facilities or in nature, during the first days of their adult life are likely to contact individuals (or host products) of the host species from which they have emerged. If parasitoids have weakly-fixed host preferences, then such initial contact with hosts of the natal species may, through learning, strengthen a preference for that host species. Parasitoids reared on alternative hosts may perform less well against the target species (van Bergeijk et al. 1989). Conversely, in some species, host preference may be sufficiently well-fixed genetically that early conditioning to alternative hosts may have little effect. Weseloh (1984), for example, found that conditioning of the tachinid *Compsilura concinnata* to rearing hosts such as *Galleria mellonella* did not change its preference for *Lymantria dispar.*

Giron (1979) proposed that successful parasitism of unparasitized hosts enable parasitoids to better distinguish parasitized and healthy hosts. This proposal was based on the observation that, following exposure to nonparasitized hosts, many species of parasitoids do change the pattern of their ovipositions to favor healthy hosts. This shift, however, is no longer interpreted as being based on an improved ability to recognize previously parasitized hosts. Superparasitism, previously thought to be uniformly maladaptive, is now recognized to be a potentially adaptive foraging strategy, particularly when patches of healthy hosts have not yet been encountered. Naive parasitioids are believed to be able to recognize parasitized hosts, but still oviposit in them because in the absence of contact with healthy hosts, it is adaptive to exploit the resource at hand, albeit a poorer quality one, than to refrain from oviposition (van Alphen et al. 1987; van Alphen and van Dijken 1988). This is adaptive in all cases for gregarious parasitoids, and for solitary parasitoids whenever second ovipositions are conspecific but not self-superparasitism and the second egg has some chance of defeating the first parasitoid.

Associative learning involves learning to link the perception of one stimulus with the presence of another object or event. Parasitoids associate a variety of secondary stimuli with hosts or food sources (Lewis and Tumlinson 1988). These secondary stimuli include characteristics of the host's habitat (form, color, or odor) or microhabitat (such as a gall or leaf mine) (Wardle and Borden 1989, 1990), the host's host plant (odor, color, form) (Kester and Barbosa 1992), the host-plant/host complex (usually odors) (Turling et al. 1991; Lewis et al. 1991), and odors associated with nectar or other food sources (Lewis and Takasu 1990). In addition, parasitoids may associate hosts with artificial stimuli, such as color (Fig. 15.10), chemicals in artificial diets (Vinson et al. 1977), or odors such as vanilla or chocolate that are used experimentally to demonstrate associative learning (Lewis and Takasu 1990). Parasitoids also may simultaneously associate two or more cues, such as odor and color, with hosts (Wäckers and Lewis 1994).

Both generalist and specialist parasitoids are able to learn (Poolman Simons et al. 1992). They may, however, employ this ability differently. Generalists adapt quickly to cues from a wider range of habitats or plants where hosts have been recently encountered. In contrast, specialists learn to search alternative habitats or plants only when these are abundant and the normal host plant or habitat is scarce. In either case, learned responses are temporary, fading after a few days (Papaj and Vet 1990; Poolman Simons et al. 1992), allowing parasitoids to continually adjust their search efforts to cues that have been profitable (i.e., have led to actual host encounters and ovipositions) in the immediate past. Young parasitoids of some species may learn or associate stimuli better than older individuals (Wardle and Borden 1985).

Learning abilities may have practical implications for several biological control activities. For example, efforts to colonize new species of natural enemies might be improved by exposing

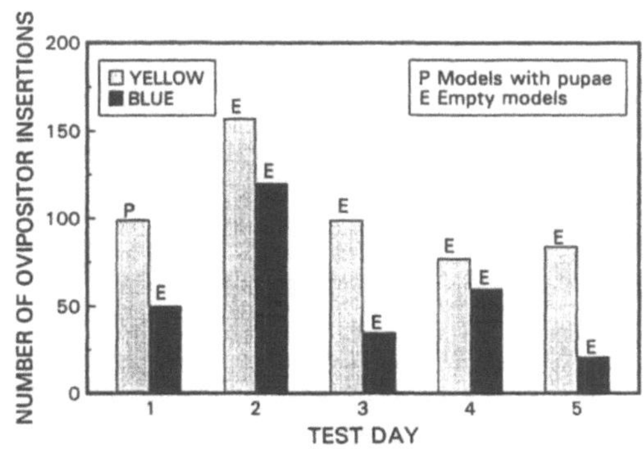

Figure 15.10. Comparison of ovipositor insertions by *Pimpla instigator* Fabricius into colored models, indicating associative learning. Wasps were offered host pupae in yellow-colored models on day one. When offered empty models subsequently, the wasps made significantly more insertions in yellow-colored models (except on day four when the difference was not significant) (after Schmidt et al. 1993). Similar associative learning behavior has been demonstrated for chemical odor-based stimuli (Lewis and Tumlinson 1988).

adult parasitoids or predators to the target prey, on the target plant, prior to release, although this has not been demonstrated. For mass-reared natural enemies, conditioning to alternative hosts or artificial rearing conditions might reduce their efficacy on the natural or target host. Exposure to the target host prior to release may be valuable in overcoming such effects (Hérard et al. 1988). Attempts to foster the field production of natural enemies in noncrop reservoirs, or on alternative hosts, may be less effective than supposed if natural enemies arising in these zones are strongly conditioned to search on the noncrop plant, or to search for the alternative host, in lieu of the pest on the crop (see Chapter 7).

EFFECTS OF PLANT FEATURES ON NATURAL ENEMIES

Characteristics of plants influence natural enemies in a wide range of ways, some of which are just beginning to be recognized (Fig. 15.11). It is common for a particular host species to be subject to varying levels of attack by natural enemies when it feeds on different plant species. Plant features influence foraging success and reproductive fitness of natural enemies. Knowledge of such interactions is important for conducting foreign exploration (to obtain species or races of natural enemies adapted to attack the target pest on the target crops); for planning conservation and augmentation programs; and for guiding changes made to crop plants through plant breeding (so that natural enemy action is enhanced rather than reduced during the creation of new crop varieties, see Boethel and Eikenbary [1986]). Strong et al. (1984) discussed relationships between plants and insects and concluded that competition for space free from natural enemy attack was important in shaping herbivore communities. Price (1986) presents a classification of how plants can affect natural enemies through either semiochemically-mediated effects, chemically-mediated effects, or physically-mediated effects.

Semiochemically-mediated effects have been discussed earlier in this chapter in relation to habitat and host finding. In some instances, plant compounds are used directly by natural enemies as cues for habitat location. In other instances, plant compounds may either be sequestered by herbivores or may be released by plants under herbivore attack, resulting in the attraction of natural enemies directly to the host. At the level of plant communities, associated plants may produce compounds that either enhance attractiveness to natural enemies or may make the detection of host plants by natural enemies more difficult. (For more on this concept see Chapter 7 on intercropping and vegetational diversity).

Chemically-mediated effects of plants on natural enemies include the use by natural enemies

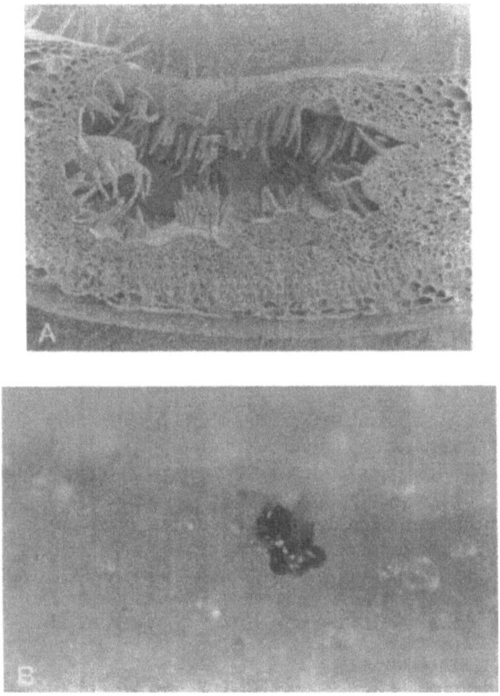

Figure 15.11. Plant structures affect natural enemies in a variety of ways, for example: (A) domatia are pit-like structures found on some plants in which phytoseiid mites harbor. Plants with such structures support larger populations of phytoseiids; (B) trichome density can affect parasitoid reproductive success by reducing walking speed. Sticky or poisonous exudates from some trichomes further impair parasitoid survival and reproduction, as seen here where an *Encarsia luteola* Howard wasp is entrapped in exudate from trichomes on the leaves of the plant *Abutilon theophrasti* (Medicus). (Photographs courtesy of D. Wallace [A] and D. Headrick [B].)

of plant products such as nectar and pollen as food sources. Nutritional qualities of plants also affect natural enemies indirectly by influencing the rate of growth and survival of herbivores which feed on them. Herbivores which develop on plants of reduced nutritional quality are likely to require a longer period to develop and may remain in stages susceptible to natural enemy attack longer than herbivores developing on more nutritious hosts. Plant qualities that cause mortality to an associated herbivore affect rates of the herbivore's survival in its various life stages. This in turn will affect the survival of immature parasitoids associated with the stage of the herbivore, and also will affect the number of hosts in subsequent stages that are available for other parasitoids or predators to attack. Finally, plant compounds may be sequestered by herbivores and used as defenses against natural enemy attack. Protection of monarch butterflies (*Danaus plexippus* [Linnaeus]) from predation by birds through sequestration by monarch larvae of plant-derived cardiac glycosides is a well-known example (Brower 1969).

Physical aspects of plants may affect natural enemies in several ways. Spatial dispersion of plants can affect the ability of herbivores and natural enemies to locate and, in some cases, successfully colonize the plants. Plant structures can provide herbivores with physical protection. Insects in the centers of large fruits or galls, for example, are less accessible to parasitoids with short ovipositors. Some plant features may directly shelter natural enemies. Domatia (pits and pockets) on leaves (Fig. 15.11), for example, are used by phytoseiids and plants with such features harbor higher numbers of these predatory mites (Walter and O'Dowd 1992; Grostal and O'Dowd 1994).

Plant features such as leaf toughness or hairiness, which in some cases defend plants against herbivores, may also affect natural enemies. Increased trichome density on leaves is associated for some parasitoids with reduced walking speed and lowered rates of foraging, making them less effective at finding hosts (Hua et al. 1987). Some predators, such as chrysopid larvae, may also experience reduced walking speed on hairy leaves (Elsey 1974), but other predators, such as larvae of the coccinellid *Adalia bipunctata*, forage for prey more efficiently on leaves with single scattered hairs (versus glabarous, waxy leaves) because hairs force more frequent turning and cause the larvae to move across the leaf surface rather than to only follow the veins and leaf edge (Shah 1982). Some phytoseiids have been shown to be more abundant on grape varieties with hairy undersurfaces, perhaps because of more favorable microclimate and protection from rain (Duso 1992). The influence of leaf pubescence on entomophagous species has been reviewed by Obrycki (1986). More broadly, the shape of plant leaves and the arrangement of branches and other plant parts affect how natural enemies structure their foraging on the plant and between sets of plants (Ayal 1987; Grevstad and Klepetka 1992). Coccinellid larvae of several species, for example, drop off leafless pea plants (*Pisum* spp.) less frequently than from normal plants because tendrils of leafless plants are easier for beetles to grasp (Kareiva and Sahakian 1990). Plant canopies in which leaves overlap are associated with increased rates of dispersal of coccinellids (in the absence of prey) than crops in which plant canopies are discrete (Kareiva and Perry 1989). Parts of plants with distinct characteristics may be searched differently by species of natural enemies. Banana plants (*Musa* spp.), for example, are searched differently for banana aphid (*Pentalonia nigronervosa* Coquillet) by *Lysiphlebus testaceipes* (Cresson), which searches open surfaces but avoids concealed areas, in contrast to *Aphidius colemani* Viereck, which searches in both zones (Stadler and Völkl 1991).

SYNCHRONY AND NATURAL ENEMY RACES

Successful natural enemies must accurately time their periods of active growth and reproduction to seasons when hosts or prey and other necessary resources are available. Some natural enemies may pass periods that are climatically unfavorable (winter cold, periods of extreme heat or dryness) or periods when hosts (or prey) are unavailable by becoming physiologically inactive. The condition of arrested physiological activity is termed diapause. For endoparasitoids, the diapause condition of the host is a source of information that can be used for seasonal synchronization, as discussed previously. In other cases, natural enemies may react directly to environmental cues to initiate or terminate diapause. The environmental signal used most commonly is photoperiod. Natural enemies with wide distributions across bands of latitude may exhibit clines with respect to the critical day length needed to induce diapause (Saunders 1966; Tauber and Tauber 1972; Beck 1980). In some species of natural enemies, diapause may be induced by low temperatures (Carton and Claret 1982) or may be absent (with metabolic rate rising and falling with temperature, without physiological arrestment). Clines of genetically-determined critical photoperiods for diapause induction may limit successful colonization of natural enemies which are moved between locations differing in latitude. Attempts to colonize populations of *Cotesia rubecula* in the eastern United States using populations from European sites at higher latitudes were impeded by a mismatch of critical photoperiod and local seasonality (Nealis 1985).

The life cycles of natural enemies may also fail to fit the seasonal pattern of a location if essential hosts are unavailable in some portion of the year. *Pediobius foveolatus*, for example, overwinters successfully in northern China in climates comparable to those of the northeastern United States. This race of the parasitoid, however, failed to establish in the northeastern United States, perhaps because the parasitoid overwinters in certain Asian epilachnines (such as

Epilachna admirabilis Crotch) which pass the winter as larvae. These hosts are absent from the United States, where the target host, *Epilachna varivestis*, overwinters as an adult (Schaefer et al. 1983).

In addition to genetic differences related to seasonal synchronization, local natural enemy populations may also differ in their adaptation to other environmental characteristics of sites, such as temperature extremes. Natural enemies which are to be colonized in multiple areas with different climates may need to be collected in several locations to obtain populations adapted to these various climatic zones. Parasitoids collected from the deserts of the Middle East have in some instances proven better adapted to California's hotter regions (such as the interior valleys), than were parasitoids from western Europe, which were suited only to cooler coastal areas (Messenger et al. 1976; see also Chapter 8).

CHAPTER

16

BIOLOGY AND DYNAMICS OF PATHOGENS

INTRODUCTION

Pathogens of arthropods include bacteria, viruses, fungi, protozoa, and nematodes. Many phyla, families, and species are involved and collectively display an enormous diversity in their biologies. These have been summarized recently by Tanada and Kaya (1993), from whom much of the information presented here has been drawn. Concepts presented here on epizootiology of disease in arthropod populations have been drawn from Fuxa and Tanada (1987).

BASIC PROCESSES IN PATHOGEN BIOLOGY

To complete their life cycles successfully, most pathogens must contact a host, gain entrance to the host's body, reproduce within one or more host tissues, and emit propagules which subsequently contact and infect new hosts. This chapter examines these components of pathogen biology, first as concepts, and then as components of the biologies of selected pathogen groups.

Host Contact

Unlike predators and parasitoids, many arthropod pathogens lack a motile stage. Therefore, host contact typically results not from active search processes but rather from chance contact following the passive dispersal of some stage of the pathogen such as fungal spores by wind, rain, or other organisms (such as parasitoids). The amount of contact between a pathogen population and its hosts is determined by the spatial pattern of the pathogen's propagules relative to the spatial distribution of the host, and the survival of these propagules over time. The occlusion bodies of nuclear polyhedrosis viruses from the cadavers of diseased gypsy moth larvae (*Lymantria dispar*), for example, are released when host cadavers rupture (Fig. 16.1). Virus occlusion bodies are initially concentrated near the site of host death, but later become distributed over nearby foliage (especially foliage directly beneath the initial point) by rain (Woods and Elkinton 1987). In agricultural systems, pathogen dispersal may also occur as the result of some irrigation practices (Young 1990). Similarly, wind acts to redistribute fungal conidia, which are produced on the cadavers of hosts, to new locations in the habitat (Fig. 16.2). Factors which affect disease transmission at the population level are discussed further in the following section section on epizootiology.

While contact between propagules of many types of pathogens and new hosts is largely a

Figure 16.1. Cadavers of virus-killed gypsy moth larvae (*Lymantria dispar* [Linnaeus]) rupture and release virus particles which contaminate bark and foliage. (Photograph courtesy of J. Cunningham, Forestry Canada.)

random process mediated by abiotic factors (horizontal transmission) (Fig. 16.3), some pathogens are transmitted between generations of hosts from mother to offspring (vertical transmission), a process that eliminates the need to contact new hosts randomly (Fig. 16.4). Additionally, a few types of pathogens, such as some nematodes and aquatic fungi, are able to move towards hosts. Some nematodes move in water between soil particles and use chemical cues such as CO_2 and host feces to aggregate near hosts (Ishibashi and Kondo 1990). Similarly, the motile spores of such aquatic fungi as *Coelomomyces* spp. and *Lagenidium* spp. move toward hosts in response to chemical cues emitted by hosts (Carruthers and Soper 1987).

Host Penetration

Once propagules of pathogens have contacted a host, the body of the host must be penetrated to reach tissues susceptible to attack. The arthropod cuticle provides protection from many

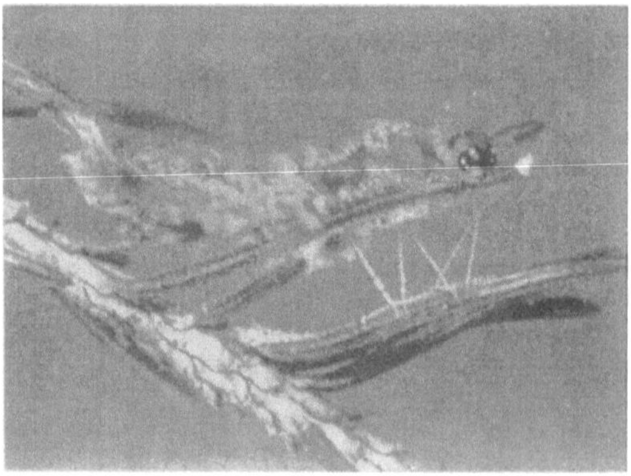

Figure 16.2. Conidia of *Zoophthora radicans* (Brefeld) Batko are produced by conidiophores which emerge from fungus-killed *Choristoneura fumiferana* (Clemens) larvae. These condia are dispersed by wind and rain. (Photograph courtesy of D. MacLeod, Forestry Canada.)

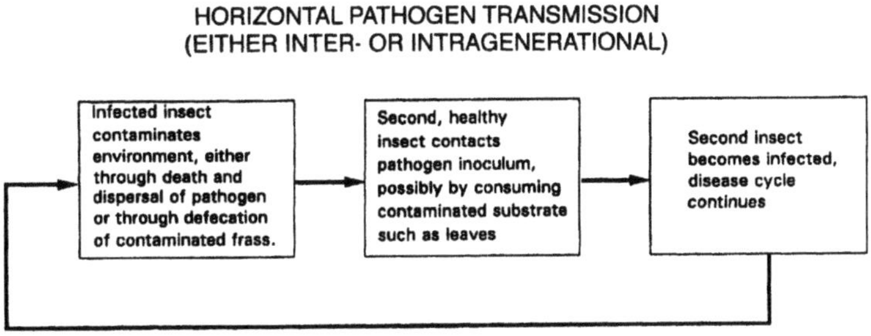

Figure 16.3. Horizontal transmission of pathogens in insect populations occurs among members of the same general life stage and usually among members of the same generation.

types of pathogens. Most bacteria, viruses, and protozoa enter arthropods through the thin (nonchitinized) wall of the midgut after being ingested. Consumption of food that is contaminated with pathogen stages is a major mechanism of contagion for chewing arthropods. Sucking arthropods, in contrast, escape exposure to such contamination by feeding on plant fluids, which are relatively free of microbial contaminants. As a consequence, sucking insects such as aphids are little affected by such pathogens as bacteria, viruses, and protozoa that normally enter hosts by ingestion. Nematodes and fungi are able to penetrate arthropod hosts through routes other than the midgut. Nematodes may enter hosts either through the midgut (after being ingested or actively entering the mouth), or may enter hosts directly through wounds or spiracles, or may mechanically penetrate intact cuticle using stylets or spears as cutting devices. The characteristic route of entry for fungi is through the cuticle, a process that is achieved by special penetration hyphae that produce enzymes capable of digesting insect cuticle (Fig. 16.5).

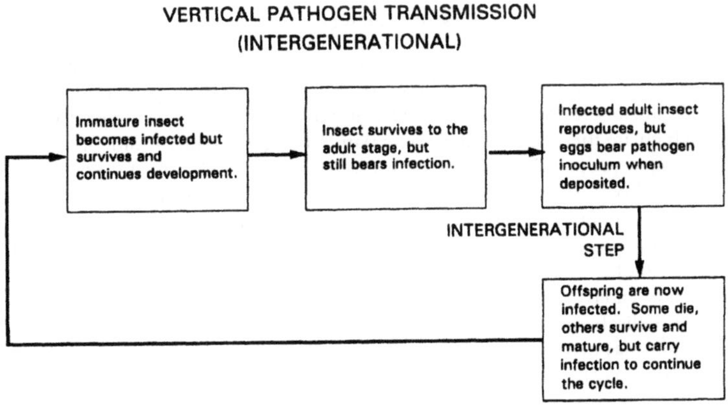

Figure 16.4. Vertical transmission of pathogens in insect populations occurs between members of two succeeding generations and involves passing the inoculum from reproducing parents to offspring.

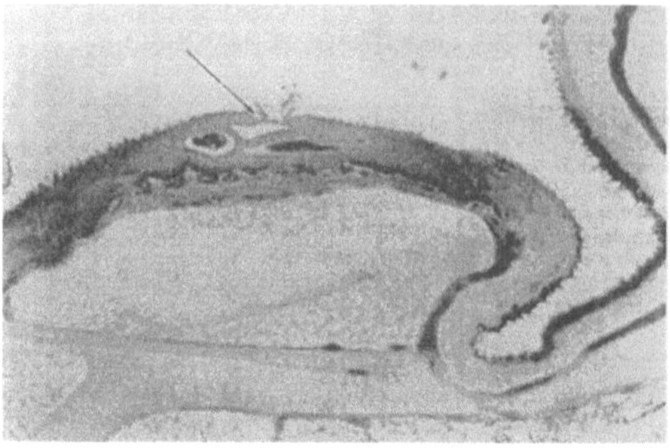

Figure 16.5. Host penetration by fungi is usually through the cuticle. Penetration hyphae are used by an *Entomophthora* species to enter (arrow) its host, the pine sawfly larva, *Diprion similus* (Hartig). (Photograph courtesy of M.G. Klein, from Klein, M. G. and H. C. Coppel. 1973, *Annals of the Entomological Society of America* 66:1178–1180, with permission.)

Reproduction in Host Tissues

Once a pathogen has penetrated its host, the pathogen typically proceeds to reproduce in one or several tissues (Fig. 4.3b,d). In some pathogen groups, reproduction may be limited to specific tissues as, for example, the nonoccluded *Oryctes* virus that reproduces principally in fat body and midgut epthelium. Others reproduce in virtually all tissues in their hosts. The range of tissues infected affects the number of pathogen propagules produced per unit weight of the host, such that pathogens producing systemic infections in all tissues may be more economical to rear than ones limited to only certain of their host's tissues. Steinernematid and heterorhabditid nematodes continue to reproduce after hosts have been killed by symbiotic bacteria. Consequently, most host tissue is available for nematode reproduction.

Exit of Pathogen Propagules from Host or Cadaver

Following reproduction of a pathogen inside the body of its host, the offspring of the pathogen must in some fashion contact other hosts to continue the transmission cycle. In groups in which vertical transmission from parent to offspring occurs, contact is achieved by contamination of eggs which are then deposited in some part of the general environment. In most cases, however, pathogen propagules are released independently back into the environment and later contact new hosts. If pathogens kill the host, release may take place as a consequence of the physical disintegration of the host cadaver, as in the case of virus diseases in which host cadavers liquify and are broken apart by abiotic factors. For fungi, release of spores may be by way of the growth of special spore-producing hyphae which reach through the cuticle to the outside of the host cadaver. Release from such structures may be passive or may be assisted by active mechanical discharge of spores. Nematodes leave hosts in several ways. In some groups, juveniles or adults may actively disperse from host cadavers in soil or water (Fig. 16.6). In other groups, nematodes may be dispersed by way of the host's reproductive tract during oviposition attempts by infected hosts. Dispersal of protozoans and bacteria from infected hosts may occur in the form of contaminated feces while the host lives, or later through breakup or consumption (by other hosts) of the host cadaver.

Figure 16.6. Emergence of the post-parasitic stage of *Romanomermis culicivorax* Ross and Smith from its mosquito host. (Photograph courtesy of J.J. Petersen.)

Spread and Persistence of Pathogen Propagules in the Environment

Following the release of infectious pathogen particles into the environment, the continuity of the pathogen's population depends on contacting new hosts. Because the host's physical and temporal occurrence may be patchy and unpredictable, pathogens require adaptations both for dispersal and persistence. Dispersal in the environment is largely accomplished, as mentioned above, by physical factors such as wind and rain, with host contact being largely a matter of chance. Transmission to new hosts by such dispersal is most likely when hosts are highly aggregated. Insects such as whiteflies, aphids, Lepidoptera, or other insects undergoing high-density population outbreaks are especially favorable for disease transmission. Insects being reared in colonies in laboratory or commercial settings, unless reared individually, are also especially susceptible to the propagation of disease.

The likelihood of a pathogen contacting a host is enhanced by a variety of life history characteristics. Some groups of pathogens are transmitted vertically from parent to offspring, such as the microsporidium *Nosema pyrausta* (Paillot) which is transmitted primarily through transovarial infection in the first generation of its host noctuid *Ostrinia nubilalis* (Siegel et al. 1988). Other pathogens are transmitted by infected hosts to the specific habitat where additional hosts are likely to be present. Some nematodes in the families Iotonchiidae and Phaenopsitylenchidae, for example, are deposited in the habitat of their hosts' larvae by infected adults (muscoid flies and wood wasps) whose ovaries are infected by the nematodes (Fig. 4.9). Because at times hosts are likely to be scarce or environmental conditions unfavorable for host infection, many bacteria, protozoa, and fungi produce resting stages (of various types) that are able to persist for long periods. The production of such structures aids host contact by permitting pathogen propagules to be conserved until conditions improve.

At the population level, two general routes of transmission can be defined. Vertical transmission occurs when pathogens are passed directly from parents to offspring (Fig. 16.3). There are two types of vertical transmission: transovarial and transovum. In transovarial transmission, the pathogens are present inside the eggs at the time they are laid due to invasion by the pathogen during egg development inside the mother. This manner of transmission predominates, for example, in the nonoccluded iridescent viruses and the microsporidia. Transovum transmission occurs when offspring acquire disease as a consequence of external contamination of their surfaces at the time of oviposition by pathogens from within the mother. Such transmission occurs, for example, in some baculoviruses and cytoplasmic polyhedrosis viruses of Lepidoptera and Hymenoptera (Andreadis 1987).

All other routes of transmission are considered to be horizontal (Fig. 16.4). One form of horizontal transmission occurs when pathogen particles emitted by an infected individual enter the environment and subsequently contact and infect another host. For example, horizontal transmission occurs when an infected caterpillar defecates pathogen-contaminated feces on foliage that is later eaten by a healthy caterpillar, which then contracts the disease. Horizontal transmission is common in baculoviruses and fungi, among other groups. Transmission of venereal diseases during copulation is also considered a form of horizontal transmission. In this process, pathogens pass directly between hosts without entering the nonhost environment.

EPIZOOTIOLOGY OF ARTHROPOD PATHOGENS

The prevalence of a disease in a system may change markedly over time. Periods when disease is particularly prevalent are termed epizootic outbreaks of a disease. Some groups of pathogens are more frequently involved in arthropod epizootics (viruses, fungi, protozoa) than others (bacteria). A number of factors can contribute to the outbreak of a disease in a particular insect population at a particular time and place. These include characteristics of both the host and the pathogen, host population density and distribution, and environmental conditions such as temperature, rainfall and humidity. The study of how such factors determine the course of disease outbreaks is termed epizootiology; Fuxa and Tanada (1987) discuss epizootiology of insect diseases.

Host Features That Influence Disease Prevalence

Among the host factors that can affect the development of an epizootic are host density, spatial distribution, health of hosts, age, moulting status, and behavior.

Because pathogen propagule density will be diluted as the cube of the distance to the next nearest host, contact rates will be highest when hosts are closest together. Disease transmission is facilitated in insect-rearing colonies with artificially high densities, or in kinds of hosts that occur in nature in colonies or groups, or have significantly aggregated spatial distributions. For chewing insects such as caterpillars, horizontal transmission is facilitated when hosts are dense because contact with feces or fragments of host cadavers is more likely to occur. For sucking insects, aggregation enhances transmission by proximity to sources of pathogens such as fungal spores released from diseased aphids within aphid colonies.

The health of hosts is also a factor affecting the rate of transmission of a pathogen because hosts that are stressed by other pathogens, poor nutrition, or adverse physical conditions are often less resistant to pathogen invasion. Similarly, age and moulting status affect susceptibility to infection. Young caterpillars are often more susceptible to the bacterium *Bacillus thuringiensis* and to viruses. Newly moulted insects, in which the cuticle is still rather thin, are more susceptible to fungal invasion than insects whose cuticle is fully hardened. When large numbers of individuals in a population are stressed, or in a susceptible age class, or moulting status, an epizootic of a disease becomes more likely because hosts are more easily infected.

A variety of host behaviors can affect the rate of transmission of a disease in an insect population. Diseased individuals may exhibit abnormal behaviors that increase the dispersal of the pathogen propagules released from the individual. In some diseases, infected individuals frequently die in relatively high positions on their food plant or habitat. Some caterpillars infected with virus, for example, migrate upward and die at the tips of branches, a behavior that positions the cadaver to contaminate foliage lower down as the cadaver disintegrates. Similarly, some flies infected with various fungal pathogens die at the tips of grass stems, such that wind dispersal of spores produced on the cadaver is facilitated. In other instances, normal behaviors such as grooming of nestmates in termites or cannibalism (in which moribund individuals may be consumed by other members of the group) may increase the spread of a disease.

Pathogen Features That Influence Disease Prevalence

Many pathogen characteristics influence the rate of disease in a host population. These include: infectivity, virulence, production of toxins, nature of the pathogen life cycle, and inoculum density, distribution, and persistence. Understanding pathogen biology and constraints on pathogen transmission in the environment is important in the development of pathogens as pest control products.

The genotype of a particular pathogen strain will strongly influence its infectivity and virulence to a particular host. Infectivity is the ability of the pathogen to gain entrance to the host's body and virulence is the ability, once inside the host, to cause disease. Pathotypes within a pathogen species vary significantly with regard to what host species can be successfully attacked. In fungi, strains may vary in the level of enzymes produced by penetration hyphae, changing their infectivity to the host. In *Bacillus thuringiensis*, isolates differ in the kinds and quantities of the toxins they produce. These differences in toxins determine which groups of hosts will be affected by different strains of the bacterium.

Pathogen life cycles vary from simple to highly complex, with the need in some cases for alternate hosts. Highly complex life cycles may place constraints on pathogen transmission if alternate hosts or special conditions are available in only some habitats or periods. The requirement for copepods or ostracods as alternate hosts by fungal pathogens in the genus *Coelomomyces*, for example, means that continuous reproduction of this pathogen following artificial application is only possible if these hosts are present (Tanada and Kaya 1993).

The density, distribution, and persistence of the inoculum of a pathogen is important in determining the normal rate of the disease in the host population and the frequency and intensity of epizootics. Inoculum alone in the absence of favorable environmental conditions is insufficient to cause epizootics. However, abundant, persistent sources of inoculum in the habitat favor the occurrence of epizootics by ensuring the presence of inoculum during periods when environmental conditions are favorable for disease initiation. Factors that influence the amount of inoculum in the local habitat include those that affect the level of production of the inoculum, dispersal in the habitat, and persistence of the inoculum in various specific parts of the habitat. Quantities of inoculum produced per host will vary greatly between types of pathogens for a variety of reasons. Some pathogens, for example, multiply in the entire host, while others may be restricted to specific tissues. Pathogens killing hosts at a early stage are likely to produce less inoculum per host than ones killing older, larger hosts.

Spread of a given pathogen in the habitat will depend on the nature of the release mechanism from the host. Wind-blown fungal spores are likely to be more widely dispersed than viruses liberated by liquification of host cadavers with local contamination of foliage in the drip zone below the cadavers. Persistence of pathogen inoculum will be strongly influenced by physical factors that destroy inoculum, particularly ultraviolet light, high temperatures, and dryness. The production of pathogen stages resistant to such degradation will be important in determining how well a given pathogen can persist at a site. Some microhabitats, especially soil and protected spaces such as bark crevices, provide physical conditions that are more favorable for the survival of pathogen inoculum. Host contact with these zones, or movement of material from these areas to areas where hosts feed, such as the movement of soil onto crop foliage by the splashing of rain droplets, can be important mechanisms for bringing inoculum back into contact with hosts.

Environmental Factors That Influence Disease Prevalence

Many abiotic and biotic factors affect the level of disease in an arthropod population. Among these are temperature, humidity, desiccation, light, soil characteristics, and other organisms that aid in pathogen transmission. A more detailed discussion of the effects of these factors on insect epizootics is given by Benz (1987).

Effects of temperature on disease rates are complex. Temperature differences may affect either the pathogen or host directly, but the effect on the disease rate can only be understood by also considering the impact on behavior, growth, movement, and other factors that taken together define the effect of temperature on the pathogen-host interaction. The route of entry of the pathogen, for example, can play an important role in this process. For organisms in which ingestion of contaminated food is the principal route of entrance, infections can only be acquired at temperatures at which hosts feed. For fungi, which enter hosts primarily through the integument rather than the mouth, infections may be acquired at temperatures below those at which the host feeds, if temperatures are favorable for fungal spore germination and hyphal growth.

Humidity, free water, and desiccating conditions are important in some situations. High humidity generally favors outbreaks of fungi, promoting both the germination of existing spores and formation of new spores on cadavers. High humidities and soil moisture levels are also important in favoring nematode epizootics. Bacterial, viral, and protozoan disease prevalences seem less influenced by these factors. Rain appears to have relatively little direct effect on disease rates and does not wash significant amounts of pathogen inoculum from plant surfaces. Desiccation, in contrast, is an important, lethal factor for many types of pathogens, including nematodes, bacteria, and protozoa. Many pathogens have special stages that can resist desiccation. These include the occlusion bodies of baculoviruses, the spores of some bacteria (*Bacillus* spp.), the cysts of amoeba and other protozoans, the resting spores of fungi, and the eggs and juvenile resting stages of some nematodes.

The deleterious effect of sunlight, especially ultraviolet light, on baculoviruses is well known. Baculoviruses deposited on upper leaf surfaces exposed to sunlight are typically inactivated in a short period, ranging from a few hours to a few days. Fungal spores are also sensitive to light, but many are protected by light-absorbing pigments. Few data are available concerning the effect of light on protozoa and nematodes.

Soil is a complex habitat with many physical and biotic components. Because soils are often moist and dark, they are a favorable location for the survival of resting stages of such pathogens as bacteria, baculoviruses, and fungi. Soil pH and organic content influence the rate of degradation of pathogens, as does the species composition and abundance of soil microorganisms.

Contamination of predators and parasitoids with viruses or other pathogens may occur when these feed on or emerge from diseased hosts. Pathogens may then be spread to new hosts by the movement and behaviors of such agents. Fuxa et al. (1993) demonstrated this process by tracking the movement of an introduced virus (the *Anticarsia gemmatalis* virus from Brazil) by such agents in a soybean (*Glycine max*) field in Louisiana (U.S.A.) following its release. Because the virus was not native to the region, all virus detected was known to have originated from the released material.

In addition to these environmental and biotic factors, several features of interacting host-pathogen systems have been identified that can cause periodic or aperiodic cycles of disease prevalence. These features are part of the natural cycling of densities between a host and its pathogen independent of changes in host susceptibility, environmental suitability, or pathogen virulence. These issues have received considerable attention in the literature on dynamics of interacting host-pathogen systems. An introduction to this topic is given in Chapter 18.

BACTERIAL PATHOGENS OF ARTHROPODS

Bacteria are unicellular organisms that have rigid cell walls. Their shapes include rods, spheres, spirals, and forms that have no fixed shape. A description of the families and genera in which arthropod bacterial pathogens are found is given in Chapter 4. The biologies of species causing disease in arthropods are discussed in detail by Tanada and Kaya (1993).

A generalized arthropod bacterial infection may be described as follows. Most pathogenic bacteria enter arthropod hosts through the mouth when contaminated food is ingested. They multiply in the gut, producing enzymes (such as lecithinase and proteinases) and toxins that damage midgut cells, aiding the bacteria to invade the hemocoel. Once they have successfully invaded the hemocoel, bacteria cause a general septicemia, during which more toxins are produced. Host death results from bacterial multiplication in the hemocoel and poisoning by bacterial toxins. Before death, hosts lose their appetite and cease feeding. Diseased hosts become diarrhetic and discharge watery feces. Hosts may vomit. Both of these processes distribute infective particles into the environment, promoting horizontal transmission to new hosts. Insects killed by bacteria darken and become soft. Tissues become viscous and have a putrid odor. The integument remains intact, while cadavers shrivel, become dry and harden.

While most arthropod bacterial pathogens enter hosts through the mouth, a few are able to penetrate the integument directly, as is the case of *Micrococcus nigrofaciens*, a pathogen of scarab beetles in the genus *Lachnosterna* (Northrup 1914). Some bacteria are transmitted vertically from parent to offspring, for example, *Serratia marcescens* Bizio in the brown locust, *Locustana pardalina* (Walker) (Prinsloo 1960). In some cases, bacteria can be introduced into the host's hemocoel directly on the ovipositors of parasitic Hymenoptera. However, such cases are of minor importance. The dominant route of host entry for arthropod bacterial pathogens is by mouth when contaminated foliage or other food is eaten.

Because the hemocoel is the characteristic site for most bacterial infections in arthropods, mechanisms to gain entrance to this region are important determinants of which groups of bacteria are effective pathogens of arthropods. Several mechanisms exist that permit bacteria to reach the hemocoel. Some species in the genus *Bacillus* produce crystalline toxic proteins. These aid the bacteria in penetrating the midgut epithelial cells. Bacteria enter the hemocoel by causing pores to open in midgut membrane cells, altering their permeability (Honée and Visser 1993). Many other groups of bacteria, however, lack such toxins. These bacteria are usually unable to gain entrance to the hemocoel and normally exist as saprophytes in the insect gut and exteriorly in other habitats. Such groups (*Proteus*, *Serratia*, *Pseudomonas*) are pathogenic if they penetrate into the hemocoel, but usually do not do so unless the host is stressed by some other factor. When the host is stressed, these bacteria are able to multiply in the gut and more effectively penetrate into the hemocoel, aided by enzymes such as chitinase. Bacteria in the genera *Photorhabdus* and *Xenorhabdus* are symbiotic with nematodes in the families Heterorhabditidae and Steinernematidae, respectively. The nematodes serve as vectors that penetrate mechanically into the insect hemocoel and deposit these bacteria there directly.

Some species of bacterial pathogens of pest arthropods, such as *Bacillus thuringiensis*, do not cause significant epizootics in nature, due to low spore production and ineffective horizontal transmission. Indeed, some evidence suggests this species may be primarily a saprophyte in nature, and only incidentally an insect pathogen (Martin 1994). Nevertheless, such species may be of importance to biological control as augmentative biological control agents (see Chapter 11). Death of hosts from *Bacillus thuringiensis* products requires delta endotoxins (for bacteria to gain access to hemolymph), but LC_{50} values of *B. thuringiensis* products are greatly reduced by the presence of live spores, demonstrating the importance of live cells in the disease process (Miyasono et al. 1994). Physiological modes of action of *B. thuringiensis* endotoxins in insects are reviewed by Gill et al. (1992). Other species, such as *Bacillus sphaericus* and *Bacillus popillae*, are more effective in horizontal transmission and do maintain continuous disease cycles in affected arthropod populations for many years.

VIRAL PATHOGENS OF ARTHROPODS

Viruses are obligate intracellular parasites consisting of a genome (either DNA or RNA) enclosed in a protective protein coat (**capsid**) and further enclosed in additional layers termed **envelopes**. The capsid plus the DNA or RNA it encloses is termed **nucleocapsid** or **virion**. Virions may be further embedded (in some viruses) in protective protein matrices termed **occlusion bodies**. Viruses replicate inside host cells using the host's protein-synthesizing metabolism and materials in the host cell (Matthews 1991). Some features of virus biology important to biological control are: virus susceptibility to abiotic factors such as ultraviolet light and high temperatures; need for a mechanism to penetrate into host cells; and the occurrence of significant virus epizootics in nature, especially those of members of the Baculoviridae.

Susceptibility to abiotic factors is reduced in the Baculoviridae, Entomopoxviridae, and Reoviridae by protein matrices (occlusion bodies) which surround and protect the virions. Occlusion bodies increase virion stability and persistence during periods spent outside the host. Other viruses, such as Rhabdoviridae, are not exposed to such abiotic factors because they are closely associated with hosts at all times, being transmitted vertically between host generations (Tanada and Kaya 1993).

Initial entrance into hosts by the Baculoviridae is by mouth when contaminated food is consumed. The high pH found in the insect midgut dissolves the protein occlusion bodies of baculoviruses, liberating virions. Virion envelopes merge with the cell membranes of gut microvilli, and nucleocapsids enter host cells. Alternatively, an enveloped virion may be envaginated in a vacuole of cell membrane and later be liberated in the cell after enzymes have dissolved the vacuole and the virion envelope. This process is call viroplexis. Once in the hemocoel, nonoccluded forms of the virus (plasma-enveloped virions) are the infective structure. Details of cytopathology and viral replication may be found in Tanada and Kaya (1993). In most Lepidoptera, baculoviruses in the nuclear polyhedrosis subgroup infect many host tissues (fat body, hypodermis, trachea, blood cells). In sawflies (Symphyta, Hymenoptera), nuclear polyhedrosis viruses infect only the midgut tissue. Such differences influence greatly the number of virions produced per infected host, affecting both the dynamics of horizontal transmission in nature and the economics of commercial virus production.

Arthropod larvae (such as those of many Lepidoptera) infected with baculoviruses lose their appetite but continue to feed at lower rates up until a few days prior to death, which may occur 5–21 days after infection. Before death, some species of infected larvae move to positions high in the plant canopy, a behavior that facilitates the horizontal transmission of viruses in nature through food contamination. Dead hosts often become flaccid and the integument ruptures, liberating occlusion bodies containing virions into the environment. Consumption of foliage contaminated by occluded virions by new hosts completes the virus transmission cycle, commonly leading to epizootics. Transmission of *Oryctes* virus (a nonoccluded virus) in rhinoceros beetle, *Oryctes rhinoceros*, occurs by means of contact with virus-contaminated feces produced by infected adult beetles that may live as long as 30 days. Adult feeding and defecation in sites also used for oviposition and larval development promote horizontal transmission of this virus.

FUNGAL PATHOGENS OF ARTHROPODS

Morphologically, fungi may occur as single cells (such as yeasts), or branched filaments (hyphae) which form mats (mycelia). Hyphae may be uninucleate, or multinucleate segments which have numerous nuclei not separated by transverse walls. Fungi may reproduce sexually, asexually, or both. Sexual reproduction involves fusion between two structures. These may be motile gametes, or one motile and one stationary gamete, or two sexually differentiated hyphae, or two nondifferentiated hyphae. Fungi may be either homothallic or heterothallic, with one fungal body producing both sexes in the former and only one sex in the latter.

Infective propagules in fungi are of several distinct types, including spores, conidia, zoospores, planonts, and ascospores, among others. While these differ in their etiology, most (with the exception of the zoospores of aquatic fungi) are functionally similar in the way in which they contact and penetrate hosts. Host entry is most often through the integument, less often through natural body openings. Transovarial transmission is extremely rare. Unlike the bacteria and viruses, most fungi do not invade hosts through the gut after being ingested. Arthropods such as aphids that feed by sucking plant fluids are often subject to fungal infections (though the integument) but relatively unaffected by bacterial or viral infections. Host ranges of fungi vary from narrow to broad, and species with broad host ranges may consist of a series of pathotypes that are relatively specific to different hosts within the overall host range.

Fungal infections (termed mycoses) begin with contact between susceptible hosts and infective particles (such as spores or conidia). Zoospores of some aquatic fungi such as the Oomycota (*Lagenidium* spp.) are mobile (by flagellae) and able to detect chemical clues from hosts and actively orient towards them. In contrast, spores of most terrestrial fungi lack this active mobility and contact hosts through random movement caused by wind and other forces.

Following host contact, the first step in fungal infection is adhesion and germination of the spore on the host's cuticle. The physical and chemical properties of a potential host's cuticle contribute to the specificity of fungal pathogens by influencing the success of adhesion and spore germination. The degree of adhesion of fungal spores is influenced by the presence of mucous materials and physical structures (appressoria). Penetration of the host's integument is accomplished by mechanical entrance of a penetration hypha (germination tube). Penetration involves both physical pressure from the penetrating hypha and degradation of the components of the cuticle by proteinases and other enzymes. Fully hardened cuticle presents a greater barrier to fungal penetration than does new cuticle following a moult. For this reason, insects are more susceptible to fungal invasion after a moult. Differences in infectivity of fungal strains may relate to variation in levels of chitin-degrading enyzmes (Gupta et al. 1994). Fungal penetration through the digestive tract (buccal cavity, esophagus, gut) has been observed in some instances.

Fungi such as the Deuteromycotina, once they have entered the hemocoel, reproduce quickly and generally kill the host. Fungal growth can be through various structures including yeast-like hyphal bodies, hyphal strands, or wall-less protoplasts. Protoplasts assist fungi in overcoming host defenses because they are not recognized by host immune systems. These yeast-like cells develop rapidly and produce toxins that help suppress the host's immune reactions. After the host dies, further fungal growth is saprotrophic. This growth leads to the development of a mycelium, which becomes a sclerotium (a durable resting structure), from which sexual reproductive spores are later produced. In addition, in many groups non-sexual hyphae emerge from the cadaver and under favorable humidities produce asexual spores. These spores disperse and are often important in horizontal transmission to new hosts, leading to epizootics of the pathogen in the host population. Dispersal of such spores may sometimes occur through movement of insects, but most often is passive by means of wind or water. Initial local dispersal may be actively facilitated by the forceful discharge of spores from spore-producing structures on the host cadaver.

Temperatures most favorable for the development of fungal infections are 20–30°C. High humidities (above 90%) are often required for spore germination and for spore production outside of hosts. Films of free water are necessary for conidial germination of some Deuteromycotina but are unfavorable for most Entomophthorales. Lower humidities (below 50%), darkness, and vibration may be needed for spore release for some groups, such as *Beauveria* and *Metarhizium*.

As with many pathogens, reproductive effort of pathogenic fungi is divided into the production of spores of different types. Under certain conditions conidial (or other type) spores are produced in large numbers, facilitating short-term horizontal transmission, often resulting in epizootics. Under less favorable conditions, thicker-walled resting spores are produced that are more resistant to adverse environmental conditions. These spores aid the pathogen in surviving periods of environmental stress or periods when suitable hosts are unavailable.

Fungal biology also strongly influences the degree to which any given group may be used in pest control through augmentation (based on commercial production of infective particles) (Chapter 11). Some fungi are able to grow and sporulate on nonliving media. Other groups, for example, many members of the Entomophthorales and the water molds in the Chytridiomycota (*Coelomomyces*), have an obligate need for live hosts to complete their life cycles and, therefore, are poorly suited for commercial production. Detailed treatments of the biologies of various groups of fungi are found in Tanada and Kaya (1993).

PROTOZOAN PATHOGENS OF ARTHROPODS

Protozoans are minute, single-celled organisms found in most habitats, except perhaps the air. They vary widely in shape, color, and morphology. They exhibit both sexual and asexual reproduction. Of the 15,000 described species about 1200 are associated with insects, some of which are pathogenic (Tanada and Kaya 1993).

Protozoan infections typically begin when infective forms are ingested and enter the host gut. A few forms such as the ciliates in the genus *Lambornella* are able to attach to and penetrate the integument. The infective stage is most often a cyst or spore, but may be an actively-living stage. Cannibalism may be an important route of horizontal transmission, as may be the consumption of infected prey of other species (for example in flagellate infections). Most of the protozoans associated with insects remain in the gut and groups such as ciliates, flagellates, and gregarines often cause little pathology. Some forms, however, such as the apicomplexans and microsporidia, are more likely to invade the hemocoel where they are pathogenic. The microsporidia are the most important group of protozoan pathogens affecting insects. Epizootics of microsporidia occur in nature. These are all obligate parasites and multiply only in living cells. Tissues affected by protozoan infections vary but may involve the gut epithelium, the Malpighian tubules, or the fat body, among others.

In addition to entering hosts through ingestion, microsporidia may be introduced into hosts on the ovipositors of Hymenoptera. Microsporidian infections can reduce the effectiveness of rearing insects for research or for commercial mass production of beneficial insects for augmentative control. Vertical transmission from parent to offspring of the host occurs in many groups of protozoa, especially the microsporidia.

Most protozoan infections are chronic infections that persist for extended periods but do not kill their hosts. Infected hosts often show few or no external signs or symptoms of infection. Toxins have not been detected in protozoan infections. Infections may be either systemic or limited to one or more tissues. Complex sexual or asexual patterns of multiplication may occur. Details concerning the reproduction cycles of individual groups are given by Tanada and Kaya (1993).

NEMATODES PATHOGENIC IN ARTHROPODS

Nematodes represent a single phylum, the Nematoda, within which about nine families occur that are parasitic on insects and have potential for use as biological control agents (see Chapter 4). Nematodes are translucent, usually elongate, and cylindrical in form. The body is covered with an elastic, noncellular cuticle, but is not segmented. Unlike bacteria, viruses, and protozoa, nematodes are multicellular animals that possess well-developed excretory, nervous, digestive, muscular, and reproductive systems. They do not have circulatory or respiratory systems. The digestive system consists of a mouth, buccal cavity, intestine, rectum, and anus. Nematode taxonomy is based largely on sexual characters of adults; consequently, immature stages are difficult or impossible to identify.

Nematodes are diverse and are found in nearly all habitats. Nematodes may be free- living or parasitic on either plants or animals. Nematode associations with insects range from phoresy to parasitism. Some nematodes, such as *Beddingia siricidicola*, have complex life histories with both parasitic and free-living cycles that may continue indefinitely.

Many nematodes have relatively simple life cycles with three life stages: eggs, juveniles, and adults. Mated female nematodes deposit eggs in the environment; the first stage juvenile usually moults inside the egg and emerges as a second stage juvenile. Most nematodes moult four times. In many groups, the third stage juvenile remains ensheathed in the cuticle of the second stage, which provides it with increased resistance to adverse conditions. This third stage form is called a dauer juvenile, dauer being the German word for durability. Moulting to the adult stage may occur inside the host or free in the environment. All nematodes stages, except the egg, are mobile. Most nematodes have separate, single-sexed individuals and mating is required.

Nematode infections usually occur in the hemocoel, but in some groups such as the Phaenopsitylenchidae (e.g., *Beddingia*) and Iotonchiidae (e.g., *Paraiotonchium*) nematodes may invade the sexual organs. Nematode infections may severely affect the host, causing debilitation, castration, or death. Most of the obligately parasitic nematodes are relatively host-specific and are associated with one or a small group of hosts. Some groups, however, such as the steinernematids and heterorhabditids, often have broad host ranges under laboratory conditions. However, such laboratory host ranges are typically broader than actual host ranges in nature because of the absence in such tests of ecological factors that restrict host contacts to species found only in certain habitats.

In nematodes, unlike most of the other pathogens discussed in this chapter, host finding may be an active process in which nematodes move towards and recognize hosts using cues such as bacterial gradients, host fecal components, or carbon dioxide (Grewal et al. 1993a). Nematode species vary in their host searching strategies, with some being ambush predators and others actively moving in search of hosts (Kaya et al. 1993). Host entrance may be a passive process, as when nematode eggs or juveniles of Tetradonematidae are ingested by larvae of sciarid flies. In most instances, however, host penetration is an active process in which juvenile nematodes penetrate hosts through the integument or natural openings (mouth, anus, spiracles). In the cases of natural openings, nematodes seeking entrance have only to move through the opening, avoiding efforts of the host to brush them aside (in the case of the mouth). Once inside the gut, nematodes use mechanical devices (stylets, spears) to puncture the gut wall and enter the hemocoel, the site of nematode infection. Stylets and spears may also be used externally to perforate the cuticle to penetrate directly to the hemocoel in some groups of nematodes. Other kinds of nematodes, such as the Sphaerulariidae, may use adhesive materials which attach the nematode to the host's cuticle to assist in perforating the cuticle with stylets.

Nematode infections produce relatively few external signs other than, in some cases, distended abdomens or changes in color. One exception to this is the formation of intercaste or intersex individuals infected by mermithids. Internal effects of infection may be profound. Sterility is induced by several groups of nematodes, including Mermithidae, Phaenopsitylenchidae, and Iotonchiidae. Moulting may be inhibited in some cases. Behaviors of nematode-infected individuals may be abnormal. Infected individuals may have difficulty walking or flying normally, or may exhibit unnatural phototropisms.

Mermithids differ from other nematodes because they leave their hosts before reaching the adult stage. Postparasitic juveniles exit from hosts and then moult to adults that mate and produce progeny as free-living stages.

Steinernematids and heterorhabditids, the groups of nematodes used most extensively in augmentative biological control, kill their hosts in 2–3 days, a much shorter time than for other groups of nematodes. This occurs because these families of nematodes have mutualistic bacteria in their guts (*Xenorhabdus* spp., *Photorhabdis* spp.) that kill hosts by septicemia. Juvenile nematodes reach the hemocoel by penetrating the midgut wall after being ingested by the host, or by penetrating the host integument. *Xenorhabdus* spp. or *Photorhabdis* spp. bacteria are then released into the host hemocoel by defecation of the juvenile nematodes. Juveniles feed saprophytically on the dead host's tissues and then mature to adults which reproduce. When a new generation of the dauer stage is attained, they leave the host cadaver.

Nematodes in the families Phaenopsitylenchidae and Iotonchiidae include both facultative and obligate parasites. The phaenopsitylenchid *Beddingia siricidicola* has two life cycles. One is free living and feeds on a fungus that is mutualistic with the insect host. This fungus is spread by the host (a siricid wood wasp) and grows in the cambium of the host tree attacked by the wasp. In the free-living cycle, juvenile nematodes feed on fungus, become adults, mate, and lay eggs. In the parasitic life cycle, adult female nematodes penetrate the cuticle of wood wasp larvae which themselves feed on the fungus. After the host insect has pupated, nematodes develop in the hemocoel and produce offspring that invade the developing eggs of the wood wasp. When the wood wasp oviposits in new fungal patches, it deposits nematode-infected eggs rather than healthy ones. The eggs are killed by the nematodes, which emerge and continue their development, either through the fungus-based life cycle or the insect-based life cycle depending on the presence or absence of insect hosts on the patch. In a similar manner, the iotonchiid *Paraiotonchium autumnale*, a parasite of *Musca autumnalis*, invades ovaries of its host and is dispersed in the habitat by the fly's oviposition attempts, as does *Paraiotonchium muscadomesticae* Coler and Nguyen, a parasitoid of *Musca domestica* (Coler and Nguyen 1994).

Further details on the biologies of specific groups of nematodes are given in Gaugler and Kaya (1990), Kaya (1993), and Tanada and Kaya (1993).

CASE HISTORIES I: TWO VIRUSES

Gypsy Moth Nuclear Polyhedrosis Virus

The gypsy moth, *Lymantria dispar*, is a Eurasian forest lepidopteran that undergoes occasional outbreaks in North America, an area to which it was introduced. Outbreaks are ended by epizootics of a nuclear polyhedrosis virus. This virus is amplified within years by horizontal transmission from young to old larvae and maintained between years in field populations by surface contamination of eggs. Egg masses become contaminated with occlusion bodies at the time of oviposition by being deposited on surfaces that have been contaminated by virus from cadavers of older larvae which died of virus. Viruses mixed into the egg mass, which is covered with setae from the moth, are protected from environmental degradation and are able to persist from late summer of one year until early spring of the following year. Infection occurs in young larvae (first and second instars) when they ingest chorion from contaminated egg masses. Cadavers of young larvae killed by virus disintegrate, contaminating adjacent foliage and causing a new set of infections in older larvae that consume this foliage (Elkinton and Liebhold 1990). Because this type of horizontal transmission becomes more efficient as host densities increase, epizootics become more likely and more intense as larval densities increase over a series of years, culminating in an epizootic that depresses the pest population to extremely low levels. This in turn reduces the effectiveness of disease transmission, reducing the prevalence of disease in the population.

Oryctes Non-Occluded Virus

In contrast to the gypsy moth virus which is able, once incorporated into a host egg mass, to survive in the habitat for an eight-month period (August-April), the nonoccluded *Oryctes* virus degrades in less than a week if left unprotected in the environment (Hochberg and Waage 1991). This virus infects the rhinoceros beetle, *Oryctes rhinoceros*, a pest of coconut palms in the southwest Pacific and southeast Asia through to east Africa. The virus has been introduced into several locations with subsequent reduction in the level of damage from the beetle (Zelazny et al. 1990). Virus persistence in the host population is the result of infected adult beetles surviving for 2–4 weeks, during which time their feces are contaminated with virus particles. During this period, beetles are capable of normal flight and feeding behaviors and visit feeding sites in the crown of palms and oviposition sites in decayed palm logs. Feces from infected beetles dropped at such sites are ingested by larvae and other adults which then acquire the disease. The disease prevalence is augmented by releases of laboratory-infected adult beetles. The dynamics of this system have been modeled by Hochberg and Waage (1991).

CASE HISTORIES II: TWO BACTERIA

Most bacterial pathogens have not been found to be important regulators of insect populations in nature. However, the biologies of several species are such that they, or their chemical products, can be used in augmentative biological control. The greatest commercial success with the augmentative use of bacteria for insect control has been with members of the genus *Bacillus*. Two species, *B. thuringiensis* and *B. popilliae*, illustrate how biological features of individual species can constrain the augmentative uses of microbes.

Bacillus thuringiensis

Bacillus thuringiensis is a bacterium that does not persist in sufficient numbers in nature in most habitats to cause epizootics, but which produces stable toxins that can be harvested and used as pest control chemicals. Commercial production of the bacterium is feasible because it will grow and produce spores and toxins in fermentation media, making inexpensive, large-scale production systems possible. Several strains which produce toxins effective against various insect groups are known. Genetically-modified strains have been constructed that combine several toxins into one bacterial strain, making it feasible to produce products that are effective against pests in three insect orders: Lepidoptera, Coleoptera, and Diptera.

Bacillus popilliae

In contrast, in North America, *B. popilliae* is able to persist in soil and affect the larval populations of its host, the immigrant scarabaeid *Popillia japonica*, for decades (Ladd and McCabe 1967) and cause a degree of population suppression. While interest also exists in using this organism for augmentative biological control, differences between its biology and that of *B. thuringiensis* have made this unsuccessful to date. Toxins play little or no role in the infection process or host death for *B. popilliae*. Rather, spores must be ingested and disease induced through bacterial cell multiplication for death to occur. While abundant vegetative growth of this pathogen can be induced on artificial media, formation of spores (the obligate infective stage) is limited and difficult to obtain apart from living hosts (Tanada and Kaya 1993). Because of the lack of efficient systems for producing spores, commercial production of spores of this bacterium is limited to the small-scale, expensive process of rearing the bacterium in live host larvae which must be collected in the field and inoculated individually by injection (Fig. 11.3). Consequently, this organism, while useful in nature for biological control through introduction, has been less successful than *B. thuringiensis* as a commercial product for augmentative control.

CASE HISTORIES III: TWO NEMATODES

Recent improvements in large scale production of nematodes in the Steinernematidae and Heterorhabditidae have led to an increasing augmentative use of several species of nematodes in these families for control of soil-dwelling pests (see Chapter 11). The role of nematodes in biological control, however, also includes the introduction of specialized nematodes for the permanent suppression of immigrant pest species. This use of nematodes is illustrated by the release of the phaenopsitylenchid *Beddingia siricidicola* in Australia for the control of a wood wasp and of the steinernematid *Steinernema scapterisci* for control of immigrant mole crickets in Florida.

Beddingia siricidicola

The siricid wood wasp *Sirex noctilio* is a pest of European origin that attacks Monterey pine (*Pinus radiata* D. Don) in Australia. The nematode *Beddingia siricidicola* was imported and released in Tasmania by placing it and an associated fungus (*Amylostereum chailletii* [Fr.] Boidin) into holes bored in trees (Bedding 1968). The nematode has two possible life histories. In one, it develops and feeds on the fungus that is vectored by the wood wasp. (The fungus spread by the wood wasp is the agent that kills the tree, making the infected tree suitable for reproduction by the wood wasp.) Also, the nematode can infect larvae of wood wasps. When it does, it defers development until the wood wasp has reached the adult stage. At this time the nematodes reproduce in the host's hemocoel and the resulting juveniles migrate into the wasp's ovaries, where they remain until the wood wasp attempts to oviposit in new breeding locations. At this point nematode-infected, rather than healthy, eggs are deposited. Nematodes emerge (the wasp egg dies), placing nematodes in locations where both new fungus and new wood wasp larval hosts are present. The nematodes then either feed on fungus or infect other wasp larvae. Using inoculations of this nematode, Bedding (1974) was able to reduce the number of trees killed per year by wood wasps in a 1000 acre test area from 200 to none over a four-year period.

Steinernema scapterisci

Scapteriscus spp. mole crickets are turf and pasture pests in Florida; all species of *Scapteriscus* in Florida are immigrant species from South America. Surveys of natural enemies of mole crickets in Uruguay revealed a species of steinernematid, subsequently described as *Steinernema scapterisci* (Nguyen and

Smart 1990). Host specificity studies of this species have shown that it is specific to certain mole crickets at moderate inoculum levels (Nguyen and Smart 1991), although a broader set of species can be attacked if higher inoculum levels are administered in laboratory petri dish assays (Grewal et al. 1993b). *Steinernema scapterisci* has persisted at release sites for up to five years and continues to infect a substantial proportion of hosts (Parkman et al. 1993). Natural spread of the nematode to additional locations has been observed and potential for control of the target pests appears promising.

These projects indicate that importing and establishing new species of nematodes is a valuable approach for control of immigrant pests.

CHAPTER
17

Biology of Weed Control Agents

INTRODUCTION

Three principal groups of organisms—herbivorous arthropods, fungal pathogens, and fish—have been used for biological control of weeds. This chapter considers key aspects of the biologies of each of these groups.

HERBIVOROUS INVERTEBRATES

The two most important features of the biology of herbivorous arthropods being considered for use as biological weed control agents are host specificity and degree of effect on plant performance and population dynamics. Agents introduced to new regions must possess sufficient specificity that risk to important economic or native plants is limited. The factors affecting breadth of the host ranges in herbivorous arthropods are important to the design and interpretation of appropriate host-range tests. Tests allow potential levels of safety or risk to be identified, which can then be judged as to their acceptability. Similarly, knowledge of plant physiology and ecology is important in understanding the biological features that determine which agents are effective in suppressing their target plants, and why they do so. Of particular importance is an appreciation of the relation between kinds and degrees of herbivory and resultant levels of plant injury. Further ecological knowledge is required to understand how given levels of plant injury affect plant populations, including effects on plant density and reproductive rates over time.

Interactions among the agent, the target plant, and the environment are often crucial in determining the success or failure of a particular agent in a given location. Besides discussing the determinants of herbivore host range, methods for its measurement, and the nature of the physiological effects of herbivores on plants, some of the interactions that shape the outcome of biological weed control projects are discussed. These include interactions between different herbivores, between target plants and their competitors, and between stresses from herbivores and abiotic factors.

Finally, efforts to integrate knowledge about agents, the weed, and its environment in ways that can guide agent selection are considered. Agent selection, while unlikely to be able to reliably identify "winners," may be useful in identifying agents with high probabilities of failure. Such winnowing of candidate agents may be required by the high cost of screening and testing biological weed control agents (Schroeder and Goeden 1986).

Biological Determinants of Host Plant Range

Why given species of herbivorous arthropods oviposit and feed on some plants and not others is a basic question of ecology that has been the focus of many studies. Estimating the degree of specificity exhibited by potential candidates for introduction as biological weed control agents is essential for planning in biological weed control. This chapter discusses factors affecting specificity of herbivores and relates these to biological control practice. Reviews of the subject include those of Courtney and Kibota (1990), Jaenike (1990), and Rausher (1992).

The inclusion of a plant species in the host range of an herbivorous arthropod involves two steps: finding and accepting the plant (usually by an ovipositing adult) (Thompson and Pellmyr 1991; Renwick and Chew 1994) and successful use of the plant as food (usually by the larvae or nymphs). Of these two steps, host finding and acceptance by adults is believed to be the factor most frequently determining the breadth of the host range of herbivorous arthropods (Courtney and Kibota 1990). A wide range of factors can affect the ability of ovipositing insects to locate particular plant species and then either to accept or reject them as oviposition sites. These factors include features of the plant and its ecological community (plant chemistry, morphology, spatial distribution, and biotic associations), and features of the insect itself (such as hunger, egg load, and previous host contacts, as well as the constraining phenotypic and genotypic limits of the herbivore) (Courtney and Kibota 1990). Collectively, these factors determine the host range in nature (the ecological host range), which often differs from the range of hosts consumed under conditions of close confinement in laboratory tests (the physiological host range). Factors shaping the ecological host range include herbivore physiology and sensory cues, developmental history and experience, and the herbivore's phylogenetic history.

Herbivore Physiology and Sensory Cues. Five factors can be recognized as the immediate causes of host choices by ovipositing herbivorous arthropods (Courtney and Kibota 1990): plant chemistry, plant morphological features, environmental variation, biotic interactions, and insect physiological factors.

Chemical features of plants affecting ovipositing adults may include specific deterrents, specific stimulants, or more general stimulants (sugars, water, etc.). Host acceptability (to adults) and host plant suitability (as food for larvae) have been found to be much less strongly linked than previously assumed. Many instances have been noted in which plants that are nutritionally suitable for larval development are rejected by ovipositing adults, and, conversely, instances in which plants chosen for oviposition are not the best species for larval development. Plant chemistry is likely to affect both acceptability and suitability, separately and not always in the same way. But because plant choices are made, most often, by ovipositing adults, it is the factors governing this step that most strongly affect the host plant range. (An exception to this would be species in which eggs are deposited away from the host and the young disperse and find their own food.) Species of herbivores with narrow host ranges are frequently adapted to respond to specific chemicals found in their host plants (Rausher 1992). Species with broad host ranges often respond more to nonspecific stimuli coupled with the absence of specific deterrent compounds.

Many morphological features of plants, including size, shape, color, texture, and general appearance can potentially affect oviposition choices of herbivores. In addition, the phenology of plant growth may be important; if oviposition occurs on certain plant structures, for example, seed heads, then only plant species that are in this stage when a particular herbivore is in the adult stage can be attacked by that species. Related plant species which flower at other times will not be used.

Many environmental factors can increase or decrease the acceptability of a plant to a herbivore. Plants growing in sunny versus shady areas, for example, may be treated differently. Plants under water stress may be either more or less attractive to ovipositing adults than unstressed plants. Plants of low nutritional value might be used in regions where the season of favorable temperature is long enough to permit completion of the developmental cycle but not in areas with shorter seasons (Scriber and Lederhouse 1992).

Biotic factors such as plant density can also affect the ovipositional responses of herbivores to plants. Specialized herbivores may be attracted to dense patches of a host plant. The presence of other plant species may increase or decrease the acceptability of a plant. A less preferred plant, for example, may be used if it occurs together with a more preferred plant.

Physiological factors such as egg load and age may influence the readiness of adults to oviposit. In some insect species, adults with high egg loads more readily accept less preferred hosts for oviposition (Jaenike 1990; Minkenberg et al. 1992). Older females of *Rhagoletis pomonella* become more likely with age to accept hosts not used by younger flies (Stanek et al. 1987).

Herbivore Developmental History and Experience. Early adult experiences of insects, such as experiences gained on the larval host plant after emergence of the adult insect, may affect host preferences. Such experiences can alter the sensitivity of peripheral sensory organs to host plant stimuli, or they alter the central nervous system's response threshold to the stimuli. These effects may occur separately or together. Insects may learn to use new host species, especially if the preferred host is rare and other potential hosts are common (Jaenike and Papaj 1992).

Herbivore Phylogenetic History. Physiological or behavioral limits, defined phenotypically or genotypically, can limit what stimuli ovipositing adults (or larvae, if they are the stage that chooses the host plant) are able to respond to and, therefore, constrain the species' host range. The ability to expand host ranges to include new species may be greatest if new hosts have chemistry or other features that are similar to old hosts. The action of such phylogenetic influence is the rationale behind the host range testing scheme that Wapshere (1974a) termed the "centrifugal phylogenetic method." This approach seeks to discover the taxonomic limits in host diet breadth by progressively testing plants less and less related to the target plant.

Host Range Testing and Field Evaluation

The goal of host range testing is to estimate the species of plants likely to be attacked by a herbivore after it is released in a new location. Methods for host range determination include laboratory tests to measure the physiological host range of the feeding stage and tests to determine which plants are acceptable for oviposition by the adult herbivore. In addition, oviposition choice tests under field conditions in the country of origin may be conducted to avoid any errors that might arise due to the artificiality of confinement that exists in laboratory tests. Baudoin et al. (1993), for example, planted artichoke and selected *Cirsium* spp. in plots of musk thistle (*Carduus thoermeri* Weinmann) in a three-year field trial to confirm the inability of the rust fungus *Puccinia carduorum* Jacky to infect these nontarget plants. Similarly, Clement and Sobhian (1991) used garden test plots in Greece to assess the host ranges of 15 species of capitulum-feeding insects, incorporating plants of the target weed, yellow starthistle (*Centaurea solstitialis*), and other related native plants or crops.

Integration of data from these various types of tests allows the host range to be estimated. A method for such integration was developed by Wapshere (1989) and termed a "reverse order sequence" for host range determination (Fig. 17.1). The sequence begins with larval feeding tests (under close confinement) and proceeds to adult oviposition tests under close and then loose confinement, followed by open field tests. Plants not accepted at each level are dropped from the estimated host range. Thus, a plant fed on by a larva under close confinement that was not subsequently also oviposited on by the adult female would be considered as outside the host range of the herbivore. Candidate plants in herbivore-host range tests may be presented either separately ("no-choice" tests) or in groups of several species ("choice" tests). In some instances, plants will be accepted for oviposition in no-choice tests that are not used when the preferred host is present (see Table 17.1, Wilson and Garcia 1992 as an example). The pondweeds *Potamogeton crispus* Linnaeus and *Potamogeton perfoliatus* Linnaeus received ovipositions of the ephydrid leaf miner *Hydrellia pakistanae* Deonier when its preferred host *Hydrilla verticillata* (L.f.) Royle was not present, but received few or no ovipositions in choice tests (Buckingham et al. 1989). The curculionid *Neohydronomus affinis* fed and oviposited on plant species in no-choice tests but then did not use those plants in choice tests which included its

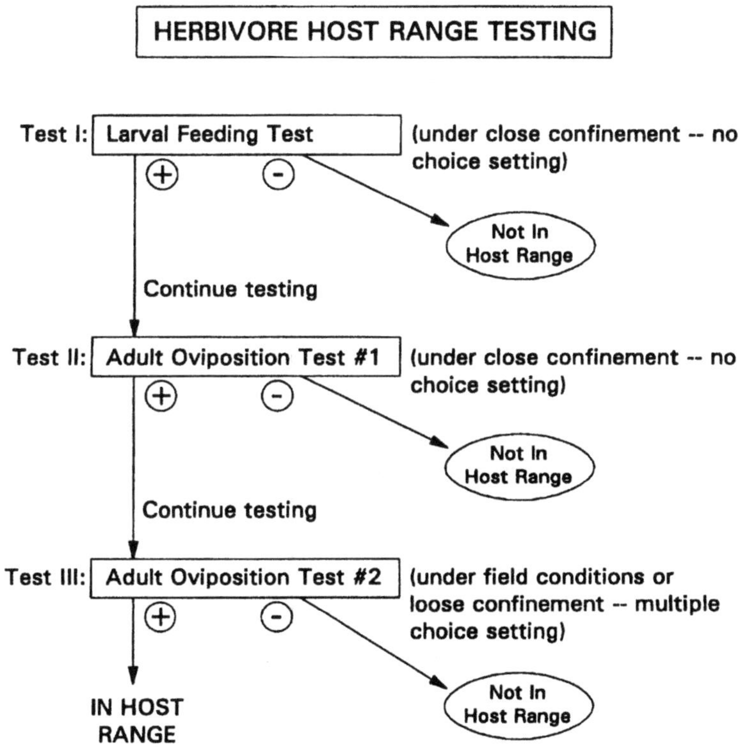

Figure 17.1. "Reverse-order" testing sequence for estimating host ranges of herbivorous arthropods being evaluated for introduction as biological control agents (after Wapshere 1989).

TABLE 17.1 "Multiple-Choice" and "No-Choice" Host Specificity Testing of *Heteropsylla* sp. for Oviposition and Adult and Nymphal Survival (From Wilson and Garcia 1992)

	No-Choice		Multiple-choice	
Plant species	No. eggs	No. adults or nymphs alive after 4 days	No. eggs	No. nymphs alive 2 days after eclosion
Delonix regia (Hooker) Rafinesque	4	0	0	0
Neonotonia wightii (Arnott) Lackey	5	0	0	0
Trifolium repens Linnaeus	1	0	0	0
Acacia angustissima (Miller) Kuntze	27	0	19	0
Acacia melanoxylon R. Brown	6	0	0	
Albizia sp.	5	0	0	0
Desmanthus virgatus von Willdenow	20	0	0	0
Dichrostachys cinerea (Linnaeus) Wright and Arnott	10	0	0	0
Leucaena collinsii Britton and Rose	4	0	0	0
Leucaena diversifolia Bentham	15	0	0	0
Leucaena leucocephala (de Lamarck) de Wit	4	0	0	0
Leucaena macrophylla Bentham	4	0	0	0
Leucaena pallida Britton and Rose	27	0	0	0
Leucaena pulverulenta (von Schlectendal) Bentham	17	0	0	0
Mimosa invisa Martius ex Colla	685	15 adults 28 nymphs	>4,000	>4,000
Mimosa pigra Linnaeus	26	0	0	0
Mimosa pudica Linnaeus	13	0	0	0
Neptunia gracilis Bentham	20	0	0	0

preferred host, *Pistia stratiotes* (Thompson and Habeck 1989). Further advice on interpreting the results of host range tests is given by Harley and Forno (1992). For insects such as aphids that are able to transmit plant pathogens, screening trials must also include tests of the species' ability to acquire and transmit a variety of plant pathogens that may be present in the country of introduction (Briese 1989). Such screening trials should focus both on assessing the ability of an agent to transmit any diseases of the target weed (a desirable ability) and diseases of other species of plants (which might be undesirable). Host range determination is as important for pathogens being considered for development as bioherbicides as it is for agents of introduction. Weidemann and Tebeest (1990) discuss host range testing in this context and stress the importance of making appropriate allowances for genetic variability both within the pathogen and the plant species being tested.

Once estimates of the host range have been made by such experiments, the degree of host specificity required for a particular situation must be determined. Few herbivores are monophagous. Typically natural enemies employed in biological weed control are oligophagous. Such agents may be judged as possessing sufficiently narrow host ranges if they do not attack any important species of economic or native plants, or if evidence suggests that any native plants that might be attacked will not be reduced at the population level. Polyphagous herbivorous invertebrates are usually not employed as biological weed control agents, but might be in some circumstances. This might occur, for example, if agents with wide host ranges were to be introduced to locations in which the target weed were the only member of the host range, or if other members of the host range in the target area were also weeds.

Following releases of biological weed control agents in new locations, field studies of the range of plant species actually attacked in the field need to be conducted. These studies provide information that allows the accuracy of the prediction about host range made from laboratory tests to be assessed. Use by introduced herbivores of additional plant species beyond those in the known host range is most likely when new plant species are chemically, morphologically, or phenologically similar to the normal host and occur in the same habitat. About 20 cases have been recognized in which herbivorous arthropods used in biological weed control have included new plant species in their host range following release in a new location (C. Moran, personal communication as cited in Dennill et al. 1993), although none of these has had important ecological or economic consequences.

Effects of Herbivory on Plants

The effects that herbivorous arthropods have on plants occurs at two levels: changes in the performance of individual plants and changes in the dynamics of the target plant population. A great deal of literature exists on herbivore effects on plant performance, while very few studies have adequately evaluated effects on plant population dynamics (Crawley 1989).

Herbivores affect plant performance in a variety of ways, including: reduced flowering and seed set; reduced post-dispersal seed survival and reduced seedling survival; loss of biomass from defoliation; reduced growth and reproduction; reduced competitiveness with other plants; reduced root reserves and increased susceptibility to abiotic stress; and increased mortality of established plants (Fig. 17.2). Many studies exist that document one or more of these effects for various insects (Crawley 1989).

By themselves, effects on plant performance are of limited interest to biological control practitioners. Rather, the critical point is how and to what degree these effects change the population dynamics of the target weed. Some of these changes include reduced seedling recruitment (from seed destruction), increased direct mortality of established plants from herbivory, and indirect mortality of established plants from a combination of herbivory and other stresses.

Reduced Recruitment to the Plant Population. Seed-feeding insects are often introduced against weeds with the intent to lower seed production, thus reducing recruitment to the seedling stage. The relationship between seed production and seedling recruitment, however, is often not linear. Seed production rates may be limiting for some kinds of plants, such as some annuals, but not for stable populations of long-lived woody perennials. If large populations of seeds accumulate and survive in soils and if favorable microsites suitable for germination and seedling establishment are scarce, then it may be the density of these germination sites, not the size of the population of seeds in the soil, that governs the rate of seedling recruitment (Andersen 1989). One experimental method to determine whether recruitment to plant populations is limited by seed density or by the number of germination and establishment sites is to add seeds to natural plant populations to see if it enhances seedling recruitment rates. Sowing 1000 seeds per m^2 of tansy ragwort, *Senecio jacobaea*, for example, did not result in any increased recruitment of seedlings, suggesting that seeds were not limiting in the test population (Crawley and Nachapong 1985). From a life table perspective, this situation is one in which the impact of mortality in one stage (here, the seed) is reduced in biological significance if mortality in a subsequent stage (here, seedling establishment) is positively density dependent. In this case, the existence of a fixed and limited supply of seedling establishment sites acts

Figure 17.2. Natural enemy impacts on plants are varied and include: (A) seed reduction by seed feeders (larva of *Botanophila* sp., Diptera: Anthomyiidae, feeding on flower head of *Senecio jacobaea* Linnaeus); (B) defoliation (adult of *Chrysolina quadrigemina* [Suffrian] feeding on *Hypericum perforatum* Linnaeus); (C) disruption of vascular systems (larva of *Longitarsus jacobaeae* [Waterhouse] feeding in root crown of *Senecio jacobaea* Linnaeus); (D) galling of various tissues (galls of *Trichilogaster acaciaelongifoliae* [Froggatt] on *Acacia longifolia* [Andrews] Willdenow); and (E) disease-induction (*Puccinia chondrillina* Bubak and Sydow infecting skeletonweed, *Chondrilla juncea* Linnaeus). (Photographs courtesy of USDA-ARS [A, B, C, E] and A. J. Gordon [D].)

as a density-dependent source of mortality once the seed supply exceeds the supply of favorable germination sites.

However, seedling recruitment of some plant species has been reduced following the introduction of seed- or flower-feeding herbivores. The apionid *Trichapion lativentre* (Bequin Billecocq), after being introduced to South Africa, reduced seed production of its host, *Ses-*

bania punicea (Cavanille) Bentham, by 98% (Hoffmann and Moran 1991). A second agent, the weevil *Rhyssomatus marginatus* Fåhraeus, was later introduced that destroyed 84% of the remaining seeds (Hoffman and Moran 1992), and the two agents together have almost completely arrested reproduction of the weed. Cessation of seed production led to a reduction in the rate of seedling establishment and a slowing of the spread of the weed to new sites. A fuller discussion of the importance of seed-attacking agents relative to various patterns of plant biology is found in Neser and Kluge (1986).

Direct Mortality of Established Plants. Herbivorous arthropods may kill their hosts outright by complete consumption of whole plants, massive disruption of key systems such as that for water transport, or defoliation coupled with destruction of carbohydrate reserves (preventing regrowth). In other cases, effects of herbivory act in concert with additional agents to cause mortality indirectly.

Indirect Mortality from Herbivory. Herbivory (defoliation, removal of plant sap by sucking insects, destruction of roots by root-feeding insects, galling of plant tissues) reduces existing biomass, reduces stored reserves in roots or other organs, and reduces the growth rate of new plant biomass. The relation of plant performance (as total growth or seed set) to injury from herbivory is usually linear (Crawley 1989). Many plants, however, are able to compensate for moderate levels of injury (Trumble et al. 1993), so that the relationship between the proportion of plants dying and the degree of injury from herbivory is often not linear, but rather is governed by a damage threshold (Harris 1986). This threshold is the amount of biomass which must be lost before plants suffer injury for which they cannot compensate and, therefore, incur an increased risk of death. Many forage grasses, for example, are able to exist in a healthy condition with annual biomass removal rates of 40%, but decline if levels exceed 50% (Harris 1986).

In addition to the amount of biomass consumed, the ability of plants to survive herbivory depends on which plant parts are destroyed relative to the organs used for storage of reserve carbohydrates and the timing of the damage relative to the growing season. For plant species in which carbohydrate reserves are stored in foliage, defoliation is more likely to result in plant death than for plant species in which reserves are stored in the roots. Such later species, in contrast to the former, often are able to recover from defoliation by use of stored reserves in their roots. Defoliation early in the year often is of less importance to the plant than midseason defoliation, because in the latter case there remains less time to regenerate new foliage and use it to produce new carbohydrate reserves. The timing of defoliation relative to plant growth is also important. Defoliation of musk thistle (*Carduus thoermeri*) at the rosette stage caused mortality, but similar levels of defoliation after bolting had little effect (Cartwright and Lok 1990).

Galling of plant parts typically has its greatest effect when tissues are attacked early in the period of their growth. The plant reacts as if the resources used for gall formation had been used for growth of normal tissue and little compensatory growth occurs (Harris 1986). Extensive galling acts as a nutrient sink, reducing vegetative growth and, if reproductive tissues are the gall site, reducing reproduction (Dennill 1988).

Methods to Measure Effects on Plant Population Dynamics. Ultimately, the effects of herbivores on plant performance must be synthesized in terms of influences on plant population dynamics. Two approaches have been used. Each depends on being able to establish plots having and plots lacking the natural enemy whose effect is to be evaluated. Native (or already widely established adventive) species can be evaluated by using insecticides to exclude the agent from some plots and comparing changes in plant numbers in these to other unsprayed plots in the vicinity. McBrien et al. (1983), for example, used this approach to evaluate the effect of grazing by *Trirhabda* spp. beetles on the goldenrod *Solidago canadensis* Linnaeus. For agents that are in the process of introduction, experimental data may be obtained by studying plant populations before and after the introduction of the new agent and comparing these differences to other locations, studied over the same time period, at which the agent is not released. Huffaker et al. (1983), for example, used this approach to evaluate the effect of two introduced weevils on puncturevine, *Tribulus terrestris*, in California. In such plots, measurements of both number and biomass of plants or their parts (flowers, leaves, stems, roots) are taken over time. Analysis of trends in these measures may be facilitated by use of Allen curves or relative-growth curves, which depict interactions between number and mass over time and allow calculation of net annual above-ground productivity (Hoffmann and Moran 1989). Laboratory tests or field cage tests (using various numbers of agents per plant) can be used to assess the ability of candidate weed control agents to affect plant performance, in terms of mortality and reduced growth and reproduction, as well as assessing the ability of the plant for compensatory growth (Julien and Bourne 1988). Principles of natural enemy evaluation are more thoroughly developed in Chapter 13.

Interactions Between Herbivory and Other Stresses

The degree of biomass or nutrient loss that plants can survive also depends on the timing and strength of other stresses that affect the plant.

Multiple Stresses From Different Herbivores. In climates that are relatively lacking in abiotic stresses for plants (no cold winters, no extremely hot or dry periods), a series of introduced herbivores that act sequentially over the year may be needed to place enough cumulative stress on plants to reduce plant survival. In parts of Hawaii, for example, control of *Lantana camara* was not achieved by the tingid *Teleonemia scrupulosa*, even though it annually defoliated the plant, until additional agents were introduced that attacked the plant in the winter months (Andres and Goeden 1971). Seed production of *Sesbania punicea* in South Africa was virtually eliminated by the combined action of two seed-feeding insects (Hoffmann and Moran 1992). Studies of effects of introduced herbivores attacking tansy ragwort (*Senecio jacobaea*) indicated that control resulted from complementary effects of two herbivores, cinnabar moth (*Tyria jacobaeae*) and ragwort flea beetle (*Longitarsus jacobaeae*) (McEvoy et al. 1990).

Herbivore Stress Followed "Soon" by Abiotic Stress. Plants defoliated (or otherwise severely injured) by herbivores tend to respond to injury by regrowth. Stored reserves of carbohydrates are used to support such regrowth, often reducing reserves to low levels. New foliage is then employed by the plant to rebuild carbohydrate reserves. Consequently, plants that have used stored reserves for regrowth are more vulnerable to abiotic stresses (such as low winter temperature or summer drought) until such time as these stores have been replaced. The impact of a defoliation-refoliation cycle, therefore, will be more severe if it occurs together with or shortly before a period of abiotic stress (winter cold, summer drought). If these additional stresses occur too late, sufficient time may be available for plants to replenish carbohydrate reserves, making them better able to survive abiotic stresses. Defoliation of tansy ragwort, *Senecio jacobaea* rosettes by cinnabar moth larvae (*Tyria jacobaeae*), for example, reduced root reserves and increased death rates from frost (Harris et al. 1978). Willis et al. (1993) found a weak, but positive, interaction between water stress and herbivory from mites or aphids for *Hypericum perforatum*.

Herbivory and Plant Diseases. Establishment of herbivores may increase plant susceptibility to disease. Dieback of *Mimosa pigra* in northern Australia caused by the pathogen *Botryodiplodia theobromae* Patouillard was believed to have been stimulated by establishment of a stem boring moth introduced against the weed (Wilson and Pitkethley 1992).

Herbivory in Combination with Competition from Other Plants. Plants compete with other species growing around them for water, space, and nutrients. Grasses, for example, are important competitors for the rangeland weed *Centaurea maculosa* Lamarck, reducing survival of rosettes to 17% of that in controls lacking grass competition (Müller-Schärer 1991). The strength of such competition from neighboring plants is an important factor increasing the effect of damage from herbivores to a target weed. Experiments incorporating both plant competition and natural enemy attack can be used to quantify the interaction between these factors. Pantone et al. (1989) quantified wheat competition with the weed *Amsinckia intermedia* von Fischer and Meyer with and without the flower gall nematode *Anguina amsinckiae* (Steiner and Scott) Thorne and found that the nematode's presence sharply increased the competitiveness of wheat with the weed.

The ability of weeds to compensate for herbivory by increased growth is greatest when competition is absent and nutrients and water are not limiting. As plant competition increases, target weeds are less able to regrow after damage from herbivores. Also, competition may act over time such that bare areas left by dead plants of the target species, if quickly filled by other plants, are not available as sites favorable for germination and establishment of the seedlings of the target weed.

Community Interactions

Weed biological control programs by design seek to change plant communities, often suppressing monospecific stands of adventive weeds and allowing the return of a more diverse vegetation. The ecological consequences of biological weed control are numerous and usually positive. Huffaker and Kennett (1959), for example, observed that diversity of native plants in Californian rangeland increased dramatically after biological control suppressed the density of St. John's wort, *Hypericum perforatum*. Coastal prairie sites in California returned to a nearly natural state after introduced insects controlled tansy ragwort (*Senecio jacobaea*) stands that formerly dominated the vegetation (Pemberton and Turner 1990). For more on the use of weed biological control for the restoration of damaged native ecosystems see Chapter 21.

Apart from effects on native, nontarget organisms, ecological ramifications of weed biological control efforts that need to be noted include the following. First, some species of newly introduced herbivores may become hosts or prey for local natural enemies, usually generalist species with broad host ranges (Cornell and Hawkins 1993). For example, the arctiid *Pareuchaetes pseudoinsulata* Rego Barros, while successful in some locations as a control agent of the weedy shrub *Chromolaena odorata* (Linnaeus) R. King and H. Robinson, typically has failed to establish in areas where it is subject to ant predation (Kluge 1991). Similarly, the skeleton weed gall midge, *Cystiphora schmidti* (Rübsaamen), is heavily parasitized in some areas by native parasitoids, such as *Mesopolobus* sp. in Washington state (Wehling and Piper 1988). Second, the void created by the successful control of the target weed is going to be filled, but the nature of the plant species that fill the vacated niche may vary. In some systems, native species may increase. In managed pastures, desirable forage species may be seeded into the available space. However, in other circumstances, elimination of the target weed may lead to increases in other weed species (Harris 1976). Third, plant species related to the target weed that are not preferred hosts may be attacked in field populations when the less preferred host and the preferred target species occur together (Goeden and Kok 1986).

Agent Selection: Synthesizing Agent and Target Biology

Accurate prediction of which of any potential list of candidate agents is the set of species that will provide control of a target weed has not proved possible. However, because of the high cost of host-specificity testing, criteria that allow placing candidate agents even approximately into such categories as "possible" and "unlikely" are valuable. Where possible, the most effective means of selecting weed control agents often is to use species that have controlled the target weed in other countries with similar climates. For example, the use of *Cyrtobagous salviniae* to control *Salvinia molesta* in South Africa was an obvious choice following its successful use earlier in Australia and Papua New Guinea (Cilliers 1991).

More generally, three approaches to agent selection have been suggested (Wapshere 1992). One is to conduct studies in the weed's native range and identify the agents having the greatest effect there, making appropriate allowances for parasitoids and other factors that might be suppressing the agent and which can be eliminated in the introduction process (Wapshere 1974b). Another is to study the historical record of past projects (Julien 1992) and identify, if possible, any features of agents, or their hosts, that were common to successful cases. Finally, agents might be scored based on possession of specific attributes which they, or their hosts, possess that seem, based on theoretical considerations, to be ones important for successful biological control (Harris 1973; Goeden 1983).

Characteristics thought to be important for weed biological control agents include the following. First, a good agent should not be self-limiting. It should be tolerant of high densities of its own species so that it will be able to become sufficiently numerous to damage its host. Second, species that are limited in their country of origin by natural enemies may be very effective, provided the natural enemies can be eliminated during the introduction process, and there are no species in the proposed area of introduction likely to take their place. Third, a good agent would be a species that will complement the stresses imposed on the weed by the other agents already in the system. While not a detailed guide to agent selection, these and other general concepts can help sort agents to some degree. For example, a consideration of the biology of the buprestid *Taphrocerus schaefferi* Nicolay and Weiss suggests that it would be unlikely to be an effective agent for yellow nutsedge, *Cyperus esculentus* Linnaeus. It is not tolerant of high densities of its own species (due to larval cannibalism), is subject to parasitism, and does little damage since most feeding occurs on senescing tissues (Story and Robinson 1979).

PLANT PATHOGENS

Plant pathogens affect the growth and survival of their hosts in a variety of ways, including necrosis of various tissues, permanent wilting from blockage of xylem or as a reaction to toxins, hypertrophy or hyperplasia, leaf abscission, etiolation, and prevention of reproduction (Kendrick 1992). Unlike arthropods, which for weed biological control have typically been employed in programs of agent introduction, plant pathogens have been used in both introduction and augmentative programs in which the agent is applied as a bioherbicide (Templeton and Trujillo 1981). Agents for these two different purposes often possess different biological characteristics. While several groups of organisms exist that are pathogenic to plants (bacteria, fungi, viruses, nematodes), most attention has been focused on fungi as plant biological control agents.

Pathogens used for introduction to new regions for biological control need not be rearable on artificial media because they must be produced only for a brief period during testing and release. For this, live plant hosts can be used. For pathogens used in introduction programs, characteristics that are important are high host specificity, high pathogenicity, and an ability to disperse readily between host patches (Hasan 1980). Rust fungi, for example, often have such characteristics. Among rusts, species requiring only one host for the entire life cycle are suitable for use. However, species that require two hosts (which typically are taxonomically widely separated) are not suitable, unless both plant species are weeds suitable as targets of biological control. Rusts have great potential as biological weed control agents, and important successes have been achieved with this group of agents, including skeleton weed in Australia (Hasan and Wapshere 1973; Hasan 1981; Delfosse et al. 1985; Watson 1991) and mist flower in Hawaii (Barreto and Evans 1988). In addition, the rust fungus *Uromyces heliotropii* Scredinski has been released in Australia against common heliotrope, *Heliotropium europaeum* Linnaeus (Delfosse 1985; Hasan et al. 1992) and control is being evaluated.

For fungi or other pathogens to be used as bioherbicides, dispersal characteristics are not particularly important because the agent will be physically applied to the plants to be killed and spread to new areas is not particularly desirable (Templeton et al. 1984). High pathogenicity and specificity are important (Charudattan 1989), and the ability to rear the agent on artificial media is crucial. Fungi such as rusts that must be reared in living hosts are usually too expensive to rear to be commercially successful as bioherbicides. Additional discussion of characteristics important for fungi used as bioherbicides is found in Chapter 11.

HERBIVOROUS FISH

Eleven species of fish are listed by Julien (1992) as agents that have been introduced for biological control of weeds (see Chapter 5). Unlike arthropods agents, which often are selected on the basis of limited host plant ranges, fish as weed biological control agents often are nonselective grazers used to suppress entire communities of aquatic macrophytes. As such, host plant range testing plays little role in the choice of agents.

Perhaps the central question concerning fish biology in the selection and manipulation of fish species for weed biological control has been their ability or inability to establish breeding populations in the release area. If mismanaged, herbivorous fish that suppress macrophytic vegetation have potential to profoundly affect aquatic systems in a variety of ways, including increased turbidity and algal growth and competition with other fish or wildlife. If the species used is native to the region or already widely established, these concerns may be minimal, and fish reproduction may be desirable, especially if the fish can be eaten. If, however, the species is not native and is not widely distributed, it may be undesirable to permit the escape of fish capable of breeding. Enclosed systems such as irrigation ponds and tanks are often the sites where weed control is desired. Physical isolation, however, is not a reliable check on the movement of fish populations, as fish may be moved by floods or deliberately transported by people between water bodies. Sterile fish, either hybrids or artificially induced polyploids (Fig. 17.3), may be used to prevent establishment of breeding populations.

Figure 17.3. Sterile tripolid grass carp, *Ctenopharyngodon idella* (Cuvier and Valenciennes), are generalist herbivores capable of suppressing macrophyte communities (Photograph courtesy of R. Stocker.)

Another aspect of fish biology that has been of importance in the choice of species is the system of brood care used and potential of a species to act as a predator of fry of other species. Three species of fish were imported, for example, to suppress aquatic vegetation in irrigation canals in California and released together. Two of the three species ultimately displaced the third, *Tilapia zillii*. This displacement was undesirable because it was ultimately learned that this third species, if the only species present, gave the best control. The displacement was a result of the ability of the other two species to mouth brood (and hence protect) their fry; since *T. zillii* does not mouth brood, its fry were exposed in a nonreciprocal manner to predation by the other species (Legner 1986a).

CHAPTER

18

Population Regulation Theory and Implications for Biological Control

INTRODUCTION

The concepts of population equilibria and population regulation lie at the heart of biological control. Characteristics of successful biological control programs through introduction are, for example, the reduction of pest populations and their maintenance, in a stable interaction with their natural enemies, about some low, nonpest level. Such outcomes have frequently been recorded (DeBach 1964a; Clausen 1978; Bellows 1993 (see also; Chapter 8)), but evidence documenting the mechanisms of population interactions which are responsible for these dynamical changes is less common (Beddington et al. 1978; Gould et al., 1992a,b). Examples tracking the decline of a population following the introduction of natural enemies are shown in Fig. 18.1, 18.2, and Fig. 8.7. Biological control programs seek to enhance such natural control of populations, and an understanding of the principles involved requires a fundamental understanding of many aspects of population ecology.

Models have played an important role in examining interactions among natural enemies and their host populations. The early work of Thompson (1924), Nicholson (1933), and Nicholson and Bailey (1935) explored the consequences of very simple interactions between hosts and parasitoids and formed the basis for development of mathematical descriptions of these interactions. Since that work, models and their underlying assumptions have become more complex to reflect ever increasingly detailed knowledge of the behavior of natural enemies and of other features of populations, including spatial dispersion of populations, nonrandom search, and density-dependent processes. Models serve to permit the consequences of each of these processes to be examined independently and in combination, and, thereby, permit us to explore the significance of each to the behavior of interacting host-parasitoid systems. In this way, models serve as generators or clarifiers of hypotheses about how populations interact, formalizing the structure of the hypotheses in terms of axioms (assumptions of the models) and results (outcomes of the models). They also serve as tools to permit us to examine the likely or possible effect of certain types of interactions more simply than conducting a large number of experiments. Models do not replace experiments, but rather indicate the potential range of outcomes for particular experiments, helping to clarify our understanding about the occasionally nonobvious impact of certain types of interactions. Models help to extend the theoretical treatment of how and why biological control works, how populations interact to achieve population balance, and what features of these interactive systems are consistent with desirable

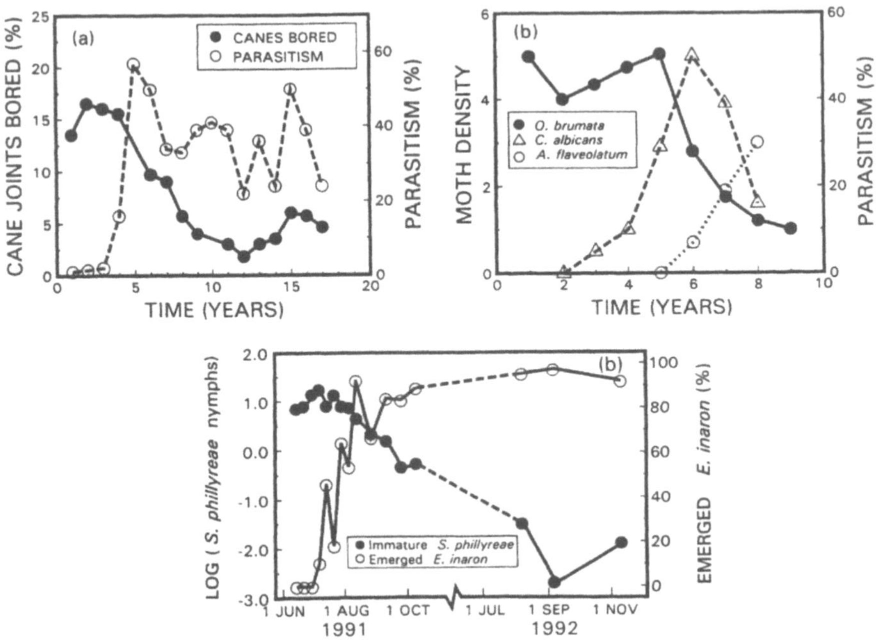

Figure 18.1. Biological control of three insect pests. (a) Sugarcane borers (*Diatraea* spp.) following introductions of the tachinid *Lixophaga diatraeae* (Townsend) and the braconid *Cotesia flavipes* (Cameron) (after Anon. 1980). (b) Winter moth (*Operophtera brumata* [Linnaeus]) in Nova Scotia, Canada, following introductions of the tachinid *Cyzenis albicans* (Fallén) and the ichneumonid *Agrypon flaveolatum* (Gravenhorst) (after Embree 1966). (c) Ash whitefly (*Siphoninus phillyreae* [Haliday]) nymphal populations in California (U.S.A.) prior to and following the introduction of the parasitic wasp *Encarsia inaron* (Walker) (after Bellows et al. 1992b).

biological control. Experiments form the basis for tests of such theories; models assist in defining the types of experiments which may shed the greatest light on pertinent questions.

This chapter considers mechanisms through which natural enemies contribute to reduction of pest population levels and to population regulation, and model frameworks which represent them. The mathematics used in this chapter is not very complicated, and does not require more than an introductory understanding of algebra. The interested reader will find the detailed treatments and analyses which lead to the conclusions discussed here in the original papers. To develop ideas relating to interacting populations, as is applicable to biological control systems, ideas are first discussed relating to single-species populations, introducing single and multiple age-class systems for both homogenous and heterogenous environments. These ideas are then extended to interspecific competition systems and also to several types of parasitoid-host interactions. Some simple pathogen-host and multispecies systems are also introduced. Although no particular system would apply uniformly to all biological control interactions, the insights gained from many of these appear to apply to many systems, and shed light on the likely mechanisms important in determining the dynamics of multitrophic level systems.

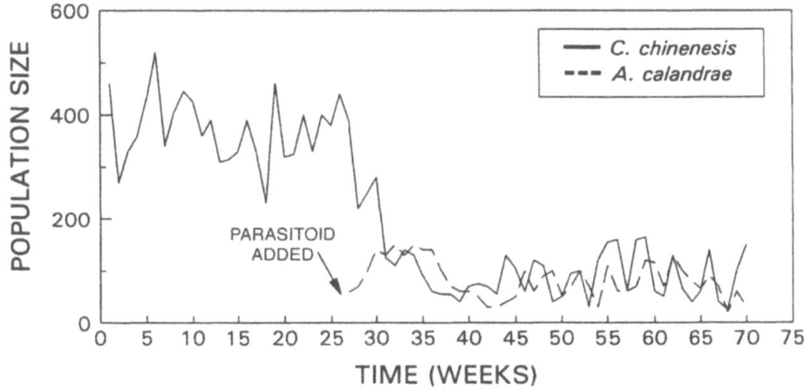

Figure 18.2. Dynamics of *Callosobruchus chinensis* (Linnaeus) and *Anisopteromalus calandrae* (Howard) populations in the laboratory (after May and Hassell 1988).

Throughout much of this development, focus is on models that are tractable analytically, that is, their properties and the implications for population dynamics and biological control systems can be determined explicitly from the models' structures. Another paradigm of modeling is that of systems or simulation modeling, where models are more complex and their properties are determined by using the models to simulate particular situations, rather than analytically from model structure. These models are briefly introduced to consider some of the more complex interactions in biological control, such as multiple age-class populations. The chapter concludes by discussing the implications of research on population dynamic theory to biological control.

Although there is abundant information on the effects of herbivory on the performance of individual plants, there are few data on the effects of insect herbivory on the dynamics of plant populations (Crawley 1983, 1989). For this reason, discussion focuses primarily on insect predator-prey, host-parasitoid and host-pathogen interactions.

SINGLE-SPECIES POPULATIONS

Homogeneous Environments

Single Age-Class Systems The study of single species population dynamics has a long history of both theoretical and empirical development both on systems in continuous time and with discrete generations. In this section, we examine the dynamics of populations with discrete generations, which is appropriate for many insect populations. The algebraic framework is straightforward:

$$N_{t+1} = Fg(N_t)N_t \tag{1}$$

Here N is the host population denoted in successive generations t and $t+1$, and $Fg(N_t)$ is the per capita net rate of increase of the population dependent on the per capita fecundity F and the relationship between density and survival g (which is density dependent for $dg/dN < 0$).

Assuming that $g(N_t)$ is density dependent, the fundamental concept embodied in equation [1] is that some resource, critical to population reproduction or survival, occurs at a finite and limiting level (when $g = 1$, there is no resource limitation and the population grows without limit). Individuals in the population compete for the limiting resource, for example, as adults for oviposition sites (Utida 1941; Bellows 1982a) or food (Nicholson 1954), or as larvae for food (Nicholson 1954, Bellows 1981).

The dynamics of such single-species systems spans the range of behavior from geometric (or unconstrained) growth (when competition does not occur), monotonic approach to a stable equilibrium, damped oscillations approaching a stable equilibrium, stable cyclic behavior, and, finally, systems characterized by aperiodic oscillations or "chaos" (Fig. 18.3) (May 1975; May and Oster 1976). Which type of behavior occurs in a particular case depends both on the reproductive rate and the strength (nonlinearity) of the density dependent competition. Thus, species showing contest competition have more stable dynamics than species showing scramble competition, which tend to show more cyclic behavior when reproductive rates are sufficiently high (May 1975; Hassell 1975).

The exact form of the function used to describe g is not particularly critical to these general conclusions and many forms have been proposed (May 1975; May and Oster 1976; Bellows 1981), although different forms may have specific attributes more applicable to particular cases (Bellows 1981). Perhaps the most flexible is that proposed by Maynard Smith and Slatkin (1973), where $g(N)$ takes the form

$$g(N) = [1 + (N/a)^b]^{-1} \tag{2}$$

Here the relationship between proportionate survival and density is defined by the two parameters a, the density at which density dependent survival is 0.5, and b, which determines

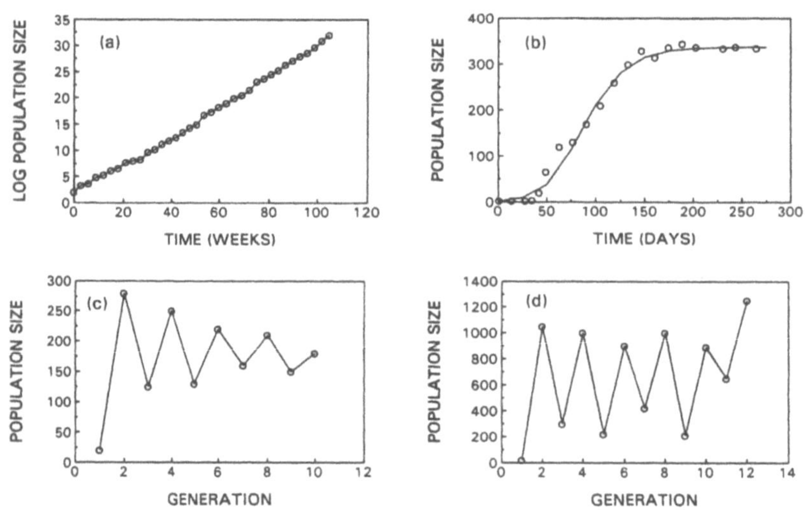

Figure 18.3. Dynamical behavior of single species populations. (a) Unrestricted growth of *Lasioderma serricorne* (Fabricius) (after Lefkovitch 1963). (b) Logistic growth of *Rhizopertha dominica* (Fabricius) (after Crombie 1945). (c) Oscillatory growth of *Callosobruchus chinensis* (Linnaeus) (after Utida 1967). (d) Cyclic dynamics of *Callosobruchus maculatus* (Fabricius) (after Fujii 1968).

the severity of the competition. As b approaches 0, competition becomes less severe until it no longer occurs ($b = 0$). When $b = 1$ density dependence results in contest competition with the number of survivors reaching a plateau as density increases. Finally, for $b > 1$ scramble competition occurs, with the number of survivors declining as the density exceeds $N = a$. Examples of different types of dynamical behavior represented by equation [1] with $g(N)$ defined by equation [2] are shown in Fig. 18.4.

Reviews of insect populations showing density dependence in natural and laboratory settings suggest that most populations exhibit monotonic or oscillatory damping towards a stable equilibrium (Hassell et al. 1976; Bellows 1981). Of natural populations, only the Colorado potato beetle, ***Leptinotarsa decimlineata*** (Harcourt 1971), was predicted to show limit cycles, and only Nicholson's (1954) blowflies (*Lucilia cuprina* Linnaeus) fell within the region of chaos (Hassell 1975; Bellows 1981). By representing the dynamics in a single-species framework, however, there is the possibility that cycles and higher order behavior arising from interactions with other species have been overlooked (Hassell et al. 1976; Godfray and Blythe 1990).

Multiple Age-Class Systems. Most populations are separable into distinct age- or stage-classes, which can bear importantly on the outcome of competition. In most insects the preimaginal stages must compete for resources for growth and survival, while adults may additionally compete for resources for egg maturation and oviposition sites. In such cases, competition

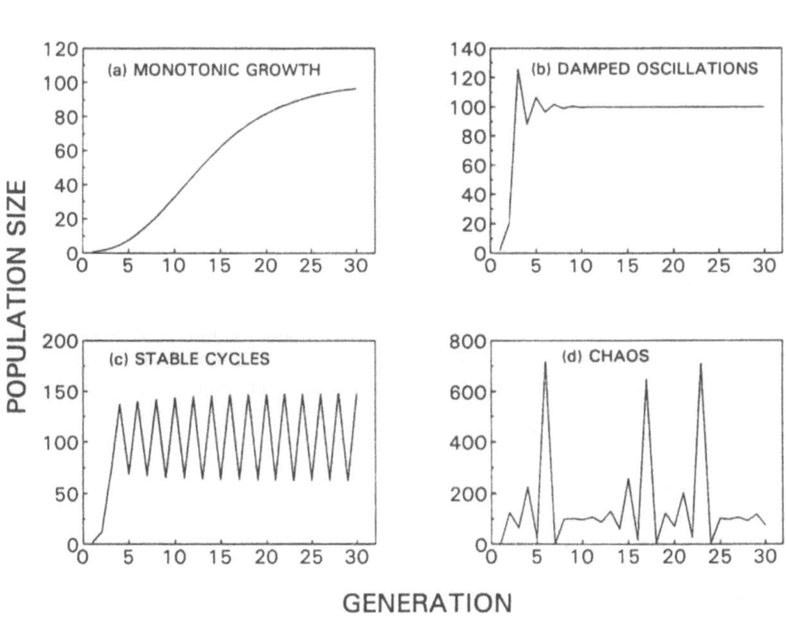

Figure 18.4. Examples of the population dynamics exhibited by the single-species model of equations 1 and 2. In all cases $N^* = 100$. (a) Monotonic population growth to stable equilibrium ($F = 5$, $b = 0.2$). (b) damped oscillations to a stable equilibrium ($F = 10$, $b = 1.7$) (c) Stable limit cycles ($F = 6$, $b = 2.5$). (d) Chaotic behavior ($F = 50$, $b = 3$).

within populations divides naturally into sequential stages. Equation [1] may now be extended to the case of two age classes (May et al. 1974) and, assuming competition occurs primarily within stages (larvae compete with larvae and adults with adults), gives:

$$A_{t+1} = g_L(L_t)L_t, \quad (3a)$$

$$L_{t+1} = Fg_A(A_t)A_t, \quad (3b)$$

where A and L denote the adult and larval populations. Here larval competition, governed by $g_L(L_t)$ in equation [3a], leads to A_t adults which compete for resources $[g_A(A_t)]$ and produce the next generation of larvae. In such multiple age-class systems, the dynamical behavior of the population is dominated by the outcome of competition in the stage in which b is nearest to one. For example, a population in which adults contest for oviposition sites while larvae scramble for food will show a monotonic approach to a stable equilibrium. More complex approaches to constructing models of age- or stage-structured single-species populations have been developed (Nisbet and Gurney 1982).

Patchy Environments

For many populations, resources are not distributed either continuously or uniformly over the environment, but rather occur in discrete units or patches. Consider an environment divided into j discrete patches (such as individual plants) which are utilized by an insect species. Adults (N) disperse and distribute their progeny among the patches. Progeny compete for resources only within their patch. The population dynamics is now dependent partly on the distribution of adults reproducing in patches, denoted by Γ, and partly on the density dependent relationship g which characterizes preimaginal competition. Population reproduction over the entire environment (summed over all patches) is characterized by the relationship (de Jong 1979)

$$N_{t+1} = jF\Sigma\,\Gamma(n_t)n_t g[Fn_t], \quad (4)$$

where n_t is the number of adults in a particular patch, $\Gamma(n_t)$ is the proportion of patches colonized by n_t adults, F is the per capita reproductive rate, and g is the density dependent survival rate.

De Jong (1979) considers four distinct distributions of adults among patches. In the case of uniform dispersion, equation [4] is equivalent to equation [1] for homogeneous environments. For three random cases, positive binomial, independent (Poisson), and negative binomial, the outcome depends somewhat on the form taken for the function g. For most reasonable forms of g, the general outcomes are a lower equilibrium population level and enhanced stability compared to a homogeneous environment with the same F and g. In addition, for a fixed amount of resource, stability tends to increase as the number of patches increases (the more finely divided the resource, the more stable the interaction). Finally, one additional feature arises in systems with overcompensatory or scramble competition: there is an optimal fecundity for maximum population density, with fecundities both above and below the optimum resulting in fewer surviving progeny.

INTERSPECIFIC COMPETITION

Although Strong et al. (1984), Lawton and Strong (1981), and Lawton and Hassell (1984) suggest that competition is not commonly a dominant force in shaping many herbivorous insect communities, it certainly is important in some communities, such as the social insects and, to a lesser extent, insects feeding on ephemeral resources (*Drosophila* spp., Atkinson and Shorrocks 1977), insect parasitoids (Luck and Podoler 1985), predatory beetles, and multispecies plant communities. The processes and outcomes of interspecific competition in insects have been studied widely in the laboratory (Crombie 1945; Fujii 1968, 1969, 1970; Bellows and Hassell 1984) (Fig. 18.5) as well as in the field (see review by Lawton and Hassell 1984).

Homogeneous Environments

Single Age-Class Systems Many of the same mechanisms implicated in intraspecific competition for resources (competition for food and oviposition sites) also occur between species (Crombie 1945, 1946; Park 1948; Fujii 1969, 1970). The dynamics of these interspecific systems can be considered in a framework very similar to that for single species populations.

Equation [1] may be extended to the case for two (or more) species by allowing the function g to depend on the density of both competing species (Hassell and Comins 1976). The reproduction of species X now depends not only on the density of X but also on the density of Y (and similarly for species Y):

$$X_{t+1} = Fg_x(X_t + \alpha Y_t)X_t, \tag{5a}$$

$$Y_{t+1} = Fg_y(Y_t + \beta X_t)Y_t. \tag{5b}$$

Here the parameters α and β are the competition coefficients reflecting the relative severity of interspecific competition with respect to intraspecific competition.

Population interactions characterized by equation [5] may have one of four possible outcomes: the two species may coexist, species X may always exclude species Y, species Y may always exclude species X, or either species may exclude the other depending on their initial relative abundance. Coexistence is possible only when the product of the interspecific competition parameters $\alpha\beta < 1$. As for single species systems, equation [5] shows a range of dynamical behavior from stable points (Fig. 18.5), to cycles and higher order behavior determined by the severity of the intraspecific competition described by g (Hassell and Comins 1976).

Multiple Age-Class Systems. Many insect populations compete in both preimaginal and adult stages, perhaps by competing as adults for oviposition sites and subsequently as larvae for food (Fujii 1968), and in some cases the superior adult competitor may be inferior in larval competi-

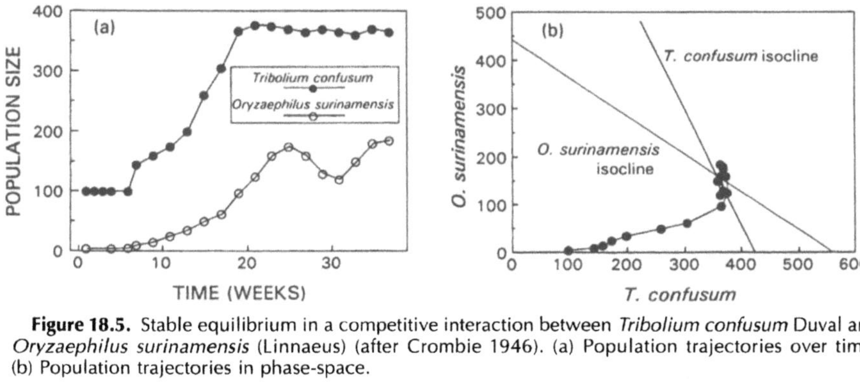

Figure 18.5. Stable equilibrium in a competitive interaction between *Tribolium confusum* Duval and *Oryzaephilus surinamensis* (Linnaeus) (after Crombie 1946). (a) Population trajectories over time. (b) Population trajectories in phase-space.

tion (Fujii 1970). The properties of such multiple age-class systems may be considered by extending equation [3] as follows (Hassell and Comins 1976):

$$X_{t+1} = x_t g_{x_L}(x_t + \alpha_L y_t) \tag{6a}$$
$$Y_{t+1} = y_t g_{y_L}(y_t + \beta_L x_t) \tag{6b}$$
$$x_{t+1} = X_t F_x g_{x_A}(X_t + \alpha_A Y_t) \tag{6c}$$
$$y_{t+1} = Y_t F_y g_{y_A}(Y_t + \beta_A X_t), \tag{6d}$$

where x and y are the preimaginal or larval stages and X and Y are the adults. Larval survival of each species is dependent on the combined larval densities and, similarly, adult reproduction of each species is dependent on the combined adult densities. Interspecific larval competition is characterized by the larval competition coefficients α_L and β_L, while adult competition is characterized by α_A and β_A.

The simple extension of competition to more than one age class has important effects on dynamics. The isoclines of zero population growth are now no longer linear (as in the case of single age-class systems, compare Figs. 18.5 and 18.6) and, in addition, multiple equilibria may now occur involving in some cases more than one stable equilibrium (Hassell and Comins 1976). Such curvilinear isoclines are in accord with those found for competing populations of *Drosophila* spp. (Ayala et al. 1974).

More complex systems can be envisaged with additional age classes and with competition between age classes (Bellows and Hassell 1984). The general conclusions from studies of these more complex systems are similar to those for the two age-class system, specifically that more complex systems have nonlinear isoclines and, consequently, may have more complicated dynamical properties. More subtle interactions may also affect the competitive outcome, such as differences in developmental time between two competitors. In the case of *Callosobruchus chinensis* (Linnaeus) and *Callosobruchus maculatus* (Fabricius), the intrinsically superior competitor (*C. maculatus*) can be outcompeted by *C. chinensis* because of its faster development which allows it earlier access to resources in succeeding generations. This earlier access

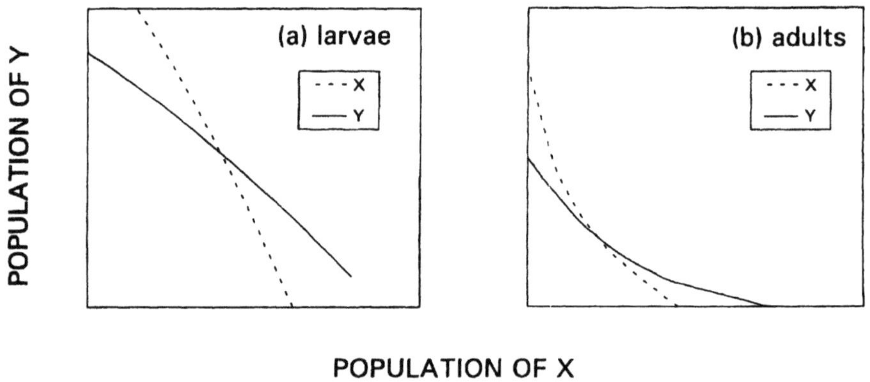

Figure 18.6. Zero-growth isoclines for populations with two age classes may not be linear, but curved to greater or lesser degree. The isoclines are differently shaped for the different age classes. In this example a single stable equilibrium is indicated at the place where the isoclines intersect. In other cases multiple intersections may occur, indicating multiple equilibria (after Hassell and Comins 1976).

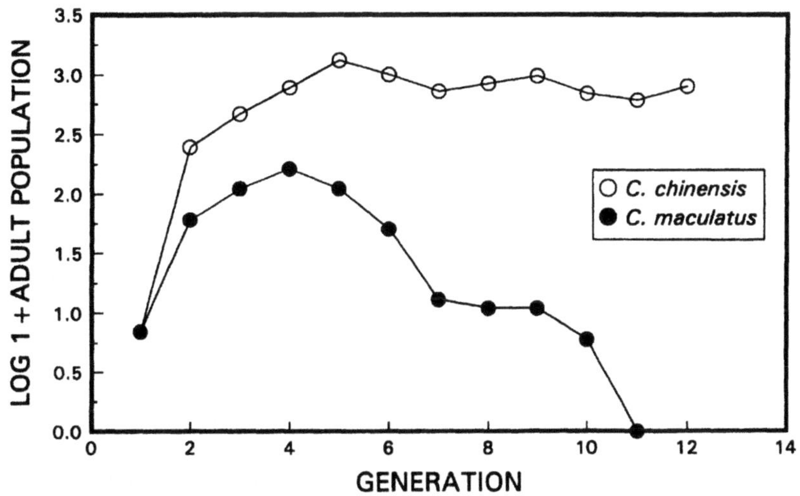

Figure 18.7. In mixed populations of *Callosobruchus chinensis* (Linnaeus) and *Callosobruchus maculatus* (Fabricius), *C. chinensis* excludes the competitively superior *C. maculatus.* This is a result of the faster developmental time of *C. chinensis,* which provides increasingly earlier access to resources in succeeding generations. This earlier access allows *C. chinensis* to successfully colonize resources (in this case, dried beans) and reproduce before *C. maculatus* can fully occupy the beans, leading eventually to complete exclusion of *C. maculatus* (after Bellows and Hassell 1984).

This earlier access confers sufficient competitive advantage to *C. chinensis* that it eventually excludes *C. maculatus* from the two-species system (Bellows and Hassell 1984) (Fig. 18.7).

Patchy Environments

Many insect populations are dependent on resources which occur in patches (fruit, fungi, dung, flowers, dead wood). Dividing the resource into discrete patches can have important effects on the conditions for coexistence of the competing species. This is well illustrated by the models of Atkinson and Shorrocks (1981), de Jong (1981), Hanski (1981), Ives and May (1985), and Comins and Hassell (1987). This body of work emphasizes the importance for coexistence of the spatial distribution of the competing species. Thus Atkinson and Shorrocks (1981) concluded that coexistence becomes more likely if the patches are more finely divided and if the competitors are aggregated in their distribution between patches independently of one another. Particularly important for coexistence is the marked aggregation of the superior competitor, thus providing more patches in which it is absent and in which the inferior competitor can survive.

HOST-PARASITOID SYSTEMS

Equation [1] for a single species in a homogeneous environment can be extended to include the additional effect of mortality caused by a natural enemy. Following previous work (Nicholson and Bailey 1935; Hassell and May 1973, 1974; Beddington et al. 1978; May et al. 1981), we will consider principally insect parasitoids. Such systems have been the focus of considerable work, both theoretical and experimental (see Hassell [1978] for a review). Continuing with the discrete-generation framework of the preceding sections and assuming a coupled, synchronized parasitoid population, we can write the following generalized model:

$$N_{t+1} = Fg[f(N_t,P_t)\cdot N_t]N_t f(N_t,P_t) \tag{7a}$$

$$P_{t+1} = cN_t\{1 - f(N_t,P_t)\}. \tag{7b}$$

Here N and P are the host and parasitoid populations. $Fg[f(N_t,P_t)\cdot N_t]$ is the per capita net rate of increase of the host population, with intraspecific competition defined as before by the function g with density dependence for $dg/dn < 0$. The function f defines the proportion of hosts which survive parasitism and embodies the functional and numerical responses of the parasitoid, and c is the average number of adult female parasitoids which emerge from each attacked host (c therefore includes the average number of eggs laid per host parasitized, the survival of these progeny, and their sex ratio). In such discrete-generation frameworks with both host density dependence (g) and parasitism (f), different dynamics can result depending on the sequence in the host's life cycle that these occur (Wang and Guttierrez 1980; May et al. 1981, Hassell and May 1986). In effect, therefore, the model represents a minimally complicated age-structured host population with pre- and post-parasitism stages. Equation [7] describes the particular case of parasitism acting first followed by the density dependence defined by g (May et al. [1981] provide a discussion of alternatives).

Within the framework of equation [7], the degree to which the parasitoid population can reduce the average host population level (leaving aside whether or not this is a stable equilibrium) can be defined by the ratio, q, of average host abundances with and without the parasitoid (i.e., $q = K/N^*$, where K is the carrying capacity of the host population in the absence of the parasitoid and N^* is the parasitoid-maintained equilibrium). The magnitude of this depression depends upon the balance between:

(1) the host's net rate of increase ($Fg[f(N_t,P_t)\cdot N_t]$ in equation [7a], and

(2) the various factors affecting overall parasitoid performance contained within the function $f(N_t,P_t)$ and the term c. These include the per capita searching efficiency and maximum attack rate of adult females, the spatial distribution of parasitism in relation to that of the host (see following), and the sex ratio and survival of parasitoid progeny.

The general framework of equation [7] has been explored with many variants for the functions f and g. The original, and most familiar, version is that of Nicholson (1933) and Nicholson and Bailey (1935), where $c = g(N_t) = 1$:

$$N_{t+1} = FN_t\exp(-aP_t), \tag{8a}$$

$$P_{t+1} = N_t\{1 - \exp(-aP_t)\}, \tag{8b}$$

Here $f(N,P)$ is represented by the zero term of the Poisson distribution, implying that each host is equally susceptible to parasitism by the P_t adult parasitoids, and a is the per capita searching efficiency of the parasitoids (the "area of discovery") that sets the proportion of hosts encountered per parasitoid per unit time. The model thus implicitly assumes a type I functional response (Holling 1959a) with no upper limit to the number of hosts that a parasitoid can successfully parasitize. Handling time is, therefore, assumed to be negligible and parasitoid egg supply unlimited. The model thus assumes: that each host is equally subject to attack (random search); that the parasitoids have a linear functional response; that each host parasitized produces one female progeny for the next generation ($c = 1$); and that the host population suffers no additional density dependence due, for instance, to resource limitation ($g = 1$). The model predicts expanding oscillations of host and parasitoid populations around an unstable equilibrium. The inclusion of a finite handling time (T_h), and thus a type II functional response, makes this instability more acute (Hassell and May 1973).

Since such unstable interactions have only been observed from a few simple laboratory experiments (for example, Burnett 1958) (Fig. 18.8), there has been much interest in factors that could be important in promoting the persistence of more stable host-parasitoid interactions. One possibility is that host-parasitoid systems persist by grace of additional density dependent factors ($g < 1$) affecting the host population, although mechanisms for this become hard to invoke at low levels of host abundance. Alternatively, the description of parasitism in equation [8] may be inadequate and parasitism itself may be a regulatory process. A number of such regulatory mechanisms have been proposed, including sigmoid functional responses (Murdoch and Oaten 1975; Nunney 1980), mutual interference between searching parasitoids (Hassell and Varley 1969; Hassell and May 1973), density-dependent sex ratios (Hassell et al. 1983; Comins and Wellings 1985), and heterogeneity in the distribution of parasitism (following and section on Spatial Patterns of Parasitism).

An example of how readily functional forms of $f(N_t,P_t)$ can be found that stabilize host-parasitoid interactions is given by May (1978). Here, the distribution of parasitoid attacks among hosts is described by a clumped distribution, the negative binomial, instead of the independently random Poisson. Host survival, f, in equation [7] is thus given by the zero term of the negative binomial distribution (May 1978):

$$f(N,P) = \left[1 + \frac{aP}{k}\right]^{-k}. \qquad (9)$$

Here k describes the contagion in the distribution of parasitoid attacks among host individuals. Contagion increases as $k \rightarrow 0$, whereas in the opposite limit of $k \rightarrow \infty$ attacks become

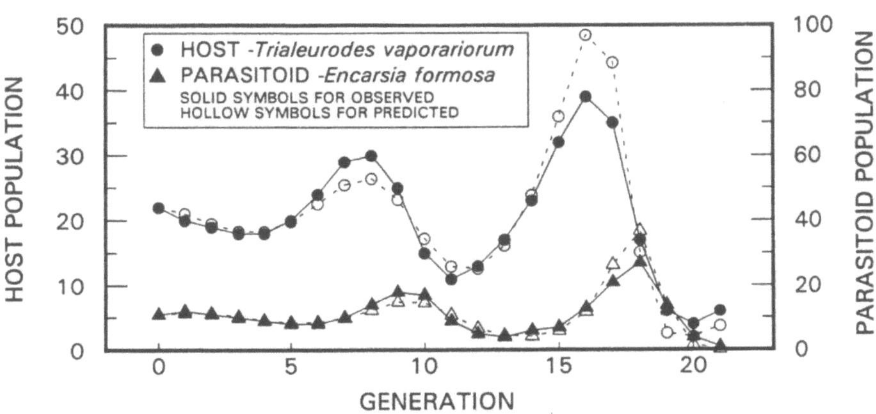

Figure 18.8. Dynamics of a simple host-parasitoid laboratory system consisting of *Trialeurodes vaporariorum* (Westwood) and *Encarsia formosa* Gahan (after Burnett 1958). Solid symbols are observed population numbers, hollow symbols are those predicted from equation (8) with $F = 2$, $a = 0.067$.

distributed independently and the Poisson distribution (equation [8]) is recovered. As May and Hassell (1988) have discussed, the outcome of a parasitoid's searching behavior cannot usually be fully characterized so simply as equation [9] (cf. Hassell and May 1974; Chesson and Murdoch 1986; Perry and Taylor 1986; Kareiva and Odell 1987). Nonetheless, the use of equation [9] with a constant k permits the dynamical effects of nonrandom or aggregated parasitoid searching behavior to be explored without introducing a large list of behavioral parameters. More complex cases, such as the value of k varying with host density, can be considered (Hassell 1980), but have little effect on the dynamical properties of the host-parasitoid interaction.

The simple change from independently-random search in equation [8] to the more general case of equation [9] has profound effects on the dynamics of the interaction. The populations are stable for $k < 1$ and show increasing oscillations for $k > 1$ (Fig. 18.9). A modification of this model in which the k is a function of average host density per generation is described by Hassell (1980) in relation to the winter moth, *Operophtera brumata* in Nova Scotia parasitized by a tachinid, *Cyzenis albicans* (cf. Embree 1966).

Host Density Dependence

The preceding discussion has assumed no host density dependence ($g = 1$). This is likely to be appropriate for many situations, particularly where biological control agents are established and populations are maintained substantially below their environmentally determined carrying capacity. In other cases, however, the relative contributions to regulation of both host density dependence and parasitism must be addressed.

The framework presented in equation [7] can be used to explore the joint effects of density dependence in the host and the action of parasitism (Beddington et al. 1975; May et al. 1981). One important feature of such discrete systems incorporating both host and parasitoid density dependence is that the outcomes of the interactions will depend on whether the parasitism acts before or after the density dependence in the host population. May et al. (1981) envisaged two general cases, the first where host density dependence acts first and the second where parasitism acts first (their models 2 and 3). They employed equation [9] for function f, with

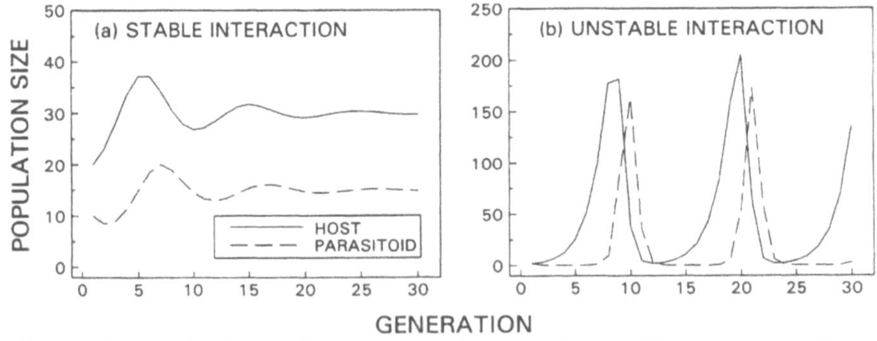

Figure 18.9. Examples of (a) stable ($F = 2$, $a = 0.1$, $k = 0.5$) and (b) unstable ($F = 2$, $a = 0.1$, $k = 1$) dynamics in the parasitoid-host model with negative binomial distribution of attacks (equations 7 and 9).

host density dependence described by $g = \exp(-\alpha N_t)$. With host density dependence acting before parasitism, we have

$$N_{t+1} = Fg(N_t)N_t f(P_t), \tag{10a}$$

$$P_{t+1} = N_t g(N_t)\{1 - f(P_t)\}; \tag{10b}$$

and with parasitism acting first:

$$N_{t+1} = Fg[f(P_t)\cdot N_t]N_t f(P_t), \tag{11a}$$

$$P_{t+1} = N_t\{1 - f(P_t)\}. \tag{11b}$$

Beddington et al. (1975) and May et al. (1981) have explored the stability properties of such interactions in terms of two biological features of the system—the host's intrinsic rate of increase ($\log F$) and the level of the host equilibrium in the presence of the parasitoid (N^*) relative to the carrying capacity of the environment (K, the host equilibrium due only to host density dependence in the absence of parasitism). This ratio between the parasitoid-induced equilibrium N^* and K is termed q; $q = N^*/K$.

When equation [9] is employed in the function f, the behavior of these systems with both density dependence and parasitism depends on three parameters: the host rate of increase ($\log F$), the degree of suppression of the host equilibrium (q), and the degree of contagion in the distribution of parasitoid attacks (k of equation [9]). The specific relationships between these parameters which determine the dynamics of the system depend on whether parasitism occurs before or after density dependence in the life cycle of the host (Fig. 18.10). In both cases, the degree of host suppression possible increases with increased contagion of attacks. The new parasitoid-caused equilibrium may be stable or unstable, and for unstable equilibria the populations may exhibit geometric increase or oscillatory or chaotic behavior (Fig. 18.11). For density dependence acting after parasitism (Fig. 18.10c, d) and for $k < 1$, any population reduction is stable. May et al. (1981) provide additional detail on the possible dynamical outcomes of such systems, but in general, much of the parameter space for both cases implies a stable reduced population whenever $k < 1$, when there is sufficient contagion of attacks.

Spatial Patterns of Parasitism

In the same way that we have treated single-species and competing species in heterogeneous or patchy environments, we can also consider host-parasitoid systems where the host populations are distributed among discrete patches. The consequences of such heterogeneous host distributions on the dynamics of the host-parasitoid system depends significantly on the variation in the distribution of parasitism among patches.

Several mechanisms exist which could lead to variable parasitism rates per patch. For example, aggregations of natural enemies in patches of relatively high prey densities may result from such patches being more easily discovered by natural enemies (Sabelis and Laane 1986), or from searching behavior changing upon discovery of a host (Hassell and May 1974; Waage 1979).

The considerable interest in the effects of heterogeneity on host-parasitoid dynamics have led many workers to record the distribution of parasitism in the field in relation to the local density of hosts per patch. Of 194 different examples listed in the reviews of Lessells (1985), Stiling (1987), and Walde and Murdoch (1988), 58 show variation in attack rates among patches depending directly on host density (Fig. 18.12a,c), 50 show inversely density dependent relationships (Fig. 18.12b), and 86 show variation uncorrelated with host density (density independent) (Fig. 18.12d,e).

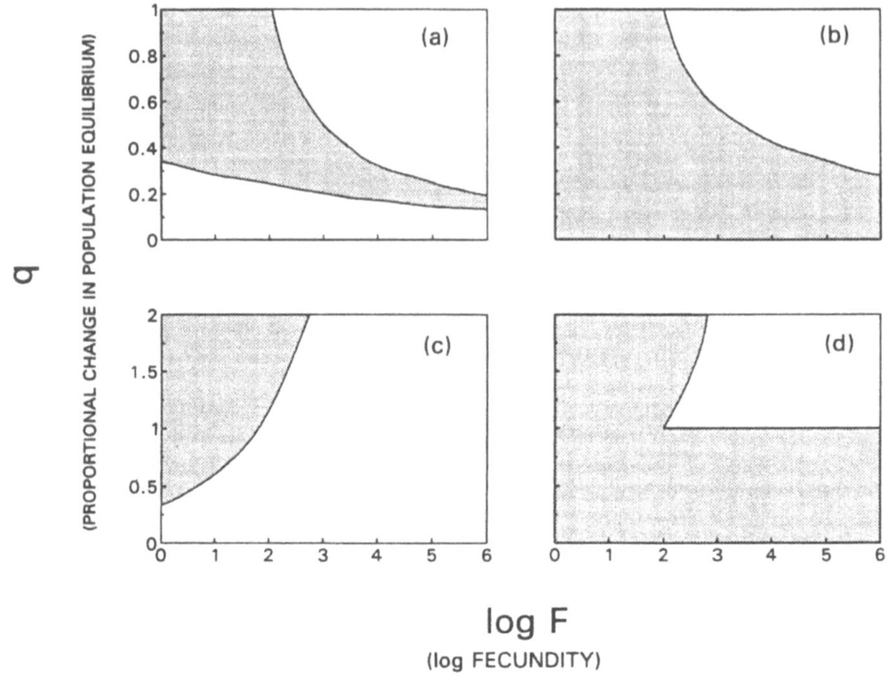

Figure 18.10. Regions of stable and unstable population dynamics for equations (10) (a, b) and (11) (c, d), where both parasitism and density dependence act on the host population. In (a) and (b) density dependence acts before parasitism, while in (c) and (d) density dependence acts after parasitism in the host life cycle. In (a) and (c), parasitoid attacks are distributed independently in the host population (Poisson distribution, with negative binomial parameter $k = \infty$. In (b) and (d), attacks are aggregated ($k = 0.5$). Stable two-species equilibria are possible only in the shaded areas. In general, stability is enhanced by aggregated attacks.

A popular interpretation of these data has been that only the direct density dependent patterns promote the stability of the interacting populations. This picture, however, is incomplete. Both inverse density dependent patterns (Hassell 1984; Walde and Murdoch 1988) and variation in parasitism that is independent of host density (Chesson and Murdoch 1986; May and Hassell 1988; Pacala et al. 1990; Hassell et al. 1991) can be just as important to population regulation. The reason is that any variation in levels of parasitism from patch to patch has the net effect of reducing the per capita parasitoid searching efficiency (measured over all hosts) as average parasitoid density increases (the "pseudo-interference" effect of Free et al. [1977]).

To explore the dynamical effects of such heterogeneous parasitism, consider a habitat which is divided into discrete patches (food plants for an herbivorous insect) among which adult insects with discrete generations distribute their eggs. The immature stages of these insects are hosts for a specialist parasitoid species whose adult females forage across the patches according to some unspecified foraging rule. We also assume that parasitism dominates host mortality such that the hosts are on average kept well below their carrying capacity.

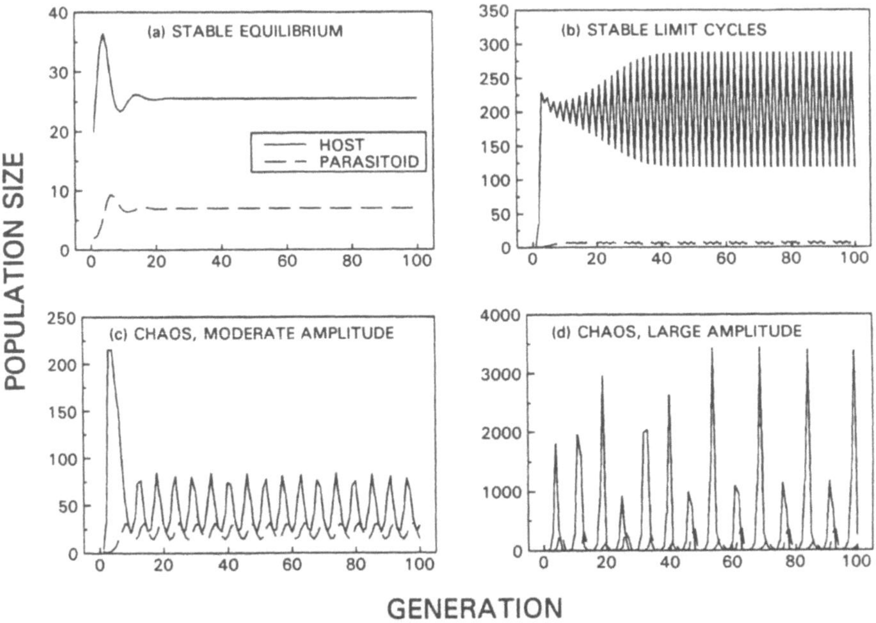

Figure 18.11. Examples of population dynamics for host-parasitoid systems with host density dependence (equation 10). In all cases $a = 0.1$. (a) stable equilibria ($F = 2$, $k = 0.5$, $b = 0.01$); (b) stable limit cycles ($F = 10$, $k = 0.2$, $b = 0.01$); (c) chaotic cycles of moderate amplitude ($F = 10$, $k = 5$, $b = 0.01$); (d) chaotic cycles of larger amplitude ($F = 10$, $k = 2$, $b = 0.001$).

With this scenario, equation [7] applies with $c = g = 1$ and $f(N_t, P_t)$ representing the average, across all patches, of the fraction of hosts escaping parasitism. The distribution of hosts in such a patchy setting can either be random or vary in some other prescribed way. Similarly, the density of searching parasitoids in each patch can either be a random variable independent of local host density or a deterministic function of local host density. Pacala et al. (1990), Hassell et al. (1991), and Pacala and Hassell (1991) call these patterns of heterogeneity in parasitoid distribution host-density dependent heterogeneity (HDD) and host-density independent heterogeneity (HDI), respectively. Comparable terms have been coined by Chesson and Murdoch (1986), who labeled models with randomly distributed parasitoids as "pure error models" and those with parasitoids responding to host density in a deterministic way as "pure regression models."

May and Hassell (1988) suggested a very simple and approximate stability condition for model (7) in a patchy environment, namely, that the populations will be stable if the distribution of parasitoids from patch to patch (measured as the square of the coefficient of variation, CV^2, is greater than unity. Pacala et al. (1990), Hassell and Pacala (1990), and Hassell et al. (1991) extended this work and showed that a very similar criterion applies across a broad range of models. Their criterion differs in that the density of searching parasitoids per patch is now weighted by the number of hosts in that patch.

The $CV^2 > 1$ rule is in terms of the distribution of searching parasitoids. Such data, however,

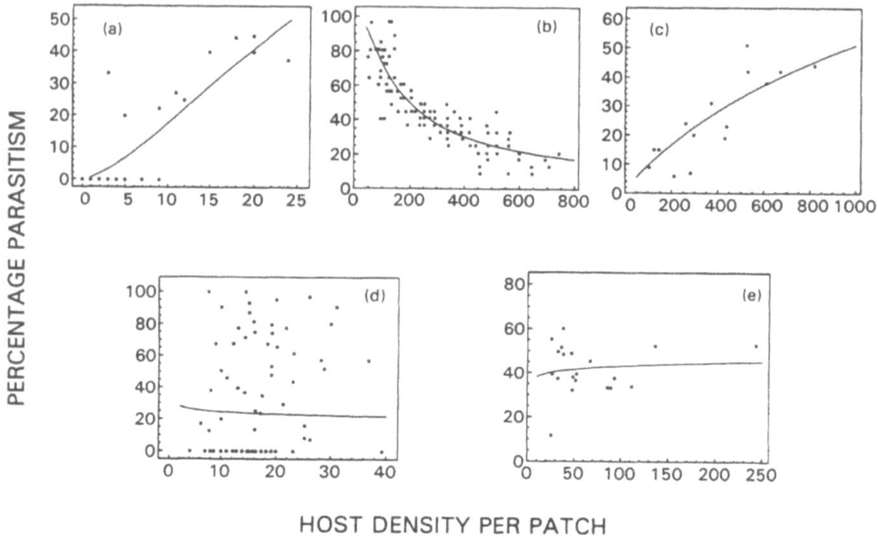

Figure 18.12. Examples of field studies showing percentage parasitism as a function of host density. (a) The eucoilid *Trybliographa rapae* (Westwood) parasitizing the anthomyiid *Delia radicum* (Linnaeus) (Jones and Hassell 1988). (b) The encyrtid *Ooencyrtus kuwanai* (Howard) parasitizing the lymantriid *Lymantria dispar* (Linnaeus) (Brown and Cameron 1979). (c) The aphelinid *Aspidiotiphagus citrinus* (Crawford) parasitizing the diaspidid *Fiorinia externa* Ferris (McClure 1977). (d) The eulophid *Tetrastichus* sp. parasitizing the cecidomyiid *Rhopalomyia californica* Felt (Ehler 1986). (e) The aphelinid *Coccophagoides utilis* Doutt parasitizing the diaspidid *Parlatoria oleae* (Colvée) (Murdoch et al. 1984). Curves are predicted values given by Hassell and Pacala (1990) from a host density-independent probabilistic model relating percentage parasitism to host density per patch (after Hassell and Pacala 1990).

are rarely available from natural populations; most of the information is in the form of relationships between percentage parasitism and host density per patch (Fig. 18.12). Pacala and Hassell (1991), however, show that it is possible to estimate values of all the parameters necessary to calculate CV^2 from such data. In particular, they show that CV^2 from this general model may be approximated as:

$$CV^2 = C_I C_D - 1 \tag{12}$$

Here $C_I = 1 + \sigma^2$ where σ^2 is the variance of a gamma-distributed random variable describing the density independent component of parasitoid distribution, and $C_D = 1 + V^2\mu^2$ where V^2 describes the degree of contagion in the host's distribution and μ describes the strength of any density dependence (positive or negative) in the parasitoids' response to host distribution. Details are given in Pacala and Hassell (1991).

Hassell and Pacala (1990) analyzed 65 examples from field studies reporting percentage parasitism versus local host density per patch, for each of which they obtained estimates of σ^2, μ and V^2, and then calculated C_I, C_D and CV^2. It was also possible in each case to predict the mean percentage parasitism in relation to host density per patch from the expression which provides the fitted lines in Fig. 18.12. Interestingly, of the five examples in Fig. 18.12, heterogeneity in parasitism is sufficient (if typical of the interactions) to stabilize the dynamics only in Fig. 18.12a and Fig. 18.12d. The inverse pattern in Fig. 18.12b would have been sufficient had the host population been more clumped in its distribution.

On surveying the results of the 65 examples, Pacala and Hassell (1991) found that in 18 of the 65 cases, $CV^2 > 1$ indicating that heterogeneity at this level ought to be sufficient to stabilize the populations. Interestingly, in 14 of these 18 cases heterogeneity in C_I alone was sufficient to make $CV^2 > 1$. This analysis suggests that density independent spatial patterns of parasitism may be more important in promoting population regulation than density dependent patterns.

The work of Pacala and Hassell (1991) shows how relatively simple models of host-parasitoid interactions can be applied to field data on levels of parasitism in a patchy environment. Such heterogeneity has often been regarded as a complicating factor in population studies, and one that rapidly leads to analytical intractability of models. Clearly, this need not necessarily be so. The $CV^2 > 1$ rule explains the consequences of heterogeneity for population dynamics in terms of a simple description of the heterogeneity itself. The rule gives a rough prediction of the effects of heterogeneity and also identifies the kinds of heterogeneity that contribute to population regulation.

Aggregation as a phenomenon affecting the dynamics and stability of model host-parasitoid systems has been addressed in a number of analytical frameworks other than the one we have explored here. Murdoch and Stewart-Oaten (1989) consider the effects of aggregation on stability in a continuous-time model and conclude that aggregation is in general not stabilizing in that model framework, and Murdoch et al. (1992) extended this analysis to a model of two populations linked by migration (a metapopulation) and found both stabilizing and destabilizing effects of aggregation. These findings were revisited by Godfray and Pacala (1992) and Rohani et al. (1994), who argued that aggregation by parasitoids is indeed stabilizing for most reasonable mechanistic models of host-parasitoid interactions. Rohani et al. (1994) demonstrated that, for a model framework somewhat hybrid between the discrete form of equation [7] and the continuous-time forms of Murdoch and Stewart-Oaten (1989), aggregation independent of host density is a more potent stabilizing influence than aggregation dependent upon host density. This finding is consistent with the analyses of data conducted by Pacala and Hassell (1991) using the general CV^2 rule. Thus aggregation and heterogeneity, while having been two of the more continuously invoked stabilizing influences in model host-parasitoid systems, also appear to lead to hypotheses that are consistent with findings of studies of field populations (Lessells [1985], Stiling [1987], and Walde and Murdoch [1988].

Age-Structured Systems and Simulation Models

Insects grow through distinct developmental stages; therefore, the concepts of age- and stage-structure are linked (more closely in some systems than in others). Many of the general models described in the previous sections take some account of developmental stages (equations [3], [10] and [11]). A more detailed treatment of age- or stage-structure, however, requires different approaches to developing population models and theories of population interaction and regulation.

One such approach has been to construct more complex models, often called system or simulation models, which incorporate more biological detail at the expense of analytical tractability. Such models can be used not only to explore population dynamics but also other features such as population developmental rate, biomass and nutrient allocation, community structure, and management of ecosystems. Here, we will be concerned only with those features of such systems which bear on population regulation in ways which are not directly addressable in the simpler analytical frameworks previously presented.

Synchrony of Parasitoid and Host Development. Insect populations in continuously favorable environments (laboratory populations, some tropical environments) may develop continuously overlapping generations. In the presence of parasitism as a major cause of mortality, such populations often exhibit more-or-less discrete generations (Tothill et al. 1930; Taylor 1937; Van der Vecht 1954; Utida 1957; Wood 1968; Hassell and Huffaker 1969; White and Huffaker 1969; Metcalfe 1971; Bigger 1976; Banerjee 1979). Godfray and Hassell (1987, 1989) develop both a simulation model and a delayed differential equation model in which they consider an insect host-parasitoid population system growing in a continuously favorable environment (with no intraspecific density-dependence) which passes through both an adult (reproductive) and preimaginal stages.

The dynamical behavior of these systems was characterized either by stable populations in which all stages were continuously present in overlapping generations, populations which were stable but which occurred in discrete cycles of approximately the generation period of the host, or by unstable populations. These dynamics were dependent principally upon two parameters, the degree of contagion in parasitoid attacks, *k*, and the relative lengths of preimaginal host and parasitoid developmental time (Fig. 18.13). Low values of *k* (strong contagion) promoted continuous, stable generations. Moderate values of *k* (less strong contagion) were accompanied by continuous generations when the parasitoid had developmental times approximately the same length as the host, approximately twice as long, or very short. When developmental times of the parasitoid were approximately half or 1.5 times that of the host, discrete generations arose. Unstable behavior arose for larger values of *k*. Therefore, degree of synchrony between host and parasitoid can be an important factor affecting the dynamical behavior of models of continuously-breeding populations, particularly for parasitoids which develop faster than their hosts.

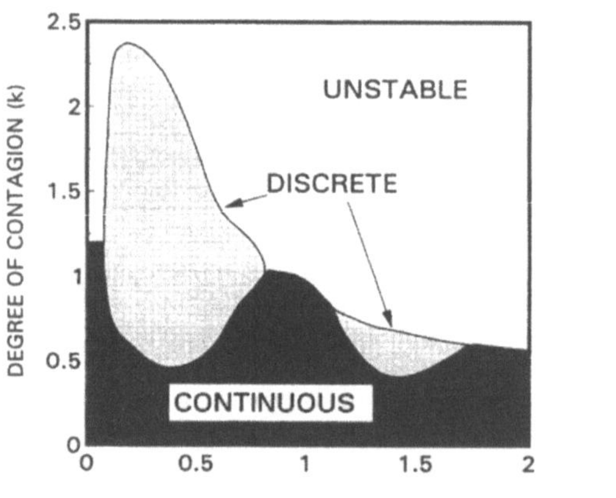

Figure 18.13. Regions of continuous reproduction, discrete generations and unstable behavior for a system with developmental asynchrony between host and parasitoid (after Godfray and Hassell 1987).

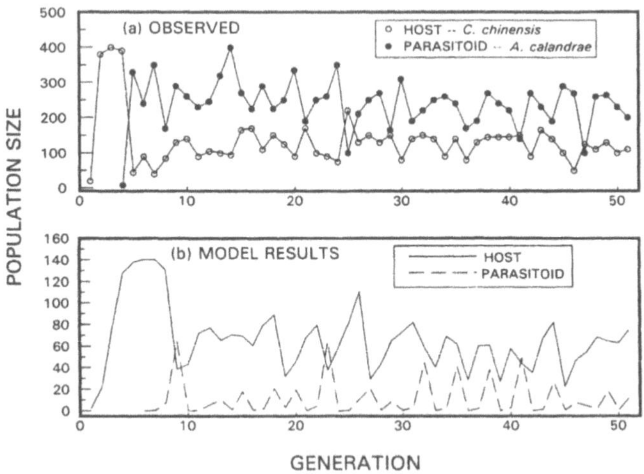

Figure 18.14. Observed (a, after Utida 1950) and simulated (b, after Bellows and Hassell 1988) dynamics for laboratory host-parasitoid systems with both developmental asynchrony and host competition. (a) *Callosobruchus chinensis* (Linnaeus) and *Anisopteromalus calandrae* (Howard). (b) *C. chinensis* and *Lariophagus distinguendus* Förster.

Parasitism and Competition in Asynchronous Systems. Utida (1950, 1957) described the dynamics of a host-parasitoid system which had unusual dynamical behavior characterized by bounded, but aperiodic, cyclic oscillations (Fig. 18.14a). The laboratory system consisted of a regularly renewed food source, a phytophagous weevil, and a hymenopteran parasitoid. Important characteristics of the system were host-parasitoid asynchrony (the parasitoid developed in two-thirds of the weevil developmental time), host density dependence (the weevil adults competed for oviposition sites and larvae for food resources), and age-specificity in the parasitoid-host relationship (parasitoids could attack and kill three larval weevil stages and pupae, but could only produce female progeny on the last larval stage and pupae).

Bellows and Hassell (1988) describe a model of a similar system which incorporated age-structured host and parasitoid populations, intraspecific competition among host larvae and adults, and age-specific interactions between host and parasitoid. The dynamics of the model (Fig. 18.14b) had characteristics similar to those shown by the experimental populations and distinct from those of any simpler model. In particular, asynchrony of host and parasitoid, the attack by the parasitoid of young hosts (on which reproduction was limited to male offspring), and intraspecific competition by the host were all important features contributing to the observed dynamics. The interaction of these three factors caused continual changes in both host density and age-class structure. In generations where parasitoid emergence was synchronized with the presence of late larval hosts, there was substantial host mortality and parasitoid reproduction. This situation produced a large parasitoid population in the succeeding generation which, emerging coincident with young host larvae, killed many host larvae but produced few female parasitoids. The reduced host larval population suffered little competition (because of reduced density) and little parasitism. This continual change in intensity of competition and parasitism contributed significantly to the cyclic behavior of the system.

Invulnerable Age-Classes. The two previous models incorporated both susceptible and unsusceptible stages, ideas which are inherent to any stage-specific model where the parasitoid attacks a specific stage such as egg, larvae, or pupae. The consequences of unsusceptible or invulnerable stages in a population has been considered by Murdoch et al. (1987) in relation to the interaction between California red scale, *Aonidiella aurantii*, and its parasitoid *Aphytis melinus*. Their models included both vulnerable and invulnerable host stages, and both juvenile and adult parasitoids; they also contained no explicit density dependence in any of the stages, but did contain time delays in the form of developmental times from juvenile to adult stages of both populations.

Murdoch et al. (1987) developed two models, one in which the adult hosts were invulnerable, and one in which the juvenile hosts were invulnerable. They found that either model can have stable equilibria (approached either monotonically or through damped oscillations), stable cyclic behavior, or chaotic behavior. The realm of parameter space which permitted stable populations was substantially larger for the model in which the adult was invulnerable than for the model when the juvenile was invulnerable. Whether or not the stabilizing effect on an invulnerable age class was sufficient to overcome the destabilizing influence of parasitoid developmental delay depended on the relative values of several parameters, but short adult parasitoid lifespan, low host fecundity, and a long adult invulnerable age class all promoted stability.

Many insect parasitoids attack only one or few stages of a host population (although predators may be more general), and, therefore, many populations contain unattacked stages. In addition, however, many insect populations are attacked by more than one natural enemy, and general statements concerning the aggregate effect of a complex of natural enemies attacking different stages of a continuously developing host population are not yet possible. Nonetheless, it appears that, at least for the California red scale and *A. melinus* system, the combination of an invulnerable adult stage and overlapping generations is likely a factor contributing to the observed stability of the system (Reeve and Murdoch 1985; Murdoch et al. 1987).

Spatial Complexity and Asynchrony

In predator-prey or parasitoid-host systems which occur in a patchy or heterogeneous environment, we may distinguish between dynamics which occur between the species within a patch and the dynamics of the regional or global system. Here we distinguish between "local" dynamics (those within a patch) and "global" dynamics (the characteristics of the system as a whole). While still interested in such dynamical behavior as stability of the equilibrium, we also seek to understand what features of the system might lead to global persistence (the maintenance of the interacting populations) in the face of unstable dynamical behavior at the local level (see review by Taylor [1988]). One set of theories concerned with the global persistence of predator-prey systems emphasizes the importance of asynchrony of local, within-patch, predator-prey cycles (Den Boer 1968; Reddingius and Den Boer 1970; Reddingius 1971; Maynard Smith 1974; Levin 1974, 1976; Crowley 1977, 1978, 1981). In this context, asynchrony between patches implies that, on a regional basis, unstable predator-prey cycles may be occurring in each patch, but out of phase with one another (prey populations may be increasing in some fraction of the environment while they are being driven to extinction by predators in another). Such asynchrony may reduce the likelihood of global extinction and promote the persistence of the populations. (This asynchrony in populations among patches is distinct from the developmental asynchrony between host and parasitoid of the three preceding systems.)

An example of such a spatially heterogenous system comes from the simulation or systems model of interacting populations of the spider mite *Tetranychus urticae* and the predatory mite

Phytoseiulus persimilis described by Sabelis and Laane (1986). This model is a regional model of a plant-phytophage-predator system that incorporates patches of plant resource which may be colonized by dispersing spider mites; colonies of spider mites may in turn be discovered by dispersing predators. The dynamics of the populations within the patch are unstable (Sabelis 1981; Sabelis et al. 1983; Sabelis and van der Meer 1986), with overexploitation of the plant by the spider mite leading to decline of the spider mite population in the absence of predators. When predators are present in a patch, they consume prey at a rate sufficient to cause local extinction of the prey and subsequent extinction of the predator.

In contrast to the local dynamics of the system, the regional or global dynamics of the system was characterized by two states. In one, the plant and spider mite coexisted but exhibited stable cycles (driven by the intraspecific depletion of plant resource in each patch and the time delay of plant regeneration). The other state was one in which all three species coexisted. This latter case was also characterized by stable cycles, but these were primarily the result of predator-prey dynamics; the average number of plant patches occupied by mites in the three-species system was less than 0.01 times the average number occupied by spider mites in the absence of predators. The overall system thus persisted despite the unstable dynamics at the patch level (Fig. 18.15).

Principal among the model's features which contributed to global persistence was asynchrony of local cycles, making it very unlikely that prey would be eliminated in all patches at the same time. This asynchrony could be disturbed if the predators became so numerous that all prey patches were likely to be simultaneously discovered, leading to global extinction of both prey and predator.

These results are broadly in accord with the experiments of Huffaker (1958) and Huffaker et al. (1963) (Fig. 18.16), who found that increasing spatial heterogeneity enhanced population persistence. Three features of those experiments were in accord with the behavior of the model of Sabelis and Laane (1986): (i) overall population numbers did not converge to an equilibrium value but oscillated with a more or less constant period and amplitude; (ii) increased prey dispersal relative to predator dispersal enhanced the persistence of the populations (Huffaker 1958); (iii) increased food availability per prey patch resulted in increased

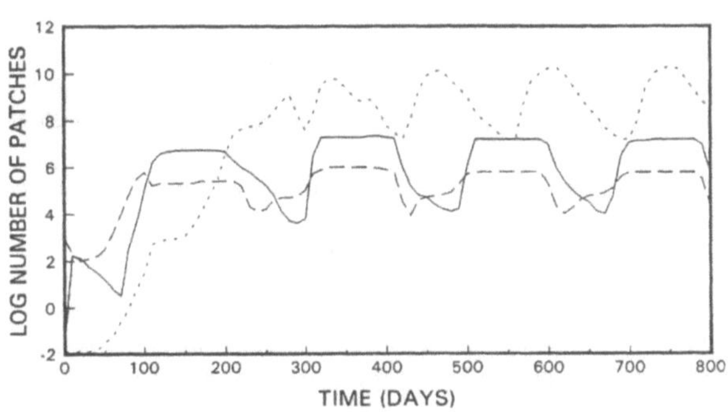

Figure 18.15. Regional dynamics of a system model for a spatially discrete predator-prey system (after Sabelis and Laane 1986). Dotted line is number of plant patches occupied by mites for system with only plants and phytophagous mites. Solid line (patches with only phytophagous mites) and dashed line (patches with either both phytophagous mites and predatory mites, or only predatory mites) are for system with plants, prey, and predator.

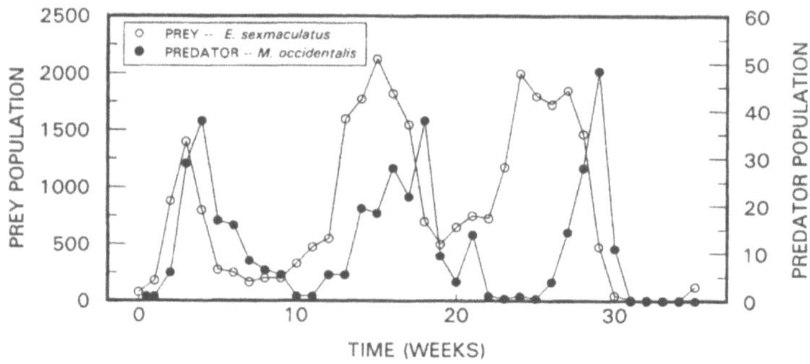

Figure 18.16. Observed dynamics of a mite predator (*Metaseiulus occidentalis* [Nesbitt])-mite prey (*Eotetranychus sexamaculatus* [Riley]) system in the laboratory (after Huffaker 1958).

predator production at times of high prey density, which in turn led to synchronization of the local cycles and global extinction (Huffaker et al. 1963).

Results reported in larger-scale systems, particularly greenhouses, include elimination of prey and, subsequently, of predator (Chant 1961; Bravenboer and Dosse 1962; Laing and Huffaker 1969; Takafuji 1977; Takafuji et al. 1981, 1983), fluctuations of varying amplitude (Hamai and Huffaker 1978), and wide fluctuations of increasing amplitude (Burnett 1979, Nachman 1981). Interpreting these results requires caution because of differences in scale, relation of the experimental period to the period of the local cycles, and relative differences in ease of prey and predator redistribution in different systems. Nonetheless, it is clear that asynchrony among local patches can play an important role in conferring global stability or persistence to a system composed of locally unstable population interactions.

Generalist Natural Enemies

The preceding discussions have focused on natural enemies that are specific to their hosts. Many species of natural enemies, however, are generalists and feed or reproduce on a variety of different hosts, making their population dynamics more-or-less independent of a particular host population.

Equations [7a] and [9] may be modified to represent a host population subject to a generalist natural enemy,

$$N_{t+1} = FN_t[\{1 + aG_t/k\}^{-k}], \tag{13}$$

where G_t is the number of generalist natural enemies attacking the N_t hosts, and the other parameters have the same meaning as before. This equation embodies a response for a generalist whose interactions with the host population may be aggregated or independently distributed (depending on the value of k).

Central to such a model are the details of the numerical response of the generalist predator which determine the values of G_t in equation [13]. Data reported in the literature tend to show a tendency for G_t to rise with increasing N_t to an upper asymptote (Holling 1959b; Mook 1963; Kowalski 1976) (Fig. 18.17). This simple relationship may be described by the following equation (Southwood and Comins 1976; Hassell and May 1986):

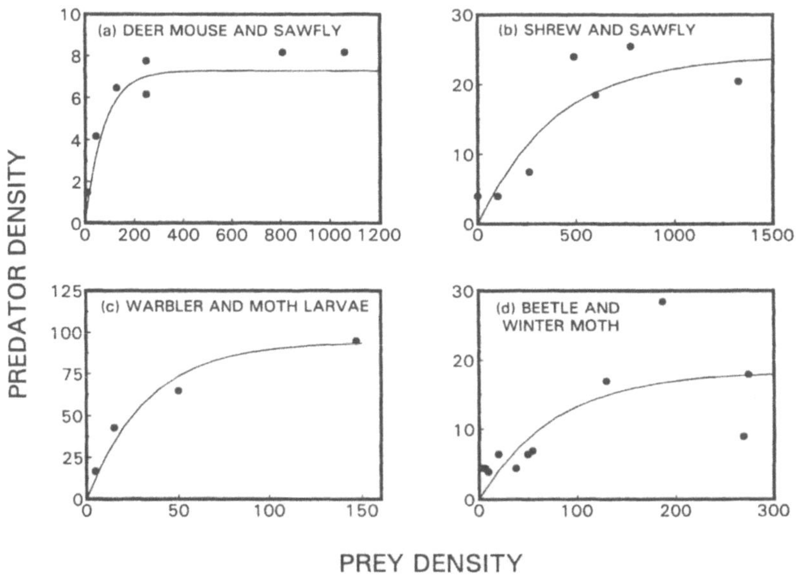

Figure 18.17. Numerical responses by generalist predators: (a) *Peromyscus maniculatus* How and Kennicott and (b) *Sorex cinereus* Kerr (both as numbers per acre) in relation to the density of larch sawfly (*Neodiprion setifer* [Geoffroy]) cocoons (thousands per acre) (after Holling 1959b); (c) the bay-breasted warbler (*Dendroica fusca* [Müller]) (nesting pairs per 100 acres) in relation to third instar larvae of the spruce budworm (*Choristoneura fumiferana* [Clemens]) (numbers per 10 ft^2 of foliage) (after Mook 1963); and (d) the staphylinid *Philonthus decorus* Stephen (pitfall trap index) in relation to winter moth (*Operophtera brumata* [Linnaeus]) larvae per m^2 (after Kowalski [1976]).

$$G_t = m[1 - \exp(-N_t/b)]. \tag{14}$$

Here m is the saturation or maximum number of predators and b determines how steeply the curve rises toward this maximum. Such a numerical response implies that the generalist predator population responds to changes in host density quickly, relative to the generation time of the host, as might occur from rapid reproduction relative to the time scale of the host or by switching from feeding on other prey to feeding more prominently on the host in question (Murdoch 1969). The complete model for this host-generalist interaction incorporating equation [14] into equation [13] now becomes

$$N_{t+1} = FN_t\left[1 + \frac{am\{1 - \exp(-N_t/b)\}}{k}\right]^{-k}. \tag{15}$$

This equation represents a reproduction curve with implicit density dependence. Hassell and May (1986) present an analysis of this interaction together with the following conclusions. First, the action of the generalist reduces the growth rate of the host population (which in the absence of the natural enemy grows without limit in this model). Whether the growth rate has been reduced sufficiently to produce a new equilibrium depends upon the attack rate a and the maximum number of generalists m being sufficiently large relative to the host fecundity F. The host equilibrium decreases as predation by the generalist becomes less clumped, as the combined effect of search efficiency and maximum number of generalists (the overall measure of natural enemy efficiency *am*) increases, and as the host fecundity (F) decreases. A new equilibrium may be stable or unstable; unstable populations will show limit cycle or chaotic dynamics. These latter persistent but nonsteady state interactions arise when the generalists cause sufficiently severe density-dependent mortality, promoted by low degrees of aggregation (high values for k), large *am*, and intermediate values of host fecundity F.

HOST-PATHOGEN SYSTEMS

Insect populations can be subject to infection by viruses, bacteria, protozoa, fungi and nematodes, the effects of which may vary from reduced fertility to death. In many cases, these have been intentionally manipulated against insect populations. Reviews of case studies have been presented by Tinsley and Entwistle (1974), Tinsley (1979), Falcon (1982), Entwistle (1983), and Tanada and Kaya (1993).

Much early work with insect pathogens was largely empirical, and a theoretical framework for interactions between insects populations and their pathogens has only recently been developed (Anderson and May 1981; Régnière 1984; May 1985; May and Hassell 1988; Hochberg 1989; Hochberg et al. 1990). Following this work, we will consider first a host population with discrete, nonoverlapping generations (for example a univoltine temperate Lepidoptera such as the gypsy moth, *Lymantria dispar*, and its nuclear polyhedrosis virus disease), affected by a lethal pathogen which is spread in an epidemic fashion through contact between infected and healthy individuals each generation prior to reproduction by the insect. We may apply a variant of equation [7] to describe the dynamics of such a population (where $g = 1$ so that there is no other density-dependent mortality):

$$N_{t+1} = FN_tS(N_t), \tag{16}$$

Here $S(N_t)$ represents the fraction escaping infection as an epidemic spreads through a population of density N_t. This fraction is given implicitly by the Kermack-McKendrick expression, $S = \exp\{-(1 - S)N_t/N_T\}$ (Kermack and McKendrick 1927), where N_T is the threshold host density (which depends on the virulence and transmissibility of the pathogen) below which the pathogen cannot maintain itself in the population. For populations of size N less than N_T the epidemic cannot spread ($S = 1$), and the population consequently grows geometrically while the infected fraction S decreases to ever smaller values. As the population continues to grow, it eventually exceeds N_T, and the epidemic can again spread. This very simple system has very complicated dynamical behavior; although completely deterministic, it has neither a stable equilibrium or stable cycles, but exhibits chaotic behavior (where the population fluctuates between relatively high and low densities) in an apparently random sequence (Fig. 18.18) (May 1985).

Hochberg et al. (1990) have explored an extension of equation [16] where transmission is via free-living stages of the pathogen (rather than direct contact between diseased and healthy individuals). They found that the ability of a pathogen to produce long-lived external stages contributed to the persistence of the pathogen, and tended to dampen the chaos described by equation [16]. However, if stages were too long-lived, then the pathogen could build up a reservoir resulting in population fluctuations of long period.

Additionally, many such populations may have generations which overlap to a sufficient degree that differential, rather than difference, equations are a more appropriate framework for their analysis. The study of many insect host-pathogen systems have thus been framed in differential equations. Following Anderson and May (1980, 1981), we first assume that the host

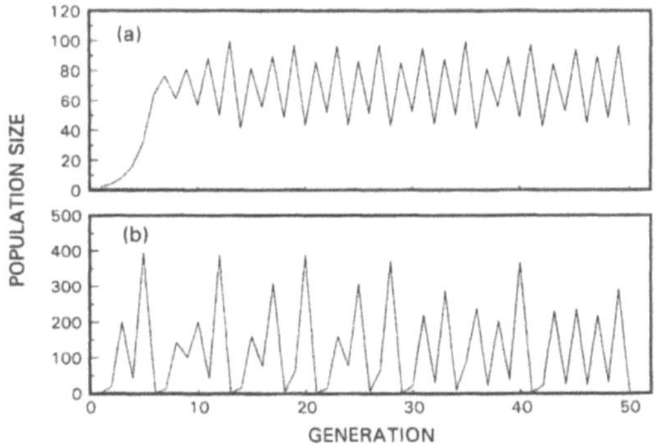

Figure 18.18. Chaotic dynamics of the deterministic host-pathogen system of equation (16). In both examples $N_T = 50$ and the starting population size was $N_1 = 2$. In (a) $F = 2$, in (b) $F = 10$.

population has constant per capita birth rate a and death rate (from sources other than the pathogen) b. We divide the host population $N(t)$ into uninfected $(X(t))$ and infected $(Y(t))$ individuals, so that $N = X + Y$. For consideration of insect systems, the model does not require the separate class of individuals which have recovered from infection and are immune, as may be required in vertebrate systems, because current evidence does not clearly indicate that insects are able to acquire immunity to infective agents. This basic model further assumes that infection is transmitted directly from infected to uninfected hosts as a rate characterized by the parameter β, so that the rate at which new infections arise is βXY (Anderson and May 1981). Infected hosts either recover at rate g or die from the disease at rate α. Both infected and healthy hosts continue to reproduce at rate a and be subject to other causes of death at rate b.

The dynamics of the infected and healthy portions of the population are now characterized by

$$dX/dt = a(X + Y) - bX - \beta XY + gY, \tag{17a}$$

$$dY/dt = \beta XY - (\alpha + b + g)Y. \tag{17b}$$

The healthy host population increases from both births and recovery of infected individuals. Infected individuals appear at rate βXY and remain infectious for average time $1/(\alpha + b + g)$ before they die from disease or other causes or recover. The dynamics of the entire population are characterized by

$$dN/dt = rN - \alpha Y, \tag{18}$$

where $r = a - b$ is the per capita growth rate of the population in the absence of the pathogen. There is no intraspecific density dependence or self-limiting feature in the host population, so that in the absence of the pathogen the population will grow exponentially at rate r.

We may now consider the consequences of introducing a few infectious individuals into a population previously free from disease. The disease will spread and establish itself provided the right-hand side of equation [17b] is positive, or more generally if the total population of the host exceeds a threshold density $N_{T'}$,

$$N_{T'} = (\alpha + b + g)/\beta. \tag{19}$$

Because the population in this simple analysis increases exponentially in the absence of the disease, the population will eventually increase beyond the threshold. In a more general situation where other density-dependent factors may regulate the population around some long-term equilibrium level K (in the absence of disease), the pathogen can only establish in the population if $K > N_T$ (Anderson and May 1981).

Once established in the host population, the disease can (in the absence of other density-dependent factors) regulate the population so long as it is sufficiently pathogenic, with $\alpha > r$. In such cases, the population of equation [17] will be regulated at a constant equilibrium level $N^* = [\alpha/(\alpha - r)]N_T$. The proportion of the host population infected is $Y^*/N^* = r/\alpha$. Hence the equilibrium fraction infected is inversely proportional to disease virulence, and so decreases with increasing virulence of the pathogen. If the disease is insufficiently pathogenic to regulate the host ($\alpha < r$), then the host population will increase exponentially at the reduced per capita rate $r' = r - \alpha$ (until other limiting factors affect the population).

The relatively simple system envisaged by equation [17] permits some additional analysis. First, in single host-single pathogen systems, pathogens cannot in general drive their hosts to extinction because the declining host populations eventually fall below the threshold density for maintenance of the pathogen. Additionally, we may consider what features of a pathogen might be implicated in maximal reduction of pest density to an equilibrium regulated by the disease; most particularly, what degree of pathogenicity produces optimal host population suppression. Pathogens with low or high virulence lead to high equilibrium host populations, while pathogens with intermediate virulence lead to optimal suppression (Anderson and May 1981). This point is important because many control programs (and indeed many genetic engineering programs) often begin with an assumption that high degrees of virulence are desirable qualities. While this may be true in some special cases of inundation, it is not true for systems which rely on any degree of perpetual host-pathogen interaction (Anderson 1982; May and Hassell 1988; Hochberg et al. 1990).

Some potentially important biological features are not considered explicitly in equation [17] (Anderson and May 1981). Several of these have fairly simple consequences for the general conclusions presented above. Pathogens may reduce the reproductive output of infected hosts prior to their death (which renders the conditions for regulation of the host population by the pathogen less restrictive). Pathogens may be transmitted between generations (vertically) from parent to unborn offspring (which reduces N_T and thus permits maintenance of the pathogen in a lower density host population). The pathogen may have a latency period where infected individuals are not yet infectious (which increases N_T and also makes population regulation by the pathogen less likely). The pathogenicity of the infection may depend on the nutritional state of the host and hence, indirectly, on host density. Under these conditions, the host population may alternate discontinuously between two stable equilibria. Anderson and May (1981) give further attention to these cases.

A more significant complication arises when the free-living transmission stage of the pathogen is long-lived relative to the host species. Such is the case with the spores of many bacteria, protozoa, and fungi, and the encapsulated forms of many viruses (Tinsley 1979). Most of the conclusions from equation [17] still hold, but the regulated state of the system may now be either a stable point or a stable cycle with a period greater than two generations. Anderson and May (1981) show that the cyclic solution is more likely for organisms of high pathogenicity (and many insect pathogens are highly pathogenic, see Anderson and May [1981] and Ewald [1987]) and which produce large numbers of long-lived infective stages. The cyclic behavior results from the time delay introduced into the system by the pool of long-lived infectious stages. Such cyclic behavior appears characteristic of populations of several forest Lepidoptera and their associated diseases (Anderson and May 1981). In one particular case where sufficient data are available to estimate the necessary parameters, there is substantial agreement between the expected and observed period of population oscillation (Anderson and May 1981; McNamee et al. 1981).

More recently, Hochberg (1989) has shown that heterogeneity in the structure of the pathogen population can have important effects on pathogen persistence and the ability of pathogens to regulate their hosts. In particular, the formation of a reservoir of long-lived stages can dampen the tendency to cycle that was identified in the models of Anderson and May (1980, 1981).

MULTISPECIES SYSTEMS

Emphasis has thus far centered on the dynamics of single- and two-species systems, divorced from the more complex webs of multispecies interactions of which they will often be an integral part. If we are to understand how population dynamics can influence the structure of simple communities, it is important to determine first how the dynamics of two interacting species are influenced by the additional linkages typically found with other species in the food web. With this in mind, the dynamics of a wide range of different three-species systems have been examined. These range from a natural enemy species (predator or pathogen) attacking competing prey or host species (Roughgarden and Feldman 1975; Comins and Hassell 1976; Anderson and May 1986), competing natural enemy species sharing a common prey or host species (May and Hassell 1981; Kakehashi et al. 1984; Anderson and May 1986; Hochberg et al. 1990), a prey species attacked by both generalist and specialist natural enemies (Hassell and May 1986), and various three trophic level interactions (Beddington and Hammond 1977; May and Hassell 1981; May and Hassell 1988). In some of these systems the dynamics are not just the expected blend of the component two-species interactions. Rather, the additional nonlinearities introduced by the third species can lead to quite unexpected dynamical properties.

We now turn to four examples illustrating how population dynamics are affected in moving from two- to three-species interactions.

Competing Natural Enemies

In many natural systems, phytophagous species are attacked by a suite of natural enemies, and plants are attended by a complex of herbivores. In biological control programs, attempts to reconstruct such multiple-species systems have often met with some debate in spite of their ubiquitous occurrence. Some workers have suggested that interspecific competition among multiple natural enemies will tend to reduce the overall level of host suppression (Turnbull and Chant 1961; Watt 1965; Kakehashi et al. 1984). Others view multiple introductions as a potential means to increase host suppression with no risk of diminished control (van den Bosch and Messenger 1973; Huffaker et al. 1971; May and Hassell 1981; Waage and Hassell 1982). The significance of this issue probably varies in different systems, but the basic principles may be addressed analytically.

The dynamics of a system with a single host and two parasitoids may be addressed by extending the single host-single parasitoid model of equation [7] (with no host density dependence, $g = 1$) to include an additional parasitoid. One possibility is the case described by May and Hassell (1981):

$$N_{t+1} = FN_t h(Q_t) f(P_t), \tag{20a}$$

$$Q_{t+1} = N_t\{1 - h(Q_t)\}, \tag{20b}$$

$$P_{t+1} = N_t h(Q_t)\{1 - f(P_t)\}. \tag{20c}$$

Here the host is attacked sequentially by parasitoids Q and P. The functions h and f represent the fractions of the host population surviving attack from Q and P, respectively, and are described by equation [9]; the distribution of attacks by one species is independent of attacks by the other. Variations on this theme have also been considered, such as when *P* and *Q* attack the same stage simultaneously (May and Hassell 1981). The general qualitative conclusions, however, remain unchanged.

Three general conclusions arise from an examination of this system. First, the coexistence of the two species of parasitoids is more likely if both contribute in some measure to the stability of the interaction (both species have values of $k < 1$ in equation [9]). Second, if we consider the system with the host and parasitoid *P* already coexisting and attempt to introduce parasitoid *Q*, then coexistence is more likely if *Q* has a searching efficiency higher than *P*. If *Q* has too low a searching efficiency, it will fail to become established, precluding coexistence. If the search efficiency of *Q* is sufficiently high, it may suppress the host population below the point at which *P* can continue to persist, thereby leading to a new single host-single parasitoid system. Apparent examples of such competitive displacement include the introductions of parasitoids against *Dacus dorsalis* Hendel in Hawaii (U.S.A.) (Bess et al. 1961, Fig. 18.19) and the displacement of *Aphytis lingnanensis* by *A. melinus* in interior southern California (U.S.A.) (Luck and Podoler 1985). Finally, the successful establishment of a second parasitoid species (*Q*) will in almost every case further reduce the equilibrium host population. For certain parameter values, it can be shown that the equilibrium might have been lower still if only the host and parasitoid *Q* were present, but this additional depression is slight. In general, the analysis points to multiple introductions as a sound biological strategy.

Kakehashi et al. (1984) have considered a case similar to equation [20], but where the distributions of attacks by the two parasitoid species are identical rather than independent. This assumes, therefore, that the two species of parasitoids respond in the same way to environmental cues involved in locating hosts. This modification does not change appreciably the stability properties of equation [20], but does change the equilibrium properties. In particular, a system with the superior parasitoid alone now can cause a greater host population depression than does the three-species system. In natural systems, however, complete covariance of parasitism between species may be less likely than more independent distributions (Hassell and Waage 1984). Nonetheless, this is one example where general, tactical predictions can be affected by changes in detailed model assumptions.

Generalist and Specialist Natural Enemies

We now turn to interactions between populations of specialist and generalist natural enemies sharing a common host species. As noted earlier, discrete systems with more than one mor-

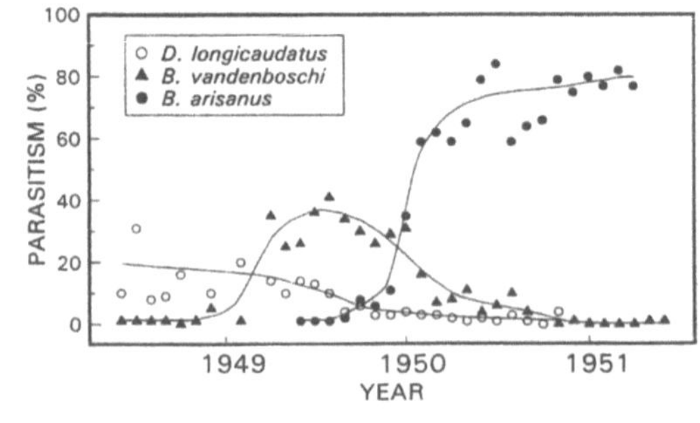

Figure 18.19. The sequential introduction of three fruit fly parasitoids (*Diachasmimorpha longicaudatus* [Ashmead], *Biosteres vandenboschi* [Fullaway] and *Biosteres arisanus* [Sonan]) into Hawaii (U.S.A.) indicates the sequential competitive replacement of one species by another together with increased overall parasitism rates (after Bess et al. 1961).

tality factor may have different dynamics depending on the sequence of mortalities in the hosts life cycle. We will consider here a situation where the specialist natural enemy acts first, followed by the generalist, both preceding reproduction of the adult host (Hassell and May 1986):

$$N_{t+1} = FN_t f(P_t) g[N_t f(P_t)], \tag{21a}$$

$$P_{t+1} = N_t\{ 1 - f(P_t)\}. \tag{21b}$$

Here $g(N_t)$ is the host survival from the generalist which incorporates a numerical response together with the negative binomial distribution of attacks:

$$g(N) = \left[1 + \frac{am\{1 - \exp(-N/b)}{k}\right]^{-k}. \tag{22}$$

The function $f(P)$ is the proportion surviving parasitism and, again assuming a negative binomial distribution of parasitoid attacks, is given by

$$f(P) = [1 + a'P/k']^{-k'}. \tag{23}$$

We now consider the conditions for generalist and specialist to coexist and examine their combined effects on the host population. In particular, a specialist natural enemy can coexist with the host and generalist most easily when the effect of the generalist is small (k and am are small, indicating low levels of highly aggregated attacks), when the efficiency of the specialist is high, and when there is low density dependence in the numerical response of the generalist (Hassell and May 1986). In simple terms, if the effect of the generalist is small, there is greater potential that the host population can support an additional natural enemy (the specialist). Conversely, if the host rate of increase F is low or the efficiency of the generalist population (am) too high, then a specialist is unlikely to be able to coexist in the host-generalist system. Further details are given in Hassell and May (1986).

Parasitoid-Pathogen-Host Systems

We now consider interactions in which a host is attacked by both a parasitoid (or predator) and a pathogen (Carpenter 1981; Anderson and May 1986; May and Hassell 1988; Hochberg et al. 1990). These interactions may be considered cases of two-species competition, where the natural enemies compete for the resource represented by the host population. As for interspecific competition by two parasitoids discussed previously, these interactions are characterized by four possible outcomes: the parasitoid and pathogen may coexist with each other and with the host; either parasitoid or pathogen may regulate the host population at a density below the threshold for maintenance of the other agent; or there may be two, alternative stable (or unstable) states (one with host and parasitoid and one with host and pathogen), where the outcome of a particular situation depends on the initial condition of the system.

Consider a population which is first attacked by a lethal pathogen (spread by direct contact) and the survivors are then attacked by parasitoids, represented by combining the models of equations [7] (with $g = 1$) and [16]:

$$N_{t+1} = FN_t S(N_t) f(P_t), \tag{24a}$$

$$P_{t+1} = cN_t S(N_t)\{1 - f(P_t)\}. \tag{24b}$$

Here $S(N)$ is the fraction surviving the epidemic given by the implicit relation $S = \exp[-(1 - S) N_t/N_T]$; f has the Nicholson-Bailey form $f(P) = \exp(-aP)$ representing independent, random search by parasitoids; F is the per capita host reproductive rate; and c is the number of parasitoids produced by a single parsitized host.

The dynamical character of this system has been summarized by May and Hassell (1988) and considered in some detail by Hochberg et al. (1990). For $acN_T(\ln F)/(F-1) < 1$ the pathogen excludes the parasitoid by maintaining the host population at levels too low to sustain the parasitoid. For parasitoids with greater searching efficiency (a), or greater degrees of gregariousness, or for systems with higher thresholds (N_T), so that $acN_T(\ln F)/(F-1) > 1$, a linear analysis suggests that the parasitoid would exclude the pathogen in a similar manner. However, the diverging oscillations of the Nicholson-Bailey system eventually lead to densities higher than N_T, and the pathogen can repeatedly invade the system as the host population cycles to high densities. The resulting dynamics (Figure 18.20) can be quite complex, even from the simple and purely deterministic interactions of equation [24]. Here the basic period of the oscillation is driven by the Nicholson-Bailey model, with the additional effects of the (chaotic) pathogen-host interaction leading to approximately stable (rather than diverging) oscillations. As May and Hassell (1988) discuss, in such complex interactions, it can be relatively meaningless to ask whether the dynamics of the system are determined mainly by the parasitoid or by the pathogen. Both contribute significantly to the dynamical behavior, the parasitoid by setting the average host abundance and the period of the oscillations, and the pathogen providing long term "stability" in the sense of limiting the amplitude of the fluctuations, and thereby preventing catastrophic overcompensation and population extinction. Extensions to the model of May and Hassell (1988), where the pathogen is capable of producing long-lived external stages, and the parasitoid and pathogen compete within host-individuals, have also been explored by Hochberg et al. (1990).

Competing Herbivores and Natural Enemies

The presence of polyphagous predators in communities of interspecific competitors can have profound effects both on the number of species in the community and on the relative roles which predation and competition play in population dynamics. Paine (1966, 1974) demonstrated that intertidal communities contain more species when subject to predation by the predatory starfish, *Pisaster ochraceus* (Brandt), than when it is absent. Since then there has been much theoretical attention to the relative roles of predation and competition in multi-

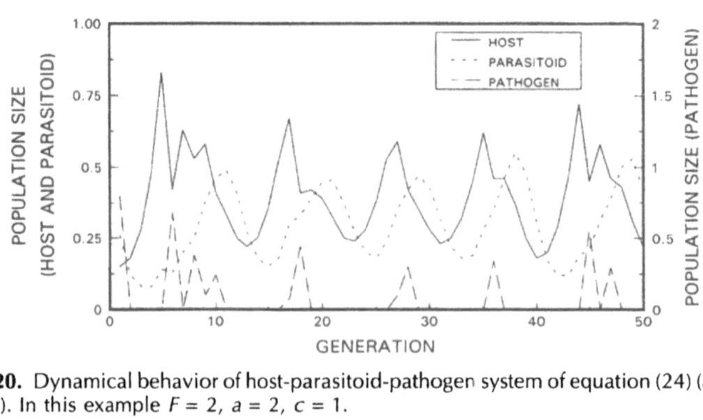

Figure 18.20. Dynamical behavior of host-parasitoid-pathogen system of equation (24) (after May and Hassell 1988). In this example $F = 2$, $a = 2$, $c = 1$.

species communities (Parrish and Saila 1970; Cramer and May 1972; Steele 1974; van Valen 1974, Murdoch and Oaten 1975; Roughgarden and Feldman 1975; Comins and Hassell 1976; Fujii 1977; Hassell 1978, 1979; Hanski 1981). One general conclusion of this work is that natural enemies can, under certain conditions, enable competing species to coexist where otherwise they could not. This effect is enhanced if the natural enemy shows some preference for the dominant competitors or switches between prey species as one becomes more abundant than the other.

This work has also been extended to the case of competing prey and natural enemies existing in a patchy environment. Comins and Hassell (1987) considered the cases of patchily-distributed, competing prey and either a generalist natural enemy (whose dynamics were unrelated to the dynamics of the prey community) or a more specialized natural enemy coupled to the competing prey in the community. In both cases, the natural enemy population could, under certain conditions, add stability to an otherwise unstable competition community. This occurred more readily with the generalist than the specialist. In all cases, aggregation by the natural enemy in patches of high prey density (which leads to a switching effect) was an important attribute contributing to stability. Predation which was independently random across patches was destabilizing for both the generalist and specialist cases. Coexistence of competing prey species was possible in this spatially heterogeneous model even when the distributions of the prey species in the environment were correlated and when interspecific competition was extreme.

IMPLICATIONS FOR BIOLOGICAL CONTROL

The models presented in this chapter are framed in general, analytical terms, and are designed to capture the outline, or general features, of interactions between natural enemies and their hosts. They are not designed to provide detailed predictions or hypotheses about the results of a particular biological control program; to model an interaction at sufficient detail to evaluate interactions among particular species usually requires much more detailed models (Bellows 1982a,b; Bellows and Hassell 1984, 1988; Godfray and Waage 1991). Rather, the intent of models such as those presented in this chapter is to represent the general features of a biological control system, and then to evaluate the consequences of changing one of those major features. This approach to exploring the dynamical features of biological control systems examines such questions as: Are generalist and specialist predators able to work together effectively against a pest? How might the presence of more than one parasitoid affect a natural enemy-host interactions? Can pathogens and parasitoids possibly work together? These questions can be addressed and answered in generalities by the analytical and system models presented here.

Many of the theoretical constructs in the foregoing sections rest on substantial experimental evidence indicating that either the mechanisms, or, in some cases, both the mechanisms and their outcomes, have been documented to occur in laboratory or natural systems. This evidence lends strength to the premise that the ideas embodied in these theories indeed apply, at least in principle, to biotic systems. Consequently, several of the more general conclusions can be used to provide a basis for ecological practice in the execution of biological control programs. These generalities have been discussed in the preceding sections, but some are reiterated collectively here.

In both homogeneous and heterogeneous environments (or for both uniformly distributed and patchily distributed populations), intraspecific competition can lead to very complicated dynamical behavior, including stable cycles and chaos. However, in most natural populations which exhibit density dependence, such density dependence appears moderate, and populations dynamics is more nearly monotonic in behavior. This is true in both single- and multiple-age class systems, and also applies generally to multispecies systems where interspecific compeition occurs (although in this latter case, there is the additional possibility of competitive exclusion of particular species). In host-parasitoid systems, density dependence in the host population can stabilize an otherwise unstable interaction, but stable host-parasitoid systems may occur at such low densities of the host that such intraspecific competition is unlikely to play a major role in the system. One important stabilizing feature in upper trophic levels appears to be spatial patterns of attack by natural enemies; many different examples of spatial heterogeneity have been described. Other stabilizing features may include invulnerable host stages.

In the practice of biological control, population dynamics theory can be applied most directly to programs which envisage the permanent altering of population interactions, such as by the introduction of new species. In cases of introductions, theory indicates that it is unlikely that biological control of a target population caused by an existing parasitoid population could be negatively affected by the introduction of additional parasitoid species. The additional natural enemies may exclude existing ones, but except in very special cases this can occur only when the newly introduced natural enemies cause total mortality greater than that caused by existing natural enemies. Therefore if the introduction of additional parasitoids is justifiable on the basis of need for additional biological control, in most cases it is not necessary to address concerns about potential competition among the candidate parasitoids. Indeed, the action of several natural enemies may prove necessary to provide adequate suppression of target populations, as has proved to be the case in some biological control programs.

Some pathogen life histories are possibly linked to chaotic dynamics, which are undesirable in pest species populations. Consequently, the evaluation of the possible effects of certain pathogen life cycles on the dynamics of populations of the target organism should be considered before such pathogens are introduced as biological control agents.

BIOLOGICAL CONTROL SYSTEMS AND ECOLOGICAL EVALUATIONS

Many biological control programs take on the character, on a very large scale, of ecological experiments. This is especially true for those programs which involve adventive species which have no or a very limited number of natural enemies in a region, and often reach pest status. This phase of a program provides scope for investigation of niche expansions, effect of addition of a competitor species into established guilds, evaluation of physiological and ecological limits of range expansion, all in the absence of a limiting upper trophic level fauna. These and other ecological consequences of introductions can be evaluated for adventive plant species, arthropods, vertebrates, and microorganisms. The consequences of such introductions are often spectacular. There are many examples (see Chapter 8; DeBach 1964a; DeBach and Rosen 1991), and both intentional and accidental introductions may be frequent. California, for example, received over 200 unintentional introductions of invertebrates from 1955 to 1988, one every 60 days, although not all of these became pests (Dowell and Gill 1989).

When an adventive species reaches pest status, a study of the upper trophic level fauna in possible areas of origin often uncovers natural enemies, and is often followed by the introduction of natural enemy species into the affected area. These introductions are also large-scale ecological experiments, with similar scope for investigation, but with additional impetus for determining whether or not any particular agent or agents might provide suitable suppression of the target pest. All of these interactions are excellent opportunities for the study of ecological mechanisms of population interactions and dynamical behavior.

SECTION

VI

Additional Topics

In this final section several additional topics are covered that bear on biological control. Chapters 19 and 20 consider social factors affecting biological control. Chapter 19 discusses the role of farmer training in promoting adoption of biological control. In Chapter 20 the effects of government policies of various types are explored with respect to how they advance or inhibit the use of biological control methods. Chapter 21 discusses biological control's role in suppressing environmental pests, reviewing principles of why some pests damage natural systems (or particular species) and how biological control can be employed to resolve such problems and protect our natural environments. In Chapter 22, we look briefly to the future.

CHAPTER

19

The Role of Grower Education in Biological Control

INTRODUCTION

To be successful, biological control programs need to be explained and demonstrated to farmers, the public, and governmental officials responsible for formulating agricultural and environmental policies (Van Driesche 1989a; Andrews et al. 1992). Agricultural extension employees and the public communication media play key roles in this process. This chapter considers: (1) the needs of farmers for concepts and specific information concerning biological control; (2) media resources used to train farmers and other groups; (3) improving knowledge of extension agents about biological control; (4) explaining biological control to politicians; and (5) communicating with environmental groups.

In many systems, biological control of a pest can function independently of controls for other pests. In other systems biological control will be a component of a larger pest management plan that may include such additional tactics as use of resistant cultivars, cultural practices or selective pesticides. The types of information needed by growers and other groups change as the degree of integration that is required between biological control and other methods increases.

TYPES OF INFORMATION NEEDED BY GROWERS

Farmers and pest control advisors need to understand the principles of how biological control works and the detailed knowledge and skills to effectively apply biological controls of particular pests in the crops they manage. This knowledge includes an ability to recognize key natural enemies and to understand their life cycles and connections to pest life histories. It is also important to understand principles of economic thresholds and pest population management and know how to monitor both pests and natural enemies. Decisions about pest control can then be made based on this knowledge. Such information is often incomplete, but will be needed to develop pest management systems based on biological control.

Concepts of Biological Control Mechanisms

Basic concepts of how biological control functions are often poorly understood by farmers, the general public, and government agricultural administrators. This lack of understanding results from a lack of information and mistaken ideas about biological control, pests, and pest management. Before biological control facts related to specific pests and their natural enemies can be assimilated, basic information must be taught. Important concepts include the following.

Not All Insects Are Pests. Farmers need to be shown beneficial species and taught about parasitism, predation, and disease as factors keeping pest populations in check. Recognition of the concept that beneficial species exist and can help combat pests is the first step in teaching biological control concepts to growers and other groups.

Density Makes the Pest. It is important to teach growers and pest control consultants that it is not mere pest presence that must be managed, but actual losses, either as yield or quality. This concept is important to farmers, who may be very insecure about potential crop losses from pests. The presence of a few pests in a crop is not a sign that chemical controls are needed. The concept of economic injury levels and sampling to determine whether pests in any given field have actually reached a damaging level need to be explained to farmers and demonstrated in crop fields. Where data are lacking to define injury levels, research will be needed for their development.

A further extension of this idea is that healthy and parasitized pests need to be considered separately in assessing pest levels. Methods to distinguish parasitized or diseased pests from healthy ones need to be demonstrated to farmers so they can modify pest counts accordingly. For example, Coop and Berry (1986) present a modified economic threshold for the variegated cutworm, *Peridroma saucia* in peppermint that recognizes the fact that parasitized, but living, cutworms will consume only about 7% of the foliage eaten by healthy cutworms. Using the prevalent rate of parasitism and relative consumption rates of healthy and parasitized larvae, they suggested that the economic threshold previously in use could be raised 34%. Similarly, Van Driesche et al. (1994) present a method for modifying the action threshold for the apple blotch leaf miner, *Phyllonorycter crataegella*, to account for reduction in population growth rate due to larval parasitism (Fig. 14.2). Hoffmann et al. (1991) present a modified threshold for *Heliothis zea* on tomatoes that accounts for egg parasitism (Fig. 13.1).

Not All Pesticide Effects Are Good. The benefits of chemical pesticides (especially insecticides, but potentially other groups, including fungicides and herbicides) are often easier to see and understand than their harmful effects. Benefits include quick pest suppression, often very effective initially, when and where pest control is needed. Growers will be familiar with these benefits. To gain a better understanding of the relative merits of pesticide-based control and biological control, growers need to be taught about (1) pesticide resistance, (2) pest resurgence due to destruction of the pest's natural enemies, (3) pest creation or enhancement (secondary pest outbreaks) due to destruction of natural enemies of minor or nonpest species (Heinrichs and Mochida 1983), (4) potential harm to human health from direct exposure to pesticides during application and later by exposure to residues in food and water, (5) potential harm to wildlife, including endangered species, and (6) other environmental damage that may arise from pesticide use such as contamination of drinking water supplies. Distinctions can also be made between broad- or narrow-spectrum pesticides, contact and stomach pesticides, and other features that may limit the undesired effects of pesticides (see Chapter 7).

The public also needs to be informed of these same issues, as do agricultural public policy makers. This is especially important, so that pesticides are not subsidized by governments (formerly a common practice in some countries). Subsidies encourage pesticide use beyond true need. It is important to impress on policy makers that pesticides should not be viewed as fertilizer, where (more or less) the more you use, the bigger the harvest, but rather that inappropriate or excessive use of pesticides can actually reduce yields, or even cause crop failure. It should also be noted that, in some cases, pesticide use on ornamental plants may lower crop quality through pesticide phytotoxicity.

Farm Practices Affect Natural Enemies. Farmers need to be taught that natural enemies of pests are a resource, present on their farms, that can be used to help meet their crop production goals. Natural control of pests by existing natural enemies has to be demonstrated to growers by direct teaching. Catching natural enemies in their fields and showing them to farmers, showing them eating pests, putting them under microscopes so farmers can see them better, are more effective teaching tools than showing pictures of natural enemies or giving talks about them (although these also have a role to play in teaching biological control to farmers). Farmers need to learn what reduces the effectiveness of natural enemies, with specific attention given to the effects of pesticides, dust, ants, and high levels of nitrogen fertilization. Farmers need to learn that some types of pesticides are less harmful to natural enemies than others. For example, a granular application of a pesticide into the soil will not affect parasitoids looking for hosts on foliage (but would affect natural enemies in the soil). Also, such stomach poisons as cryolite and *Bacillus thuringiensis* may affect natural enemies less than such contact poisons as organophosphate or carbamate pesticides (for more, see Chapter 7). Demonstrations can be devised that can illustrate these ideas and can be performed on growers' farms.

Farmers also need to learn from extension agents methods to enhance colonization, retention, reproduction, and longevity of natural enemies in their crops. Such techniques as strip harvesting, cover crops, natural enemy refuges, overwintering locations, provision of carbohydrate sources for natural enemies, and other methods which enhance natural enemies need to be developed and taught to farmers. Demonstration farms can be used to show the effects of these practices.

New Pests Need New Natural Enemies. Farmers are not likely to know whether or not a particular pest is native to their country unless the pest's arrival was a spectacular pest invasion that occurred during their lifetimes. All earlier invasions will have become the accepted pest background and will be considered native. Organic farming groups and other farmers who have learned that natural enemies are important and should be conserved may have unrealistic hopes that native natural enemies will control all pests. It needs to be explained that some pest species may not have natural enemies anywhere that are sufficiently effective to produce effective pest control (possible examples include direct pests of fruit, pests found deep inside plant tissues, and pests in the soil) and that pest species not native to a region are not likely to be controlled in that region by local natural enemies. Nonnative pest species require the introduction of new species of natural enemies, using species from the pest's native home that are specialized to attack the pest. This concept is a critical foundation for much of biological control, and it is very important that this idea be widely and accurately understood, not just by farmers, but by the public in general, government officials in charge of agricultural policy, and conservation groups.

It is especially important to enlist the active support of conservation groups by ensuring that they understand the distinction between invasions of immigrant pests (which are damaging) and planned introductions of specific natural enemies which can be made safely and which are intended to correct the effects of pest invasions. Without such understanding, conservation groups may oppose biological control activities out of fear that they will pose risks to native species. Conservation groups must also be educated about the great potential that biological control introductions have to help reduce the negative impacts of adventive species, particularly aggressive plants, on conservation areas (see Chapter 21) and to reduce the need for insecticides.

Government officials need to have a clear understanding of the adventive organism problem, including why introductions of new natural enemies may be needed and how such introductions are actually achieved. Officials need this information to establish plant quarantine laws to help reduce the rate of pest invasions. In addition, they need to understand the necessity of government-supported quarantine laboratories and reasonable regulations governing natural enemy importation so that desired biological control introductions can be made efficiently, legally, and safely.

Pathogen-Based Pesticides Act Slowly. Farmers accustomed to using conventional pesticides will be accustomed to seeing rapid kill of pests (insects, weeds in some cases) within hours of application. Biological control implementation sometimes involves the substitution of pathogen-based or pathogen-derived pesticides for conventional products. These products act more slowly than the synthetic pesticides. For farmers to accept pathogen-based materials and use them effectively, they must be trained in their use. First, they must be taught that these materials act over several days, not hours. Second, they must be instructed that the timing of application of these materials is more important than timing of conventional insecticides, as they are often more effective against younger pest stages than older ones (Vail et al. 1991). Timing of applications may require careful monitoring of the time of pest colonization, hatch, or development in individual fields (Ferro and Lyon 1991; Zehnder et al. 1992).

To teach farmers about the value of mortality from natural pathogens (those not applied as microbial pesticides) is an even more difficult task because this mortality may occur slowly over long periods. Rather than a short-term dramatic decline in pest numbers that, while slower to act than chemical pesticides, can still be seen, unmanipulated natural populations of pathogens act almost invisibly by reducing the rate of pest population growth (Ekbom and Pickering 1990). Use of computer models can help demonstrate the significance of higher or lower population growth rates (caused by pathogens or other factors) for specific pests of local interest.

For Insectary-Reared Natural Enemies, Quality, Quantity, and Cost Are Critical to Success. Farmers may be familiar with natural enemies reared commercially for biological control. The gardening public in many countries is also aware of the possibility of purchasing natural enemies due to their contact with advertising by vendors of such products. Some individuals, however, may be limited in their ability or experience to judge the value they obtain from such purchased natural enemies because pests may or may not actually occur in a particular plot the year of release, and results are often hard to measure. Extension agents need to stress that these are living organisms that need to be properly handled and released according to instructions (see Chapter 10).

Quality implies that the natural enemy sold for control of a particular pest has the biological capacity to be effective. That is, it must be the right species for the job, and the strain, as actually produced, must still have this capacity, given the rearing methods and other conditions employed in the production of the natural enemy. Also, the rearing conditions must be such that the agents reared have a high proportion of females (especially for parasitoids). Shipping arrangements must permit the agents to arrive alive and on time. Users should be shown what visible signs (such as black parasitized stages for whiteflies, darkened *Heliothis* eggs, or aphid mummies) would indicate reproduction of the natural enemy at the release site, so that users can evaluate their releases.

Quantity is important: the numbers released must be adequate for control. This may be a very large number per ha. Releases of small numbers may be of little value. Before release levels can be recommended, various release rates must be tested by agricultural scientists not employed by the firm selling the product. Finally, costs of using reared natural enemies must be reasonably similar to costs of other forms of control. Some increases of cost may be offset by reduced problems from pesticides, but costs must be appropriate in view of other crop production costs and potential sale value. In general, reared natural enemies may be expensive because of the quantity of labor needed to produce them.

There Are Many Kinds of Natural Enemies. Farmers and the general public are likely to vastly underestimate the numbers of kinds of natural enemies of insects and plants that exist in nature. Accustomed to thinking in terms of dozens or perhaps hundreds of types of other organisms (birds, mammals, plants), most nonentomologists are unprepared to grasp the numbers of species of arthropods, with perhaps 100,000 to 1,000,000 or more species of parasitoids and an equal or larger number of predators. Without this perspective, farmers and the public may consider whole families of natural enemies to be more or less the same. All coccinellids are ladybugs; all carabids, ground beetles; all *Trichogramma*, parasitoids of moth eggs. Without some understanding of important differences at the species level, farmers will be unable to manipulate natural enemies effectively. The solution to this problem is first to make the numbers of species clear to farmers, and to stress that differences between species and even populations within species are often critical for success. Then, recognition of particular species can be taught as necessary on a crop-by-crop, pest-by-pest basis for each region. Potato farmers in the northeastern United States, for example, need to be able to differentiate between *Coleomegilla maculata*, which feeds on eggs of the Colorado potato beetle, *Leptinotarsa decemlineata*, and other coccinelliid species possibly found in the crop which do not feed on this pest.

Details of Natural Enemy Biology Are Important. Closely related natural enemies may differ in aspects of their biologies in ways that make large differences to their effectiveness as pest control agents in a particular region or country. To effectively train farmers in the use of biological control agents, the subject must be taught at a precise enough level to make all the necessary distinctions that allow effective, important natural enemies in a given crop-pest system to be recognized. The difference between a parasitoid strain well-adapted to the local climate and one that is poorly adapted, the difference between one species of *Trichogramma* wasp and another, the difference between one ladybird beetle and another, the difference between a fungal and bacterial pathogen, all may be critical to the outcome, and these differences must be communicated to farmers if they are to effectively employ biological control.

Conversion from Pesticides to Biological Control Takes Time. Farmers who seek to reduce their use of pesticides and develop methods to enhance the role of biological control agents in their fields need to be told that this process takes time and that various difficulties are likely to arise during the transition period. To redevelop healthy natural enemy faunas in agricultural fields that have received regular pesticide applications in the past will very likely take several years if not longer. Farmers should expect the process to take three or more years and to require changes in their management practices.

Biological Controls Are Sustainable and Safe. Farmers and agricultural policymakers both have a need to identify farming methods that can be employed indefinitely without important damage to the general environment or wildlife of the region. Biological control, if properly conducted, is sustainable and safe (see sections on safety in Chapters 8, 10, and 11 for more details). Biological controls are relatively free from problems of pest resistance and are extremely safe to human health.

Biological Control Increases Profits. Crop forecasting models need to incorporate and quantify the value of pest suppression contributed by natural enemies so that farmers can determine the dollar value of conserving existing natural enemies. In cotton in east Texas, for example, crop models indicated that generalist predators contributed from $0.87 to $15.50 per acre (0.4 ha) of value (Sterling et al. 1992).

Natural Enemy Recognition Skills

Extension agents need to teach farmers to recognize the important natural enemies found in their crops and to relate them to the particular pests the natural enemies attack. This type of training occurs at two levels. The first level is to develop the ability of farmers to recognize families and general types of natural enemies. For example, Raupp et al. (1993) provide information for recognizing the principal families of natural enemies of insect pests affecting woody ornamental plants in the northeastern United States. Such training provides growers and other pest managers with a general understanding of the kinds of natural enemies they are likely to encounter. This understanding makes it easier for them to remember information about individual species of natural enemies of specific pests found on their crops.

The second level, recognizing specific natural enemies, begins with extension agents capturing natural enemies in the farmers' own fields and showing them to the farmers, using hand lenses for a close-up view, and explaining what pests they attack. These specimens can then be taken indoors where microscopes are available for a more detailed examination. Such exercises can also include special demonstrations on the biology of the natural enemies, as well as simple experiments such as seeing natural enemies eat specific pests, or seeing the effects of different pesticides on natural enemies. Once farmers have had the experiences of seeing live specimens of the important natural enemies and working with them in simple experiments, this familiarity can be enhanced by giving them written or visual materials (video tapes and illustrated booklets) that outline the natural enemies' biologies and their effects on the pests of the crop. It is very important that publications include good quality photographs of the natural enemies (Van Driesche and Ferro 1989; Van Driesche et al. 1989b). Also helpful are drawings illustrating life cycles, methods of attack on the pest, and management techniques to conserve the natural enemies.

Biology of Natural Enemies

Building on the recognition skills described above, farmers need to be informed about the biologies of the various natural enemies. Using the most important natural enemies in the crop, life cycles should be taught and the importance of biological features explained. For example, to effectively conserve overwintering parasitoids, farmers need to be able to see parasitoid cocoons on crop foliage or see mummified aphids, and recognize them for what they are and understand how they fit into the natural enemies' life cycles. Such biological features as host feeding by parasitoids, the effect of host size on natural enemy sex ratio, carabid species that climb into foliage to feed versus ones that stay strictly on the ground, and parasitoids that tunnel in manure to find fly pupae versus ones that attack fly pupae only on the surface of the manure pile, all are examples of types of biological features that farmers may need to understand to be able to use natural enemies effectively.

Sampling Natural Enemies

Ultimately, use of natural enemies requires sampling of natural enemies throughout the growing season to determine whether levels of natural enemies are sufficient to provide effective control (see Chapter 13). In addition to monitoring the densities of the principal crop pests, the densities of key natural enemies must also be determined. In apples, for example, counts of mites on foliage include counts of both damaging herbivorous mites and beneficial predacious mites. Ratios of these two classes of mites can be used to forecast the trend in mite populations over a short time span and to decide whether biological control can be relied on to control mites or if chemical controls will be needed (Nyrop 1988). In the same manner, parasitism of eggs of moth pests in such crops as tomatoes can be monitored by determining ratios of healthy (white) and parasitized (black) eggs on foliage (Hoffmann et al. 1991).

Adult natural enemies may also be monitored in visual, odor, or other kinds of traps. Parasitoids of some scales, for example, are attracted to the pheromones of their scale hosts (Grout and Richards 1991). Sticky-traps, baited with pheromones, catch adult parasitoids and counts of these can provide information on the timing and abundance of parasitoids in individual orchards.

Additional information on the effects of parasitoids may be obtained by sampling various life stages of the pest and rearing or dissecting the samples to determine the level of parasitism. Techniques for monitoring natural enemies are currently available only in a few instances and will need to be developed in others before this approach can be widely implemented.

Basing Pest Management Decisions on Natural Enemy Monitoring

Information on natural enemy biology and abundance is integrated into pest management systems through decision-making rules that allow farmers to base pest control decisions on counts of the natural enemies and the pest, as well as the crop growth stage. Predator to prey ratios of mites or white to black (unparasitized to parasitized) ratios of pest moth eggs, once counted, must be interpreted by reference to values that imply effective biological control or inadequate control. Levels of parasitism seen in samples of pests can be used to modify treatment or action thresholds. Treatment thresholds for *Phyllonorycter crataegella*, for example, can be raised as parasitism increases in apple orchards (Van Driesche et al. 1994).

Action thresholds based on both pest and natural enemy numbers are available in very few cases. To develop pest management programs based on biological control agents will require applied studies of natural enemy action under local crop conditions. This effort is equal to or larger than that of defining the damage levels for the pests themselves. However, without this information, extension agents will lack an adequate basis for recommending to growers which forms of biological control can provide reliable, effective pest control. If growers sense that biological control methods are unreliable, they are likely to continue using the chemical pesticides to which they are accustomed to minimize potential crop losses. Extension agents will need to work closely with crop research specialists to carry out studies to evaluate biological control methods. These are likely to have to be repeated in each major climate zone, to a much greater degree than has been true for testing of chemical pesticides.

TRAINING MEDIA

All traditional extension methods can be used to demonstrate and explain biological controls to growers, the public, and agricultural and environmental policy makers. These include: on-farm demonstration plots; talks; written educational materials; slide sets and video tapes; and "hands on" demonstrations of live natural enemies, especially where microscopes can be provided for a close look.

On-farm demonstrations are the hardest to organize, yet the most persuasive. Crop fields where farmers can inspect plots with and without some form of biological control agent or management and see the difference in terms of pest numbers, damage, and yield do more to convince farmers of a method's effectiveness than any other approach. Rice plots in Indonesia grown with and without pesticides, for example, demonstrated to rice farmers the control of rice brown planthopper, *Nilaparvata lugens*, by native predators and the lack of need for pesticides. Demonstration plots, however, are difficult to arrange, must be replicated in each zone where the information is to be communicated, and may be feasible only in certain seasons. Also, some forms of biological control may be hard to demonstrate at a single location, for example, the value of new natural enemies that have been recently introduced (unless with and without plots can be created by some form of caging or exclusion).

To supplement demonstration plots and to reach people who are not farmers, other forms of communication are needed. The most widely used methods will be talks and written materials. These can be either technical materials aimed at farmers, or materials designed to teach concepts of biological control to the public, environmental groups, and political decision makers. Talks will be most convincing if they are given by people who have actually done some of the work they are presenting. Extension agents who are participants in biological control projects are better able to present talks on such efforts because of their first-hand knowledge. To be useful, talks must have important, new content that provides farmers with information that they can apply to their pest control problems. High quality slides or video presentations about natural enemies help audiences visualize the organisms discussed.

Talks should also be given on the results of biological control projects to groups likely to be influential in determining public attitude about biological control in general and about public funding for new projects. Specific groups that should be considered include gardening and botanical clubs, environmental groups, and politicians. Environmental groups are important because they can either endorse or oppose biological control efforts, particularly importation of new species of natural enemies, depending on their views. It is important that the benefits of biological control to nature conservation be emphasized to environmental groups, both in terms of reduced pesticide use that results from the adoption of biological controls for pests of resources (farm products, rangeland, forestry), and in terms of the use of biological control against environmental pests such as aggressive adventive plants and other organisms that damage conservation interests by overrunning natural ecosystems or attacking rare or important species. Talks on biological control projects in a local region, state or country should be given to politicians who make the public policies that affect biological control (subsidies for pesticides, construction of biological control quarantine laboratories) and who provide funding for biological control research and implementation. Talks should be presented when and where they will be most convenient for the intended audience of politicians to attend and should be summarized in written form and provided in advance to participants.

Written information on biological control methods needs to be current, detailed, and illustrated with photographs of both pests and key natural enemies. Until recently, extension materials in the United States on pest control said very little about biological control. Typically, bulletins consisted primarily of descriptions of the pests, their damage, and chemicals for their control, with a few lines on one of the interior pages devoted to biological control. The message from such bulletins, in spite of statements about the importance of conserving natural enemies, was that biological controls were not serious forms of pest control and that chemicals should be used instead. Materials are needed that provide details on the recognition, importance, and, when available, monitoring of key natural enemies in crops and how to use this information to make pest management decisions. Biological control bulletins need to describe practices farmers can employ to conserve and enhance the action of the important natural enemies in specific crops. Such information is highly crop- and site-specific and must be developed locally. Bulletins describing introduction programs are also needed to promote recognition of such projects by farmers.

Examples of bulletins on biological control include the *Integrated Pest Management Manual* series from California; the *Using Biological Control in Massachusetts* series (Fig. 19.1a) (including fact sheets on mites and leaf miners of apples, cole crop Lepidoptera, and the Colorado potato beetle); Raupp et al. (1993) on biological controls for pests of woody ornamentals; Hoffmann and Frodsham (1993) on natural enemies of vegetable pests; Frank and Slosser (1991) on biological control of pests of wheat, cotton, and alfalfa in Texas; Henderson and Raworth (1990) on beneficial insects in strawberry and raspberry (*Rubus idaeus* Linnaeus) crops in Canada; and a general bulletin on biological control agents from Illinois by Henn and Weinzierl (1990).

TRAINING EXTENSION AGENTS

To be able to provide the previously described information to growers and other audiences, extension agents must be well-trained in biological control. To maximize the ability of extension personnel to understand biological control and train growers in its use, a national or provincial level system of extension biological control is needed (see Van Driesche 1989b). This system should perform five functions.

(1) **Provide leadership**. Many countries and provinces have decentralized systems for development and delivery of biological control programs. This system has the advantage of permitting researchers considerable freedom to develop projects or methods they believe are likely to be most effective. Some level of coordination, however, is also valuable. If within a country or province there is a biological control leadership position (biological control coordinator), extension personnel have someone with whom they can discuss their needs and interests concerning biological control. A biological control coordinator can be responsible, among other things, to organize the training and continued education of extension agents.

In addition, this coordinator can form links to government agencies to keep abreast of new developments in law, regulation, or policy that may affect biological control locally and help make government officials at the national or provincial levels aware of regional needs or interests concerning biological control. Van Driesche (1989b) provides a broader discussion on the concept of a biological control coordinator as a leadership position that

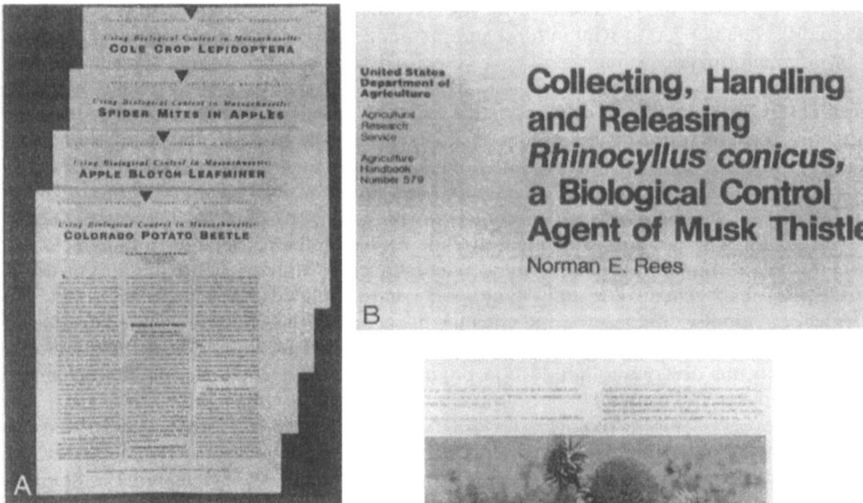

Figure 19.1. Fact sheets on biological controls for particular crops or pests are important tools to help farmers understand how to use natural enemies. (A) fact sheets of Massachusetts Extension Service on using biological control in cole crops, apple and potato; (B,C) a United States Department of Agriculture pamphlet on biological control of a pest thistle.

can be helpful to a country or province in developing a strong program of biological control.

(2) **Train Extension Agents in Biological Control**. For biological controls to become a major focus of extension pest control recommendations to farmers, extension agents will have to first educate themselves about the subject. This requires some central structure to provide the opportunities for extension personnel to acquire this training. Where a leadership position exists, the biological control coordinator, in conjunction with crop specialists, can organize training days for extension agents. Training can be a series of short events or a longer program (a short course) that might last a week or more. This training would cover all the material described in this chapter that growers need to know, from biological control concepts to natural enemy recognition, biology, monitoring, and incorporation of natural enemy levels into pest management decision-making. Extension agents would be taught

Figure 19.2. Training meetings provide opportunities for extension agents to study biological control concepts and applications for their crops.

by specialists in biological control through lectures, laboratory exercises, and field visits (Fig. 19.2).

(3) **Provide Regular Information Updates**. Once extension personnel have received their initial training, they will continue to require new information. They will need to receive regular updates on new developments, such as new natural enemy introductions or research findings, new products, news of such events as training days or farm demonstration days, and legal developments affecting biological control. One method to fill the need is to have a specialized newsletter, covering all crops and pests, that draws together all the news about biological control pertinent to the region, state, or country in a form that is appropriate for extension agent use and is sent to a specified set of people on a regular basis. In Massachusetts, for example, the newsletter *Biocontrol Flash* serves this purpose. In Canada, at the national level, the newsletter *Biocontrol News* is a valuable source of such information. In the United States, an electronic bulletin board on biological control news has been organized by the National Biological Control Institute within USDA/APHIS. Internationally, newsletters of regional sections of IOBC (International Organization for Biological Control) or national biological control programs (as in India) provide a means of exchanging information. Additionally, articles on biological control topics of local interest could be incorporated in other, crop-oriented newsletters or magazines.

If a biological control coordinator or other leadership position exists, this individual can be responsible to prepare information on biological control for extension personnel. The coordinator can serve as a link between extension agents and academic sources of information on biological control. This information is often not easily available to extension agents because they are commonly located at a distance from universities with the necessary libraries. The coordinator can read widely and pass on condensed versions of relevant material in the regional extension biological control newsletter. Abstracting journals provide one source of such information for the coordinator to follow. Especially helpful is the Commonwealth Agricultural Bureaux's publication *Biocontrol News and Information*, published quarterly, which covers the biological control literature worldwide, including items such as conference proceedings and research station annual reports, items often not carried by libraries.

In addition to this type of information exchange, extension agents need access to biological control specialists who can help with specific questions they may have. Again, this source could be the biological control coordinator in areas that have such a position. The coordinator can maintain a library of books and research articles on biological control and develop a wide set of contacts with biological control workers and policy makers at other levels of government and in other countries, all of which will help the coordinator be better able to answer agents' questions.

(4) **Provide Materials to Train Growers.** To train growers effectively in the use of biological controls, extension agents need training materials, including fact sheets, pamphlets, and slide sets on the various biological control subjects, pests, or crops. Extension agents will benefit from access to prepared materials that they can use in programs for growers. Some of these materials can be developed by crop pest management specialists, with biological control information integrated into a description of the larger crop management program. Other materials can best be prepared by biological control specialists. These might include pamphlets on biological control concepts, recognition of types of natural enemies, and descriptions of specific biological control projects. In addition, extension agents should be provided with visual training aids, including slides, slide and tape sets, and video tapes on biological control concepts and projects. These written and visual materials, combined with attendance at training events and participation in biological control projects, will give agents the materials and experience needed to train growers.

(5) **Provide Opportunities For Extension Agents to Participate in Biological Control Research.** Extension agents can teach best what they have done, at least in part, with their own hands and have seen with their own eyes. Agents' abilities to train growers in use of biological controls will, therefore, be stronger if they have opportunities to participate in biological control projects. Making these opportunities available involves creating the attitude that such participation has benefits both for the researcher and the extension agent, and then developing institutional mechanisms to allow these joint efforts to be carried out.

Most biological control projects start out primarily as research and end up primarily as extension (Fig. 19.3). Even so, there are extension aspects in even the earliest phases of work, and agent participation in all of these steps is an excellent way to improve the understanding of agents of the biological control research process and the underlying concepts. Agents, for example, can help researchers and policy makers choose good targets for new projects through their knowledge of pest concerns on local farms. They can, in some cases, help collect natural enemies to be moved into the region (if the natural enemy collection area is not too far away or extension travel funds are available), and they can contribute ideas for designing experiments on farming methods to conserve natural

Research

1. Target Identification
2. Foreign Exploration
3. Introduction and Establishment
4. Intensive Evaluation
5. Widespread Redistribution
6. Extensive Evaluation
7. Grower Training
8. General Publicity

Extension

Figure 19.3. Development of a natural enemy introduction biological control project, from initial problem definition and research through solution discovery, evaluation, and grower training, can involve extension agents at all phases. Direct participation of extension staff provides valuable, increased understanding and insight to both extension agents and researchers.

enemies. They can help with the colonization and redistribution of new agents, especially through finding cooperative growers who are willing to have their farms used as release sites and identifying growers interested in having their farms used as test locations for on-farm biological control trials. Extension agents can be effective in explaining new biological control developments or introductions to growers and the public, coordinating the personal contacts required in projects, and managing the necessary news releases. Through such participation, agents receive highly valuable training on biological control, become familiar with the biologies of the natural enemies involved, and become aware of the problems that must be overcome to make biological control agents or management methods successful. Researchers benefit from agents' ideas, time and effort, grower contacts, and communication skills.

TEACHING BIOLOGICAL CONTROL TO POLITICIANS

For a national program of biological control to prosper, capable biological control researchers, well-trained extension agents, and willing growers are not enough. Politicians who make government policies on agriculture, environment, and trade must also understand at least some basic concepts about biological control, pest management, and pesticides. If they do not, governments may adopt policies that actively discourage the use of biological control. Areas of critical importance include policies on regulation of and subsidies for pesticides, nonnative organism quarantine, biodiversity protection, food grading standards and international trade in agricultural products, financial support for integrated and biological pest controls, and creation and funding of biological control research institutions and quarantine facilities for natural enemy introductions (see Chapter 20).

Given the critical role of politicians in shaping the possibilities to conduct biological control in any given country, it is important that extension agents and biological control researchers develop methods to communicate with political decision-makers about the importance of and methods for biological control. The focus of such communication should be not on biological details behind particular biological controls, but rather, on how biological control relates to the concerns that politicians feel are important to voters or for the economic development or protection of their country. These issues are likely to include: stability of the level of food production; ability to export agricultural products to markets that impose specific, usually high, quality standards; moderating or reducing harm caused by pesticides to farmer health; lowering pesticide residues in food eaten by consumers; protecting water supplies; reducing damage to nature caused by pesticides; and protecting biodiversity. These issues are arranged in what often is the historical progression of awareness of these issues in a given country, with basic sustenance coming first, followed by economic issues, then issues of human health, then food and water purity, and finally protecting nature.

Efforts, then, to reach politicians with messages about the value of biological control should seek to relate biological control's achievements to these political concerns. Wherever possible, the economic savings from biological control should be emphasized, as well as improvements in human health and the environment. Messages should consist of key statements of fact, descriptions of current circumstances, needed actions, and references to sources of further information.

Various techniques can be used to reach political audiences:

- Place key political figures on mailing lists for biological control newsletters. Even if only some are read, this is an easy way to get some messages across.

- Hold annual meetings in the capital expressly for key politicians. Precede the meeting with mailings stressing highlights of the event to encourage attendance. Make the meetings short and conveniently located. Provide written summaries of the information presented because many politicians will be unable to attend but will send representatives.
- Develop well-coordinated messages and repeat the same messages wherever a forum is available to express them.
- Use grower organizations. Communicate key messages clearly to the heads of such organizations and rely on them to pass the messages on at appropriate times.
- Use mass media to focus attention on biological control events. Remember to tie the particular event clearly and strongly with the larger issues described in the previous paragraph. Stress the need for future action.

COMMUNICATING WITH ENVIRONMENTAL GROUPS

Environmental groups are the third audience that extension personnel and biological control researchers must reach. Biological control introductions may be linked in the mind of many environmentalists with the problems caused by adventive animals and plants whose origins are unrelated to biological control (goats, rats, crop or ornamental plants that become weeds, sport or hobby fish). The effect of such adventive species on local ecosystems and rare species is an important environmental concern that will receive increasing attention as efforts to preserve biodiversity increase.

To forge positive, mutually supportive relations with the environmental community, biological control practitioners must inform themselves about the risks associated with biological control introductions and other activities (see Safety sections in Chapters 8, 9, 10 and 11 and also see Chapter 21). It is important both to recognize that some mistakes were made in the past, but also that on the whole biological control has an excellent safety record.

Every opportunity needs to be taken to stress the positive applications of biological control for nature conservation (see Chapter 21). Whenever a new biological control project has conservation benefits, special efforts should be made to communicate them to conservation groups. Talks should be given to meetings of environmental groups about biological control. These talks can cover concepts of biological control and details of individual projects. Benefits worthy of emphasis might include pesticide reductions (from projects aimed at agricultural, range or forestry pests) and suppression of aggressive plants or other adventive organisms damaging to native ecosystems or individual native species. When projects are carried out with conservation benefits, it is important to get environmental groups to participate in the work. Control of introduced plants that have become weeds in native tropical forests of Hawaii is an excellent example of such efforts.

CHAPTER
20

GOVERNMENTAL POLICY AND BIOLOGICAL CONTROL

INTRODUCTION

Biological control embraces activities undertaken by farmers, businesses, researchers, and government employees. Governmental policies in several areas influence the breadth, efficiency, and effectiveness of the application of biological control to the pest problems of a nation or region. This chapter reviews those policies that directly affect biological control, as well as considers the indirect effects of such other policies as those on the use of pesticides, pest eradication, crop exportation standards, and cosmetic grade standards. Specific areas addressed include: (1) pesticide use, including applicator training and licensing, product safety testing, subsidies, taxes, and commercial development of pathogens as pesticides; (2) quarantine and eradication; (3) agricultural development; (4) pest management; (5) biological control, including introduction of new species of natural enemies; and (6) use of genetically-modified organisms.

PESTICIDE POLICIES

Chemical pesticides are the principal form of pest control against which biological control must compete for farmer adoption and governmental resources. Policies concerning pesticides should be clearly thought out and supportive of biological control (see Higley et al. [1992] for a broad discussion of pesticide policy issues and Dahlberg [1993] for comments on policies that encourage increased use of pesticides in the United States). The availability and cost of pesticides to the user are affected by a series of governmental decisions, including those concerning product safety, regulation of pesticide users, registration fees and taxes, government subsidies for pesticide prices, and regulation of pesticides containing pathogens (a form of biological control).

Product Safety

For a pesticide to be used safely, extensive information must be available about its chemistry, toxicology, persistence, degradation, and environmental effects. Governments typically require that companies wishing to sell pesticides submit such information as a prerequisite for permission to sell their products (a process termed "product registration"). This procedure gives governments a mechanism to exclude products that do not meet registration standards. Registration standards typically are designed to ensure safety to humans and selected components of the environment. Such standards do not, however, typically require that products be safe for natural enemies of pests in the crops to which the pesticides are going to be applied. An exception to this exists in Europe where members of the European Community require that any new pesticide be tested against a prescribed set of eight species of natural enemies, identified on the basis of importance in particular crops (Flick 1987).

Funds to develop information needed for product registration may come either from the business that wishes to obtain permission to market a chemical as a pesticide or from tax money used to support research on pesticides at public institutions. Once a pesticide has been registered in Europe, the United States, or Japan, other countries are likely to accept the product as safe for use based on experience in those countries. In some instances, however, local climate, nontarget organisms, or other factors may vary enough to dictate the need for additional local testing. Or, a product may simply be allowed to be used, with minimal consideration of safety issues. In such instances, problems such as high levels of pesticide residues in crops, water, or wildlife may be recognized later, forcing the expenditure of public funds to address the problems. All such public expenditures made to determine facts about pesticides, including how to use them effectively or to determine their environmental effects, constitute a subsidy for pesticides.

Governmental Regulation of Pesticide Users

Another subsidy commonly extended to pesticides is the cost of regulation of the actual use of products by applicators and the correction of pesticide-caused problems. Government legal powers and financial resources are often employed to control how pesticides are used and who is entitled to apply them. Government activities may include testing applicants for licenses to apply pesticides, associated record keeping, hiring technical personnel to provide assistance during pesticide emergencies such as spills and fires, sampling agricultural produce to measure pesticide levels for enforcement of residue limits, and investigation of damage to crops, wildlife, drinking water, or other resources. The costs of these activities are substantial and are typically not borne by the companies that sell the pesticides but rather are paid by the public through tax revenues used to support government regulatory institutions that respond to these concerns. The choice of whether or not to impose taxes on pesticides sufficient to cover these costs is an important public policy issue.

Registration Fees and Pesticide Taxes

Fees are usually charged by governments to register pesticide products. Setting the amount of these fees is an important public policy, because the size of the fee is determined by what costs the fee is seen as offsetting. Minimal fees may be imposed if the fee is envisioned as covering only the cost of administering the record keeping activities of registration. Larger product registration fees may be necessary if such revenues are to cover adequately the full range of costs (studies on pesticide effectiveness and environmental fate, applicator training, pollution clean-up, investigations of misuse, and regulation enforcement) created by pesticide use.

Another means of collecting revenue to cover these social costs of pesticide use is to impose a tax per unit weight of pesticide production or use. The advantage of this system is that revenues increase in direct proportion to the volume of pesticide use and thus are more likely to be adequate to meet the size of the social problems caused by pesticide use.

Additionally, a portion of such taxes on pesticides may be reserved for research on and implementation of alternative forms of pest control, such as biological control.

Government Subsidy of Pesticide Prices for Consumers

In addition to the indirect subsidies of pesticide costs discussed above, some countries directly subsidize the purchase price of pesticides paid by farmers. Such subsidies have occasionally amounted to as much as 90% of the cost of the pesticides. Such pesticide subsidies have been part of some countries' agricultural development policies, based on the mistaken notion that pesticides are simply another form of agricultural input and that increasing usage increases crop yield in a simple, direct, and unlimited fashion. In fact, pesticide use bears a much more complex relationship to yield and may cause yields to decline if pesticide use stimulates rapid pest resurgence or secondary pest outbreaks (see Chapter 7). Pesticide subsidies distort the decision-making processes of farmers by undervaluing the cost of pesticides, causing pesticide use to occur without sufficient economic justification. Recognition of this fact has prompted reversal of such policies in some countries (Oka 1991).

Regulation of Pesticides Containing Pathogens

Pest control products that can potentially reduce the need for nonselective chemical pesticides include those containing selective pathogenic organisms, either of plants or arthropods. Such products must, however, compete economically with chemical pesticides and often suffer in this regard from smaller markets (due to greater product specificity) and higher production costs (see Chapter 11). Given these constraints, governments can foster use of such products by reducing the cost of registration and by limiting the scope of safety testing to those tests appropriate for pathogens and not imposing the full range of tests required of synthetic chemical pesticides. Betz et al. (1990) lists the requirements currently in force for registering products containing pathogenic microbes in the United States and these are discussed in Chapter 11.

QUARANTINE AND ERADICATION POLICIES

Invasions of adventive plants, arthropods, and other organisms are a major source of new pests (U.S. Congress, OTA 1993). Biological control efforts seek to suppress many such pests. Prevention of invasion through promulgation of quarantine laws and their effective enforcement complements biological control by reducing the rate of creation of new pest problems. To be effective, quarantine systems need to have broad scope as to the kinds of organisms that are covered and must be well implemented (Fig. 20.1). Most commonly, quarantine laws are intended only to control the spread of weeds, diseases of domestic animals, and arthropods and diseases that attack economic plants. Many countries have quarantine laws covering such organisms.

Organisms that are less well regulated in many countries are those that have both desirable traits and the potential to become pests. Some species valued as ornamental plants or as forage grasses, for example, have potential to spread aggressively and become weeds in some habitats (see Chapter 21 for examples). Proposals in Australia are currently under consideration to further restrict the importation of some such species pending review of their potential to become pests (Hazard 1988). Similarly, the kinds of nonindigenous animals sold as pets include a wide range of creatures, some of which have been released into the wild and some have

Figure 20.1. Inspections at borders and ports of entry are intended to intercept potentially harmful adventive species. (Photograph courtesy of USDA-APHIS.)

formed breeding populations. For example, over 45 species of introduced fishes have become established in the United States as a consequence of releases or escapes of fish sold or reared for the tropical fish trade (Courtenay 1993). These and other introduced fishes have had significant harmful effects on native fishes (Taylor et al. 1984).

To effectively slow the invasion of new pests, quarantine laws should apply broadly to all nonindigenous organisms, as well as to such bulk materials as soil, ballast water in ships, and raw logs, which are likely to transport pests. Invasion of North American fresh waters by the pest mollusc *Dreissena polymorpha* Pallas (zebra mussel) has prompted adoption of rules controlling the salinity of ballast waters of ships entering the Great Lakes through the St. Lawrence Seaway (Canada, United States) in an effort to prevent further introductions by this route (Garton et al. 1993).

Given appropriately broad taxonomic coverage, effective quarantine depends on the degree of physical isolation of a country, the number and kinds of ports of entry, and the establishment of effective inspection services at these ports of entry to detect and exclude pests. Where ports of entry are limited, as for example on islands, quarantine can be an effective force in reducing the rate of invasion of immigrant species. In areas with more numerous points of entry and higher volumes of movement of people and goods, effective quarantines require proportionately greater effort. However, even carefully planned and executed detection methods may not intercept every pest. As a result, pest invasions have been associated with trade in plants and will likely continue to be in the future (Minkenberg 1988).

Once pests have eluded quarantine barriers and established breeding populations in a new country, governments can either attempt to eradicate the pest or begin the process of instituting biological control. Feasibility of eradication will depend on the extent of the infestation, the biology of the organisms involved, and the resources available. An effective decision between eradication and biological control will depend on the ability to assess realistically how early an invading species has been detected, how widespread the infestation is, and what technologies are available to implement an eradication program. If eradication campaigns are mounted against a well established species or one occupying a wide geographic area, such programs' pesticide use may be damaging to the biological control agents of other pest species in the region, as well as consuming resources that may be better spent on initiating a program of biological control against the pest. The effects of eradication programs on biological control agents of other pests should be assessed in making decisions about when, where, and how such eradication efforts should be conducted (Hoelmer and Dahlsten 1993).

Another aspect of public quarantine policy that has a decisive impact on the feasibility of applying biological control is the policy governing the importation of species of natural enemies into a country for the suppression of pests and the availability of facilities to make such introductions safely. Laws and regulations are needed to provide clear guidance as to what introductions can be made legally and to define procedures to resolve any conflicts of interests that may arise. Procedures and facilities must be available to support the safe introductions of needed natural enemies. A general policy statement concerning such importations has been developed for the international community by the Food and Agriculture Organization of the United Nations (Anon. 1992).

AGRICULTURAL DEVELOPMENT POLICIES

National or regional agricultural development policies can also affect pest populations and influence opportunities for use of biological controls in some crops. Such policies often affect pest control in a variety of ways, such as creating large concentrations of crops in specific areas, affecting the timing or manner of crop production, or establishing grading standards that determine the level of pest control required for access to a market.

In cotton production, for example, government-supplied irrigation may stimulate the development of large concentrations of the crop. Participation in such enterprises may require that farmers adhere to policies on such issues as crop variety and planting date, dates for destruction of crop residues, restrictions on movement of planting materials and farm machinery from outside the area into the planting zone, and mandatory pesticide applications, determined either by prescription or monitoring. Such practices can potentially affect biological control by modifying the risk of invasions of immigrant pests, or by influencing the degree to which existing natural enemies are conserved. Planning for such projects should include specific consideration of how project management policies will affect the balance between chemical and biological pest control. In some countries or provinces, laws for the control of noxious weeds mandate herbicidal controls for weeds where encountered. For weeds that are targets of biological control projects, such mandates become impediments and need modification to permit weeds to go untreated in zones where biological control agents exist or are being released.

Market standards also strongly affect the degree to which biological controls may be used against pests in some crops, especially fruits and vegetables. Standards can operate at several levels, with various degrees of formality. Informal cosmetic standards probably operate continuously in the marketing of fresh fruits and vegetables at all levels of the sales process, with blemished produce being more difficult to sell, or perhaps completely unmarketable. To some degree, education of consumers and distributors can influence this process by providing information on the significance of various kinds of blemishes or traces of pests on the taste or storage properties of the produce. In some instances blemishes may indeed reduce quality. In other cases, they may have little or no importance and may be acceptable to consumers if the trade-offs between such blemishes and reduced pesticide use are adequately explained.

At a more formal level, large processors, exporters, or government food purity agencies may have explicit grading standards that producers or sellers are required to meet. These restraints may increase the level of pesticide use in some crops significantly. Standards for food in the United States, for example, are issued as "defect action levels" set by the Food and Drug Administration. These standards define the maximum numbers of insects that are legally permitted to be present in various foods. United States levels were set historically to reflect the level of control that was feasible at the time the rules were developed. Many standards set in the 1930s, for example, were changed in later decades as pesticides made it possible to suppress insects such as aphids and leaf miners to lower levels (Pimentel et al. 1993). In some cases, permitted insect levels were lowered as much as 50 to 88% from earlier standards. Such standards sometimes force producers to suppress pests to much lower levels than are needed to protect crop yield or product quality.

Where feasible, grading rules of this sort should be modified to reflect levels attainable by biological control methods and levels of insects actually reducing product quality (in ways other than the mere presence of small insects or mites in a raw product). Product quality can further be protected in the packaging process through redesign of the methods used to clean the raw product prior to packaging. Agencies charged with setting food standards should be mandated to consider the trade-offs between pesticide levels in foods and levels of minor insect contaminants when adopting standards governing insect levels permitted in food. Such standards should encourage reduction in pesticide use rather than force increases in pesticide use.

Standards imposed on export crops can be a significant factor influencing the level of pesticide use in such crops as apples, citrus, mangoes (*Mangifera indica* Linnaeus), and other produce. If exports are destined for countries with similar climates and crops, importing countries may impose standards intended to protect against the risk of importation of known pests that do not occur in the importing country. Risk of importation of apple maggot, *Rhagoletis pomonella*, to areas free of this pest is an example. In such cases, risk may be managed in several different ways, which have very different implications for the balance between chemical and biological control used in the area of production. One strategy is to impose pesticide use requirements on production areas to keep them at or near a pest-free status for the pest species. A variation on this approach incorporates the use of natural enemy liberations in pest reservoirs (for example, fruit trees in urban areas) which are outside the production areas coupled with reduced levels of pesticide use in the crop itself. Use of biological control in reservoir areas can reduce pest immigration into the production zones which must be kept pest-free for access to export markets. This combined strategy of biological control and pesticide use lowers overall pesticide use. Such an approach is employed, for example, against the Caribbean fruit fly, *Anastrepha suspensa* (Loew), in Florida (Baranowski et al. 1993).

These constraints on export produce require such crops be kept as close to pest-free as possible, entailing regular inspection and pesticide use. Ideally, a pest management approach would be applied and the resultant small portion of the crop bearing live pests treated in some fashion before export to reach the zero pest level. If the pest is visible externally, hand sorting of the produce may be feasible. If not, produce may be treated to kill the pests. Methods to achieve this vary with the pest species and the ability of the produce to withstand treatments of various kinds. Potential treatment methods include chemical fumigation, hot water dips, and cold treatment (Jessup 1994).

PEST MANAGEMENT POLICIES

Governmental agencies and other public and private institutions involved in setting policies for agricultural or social development need to consider the impact of such policies on the methods of pest control selected and employed in their countries or projects. The Philippines and Indonesia (Oka 1991), for example, both have national policies on agricultural pest control that seek to promote reliance on nonpesticidal means of control such as plant breeding and biological control, combined in some rational fashion with other controls as needed. Most countries lack explicit national pest management policies, although the matter may be addressed to some degree by individual government agencies.

Similarly, international agencies involved in financing developmental efforts should be concerned with the environmental consequences of the projects they promote. One aspect of such concern is attention to the nature of pest control used in development projects. The World Bank, for example, has such a policy aimed at promoting conservation of existing natural enemies of pests found in the crops affected by development projects they fund.

The objective of governmental policies should be to encourage the development and use of nonchemical pest control methods, in rational combination with limited pesticide use, as a national or institutional goal. Impediments to implementing these policies can then be identified and plans developed to overcome these limitations, whether deficiencies in scientific knowledge, facilities, human resources, or farmer training. To be effective, policy should be adopted at the highest level, together with oversight authority to review policies and practices of other branches of the government or other institutions to ensure that these promote the goals of the overall pest management policy. All issues discussed in this chapter could potentially be matters reviewable with respect to such a policy, including pesticide registration policy, research funding for scientists involved in pest management studies, provision of public sector quarantine facilities for the importation of natural enemies, and training of extension agents and farmers in use of biological control.

BIOLOGICAL CONTROL POLICIES

Within the framework of a national pest management policy, specific policies for the promotion of biological control can be developed (Van Driesche et al. 1990). Four areas are critical for the development of an effective biological control implementation policy and need support at the highest levels within a country: (1) recognition of the importance of biological control in pest management; (2) coordination of government policies on pesticides and other agricultural matters affecting biological control; (3) development of strong leadership for biological control programs, with adequate resources and clear lines of authority; and (4) good planning, communication, and project evaluation.

For governments to foster effective programs of pest management, the fundamental role played by biological control in many pest control systems must be recognized. Underlying many outdoor pest control activities is the control exerted by existing natural enemies. Recognizing the value of this force and the importance of enhancing it and extending it (through natural enemy introductions) to cover new pest species as they invade a region is the ecological basis of pest management.

Government leaders must recognize that biological control policy cannot be effective in isolation, but will be influenced strongly by other government policies such as those concerning pesticides, quarantine and eradication, and agricultural development. Methods must be developed to coordinate these various policies to promote the use of biological control and discourage the use of pesticides.

Once a biological control policy has been developed which identifies the goals for biological control in a country or province, governments must create effective organizations for their implementation. Implementation may include developing centers for biological control research or support for biological control activities within universities, departments of provincial governments, or agricultural experiment stations. Leadership at the national level should exist to address national-level problems (such as biological control introductions against target pests which occupy large, interregional, or multicountry areas, regulation, and national funding of biological control), but will also be needed at the provincial level to identify and implement projects of more regional interest. National and provincial institutions engaged in biological control should be provided with appropriate resources, in terms of numbers and quality of scientific personnel, the level of salary support for scientists and other staff, quarantine laboratories, and other physical facilities.

Basic to the creation of an effective national or provincial biological control program are policies that establish a structure for planning, communication, and project evaluation, as these provide guidance to biological control efforts over extended periods of time and across multiple biological control projects. A basic part of such policies is the establishment of methods to identify appropriate biological control targets. Depending on the size of area involved, reviews may be conducted nationally, or for large countries, regionally. Mechanisms must be established to collect information, especially the published scientific literature, about the pests of importance and their natural enemies. In addition to this biological information, additional information about the economic losses caused by each candidate species is needed. The process of target selection should use defined procedures, such as Van Driesche and Carey (1987), McClay (1989), or Barbosa and Segarra-Carmona (1993), rather than opinion or public concern about particular pests, which may or may not be appropriate targets for biological control. Procedures should exist for groups with differing ideas about the pest status of proposed target species to present their concerns so that the public interest can be determined (for Australia, see Cullen and Delfosse 1985).

Once accomplished, such a pest-by-pest, crop-by-crop review leads to a written plan which formally identifies important targets for biological control projects in an area, against which progress can be measured (for Massachusetts, see Van Driesche and Carey 1987). The list of target species should be reviewed periodically to include consideration of new pest invasions or other developments. (Chapter 8 discusses the process of selecting targets for biological control programs).

Once a national or provincial program of biological control has been developed and its goals established, a communication and coordination system must be organized among those working in the various parts of the system. Among the many benefits that derive from scientist-to-scientist communication are exchange of information (as for example on rearing methods, new projects, or newly imported species of natural enemies), exchange of natural enemies either from laboratory colonies or field collections, and the development of transfer projects, in which a successful project achieved in one area is copied in other locations at reduced cost.

Methods of communication useful for promoting such contact among biological control workers include newsletters, annual meetings, and participation in international biological control organizations. Newsletters aimed at agricultural extension personnel can transmit news of biological control activities to the local level and, ultimately, to farmers. At the international level, major organizations involved in the exchange of information on biological control include the International Organization for Biological Control (IOBC) and the International Institute for Biological Control (IIBC). Among the activities of IOBC are the promotion of scientist "working groups" on biological control of specific pests, or on specific issues, such as developing standard methods for testing the effects of pesticides on natural enemies. IIBC both collects and publishes information about biological control and conducts implementation projects on a contract basis. *Biocontrol News and Information* is an IIBC publication that collects the world literature on biological control and publishes abstracts of new work on a quarterly basis. In the United States, the National Biological Control Institute (USDA-APHIS) maintains an electronic bulletin board on biological control topics. Additionally, some countries maintain centralized national records on releases and recoveries of natural enemies (Coulson 1992).

Program evaluation is an additional area where clear policies are needed in order to give long-term guidance to national programs of biological control. Evaluation is needed at all levels, from the evaluation of the biological and economic success of individual projects to the periodic evaluation of the national program as a whole (see Chapter 13). National policy on evaluation should include at a minimum that all projects will be evaluated, should give some guidance as to what type of information must be included in such evaluations, and should indicate how the results of project evaluations will be documented (Coulson 1992), published, and used to provide ongoing revision of the goals and methods of the national program as a whole.

USE OF GENETICALLY-MODIFIED ORGANISMS

Genetically-modified microorganisms are likely to play an important role in the future of biological control. Government policy on the testing, registration, and use of these organisms will influence the extent and speed of the development of such agents. Central to these policies are the development of concepts and procedures for assessment of risks from recombinant microorganisms. Studies of nonrecombinant agents in current use may be helpful in formulating such policies (Fuxa 1989; Wilson and Lindow 1993) (See Chapter 11). Similar issues arise with genetically-modified arthropods or other multicellular species (Hoy 1992).

CHAPTER

21

Biological Control in Support of Nature Conservation

INTRODUCTION

Natural ecosystems, biological reserves, and other undeveloped land and water of conservation importance are regularly invaded by nonindigenous plant and animal species. Such invasions reflect, in part, the level of human activity in the region, with the consequent human-assisted movement of adventive species by commerce (Fig. 1.1). Some adventive species threaten the continued existence of native flora or fauna, by competition or direct attack. Other adventive species alter basic properties of the ecosystems they invade, rendering them less suitable for the continued existence of broad sets of native species. Efforts to combat such adventive species by chemical or mechanical means are often unsatisfactory, the former because of the risk of chemical pollution, damage to nontarget species, and cost, and the latter because of cost and the difficulty of applying mechanical remedies to any large region. Biological control, through the introduction of natural enemies specialized to attack the undesired adventive species, offers a method to solve this problem in many cases. This chapter discusses the principles behind the use of biological control for conservation of natural areas, the past achievements of such efforts, and future prospects for additional applications. Means to ensure that biological control agents themselves do not become environmental pests are discussed in the safety sections in Chapters 8, 9, 10, and 11.

CONCEPT OF RESTORATIVE ECOLOGY THROUGH BIOLOGICAL CONTROL

Nature reserves are areas of land or water set aside with the intention of permanently preserving certain kinds of ecosystems or critical habitat for individual rare or significant species. Lands and waters not formally designated for nature conservation (private lands left in a relatively natural state, government or communal lands either not, or only lightly, exploited) can also be important conservation areas.

One of the possible threats to the continued ability of such areas to provide refuge to their resident species or to remain ecosystems of their original composition is the invasion and colonization of such areas by nonnative plants and animals (Cock 1985; Howarth 1985; Ashton et al. 1986; Breytenbach 1986; Macdonald 1988; Case and Bolger 1991; de Polania and Wilches 1992; McKnight 1993).

Virtually all areas are subject to repeated contact with immigrating or introduced plants and animals. Range extensions of species from contiguous adjacent areas, however, are qualitatively different in their potential impact from long distance invasions by species from distant, disjunct areas. In the latter case, immigrant species may arrive in the absence of their more specialized natural enemies and consequently be able to achieve vastly greater densities in the new region. In contrast, species that arrive in new areas through continuous range expansion will often bring with them the natural enemies that affect them in their original home. Such natural enemies, in the normal course of exploiting their host, expand their ranges into new areas concurrent with the range expansion of their hosts. Parasitoids attacking the gypsy moth (*Lymantria dispar*), for example, have moved (with a time lag) south and west in the United States with their host as it has expanded its range (Reardon and Podgwaite 1976). This does not happen, however, in cases where disjunct gypsy moth populations are created by long distant transport (for example, when egg masses are moved by attachment to vehicles) to zones beyond the dispersal powers of its natural enemies. In such cases, natural enemies of the transported stage may or may not accompany their host to the new location, but specialized natural enemies of other life stages (for gypsy moth, larval and pupal parasitoids and predators) will be left behind.

Biological control is a powerful method to control adventive species that invade areas disjunct from their main range and leave behind their natural enemies. We use the term **environmental pest** to describe adventive species which damage conservation interests in that the motivation for their control is primarily the protection of nature and natural systems, rather than protection of agriculture or forestry (Van Driesche 1994). Of the various kinds of organisms that might invade native systems, biological control has been applied most often to vascular plants and nonmarine arthropods. In general, mammals, birds, and other vertebrates have been less frequently targeted. Possibilities also exist for biological control of invertebrates other than arthropods and for nonvascular plants, but to date relatively few attempts have been made to control adventive pests in these groups through natural enemy introduction.

The concept of ecological restoration through the use of biological control begins with the fact that adventive species often invade native systems. Some such species damage individual native species or key ecosystem characteristics to such an extent that remedial action is desirable. Adventive species, however, do not invariably harm native systems. Many become integrated into the local fauna and flora simply as additional species that are present but are not damaging or excessively numerous. In part, this outcome is the result of attack by local species of natural enemies on the new immigrant species as, for example, limitations placed on the giant African land snail, *Achatina fulica*, on Christmas Island by predation from red land crabs, *Gecarcoidea natalis* Pocock (Lake and O'Dowd 1991), or the parasitism of the black Portuguese milliped, *Ommatoiulus moreletii*, in Australia by a native rhabditid nematode, *Rhabditis necromena* Sudhaus and Schulte (McKillup et al. 1988, McKillup and Bailey 1990). The effects of such local natural enemies may also be dependent on other ecological factors such as the degree of habitat disturbance (Lake and O'Dowd 1991). Native natural enemies, however, are commonly ineffective in suppressing population increases of adventive species. In some cases, adventive species are held in check by competitive pressures of native organisms in the same trophic level. Many introduced plants, for example, fail to become pests because of competition by native vegetation. In cases, however, where neither local natural enemies nor competition from native species are effective, adventive species may become serious environmental pests.

Where adventive species have become environmental pests, restoration of the native system, or reduction of competition or predation pressure exerted by adventive organisms on key native species, may be attempted through the introduction of the missing natural enemies of the adventive species (Moran et al. 1986). Successful biological control in these situations depends on four conditions: (1) the adventive species must have natural enemies where it came from that can contribute to suppressing its numbers; (2) these natural enemies must be missing in the recently invaded area; (3) these natural enemies must not be harmful to other important species in the target release areas (see the section on safety in Chapter 8); and (4) the natural enemies must reduce the survival or reproduction (or for plants, competitiveness) of the target adventive species and thus contribute to lowering its density in the target area (see Chapters 13 and 17). The first three conditions need to be assessed before natural enemies are introduced, and the last is determined in the course of subsequent field studies to evaluate the consequences of introduction of one or more natural enemies.

FACTORS PROMOTING ADVENTIVE SPECIES INVASIONS

Before addressing the effects of adventive species on conservation systems and methods to reduce such damage through biological control, processes facilitating species invasion and subsequent establishment need to be considered, including: human movement of plants and goods; habitat disturbance; altering frequency or intensity of events such as fires or floods; and prior invasions of other non-native species.

Human Movement of Plants and Goods

Movement of plant and animal populations between disjunct portions of habitat is a naturally occurring process that is an important source of colonization of new habitats. Examples of recent disjunct invasions not directly facilitated by man include the colonization of the Americas by the African cattle egret, *Bulbulcus ibis* Bonaparte, and the arrival in the Caribbean from Africa of the desert locust, *Schistocerca gregaria* (Anon. 1988b). However, human movement of plants and animals, either through deliberate choice (many birds, feral domestic animals of many kinds, many species of deer and other game animals, ornamental and edible plants, many species of fish) or by accident (weedy plants, many herbivorous arthropods associated with plants being moved between regions, many pest arthropods and some plant pathogens in such commodities as soil, logs, ballast water, or other materials moved between regions, and such vertebrates as rats and mice), contributes greatly to the rate at which new species enter local ecosystems. While most adventive arthropod species that become pests do so following accidental invasions, some important weeds were deliberately introduced as ornamentals or for other reasons and later became weeds due to excessive power of reproduction in the new environment (Patterson 1976; Neel and Will 1978; Hardt 1986; Hazard 1988). For example, Lehmann lovegrass, *Eragrostis lehmanniana* Nees, a forage species introduced from South Africa to Arizona in 1932, has occupied over 145,000 ha of semidesert rangeland, displacing native grasses (Anable et al. 1992).

The rapidity of air travel makes it likely that stowaway organisms will survive long enough to reach distant locations. For example, between 1962 and 1985 an average of 19–20 immigrant species (species not intentionally introduced) per year gained entrance to Hawaii (U.S.A.) and became established there (Funasaki et al. 1988a, Beardsley 1991). Given the extreme isolation of these islands, nearly all of these invasions were likely related to human movement of plants and other goods. Of these immigrants, 3.5 species (16%) per year were considered pests (Beardsley 1991). Between 1955 and 1988, an average of six immigrant species of insects per year entered California (Dowell and Gill 1989), although not all became established. Clearly, invasion by nonnative organisms is a major conservation challenge which can only continue to increase in direct proportion to the volume and rapidity of international commerce (especially trade in plants and pet organisms) and tourist travel (Soulé 1990) and argues strongly for continued strict quarantine measures against immigrant species.

Habitat Disturbance

Habitat disturbance increases the probability that founder organisms of adventive species which reach new locations will survive as populations and become established (Orians 1986; den Hartog and van der Velde 1987). Disturbed soil (from agriculture, excessive grazing, logging, fire, flood, or excavation), for example, provides growth sites for newly arrived plant species where competition pressures from native or previously established plants will be reduced, at least temporarily (see examples cited in Merlin and Juvik 1992). Widespread disturbances of these sorts thus favor colonization either of immigrant plants or of descendants from domesticated populations of introduced plant species that move into natural areas (for example, purple loosestrife, *Lythrum salicaria* [Linnaeus], in North America; *Opuntia* spp. cacti in Australia and South Africa). Overgrazed lands are particularly susceptible to invasion by aggressive adventive weeds, especially if these are unpalatable due to thorns or toxins, because grazing will both physically open up space in the habitat and will confer a competitive advantage on the uneaten weeds. North American range lands have been invaded by many such aggressive species, including: Klamath weed, *Hypericum perforatum*; diffuse knapweed, *Centaurea diffusa* Lamarck; spotted knapweed, *Centaurea maculosa*; tansy ragwort, *Senecio jacobaea*; and thistles of various species (see Julien 1992 for more examples of adventive, aggressive plant species). In some cases, termination of habitat disturbance alone may be sufficient to reduce the competitive advantage of the adventive species enough to limit its numbers and hence the damage it causes. In other cases, biological control methods may need to be combined with reduction of habitat disturbance. Integration of these two approaches is valuable because upper trophic level natural enemies are assisted in suppressing the pest by the competition pest plant species experience from native plants (see Chapter 17).

The role of disturbance in facilitating animal invasion lies, at least in part, in reducing predation or other forms of mortality from native natural enemies in disturbed areas. For example, levels of predation by red land crabs (*Gecarcoidea natalis*) on giant African land snails (*Achatina fulica*) on Christmas Island were lower in disturbed areas compared to areas of intact rain forests (Lake and O'Dowd 1991). Another type of disturbance that can aid the establishment of nonnative birds and other animals is resource enhancement, such as the increased food available to some types of birds in urban environments (Orians 1986). The role of disturbance (type and frequency) in maintaining communities of native species or facilitating invasions of adventive species is further discussed by Hobbs and Huenneke (1992).

Alteration of Frequency or Intensity of Fires or Floods

In some systems, changing the frequency or intensity of a regularly occurring event can in itself be a disturbance. Periodic events such as fires and floods can act as screens that exclude some immigrant species from establishing in an ecosystem. If these events are controlled by man through dams or fire suppression programs, competitive balances may shift in favor of species previously kept out of the habitat by these periodic events. Invasion of grasslands by woody species is a typical consequence of fire suppression in some habitats. Reduced seasonal flooding in the Florida Everglades (in the United States), combined with elevated soils associated with drainage canals, has facilitated the invasion of the region by the Brazilian peppertree,

Schinus terebinthifolius Raddi, and the Australian broad-leaved paperbark tree, *Melaleuca quinquenervia* (Ewel 1986).

Prior Invasions of Other Nonnative Species

In some cases, prior invasion of an ecosystem by an adventive organism may alter basic properties of the ecosystem, causing the system to diverge progressively from its original condition and become increasingly vulnerable to invasion by additional species. The introduction of deep-rooted woody species, for example, into riparian habitats may lower the water table, resulting in reduced soil moisture and leading to possible further invasions by species adapted to drier soils. The invasion of nitrogen-fixing plants can increase fertility of poor-quality soils, possibly permitting additional invasions by plants previously excluded because of low nitrogen levels. The introduction of the myxoma virus to Europe greatly reduced rabbit populations and subsequently the level of rabbit grazing. In dune systems, reduced grazing allowed the invasion of woody species, with the resulting soil development favoring more mesic-adapted plants over the xeric-adapted dune plants (Hodgkin 1984).

EFFECTS OF ADVENTIVE SPECIES

In most natural systems around the world, adventive species are present. In some cases, the numbers of such species can be a significant proportion of the total species present in the area (from 3 to 66% of the flora in various national parks of the United States, Loope 1992). Most such species, however, occur in low numbers, are concentrated in such disturbed areas as roadsides and agricultural areas, and do little damage to native species. Adventive species become pests in natural ecosystems when they threaten the continued existence of specific indigenous species or groups of species by direct attack, or they alter ecosystems on a broad scale in such important ways that they threaten the structure or existence of whole communities of plants and their associated animals.

Effects on Particular Species or Groups of Species

Most adventive species of concern to conservationists are not species that directly attack individual native species. Rather, native species are typically threatened by adventive organisms because of some broad effect (such as crowding or resource preemption) that the adventive species has on the native ecoystem. However, in some cases, adventive species may be able to include native species in their diet or host range. In such cases, the adventive species may severely reduce the density of a particular indigenous species. For example, the scales *Carulaspis minima* (Targioni-Tozzetti) and *Insulaspis pallida* (Maskell) invaded Bermuda in the 1940s, and nearly exterminated the precinctive Bermuda cedar, *Juniperus bermudiana* Linnaeus (Cock 1985) (Fig. 21.1). In Florida indigenous bromeliads (Fig. 21.2) are currently threatened by the arrival in Florida (in imported bromeliads from Central America or Mexico) of weevils in the genus *Metamasius* (Frank and Thomas 1994). Immigrant pathogens may attack indigenous species. The fungus *Cryphonectria parasitica*, for example, invaded North America where it caused a blight of the native American chestnut, *Castanea dentata*, decimating the species. Similarly, some adventive species' habitat requirements may overlap strongly with a particular native species or set of native species with a precise ecological need. Starlings, *Sturnus vulgaris* Linnaeus, for example, when introduced to North America drastically reduced populations of the native eastern bluebird, *Sialia sialis* Linnaeus, by competing for the limited supply of natural nest cavities.

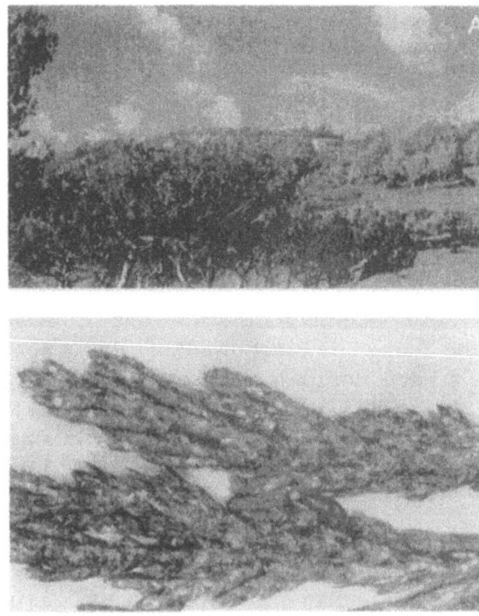

Figure 21.1. Native plants may be at risk from immigrant arthropods, as in the case of Bermuda cedar, *Juniperus bermudiana* Linnaeus (A), attacked and killed by the immigrant scales *Insulaspis pallida* (Maskell) and (B) *Carulaspis minima* (Targioni-Tozzetti). (Photographs courtesy of F. Bennett.)

Alteration of Ecosystem Properties

More commonly than attacking specific indigenous species, adventive organisms damage natural areas by changing and dominating such systems, making them less suitable for whole communities of native species that formerly thrived in the location (Vitousek 1992). Alterations may be caused in several ways (Table 21.1), among which are: changes in water table depths; changes in soil fertility or chemistry; crowding and smothering; altered flammability of the habitat; enhanced predation or competition; and enhanced disease.

Altered Water Tables. Some adventive plant species can modify local water tables. Among these are several species of salt cedar (*Tamarix* spp. from Asia) which were introduced into the

Figure 21.2. Another example of native plants currently threatened by adventive insect pests is that of Floridian (U.S.A.) bromeliads being killed by *Metamasius* weevil species from Latin America. (Photograph courtesy of H. Frank.)

TABLE 21.1 Ways in Which Immigrant Species Can Potentially Alter Native Ecosystems

1.	Changes in water tables
2.	Changes in soil fertility or chemistry
3.	Crowding and smothering
4.	Altered fire frequency or intensity
5.	Altered rates of predation, parasitism, or disease

southwestern United States as ornamentals and to stabilize blowing sand along roads and railroads. These plants have, however, invaded natural riparian habitats where they form extensive thickets (Fig. 21.3). Because these plants are deep-rooted and do not restrict their water losses from evapotranspiration, they cause water tables to drop (Neill 1983) which has led to the drying up of wet meadow habitats in riparian areas with dense salt cedar stands (Vitousek 1986). Soil dessication eliminates many indigenous species with short roots that depend on locally high water tables found under these wet meadows. The drying of the soil makes such areas more easily colonized by various adventive weeds that are not competitive under wet meadow conditions.

Altered Soil Fertility or Soil Chemistry. Adventive plants can alter local soil conditions in various ways, such as enhancing fertility or adding salt or other plant poisons. Some species restricted to low fertility soils are poor competitors under higher fertility conditions, but survive because their potential competitors are unable to grow well on low nutrient soils. In such cases, soil fertility enhancement can cause the low nutrient specialists to be lost from the ecosystem. For example, nutrient-poor soils in volcanic caldera in Hawaii support rare native plants. Recently these areas have been invaded by the Atlantic shrub *Myrica faya* Aiton which is a nitrogen-fixing species, resulting in a fourfold increase in the rate of soil nitrogen fixation (Vitousek 1990). Such alteration might lead to loss of competitiveness for species restricted to nutrient poor soils and possibly open these areas to invasion by plant species that were formerly excluded by low nitrogen levels. Iceplant, *Mesembryanthemum crystallinum* Linnaeus, which has been introduced to California for use as an ornamental, invades coastal grasslands. The plant accumulates salt which is released upon its death. Important native species in these dune communities are less salt tolerant than iceplant and, by this mechanism,

Figure 21.3. Adventive species may affect native communities by altering habitat characteristics, as, for example, the lowering of water tables in desert riparian areas in the United States by introduced salt cedars (*Tamarix* spp.), leading to loss of native shallow-rooted plant species. (Photograph courtesy of J. DeLoach.)

iceplant benefits from altered soil chemistry to invade and hold ground in competition with native species (Vivrette and Muller 1977; Kloot 1983).

Crowding and Smothering. Many adventive plants, in the absence of specialized arthropods or pathogens to reduce their growth rates, form dense stands that physically displace most or all native vegetation in the areas they invade. To correct such damage it is necessary to reduce the density of the adventive plant, but the species need not be eliminated.

Such dense stands of weeds are threats to native species and ecosystems in several ways. In unmanaged conservation areas, dense weed stands preclude the normal growth and existence of many native plants by physically preempting the available space and resources in the habitat (for example, *Hypericum perforatum* in grasslands of the western United States [Huffaker and Kennett 1959] and *Melaleuca quinquenervia* in the Florida Everglades [Ewel 1986; LaRosa et al. 1992]). Harris (1988) notes several examples in which adventive plants pose threats to endangered species. Among these are the introduced garden flower purple loosestrife, *Lythrum salicaria*, which forms dense stands in wetlands in eastern and central North America. This species is believed to have been partially responsible for the decline of several rare plants, including the threatened bulrush *Scirpus longii* (Fernald) in Massachusetts and the rare dwarf spike rush *Eleocharis parvula* (Rom and J. A. Schultes) in New York (U.S.A.) (Hight and Drea 1991). Also, in Canada, Harris (1988) notes that the immigrant species leafy spurge, *Euphorbia esula* Linnaeus, is eliminating the vegetation that forms the required habitat for the threatened northern prairie skink, *Eumeces septentrionalis septentrionalis* (Baird). In addition, threatened species may be damaged when other forms of control (mechanical or chemical) are directed at dense stands of adventive weeds. For example, the western prairie fringed orchid *Plantathera praeclara* (Cheviak and Bowles) is threatened in the Sheyenne National Grassland of North Dakota (U.S.A.) by herbicides applied against dense stands of leafy spurge.

There are numerous examples of smothering and crowding by adventive weeds in both aquatic and terrestrial habitats. Several species of floating and submerged water weeds have invaded aquatic systems throughout the tropics, including the floating fern, *Salvinia molesta* which has invaded parts of Africa, India, northern Australia, and Papua New Guinea among other locations (Forno et al. 1983; see also Chapter 8). The effect of this weed on the Sepik River system in Papua New Guinea was especially rapid and dramatic (Mitchell et al. 1980). Within ten years of its first occurrence in that drainage system, over 20% of the surface area of all the adjacent oxbow lakes were covered by solid mats of *S. molesta* (a total of 80 km^2 of mats). Mats of *S. molesta* reduced water oxygen levels by up to 37% and killed submerged vegetation, leading to further deoxygenation. Thomas (1981) reports that following *S. molesta* invasions of water bodies in Kerala, India, abundances of indigenous aquatic plants were greatly reduced, with some species disappearing altogether. Due to rapid accumulation of organic matter from *S. molesta* mats and increased evapotranspiration, water bodies declined in volume and extent and some portions were transformed into arable land.

Adventive aquatic weeds that smother or crowd native aquatic plants in tropical and subtropical areas include alligator weed (*Alternanthera philoxeroides*) and water hyacinth (*Eichhornia crassipes*). Aquatic weeds are also of concern in temperate areas. One of the most notable is Eurasian watermilfoil, *Myriophyllum spicatum* Linnaeus. In Lake George, New York, watermilfoil beds (which grow submerged on the bottoms of water bodies) caused a 50% decline in the number of species of indigenous aquatic macrophytes (Madsen et al. 1991). Eurasian watermilfoil was documented by Madsen et al. (1991) to be able to invade intact, undisturbed indigenous littoral plant communities, leading to their degradation over several years and substantial replacement by watermilfoil mats. Purple loosestrife, *Lythrum salicaria*, is an escaped garden flower that is an emergent wetlands plant. It invades wet pastures, marshes, and shallow water bodies, where it displaces cattails and other native vegetation, forming dense monocultures. In addition to displacement of native plants through crowding, it reduces cattail (*Typha* spp.) marsh quality as habitat for some native mammals and waterfowl. This plant is now widespread over the northeastern United States and is spreading into marshes and prairie potholes of the northern plains region, one of the major waterfowl nesting areas in North America.

Crowding and smothering of native plant communities is also caused by terrestrial adventive plants. Kudzu, *Pueraria lobata* (Willdenow) Ohwi, introduced to the southern United States from Asia for soil stabilization, grows rapidly into extremely dense mats that envelop and kill trees and exclude much native vegetation (Sweet and Schaefer 1985; Tayutivutikul and Yano 1989). Coastal areas of northern Australia are currently being transformed by the woody South American shrub *Mimosa pigra* which forms dense stands on seasonally inundated floodplains (Braithwaite et al. 1989). Monospecific stands of this shrub progressively invade and replace the sedgeland, then riparian, aquatic, paperbark, and monsoon forest communities. This invasion has led to the complete alteration of what formerly was a largely intact natural landscape. In addition to the dramatic effects on indigenous plants, this change of vegetation has caused severe damage to the native fauna, greater than that from replacing native forest with plantations of introduced pines. This same shrub is also spreading in Burma, Laos, Kampuchea, Vietnam, and Malaysia (Kassulke et al. 1990). In New Zealand, consideration is currently being given to biological control of heather (*Calluna vulgaris* [Linnaeus]) in national parks, where it was planted in earlier years in an effort to transform native plant communities into habitat reminiscent of Scottish moorlands, suitable for grouse hunting (Syrett 1990).

Alteration of Fire Frequency or Intensity. The fire characteristics of a habitat are important in shaping the species composition of some plant communities. Adventive species may alter these characteristics enough to affect other species in the community. For example, invasion of fynbos communities in South Africa by the woody shrub *Hakea sericea* (Fig. 21.4) caused a 60% increase in fuel loading at some sites (Versfeld and van Wilgen 1986). This added fuel load increased fire intensity, and perhaps fire frequency.

Figure 21.4. Adventive species may affect native communities by changing the frequency or intensity of fire as, for example, *Hakea sericea* Schrader, a plant introduced to South Africa which has increased fuel loading in native fynbos communities. (Photograph courtesy of A.J. Gordon.)

Addition or Deletion of Dominant Species. Adventive animal species that exhibit high densities and broad diets can be important influences on the species composition of native communities, either directly through predation, or indirectly through competition for resources. Invasions of fynbos communities in South Africa by the Argentine ant, *Linepithema humile*, for example, displaced most of the indigenous ant fauna (Breytenbach 1986). In a high-elevation shrubland in Hawaii, densities of several indigenous insects, including important pollinators, were lower in areas with Argentine ants, compared to adjacent areas not yet colonized by this species (Cole et al. 1992). In Colombia, deliberate introduction from Brazil of the adventive ant *Paratrechina fulva* (Mayr) caused a 95% reduction in native ants (only 2 of 38 native species persisted in invaded areas). In addition, one species of snake and three species of lizards disappeared from the area occupied by *P. fulva* (de Polania and Wilches 1992). Williams (1994) gives a broader treatment of the effects of adventive ant species. Introduction of the predacious Nile perch (*Lates* sp.) to Lake Victoria in Africa caused extinctions of many fish species and made many other once-common species rare (Goldschmidt et al. 1993). Invasion of Guam in the South Pacific by the brown tree snake (*Boiga irregularis* Fitzinger) virtually eliminated the entire native, forest-dwelling avifauna (McCoid 1991). The effect of the converse action, removal of a dominant predator, can also lead to loss of other species in a community as demonstrated by exclusion experiments in marine intertidal areas (Paine 1980; Hall et al. 1990) and in agricultural insect systems (Luck et al. 1988). Predator removal at the population level might occur if, for example, an adventive disease or parasitoid caused a major predator species to be decimated.

Altered Rates of Disease. Diseases can also play a major role in shaping ecosystems. Introduction of avian malaria to Hawaii, for example, is believed to be a significant factor in the narrowing of the distributions of native birds (van Riper et al. 1986).

BIOLOGICAL CONTROL FOR CONSERVATION

Biological control of weed pests of terrestrial and aquatic environments in the United States and Canada has been reviewed by DeLoach (1991) and Gallagher and Haller (1990). In the following sections, we emphasize applications of biological control to several groups important as environmental pests.

Use Against Aquatic Plants

Several species of tropical or subtropical plants growing in freshwater habitats, and a few temperate species, have become widely established around the world. Some of these species have become very damaging to native aquatic plant communities through the formation of mats or beds of such size and thickness as to alter lighting, water oxygen levels, and physically occupy available space. Native plants are often greatly reduced in density or completely excluded. Biological control has been used successfully against several such weeds and work on other species is actively being pursued. Some important examples, in addition to salvinia (discussed in Chapter 8), follow.

Water Hyacinth. *Eichhornia crassipes* is a floating freshwater plant of worldwide concern in tropical and subtropical areas that originated in Brazil but has been moved about widely as an ornamental plant for fish ponds (Gopal 1987) (Fig. 21.5a). Flooding and other mechanisms have

Figure 21.5. Impacts of adventive plants species on native communities can be significantly reduced by the introduction of herbivorous arthropods, as in the case of water hyacinth (*Eichhornia crassipes* [Martinus] Solms-Laubach). (A), a water weed of worldwide significance controlled by three natural enemies, including the curculionid *Neochetina bruchi* Hustache (B). Uncontrolled growth of the plant can entirely cover waterways (C), destroying ecosystem processes; while addition of natural enemies reduces plant growth and density and restores waterways to a near-natural state (D). (Photographs courtesy T. Center.)

caused releases of the plant into natural water bodies in various countries, with the development of extensive weed mats in slow moving waters. So extensive are these infestations that water hyacinth is regarded as the world's worst water weed Mats are so thick that, in addition to major economic damage to dams, navigation, and fisheries, the native fauna and flora of infested areas are often severely affected. Water oxygenation is reduced, in some cases leading to fish kills. Eutrophication is increased, and, in some cases, populations of waterfowl and other wildlife are reduced (Harley 1990). Siltation is increased. Evapotranspiration from weed mats is increased 3.5 times over evaporation rates from open water, leading to filling and shrinking water bodies. For example, it has been estimated that infestation of this weed in the Nile River system has reduced the river's flow by 10% (Hamdoun and Tigani 1977).

Many biological control agents have been assessed and the weevils *Neochetina bruchi* Hustache (Fig 21.5b) and *Neochetina eichhorniae* Warner and the moth *Sameodes albiguttalis* Warner have been found to be the most effective (Harley 1990). These and other agents have been released in numerous regions around the world. Very effective control has been achieved in many areas, including the Sudan, India, Australia, and the United States (Fig. 21.5c,d). In the United States, for example, the infested area in Louisiana was reduced from 500,000 ha in 1974 to 150,000 ha in 1980, eight years after initial releases (Cofrancesco et al. 1985). Control in other regions has also been substantial, but less well-documented. Elimination of extensive water hyacinth mats has allowed physical conditions of affected water bodies to revert to those of preinfestation times, to a substantial degree. This in turn has stabilized the habitat conditions for native species and reduced competitive pressure for space in the habitat.

Alligator Weed. *Alternanthera philoxeroides* is a South American freshwater weed that is established in parts of the United States, Australia, and other warm regions of the world.

Figure 21.5. (*Continued*).

Biological control efforts were begun in the United States in 1963. At that time, over 65,700 ha of water surface were infested (Schroeder 1983). Several agents have been identified and introduced. The first and most effective of these was the chrysomelid beetle *Agasicles hygrophila*, which provided substantial control in coastal areas in the United States (Coulson 1977) and later also controlled alligator weed in parts of Australia (Room 1986).

Other Aquatic Weeds. Other introduced freshwater plant species, such as water lettuce, *Pistia stratiotes*, and Eurasian watermilfoil, *Myriophllum spicatum*, are also of ecological importance and have been targets for biological control (Harley et al. 1990; Madsen et al. 1991).

Plants in saltwater systems have not yet been the focus of biological control efforts, but some species have invaded new regions and could logically become of conservation concern if resultant populations are large enough or sufficiently damaging to local ecosystems. For example, saltmarsh grass (*Spartina* sp.) from the eastern United States has invaded coastal mudflat ecosystems of Washington state on the Pacific Ocean. Because this plant's growth form is unlike any other plant native to this mudflat ecosystem, Spartina has the potential to radically alter ecological conditions. Marine algae may also be moved internationally, associated with shipments of fresh shell fish, and establishment may occur if shells are discarded into marine waters.

Use Against Terrestrial Plants

At least 116 species of weeds have been targets for biological control through natural enemy introductions (Julien 1992). Most of these were terrestrial species targeted because of the damage they caused to economic resources such as grazing and forestry lands. However, many

Figure 21.6. Dense stands of adventive terrestrial plants can reduce numbers of native plants, as in the case of the effect of the European species *Senecio jacobaea* Linnaeus (A) on grasslands in the western United States (B). This plant was subsequently controlled (C) by introduced herbivorous natural enemies. (See p. 438) (Photographs courtesy of P. McEvoy.)

of these plants also invaded extensive areas of natural habitat, valuable for the conservation of native species and, through the formation of dense, highly competitive stands, reduced the value of these areas for nature conservation. Thus many projects which were undertaken for economic purposes have also benefited natural systems. More recently, projects of biological weed control have been initiated specifically for ecological restoration of conservation areas as, for example, in native forests in Hawaii (Markin 1982; Markin et al. 1992). Invasive adventive plants are problems around the world, in both temperate and tropical regions.

Banana poka. *Passiflora mollissima* (Humbolt, Bonpland and Kunth) Bailey is a South American vine which threatens native koa forests in Hawaii by smothering trees and reducing light penetration (Waage et al. 1981). Surveys to discover herbivores capable of attacking this weed were conducted in South America (Pemberton 1989). Several species, including the notodontid *Cyanotricha necyria* (Felder and Rogenhofer) and the pyralid *Pyrausta perelegana* Hampson, were imported, determined to be sufficiently host specific for safety to other plants in Hawaii, and released (Markin et al. 1989). Of these, the latter species established (Campbell et al. 1994) and caused a 5–10% level of bud destruction.

Giant Sensitive Plant. *Mimosa pigra* is a woody shrub native to North and South America that has invaded Australia and southeast Asia and has formed monospecific stands over hundreds of square kilometers in native wetlands causing serious damage to native flora and fauna (Braithwaite et al. 1989). The seed-feeding bruchid beetles *Acanthoscelides quadridentatus* (Schaeffer) and *Acanthoscelides puniceus* Johnson from Mexico were imported, screened for specificity, and released against this weed in 1983 in Australia (Kassulke et al. 1990). They have become established but have not yet had any effect on the plant's density (Wilson and Flanagan 1991).

Gorse. *Ulex europaeus* is a woody shrub from Europe that was introduced as a hedge plant into various temperate countries on other continents. In some areas such as New Zealand (MacCarter and Gaynor 1980), Hawaii (Markin et al. 1988), and Chile (Norambuena et al. 1986) gorse proved excessively vigorous, spreading into natural grasslands and forests, to form dense monospecific thickets. As with many aggressive weeds, this degree of crowding caused both losses to economic resources and reduced space available for native plant communities. A seed weevil, *Apion ulicis* Forster, was introduced to both New Zealand (MacCarter and Gaynor 1980) and Chile (Norambuena et al. 1986) where it became established but did not control the weed. A tetranychid mite, *Tetranychus lintearius*, has also been released and established (Hill et al. 1991), and its effect on the plant in New Zealand is under evaluation. In some locations in New Zealand, gorse thickets have conservation value as habitat for endangered orthopterans (species of weta), because thickets provide both food and protection from predation by adventive species of rodents.

Kudzu. *Pueraria lobata* was introduced to the southeastern United States to control soil erosion along roadsides. The plant spread extensively and now covers large areas with a solid blanket of vines, in some cases enveloping entire tree canopies and killing trees. Its biological control is being contemplated (Tayutivutikul and Yano 1989).

Red Quinine-Tree in the Galapagos. *Cinchona succirubra* Pavon ex Klotzsch was brought to Santa Cruz Island in the Galapagos Islands (Ecuador) as a forestry and shade tree. It has multiplied extensively and invaded native forest/shrub communities which it outcompetes. It now infests over 4000 ha on the island. An extensive mechanical and herbicidal removal program has been initiated, but costs and the plant's ability to resprout has limited the application of the program to only a small portion of the infested area (Macdonald et al. 1988). This experience has prompted consideration of biological control.

Sesbania punicea. This species is a short-lived leguminous tree from South America that was intentionally introduced to South Africa, where it has invaded riparian wetlands, forming dense thickets (Hoffmann and Moran 1991). Two introduced insects have provided substantial control of this species. A bud-feeding weevil, *Trichapion lativentre*, has established and reduced seed production by 98%. Seed reduction has not affected biomass of existing stands, but has greatly slowed the invasion of new areas. A stem-boring weevil, *Neodiplogrammus quadrivitattus* (Olivier), has also become established and in locations where it occurs, the pest weed stands are declining.

Klamath Weed (St. John's Wort). *Hypericum perforatum* is a toxic European weed that invades grasslands in temperate areas with a Mediterranean climate (winter rains, summer drought). Extensive infestations developed in various regions, including most noticeably California, British Columbia (Canada), and parts of Australia. Prior to initiation of biological control in northern California, St. John's wort formed dense stands over more than 1,000,000 ha. The introduction of several herbivorous insects, most important of which were the leaf feeding beetles *Chrysolina hyperici* (Forster) and *Chrysolina quadrigemina* (Suffrian), reduced this infestation by 99% (Andres et al. 1976). Space formerly occupied by St. John's wort, over a ten-year period after the decline of weed stands, was occupied by a mixture of native forbes and grasses, as well as introduced grasses (Huffaker and Kennett 1959).

Opuntia *spp. Cacti.* *Opuntia stricta* was introduced as an ornamental plant to Queensland, Australia during European settlement and spread to occupy over 25,000,000 ha of forest and grassland, forming dense stands that excluded other vegetation. Introduction of the moth *Cactoblastis cactorum* (Fig. 5.5) from Argentina resulted in successful biological control of the cactus, allowing the return of previous vegetation, or conversion for economic use (DeBach et al. 1976). Other species of jointed cacti have also become pests in various countries around the world, for example, in South Africa, where *Opuntia aurantiaca* Lindley occupies over 830,000 ha (Zimmermann et al. 1986). Dense infestation of vast expanses of range and forest lands by adventive plants such as these *Opuntia* spp. is a classic example of the effects that nonindigenous species can have on native ecosystems through the physical occupation of the ground and consequent exclusion of many native species from access to water, light, and nutrients.

Tansy Ragwort. *Senecio jacobaea* is a toxic weed that invaded pastures and natural grasslands in Oregon (U.S.A.) and other parts of the world. A chrysomelid beetle (*Longitarsus jacobaeae*, Fig. 5.3) and an arctiid moth (*Tyria jacobaeae*, Fig. 5.6) were introduced against the weed and reduced its density by 89% (Fig. 21.6, see p. 436) (McEvoy et al. 1990; McEvoy and Cox 1991). This reduction has allowed for recovery of native coastal grasslands in some areas, with a beneficial effect on rare native plants.

Use Against Herbivorous Arthropods

Herbivorous arthropods have been the most common type of organism against which biological control has been employed. However, nearly all such efforts have been directed against pests of economic resources, rather than species adversely affecting conservation interests (Laing and Hamai 1976; Clausen 1978). Apart from financial and other organizational reasons, one explanation for this limited use may be that immigrant insects are less likely to attack indigenous plants (if these are distinct from those in their native homes), or if they do attack indigenous plants, are not excessively damaging and thus are ignored. Exceptions do arise, however, such as the effect of immigrant scales on the native Bermuda cedar, *Juniperus bermudiana*, discussed earlier (Cock 1985). Efforts to control these scales included the introduction of the aphelinid *Encarsia lounsburyi* (Berlese and Paoli) and two coccinellids, *Rhyzobius lophanthae* (Blaisdell) and *Microweisea suturalis* Schwarz. These reduced the scales's densities, but did not halt tree loss. So severe was the tree's physiological reaction to scale feeding that even low to moderate levels of scale caused tree decline, often resulting in tree death. It appears, however, that trees which survived the initial die-off are more tolerant of scale feeding and some regeneration of cedar has occurred. Howarth (1985) records cases of herbivorous arthropods which have invaded Hawaii and which may be having enough effect on native plants to merit consideration as possible targets for biological control introductions.

Frank (1994) is currently seeking to locate natural enemies to import into Florida to control an immigrant Mexican weevil, *Metamasius callizona* (Chevrolat), which invaded Florida through commercial trade in Mexican bromeliads. This weevil is currently destroying native Floridian bromeliads.

Use Against Predacious or Parasitic Arthropods

Some species of social Hymenoptera, such as the Argentine ant (*Linepithema humile*), several fire ants (*Solenopsis* spp.), and social wasps in the genus *Vespula*, have invaded new regions and become very abundant (Fig. 21.7a). Because of their high densities and the wide breadth of their diets, such species have in some instances had significant effects on indigenous invertebrates and vertebrates (Gagné and Howarth 1985; Howarth 1985; Breytenbach 1986; Gambino et al. 1990; Beggs and Wilson 1991). For such species, in forests or conservation lands, biological control may be a potential solution (Grossman 1990). The ichneumonid wasp *Sphecophaga vesparum* (Curtis) *vesparum* (Fig. 21.7b) has been found to be sufficiently host specific for release against the immigrant species *Vespula germanica* (Fabricius) and *Vespula vulgaris* (Linnaeus) (Field and Darby 1991). Releases in New Zealand have resulted in establishment, but the degree of control is still being evaluated (Moller et al. 1991).

Use Against Other Invertebrates

Biological control of invertebrates other than arthropods has been almost exclusively attempts to control molluscs (slugs and snails) that are either herbivorous terrestrial species that feed on

Figure 21.7. Invasion of native communities by vespid wasps (*Vespula* spp. [A]) has caused ecological damage in some locations such as New Zealand. Introduction of natural enemies of these pests, such as parasitic Hymenoptera (*Sphecophaga vesparum* [Curtis] *vesparum*) (B), may reduce pest populations and help provide control. (Photographs courtesy of T.E. Fitzgerald [A] and J.S. Rees [B].)

agricultural crops or aquatic species that are intermediate hosts of liver flukes that attack man or domestic animals. (See Chapter 2 for more on molluscs as targets for biological control.) No projects have yet been attempted solely for conservation purposes, but some potential needs do exist, as some adventive molluscs are damaging to conservation interests. Examples include the predatory snail *Euglandia rosea*, which is a deliberately introduced species that threatens native land snails in such locations as Moorea in French Polynesia in the Pacific (Clark et al. 1984). In North America, freshwater benthic species have been affected by immigrant molluscs in several locations. In the Great Lakes of North America and associated rivers, native benthic molluscs and other species have been reduced considerably by the smothering and competition that has resulted from high population densities of the zebra mussel, *Dreissena polymorpha*, an Asian species transported in ballast water of ships (Garton et al. 1993). A species of Asian clam in the genus *Potamocorbula* has recently invaded San Francisco Bay in California which has reached such high population levels that it is reducing phytoplankton densities, affecting the entire ecosystem through competition for the basic energy resource of the aquatic foodchain in the bay (Dietel 1991). European green crab, *Carcinus maenas* (Linnaeus), (Fig. 21.8) invaded the west coast of North America (San Francisco Bay, in California) in 1990 and has reached damaging densities. The introduction of a highly specific parasitic barnacle *Sacculina carcini* Thompson has been suggested for biological control of this adventive crab.

Use Against Vertebrates

Non-native vertebrates, including feral populations of domestic animals, have caused significant damage to indigenous species (especially plants and ground nesting birds) in many locations, and their suppression on many oceanic islands is an environmental priority (Chapuis et al. 1994). Vertebrates that prey on other vertebrates generally are inappropriate for introduction outside of their historical range as the specificity of this class of agents is usually not sufficient to limit their effects to the target pest, and such agents may pose dangers to other indigenous vertebrate species. Habitat modification that favors the action of native vertebrate predators, however, may be effective in some instances and prove safe to native species. Densities of predatory birds such as owls and hawks and mammalian species such as foxes (*Pseudalopex* spp.) may be enhanced, for example, though provision of nesting structures (for birds) and vegetation modification to increase predation rates (by clearing strips to increase prey visibility) on prey such as rabbits and rodents (Muñoz and Murúa 1990).

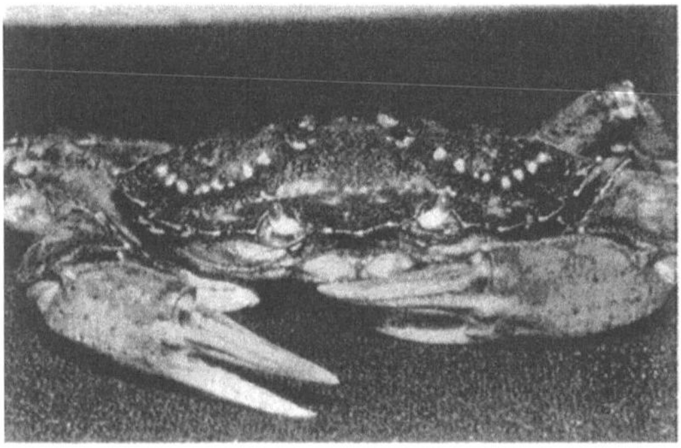

Figure 21.8. Adventive marine species can establish and increase in non-native locations, such as the European green crab (*Carcinus maenas* [Linnaeus]), and can affect fisheries and other marine ecosystem properties. The introduction of the parasitic barnacle *Sacculina carcini* Thompson has been suggested for biological control of this pest crab. (Photograph courtesy of T. Huspeni.)

Pathogens have in a few cases been used successfully for biological control of vertebrates. The most important example is the introduction of the myxoma virus of rabbits to Australia and, later, to Europe for control of the European rabbit, *Oryctolagus cuniculus*, with dramatic results (Ross and Tittensor 1986). Rabbit haemorrhagic disease, a second rabbit pathogen, has been introduced to Macquaries Island, between Tasmania and Antarctica (Anon. 1994b).

Another example of successful use of biological control against a pest vertebrate is the introduction of the feline panleucopaenia virus of domestic cats (*Felis cattus*) into a population of feral cats on Marion Island in South Africa (Fig. 21.9). This was done to reduce the killing by cats of up to 450,000 seabirds per year in a nesting colony on the island (van Rensburg et al. 1987). Initial results were successful, with a reduction from an estimated 3409 cats to 615 in the first five years.

Clearing oceanic islands of introduced herbivorous mammals such as goats is critical for the regeneration of indigenous plant communities (North et al. 1994). Dobson (1988) suggested that opportunities may exist to use pathogens against such pest vertebrates on oceanic islands where such vertebrates have been particularly destructive to native ecosystems and rare species. Dobson notes that island populations of many feral mammals have fewer species of parasites and pathogens attacking them than mainland populations. Some of these, such as the sexually transmitted protozoan *Trichomonas foetus* that affects goats, may be sufficiently specific that they could be safely used to reduce population reproduction or survival rates. This avenue would be especially valuable because attempts at eradicating such pests through hunting have often been unsuccessful due to the near impossibility of finding and killing the last 1 or 2% of a population, especially in rugged terrain.

SUPPRESSING ADVENTIVE SPECIES

Depending on the physical extent of the area affected by an adventive species, mechanical, chemical, or biological controls may be employed successfully. When areas infested are small, it may be feasible to destroy the colonizing organisms mechanically or chemically (especially larger plants of species that have limited seed production) (Santos et al. 1992). These controls are likely to require periodic repetition, increasing their cost and limiting the size of area which may be cleared by such methods. When large areas are infested, or extensive seed banks exists, or the organism is too small or hidden to attack through mechanical or chemical

Figure 21.9. The presence of feral domestic cats (*Felis cattus* Linnaeus) on oceanic islands has threatened nesting seabirds. Control of such cat populations has been achieved through a combination of introduction of the parvo virus (the causative agent of feline panleucopaenia) together with other eradication methods. (Photograph courtesy of P.J.J. van Rensburg.)

methods, biological control should be considered. The role of biological control as a management tool in national parks in the United States has been reviewed by Gardner (1990).

Use of biological control against environmental pests is limited because developmental costs for new projects, especially against plants, are large and most expenses occur at the beginning of the project during exploration for and host-specificity testing of new agents (Markin et al. 1992). Harris (1979), for example, estimates that in Canada a complete biological control program for one weed species could cost as much as $1.5 million (Can.). These costs, however, are low compared to long-term herbicide use for weed suppression on large acreages. Also, in cases in which a successful agent of a given weed has already been employed in another location, repeating the projects in other countries can be done at much lower cost.

Another important feature of biological control is that the release of biological control agents is a permanent change. Sufficient prerelease screening must, therefore, be done to ensure that only suitable organisms are released (see Chapter 8). Within these constraints, biological control has great application for the control of widespread environmental pests because effective natural enemies, once released, remain effective permanently at no further annual cost. Because natural enemies do not need to be released repeatedly and because they are able to spread to additional areas, the method can be applied even when infested areas are large (thousands to millions of hectares).

SAFETY

Introducing new species of herbivores, parasitoids, parasites, predators, or pathogens to suppress adventive environmental pests is not without potential consequences to indigenous species. Beneficial consequences include curtailing the harm caused to natural systems by the pest species the program is targeted against. Detrimental consequences may include mortality to non-target species in the area of introduction. Each proposed introduction needs careful evaluation to ensure that it is well conceived and executed to prevent or minimize any detrimental consequences. (See Chapter 9 for more on the technical aspects of safely introducing new natural enemies.) Well-conceived programs are those in which the damage from the target pest is serious and natural enemies chosen for introduction are sufficiently host specific that most of their effect will fall on the target and not on desirable indigenous species.

Specificity is rarely absolute, as few organisms are truly monophagous. The degree of specificity required for safety will depend on the characteristics of both the location in which the project is to be conducted, and the taxonomic groups of organisms involved (see Chapter 8). Each case must be considered individually because actions that may be safe in one habitat may be undesirable in another (Clarke et al. 1984; Harris 1990; Howarth 1991). Introduction of snake pathogens to Guam (where there are no native snakes) in the South Pacific to control the adventive brown tree snake, which attacks native birds, for example, could be done with greater ease and safety than the same action in areas with a diverse native snake fauna. Introductions of pathogens of domestic vertebrates could be safely undertaken on uninhabited oceanic islands to suppress feral goats, pigs, or cats, but the same action in populated areas might be less acceptable because of conflicts with owners of domestic populations of the same mammals.

The taxonomic groups involved also affect the degree of safety involved in natural enemy introductions. Different types of natural enemies vary in the degree of risk that they may pose, and some groups such as vertebrates are generally unsuited for introduction beyond their historical ranges without compelling reasons (Legner 1986a). Specialized parasitoids of arthropods are often suitable for introduction, whereas generalist invertebrate predators, social invertebrate predators (ants, vespids), and generalist predacious snails are all groups that are potentially damaging to indigenous species and should not be introduced (Nafus 1993). Highly polyphagous parasitoids, such as some tachinids, may also pose risks to native species (Tothill et al. 1930). Herbivores introduced for control of adventive plants must be carefully screened to determine their hosts ranges (see Chapter 17).

Programs of biological control must balance benefits against potential risks and pursue actions likely to enhance the overall health of the ecosystems into which introductions are being made. While some environmental risks may remain in any natural enemy introduction program, even following careful host-range testing, these small, often conjectural, risks must be balanced against the risk of doing nothing. In many cases, uncontrolled populations of environmental pests have important, serious effects on native species and ecosystems, including threats to endangered species that might contribute to their extinction.

Risk and benefit assessments should occur in open forums and be based on scientific facts related to biologies of the organisms involved. A wide knowledge of the ecosystems into which introductions are being made and of the specific taxa found within them is important in making such judgments and should be actively sought (Samways 1994). Rigorous efforts should be made to identify local biological resources of special significance so that these rare, local, or unusual life forms are not threatened. Quarantine measures should be established to slow the invasion of adventive species, and biological control remedies should be vigorously pursued for those species that become established and which cause widespread, significant environmental problems.

CHAPTER 22

FUTURE PROSPECTS

The future of biological control holds promise to provide solutions both to pest problems affecting agricultural production and to needs for environmental protection of natural ecosystems. The ecological principles which underlie the contributions that biological control makes to both managed and natural ecosystems do not change with the passage of time; they are basic to interactions among species and are inherent in the structure of ecosystems. Consequently, employing biological control principles of population management and trophic-level interactions will continue to provide productive, efficient, and economical solutions to pest problems.

New biological control efforts are currently needed for many existing pest problems, both for programs targeted against introduced pests and for additional work toward natural enemy conservation in pest management systems. Continued international man-assisted movement of plant material and associated insects will likely cause the unintentional creation of new pests by separating species from their natural enemies. Such pests will affect both natural and agricultural communities and will require research and action to locate and introduce natural enemies suitable for limiting the pests in their new environments. The continued dependence of human societies on agriculture, in both developed and developing countries, dictates the need for integrated management systems for efficient and economical production of food and fiber. These integrated systems should increasingly be based on natural enemies to reduce both costs and pesticide use. In many of these systems, emphasis will be placed on conserving native or previously introduced natural enemies.

There are also opportunities for use of biological control in areas which have, in recent history, received little attention. These include, for example, pest problems in urban environments. Ornamental landscape and interior plants in many countries, for example, are introduced from other parts of the world, and often are attacked by arthropods native to the plants' original homes. Introduction of natural enemies is often the only viable pest management solution for these landscape pests, in part because the cost of other solutions is larger than the value of the plants, and in part because of growing concern about using insecticides in or near homes and commercial buildings.

Similarly, most pests of buildings are introduced; one example is the list of pest cockroaches in North America, all of which are introduced species. Little biological control research has been done on such structural pests relative to the attention received by agricultural pests. This is, in part, due to the lack of organized funding directed at such solutions for homeowners; each homeowner is generally responsible for his own pest management, rather than belonging to a large cooperative research organization as is the case for many agricultural commodities. It is also due, in part, to a misconception that biological control of household pests is fundamentally flawed because homeowners will not tolerate the presence of natural enemies any more than they will tolerate pests. This misconception fails to recognize that pest and natural enemy populations together are usually present in only one hundredth or one thousandth the numbers of pests in the absence of natural enemies. Therefore, insect natural enemies and household pests together would likely be much rarer than the pests the natural enemies are introduced to control. In addition, education regarding "good" and "bad" insects is already affecting public perception of arthropods (in the garden, ladybird beetles are appreciated, while pest insects are not), and additional education regarding insects beneficial to households would naturally become a part of any biological control program aimed at such pests.

Amid a growing awareness worldwide of the significance of natural enemies in limiting damage from arthropod pests, weeds, and diseases, there are some important challenges to the continued development of biological control. One of these is the ubiquitous nature of its potential application, in that it can and should be applied in every setting where population management is a crucial part of a natural or agricultural ecosystem. This pervasive need prompts the attention, however fleeting, of trained entomologists, pathologists, and other agricultural professionals. Such attention can give the impression that biological control research is indeed already pervasive and widely applied. Rather, successful development of biological control programs generally requires extended commitment of professional, full-time career biological control scientists. This is largely because of the long-term nature of biological control programs, which require an understanding of the population-level processes, taxonomic issues, and ecological functioning of a particular system. This is clear in the case of conservation and augmentation research, where a sound knowledge of the structure and function of the natural enemy and pest fauna is critical to making informed decisions on modifications to the system. Similarly, but for different reasons, long-term commitment is generally necessary to the effective functioning of a program designed at introducing natural enemies of an adventive pest. This commitment is a matter of professional focus, where the time given to the study of natural enemy groups over the years of professional development provide the research scientist a worldwide perspective on particular fauna and flora. In part it also a matter of personal commitment, because travel necessary to secure natural enemies reduces research on other topics by the explorer, and places the explorer at increased risk from accidents and diseases. Finally, quantifying the effects of any changes to a natural or agricultural system, such as the introduction or conservation of particular natural enemies, requires special applications of ecological principles and research techniques.

While none of these activities is beyond the scope of a trained scientist, they together form a group of talents and training that, to perform effectively, demand nearly full-time professional attention. Alternatively, *ad hoc* attempts at biological control solutions are rarely successful, perhaps because the knowledge necessary for successful completion of a program is lacking and because there is no long-term commitment to a biological solution, which may (but not always!) take several years to identify and put into place.

In this context, training future scientists and developing research positions specifically for biological control will continue to be necessary for biological control to fulfill its promise. We expect that as increasing numbers of scientists obtain training and experience in biological control in all its facets, there will be worldwide increase in biological control programs and in public awareness of their value. It is likely that public funding will, in the future as in the past, be a vital part of the support for such programs.

Finally, it is likely that scientific cooperation in biological control programs will continue to increase, as it has in the past several decades. Biological control programs are strengthened by the joint efforts of scientists who can bring complementary expertise, as in the case of systematists, ecologists, and entomologists evaluating a system for conservation by surveying existing faunas and defining host relationships, and where pathologists, geneticists, and physiologists collaborate to explore the possibilities of biological suppression of soil-dwelling pathogens. Similarly, international programs benefit from the assistance of researchers around the world in locating, collecting, and shipping natural enemies, and in conducting faunal surveys. This cooperation is one of the fruits of an increasing population of scientists and researchers working in biological control, and one which presages increasing impact of biological control in pest management and environmental protection in the future.

REFERENCES

Ables, J.R. 1979. Methods for the field release of insect parasites and predators. *Transactions American Society Agricultural Engineers USDA*, **22**: 59–62.

Ables, J.R., S.B. Vinson, and J.S. Ellis. 1981. Host discrimination by *Chelonus insularis* [Hym.: Braconidae], *Telenomus heliothidis* [Hym.: Scelionidae], and *Trichogramma pretiosum* [Hym.: Trichogrammatidae]. *Entomophaga*, **26**: 149–156.

Abou-Awad, B.A., and E.M. El-Banhawy. 1986. Biological studies of *Amblyseius olivi*, a new predator of eriophyid mites infesting olive trees in Egypt (Acari: Phytoseiidae). *Entomophaga*, **31**: 99–103.

Abou-Awad, B.A., A.S. Reda, and S.A. Elsawi. 1992. Effects of artificial and natural diets on the development and reproduction of two phytoseiid mites *Amblyseius gossipi* and *Amblyseius swirskii* (Acari: Phytoseiidae). *Insect Science and Its Application*, **13**: 441–445.

Abraham, Y.J., D. Moore, and G. Godwin. 1990. Rearing and aspects of biology of *Cephalonomia stephanoderis* and *Prorops nasuta* (Hymenoptera: Bethylidae) parasitoids of the coffee berry borer, *Hypothenemus hampei* (Coleoptera: Scolytidae). *Bulletin of Entomological Research*, **80**: 121–128.

Adams, P.B. 1990. The potential of mycoparasites for biological control of plant diseases. *Annual Review of Phytopathology*, **28**: 59–72.

Adlung, K. 1966. A critical evaluation of the European research on use of red wood ants (*Formica rufa* Group) for the protection of forests against harmful insects. *Zeitschrift für Angewandte Entomologie*, **57**: 167–189.

Agricola, U., D. Agounké, H.U. Fischer, and D. Moore. 1989. The control of *Rastrococcus invadens* Williams (Hemiptera: Pseudococcidae) in Togo by the introduction of *Gyranusoidea tebygi* Noyes (Hymenoptera: Encyrtidae). *Bulletin of Entomological Research*, **79**: 671–678.

Agrios, G.N. 1988. *Plant Pathology*. Academic Press, New York.

Ahmed, S.S., A.L. Linden, and J.J. Cech, Jr. 1988. A rating system and annotated bibliography for the selection of appropriate, indigenous fish species for mosquito and weed control. *Bulletin of the Society of Vector Ecologists*, **13**: 1–59.

Ainsworth, G.C., (ed.). 1971. *Ainsworth and Bisby's Dictionary of the Fungi* (6th ed.). Commonwealth Mycological Institute, Kew, Surrey, U.K.

Ainsworth, G.C., F.K. Sparrow, and A.S. Sussman, (eds.). 1973. *The Fungi. An Advanced Treatise, Vols. 4a and b.* Academic Press, New York.

Aizawa, K. 1987. Strain improvement of insect pathogens, pp. 3–11. *In* Maramorosch, K., (ed.). *Biotechnology in Invertebrate Pathology and Cell Culture.* Academic Press, New York.

Aizawa, K. 1990. Registration requirements and safety considerations for microbial pest control agents in Japan, pp. 31–39. *In* Laird, M., L.A Lacey, and E.W. Davidson, (eds.). *Safety of Microbial Insecticides.* CRC Press, Inc., Boca Raton, Florida, U.S.A.

Akhurst, R.J. 1990. Safety to nontarget invertebrates of nematodes of economically important pests, pp. 233–240. *In* Laird, M., L.A Lacey, and E.W. Davidson, (eds.). *Safety of Microbial Insecticides.* CRC Press, Inc., Boca Raton, Florida, U.S.A.

Akre, R.D., A. Greene, J.F. MacDonald, P.J. Landolt, and H.G. Davis. 1980. *Yellow Jackets of America North of Mexico.* USDA Agricultural Handbook No. 552.

Alabouvette, C., F. Rouxel, and J. Louvet. 1979. Characteristics of *Fusarium* wilt-suppressive soils and proposals for their utilization in biological control, pp. 165–182. *In* B. Schippers and W. Gams, (eds.). *Soil-borne plant pathogens.* Academic Press, London.

Alam, M.M., F.D. Bennett, and K.P. Carl. 1971. Biological control of *Diatraea saccharalis* (F.) in Barbados by *Apanteles flavipes* Cam. and *Lixophaga diatraeae* T.T. *Entomophaga*, **16**: 151–158.

Aldrich, J.R., J.P. Kochansky, and C.B. Abrams. 1984. Attractant for a beneficial insect and its parasitoids: pheromone of the predatory spined soldier bug, *Podisus maculiventris* (Hemiptera: Pentatomidae). *Environmental Entomology*, **13**: 1031–1036.

Ali, A.D. and T.E. Reagan. 1985. Vegetation manipulation impact on predator and prey populations in Louisiana sugarcane ecosystems. *Journal of Economic Entomology*, **78**: 1409–1414.

Ali, M.I. and M.A. Karim. 1990. Threshold sprays of insecticides: its advantages on conservation of arthropod predators and parasites in cotton ecosystem in Bangladesh. *Bangladesh Journal of Zoology*, **18**: 17–22.

Allan, D.J. and M.G. Hill. 1984. Rearing and release of *Bonnetia comta* (Diptera: Tachinidae), a parasite of *Agrotis ipsilon* (Lepidoptera: Noctuidae). *New Zealand Entomologist*, **8**: 71–74.

Allard, G.B., C.A. Chase, J.B. Heale, J.E. Isaac, and C. Prior. 1990. Field evaluation of *Metarhizium anisopliae* (Deuteromycotina: Hyphomycetes) as a mycoinsecticide for control of sugarcane froghopper, *Aeneolamia varia saccharina* (Hemiptera: Cercopidae). *Journal of Invertebrate Pathology*, **55**: 41–46.

Altieri, M.A. and W.H. Whitcomb. 1979. The potential use of weeds in the manipulation of beneficial insects. *Hortscience*, **14**(1): 12–18.

Altieri, M.A., J. Trujillo, L. Campos, C. Klein-Koch, C.S. Gold, and J.R. Quezada. 1989. Classical biological control in Latin America in its historical context. *Manejo Integrado de Plagas No. 12.* University of California Press, Berkeley. (in Spanish)

Anable, M.E., M.P. McClaran, and G.B. Ruyle. 1992. Spread of introduced Lehmann lovegrass *Eragrostis lehmanniana* Nees. in southern Arizona, USA. *Biological Conservation*, **61**: 181–188.

Andersen, A.N. 1989. How important is seed predation to recruitment in stable populations of long-lived perennials? *Oecologia*, **81**: 310–315.

Anderson, J.R. 1978. Pesticide effects on non-target soil microorganisms, pp. 313–533. *In* Hill, I.R., and S.J.L. Wright, (eds.). *Pesticide Microbiology.* Academic Press, London.

Anderson, R.M. 1982. Theoretical basis for the use of pathogens as biological control agents of pest species. *Parasitology*, **84**: 3–33.

Anderson, R.M. and R.M. May 1980. Infectious diseases and population cycles of forest insects. *Science*, **210**: 658–661.

Anderson, R.M. and R.M. May. 1981. The population dynamics of microparasites and their invertebrate hosts. *Philosophical Transactions of the Royal Society of London, Series B*, **291**: 451–524.

Anderson, R.M. and R.M. May. 1986 . The invasion, persistence and spread of infectious diseases within animal and plant communities. *Philosophical Transactions of the Royal Society of London, Series B*, **314**: 533–570.

Andow, D.A. 1986. Plant diversification and insect population control in agroecosystems, pp. 277–386. *In* Pimentel, D., (ed.). *Some Aspects of Integrated Pest Management.* Cornell University Press, Ithaca, N.Y., U.S.A.

Andow, D.A. 1988. Management of weeds for insect manipulation in agroecosystems, pp. 265–301. *In* Altieri, M.A., and M. Liebman, (eds.). *Weed Management in Agroecosystems: Ecological Approaches.* CRC Press, Inc., Boca Raton, Florida, U.S.A.

Andow, D.A. 1990. Population dynamics of an insect herbivore in simple and diverse habitats. *Ecology*, **71**: 1006–1017.

Andow, D.A. 1991a. Control of arthropods using crop diversity, pp. 257–284. *In* Pimentel, D.P. [ed.] CRC *Handbook of Pest Management in Agriculture* (2nd ed.), Volume I. CRC Press, Inc , Boca Raton, Florida, U.S.A.

Andow, D.A. 1991b. Yield loss to arthropods in vegetationally diverse agroecosystems. *Environmental Entomology*, **20**: 1228–1235.

Andow, D.A. and D.R. Prokrym. 1991. Release density, efficiency and disappearance of *Trichogramma nubilale* for control of European corn borer. *Entomophaga*, **36**: 105–113.

Andreadis, T.G. 1987. Transmission, pp. 159–176. *In* Fuxa, J.R., and Y. Tanada, (eds.). *Epizootiology of Insect Diseases.* John Wiley and Sons, New York.

Andres, L.A. 1976. The economics of biological control of weeds. *Aquatic Biology*, **3**: 111–123.

Andres, L.A. and C.J. Davis. 1973. The biological control of weeds with insects in the United States. Proceedings of the 2nd International Symposium on Biological Control of Weeds, *Commonwealth Institute of Biological Control, Miscellaneous Publications No. 6*: 11–28.

Andres, L.A. and R.D. Goeden. 1971. The biological control of weeds by introduced natural enemies, pp. 143–164. *In* Huffaker, C.B., (ed.). *Biological Control.* Plenum Press, New York.

Andres, L.A., C. J. Davis, P. Harris, and A.J. Wapshere. 1976. Biological control of weeds, pp. 481–499. In Huffaker, C.B. and P.S. Messenger, (eds.). Theory and Practice of *Biological Control.* Academic Press, New York.

Andrews, J.H. 1992. Biological control in the phyllosphere. *Annual Review of Phytopathology*, **30**: 603–635.

Andrews, J.H. and R.F. Harris. 1985. *r*- and *K*-selection and microbial ecology. *Advances in Microbial Ecology*, **9**: 99–147.

Andrews, K.L., J.W. Bentley, and R.D. Cave. 1992. Enhancing biological control's contributions to integrated pest management through appropriate levels of farmer participation. *Florida Entomologist*, **75**: 429–439.

Angus, T.A. and P. Luthy. 1971. Formulation of microbial insecticides, pp. 623–638. *In* Burges, H.D. and N.W. Hussey, (eds.). *Microbial Control of Insects and Mites.* Academic Press, London.

Anon. 1971. *Biological Control Programmes Against Insects and Weeds in Canada, 1959–1968. Commonwealth Agricultural Bureaux Technical Communication. No. 4.*, Lamport Gilbert Printers, Reading, U.K.

Anon. 1980. *Biological Control Service. 25 Years of Achievement.* Commonwealth Agricultural Bureaux, Slough, U.K.

Anon. 1987. Early warning system. *Citrograph,* **72**(9): 182,184.

Anon. 1988a. The maize pyralid. *Phytoma,* **403**: 54. (in French)

Anon. 1988b. Locusts blow from Africa to Caribbean. *The Miami Herald,* October 21, 1988, p. 19A.

Anon. 1992. *Expert Consultation on Guidelines for Introduction of Biological Control Agents.* FAO. Rome, Italy. 17–19 September, 1991

Anon. 1994a. 1995 directory of least-toxic pest control products. *The IPM Practitioner,* **16**: 1–34.

Anon. 1994b. Rabbit control on Macquaries Island. *Oryx,* **28**(1): 13.

Antía-Londoño, O.P., F. Posada-Florez, A.E. Bustillo-Pardey, and M.T. González-Garciá. 1992. *Produccion en finca del hongo Beauveria bassiana para el control de la broca del cafe.* No. 182, pub. by Cenicafe, Chinchiná, Caldas, Colombia, Octubre, 1992. (in Spanish)

Antolin, M.F. 1989. Genetic considerations in the study of attack behavior of parasitoids, with reference to *Muscidifurax raptor* (Hymenoptera: Pteromalidae). *Florida Entomologist,* **72**: 15–32.

Arakaki, N. 1990. Phoresy of *Telenomus* sp. (Scelionidae: Hymenoptera), an egg parasitoid of the tussock moth *Euproctis taiwana. Journal of Ethology,* **8**: 1–3.

Argov, Y. and Y. Rossler. 1988. Introduction of beneficial insects into Israel for the control of insect pests. *Phytoparasitica,* **16**: 303–316.

Arnett, R.H. 1968. *The Beetles of the United States (a Manual for Identification).* The American Entomological Institute, 5950 Warren Road, Ann Arbor, Michigan, U.S.A.

Arthington, A.H. and L.N. Lloyd. 1989. Introduced poeciliids in Australia and New Zealand, pp. 333–348. *In* Meffe, G.K. and F.F. Snelson, Jr., (eds.). *Ecology and Evolution of Livebearing Fishes (Poecilidae).* Prentice Hall, Englewood Cliffs, New Jersey, U.S.A.

Arthur, A.P. 1966. Associative learning in *Itoplectis conquisitor* (Say) (Hymenoptera: Ichneumonidae). *Canadian Entomologist,* **98**: 213–223.

Ashton, P.J., C.C. Appleton, and P.B.N. Jackson. 1986. Ecological impacts and economic consequences of alien invasive organisms in southern African aquatic ecosystems, pp. 247–257. *In* Macdonald, I.A.W., F.J. Kruger, and A.A. Ferrar, (eds.). *The Ecology and Management of Biological Invasions in Southern Africa; Proceedings of the National Synthesis Symposium on the Ecology of Biological Invasions.* Oxford University Press, Cape Town, South Africa.

Askew, R.R. 1971. *Parasitic Insects.* American Elsevier Pub. Co., New York.

Askew, R.R. and M.R. Shaw. 1986. Parasitoid communities: their size, structure and development, pp. 225–264. *In* Waage, J. and D. Greathead, (eds.). *Insect Parasitoids.* Academic Press, London.

Atkinson, W.D. and B. Shorrocks. 1977. Breeding site specificity in the domestic species of *Drosophila. Oecologia,* **29**: 223–232.

Atkinson, W.D. and B. Shorrocks. 1981. Competition on a divided and ephemeral resource: A simulation model. *Journal of Animal Ecology,* **50**: 461–471.

Atlegrim, O. 1989. Exclusion of birds from bilberry stands: impact on insect larval density and damage to the bilberry. *Oecologia,* **79**: 136–139.

Auger, J., C. Lecomte, J. Paris, and E. Thibout. 1989. Identification of leek-moth and diamondback-moth frass volatiles that stimulate parasitoid *Diadromus pulchellus. Journal of Chemical Ecology,* **15**: 1391–1398.

Avilla, J. and R. Albajes. 1984. The influence of female age and host size on the sex ratio of the parasitoid *Opius concolor. Entomologia Experimentalis et Applicata,* **35**: 43–47.

Axtell, R.C. 1963. Effect of *Macrochelidae* (Acarina: Mesostigmata) on house fly production from dairy cattle manure. *Journal of Economic Entomology,* **56**: 317–321.

Axtell, R.C. 1981. Use of predators and parasites in filth fly IPM programs in poultry housing, pp. 26–43. *In*: *Status of Biological Control of Filth Flies.* Proceedings of a Workshop, February 4–5, 1981, University of Florida, Gainesville, Published by USDA, SEA.

Ayal, Y. 1987. The foraging strategy of *Diaeretiella rapae.* I. The concept of the elementary unit of foraging. *Journal of Animal Ecology,* **56**: 1057–1068.

Ayala, F.J., M.E. Gilpin, and J.G. Ehrenfield. 1974. Competition between species: theoretical models and experimental tests. *Theoretical Population Biology,* **4**: 331–356.

Ayers, W.A. and P.B. Adams. 1981. Mycoparasitism and its application to biological control of plant diseases, pp. 91–105. *In* Papavizas, G.C., (ed.). *Biological Control in Crop Production.* Beltsville Symposium in Agricultural Research 5, Allenheld, Osmum Publishing, Totowa, New Jersey, U.S.A.

Bach, C.E. 1991. Direct and indirect interactions between ants (*Pheidole megacephala*), scales (*Coccus viridis*) and plants (*Pluchea indica*). *Oecologia,* **87**: 233–239.

Bai, B. and M. Mackauer. 1991 Recognition of heterospecific parasitism: competition between aphidiid (*Aphidius ervi*) and aphelinid (*Aphelinus asychis*) parasitoids of aphids (Hymenoptera: Aphidiidae; Aphelinidae). *Journal of Insect Behavior*, **4**: 333–345.

Bai, B. and S.M. Smith. 1994. Patterns of host exploitation by the parasitoid wasp *Trichogramma minutum* (Hymenoptera: Trichogrammidae) when attacking eggs of the spruce budworm (Lepidoptera: Tortricidae) in Canadian forests. *Annals of the Entomological Society of America*, **87**: 546–553.

Bailey, J.A., (ed.). 1985. *Biology and Molecular Biology of Plant-Pathogen Interactions*. Springer-Verlag, Berlin, Germany.

Bailey, P.T. 1989. The milipede parasitoid *Pelidnoptera nigripennis* (F.) (Diptera: Sciomyzidae) for the biological control of the millipede *Ommatoiulus moreleti* (Lucas) (Diplopoda: Julida: Julidae) in Australia. *Bulletin of Entomological Research*, **79**: 381–391.

Baker, R. 1985. Biological control of plant pathogens: definitions, pp. 25–39. *In* Hoy, M.A. and D.C. Herzog. *Biological Control in Agricultural IPM Systems*. Academic Press, Orlando, Florida, U.S.A.

Balasubramanian, S. and A.D. Pawar. 1990. Village level mass-rearing of egg parasites for the control of lepidopterous pests of rice. *Plant Protection Bulletin (Faridabad)*, **42**(3;4): 14–16.

Balch, R.E. and F.T. Bird. 1944. A disease of the European spruce sawfly, *Gilpinia hercyniae* (Htg.) and its place in natural control. *Science in Agriculture*, **25**: 65–80.

Balciunas, J.K. and D.W. Burrows. 1993. The rapid suppression of the growth of *Melaleuca quinquenervia* saplings in Australia by insects. *Journal of Aquatic Plant Management*, **31**: 265–270.

Balduf, W.V. 1931. Carnivorous moths and butterflies. *Transactions of the Illinois State Academy of Science*, **24**: 156–164.

Banerjee, B. 1979. A key-factor analysis of population fluctuation in *Andraca bipunctata* (Walker) (Lepidoptera: Bombycidae). *Bulletin of Entomological Research*, **69**: 195–201.

Banks, C.J. and E.D.M. Macaulay. 1967. Effects of *Aphis fabae* Scop. and its attendant ants and insect predators on yields of field beans (*Vicia faba* L.). *Annals of Applied Biology*, **60**: 445–453.

Baranowski, R., H. Glenn, and J. Sivinski. 1993. Biological control of the Caribbean fruit fly (Diptera: Tephritidae). *Florida Entomologist*, **76**: 245–251.

Barbosa, P. and A. Segarra-Carmona. 1993. Criteria for the selection of pest arthropod species as candidates for biological control, pp. 5–23. *In* Van Driesche, R.G. and T.S. Bellows, Jr., (eds.). *Steps in Classical Arthropod Biological Control*. Thomas Say Publications in Entomology, Entomological Society of America, Lanham, Maryland, U.S.A.

Bari, M.A. and H.K. Kaya. 1984. Evaluation of the entomogenous nematode *Neoplectana carpocapsae* (= *Steinernema feltiae*) Weiser (Rhabditida: Steinernematidae) and the bacterium *Bacillus thuringiensis* Berliner var. ***kurstaki*** for suppression of the artichoke plume moth (Lepidoptera: Pterophoridae). *Journal of Economic Entomology*, **77**: 225–229.

Barlow, N.D. 1994. Predicting the effect of a novel vertebrate biocontrol agent: a model for viral-vectored immunocontraception of New Zealand possums. *Journal of Applied Ecology*, **31**: 454–462.

Barnes, H.F. 1929. Gall midges as enemies of aphids. *Bulletin of Entomological Research*, **20**: 433–442.

Barreto, R.W. and H.C. Evans. 1988. Taxonomy of a fungus introduced into Hawaii for biological control of *Ageratina riparia* (Eupatorieae; Compositae), with observations on related weed pathogens. *Transactions of the British Mycological Society*, **91**: 81–97.

Barron, G.L. 1977. *The Nematode-Destroying Fungi. Topics in Mycobiology, Vol. 1*. Canadian Biological Publications, Guelph, Ontario, Canada.

Bartlett, B.R. 1951. The action of certain "inert" dust materials on parasitic Hymenoptera. *Journal of Economic Entomology*, **44**: 891–896.

Bartlett, B.R. 1953. Retentive toxicity on field-weathered insecticide residues to entomophagous insects associated with citrus pests in California. *Journal of Economic Entomology*, **46**: 565–569.

Bartlett, B.R. 1963. The contact toxicity of some pesticide residues to hymenopterous parasites and coccinellid predators. *Journal of Economic Entomology*, **56**: 694–698.

Bartlett, B.R. 1964a. Patterns in the host-feeding habit of adult parasitic Hymenoptera. *Annals of the Entomological Society of America*, **57**: 344–350.

Bartlett, B.R. 1964b. The toxicity of some pesticide residues to adult *Amblyseuis hibisci*, with a compilation of the effects of pesticides upon phytoseiid mites. *Journal of Economic Entomology*, **57**: 559–563.

Bartlett, B.R. 1966. Toxicity and acceptance of some pesticides fed to parasitic Hymenoptera and predatory coccinellids. *Journal of Economic Entomology*, **59**: 1142–1149.

Bartlett, B.R. 1978. Margarodidae, pp. 132–136. *In* Clausen, C.P., (ed.). *Introduced Parasites and Predators of Arthropod Pests and Weeds: a World Review*. USDA Agricultural Handbook No. 480, Washington, D.C., U.S.A.

Bartlett, B.R. and R. van den Bosch. 1964. Foreign exploration for beneficial organisms, pp. 283–304. *In* DeBach, P., (ed.). *Biological Control of Insect Pests and Weeds*. Reinhold, New York.

Bartlett, M.C. and S.T. Jaronski. 1988. Mass production of entomogenous fungi for biological control of insects, pp. 61–85. *In* Burge, M.N., (ed.). *Fungi in Biological Control Systems.* Manchester Press, Manchester, U.K.

Barton, L.C. and F.W. Stehr. 1970. Normal development of Anaphes flavipes in cereal beetle eggs killed with x-radiation and potential field use. *Journal of Economic Entomology*, **63**: 128–130.

Bateman, R.P. 1992. Controlled droplet application of mycoinsecticides: an environmentally friendly way to control locusts. *Antenna*, **16**(1): 6–13.

Bateman, R.P., M. Carey, D. Moore, and C. Prior. 1993. The enhanced infectivity of *Metarhizium flavoviride* in oil formulations to desert locusts at low humidities. *Annals of Applied Biology*, **122**: 145–152.

Baudoin, A.B.A.M., R.G. Abad, L.T. Kok, and W.L. Bruckart. 1993. Field evaluation of *Puccinia carduorum* for biological control of musk thistle. *Biological Control*, **3**: 53–60.

Baumann, L. and P. Baumann. 1989. Expression in *Bacillus subtilis* of the 51- and 42-kilodalton mosquitocidal toxin genes of *Bacillus sphaericus. Applied Environmental Microbiology*, **55**: 252–253.

Baumann, L., A.H. Broadwell, and P. Baumann. 1988. Sequence analysis of the mosquitocidal toxin genes encoding 51.4- and 41.9-kilodalton proteins from *Bacillus sphaericus* 2362 and 2297. *Journal of Bacteriology*, **170**: 2045–2050.

Baumann, P., L. Baumann, R.D. Bowditch, and A.H. Broadwell. 1987. Cloning of the gene for the larvicidal toxin of *Bacillus sphaericus* 2362: Evidence for a family of related sequences. *Journal of Bacteriology*, **169**: 4061–4067.

Baumann, P., M.A. Clark, L. Baumann, and A.H. Broadwell. 1991. *Bacillus sphaericus* as a mosquito pathogen: Properties of the organism and its toxins. *Microbiology Reviews*, **55**: 425–436.

Bay, E.C., C.O. Berg, H.C. Chapman, and E.F. Legner. 1976. Biological control of medical and veterinary pests, pp. 457–479. *In* Huffaker, C.B. and P.S. Messenger, (eds.). *Theory and Practice of Biological Control.* Academic Press, New York.

Beard, R.L. 1940. Parasite castration of *Anasa tristis* DeG. by *Trichopoda pennipes* Fab., and its effect on reproduction. *Journal of Economic Entomology*, **33**: 269–272.

Beardsley, J.W. 1991. Introduction of arthropod pests into the Hawaiian Islands. *Micronesica Supplement*, **3**: 1–4.

Bechinski, E.J. and L.P. Pedigo. 1981. Ecology of predaceous arthropods in Iowa soybean agroecosystems. *Environmental Entomology*, **10**: 771–778.

Beck, S.D. 1980. *Insect Photoperiodism*, 2nd ed. Academic Press, London.

Beckage, N.E. 1985. Endocrine interactions between endoparasitic insects and their hosts. *Annual Review of Entomology*, **30**: 371–413.

Becker, N., S. Djakaria, A. Kaiser, O. Zulhasrii, and H.W. Ludwig. 1991. Efficacy of a new tablet formulation of an asporogenous strain of *Bacillus thuringiensis israelensis* against larvae of *Aedes aegypti. Bulletin of the Society of Vector Ecologists*, **16**: 176–182.

Bedding, R.A. 1968. *Deladenus wilsoni* n. sp. and *D. siricidicola* n. sp. (Neotylenchidae), entomophagous-mycetophagous nematodes parasitic in siricid woodwasps. *Nematologica*, **14**: 515–525.

Bedding, R.A. 1974. Five new species of *Deladenus* (Neotylenchidae) entomophagous-mycetophagous nematodes parasitic in siricid woodwasps. *Nematologica*, **20**: 204–225.

Bedding, R.A. 1984. Nematode parasites of Hymenoptera, pp. 755–795. *In* Nickle, W.R., (ed.). *Plant and Insect Nematodes.* Marcel Dekker, Inc., New York.

Bedding, R.A. and L.A. Miller. 1981. Disinfesting blackcurrant cuttings of *Synanthedon tipuliformis*, using the insect parasitic nematode, *Neoplectana bibionis. Environmental Entomology*, **10**: 449–453.

Beddington, J.R. 1975. Mutual interference between parasites or predators and its effect on searching efficiency. *Journal of Animal Ecology*, **44**: 331–340.

Beddington, J.R. and P.S. Hammond. 1977. On the dynamics of host-parasite-hyperparasite interactions. *Journal of Animal Ecology*, **46**: 811–821.

Beddington, J.R., C.A. Free, and J.H. Lawton. 1975. Dynamic complexity in predator-prey models framed in difference equations. *Nature*, **255**: 58–60.

Beddington, J.R., C.A. Free, and J.H. Lawton. 1978. Characteristics of successful natural enemies in models of biological control of insect pests. *Nature*, **273**: 513–519.

Bedford, E.C.G. 1989. The biological control of the circular purple scale, *Chrysomphalus aonidum* (L.), on citrus in South Africa. *Technical Communication, Department of Agriculture and Water Supply, South Africa*, No. 218, pp. 16.

Bedford, G.O. 1980. Biology, ecology, and control of palm rhinoceros beetles. *Annual Review of Entomology*, **35**: 309–339.

Bedford, G.O. 1986. Biological control of the rhinoceros beetle (*Oryctes rhinoceros*) in the South Pacific by baculovirus. *Agriculture, Ecosystems and Environment*, **15**: 141–147.

Beegle, C.C. and T. Yamamoto. 1992. Invitation paper (C.P. Alexander Fund): history of *Bacillus thuringiensis* Berliner research and development. *Canadian Entomologist*, **124**: 587–616.

Beer, S.V., J.R. Rundle, and J.L. Norielli. 1984. Recent progress in the development of biological control for fire blight. *Acta Horticulturae*, **151**: 195–201.

Beevers, M., W. Joe Lewis, H.R. Gross, Jr., and D.A. Nordlund. 1981. Kairomones and their use for management of entomophagous insects: X. Laboratory studies on manipulation of host-finding behavior of *Trichogramma pretiosum* Riley with a kairomone extracted from *Heliothis zea* (Boddie) moth scales. *Journal of Chemical Ecology*, **7**: 635–648.

Beggs, J.R. and P.R. Wilson. 1991. The kaka *Nestor meridionalis*, a New Zealand parrot endangered by introduced wasps and mammals. *Biological Conservation* **6**: 23–38.

Begon, M., J.L. Harper, and C.R. Townsend. 1986. *Ecology: Individuals, Populations and Communities*. Blackwell Scientific Publications, Oxford, U.K.

Beirne, B.P. 1975. Biological control attempts by introductions against pest insects in the field in Canada. *Canadian Entomologist*, **107**: 225–236.

Beirne, B.P. 1984. Biological control of the European fruit lecanium, *Lecanium tiliae* (Homoptera: Coccidae), in British Columbia. *Journal of the Entomological Society of British Columbia*, **81**: 28.

Bell, M.R. 1991. *In vivo* production of a nuclear polyhedrosis virus utilizing tobacco budworm and a multicellular larval rearing container. *Journal of Entomological Science*, **26**: 69–75.

Bell, W.J. 1990. Searching behavior patterns in insects. *Annual Review of Entomology*, **35**: 447–467.

Bellotti, A. and B. Arias. 1977. Biology, ecology and biological control of the cassava hornworm (*Erinnyis ello*), pp. 227–232. *Proceedings of the Cassava Protection Workshop, Cali, Colombia*, November, 1977.

Bellotti, A.C., N. Mesa, M. Serrano, J.M. Guerrero, and C.J. Herrera. 1987. Taxonomic inventory and survey activity for natural enemies of cassava green mites in the Americas. *Insect Science and Its Application*, **8**: 845–849.

Bellows, T.S., Jr. 1981. The descriptive properties of some models for density dependence. *Journal of Animal Ecology*, **50**: 139–156.

Bellows, T.S., Jr. 1982a. Analytical models for laboratory populations of *Callosobruchus chinensis* and *C. maculatus* (Coleoptera, Bruchidae). *Journal of Animal Ecology*, **51**: 263–287.

Bellows, T.S., Jr. 1982b. Simulation models for laboratory populations of *Callosobruchus chinensis* and *C. maculatus*. *Journal of Animal Ecology*, **51**: 597–623.

Bellows, T.S., Jr. 1993. Introduction of natural enemies for suppression of arthropod pests, pp. 82–89. *In* Lumsden, R. and J. Vaughn, (eds.). *Pest Management: Biologically Based Technologies*. American Chemical Society, Washington, D.C., U.S.A.

Bellows, T.S., Jr. and M.H. Birley. 1981. Estimating developmental and mortality rates and stage recruitment from insect stage frequency data. *Researches on Population Ecology*, **23**: 232–244.

Bellows, T.S., Jr. and M.P. Hassell. 1984. Models for interspecific competition in laboratory populations of *Callosobruchus* spp. *Journal of Animal Ecology*, **53**: 831–848.

Bellows, T.S., Jr. and M.P. Hassell. 1988. The dynamics of age– structured host–parasitoid interactions. *Journal of Animal Ecology*, **57**: 259–268.

Bellows, T.S., Jr. and E.F. Legner. 1993. Foreign Exploration, pp. 25–41. *In* Van Driesche, R.G. and T.S. Bellows, Jr., (eds.). *Steps in Classical Arthropod Biological Control*. Thomas Say Publications in Entomology, Entomological Society of America, Lanham, Maryland, U.S.A.

Bellows, T.S., Jr. and J.G. Morse. 1988. Residual toxicity following dilute or low-volume applications of insecticides used for control of California red scale (Homoptera: Diaspididae) to four beneficial species in a citrus agroecosystem. *Journal of Economic Entomology*, **81**: 892–898.

Bellows, T.S., Jr. and J.G. Morse. 1993. Toxicity of pesticides used in citrus to *Aphytis melinus* DeBach (Hymenoptera: Aphelinidae) and *Rhizobius lophanthae* (Blaisd.) (Coleoptera: Coccinellidae). *Canadian Entomologist*, **125**: 987–994.

Bellows, T.S., Jr., J.C. Owens, and E.W. Huddleston. 1982a. Predation of range caterpillar, *Hemileuca oliviae* (Lepidoptera: Saturniidae) at various stages of development by different species of rodents in New Mexico during 1980. *Environmental Entomology*, **11**: 1211–1215.

Bellows, T.S., Jr., M. Ortiz, J.C. Owens, and E.W. Huddleston. 1982b. A model for analyzing insect stage-frequency data when mortality varies with time. *Researches on Population Ecology*, **24**: 142–156.

Bellows, T.S., Jr., J.G. Morse, D.G. Hadjidemetriou, and Y. Iwata. 1985. Residual toxicity of four insecticides used for control of citrus thrips (Thysanoptera: Thripidae) on three beneficial species in a citrus agroecosystem. *Journal of Economic Entomology*, **78**: 681–686.

Bellows, T.S., Jr., J.G. Morse, L.K. Gaston, and J.B. Bailey. 1988. The fate of two systemic insecticides and their impact on two phytophagous and a beneficial arthropod in a citrus agroecosystem. *Journal of Economic Entomology*, **81**: 899–904.

Bellows, T.S., Jr., R.G. Van Driesche, and J.S. Elkinton. 1989. Extensions to Southwood and Jepson's graphical method of estimating numbers entering a stage for calculating mortality due to parasitism. *Researches on Population Ecology*, **31**: 169–184.

Bellows, T.S., Jr., T.D. Paine, K.Y. Arakawa, C. Meisenbacher, P. Leddy, and J. Kabashima. 1990. Biological control sought for ash whitefly. *California Agriculture*, **44**(1): 4–6.

Bellows, T.S., Jr., J.G. Morse, and L.K. Gaston. 1992a. Residual toxicity of pesticides used for control of lepidopteran insects in citrus to the predaceous mite *Euseius stipulatus* Athias-Henriot (Acarina: Phytoseiidae). *Journal of Applied Entomology*, **113**: 493–501.

Bellows, T.S., Jr., T.D. Paine, J.R. Gould, L.G. Bezark, J.C. Ball, W. Bentley, R. Coviello, J. Downer, P. Elam, D. Flaherty, P. Gouveia, K. Koehler, R. Molinar, N. O'Connell, E. Perry, and G. Vogel. 1992b. Biological control of ash whitefly: a success in progress. *California Agriculture*, **46**(1): 24, 27–28.

Bellows, T.S., Jr., R.G. Van Driesche, and J.S. Elkinton. 1992c. Life-table construction and analysis in the evaluation of natural enemies. *Annual Review of Entomology*, **37**: 587–614.

Bellows, T.S., Jr., J.G. Morse, and L.K. Gaston. 1993. Residual toxicity of pesticides used for lepidopteran insect control on citrus to *Aphytis melinus* DeBach (Hymenoptera: Aphelinidae). *Canadian Entomologist*, **125**: 995–1001.

Ben-Ze'ev, I., R.G. Kenneth, and S. Bitton. 1981. The Entomophthorales of Israel and their arthropod hosts. *Phytoparasitica*, **9**: 43–50.

Bennett, F.D. 1971. Current status of biological control of the small moth borers of sugarcane *Diatraea* spp. (Lep., Pyralidae). *Entomophaga*, **16**: 111–124.

Bennett, F.D. and I.W. Hughes. 1959. Biological control of insect pests in Bermuda. *Bulletin of Entomological Research*, **50**: 423–436.

Bennett, F.D., P. Cochereau, D. Rosen, and B.J. Wood. 1976. Biological control of pests of tropical fruits and nuts, pp. 359–395. *In* Huffaker, C.B. and P.S. Messenger, (eds.). *Theory and Practice of Biological Control.* Academic Press, New York.

Bennison, J.A. 1992. Biological control of aphids on cucumbers, use of open rearing systems or "banker plants" to aid establishment of *Aphidius matricariae* and *Aphidoletes aphidimyza. Mededelingen van de Faculteit Landbouwwetenschappen, Universiteit Gent*, **57**: 457–466.

Benz, G. 1987. Environment, pp. 177–214. *In* Fuxa, J.R. and Y. Tanada, (eds.). *Epizootiology of Insect Diseases.* John Wiley and Sons, New York.

Berebaum, M.R. and E. Miliczky. 1984. Mantids and milkweed bugs: Efficacy of aposematic coloration against invertebrate predators. *American Midland Naturalist*, **111**: 64–68.

Berg, G.N., P. Williams, R.A. Bedding, and R.J. Akhurst. 1987. A commercial method of application of entomopathogenic nematodes to pasture for controlling subterranean insect pests. *Plant Protection Quarterly*, **2**(4): 174–177.

Bergman, J.M. and W.M. Tingey. 1979. Aspects of interaction between plant genotypes and biological control. *Bulletin of the Entomological Society of America*, **25**: 275–279.

Berliner, E. 1911. Uber die Schlaffsucht der Mehlmottneraupe. *Zeitschrift für das Gesamte Getreidewesen*, **3**: 63–70.

Berlocher, S.H. 1979. Biochemical approaches to strain, race, and species discriminations, pp. 137–144. *In* Hoy, M.A. and J.J. McKelvey Jr., (eds.). *Genetics in Relation to Insect Management.* Rockefeller Foundation, New York.

Bernal, J. and D. González. 1993. Temperature requirements of four parasites of the Russian wheat aphid *Diuraphis noxia. Entomologia Experimentalis et Applicata*, **69**: 173–182.

Berry, J.S., T.O. Holtzer, and J.M. Norman. 1991. Experiments using a simulation model of the Banks grass mite (Acari: Tetranychidae) and the predatory mite *Neoseiulus fallacis* (Acari: Phytoseiidae) in a corn microenvironment. *Environmental Entomology*, **20**: 1074–1078.

Bess, H.A., R. van den Bosch, and F.H. Haramoto. 1961. Fruit fly parasites and their activities in Hawaii. *Proceedings of the Hawaiian Entomological Society*, **17**: 367–378.

Betke, P., T. Hiepe, P. Muller, R. Ribbeck, H. Schultka, and H. Schumann. 1989. Biological control of Musca domestica with *Ophyra aenescens* on pig production enterprises. *Monatshefte für Veterinarmedizin*, **44**(23): 842–844. (in German)

Betz, F.S. 1986. Registration of baculoviruses as pesticides, pp. 203–222. *In* Granados, R.R. and B.A. Federici, (eds.). *The Biology of Baculoviruses: Volume II. Practical Application for Insect Control.* CRC Press, Inc., Boca Raton, Florida, U.S.A.

Betz, F.S., S.F. Forsyth, and W.E. Stewart. 1990. Registration requirements and safety considerations for microbial pest control agents in North America, pp. 3–10. *In* Laird, M., L.A. Lacey, and E.W. Davidson, (eds.). *Safety of Microbial Insecticides.* CRC Press, Inc., Boca Raton, Florida, U.S.A.

Bhumiratana, A. 1990. Local production of *Bacillus sphaericus*, pp. 272–283. *In* de Barjac, H. and D.J. Sutherland, (eds.). *Bacterial Control of Mosquitoes and Black Flies: Biochemistry, Genetics and Application of Bacillus thuringiensis israelensis and Bacillus sphaericus.* Rutgers University Press, New Brunswick, New Jersey, U.S.A.

Bieri, M., F. Zwygart, G. Tognina, and G. Stadler. 1989. The importance of soil water content for the biological control of

Thrips tabaci Lind. on cucumbers in the greenhouse. *Mitteilungen der Schweizerischen Entomologischen Gesellschaft*, **62**: 28. (in German)

Bigger, M. 1976. Oscillation of tropical insect populations. *Nature*, **259**: 207–209.

Bigler, F. 1986. Mass production of *Trichogramma maidis* Pint. et Voeg. and its field application against *Ostrinia nubilalis* Hbn. in Switzerland. *Zeitschrift für Angewandte Entomologie*, **101**: 23–29.

Bigler, F., (ed.). 1991. Fifth workshop of the IOBC global working group "Quality control of mass reared arthropods". Wageningen, The Netherlands, 25–28, March, 1991.

Bigler, F. 1994. Quality control in *Trichogramma* production, pp. 93–144. *In* Wajnberg, E. and S.A. Hassan, (eds.). *Biological Control With Egg Parasitoids*. Commonwealth Agricultural Bureaux, Wallingford, U.K.

Bigler, F., A. Meyer, and S. Bosshart. 1987. Quality assessment in *Trichogramma maidis* Pintureau et Voegele reared from eggs of the factitious hosts *Ephestia kuehniella* Zell. and *Sitotroga cerealella* (Olivier). *Journal of Applied Entomology*, **104**: 340–353.

Bigler, F., M. Bieri, A. Fritschy, and K. Seidel. 1988. Variation in locomotion between laboratory strains of *Trichogramma maidis* and its impact on parasitism of eggs of *Ostrinia nubilalis* in the field. *Entomologia Experimentalis et Applicata*, **49**: 283–290.

Bigler, F., S. Bosshard, and M. Waldburger. 1989. Bisherige und neue Entwicklungen bei der biologischen Bekämpfung des Maizünslers, *Ostrinia nubilalis* Hbn., mit *Trichogramma maidis* Pint. et Voeg. in der Schweiz. *Landwirtschaft Schweiz Band*, **2**: 37–43.

Bilimoria, S.L. 1986. Taxonomy and identification of baculoviruse, pp. 37–59. *In* Granados, R.R. and B.A. Federici, (eds.). *The Biology of Baculoviruses: Volume I Biological Properties and Molecular Biology*. CRC Press, Inc., Boca Raton, Florida, U.S.A.

Bishop, L. and S.E. Riechert. 1990. Spider colonization of agroecosystems: mode and source. *Environmental Entomology*, **19**: 1738–1745.

Blackshaw, R.P. 1988. A survey of insect parasitic nematodes in Northern Ireland. *Annals of Applied Biology*, **113**: 561–565.

Blakeman, J.P. 1988. Competitive antagonism of air-borne fungal pathogens, pp. 141–160. *In* Burge, M.N., (ed.). *Fungi in Biological Control Systems*. Manchester University Press, Manchester, U.K.

Blakeman, J.P. and I.D.S. Brodie. 1977. Competition for nutrients between epiphytic micro-organisms and germination of spores of plant pathogens on beetroot leaves. *Physiological Plant Pathology*, **10**: 29–42.

Blaustein, L. 1992. Larvivorous fishes fail to control mosquitoes in experimental rice plots. *Hydrobiologia*, **232**: 219–232.

Blissard, G.W. and G.F. Rohrmann. 1990. Baculovirus diversity and molecular biology. *Annual Review of Entomology*, **35**: 127–155.

Blumberg, D. 1991. Seasonal variations in the encapsulation of eggs of the encyrtid parasitoid *Metaphycus stanleyi* by the pyriform scale, *Protopulvinaria pyriformis*. *Entomologia Experimentalis et Applicata*, **58**: 231–237.

Blumberg, D. and P. DeBach. 1981. Effects of temperature and host age upon the encapsulation of *Metaphycus stanleyi* and *Metaphycus helvolus* eggs by brown soft scale *Coccus hesperidum*. *Journal of Invertebrate Pathology*, **37**: 73–79.

Blumberg, D. and R.F. Luck. 1990. Differences in the rates of superparasitism between two strains of *Comperiella bifasciata* (Howard) (Hymenoptera: Encyrtidae) parasitizing California red scale (Homoptera: Diaspididae): An adaption to circumvent encapsulation? *Annals of the Entomological Society of America*, **83**: 591–597.

Bodenheimer, F.S. 1931. Zur Fruhgeschichte der Enforschung des Insektenparasitismus. (To the early history of the study of insect parasitism). *Archieves der Geschichten Mathematik Naturwissenschaften Technologie*, **13**: 402–416. (in German).

Boethel, D.J. and R.D. Eikenbary, (eds.). 1986. *Interactions of Plant Resistance and Parasitoids and Predators of Insects*. Ellis Horwood Limited, John Wiley and Sons, New York.

Boldt, P.E. and J.J. Drea. 1980. Packaging and shipping beneficial insects for biological control. *FAO Plant Protection Bulletin*, **28**(2): 64–71.

Boller, E. 1972. Behavioral aspects of mass-rearing of insects. *Entomophaga*, **17**: 9–25.

Boller, E.F., E. Janser, and C. Potter. 1984. Testing of the side-effects of herbicides used in viticulture on the common spider mite *Tetranychus urticae* and the predacious mite *Typhlodromus pyri* under laboratory and semi-field conditions. *Zeitschrift für Pflanzenkrankheiten und Pflanzenschutz*, **91**: 561–568.

Boller, E.F., U. Remund, and M.P. Candolfi. 1988. Hedges as potential sources of *Typhlodromus pyri*, the most important predatory mite in vineyards of northern Switzerland. *Entomophaga*, **33**: 249–255.

Bomar, C.R., J.A. Lockwood, M.A. Pomerinke, and J.D. French. 1993. Multiyear evaluation of the effects of *Nosema locustae* (Microsproidia: Nosematidae) on rangeland grasshoppers (Orthoptera: Acrididae) population density and natural biological controls. *Environmental Entomology*, **22**: 489–497.

Boot, W.J., O.P.J.M. Minkenberg, R. Rabbinge, and G.H. de Moed. 1992. Biological control of the leafminer *Liriomyza bryoniae* by seasonal inoculative releases of Diglyphus isaea: simulation of a parasitoid-host system. *Netherlands Journal of Plant Pathology*, **98**: 203–212.

Borror, D.J., C.A. Triplehorn, and N.F. Johnson. 1989. *An Introduction to the Study of Insects (6th ed.).* Holt, Rinehart and Winston, Inc., Orlando, Florida, U.S.A.

Bottrell, D.G. and P.L. Adkisson. 1977. Cotton insect pest management. *Annual Review of Entomology*, **22**: 451–481.

Boyette, C.D., P.C. Quimby, Jr., W.J. Connick, Jr., D.J. Daigle, and F.E. Fulgham. 1991. Progress in the production, formulation, and application of mycoherbicides, pp. 209–222. *In* TeBeest, D.O., (ed.). *Microbial Control of Weeds.* Chapman and Hall, New York, 284 pp.

Boykin, L.S. and M.V. Campbell. 1982. Rate of population increase of the two-spotted spider mite (Acari: Tetranychidae) on peanut leaves treated with pesticides. *Journal of Economic Entomology*, **75**: 966–971.

Brady, B.L. 1981. Fungi as parasites of insects and mites. *Biocontrol News and Information*, **2**: 281–296.

Braithwaite, R.W., W.M. Lonsdale, and J.S. Estbergs. 1989. Alien vegetation and native biota in tropical Australia: The impact of Mimosa pigra. *Biological Conservation* **48**: 189–210.

Bratti, A. 1991. New research trends in the field of biological control: in vitro rearing of the larval stages of parasitoids. *Difesa delle Piante*, **14**(1): 23–37. (in Italian)

Braun, A.R., J.M. Guerrero, A.C. Bellotti, and L.T. Wilson. 1987a. Relative toxicity of permethrin to *Mononychellus progresivus* Doreste and *Tetranychus urticae* Koch (Acari: Tetranychidae) and their predators *Amblyseius limonicus* Garman and McGregor (Acari: Phytoseiidae) and *Oligota minuta* Cameron (Coleoptera: Staphylinidae): bioassays and field validation. *Environmental Entomology*, **16**: 545–550.

Braun, A.R., J.M. Guerrero, A.C. Bellotti, and L.T. Wilson. 1987b. Evaluation of possible nonlethal side effects of permethrin used in predator exclusion experiments to evaluate *Amblyseius limonicus* (Acari: Phytoseiidae) in biological control of cassava mites (Acari: Tetranychidae). *Environmental Entomology*, **16**: 1012–1018.

Braun, A.R., A.C. Bellotti, J.M. Guerrero, and L.T. Wilson. 1989. Effect of predator exclusion on cassava infested with tetranychid mites (Acari: Tetranychidae). *Environmental Entomology*, **18**: 711–714.

Bravenboer, L. and G. Dosse. 1962. *Phytoseiulus riegeli* Dosse als Prädator einiger Schadmilben aus der *Tetranychus urticae*-gruppe. *Entomologia Experimentalis et Applicata*, **5**: 291–304.

Breene, R.G., A.W. Hartstack, W.L. Sterling, and M. Nyffeler. 1989. Natural control of the cotton fleahopper, *Pseudatomoscelis seriatus* (Reuter) (Hemiptera, Miridae), in Texas. *Journal of Applied Entomology*, **108**: 298–305.

Brent, K.J. 1987. Fungicide resistance in crops—its practical significance and management, pp. 137–151. *In* Brent, K.J. and R.K. Atkin, (eds.). *Rational Pesticide Use, Proceedings of the Ninth Long Ashton Symposium.* Cambridge University Press, Cambridge, U.K.

Breytenbach, G.J. 1986. Impacts of alien organisms on terrestrial communities with emphasis on communities of the southwestern Cape, pp. 229–238. *In* Macdonald, I.A.W., F.J. Kruger, and A.A. Ferrar, (eds.). *The Ecology and Management of Biological Invasions in Southern Africa, Proceedings of the National Synthesis Symposium on the Ecology of Biological Invasions.* Oxford University Press, Cape Town, South Africa.

Briese, D.T. 1989. Host-specificity and virus-vector potential of *Aphis chloris* (Hemiptera: Aphididae), a biological control agent for St. John's wort in Australia. *Entomophaga*, **34**: 247–264.

Briese, D.T. and R.J. Milner 1986. Effect of the microsporidian *Pleistophora schubergi* on *Anaitis efformata* (Lepidoptera: Geometridae) and its elimination from a laboratory colony. *Journal of Invertebrate Pathology*, **48**: 107–116.

Brinkley, C.K., R.T. Ervin, and W.L. Sterling. 1991. Potential beneficial impact of red imported fire ant to Texas cotton production. *Biological Agriculture and Horticulture*, **8**: 145–152.

Brodeur, J. and J.N. McNeil. 1989. Seasonal microhabitat selection by an endoparasitoid through adaptive modification of host behavior. *Science*, **244**: 226–228.

Brodeur, J. and J.N. McNeil. 1992. Host behaviour modification by the endoparasitoid *Aphidius nigripes*: A strategy to reduce hyperparasitism. *Ecological Entomology*, **17**: 97–104.

Brooks, W.M. 1980. Production and efficacy of protozoa. *Biotechnology and Bioengineering*, **22**: 1415–1440.

Brooks, W.M. 1988. Entomogenous protozoa, pp. 1–149. *In* Ignoffo, C.M., (ed.). *Handbook of Natural Pesticides Volume V, Microbial Insecticides, Part A, Entomogenous Protozoa and Fungi.* CRC Press, Inc., Boca Raton, Florida, U.S.A.

Brooks, W.M. and J.J. Jackson. 1990. Eugregarines: current status as pathogens, illustrated in corn rootworms, pp. 512–515. *In* Pinnock, D.E., (ed.). *Vth International Colloquium on Invertebrate Pathology and Microbial Control.* Adelaide, Australia, 20–24 August, 1990.

Brower, J.H. 1982. Parasitization of irradiated eggs and eggs from irradiated adults of the Indian meal moth (Lepidoptera: Pyralidae) by *Trichogramma pretiosum* (Hymenoptera: Trichogrammatidae). *Journal of Economic Entomology*, **75**: 939–944.

Brower, L.P. 1969. Ecological chemistry. *Scientific American*, **220**(2): 22–29.

Brown, D.W. and R.A. Goyer. 1982. Effects of a predator complex on lepidopterous defoliators of soybean. *Environmental Entomology*, **11**: 385–389.

Brown, K.C. 1989. The design of experiments to assess the effects of pesticides on beneficial arthropods in orchards: replication versus plot size, pp. 71–80. *In* Jepson, P.C., (ed.). *Pesticides and Non-Target Invertebrates.* Intercept, Wimborne, Dorset, U.K.

Brown, M.W. and E.A. Cameron. 1979. Effects of dispalure and egg mass size on parasitism by the gypsy moth egg parasite, *Ooencyrtus kuwanai. Environmental Entomology,* **8**: 77–80.

Broza, M., M. Brownbridge, and B. Sneh. 1991. Monitoring secondary outbreaks of the African armyworm in Kenya using pheromone traps for timing *Bacillus thuringiensis* application. *Crop Protection,* **10**: 229–233.

Bruckart, W.L. and W.M. Dowler. 1986. Evaluation of exotic rust fungi in the United States for classical biological control of weeds. *Weed Science,* **34** (Supplement 1): 11–14.

Bruckart, W.L. and S. Hassan. 1991. Options with plant pathogens intended for classical control of range and pasteur weeds, pp. 69–81. *In* TeBeest, D.O., (ed.). *Microbial Control of Weeas.* Chapman and Hall, New York.

Bruns, H. 1960. The economic importance of birds in forests. *Bird Study,* **7**: 193–208.

Brust, G.E. 1991. Augmentation of an endemic entomogenous nematode by agroecosystem manipulation for the control of a soil pest. *Agriculture, Ecosystems and Environment,* **36**: 175–184.

Bruwer, I.J. and E.A. de Villiers. 1988. Biological control of the red scale, *Aonidiella aurantii,* on the Messina Experimental Farm. *Information Bulletin, Citrus and Subtropical Fruit Research Institute, South Africa,* No. 186, 12–16. (in Afrikaans)

Buchanan, R.E. and N.E. Gibbons. 1974. *Bergey's Manual of Determinative Bacteriology (8th ed.).* Williams and Wilkins, Baltimore, Maryland, U.S.A.

Buckingham, G.R., E.A. Okrah, and M.C. Thomas. 1989. Laboratory host range tests with *Hydrellia pakistanae* (Diptera: Ephydridae), an agent for biological control of *Hydrilla verticillata* (Hydrocharitaceae). *Environmental Entomology,* **18**: 164–171.220.

Bulmer, M.G. 1975. The statistical analysis of density dependence. *Biometrics,* **31**: 901–911.

Burbutis, P.P., N. Erwin, and L.R. Ertle. 1981. Reintroduction and establishment of *Lydella thompsoni* and notes on other parasites of the European corn borer in Delaware. *Environmental Entomology,* **10**: 779–781.

Burdon, J.J. and D.R. Marshall. 1981. Biological control and the reproductive mode of weeds. *Journal of Applied Ecology,* **18**: 649–658.

Burge, M.N. [ed.]. 1988. *Fungi in Biological Control Systems.* Manchester University Press, Manchester, U.K.

Burges, H.D. [ed.]. 1981a. *Microbial Control of Pests and Plant Diseases.* Academic Press, London.

Burges, H.D. 1981b. Safety, safety testing and quality control of microbial pesticides, pp. 738–769. *In* Burges, H.D., (ed.). *Microbial Control of Pests and Plant Diseases.* Academic Press, London.

Burges, H.D., G. Croizier, and J. Huber. 1980a. A review of safety tests on baculoviruses. *Entomophaga,* **25**: 329–340.

Burges, H.D., J. Huber, and G. Croizier. 1980b. Guidelines for safety tests on insect viruses. *Entomophaga,* **25**: 341–348.

Burman, M., K. Abrahamsson, J. Ascard, A. Sjoberg, and B. Erikson. 1986. Distribution of insect parasitic nematodes in Sweden, p. 312. *In* Samson, R.A., J.M. Vlak, and D. Peters, (eds.). *Fundamentals and Applied Aspects of Invertebrate Pathology.* Foundation 4th International Colloquium on Invertebrate Pathology, Wageningen, the Netherlands.

Burnett, T. 1958. A model of host-parasite interaction. *Proceedings of the Xth International Congress of Entomology,* **2**: 679–686.

Burnett, T. 1979. An acarine predator-prey population infesting roses. *Researches on Population Ecology,* **20**: 227–234.

Buschman, L.L. and L.J. DePew. 1990. Outbreaks of Banks grass mite (Acari: Tetranychidae) in grain sorghum following insecticide application. *Journal of Economic Entomology,* **83**: 1570–1574.

Bustillo, A.E. and A.T. Drooz. 1977. Cooperative establishment of a Virginia (USA) strain of *Telenomus alsophilae* on *Oxydia trychiata* in Colombia. *Journal of Economic Entomology,* **70**: 767–770.

Byrne, D.N., T.S. Bellows, Jr., and M.P. Parrella. 1990. Whiteflies in agricultural systems, pp. 227–262. *In* Gerling, D., (ed.). *Whiteflies: Their Bionomics, Pest Status and Management.* Intercept Pub., Andover, U.K.

Cabanillas, E., K.R. Barker, and L.A. Nelson. 1989. Growth of isolates of *Paecilomyces lilacinus* and their efficacy in biocontrol of *Meloidogyne incognita* on tomato. *Journal of Nematology,* **21**: 164–172.

Cabanillas, H.E. and J.R. Raulston. 1994. Evaluation of the spatial pattern of *Steinernema riobravis* in corn plots. *Journal of Nematology,* **26**: 25–31.

Caccia, R., M. Baillod, E. Guignard, and S. Kreiter. 1985. Introduction d'une souche de Amblyseius andersoni Chant (Acari, Phytoseiidae) resistant a l'azinphos, dans la lutte contre les acariens phytophages en viticulture. *Revue Suisse Viticulture, Arboriculture, et Horticulture,* **17**: 285–290.

Cade, W. 1975. Acoustically orienting parasitoids: fly phonotaxis to cricket song. *Science,* **190**: 1312–1313.

Cadogan, G.L. and R.D. Scharbach. 1993. Efficacy of Foray 48B (*Bacillus thuringiensis* Berliner) applications against the spruce budworm, *Choristoneura fumiferana* (Clemens) (Lepidoptera: Tortricidae), timed for phenological development of balsam fir and black spruce. *Canadian Entomologist,* **125**: 479–488.

Caltagirone, L.E. 1985. Identifying and discriminating among biotypes of parasites and predators, pp. 189–200. *In* Hoy, M.A. and D.C. Herzog, (eds.). *Biological Control in Agricultural IPM Systems.* Academic Press, New York.

Caltagirone, L.E. and R.L. Doutt. 1989. The history of the vedalia beetle importation to California and its impact on the development of biological control. *Annual Review of Entomology*, **34**: 1–16.

Cameron, P.J., R.L. Hill, J. Bain, and W.P. Thomas. 1989. *A Review of Biological Control of Invertebrate Pests and Weeds in New Zealand 1847–1987.* Commonwealth Agricultural Bureaux Institute of Biological Control, Technical Communication No. 10, Wallingford, U.K.

Cameron, P.J., W. Powell, and H.D. Loxdale. 1984. Reservoirs for *Aphidius ervi* Haliday (Hymenoptera: Aphidiidae), a polyphagous parasitoid of cereal aphids (Hemiptera: Aphididae). *Bulletin of Entomological Research*, **74**: 647–656.

Campbell, A. and M. Mackauer. 1975. The effect of parasitism by *Aphidius smithi* (Hymenoptera: Aphidiidae) on reproduction and population growth of the pea aphid (Homoptera: Aphididae). *Canadian Entomologist*, **107**: 919–926.

Campbell, C.L. and D.C. Sands. 1992. Testing the effects of microbial agents on plants, pp. 689–705. *In* Levin, M.A., R.J. Seidler, and M. Rogul, (eds.). *Microbial Ecology, Principles, Methods and Applications.* McGraw-Hill, New York.

Campbell, C.L., G.P. Markin, and M.W. Johnson. 1994. Determination of the fate of the biological control agents, *Cyanotricha necyria* (Lepidoptera: Notodontidae [Dioptinae]) and *Pyrausta perelegans* (Lepidoptera: Pyralidae), on banana poka (*Passiflora mollissima*) on the Big Island of Hawaii. *Proceedings of the Hawaiian Entomological Society*, **32**: 123–130.

Campbell, R. 1985. *Plant Microbiology.* Edward Arnold, London.

Campbell, R. 1989. *Biological Control of Microbial Plant Pathogens.* Cambridge University Press, Cambridge, U.K.

Campbell, R.W. and T.R. Torgersen. 1983. Compensatory mortality in defoliator population dynamics. *Environmental Entomology*, **12**: 630–632.

Canto-Saenz, M. and R. Kaltenbach. 1984. Effect of some fungicides on *Paecilomyces lilacinus* (Thom) Samson. *Proceedings of the First International Congress of Nematology, Guelph, Canada*, **1**: 13–14.

Cantwell, G.E. and T. Lehnert. 1979. Lack of effect of certain microbial insecticides on the honeybee. *Journal of Invertebrate Pathology*, **33**: 381–382.

Capinera, J.L., S.L. Blue, and G.S. Wheeler. 1982. Survival of earthworms exposed to *Neoaplectana carpocapsae* nematodes. *Journal of Invertebrate Pathology*, **39**: 419–421.

Capinera, J.L., D. Pelissier, G.S. Menout, and N.D. Epsky. 1988. Control of black cutworm, *Agrotis ipsilon* (Lepidoptera: Noctuidae), with entomogenous nematodes (Nematoda: Steinernematidae, Heterorhabditidae). *Journal of Invertebrate Pathology*, **52**: 427–435.

Capper, A.L. and R. Campbell. 1986. The effect of artificially inoculated antagonist bacteria on the prevalence of take-all disease of wheat in field experiments. *Journal of Applied Bacteriology*, **60**: 155–160.

Caprio, M.A., M.A. Hoy, and B.E. Tabashnik. 1991. Model for implementing a genetically improved strain of a parasitoid. *American Entomologist*, **37**: 232–239.

Cardé, R.T. and Hai-Poong Lee. 1989. Effect of experience on the responses of the parasitoid *Brachymeria intermedia* (Hymenoptera: Chalcididae) to its host, *Lymantria dispar* (Lepidoptera: Lymantriidae) and to kairomone. *Annals of the Entomological Society of America*, **82**: 653–657.

Carey, J.R. 1993. *Applied Demography for Biologists, with Special Emphasis on Insects.* Oxford University Press, New York.

Carey, J.R. 1989. The multiple decrement life table: a unifying framework for cause-of-death analysis in ecology. *Oecologia*, **78**: 131–137.

Carlton, B.C., C. Gawron-Burke, and T.B. Johnson. 1990. Exploiting the genetic diversity of *Bacillus thuringiensis* for the creation of new bioinsecticides, pp. 18–22. *In* Vth International Colloquium on Invertebrate Pathology and Microbial Control. Adelaide, Australia, 20–24 August, 1990.

Carpenter, S.R. 1981. Effect of control measures on pest populations subject to regulation by parasites and pathogens. *Journal of Theoretical Biology*, **92**: 181–184.

Carruthers, R.I. and K. Hural. 1990. Fungi as naturally occurring entomopathogens, pp. 115–138. *In* Baker, R.R. and P.E. Dunn, (eds.). *New Directions in Biological Control, Alternatives for Suppressing Agricultural Pests and Diseases.* Alan R. Liss, Inc., New York.

Carruthers, R.I. and R.S. Soper. 1987. Fungal diseases, pp. 357–416. *In* Fuxa, J.R. and Y. Tanada. *Epizootiology of Insect Diseases.* John Wiley and Sons, New York.

Carter, M.C. and A.F.G. Dixon. 1984a. Honeydew: an arrestant stimulus for coccinellids. *Ecological Entomology*, **9**: 383–387.

Carter, M.C. and A.F.G. Dixon. 1984b. Foraging behaviour of coccinellid larvae: Duration of intensive search. *Entomologia Experimentalis et Applicata*, **36**: 133–136.

Carter, N. 1987. Management of cereal aphid (Hemiptera: Aphididae) populations and their natural enemies in winter wheat by alternate strip spraying with a selective insecticide. *Bulletin of Entomological Research*, **77**: 677–682.

Carter, N., D.P. Aikman, and A.F.G. Dixon. 1978. An appraisal of Hughes' time-specific life table analysis for determining aphid reproductive and mortality rates. *Journal of Animal Ecology*, **47**: 677–687.

Carton, Y. and J. Claret. 1982. Adaptative significance of a temperature induced diapause in a cosmopolitan parasitoid of *Drosophila. Ecological Entomology*, **7**: 239–247.

Cartwright, B. and Loke T. Kok. 1990. Feeding by *Cassida rubiginosa* (Coleoptera: Chrysomelidae) and the effects of defoliation on growth of musk thistles. *Journal of Entomological Science*, **25**: 538–547.

Casas, J. 1989. Foraging behaviour of a leafminer parasitoid in the field. *Ecological Entomology*, **14**: 257–265.

Case, T.J. and D.T. Bolger. 1991. The role of introduced species in shaping the distribution and abundance of island reptiles. *Evolutionary Ecology*, **5**: 272–290.

Cate, J.R. and M.K. Hinkle. 1993. Integrated pest management: the path of a paradigm. A Special Report of the National Audubon Society, Washington, D. C., U.S.A.

Cayrol, J.C. 1983. Lutte biologique contre les *Meloidogyne* au moyen d'*Arthrobotrys irregularis. Revue de Nématologie*, **6**: 265–73.

Cayrol, J.C. and J.P. Frankowski. 1979. Une methode de lutte biologique contre les nematodes a galles des racines appartenant au genre *Meloidogyne. Revue Horticole*, **193**: 15–23.

Cayrol, J.C., J.P. Frankowski, A. Laniece, G. d'Hardemare, and J.P. Talon. 1978. Contre les nematodes en champignonniere. Mise au point d'une methode de lutte bologique a l'aide d'un hyphomycete predateur: Arthobotrys robusta souche antipolis (Royal 300). *Revue Horticole*, **184**: 23–30.

Cederberg, B. 1983. The role of trail pheromones in host selection by *Psithyrus rupestris* (Hymenoptera, Apidae). *Annals Entomologici Fennici*, **49**: 11–16.

Chabora, P.C. and A.J. Chabora. 1971. Effects of an interpopulation hybrid host on parasite population dynamics. *Annals of the Entomological Society of America*, **64**: 558–562.

Chambers, D.L. 1977. Quality control in mass rearing. *Annual Review of Entomology*, **22**: 289–308.

Chandler, L.D., F.E. Gilstrap, and H.W. Browning. 1988. Evaluation of the within-field mortality of *Liriomyza trifolii* (Diptera: Agromyzidae) on bell pepper. *Journal of Economic Entomology*, **81**: 1089–1096.

Chant, D.A. 1961. An experiment in biological control of *Tetranychus telarius* (L.) (Acarina: Tetranychidae) in a greenhouse using the predaceous mite *Phytoseiulus persimilis* Athias–Henriot (Phytoseiidae). *Canadian Entomologist*, **93**: 437–443.

Chapuis, J.L., P. Boussés, and G. Barnaud. 1994. Alien mammals, impact and management in the French subantarctic islands. *Biological Conservation*, **67**: 97–104.

Charnov, E.L. 1976. Optimal foraging, the marginal value theorem. *Theoretical Population Biology*, **9**: 129–136.

Charnov, E.L. 1979. The genetical evolution of patterns of sexuality: Darwinian fitness. *American Naturalist*, **113**: 465–480.

Charudattan, R. 1989. Assessment of efficacy of mycoherbicide candidates, pp. 455–464. *In* Delfosse, E.S., (ed.). *Proceedings of the VIIth International Symposium on Biological Control of Weeds*, held 6–11 March, 1988, Rome, Italy.

Charudattan, R. 1990. Release of fungi: Large-scale use of fungi as biological weed control agents, pp. 70–84. *Risk Assessment in Agricultural Biotechnology: Proceedings of the International Conference.* Davis, California, U.S.A., August, 1988.

Charudattan, R. 1991. The mycoherbicide approach with plant pathogens, pp. 24–57. *In* TeBeest, D.O., (ed.). *Microbial Control of Weeds.* Chapman and Hall, New York.

Chatelain, M.P. and J.A. Schenk. 1984. Evaluation of frontalin and exo-brevicomin as kairomones to control mountain pine beetle (Coleoptera: Scolytidae) in lodgepole pine. *Environmental Entomology*, **13**: 1666–1674.

Chelliah, S., L.T. Fabellar, and E.A. Heinrichs. 1980. Effects of sub-lethal doses of three insecticides on the reproductive rate of the brown planthopper, *Nilaparvata lugens*, on rice. *Environmental Entomology*, **9**: 778–780.

Chen, B.H., J.E. Foster, J.E. Araya, and P.L. Taylor. 1991. Parasitism of *Mayetiola destructor* (Diptera: Cecidomyiidae) by *Platygaster hiemalis* (Hymenoptera: Platygasteridae) on Hessian fly-resistant wheats. *Journal of Entomological Science*, **26**: 237–243.

Chen, C.N. and S.C. Chiu, (eds.). 1986. Proceedings of a symposium on biological control of crop pests at Taiwan Agriculture Research Institute, 16 November, 1984. Published by Taiwan Agricultural Research Institute and Plant Protection Society of the Republic of China, Taiwan.

Cherwonogrodsky, J.W. 1980. Microbial agents as insecticides. *Residue Reviewss*, **76**: 73–96.

Chesson, P.L. and W.W. Murdoch. 1986. Aggregation of risk: relationships among host-parasitoid models. *American Naturalist*, **127**: 696–715.

Chet, I. and Y. Henis. 1985. Trichoderma as a biocontrol agent against soilborne root pathogens, pp. 110–112. *In* Parker, C.A. et al., (eds.). *Ecology and Management of Soilborne Plant Pathogens.* American Phytopathological Society, St. Paul, Minnesota, U.S.A.

Chiang, H.C. 1970. Effects of manure applications and mite predation on corn rootworm populations in Minnesota. *Journal of Economic Entomology*, **63**: 934–936.

Chiverton, P.A. 1986. Predator density manipulation and its effects on populations of *Rhopalosiphum padi* (Hom.: Aphididae) in spring barley. *Annals of Applied Biology*, **109**: 49–60.

Chiverton, P.A. 1987. Effects of exclusion barriers and inclusion trenches on polyphagous and aphid specific predators in spring barley. *Journal of Applied Entomology*, **103**: 193–203.

Choi, K.M. and M.H. Lee. 1990. Use of natural enemies for controlling agricultural pests in Korea. *In* FFTC-NARC International Seminar on "The use of parasitoids and predators to control agricultural pests." Tukuba Science City, Ibaraki-ken, 305, Japan, October 2–7, 1989.

Cilliers, C.J. 1991. Biological control of water fern, Salvina molesta (Salviniaceae), in South Africa. *Agriculture, Ecosystems and Environment*, **37**: 219–224.

Clarke, B., J. Murray, and M.S. Johnson. 1984. The extinction of endemic species by a program of biological control. *Pacific Science*, **38**: 97–104.

Clarke, R.D. and P.R. Grant. 1968. An experimental study of the role of spiders as predators in a forest litter community. Part I. *Ecology*, **49**: 1152–1154.

Clausen, C.P. 1940. *Entomophagous Insects.* McGraw-Hill Book Co., New York.

Clausen, C.P. (ed.). 1978. *Introduced Parasites and Predators of Arthropod Pests and Weeds: A World Review.* Agricultural Handbook No. 480. U.S. Department of Agriculture, Washington, D.C., U.S.A.

Clement, S. L. and R. Sobhian. 1991. Host-use patterns of capitulum-feeding insects of yellow starthistle: Results from a garden plot in Greece. *Environmental Entomology*, **20**: 724–730.

Cloutier, C. and S.G. Johnson. 1993. Predation by *Orius tristicolor* (Hemiptera: Anthocoridae) on *Phytoseiulus persimilis* (Acarina: Phytoseiidae): testing for compatibility between biocontrol agents. *Environmental Entomology*, **22**: 477–482.

Cloutier, C., C.A. Lévesque, D.M. Eaves, and M. Mackauer. 1991. Maternal adjustment of sex ratio in response to host size in the aphid parasitoid *Ephedrus californicus. Canadian Journal of Zoology*, **69**: 1489–1495.

Cock, M.J.W., (ed.). 1985. *A Review of Biological Control of Pests in the Commonwealth Caribbean and Bermuda up to 1982.* Commonwealth Agricultural Bureaux Technical Communication No. 9, Unwin Brothers Limited, Surrey, U.K.

Cofrancesco, A.F., R.M. Stewart, and D.R. Sanders. 1985. The impact of *Neochetina eichhorniae* (Coleoptera: Curculionidae) on waterhyacinth in Louisiana, pp. 525–535. *In* Delfosse, E.S., (ed.). *Proceedings of the VI International Symposium on Biological Control of Weeds.* 19–25 August 1984, Vancouver, Canada. Agriculture Canada, Ottawa, Canada.

Cohen, A.C. 1985. Simple methods for rearing the insect predator *Geocoris punctipes* (Heteroptera: Lygaeidae) on a meat diet. *Journal of Economic Entomology*, **78**: 1173–1175.

Cohen, A.C. and C.G. Jackson. 1989. Using rubidium to mark a predator, *Geocoris punctipes* (Hemiptera: Lygaeidae). *Journal of Entomological Science*, **24**: 57–61.

Cole, F.R., A.C. Medeiros, L.L. Loope, and W.W. Zuehlke. 1992. Effects of the Argentine ant on arthropod fauna of Hawaiian high-elevation shrubland. *Ecology*, **73**: 1313–1322.

Coler, R.R. and K.B. Nguyen. 1994. *Paraiotonchium muscadomesticae* n. sp. (Tylenchida: Iotonchiidae) a parasite of the house fly (*Musca domestica*) in Brazil and a key to species of the genus *Paraiotonchium. Journal of Nematology*, **26**: 392–401.

Coll, M. and D.G. Bottrell. 1991. Microhabitat and resource selection of the European corn borer (Lepidoptera: Pyralidae) and its natural enemies in Maryland field corn. *Environmental Entomology*, **20**: 526–533.

Coll, M. and D.G. Bottrell. 1992. Mortality of European corn borer larvae by natural enemies in different corn microhabitats. *Biological Control*, **2**: 95–103.

Collier, T.R., W.W. Murdoch, and R.M. Nisbet. 1994. Egg load and the decision to host feed in the parasitoid *Aphytis melinus. Journal of Animal Ecology*, **63**: 299–306.

Comins, H.N. and M.P. Hassell. 1976. Predation in multi-prey communities. *Journal of Theoretical Biology*, **62**: 93–114.

Comins, H.N. and M.P. Hassell. 1987. The dynamics of predation and competition in patchy environments. *Theoretical Population Biology*, **31**: 393–421.

Comins, H.N. and P.W. Wellings. 1985. Density-related parasitoid sex ratios: influence on host-parasitoid dynamics. *Journal of Animal Ecology*, **54**: 583–594.

Connick, W.J., Jr., J.A. Lewis, and P.C. Quimby, Jr. 1990. Formulation of biocontrol agents for use in plant pathology, pp. 345–372. *In* Baker, R.R. and P.E. Dunn, (eds.). *New Directions in Biological Control.* UCLA Symposium, Alan Liss Pub., New York.

Connick, W.J., Jr., D.J. Daigle, and P.C. Quimby, Jr. 1991. An improved invert emulsion with high water retention for mycoherbicide delivery. *Weed Technology*, **5**: 442–444.

Cook, R.J. 1993. Making greater use of introduced microorganisms for biological control of plant pathogens. *Annual Review of Phytopathology*, **31**: 53–80.

Cook, R.J. and K.F. Baker. 1983. *The Nature and Practice of Biological Control of Plant Pathogens.* American Phytopathological Society, St. Paul, Minnesota, U.S.A.

Cook, R.M. and S.F. Hubbard. 1977. Adaptive searching strategies in insect parasites. *Journal of Animal Ecology*, **46**: 115–125.

Cooke, R.C. 1968. Relationships between nematode-destroying fungi and soil-borne phytonematodes. *Phytopathology*, **58**: 909–913.

Cooke, R.C. and V.E. Satchuthananthavale. 1968. Sensitivity to mycostasis of nematode-trapping Hyphomycetes. *Transactions of the British Mycological Society*, **51**: 555–561.

Coop, L.B. and R.E. Berry. 1986. Reduction in variegated cutworm (Lepidoptera: Noctuidae) injury in peppermint by larval parasitoids. *Journal of Economic Entomology*, **79**: 1244–1248.

Coppel, H.C. and J.W. Mertins. 1977. *Biological Insect Pest Suppression*. Springer-Verlag, New York.

Corbett, A., T.F. Leigh, and L.T. Wilson. 1991. Interplanting alfalfa as a source of *Metaseiulus occidentalis* (Acari: Phytoseiidae) for managing spider mites in cotton. *Biological Control*, **1**: 188–196.

Cornell, H.V. and B.A. Hawkins. 1993. Accumulation of native parasitoid species on introduced herbivores: A comparison of hosts as natives and hosts as invaders. *American Naturalist*, **141**: 847–865.

Cory, J.S. and P.F. Entwistle. 1990. Assessing the risks of releasing genetically manipulated baculoviruses. *Aspects of Applied Biology*, **24**: 187–194.

Coulson, J.R. 1977. *Biological control of allligatorweed, 1959–1972.: A review and evaluation*. U.S. Department of Agriculture Technical Bulletin 1547, 98 pp.

Coulson, J.R. 1992. Documentation of classical biological control introductions. *Crop Protection*, **11**: 195–205.

Coulson, J.R. and R.S. Soper. 1989. Protocols for the introduction of biological control agents in the U.S., pp. 1–35. *In* Kahn, R.P., (ed.). *Plant Protection and Quarantine. Vol. III. Special Topics*. CRC Press, Inc., Boca Raton, Florida, U.S.A.

Coulson, J.R., W. Klaasen, R.J. Cook, E.G. King, H.C. Chiang, K.S. Hagen, and W.G. Yendol. 1982. Notes on biological control of pests in China, 1979. *In*: *Biological Control of Pests in China*. United States Department of Agriculture, Washington, D.C., U.S.A.

Coulson, J.R., R.S. Soper, and D.W. Williams, (eds.). 1991. *Biological Control Quarantine: Needs and Procedures, Appendix III Proposed ARS Guidelines for Introduction and Release of Exotic Organisms for Biological Control*. Proceedings of a workshop, United States Department of Agriculture, Agricultural Research Service-99.

Courtenay, W.R., Jr. 1993. Biological pollution through fish introductions, pp. 35–61. *In* McKnight, B.N., (ed.). *Biological Pollution: the Control and Impact of Invasive Exotic Species*. Indiana Academy of Sciences, Indianapolis, Indiana, U.S.A.

Courtenay, W.R., Jr. and G.K. Meffe. 1989. Small fishes in strange places: a review of introduced poeciliids, pp. 319–331. *In* Meffe, G.K. and F.F. Snelson, Jr., (eds.). *Ecology and Evolution of Livebearing Fishes (Poeciliidae)*. Prentice-Hall, Englewood Cliffs, New Jersey, U.S.A.

Courtney, S.P. and T.T. Kibota. 1990. Mother doesn't know best: Selection of hosts by ovipositing insects, pp. 161–188. *In* Bernays, E., *Insect-Plant Interactions, Volume II*, CRC Press, Inc., Boca Raton, Florida, U.S.A.

Cox, C. 1990. Biological control of stored grain pests: A small business vs. government regulations. *Journal of Pesticide Reform*, **10**(3): 18–19.

Cramer, N.F. and R.M. May. 1972. Interspecific competition, predation and species diversity: A comment. *Journal of Theoretical Biology*, **34**: 289–293.

Crawford, H.S. and D.T. Jennings. 1989. Predation by birds on spruce budworm *Choristoneura fumiferana*: Functional, numerical and total responses. *Ecology*, **70**: 152–163.

Crawley, M.J. 1983. *Herbivory. The Dynamics of Animal-Plant Interactions*. Blackwell Science Pub., Oxford, U.K.

Crawley, M.J. 1989. Insect herbivores and plant population dynamics. *Annual Review of Entomology*, **34**: 531–564.

Crawley, M.J. and M. Nachapong. 1985. The establishment of seedlings from primary and regrowth seeds of ragwort (*Senecio jacobaea*). *Journal of Ecology*, **73**: 255–261.

Crickmore, N., C. Nicholls, D.J. Earp, T.C. Hodgman, and D.J. Ellar. 1990. The construction of *Bacillus thuringiensis* strains expressing novel entomocidal delta endotoxin combinations. *Biochemical Journal*, **270**: 133–136.

Croft, B.A. 1976. Establishing insecticide-resistant phytoseiid mite predators in deciduous tree fruit orchards. *Entomophaga*, **21**: 383–399.

Croft, B.A. 1990. *Arthropod Biological Control Agents and Pesticides*. John Wiley and Sons, New York.

Croft, B.A. and M.M. Barnes. 1971. Comparative studies on four strains of Typhlodromus occidentalis. III. Evaluations of releases of insecticide-resistant strains into an apple orchard ecosystem. *Journal of Economic Entomology*, **64**: 845–850.

Crombie, A.D. 1945. On competition between different species of graminivorous insects. *Proceedings of the Royal Society of London, Series B*, **132**: 362–395.

Crombie, A.D. 1946. Further experiments on insect competition. *Proceedings of the Royal Society of London, Series B*, **133**: 76–109.

Crowley, P.H. 1977. Spatially distributed stochasticity and the constancy of ecosystems. *Bulletin of Mathematical Biology*, **39**: 157–166.

Crowley, P.H. 1978. Effective size and the persistence of ecosystems. *Oecologia*, **35**: 185–195.

Crowley, P.H. 1981. Dispersal and the stability of predator-prey interactions. *American Naturalist*, **118**: 673–701.

Crowley, P.H. 1992. Density dependence, boundedness and attraction: detecting stability in stochastic systems. *Oecologia*, **90**: 246–254.

Crute, S. and K. Day. 1990. Understanding the impact of natural enemies on spruce aphid populations through simulation modelling, pp. 329–337. *In* Watt, A.D., S.R. Leather, M.D. Hunter, and N.A C. Kidd, (eds.). *Population Dynamics of Forest Insects*. Intercept, Andover, U.K.

CSIRO. 1970. *The Insects of Australia. A Textbook for Students and Research Workers*. Melbourne University Press, Carlton, Victoria, Australia.

Cudjoe, A.R., P. Neuenschwander, and M.J.W. Copland. 1993. Interference by ants in biological control of the cassava mealybug *Phenacoccus manihoti* (Hemiptera: Pseudococcidae) in Ghana. *Bulletin of Entomological Research*, **83**: 15–22.

Cullen, J.M. 1978. Evaluating the success of the programme for the biological control of *Chondrilla juncea* L. Proceedings of the 4th International Symposium on *Biological Control* of Weeds, pp. 117–121.

Cullen, J.M. 1985. Bringing the cost benefit analysis of biological control of *Chondrilla juncea* up to date, pp. 145–152. *In* Delfosse, E.S., (ed.). *Proceedings of the VI International Symposium on Biological Control of Weeds*. 19–25 August 1984, Vancouver, Canada. Agriculture Canada, Ottawa, Canada.

Cullen, J.M. 1990. Current problems in host-specificity screening, pp. 27–36. *In* Proceedings of the VIIth International Symposium on *Biological Control* of Weeds. Rome, Italy.

Cullen, J.M. and E.S. Delfosse. 1985. *Echium plantagineum*: Catalyst for conflict and change in Australia, pp. 249–292. *In* Delfosse, E.S., (ed.). *Proceedings of the VI International Symposium on Biological Control of Weeds*. 19–25 August 1984, Vancouver, Canada. Agriculture Canada, Ottawa, Canada.

Cunningham, J.C. 1988. Baculoviruses: their status compared to Bacillus thuringiensis as microbial insecticides. *Outlook on Agriculture*, **17**(1):10–17.

Curl, G.D. and P.P. Burbutis. 1978. Host-preference studies with *Trichogramma nubilale*. *Environmental Entomology*, **7**: 541–543.

Dahlberg, K.A. 1993. Government policies that encourage pesticide use in the United States, pp. 281–306. *In* Pimentel, D. and H. Lehman, (eds.). *The Pesticide Question: Environment, Economics and Ethics*. Chapman and Hall, New York.

Dahlman, D.L. 1991. Teratocytes and host/parasitoid interactions. *Biological Control*, **1**: 118–126.

Dai, K.J., L.W. Zhang, Z.J. Ma, L.S. Zhong, Q.X. Zhang, A.H. Cao, K.J. Xu, Q.Li, and Y.G. Gao. 1988. Research and utilization of artificial egg for propagation of parasitoid *Trichogramma*. *Colloques de l'INRA*, **43**: 311–318.

Dalgarno, W.T. 1935. Notes on the biological control of insect pests in the Bahamas. *Tropical Agriculture*, **12**: 78.

Davies, D.H. and M.T. Siva-Jothy. 1991. Encapsulation in insects: Polydnaviruses and encapsulation-promoting factors, pp. 119–132. *In* Gupta, A.P., (ed.). *Immunology of Insects and other Arthropods*. CRC Press, Inc., Boca Raton, Florida, U.S.A.

Davis, D.E., K. Myers, and J.B. Hoy. 1976. Biological control among vertebrates, pp. 501–519. *In* Huffaker, C.B. and P.S. Messenger, (eds.). 1976. *Theory and Practice of Biological Control*. Academic Press, New York.

Day, W.H. 1970. The survival value of its jumping cocoons to *Bathyplectes anurus*, a parasite of the alfalfa weevil. *Journal of Economic Entomology*, **63**: 586–589.

Day, W.H. 1981. Biological control of the alfalfa weevil in the northeastern United States, pp. 361–374. *In* Papavizas, G.C., (ed.). *Biological Control in Crop Production*. BARC Symposium No. 5, Allanheld, Osmun, Totowa, New Jersey, U.S.A.

Day, W.H., R.C. Hedlund, L.B. Saunders, and D. Coutinot. 1990. Establishment of *Peristenus digoneutis* (Hymenoptera: Braconidae), a parasite of the tarnished plant bug (Hemiptera: Miridae), in the United States. *Environmental Entomology*, **19**: 1528–1533.

de Barjac, H. 1978. Un nouveau candidat a la lutte biologique contre les moustiques: *Bacillus thuringiensis* var. *israelensis*. *Entomophaga*, **23**: 309–319.

de Barjac, H. and E. Frachon. 1990. Classification of *Bacillus thuringiensis* strains. *Entomophaga*, **35**: 233–240.

De Clercq, P. and D. DeGheele. 1993. Quality assessment of the predatory bugs *Podisus maculiventris* (Say) and *Podisus sagitta* (Fab.) (Heteroptera: Pentatomidae) after prolonged rearing on a meat-based artificial diet. *Biocontrol Science and Technology*, **3**: 133–139.

de Hoog, G.S. 1972. The genera *Beauveria, Isaria, Tritirachium*, and *Acrodontium* gen. nov. *Studies in Mycology*, **1**: 1–41.

de Jong, G. 1979. The influence of the distribution of juveniles over patches of food on the dynamics of a population. *Netherlands Journal of Zoology*, **29**: 33–51.

de Jong, G. 1981. The influence of dispersal pattern on the evolution of fecundity. *Netherlands Journal of Zoology*, **32**: 1–30.

de Jong, M.D. 1988. Risk to fruit trees and native trees due to control of black cherry (*Prunus serotina*) by silverleaf fungus (*Chondrostereum purpureum*). Dissertation, abstract in English, Landbouwuniversiteit te Wageningen, The Netherlands.

De Klerk, M.-L. and P.M.J. Ramakers. 1986. Monitoring population densities of the phytoseiid predator *Amblyseius cucumeris* and its prey after large scale introductions to control *Thrips tabaci* on sweet pepper. *Mededelingen van de Faculteit Landbouwwetenschappen, Rijksuniversiteit Gent*, **51**: 1045–1048.

De Leij, F.A.A.M. and B.R. Kerry. 1991. The nematophagous fungus, *Verticillium chlamydosporium*, as a potential biological control agent for *Meloidogyne arenaria*. *Revue de Nématologie*, **14**: 157–164.

Dean, R.A. and J. Kúc. 1986. Induced systemic protection in cucumber: time of production and movement of the signal. *Phytopathology*, **76**: 966–970.

DeBach, P. 1958. Application of ecological information to control of citrus pests in California. *Proceedings of the Xth International Congress of Entomology*, **3**: 187–194.

DeBach, P., (ed.). 1964a. *Biological Control of Insect Pests and Weeds*. Reinhold, New York.

DeBach, P. 1964b. Successes, trends, and future possibilities, pp. 673–713. *In* DeBach, P., (ed.). *Biological Control of Insect Pests and Weeds*. Reinhold, New York.

DeBach, P. 1974. *Biological Control by Natural Enemies*. Cambridge University Press, London (See pp. 8, 9).

DeBach, P. and C.B. Huffaker. 1971. Experimental techniques for evaluation of the effectiveness of natural enemies, pp. 113–140. *In* Huffaker, C.B., (ed.). *Biological Control*. Plenum, New York.

DeBach, P. and D. Rosen. 1991. *Biological Control by Natural Enemies (2nd ed.)*. Cambridge University Press, Cambridge, U.K.

DeBach, P., C.A. Fleschner, and E.J. Dietrick. 1951. A biological check method for evaluating the effectiveness of entomophagous insects in the field. *Journal of Economic Entomology*, **44**: 763–766.

DeBach, P., D. Rosen, and C.E. Kennett. 1971. Biological control of coccids by introduced natural enemies, pp. 165–194. *In* Huffaker, C.B., (ed.). *Biological Control*. Academic Press, New York.

DeBach, P., C.B. Huffaker, and A.W. MacPhee. 1976. Evaluation of the impact of natural enemies, pp. 255–285. *In* Huffaker, C.B. and P.S. Messenger, (eds.). *Theory and Practice of Biological Control*. Academic Press, New York.

Debolt, J.W. 1991. Behavioral avoidance of encapsulation by *Leiophron uniformis* (Hymenoptera: Braconidae), a parasitoid of *Lygus* spp. (Hemiptera: Miridae): Relationship between host age, encapsulating ability, and host acceptance. *Annals of the Entomological Society of America*, **84**: 444–446.

Deeker, W. 1992. *New Rabbit Biological Control Strategies for the 90s in Australia*. Vertebrate Biocontrol Centre Paper No. 1, CSIRO Pub., East Melbourne, Australia.

Delfosse, E.S. 1985. Re-evaluation of the biological control program for *Heliotropium europaeum* in Australia, pp. 735–742. *In* Delfosse, E.S., (ed.). *Proceedings of the VI International Symposium on Biological Control of Weeds*. 19–25 August 1984, Vancouver, Canada. Agriculture Canada, Ottawa, Canada.

Delfosse, E.S. 1992. The biological control regulatory process in Australia, pp. 135–141. *In* Charudattan, R. and H.W. Browning, (eds.). *Regulations and Guidelines: Critical Issues in Biological Control*. Proceedings of a USDA/CSRS National Workshop. Institute of Food and Agriculture Sciences, University of Florida, Gainesville, Florida, U.S.A.

Delfosse, E.S., S. Hasan, J.M. Cullen, and A.J. Wapshere. 1985. Beneficial use of an exotic phytopathogen, *Puccinia chondrillina*, as a biological control agent for skeleton weed, *Chondrilla juncea*, in Australia, pp. 171–177. *In* Gibbs, A.J. and H.R.C. Meischke, (eds.). *Pests and Parasites as Migrants: An Australian Perspective*. Australian Academy of science, Camberra, Australia.

DeLoach, C.J. 1991. Past successes and current prospects in biological control of weeds in the United States and Canada. *Natural Areas Journal*, **11**: 129–142.

Dempster, J.P. 1956. The estimation of the numbers of individuals entering each stage during the development of one generation of an insect population. *Journal of Animal Ecology*, **25**: 1–5.

Dempster, J.P. 1967. The control of *Pieris rapae* with DDT, I. The natural mortality of the young stages of *Pieris*. *Journal of Applied Ecology*, **4**: 485–500.

Dempster, J.P. 1987. Effects of pesticides on wildlife and priorities in future studies, pp. 17–25. *In* Brent, K.J. and R.K. Atkin, (eds.). *Rational Pesticide Use, Proceedings of the Ninth Long Ashton Symposium*. Cambridge University Press, Cambridge, U.K.

Den Boer, P.J. 1968. Spreading of risk and stabilization of animal numbers. *Acta Biotheoretica*, **18**: 165–194.

Den Boer, P.J., (ed.). 1971. Dispersal and Dispersal Power of Carabid Beetles. Miscellaneous paper #8, Landbouwhogeschool, Wageningen, The Netherlands.

Den Boer, P.J., H.U. Thiele, and F. Weber, (eds.). 1979. On the Evolution of Behaviour in Carabid Beetles. Miscellaneous Paper #18, Agricultural University of Wageningen, The Netherlands.

den Hartog, C. and G. van der Velde. 1987. Invasions by plants and animals into coastal, brackish and fresh water of the Netherlands. *Proceedings of the Koninklijke Nederlandse Akademie van Wetenschappen, Series C*, **90**: 31–37.

Deng, G.R., H.H. Yang, and M.X. Jin. 1987. Augmentation of coccinellid beetles for controlling sugarcane woolly aphid. *Chinese Journal of Biological Control*, **3**: 166–168. (in Chinese)

Deng, X., Z.Q. Zheng, N.X. Zhang, and X.F. Jia. 1988. Methods of increasing the winter-survival of *Metaseiulus occidentalis* (Acari: Phytoseiidae) in northwest China. *Chinese Journal of Biological Control*, **4**: 97–101.

Dennill, G.B. 1988. Why a gall former can be a good biocontrol agent: the gall wasp *Trichilogaster acaciaelongifoliae* and the weed *Acacia longifolia*. *Ecological Entomology*, **13**: 1–9.

Dennill, G.B. and V.C. Moran. 1989. On insect-plant associations in agriculture and the selection of agents for weed biocontrol. *Annals of Applied Biology*, **114**: 157–166.

Dennill, G.B. and D. Donnelly. 1991. Biological control of *Acacia longifolia* and related weed species (*Fabaceae*) in South Africa. *Agriculture, Ecosystems and Environment*, **37**: 115–135.

Dennill, G.B., D. Donnelly, and S.L. Chown. 1993. Expansion of host-plant range of a biocontrol agent *Trichilogaster acaciaelongifoliae* (Pteromalidae) released against the weed *Acacia longifolia* in South Africa. *Agriculture, Ecosystems and Environment*, **43**: 1–10.

Dennis, B. and M.L. Taper. 1994. Density dependence in time series observations of natural populations: estimation and testing. *Ecological Monographs*, **64**: 205–224.

Dennis, C., (ed.). 1983. *Post-Harvest Pathology of Fruits and Vegetables*. Academic Press, London.

Dennis, P. and G.L.A. Fry. 1992. Field margins: can they enhance natural enemy population densities and general arthropod diversity on farmland? *Agriculture, Ecosystems and Environment*, **40**: 95–115.

Dennis, P., M.B. Thomas, and N.W. Sotherton. 1994. Structural features of field boundaries which influence the overwintering densities of beneficial arthropod predators. *Journal of Applied Ecology*, **31**: 361–370.

Dennis, P. and S.D. Wratten. 1991. Field manipulation of populations of individual staphylinid species in cereals and their impact on aphid populations. *Ecological Entomology*, **16**: 17–24.

de Polania, I.Z. and O.M. Wilches. 1992. Impacto ecologico de la hormiga loca, *Paratrechina fulva* (Mayr), en el municipio de *Cimitarra* (Santander). *Revista Colombiana de Entomologia*, **18**: 14–22.

Deseo, K.V., P. Fantoni, and G.L. Lazzari. 1988. Presenza di nematodi entomopatogeni (*Steinernema* spp., *Heterorhabditis* spp.) nei terreni agricoli in Italia. *Atti Giornate Fitopatologia*, **2**: 269–280. (in Italian)

Dicke, M. 1988. Microbial allelochemicals affecting the behavior of insects, mites, nematodes, and protozoa in different trophic levels, pp. 125–163. *In* Barbosa, P. and D.K. Letourneau, (eds.). *Novel Aspects of Insect-Plant Interactions*. John Wiley and Sons, New York.

Dicke, M., J.C. van Lenteren, G.J.F. Boskamp, and E. van Dongen-Van Leeuwen. 1984. Chemical stimuli in host-habitat location by *Leptopilina heterotoma* (Thompson) (Hymenoptera: Eucolidae), a parasite of *Drosophila*. *Journal of Chemical Ecology*, **10**: 695–712.

Dicke, M., J.C. van Lenteren, G.J.F. Boskamp, and R. van Voorst. 1985. Intensification and prolongation of host searching in *Leptopilina heterotoma* (Thompson) (Hymenoptera: Eucoilidae) through a kairomone produced by *Drosophila melanogaster*. *Journal of Chemical Ecology*, **11**: 125–136.

Dicke, M., M de Jong, M.P.T. Alers, F.C.T. Stelder, R. Wunderink, and J. Post. 1989. Quality control of mass-reared arthropods: Nutritional effects on performance of predatory mites. *Journal of Applied Entomology*, **108**: 462–475.

Dicke, M., M.W. Sabelis, J. Takabayashi, J. Bruin, and M.A. Posthumus. 1990a. Plant strategies of manipulating predator-prey interactions through allelochemicals: Prospects for application in pest control. *Journal of Chemical Ecology*, **16**: 3091–3118.

Dicke, M., T.A. Van Beek, M.A. Posthumus, N. Ben Dom, H. Van Bokhoven, and A.E. de Groot. 1990b. Isolation and identification of volatile kairomone that affects acarine predator-prey interactions. Involvement of host plant in its production. *Journal of Chemical Ecology*, **16**: 381–396.

Dietel, C. 1991. Asian clam invades San Francisco Bay. *Outdoor California*, **52**: 1–4.

Dobbs, C.G. and W.H. Hinson. 1953. A widespread fungistasis in soils. *Nature*, **172**: 197–199.

Dobson, A.P. 1988. Restoring island ecosystems: the potential of parasites to control introduced mammals. *Conservation Biology*, **2**: 31–39.

Dong, H.F. and L.P. Niu. 1988. Effect of four fungicides on the establishment and reproduction of *Phytoseiulus persimilis* (Acar.: Phytoseiidae). *Chinese Journal of Biological Control*, **4**: 1–5.

Donisthorpe, H.St.J.K. 1927. *The Guests of British Ants, Their Habits and Life-Histories*. Routledge, London.

Doucet, M.M.A. de and M.E. Doucet. 1990. *Steinernema ritteri* (Nematoda: Steinernematidae) with a key to the species of the genus. *Nematologia*, **36**: 257–265.

Doutt, R.L. 1959. The biology of parasitic Hymenoptera. *Annual Review of Entomology*, **3**: 161–182.

Doutt, R.L. and J. Nakata. 1973. The *Rubus* leafhopper and its egg parasitoid: an endemic biotic system useful in grape-pest management. *Environmental Entomology*, **3**: 381–386.

Doutt, R.L., D.P. Annecke, and E. Tremblay. 1976. Biology and host relationships of parasitoids, pp. 143–168. *In* Huffaker, C.B. and P.S. Messenger, (eds.). *Theory and Practice of Biological Control.* Academic Press, New York.

Dover, B.A. and S.B. Vinson. 1989. The role of parasitoid polydnaviruses in successful parasitism of their hosts, pp. 14–18. *In* Anon. *Symposium on Biological and Integrated Pest Management.* 13–14 December, 1988, Clemson, South Carolina, U.S.A.

Dowell, R.V. and R. Gill. 1989. Exotic invertebrates and their effects on Calfiornia. *Pan-Pacific Entomologist,* **65**: 132–145.

Dowell, R.V., G.E. Fitzpatrick, and J.A. Reinert. 1979. Biological control of citrus blackfly in southern Florida. *Environmental Entomology,* **8**: 595–597.

Downing, R.S. and T.K. Moilliet. 1972. Replacement of Typhlodromus occidentalis by *T. caudiglans* and *T. pyri* (Acarina: Phytoseiidae) after cessation of sprays on apple trees. *Canadian Entomologist,* **104**: 937–940.

Dray, F.A., Jr., T.D. Center, D.H. Habeck, C.R. Thompson, A.F. Cofrancesco, and J.K. Balciunas. 1990. Release and establishment in the southeastern United States of *Neohydronomus affinis* (Coleoptera: Curculionidae), an herbivore of water lettuce. *Environmental Entomology,* **19**: 799–802.

Drea, J.J. and R.W. Carlson. 1990. Establishment of *Cybocephalus* sp. (Coleoptera: Nitidulidae) from Korea on *Unaspis euonymi* (Homoptera: Diaspididae) in the eastern United States. *Proceedings of the Entomological Society of Washington,* **90**: 307–309.

Drooz, A.T, A.E. Bustillo, G.F. Fedde, and V.H. Fedde. 1977. North American egg parasite successfully controls a different host in South America. *Science,* **197**: 390–391.

Drukker, B., J.S. Yaninek, and H.R. Herren. 1993. A packaging and delivery system for aerial release of *Phytoseiidae* for biological control. *Experimental and Applied Acarology,* **17**: 129–143.

Duddington, C.L., F.G.W. Jones, and F. Moriarty. 1956a. The effect of prectaceous fungus and organic matter upon the soil population of beet eelworm, *Heterodera schachtii* Schm. *Nematologica,* **1**: 344–348.

Duddington, C.L., F.G.W. Jones, and T.D. Williams. 1956b. An experiment on the effect of a predaceous fungus upon the soil population of potato root eelworm, *Heterodera rostochiensis* Woll. *Nematologica,* **1**: 341–343.

Duddington, C.L., C.O.R. Everard, and C.M.G. Duthoit. 1961. Effect of green manuring and a predaceous fungus on cereal root eelworm on oats. *Plant Pathology,* **10**: 108–109.

Duffey, S.S. 1980. Sequestration of plant natural products by insects. *Annual Review of Entomology,* **25**: 447–477.

Dulmage, H.T. and R.A. Rhodes. 1971. Production of pathogens in artificial media, pp. 507–540. *In* Burges, H.D. and N.W. Hussey, (eds.). *Microbial Control of Insects and Mites.* Academic Press, New York.

Dunn, M.T. 1983. *Paecilomyces nostocides,* a new hyphomycete isolated from cysts of *Heterodera zeae. Mycologia,* **75**: 179–182.

Duso, C. 1989. Role of the predatory mites *Amblyseius aberrans* (Oud.), *Typhlodromus pyri* Scheuten and *Amblyseius andersoni* (Chant) (Acari, Phytoseiidae) in vineyards. 1. The effects of single or mixed phytoseiid population releases on spider mite densities (Acari, Tetranychidae). *Journal of Applied Entomology,* **107**: 474–492.

Duso, C. 1992. Role of *Amblyseius aberrans* (Oud.), *Typhlodromus pyri* Scheuten and *Amblyseius andersoni* (Chant) (Acari, Phytoseiidae) in vineyards. *Journal of Applied Entomology,* **114**: 455–462.

Dutky, S.R., J.V. Thompson, and G.E. Cantwell. 1964. A technique for mass propagation of the DD-136 nematode. *Journal of Insect Pathology,* **6**: 417–422.

Dysart, R.J. 1991. Biological notes on two chloropid flies (Diptera: Chloropidae), predaceous on grasshopper eggs (Orthoptera: Acrididae). *Journal of the Kansas Entomological Society* **64**: 225–230.

Dysart, R.J., H. L. Maltby, and M.H. Brunson. 1973. Larval parasites of *Oulema melanopus* in Europe and their colonization in the United States. *Entomophaga,* **18**: 133–167.

Edson, K.M., S.B. Vinson, D.B. Stoltz, and M.D. Summers. 1981. Virus in a parasitoid wasp: Suppression of the cellular immune response in the parasitoid's host. *Science,* **211**: 582–583.

Edwards, W.H. 1936. Pests attacking citrus in Jamaica. *Bulletin of Entomological Research,* **21**: 335–337.

Egley, G.H., J.E. Hanks, and C.D. Boyette. 1993. Invert emulsions droplet size and mycoherbicidal activity of *Colletotrichum truncatum. Weed Technology,* **7**: 417–424.

Ehler, L.E. 1976. The relationship between theory and practice in biological control. *Bulletin of the Entomological Society of America,* **22**: 319–21.

Ehler, L.E. 1979. Assessing competitive interactions in parasite guilds prior to introduction. *Environmental Entomology,* **8**: 558–560.

Ehler, L.E. 1982. Foreign exploration in California. *Environmental Entomology,* **11**: 525–530.

Ehler, L.E. 1986. Distribution of progeny in two ineffective parasites of a gall midge (Diptera: Cecidomyiidae). *Environmental Entomology,* **15**: 1268–1271.

Ehler, L.E. 1990. Introduction strategies in biological control of insects, pp. 111–134. *In* Mackauer, M., L.E. Ehler, and J. Roland, (eds.). *Critical Issues in Biological Control.* Intercept, Andover, U.K.

Eibl-Eibesfeldt, J. 1967. On the guarding of leafcutter ants by minima-workers. *Naturwissenschaften*, **54**: 346.

Eichers, T.R. 1981. Use of pesticides by farmers, pp. 3–54. *In* Pimentel, D., (ed.). *CRC Handbook of Pest Management in Agriculture, Volume II.* CRC Press, Inc., Boca Raton, Florida, U.S.A.

Eikenbary, R.D. and C.E. Rogers. 1974. Importance of alternate hosts in establishment of introduced parasites. *Proceedings of the Tall Timbers Conference on Ecological Animal Control by Habitat Management*, **5**: 119–133.

Eisner, T., E. van Tassell, J.E. Carrel. 1967. Defensive use of a "fecal shield" by a beetle larva. *Science*, **158**: 1471–1473.

Ekbom, B.S. and J. Pickering. 1990. Pathogenic fungal dynamics in a fall population of the blackmargined aphid (*Monellia caryella*). *Entomologia Experimentalis et Applicata*, **57**: 29–37.

Elad, Y. 1986. Mechanisms of interactions between rhizosphere micro-organisms and soil-borne plant pathogens, pp. 49–60. *In* Jensen, V., A. Kjøller, and L.H. Sorensen, (eds.). *Microbial Communities in Soil.* Elsevier, New York.

Elkinton, J.S. and A.M. Liebhold. 1990. Population dynamics of gypsy moth in North America. *Annual Review of Entomology*, **35**: 571–596.

Elkinton, J.S., J.R. Gould, C.S. Ferguson, A.M. Liebhold, and W.E. Wallner. 1990. Experimental manipulation of gypsy moth density to assess impact of natural enemies, pp. 275–287. *In* Watt, A.D., S.R. Leather, M.D. Hunter, and N.A.C. Kidd, (eds.). *Population Dynamics of Forest Insects.* Intercept, Andover, U.K.

Elkinton, J.S., J.P. Buonaccorsi, T.S. Bellows, Jr., and R.G. Van Driesche. 1992. Marginal attack rate, *k*-values and density dependence in the analysis of contemporaneous mortality factors. *Researches on Population Ecology*, **34**: 29–44.

Elliott, N.C., R.W. Kieckhefer, and W.C. Kauffman. 1991. Estimating adult coccinellid populations in wheat fields by removal, sweepnet, and visual count sampling. *Canadian Entomologist*, **123**: 13–22.

Ellis, C.R., B. Kormos, and J.C. Guppy. 1988. Absence of parasitism in an outbreak of the cereal leaf beetle, *Oulema melanopus* (Coleoptera: Chrysomelidae), in the central tobacco growing area of Ontario. *Proceedings of the Entomological Society of Ontario*, **119**: 43–46.

Elsey, K.D. 1974. Influence of plant host on searching speed of two predators. *Entomophaga*, **19**: 3–6.

Elvin, M.K., J.L. Stimac, and W.H. Whitcomb. 1983. Estimating rates of arthropod predation on velvetbean caterpillar larvae in soybeans. *Florida Entomologist*, **66**: 319–330.

Elzen, G.W., H.J. Williams, and S.B. Vinson. 1986. Wind tunnel flight responses by hymenopterous parasitoid *Campoletis sonorensis* to cotton cultivars and lines. *Entomologia Experimentalis et Applicata*, **42**: 285–289.

Embree, D.G. 1966. The role of introduced parasites in the control of the winter moth in Nova Scotia. *Canadian Entomologist*, **98**: 1159–1168.

Embree, D.G. 1971. The biological control of the winter moth in eastern Canada by introduced parasites, pp. 217–226. *In* Huffaker, C.B., (ed.). *Biological Control.* Plenum Press, New York.

Emlen, J.M. 1966. The role of time and energy in food preference. *American Naturalist*, **100**: 611–617.

Entwistle, P.F. 1983. Control of insects by virus diseases. *Biocontrol News and Information*, **4**(3): 203–225.

Entwistle, P.F., J.S. Cory, M.J. Bailey, and S. Higgs, (eds.). 1993. *Bacillus thuringiensis*, an Environmental Pesticide: Theory and Practice. John Wiley and Sons, New York.

Environmental Protection Agency. 1983. *Title 40, Protection of Environment, Chapter I, Environmental Protection Agency, Subchapter E, Pesticide Programs (OPP-30063A), Part 158, Data Requirements for Pesticide Registration.* Environmental Protection Agency, Washington, D.C.

Epton, H.A.S., M. Wilson, S.L. Nicholson, and D.C. Sigee. 1994. Biological control of *Erwinia amylovora* with *Erwinia herbicola*, pp. 335–352. *In* Blakeman, J.P. and B. Williamson, (eds.). *Ecology of Plant Pathogens.* CAB International, Wallingford, U.K.

Ervin, R.T., L.J. Moffitt, and D.E. Meyerdirk. 1983. Comstock mealybug (Homoptera: Pseudococcidae): cost analysis of a biological control program in California. *Journal of Economic Entomology*, **76**: 605–609.

Erwin, T.L., G.E. Ball, D.R. Whitehead, and A.L. Halpern. 1979. Carabid Beetles: Their Evolution, Natural History and Classification. *Proceedings of the First International Symposium of Carabidology, Smithsonian Institution, Washington, D.C., U.S.A., August 21, 23, and 25, 1976.* W. Junk, the Hague, The Netherlands.

Etienne, J. 1973. Consequences de l'elevage continu de *Lixophaga diatraeae* (Dipt., Tachinidae) sur l'hôte de remplacement: *Galleria mellonella* (Lep., Galleriidae). *Entomophaga*, **18**: 193–203. (in French)

Etzel, L.K., S.O. Levinson, and L.A. Andres. 1981. Elimination of *Nosema* in *Galeruca rufa*, a potential biological control agent for field bindweed. *Environmental Entomology*, **10**: 143–146.

Evans, W.G. 1984. Odor-mediated responses of *Bembidion obtusidens* (Coleoptera: Carabidae) in a wind tunnel. *Canadian Entomologist*, **116**: 1653–1658.

Ewald, P.W. 1987. Pathogen-induced cycling of outbreak insect populations, pp. 269–286. *In* Barbosa, P. and J.C. Schultz, (eds.). *Insect Outbreaks.* Academic Press, San Diego, California, U.S.A.

Ewel, J.J. 1986. Invasibility: lessons from South Florida, pp. 214–231. *In* Mooney, H.A. and J.A. Drake, (eds.). *Ecology of Biological Invasions of North America and Hawaii.* Springer-Verlag, New York.

Fabritius, K. 1984. Investigations on inbreeding of *Muscidifurax raptor* under laboratory conditions (Hymenoptera: Pteromalidae). *Entomologia Generalis*, **9**: 237–241. (in German)

Faeth, S.H. 1990. Structural damage to oak leaves alters natural enemy attack on a leafminer. *Entomologia Experimentalis et Applicata*, **57**: 57–63.

Faeth, S.H. and D. Simberloff. 1981. Population regulation of a leaf-mining insect, *Cameraria* sp. nov., at increased field densities. *Ecology*, **62**: 620–624.

Falcon, L.A. 1976. Problems associated with the use of arthropod viruses in pest control. *Annual Review of Entomology*, **21**: 305–324.

Falcon, L.A. 1982. Use of pathogenic viruses as agents for the biological control of insect pests, pp. 191–210. *In* Anderson, R.M. and R.M. May, (eds.). *Population Biology of Infectious Diseases*. Springer-Verlag, New York.

Falcon, L.A. 1985. Development and use of microbial insecticides, pp. 229–242. *In* Hoy, M.A. and D.C. Herzog, (eds.). *Biological Control in Agricultural IPM Systems*. Academic Press, New York.

Farrar, R.R. and G. Kennedy. 1991. Inhibition of *Telenomus sphingis* an egg parasitoid of *Manduca* spp. by trichome/2-tridecanone-based host plant resistance in tomato. *Entomologia Experimentalis et Applicata*, **60**: 157–166.

Farrar, R.R., G.G. Kennedy, and R.K. Kashyap. 1992. Influence of life history differences of two tachinid parasitoids of *Helicoverpa zea* (Boddie) (Lepidoptera: Noctuidae) on their interactions with glandular trichome/methyl ketone-based insect resistance in tomato. *Journal of Chemical Ecology*, **18**: 499–515.

Faull, J.L. 1988. Competitive antagonism of soil-borne plant pathogens, pp. 125–140. *In* Burge, M.N., (ed.). *Fungi in Biological Control Systems*. Manchester University Press, Manchester, U.K.

Fedde, G.F, V.H. Fedde, and A.T. Drooz. 1976. Biological control prospects of an egg parasite, *Telenomus alsophilae* Viereck, pp. 123–27. *In* Waters, W.E., (ed.). *Current Topics in Forest Entomology*. Selected papers from XV International Congress of Entomology, United States Department of Agriculture, Forest Service General Technical Report, WO-8.

Federici, B.A. 1991. Viewing polydnaviruses as gene vectors of endoparasitic Hymenoptera. *Redia*, **74**: 387–392.

Feener, D.H., Jr. and B.V. Brown. 1992. Reduced foraging of *Solenopsis geminata* (Hymenoptera: Formicidae) in the presence of parasitic *Pseudacteon* spp. (Diptera: Phoridae). *Annals of the Entomological Society of America*, **85**: 80–84.

Feitelson, J.S., J. Payne, and L. Kim. 1992. *Bacillus thuringiensis*: Insects and beyond. *Bio/technology*, **10**: 271–275.

Feng, J.G., Y. Zhang, X. Tao, and X.L. Chen. 1988. Use of radioisotope P^{32} to evaluate the parasitization of *Adoxophyes orana* (Lep.: Tortricidae) by mass released *Trichogramma dendrolimi* (Hymen.: Trichogrammatidae) in an apple orchard. *Chinese Journal of Biological Control*, **4**: 152–154. (in Chinese)

Feng, M.G., T.J. Poprawski, and G.G. Khachatourians. 1994. Production, formulation and application of the entomopathogenic fungus *Beauveria bassiana* for insect control: Current status. *Biocontrol Science and Technology*, **4**: 3–34.

Fenner, F. and I.D. Marshall. 1957. A comparison of the virulence for European rabbits (*Oryctolagus cuniculus*) of strains of myxoma virus recovered in the field in Australia, Europe and America. *Journal of Hygiene*, **55**: 149–191.

Fenner, F. and F.N. Ratcliffe. 1965. *Myxomatosis*. Cambridge University Press, Cambridge, U.K.

Ferris, V.R. and J.M. Ferris. 1989. Why ecologists need systematists: importance of systematics to ecological research. *Journal of Nematology*, **21**: 308–314.

Ferro, D.N. and S.M. Lyon. 1991. Colorado potato beetle (Coleoptera: Chrysomelidae) larval mortality: Operative effects of *Bacillus thuringiensis* subsp. san diego. *Journal of Economic Entomology*, **84**: 806–809.

Ferron, P. 1978. Biological control of insect pests by entomogenous fungi. *Annual Review of Entomology*, **23**: 409–442.

Field, R.P. and M.A. Hoy. 1985. Diapause behavior of genetically-improved strains of the spider mite predator *Metaseiulus occidentalis* (Acarina: Phytoseiidae). *Entomologia Experimentalis et Applicata*, **38**: 113–120.

Field, R.P. and S.M. Darby. 1991. Host specificity of the parasitoid, *Sphecophaga vesparum* (Curtis) (Hymenoptera: Ichneumonidae), a potential biological control agent of the social wasps, *Vespula germanica* (Fabricius) and *V. vulgaris* (Linnaeus) (Hymenoptera: Vespidae) in Australia. *New Zealand Journal of Zoology*, **18**: 193–197.

Fillman, D.A. and W.L. Sterling. 1983. Killing power of the red imported fire ant (Hymen.: Formicidae): a key predator of the boll weevil (Col.: Curculionidae). *Entomophaga*, **28**: 339–334.

Fisher, R.A. 1930. *The Genetical Theory of Natural Selection*. Oxford University Press, Oxford, U.K.

Fisher, S.W. and J.D. Briggs. 1992. Testing of microbial pest control agents in nontarget insects and acari, pp. 761–777. *In* Levin, M.A., R.J. Seidler, and M. Rogul, (eds.). *Microbial Ecology, Principles, Methods and Applications*. McGraw-Hill, New York.

Fisher, T.W. 1978. University of California quarantine facility, Riverside, pp. 56–60. *In* Leppla, N.C. and T.R. Ashley, (eds.). *Facilities for Insect Research and Production*. United States Department of Agriculture Technical Bulletin 1576.

Fisher, T.W. and R.E. Orth. 1985. Biological control of snails. *Occasional Papers*, 1. Department of Entomology, University of California, Riverside, California, U.S.A.

Flaherty, D.L. and C.B. Huffaker. 1970. Biological control of Pacific mites and Willamette mites in the San Joaquin Valley vineyard. Part I. Role of *Metaseiulus occidentalis*. Part II. Influence of dispersion patterns of *Metaseiulus occidentalis*. *Hilgardia*, **40**: 267–330.

Fleming, J.-A. and G.W. Fleming. 1992. Polydnaviruses: Mutualists and pathogens. *Annual Review of Entomology*, **37**: 401–425.

Fleming, R.A. 1988. Difficulties implementing a modelling-based integrated pest management program for alfalfa. *Memoirs of the Entomological Society of Canada*, **143**: 47–59.

Fleschner, C.A. 1958. The effect of orchard dust on the biological control of avocado pests. *California Avocado Society Yearbook*, **42**: 94–98.

Fleschner, C.A., J.C. Hall, and D.W. Ricker. 1955. Natural balance of mite pests in an avocado grove. *California Avocado Society Yearbook*, **39**: 155–162.

Flexner, J.L., B. Lighthart, and B.A. Croft. 1986. The effects of microbial pesticides on non-target, beneficial arthropods. *Agriculture, Ecosystems and Environment*, **16**: 203–254.

Flick, G. 1987. The new Plant Protection Act in the German Federal Republic and its effects. *Mededelingen van de Faculteit Landbouwwetenschappen, Rijksuniversiteit Gent*, **52**: 349–352. (in German)

Foelix, R.F. 1982. *Biology of Spiders*. Harvard University Press, Cambridge, Massachusetts, U.S.A.

Fokkema, N.J. and J. van den Heuvel. 1986. *Microbiology of the Phyllosphere*. Cambridge University Press, Cambridge, U.K.

Fokkema, N.J. and M.P. de Nooij. 1981. The effect of fungicides on the microbial balance in the phyllosphere. *European and Mediterranean Plant Protection Organization Bulletin*, **11**: 303–310.

Forno, I.W. 1987. Biological control of the floating fern *Salvinia molesta* in north-eastern Australia: Plant-herbivore interactions. *Bulletin of Entomological Research*, **77**: 9–17.

Forno, I.W., D.P.A. Sands, and W. Sexton. 1983. Distribution, biology and host specificity of *Cyrtobagous singularis* Hustache (Coleoptera: Curculionidae) for the biological control of *Salvinia molesta*. *Bulletin of Entomological Research*, **73**: 85–95.

Forschler, B.T., J.N. All, and W.A. Gardner. 1990. *Steinernema feltiae* activity and infectivity in response to herbicide exposure in aqueous and soil environments. *Journal of Invertebrate Pathology*, **55**: 375–379.

Fournier, D., P. Millot, and M. Pralavorio. 1985. Rearing and mass production of the predatory mite *Phytoseiulus persimilis*. *Entomologia Experimentalis et Applicata*, **38**: 97–100.

Fournier, D., M. Pralavorio, J. Coulon, and J.B. Berge. 1988. Fitness comparison in *Phytoseiulus persimilis* strains resistant and susceptible to methidathion. *Experimental and Applied Acarology*, **5**: 55–64.

Foy, C.L., D.R. Forney, and W.E. Cooley. 1983. History of weed introductions, pp. 65–92. *In* Wilson, C.L. and C.L. Graham, (eds.). *Exotic Plant Pests and North American Agriculture*. Academic Press, New York.

Francki, R.I.B., C.M. Fauquet, D.L. Knudson, and F. Brown. 1991. Classification and nomenclature of viruses. Fifth report of the International Committee on Taxonomy of Viruses. *Archives of Virology, Supplement 2*, Springer-Verlag, Wien, Austria.

Frank, J.H. 1967. The insect predators of the pupal stage of the winter moth, *Operophtera brumata* (L.) (Lepidoptera: Hydriomenidae). *Journal of Animal Ecology*, **36**: 375–389.

Frank, J.H. and E.D. McCoy. 1990. Endemics and epidemics of shibboleths and other things causing chaos. *Florida Entomologist*, **73**: 1–9.

Frank, J.H. and E.D. McCoy. 1994. Commercial importation into Florida of invertebrate animals as biological control agents. *Florida Entomologist*, **77**: 1–20.

Frank, J.H. and M.C. Thomas. 1994. *Metamasius callizona* (Chevrolat) (Coleoptera: Curculionidae), an immigrant pest, destroys bromeliads in Florida. *Canadian Entomologist*, **126**: 673–682.

Frank, W.A. and J.E. Slosser. 1991. *An Illustrated Guide to the Predaceous Insects of the Northern Texas Rolling Plains*. Miscellaneous Publications of the Texas Experiment Station No. MP-1718.

Franz, J.M. 1961a. Biological control of pest insects in Europe. *Annual Review of Entomology*, **6**: 183–200.

Franz, J.M. 1961b. Biologische Schadlingsbekampfung, pp. 1–302. *In* Sorauer, P., (ed.). *Handbuch der Pflanzenkrankheiten*. Band VI. Paul Parey Verlag, Berlin, Germany.

Franz, J.M. and A. Krieg. 1982. *Biologische Schadlingsbekampfung, 3 Auflage*. Verlag Paul Parey, Berlin and Hamburg, Germany.

Fravel, D.R.J.J. Marios, R.D. Lumsden, and W.J. Connick, Jr. 1985. Encapsulation of potential biocontrol agents in an alginate-clay matrix. *Phytopathology*, **75**: 774–777.

Frazer, B.D. and D.A. Raworth. 1985. Sampling for adult coccinellids and their numerical response to strawberry aphids (Coleoptera: Coccinellidae, Homoptera: Aphididae). *Canadian Entomologist*, **117**: 153–161.

Frazer, B.D., N. Gilbert, V. Nealis, and D.A. Raworth. 1981. Control of aphid density by a complex of predators. *Canadian Entomologist*, **113**: 1035–1041.

Free, C.A., J.R. Beddington, and J.H. Lawton. 1977. On the inadequacy of simple models of mutual interference for parasitism and predation. *Journal of Animal Ecology*, **46**: 543–554.
Frick, K.E. 1974. Biological control of weeds: Introduction, history, theoretical and practical applications, pp. 204–223. *In* Maxwell, F.G. and F.A. Harris, (eds.). *Proceedings of the Summer Institute on Biological Control of Plant Insects and Diseases*. University Press of Mississippi, Jackson, Mississippi, U.S.A.
Friedlander, T.P. (1985, pub. 1986). Egg mass design relative to surface-parasitizing parasitoids, with notes on *Asterocampa clyton* (Lepidoptera: Nymphalidae). *Journal of Research on the Lepidoptera*, **24**: 250–257.
Friedman, M.J. 1990. Commercial production and development, pp. 153–172. *In* Gaugler, R. and H.K. Kaya, (eds.). *Entomopathogenic Nematodes in Biological Control*. CRC Press, Inc., Boca Raton, Florida, U.S.A.
Friese, D.D., B. Megevand, and J.S. Yaninek. 1987. Culture maintenance and mass production of exotic phytoseiids. *Insect Science and Its Application*, **8**: 875–878.
Fry, J.M., (Compiler). 1989. *Natural Enemy Databank, 1987. A Catalogue of Natural Enemies of Arthropods Derived from Records in the CIBC Natural Enemy Databank*. Commonwealth Agricultural Bureaux, Wallingford, U.K.
Fujii, K. 1968. Studies on interspecies competition between the azuki bean weevil and the southern cowpea weevil. III. Some characteristics of strains of two species. *Researches on Population Ecology*, **10**: 87–98.
Fujii, K. 1969. Studies on the interspecies competition between the azuki bean weevil and the southern cowpea weevil. IV. Competition between strains. *Researches on Population Ecology*, **11**: 84–91.
Fujii, K. 1970. Studies on the interspecies competition between the azuki bean weevil, *Callosobruchus chinensis*, and the southern cowpea weevil, *C. maculatus*. V. The role of adult behavior in competition. *Researches on Population Ecology*, **12**: 233–242.
Fujii, K. 1977. Complexity-stability relationships of two-prey-one-predator species systems model: Local and global stability. *Journal of Theoretical Biology*, **69**: 613–623.
Funasaki, G.Y., P.-Y. Lai, L.M. Nakahara, J.H. Beardsley, and A.K. Ota. 1988a. A review of biological control introductions in Hawaii: 1890 to 1985. *Proceedings of the Hawaiian Entomological Society*, **28**: 105–160.
Funasaki, G.Y., I.M. Nakahara, and B.R. Kumashiro. 1988b. Introductions for biological control in Hawaii: 1985 and 1986. *Proceedings of the Hawaiian Entomological Society*, **28**: 101–104.
Fuxa, J.R. 1989. Fate of released entomopathogens with reference to risk assessment of genetically engineered microorganisms. *Bulletin of the Entomological Society of America*, **35**: 12–24.
Fuxa, J.R. 1990a. New directions for insect control with baculoviruses, pp. 97–113. *In* Baker, R.R. and P.E. Dunn, (eds.). *New Directions in Biological Control: Alternatives for Suppressing Agricultural Pests and Diseases*. Alan R. Liss, Inc., New York.
Fuxa, J.R. 1990b. Environmental risks of genetically engineered entomopathogens, pp. 203–207. *In* Laird, M., L.A. Lacey, and E.W. Davidson, (eds.). *Safety of Microbial Insecticides*. CRC Press, Inc., Boca Raton, Florida, U.S.A.
Fuxa, J.R. and Y. Tanada, (eds.). 1987. *Epizootiology of Insect Diseases*. John Wiley and Sons, New York.
Fuxa, J.R., A. R. Richter, and M.S. Strother. 1993. Detection of Anticarsia gemmatalis nuclear polyhedrosis virus in predatory arthropods and parasitoids after viral release in Louisiana soybean. *Journal of Entomological Science*, **28**: 51–60.
Gage, S.H. and D.L. Haynes. 1975. Emergence under natural and manipulated conditions of *Tetrastichus julis*, an introduced larval parasite of the cereal leaf beetle, with reference to regional population management. *Environmental Entomology*, **4**: 425–434.
Gagné, W.C. and F.G. Howarth. 1985. Conservation status of endemic Hawaiian Lepidoptera, pp. 74–84. Proceedings of the 3rd Congress on European Lepidoptera, 1982. Cambridge, U.K.
Gair, R., P.L. Mathias, and P.N. Harvey. 1969. Studies of cereal nematode populations and cereal yields under continuous or intensive culture. *Annals of Applied Biology*, **63**: 503–512.
Gallagher, J.E. and W.T. Haller. 1990. History and development of aquatic weed control in the United States. *Reviews of Weed Science*, **5**: 115–192.
Galper, S., E. Cohn, Y. Spiegel, and I. Chet. 1991. A collagenolytic fungus, Cunninghamella elegans, for biological control of plant parasitic nematodes. *Journal of Nematology*, **23**: 269–274.
Gambaro, R.I. 1988. Natural alternative food for *Amblyseius andersoni* (Chant) (Acarina: Phytoseiidae) on plants without prey. Long-term research in orchards with a prey-predator equilibrium. *Redia*, **71**: 161–171.
Gambino, P., A.C. Medeiros, and L.L. Loope. 1990. Invasion and colonization of upper elevations on east Maui (Hawaii) by *Vespula pensylvanica* (Hymenoptera: Vespidae). *Annals of the Entomological Society of America*, **83**: 1088–1095.
Gao, R.X., Z. Ouyang, Z.X. Gao, and J.X. Zheng. 1985. A preliminary report on citrus white fly control with *Aschersonia aleyrodis*. *Journal of the Fujian Agricultural College*, **14**(2): 127–133. (in Chinese)
Gaponyuk, I.L. and E.A. Asriev. 1986. Metaseiulus occidentalis in vineyards. *Zashchita Rastenii*, **8**: 22–23. (in Russian)
Gardner, D.E. 1990. *Role of Biological Control as a Management Tool in National Parks and Other Natural Areas*. National Park Service, United States Department of Interior, Cooperative Park Studies Unit, Dept. Botany, University of Hawaii at Manoa, Honolulu, U.S.A.

Gargiulo, G., C. Malva, F. Pennacchio, and E. Tremblay. 1988. Structure of *Aphidius Nees* (Hymenoptera, Braconidae) rDNA: A molecular tool in biosystematic research. *Bolletino del Laboratorio di Entomologia Agraria "Filippo Silvestri" di Portici*, **45**: 202–219.

Garton, D.W., D.J. Berg, A.M. Stoeckmann, and W.R. Haag. 1993. Biology of recent invertebrate invading species in the Great Lakes: The spiny water flea, *Bythotrephes cederstroemi*, and the zebra mussel, *Dreissena polymorpha*, pp. 63–84. In McKnight, B.N., (ed.). *Biological Pollution: The Control and Impact of Invasive Exotic Species.* Indiana Academy of Science, Indianapolis, Indiana, U.S.A.

Gaston, K.J. and J.H. Lawton. 1987. A test of statistical techniques for detecting density dependence in sequential censuses of animal populations. *Oecologia* **74**: 404–410.

Gaugler, R. and G.M. Boush. 1978. Effects of ultraviolet radiation and sunlight on the nematode *Neoaplectana carpocapsae. Journal of Invertebrate Pathology*, **32**: 291–296.

Gaugler, R. and G.M. Boush. 1979. Nonsusceptibility of rats to the entomogenous nematode, *Neoaplectana carpocapsae. Environmental Entomology*, **8**: 658–660.

Gaugler, R. and R. Georgis. 1991. Culture method and efficacy of entomopathogenic nematodes (Rhabditida: Steinernematidae and Heterorhabditidae). *Biological Control*, **1**: 269–274.

Gaugler, R. and H.K. Kaya, (eds.). 1990. *Entomopathogenic Nematodes in Biological Control.* CRC Press, Inc., Boca Raton, Florida, U.S.A.

Gaugler, R., J.F. Campbell, and T.R. McGuire. 1989. Selection for host-finding in *Steinernema feltiae. Journal of Invertebrate Pathology*, **54**: 363–372.

Gaugler, R., J.F. Campbell, and P. Gupta. 1991. Characterization and basis of enhanced host-finding in a genetically improved strain of *Steinernema carpocapsae. Journal of Invertebrate Pathology*, **57**: 234–241.

Gaugler, R., A. Bednarek, and J.F. Campbell. 1992. Ultraviolet inactivation of heterorhabditid and steinernematid nematodes. *Journal of Invertebrate Pathology*, **59**: 155–160.

Gauld, I. and B. Bolton, (eds.). 1988. *The Hymenoptera.* Oxford University Press, Oxford, U.K.

Geden, C.J., L. Smith, S.J. Long, and D.A. Rutz. 1992. Rapid deterioration of searching behavior, host destruction and fecundity of the parasitoid *Muscidifurax raptor* (Hymenoptera: Pteromalidae) in culture. *Annals of the Entomological Society of America*, **85**: 179–187.

Gelernter, W.D. 1990. MVP-Bioinsecticidet: A bioengineered, bioencapsulated product for control of lepidopteran larvae, p. 14. *In* Pinnock, D.E., (ed.). *Vth International Colloquium on Invertebrate Pathology and Microbial Control.* Adelaide, Australia, 20–24, August, 1990.

Gelernter, W.D. 1992. Application of biotechnology for improvement of *Bacillus thuringiensis* based products and their use for control of lepidopteran pests in the Caribbean. *Florida Entomologist*, **75**: 484–493.

Genini, M. and M. Baillod. 1987. Introduction de souches resistantes de *Typhlodromus pyri* (Scheuten) et *Amblyseius andersoni* Chant (Acari: Phytoseiidae) en vergers de pommiers. *Revue Suisse Viticulture, Arboriculture, et Horticulture*, **19**: 115–123.

Georghiou, G. and A. Legunes-Tejeda. 1991. *The Occurrence of Resistance to Pesticides in Arthropods.* Food and Agriculture Organization of the United Nations, Rome, Italy.

Georgis, R. 1990. Formulation and application technology, pp. 173–191. *In* Gaugler, R. and H.K. Kaya, (eds.). *Entomopathogenic Nematodes in Biological Control.* CRC Press, Inc., Boca Raton, Florida, U.S.A.

Georgis, R. and N.G.M. Hague. 1991. Nematodes as biological insecticides. *Pesticide Outlook*, **2**: 29–32.

Georgis, R., H.K. Kaya, and R. Gaugler. 1991. Effect of steinernematid and heterorhabditid nematodes (Rhabditida: Steinernematidae and Heterorhabditidae) on nontarget arthropods. *Environmental Entomology*, **20**: 815–822.

Gerling, D., B.D. Roitberg, and M. Mackauer. 1988. Behavioral defense mechanisms of the pea aphid, pp. 55–56. *In* Anon. *Parasitoid Insects, European Workshop, Colloques de l'INRA No. 48.* Lyon, France,7–10 September 1987.

Gerson, U. 1992. Perspectives of non-phytoseiid predators for the biological control of plant pests. *Experimental and Applied Acarology*, **14**: 383–391.

Gerson, U. and E. Cohen. 1989. Resurgences of spider mites (Acari: Tetranychidae) induced by synthetic pyrethroids. *Experimental and Applied Acarology*, **6**: 29–46.

Gerson, U. and R.L. Smiley. 1990. *Acarine Biocontrol Agents, an Illustrated Key and Manual.* Chapman and Hall, New York.

Ghee, K.S. 1990. Use of natural enemies for controlling agricultural pests in Malaysia. *In* FFTC-NARC International Seminar on "The use of parasitoids and predators to control agricultural pests." Tukuba Science City, Ibaraki–ken, 305, Japan, October 2–7, 1989.

Gibb, J.A. 1962. L. Tinbergen's hypothesis of the role of specific searching images. *Ibis*, **104**: 106–111.

Gilkeson, L.A. 1990. Cold storage of the predatory midge *Aphidoletes aphidimyza* (Diptera: Cecidomyiidae). *Journal of Economic Entomology*, **83**: 965–970.

Gilkeson, L.A. 1991. State of the art: Biological control in greenhouses, pp. 3–8. *In* McClay, A.S., (ed.). *Proceedings of the Workshop on Biological Control of Pests in Canada.* October 11–12, 1990, Calgary, Alberta, Alberta Environmental Centre, Vegreville, Alberta, Canada, AECV91–P1.

Gilkeson, L.A. 1992. Mass rearing of phytoseiid mites for testing and commercial application, pp. 489–506. *In* Anderson, T.E. and N.C. Leppla, (eds.). *Advances in Insect Rearing for Research and Pest Management*. Westview Press, Boulder, Colorado, U.S.A.

Gilkeson, L.A. and S.B. Hill. 1986. Genetic selection for and evaluation of nondiapause lines of the predatory midge, *Aphidoletes aphidimyza* (Rondani) (Diptera: Cecidomyiidae). *Canadian Entomologist*, **118**: 867–879.

Gilkeson, L.A., J.P. McLean, and P. Dessart. 1993. *Aphanogmus fulmeki* Ashmead (Hymenoptera: Ceraphronidae), a parasitoid of *Aphidoletes aphidimyza* Rondani (Diptera: Cecidomyiidae). *Canadian Entomologist*, **125**: 161–162.

Gill, S.S., E.A. Cowles, and P.V. Pietrantonio. 1992. The mode of action of *Bacillus thuringiensis* endotoxins. *Annual Review of Entomology*, **37**: 615–636.

Gillaspy, J.E. 1971. Papernest wasps (*Polistes*): Observations and study methods. *Annals of the Entomological Society of America*, **64**: 1357–1361.

Gillespie, A.T. 1988. Use of fungi to control pests of agricultural importance, pp. 37–60. *In* Burge, M.N., (ed.). *Fungi in Biological Control Systems*. Manchester University Press, Manchester, U.K.

Gillespie, D.R. and C.A. Ramey. 1988. Life history and cold storage of *Amblyseius cucumeris* (Acarina: Phytoseiidae). *Journal of the Entomological Society of British Columbia*, **85**: 71–76.

Gillman, M., J.M. Bullock, J. Silvertown, and B. Clear Hill. 1993. A density-dependent model of *Cirsium vulgare* population dynamics using field-estimated parameter values. *Oecologia*, **96**: 282–289.

Gilreath, M.E. and J.W. Smith, Jr. 1988. Natural enemies of *Dactylopius confusus* (Homoptera: Dactylopiidae): Exclusion and subsequent impact on *Opuntia* (Cactaceae). *Environmental Entomology*, **17**: 730–738.

Gilstrap, F.E. 1988. Sorghum-corn-Johnsongrass and Banks grass mite: a model for biological control in field crops, pp. 141–159. *In* Harris, M.K. and C.E. Rogers, (eds.). *The Entomology of Indigenous and Naturalized Systems in Agriculture*. Westview Press, Boulder, Colorado, U.S.A.

Giron, A. 1979. Host discrimination and host acceptance behavior of Gelis tenellus, a hyperparasite of *Apanteles melanoscelus*. *Environmental Entomology*, **8**: 1029–1031.

Giuma, A.Y., A.M. Hackett, and R.C. Cooke. 1973. Thermostable nematoxins produced by germinating conidia of some endozoic fungi. *Transactions of the British Mycological Society*, **60**: 49–56.

Glaser, R.W., E.E. McCoy, and H.B. Girth. 1940. The biology and economic importance of a nematode parasitic in insects. *Journal of Parasitology*, **26**: 479–495.

Glazer, I., M. Klein, A. Navon, and Y. Nakache. 1992. Comparison of efficacy of entomopathogenic nematodes combined with antidesiccants applied by canopy sprays against three cotton pests (Lepidoptera: Noctuidae). *Journal of Economic Entomology*, **85**: 1636–1641.

Godfray, H.C.J. 1994. *Parasitoids: Behavioral and Evolutionary Ecology*. Princeton University Press, Princeton, New Jersey, U.S.A.

Godfray, H.C.J. and S.P. Blythe. 1990. Complex dynamics in multispecies communities. *Philosophical Transactions of the Royal Society of London, Series B*, **330**: 221–233.

Godfray, H.C.J. and M.S. Chan. 1990. How insecticides trigger single-stage outbreaks in tropical pests. *Functional Ecology*, **4**: 329–337.

Godfray, H.C.J. and S.W. Pacala. 1992. Aggregation and the population dynamics of parasitoids and predators. *American Naturalist* **140**: 30–40.

Godfray, H.C.J. and M.P. Hassell. 1987. Natural enemies may be a cause of discrete generations in tropical insects. *Nature*, **327**: 144–147.

Godfray, H.C.J. and M.P. Hassell. 1988. The population biology of insect parasitoids. *Science Progress*, **72**: 531–548.

Godfray, H.C.J. and M.P. Hassell. 1989. Discrete and continuous insect populations in tropical environments. *Journal of Animal Ecology*, **58**: 153–174.

Godfray, H.C.J. and J.K. Waage. 1991. Predictive modeling in biological control: the mango mealy bug (*Rastrococcus invadens*) and its parasitoids. *Journal of Applied Ecology*, **28**: 434–453.

Godfray, H.C.J., M.J.W. Cook, and J.D. Holloway. 1987. An introduction to the Limacodidae and their bionomics, pp. 1–8. *In* Cook, M.J.W., H.C.J. Godfray, and J.D. Holloway, (eds.). *Slug and Nettle Caterpillars: The Biology, Taxonomy and Control of the Limacodidae of Economic Importance on Palms in South-east Asia*. Commonwealth Agricultural Bureaux International, Wallingford, U.K.

Godfrey, K.E., W.H. Whitcomb, and J.L. Stimac. 1989. Arthropod predators of velvetbean caterpillar, *Anticarsia gemmatalis* Hübner (Lepidoptera: Noctuidae), eggs and larvae. *Environmental Entomology*, **18**: 118–123.

Godoy, G., R. Rodriguez-Kabana, and G. Morgan-Jones. 1983. Fungal parasites of *Meloidogyne arenaria* in eggs in an Alabama soil. A mycological survey and greenhouse studies. *Nematropica*, **13**: 201–213.

Goedaert, J. 1662. *Metamorphosis et historia naturales insectorum I*, pp 175–178. Jacobum Fierensium, Medisburg.

Goeden, R.D. 1978. Part II. Biological control of weeds, pp. 357–414. *In* Clausen, C.P., (ed.). *Introduced Parasites and Predators of Arthropod Pests and Weeds: A World Review*. Agricultural Handbook No. 480. United States Department of Agriculture, Washington, D.C., U.S.A.

Goeden, R.D. 1983. Critique and revision of Harris' scoring system for selection of insect agents in biological control of weeds. *Protection Ecology*, **5**: 287–301.

Goeden, R.D. and L.T. Kok. 1986. Comments on a proposed "new" approach for selecting agents for the biological control of weeds. *Canadian Entomologist*, **118**: 51–58.

Goeden, R.D., C.A. Fleschner, and D.W. Ricker. 1967. Biological control of prickly pear cacti on Santa Cruz Island, California. *Hilgardia*, **38**: 579–606.

Goettel, M.S., R.J.S. Leger, S. Bhairi, M.K. Jung, B.R. Oakley, D.W. Roberts, and R.C. Staples. 1990. Pathogenicity and growth of Metarhizium anisopliae stably transformed to benomyl resistance. *Current Genetics*, **17**: 129–132.

Goh, K.S., R.C. Berberet, L.J. Young, and K.E. Conway. 1989. Mortality of *Hypera postica* (Coleoptera: Curculionidae) in Oklahoma caused by *Erynia phytonomi* (Zygomycetes: Entomophthorales). *Environmental Entomology*, **18**: 964–969.

Goldschmidt, T., F. Witte, and J. Wanink. 1993. Cascading effects of the introduced Nile perch on the detritivorous/phytoplanktivorous species in the sublittoral areas of Lake Victoria. *Conservation Biology*, **7**: 686–700.

Goldstein, L.F., P.P. Burbutis, and D.G. Ward. 1983. Rearing *Trichogramma nubilale* (Hymenoptera: Trichogrammatidae) on ultraviolet-irradiated eggs of the European corn borer (Lepidoptera: Pyralidae). *Journal of Economic Entomology*, **76**: 969–971.

González, D. 1988. Biotypes in biological control—Examples with populations of *Aphidius ervi*, *Trichogramma pretiosum* and *Anagrus epos* (parasitic Hymenoptera). *Advances in Parasitic Hymenoptera Research*, **1988**: 475–482.

González, D., G. Gordh, S.N. Thompson, and J. Adler. 1979. Biotype discrimination and its importance to biological control, pp. 129–136. *In* Hoy, M.A. and J.J. McKelvey Jr., (eds.). *Genetics in Relation to Insect Management*. Rockefeller Foundation, New York.

Gopal, B. 1987. *Water Hyacinth*. Elsevier Science Publishers, Amsterdam, the Netherlands.

Gordon, R.D. 1985. The Coccinellidae (Coleoptera) of America north of Mexico. *Journal of the New York Entomological Society*, **93**: 1–912.

Gormally, M. J. 1988. Studies on the oviposition and longevity of *Ilione albiseta* (Dipt.: Sciomyzidae)—Potential biological control agent of liver fluke. *Entomophaga*, **33**: 387–395.

Gould, F. 1988. Evolutionary biology and genetically engineered crops, consideration of evolutionary theory can aid in crop design. *BioScience*, **38**: 26–33.

Gould, F., G.G. Kennedy, and M.T. Johnson. 1991. Effects of natural enemies on the rate of herbivore adaptation to resistant host plants. *Entomologia Experimentalis et Applicata*, **58**: 1–14.

Gould, J.R., J.S. Elkinton, and W.E. Wallner. 1990. Density-dependent suppression of experimentally created gypsy moth, *Lymantria dispar* (Lepidoptera: Lymantriidae), populations by natural enemies. *Journal of Animal Ecology*, **59**: 213–233.

Gould, J.R., T.S. Bellows, Jr., and T.D. Paine. 1992a. Population dynamics of *Siphoninus phillyreae* in California in the presence and absence of a parasitoid, *Encarsia partenopea*. *Ecological Entomology*, **17**: 127–134.

Gould, J.R., T.S. Bellows, Jr., and T.D. Paine. 1992b. Evaluation of biological control of *Siphoninus phillyreae* (Haliday) by the parasitoid *Encarsia partenopea* (Walker), using life table analysis. *Biological Control*, **2**: 257–265.

Gowling, G.R. 1988. Interaction of partial plant resistance and biological control, p. 253. *In* Anon. *Aspects of Applied Biology Part, 17, 1988. Part 2, Environmental Aspects of Applied Biology*. Association of Applied Biologists, University of York, U.K.

Grafton-Cardwell, E.E. and M.A. Hoy. 1986. Genetic improvement of common green lacewing, *Chrysoperla carnea* (Neuroptera: Chrysopidae): Selection for carbaryl resistance. *Environmental Entomology*, **15**: 1130–1136.

Graham, F., Jr. 1970. *Since Silent Spring*. Houghton Mifflin Co., Boston, Massachusetts, U.S.A.

Granados, R.R. and B.A. Federici, (eds.). 1986. *The Biology of Baculoviruses: Volume I. Biological Properties and Molecular Biology and Volume II. Practical Application for Insect Control*. CRC Press, Inc., Boca Raton, Florida, U.S.A.; Vol. I, 304 pp.; Vol. II, 324 pp.

Granados, R.R., K.G. Dwyer, and A.C.G. Derksen. 1987. Production of viral agents in invertebrate cell cultures, pp. 167–181. *In* Maramorosch, K., (ed.). *Biotechnology in Invertebrate Pathology and Cell Culture*. Academic Press, San Diego, California, U.S.A.

Gravena, S. and H.F. Da Cunha. 1991. Predation of cotton leafworm first instar larvae, *Alabama argillacea* (Lep.: Noctuidae). *Entomophaga*, **36**: 481–491.

Gravena, S. and W.L. Sterling. 1983. Natural predation on the cotton leafworm (Lepidoptera: Noctuidae). *Journal of Economic Entomology*, **76**: 779–784.

Gray, N.F. 1987. Nematophagous fungi with particular reference to their ecology. *Biological Reviews of the Cambridge Philosophical Society*, **62**: 245–304.

Gray, N.F. 1988. Fungi attacking vermiform nematodes, pp. 3–38. *In* Poinar, G.O., and H.-B. Jansson, (eds.). *Diseases of Nematodes, Vol. II*. CRC Press, Inc., Boca Raton, Florida, U.S.A.

Greany, P.D. and E.R. Oatman. 1972. Demonstration of host discrimination in the parasite *Orgilus lepidus* (Hymenoptera: Braconidae). *Annals of the Entomological Society of America*, **65**: 375–376.

Greany, P.D., J.H. Tumlinson, D.L. Chambers, and G.M. Boush. 1977. Chemically-mediated host finding by *Biosteres (Opius) longicaudatus*, a parasitoid of tephritid fruit fly larvae. *Journal of Chemical Ecology*, **3**: 189–195.

Greathead, D.J. 1971. *A Review of Biological Control in the Ethiopian Region.* Commonwealth Agricultural Bureaux Technical Communication No. 5. Lamport Gilbert Printers, Reading, U.K.

Greathead, D.J. 1973. Progress in the biological control of Lantana camara in East Africa and discussion of problems raised by the unexpected reaction of some of the more promising insects to Sesamum indicum, pp. 89–92. *In* Dunn, P.H., (ed.). *Proceedings of the 2nd International Symposium on Biological Control of Weeds.* Commonwealth Institute of Biological Control, Miscellaneous Publications 6, Farnham Royal, U.K.

Greathead, D.J., (ed.). 1976. *A Review of Biological Control in Western and Southern Europe.* Commonwealth Agricultural Bureaux Technical Communication No. 7, Farnham Royal, Slough, U.K.

Greathead, D.J. 1980. Arthropod natural enemies of bilharzia snails and the possibilities for biological control. *Biocontrol News and Infromation*, **1**(2): 197–202.

Greathead, D.J. 1986a. Parasitoids in classical biological control. pp. 289–318. *In* Waage, J. and D. Greathead, (eds.). *Insect Parasitoids.* 13th Symposium of the Royal Entomological Society of London, 18–19, Sept. 1985. Academic Press, London.

Greathead, D.J. 1986b. Opportunities for biological control of insect pests in tropical Africa. *Revue d'Zoologie Africaine*, **100**: 85–96.

Greathead, D.J. and A.H. Greathead. 1992. Biological control of insect pests by parasitoids and predators: The BIOCAT database. *Biocontrol News and Information*, **13**(4): 61N–68N.

Green, M., M. Heumann, R. Sokolow, L.R. Foster, R. Bryant, and M. Skeels. 1990. Public health implications of the microbial pesticide *Bacillus thuringiensis*: An epidemiological study, Oregon, 1985–1986. *American Journal of Public Health*, **80**: 848–852.

Greenstone, M.H. and J.H. Hunt. 1993. Determination of prey antigen half-life in *Polistes metricus* using a monoclonal antibody-based immunodot assay. *Entomologia Experimentalis et Applicata*, **68**: 1–7.

Grenier, S. 1988. Applied biological control with tachinid flies (Diptera, Tachinidae): A review. *Anzeiger für Schädlingskunde, Pflanzenschutz, Umweltschutz*, **61**: 49–56.

Grevstad, F.S. and B.W. Klepetka. 1992. The influence of plant architecture on the foraging efficiencies of a suite of ladybird beetles feeding on aphids. *Oecologia*, **92**: 399-404.

Grewal, P.S., R. Gaugler, and S. Selvan. 1993a. Host recognition by entomopathogenic nematodes: Behavioral response to contact with host feces. *Journal of Chemical Ecology*, **19**: 1219–1231.

Grewal, P.S., R. Gaugler, H.K. Kaya, and M. Wusaty. 1993b. Infectivity of the entomopathogenic nematode *Steinernema scapterisci* (Nematoda: Steinernematidae). *Journal of Invertebrate Pathology*, **62**: 22–28.

Grissell, E.E. and M.E. Schauff. 1990. *A Handbook of the Families of Nearctic Chalcidoidea (Hymenoptera).* Entomological Society of Washington, Washington, D.C., U.S.A.

Gross, H.R., Jr., and R. Johnson. 1985. *Archytas marmoratus* (Diptera:Tachinidae): Advances in large-scale rearing and associated biological studies. *Journal of Economic Entomology*, **78**: 1350–1353.

Gross, H.R., Jr., W.J. Lewis, and D.A. Nordlund. 1981. *Trichogramma pretiosum*: Effect of prerelease parasitization experience on retention in release areas and efficiency. *Environmental Entomology*, **10**: 554–556.

Gross, H.R., Jr., W.J. Lewis, M. Beevers, and D.N. Nordlund. 1984. *Trichogramma pretiosum* (Hymenoptera: Trichogrammatidae): Effects of augmented densities and distributions of *Heliothis zea* (Lepidoptera: Noctuidae) host eggs and kairomones on field performance. *Environmental Entomology*, **13**: 981–985.

Gross, H.R., Jr., S.D. Pair, and R.D. Jackson. 1985. Behavioral responses of primary entomophagous predators to larval homogenates of *Heliothis zea* and *Spodoptera frugiperda* (Lepidoptera: Noctuidae) in whorl-stage corn. *Environmental Entomology*, **14**: 360–364.

Gross, H.R., J.J. Hamm, and J.E. Carpenter. 1994. Design and application of a hive-mounted device that uses honey bees (Hymenoptera: Apidae) to disseminate *Heliothis* nuclear polyhedrosis virus. *Environmental Entomology*, **23**: 492–501.

Gross, P. 1993. Insect behavioral and morphological defenses against parasitoids. *Annual Review of Entomology*, **38**: 251–273.

Grossman, J. 1990. Update: Yellowjacket biological control in New Zealand. *IPM Practitioner*, **12**(3): 1–5.

Grostal, P. and D.J. O'Dowd. 1994. Plants, mites and mutualism: Leaf domatia and the abundance and reproduction of mites on *Viburnum tinus* (Caprifoliaceae). *Oecologia*, **97**: 308–315.

Grout, T.G. and G.I. Richards. 1991. Value of pheromone traps for predicting infestations of red scale, *Aonidiella aurantii* (Maskell) (Hom., Diaspididae), limited by natural enemy activity and insecticides used to control citrus thrips, *Scirtothrips aurantii* Faure (Thys., Thripidae). *Journal of Applied Entomology*, **111**: 20–27.

Gruys, P. 1982. Hits and misses: The ecological approach to pest control in orchards. *Entomologia Experimentalis et Applicata*, **31**: 70–87.

Gu, D.J. and J.K. Waage. 1990. The effect of insecticides on the distribution of foraging parasitoids, *Diaeretierlla rapae* (Hym.: Braconidae), on plants. *Entomophaga*, **35**: 49–56.

Guerra, A.A., K.M. Robacker, and S. Martinez. 1993. *In vitro* rearing of *Bracon mellitor* and *Catolaccus grandis* with artificial diets devoid of insect components. *Entomologia Experimentalis et Applicata*, **68**: 303–307.

Guerra, G.P. and M. Kosztarab. 1992. *Biosytematics of the Family Dactylopiidae (Homoptera: Coccinea) with Emphasis on the Life Cycle of Dactylopius coccus Costa.* Bulletin 92–1, Virginia Agricultural Experiment Station, Virginia Polytechnic Institute and State University, Blacksburg, Virginia, U.S.A.

Gumovskaya, G.N. 1985. The coccinellid fauna. *Zashchita Rastenii*, **11**: 43. (in Russian)

Gupta, S.C., T.D. Leathers, G.N. El-Sayed, and C.M. Ignoffo. 1994. Relationships among enzyme activities and virulence parameters in *Beauveria bassiana* infections of *Galleria mellonella* and *Trichoplusia ni*. *Journal of Invertebrate Pathology*, **64**: 13–17.

Gutierrez, A.P., P. Neuenschwander and J.J.M. van Alphen. 1993. Factors affecting biological control of cassava mealybug by exotic parasitoids: a ratio-dependent supply-demand driven model. *Journal of Applied Ecology*, **30**: 706–721.

Habeck, D.H., S.B. Lovejoy, and J.G. Lee. 1993. When does investing in classical biological control research make economic sense? *Florida Entomologist*, **76**: 96–101.

Haccou, P., S.J. de Vlas, J.J.M. van Alphen, and M.E. Visser. 1991. Information processing by foragers: Effects of intra-patch experience on the leaving tendency of *Leptopilina heterotoma*. *Journal of Animal Ecology*, **60**: 93–106.

Hagen, K.S. and J.M. Franz. 1973. A history of biological control, pp. 433–476. *In* Smith, R.F., T.E. Mittler, and C.N. Smith, (eds.). *A History of Entomology.* Annual Review Inc., Palo Alto, California, U.S.A.

Hagen, K.S. and R. van den Bosch. 1968. Impact of pathogens, parasites and predators on aphids. *Annual Review of Entomology*, **13**: 325–384.

Hagen, K.S., E.F. Sawall, Jr., and R.L. Tassen. 1970. The use of food sprays to increase effectiveness of entomophagous insects. *Proceedings of the Tall Timbers Conference on Ecological Animal Control by Habitat Management*, **2**: 59–81.

Hagen, K.S., S. Bombosch, and J.A. McMurtry. 1976a. The biology and impact of predators, pp. 93–142. *In* Huffaker, C. and P.S. Messenger, (eds.). *Theory and Practice of Biological Control.* Academic Press, New York.

Hagen, K.S., G.A. Viktorov, K. Yamumatsu, and M.F. Schuster. 1976b. Biological control of pests of range, forage and grain crops, pp. 397–442. *In* Huffaker, C.B. and P.S. Messenger, (eds.). *Theory and Practice of Biological Control.* Academic Press, New York.

Hagler, J.R. and A.C. Cohen. 1991. Prey selection by *in vitro*- and field-reared Geocoris punctipes. *Entomologia Experimentalis et Applicata*, **59**: 201–205.

Hagler, J.R., S.E. Naranjo, D. Bradley-Dunlop, F.J. Enriquez, and T.J. Henneberry. 1994. A monoclonal antibody to pink bollworm (Lepidoptera: Gelechiidae) egg antigen: a tool for predator gut analysis. *Annals of the Entomological Society of America*, **87**: 85–90.

Hagley, E.A.C. 1989. Release of *Chrysoperla carnea* Stephens (Neuroptera: Chrysopidae) for control of the green apple aphid, *Aphis pomi* De Geer (Homoptera: Aphididae). *Canadian Entomologist*, **121**: 309–314.

Hagley, E.A.C. and C.M. Simpson. 1981. Effect of food sprays on numbers of predators in an apple orchard. *Canadian Entomologist*, **113**: 75–77.

Hågvar, E.B. and T. Hofsvang. 1989. Effect of honeydew and hosts on plant colonization by the aphid parasitoid *Ephedrus cerasicola*. *Entomophaga*, **34**: 495–501.

Hajek, A.E., R.A. Humber, and M.H. Griggs. 1990. Decline in virulence of *Entomophaga, maimaiga* (Zygomycetes: Entomophthorales) with repeated in vitro subculture. *Journal of Invertebrate Pathology*, **56**: 91–97.

Hall, R.A. 1985. Whitefly control by fungi, pp. 116–124. *In* Hussey, N.W. and N. Scopes, (eds.). *Biological Pest Control, the Glasshouse Experience.* Cornell University Press, Ithaca, New York, U.S.A.

Hall, R.W. and L.E. Ehler. 1979. Rate of establishment of natural enemies in classical biological control. *Bulletin of the Entomological Society of America*, **25**: 280–282.

Hall, R.A. and B. Papierok. 1982. Fungi as biological control agents of arthropods of agricultural and medical importance. *Parasitology*, **84**: 205–240.

Hall, R.W., L.E. Ehler, and B. Bisabri-Ershadi. 1980. Rate of success in classical biological control of arthropods. *Bulletin of the Entomological Society of America*, **26**: 111–114.

Hall, S.J., D. Raffaelli, and W.R. Turrell. 1990. Predator-caging experiments in marine systems: A reexamination of their value. *American Naturalist*, **136**: 657–672.

Halley, S. and E.J. Hogue. 1990. Ground cover influence on apple aphids, *Aphis pomi* De Geer (Homoptera: Aphididae), and its predators in a young apple orchard. *Crop Protection*, **9**: 221–230.

Hamai, J. and C.B. Huffaker. 1978. Potential of predation by *Metaseiulus occidentalis* in compensating for increased, nutritionally induced, power of increase of *Tetranychus urticae*. *Entomophaga*, **23**: 225–237.

Hamburg, H. Van and M.P. Hassell. 1984. Density dependence and the augmentative release of egg parasitoids against graminaceous stalk borers. *Ecological Entomology*, **9**: 101–108.

Hamdoun, A.M. and K.B. El Tigani. 1977. Weed problems in the Sudan *Pest Articles News Summary*, **23**: 190–194.

Hamilton, W.D. 1967. Extraordinary sex ratios. *Science*, **156**: 477–488.

Hamilton, W.D. 1979. Wingless and fighting males in fig wasps and other insects, pp. 167–220. *In* Blum, M.S. and N.A. Blum, (eds.). *Sexual Selection and Reproductive Competition in Insects*. Academic Press, New York.

Hance, Th. and C. Gregoire-Wibo. 1987. Effect of agricultural practices on carabid populations. *Acta Phytopathogica et Entomologica Hungarica*, **22**: 147–160.

Haney, P.B., R.F. Luck, and D.S. Moreno. 1987. Increases in densities of the citrus red mite, *Panonychus citri* (Acarina: Tetranychidae), in association with the Argentine ant, *Iridomyrmex humilis* (Hymenoptera: Formicidae), in southern California citrus. *Entomophaga*, **32**: 49–57.

Hanski, I. 1981. Coexistence of competitors in patchy environment with and without predation. *Oikos*, **37**: 306–312.

Hanski, I. 1990. Small mammal predation and the population dynamics of *Neodiprion sertifer*, pp. 253–264. *In* Watt, A.D., S.R. Leather, M.D. Hunter, and N.A.C. Kidd, (eds.). *Population Dynamics of Forest Insects*. Intercept, Andover, U.K.

Hanski, I., I. Woiwood, and J. Perry. 1993. Density dependence, population persistence, and largely futile arguments. *Oecologia* **95**: 595–598.

Hara, A.H., R. Gaugler, H.K. Kaya, and L.M. LeBeck. 1991. Natural populations of entomopathogenic nematodes (Rhabditida: Heterorhabditidae, Steinernematidae) from the Hawaiian Islands. *Environmental Entomology*, **20**: 211–216.

Harcourt, D.G. 1971. Population dynamics of *Leptinotarsa decemlineata* (Say) in eastern Ontario. III. Major population processes. *Canadian Entomologist*, **103**: 1049–1061.

Hardt, R.A. 1986. Japanese honeysuckle: from "one of the best" to ruthless pest. *Arnoldia*, **46**: 27–34.

Hardy, I.C. 1994. Sex ratio and mating structure in the parasitoid Hymenoptera. *Oikos*, **69**: 3–20.

Harley, K.L.S. 1990. The role of biological control in the management of water hyacinth, *Eichhornia crassipes*. *Biocontrol News and Information*, **11**(1): 11–22.

Harley, K.L.S. and I.W. Forno. 1992. *Biological Control of Weeds, a Handbook for Practitioners and Students*. Inkata Press, Melbourne, Australia.

Harley, K.L.S., R. C. Kassulke, D.P.A. Sands, and M.D. Day. 1990. Biological control of water lettuce *Pistia stratiotes* (Araceae) by *Neohydronomos affinis* (Coleoptera: Curculionidae). *Entomophaga*, **35**: 363–374.

Harper, J.D. 1987. Applied epizootiology: Microbial control of insects, pp. 473–496. *In* Fuxa, J.R. and Y. Tanada, (eds.). *Epizootiology of Insect Diseases*. John Wiley and Sons, New York.

Harris, P. 1973. The selection of effective agents for the biological control of weeds. *Canadian Entomologist*, **105**: 1495–1503.

Harris, P. 1976. Biological control of weeds: from art to science. *Proceedings of the XV International Congress of Entomology*, Washington, D.C., U.S.A.

Harris, P. 1979. Cost of biological control of weeds by insects in Canada. *Weed Science*, **27**: 242–250.

Harris, P. 1984. *Carduus nutans* L., nodding thistle, and *C. acanthoides* L., plumeless thistle (Compositae), pp. 115–126. *In* Kelleher, J.S. and M.A. Hulme, (eds.). *Biological Control Programmes Against Insects and Weeds in Canada 1969–1980*. Commonwealth Agricultural Bureaux International, London.

Harris, P. 1986. Biological control of weeds, pp. 123–138. *In* Franz, J.M., (ed.). *Biological Plant and Health Protection, Biological Control of Plant Pests and of Vectors of Human and Animal Diseases*. International symposium of the Akademie der Wissenschaften und der Literatur, Mainz, November 15th–17th 1984 at Mainz and Darmstadt, Germany.

Harris, P. 1988. Environmental impact of weed-control insects. *BioScience*, **38**: 542–548.

Harris, P. 1990. Environmental impact of introduced biological control agents, pp. 289–300. *In* Mackauer, M., L.E. Ehler, and J. Roland (eds). *Critical Issues in Biological Control*. Intercept Ltd., Hants, U.K.

Harris, P. 1991. Classical biocontrol of weeds: its definition, selection of effective agents, and administrative-political problems. *Canadian Entomologist*, **123**: 827–849.

Harris, P., A.T.S. Wilkinson, L.S. Thompson, and M. Neary. 1978. Interaction between the cinnabar moth, *Tyria jacobaeae* L. (Lep.: Arctiidae) and ragwort *Senecio jacobaea* L. (Compositae) in Canada, pp. 174–180. *In* Anon. *Proceedings of the 4th International Symposium on Biological Control of Weeds*. Gainesville, Florida, U.S.A.

Harris, V.E. and J.W. Todd. 1980. Male-mediated aggregation of male, female and 5th-instar southern green stink bugs and concomitant attraction of a tachinid parasite, *Trichopoda pennipes*. *Entomologia Experimentalis et Applicata*, **27**: 117–126.

Hartley, C. 1921. Damping-off in forest nurseries. *USDA Bulletin*, **934**: 1–99.

Hasan, S. 1980. Plant pathogens and biological control of weeds. *Reviews in Plant Pathology*, **59**: 349–356.

Hasan, S. 1981. A new strain of the rust fungus *Puccinia chondrillina* for biological control of skeleton weed in Australia. *Annals of Applied Biology*, **99**: 119–124.

Hasan, S. and A.J. Wapshere. 1973. The biology of *Puccinia chondrillina* a potential biological control agent of skeleton weed. *Annals of Applied Biology*, **74**: 325–332.

Hasan, S., E.S. Delfosse, E. Aracil, and R.C. Lewis. 1992. Host-specificity of *Uromyces heliotropii*, a fungal agent for the biological control of common heliotrope (*Heliotropium europaeum*) in Australia. *Annals of Applied Biology*, **121**: 697–705.

Hashimoto, S., T. Kashio, and T. Tsutsumi. 1989. A study on the biological control of the whitespotted longicorn beetle, *Anoplophora malasiaca*, by an entomogenous fungus, *Beauveria brongniartii*. II. Field evaluation of the banding of a polyurethane foam sheet with *B. brongniartii* conidia, around the trunk of citrus trees, for the control of adult beetles. *Proceedings of the Association for Plant Protection, Kyushu*, **35**: 129–133. (in Japanese)

Hassan, S.A. 1976/77. The use of the lacewing *Chrysopa carnea* Steph. (Neuroptera, Chrysopidae) to control the green peach aphid *Myzus persicae* (Sulzer) on glasshouse red pepper. *Zeitschrift für Angewandte Entomologie*, **82**: 243–239.

Hassan, S.A. 1977. Standardized techniques for testing side-effects of pesticides on beneficial arthropods in the laboratory. *Zeitschrift für Pflanzenkrankheiten und Pflanzenschutz*, **84**: 158–163.

Hassan, S.A. 1978. Releases of *Chrysopa carnea* Steph. to control *Myzus persicae* (Sulzer) on eggplant in small greenhouse plots. *Journal of Plant Diseases and Protection*, **85**: 118–123.

Hassan, S.A. 1980. A standard laboratory method to test the duration of harmful effects of pesticides on egg parasites of the genus *Trichogramma* (Hymenoptera: Trichogrammatidae). *Zeitschrift für Angewandte Entomologie*, **89**: 282–289.

Hassan, S.A. 1985. Standard methods to test the side-effects of pesticides on natural enemies of insects and mites developed by the IOBC/WPRS Working Group "Pesticides and Beneficial Organisms". *Bull. OEPP/EPPO*, **15**: 214–255.

Hassan, S.A. 1989a. Testing methodology and the concept of the IOBC/WPRS working group, pp. 1–18. *In* Jepson, P.C., (ed.). *Pesticides and Non-Target Invertebrates.* Intercept, Wimborne, Dorset, U.K.

Hassan, S.A. 1989b. Selection of suitable *Trichogramma* strains to control the codling moth *Cydia pomonella* and the two summer fruit tortrix moths *Adoxophyes orana*, *Pandemis heparana* (Lep.: Tortricidae). *Entomophaga*, **34**: 19–27.

Hassan, S.A. 1994. Strategies to select *Trichogramma* species for use in biological control, pp. 55–71. *In* Wajnberg, E. and S.A. Hassan, (eds.). *Biological Control With Egg Parasitoids.* Commonwealth Agricultural Bureaux, Wallingford, U.K.

Hassan, S.A., E. Stein, K. Dannemann, and W. Reichel. 1986. Massenproduktion und Anwendung von *Trichogramma*: 8 Optimierung des Einsatzes zur Bekämpfung des Maiszünslers *Ostrinia nublilalis* Hbn. *Journal of Applied Entomology*, **101**: 508–515.

Hassan, S.A., R. Albert, F. Bigler, P. Blaisinger, G. Bogenschutz, et al. 1987. Results of the third joint pesticide testing programme by the IOBC/WPRS Working Group "Pesticides and Beneficial Organisms." *Journal of Applied Entomology*, **103**: 92–107.

Hassan, S.A., E. Kohler, and W.M. Rost. 1988. Mass production and utilization of *Trichogramma*: 10. Control of the codling moth, *Cydia pomonella*, and the summer fruit tortrix moth, *Adoxophyes orana* (Lep.: Tortricidae). *Entomophaga*, **33**: 413–420.

Hassell, M.P. 1968. The behavioural response of a tachinid fly (*Cyzenis albicans* [Fall.]) to its host, the winter moth (*Operophtera brumata* [L.]). *Journal of Animal Ecology*, **37**: 627–639.

Hassell, M.P. 1971. Mutual interference between searching insect parasites. *Journal of Animal Ecology*, **40**: 473–486.

Hassel, M.P. 1975. Density dependence in single-species populations. *Journal of Animal Ecology*, **42**: 693–726.

Hassell, M.P. 1978. *The Dynamics of Arthropod Predator-Prey Systems.* Princeton University Press, Princeton, New Jersey, U.S.A.

Hassell, M.P. 1979. The dynamics of predator-prey interactions: Polyphagous predators, competing predators and hyperparasitoids, pp. 283–306. *In* Anderson, R.M., B.D. Turner, and L.R. Taylor, (eds.). *Population Dynamics.* Blackwell Scientific Publications, Oxford, U.K.

Hassell, M.P. 1980. Foraging strategies, population models and biological control: a case study. *Journal of Animal Ecology*, **49**: 603–628.

Hassell, M.P. 1984. Insecticides in host-parasitoid interactions. *Theoretical Population Biology*, **25**: 378–386.

Hassell, M.P. and H.N. Comins. 1976. Discrete time models for two-species competition. *Theoretical Population Biology*, **9**: 202–221.

Hassell, M.P. and C.B. Huffaker 1969. Regulatory processes and population cyclicity in laboratory populations of *Anagasta kühniella* (Zeller) (Lepidoptera: Phycitidae). III. The development of population models. *Researches on Population Ecology*, **11**: 186–210.

Hassell, M.P. and R.M. May. 1973. Stability in insect host-parasite models. *Journal of Animal Ecology*, **42**: 693–726.

Hassell, M.P. and R.M. May. 1974. Aggregation of predators and insect parasites and its effect on stability. *Journal of Animal Ecology*, **43**: 567–594.

Hassell, M.P. and R.M. May. 1986. Generalist and specialist natural enemies in insect predator-prey interactions. *Journal of Animal Ecology*, **55**: 923–940.

Hassell, M.P. and S.W. Pacala. 1990. Heterogeneity and the dynamics of host-parasitoid interactions. *Philosophical Transactions of the Royal Society of London, Series B*, **330**: 203–220.

Hassell, M.P. and G.C. Varley. 1969. New inductive population model and its bearing on biological control. *Nature*, **223**: 1133–1137.

Hassell, M.P. and J.K. Waage. 1984. Host-parasitoid population interactions. *Annual Review of Entomology*, **29**: 89–114.

Hassell, M.P., J.H. Lawton, and R.M. May. 1976. Patterns of dynamical behavior in single species populations. *Journal of Animal Ecology*, **45**: 471–486.

Hassell, M.P., J.K. Waage, and R.M. May. 1983. Variable parasitoid sex ratios and their effect on host-parasitoid dynamics. *Journal of Animal Ecology*, **52**: 889–904.

Hassell, M.P., R.M. May, S.W. Pacala, and P.L. Chesson. 1991. The persistence of host-parasitoid associations in patchy environments. I. A general criterion. *American Naturalist*, **138**: 568–583.

Haugen, D.A. and M.G. Underdown. 1991. Woodchip sampling for the nematode *Deladenus siricidicola* and the relationship with the percentage of *Sirex noctilio* infected. *Australian Forestry*, **54**(1–2): 3–8.

Havelka, J. and R. Zemek. 1988. Intraspecific variability of aphidophagous gall midge *Aphidoletes aphidimyza* (Rondani) (Dipt., Cecidomyiidae) and its importance for biological control of aphids. 1. Ecological and morphological characteristics of populations. *Journal of Applied Entomology*, **105**: 280–288.

Havron, A., D. Rosen, H. Prag, and Y. Rossler. 1991. Selection for pesticide resistance in *Aphytis*. I. *A. holoxanthus*, a parasite of the Florida red scale. *Entomologia Experimentalis et Applicata*, **61**: 221–228.

Hawkins, B.A. 1990. Global patterns of parasitoid assemblage size. *Journal of Animal Ecology*, **59**: 57–72.

Hays, D.B. and S.B. Vinson. 1971. Acceptance of *Heliothis virescens* (F.) as a host by the parasite *Cardiochiles nigriceps* Viereck (Hymenoptera, Braconidae). *Animal Behavior*, **19**: 344–352.

Hazard, W.H.L. 1988. Introducing crop, pasture and ornamental species into Australia: The risk of introducing new weeds. *Australian Plant Introduction Review*, 19–36

Hazzard, R.V., D.N. Ferro, R.G. Van Driesche, and A.F. Tuttle. 1991. Mortality of eggs of Colorado potato beetle (Coleoptera: Chrysomelidae) from predation by *Coleomegilla maculata* (Coleoptera: Coccinellidae). *Environmental Entomology*, **20**: 841–848.

Heads, P.A. and J.H. Lawton. 1983. Studies on the natural enemy complex of the holly leaf-miner: The effects of scale on the detection of aggregative responses and the implication for biological control. *Oikos*, **40**: 267–276.

Hegedus, D.D. and G.G. Khachatourians. 1993. Construction of cloned DNA probes for the specific detection of the entomopathogenic fungus *Beauveria bassiana* in grasshoppers. *Journal of Invertebrate Pathology*, **62**: 233–240.

Heidger, C. and W. Nentwig. 1989. Augmentation of beneficial arthropods by strip-management. 3. Artificial introduction of a spider species which preys on wheat pest insects. *Entomophaga*, **34**: 511–522.

Heimbach, U. and C. Abel. 1991. Side effects of soil insecticides in different formulations on some beneficial arthropods. *Verhandlung der Gesellschaft für Ökologie*, **19**: 163–170. (in German)

Heinrichs, E.A. and O. Mochida. 1983. From secondary to major pest status: the case of insecticide-induced rice brown planthopper, *Nilaparvata lugens*, resurgence. Paper presented at the Rice Pest Management Symposium, Section J, Entomology, of the XV Pacific Science Congress, February 1–11, 1983, Dunedin, New Zealand.

Heinrichs, E.A., G.B. Aquino, S. Chelliah, S.L. Valencia, and W.H. Reissig. 1982. Resurgence of *Nilaparvata lugens* (Stål) populations as influenced by method and timing of insecticide applications in lowland rice. *Environmental Entomology*, **11**: 78–84.

Heiny, D.K.and G.E. Templeton. 1993. Economic comparisons of mycoherbicides to conventional herbicides, pp. 395–408. *In* Altman, J., (ed.). *Pesticide Interactions in Crop Production, Beneficial and Deleterious Effects*. CRC Press, Inc., Boca Raton, Florida, U.S.A.

Heinz, K.M. and M.P. Parrella. 1990. The influence of host size on sex ratios in the parasitoid *Diglyphus begini* (Hymenoptera: Eulophidae). *Ecological Entomology*, **15**: 391–399.

Heirbaut, M. and P. van Damme. 1992 The use of artificial nests to establish colonies of the black cocoa ant (*Dolichoderus thoracicus* Smith) used for biological control of *Helopeltis theobromae* Mill. in Malaysia. *Mededelingen van de Faculteit Landbouwwetenschappen*, **57**: 533–542.

Helgesen, R.G. and M.J. Tauber. 1974. Biological control of greenhouse whitefly, *Trialeurodes vaporariorum* (Aleyrodidae: Homptera), on short-term crops by manipulating biotic and abiotic factors. *Canadian Entomologist*, **106**: 1175–1188.

Helmuth. R. 1988. *Bacillus thuringiensis*—Health aspects for man and animals. *Mitteilungen aus der Biologischen Bundesanstalt fur Land-und Forstwirschaft Berlin-Dahlem*, **246**: 95–101. (in German)

Hemerik, L., G. Driessen, and P. Haccou. 1993. Effects of intra-patch experiences on patch time, search time and searching efficiency of the parasitoid *Leptopilina clavipes*. *Journal of Animal Ecology*, **62**: 33–44.

Hemptinne, J.-L. 1988. Ecological requirements for hibernating *Propylea quatuordecimpunctata* (L.) and *Coccinella septempunctata* (Col.: Coccinellidae). *Entomophaga*, **33**: 505–515.

Hemptinne, J.-L., A.F.G. Dixon, and J. Coffin. 1992. Attack strategy of ladybird beetles (Coccinellidae): factors shaping their numerical response. *Oecologia*, **90**: 238–245.

Henderson, D.E. and D.A. Raworth. 1990. *Beneficial insects and common pests on strawberry and raspberry crops*. Agriculture Canada Publication 1863/E, Ottawa, Canada.

Hendrickson, R.M., Jr., S.E. Barth, and L.R. Ertle. 1987. Control of relative humidity during shipment of parasitic insects. *Journal of Economic Entomology*, **80**: 537–539.

Henn, T. and R. Weinzierl. 1990. *Alternatives in Insect Management: Beneficial Insects and Mites*. Circular 1298, University of Illinois Cooperative Extension Service, Urbana-Champaign, Illinois, U.S.A.

Henriquez, N.P. and J.R. Spence. 1993. Host location by the gerrid egg parasitoid *Tiphodytes gerriphagus* (Marchal) (Hymenoptera: Scelionidae). *Journal of Insect Behavior*, **6**: 455–466.

Henry, A.W. 1931. The natural microflora of the soil in relation to the foot rot problem of wheat. *Canadian Journal of Research*, **4**: 69–77.

Henry, J.E. 1990. Control of insect by protozoa, pp. 161–176. *In* Baker, R.R. and P.E. Dunn, (eds.). *New Directions in Biological Control: Alternatives for Suppressing Agricultural Pests and Diseases*. Alan R. Liss, Inc., New York.

Henry, J.E. and J.E. Onsager. 1982. Large-scale test of control of grasshoppers on rangeland with *Nosema locustae*. *Journal of Economic Entomology*, **75**: 31–35.

Henry, J.E., E.A. Oma, and J.A. Onsager. 1978. Relative effectiveness of ULV spray applications of spores of *Nosema locustae* against grasshoppers. *Journal of Economic Entomology*, **71**: 629–632.

Hérard, F., M.A. Keller, W.J. Lewis, and J.H. Tumlinson. 1988. Beneficial arthropod behavior mediated by airborne semiochemicals. IV. Influence of host diet on host-orientated flight chamber responses of *Microplitis demolitor* Wilkinson. *Journal of Chemical Ecology*, **14**: 1597–1606.

Herren, H.R. 1987. Africa-wide biological control project of cassava pests. A review of objectives and achievements. *Insect Science and Its Application*, **8**: 837–840.

Herren, H.R. and P. Neuenschwander. 1991. Biological control of cassava pests in Africa. *Annual Review of Entomology*, **36**: 257–284.

Herren, H.R., T. J.Bird, and D.J. Nadel. 1987. Technology for automated aerial release of natural enemies of the cassava mealybug and cassava green mite. *Insect Science and Its Application*, **8**: 883–885.

Herrnstadt, C., F. Gaertner, W. Gelernter, and D.L. Edwards. 1987. *Bacillus thuringiensis* isolate with activity against Coleoptera, pp. 101–113. *In* Maramorosch, K. (ed.). *Biotechnology in Invertebrate Pathology and Cell Culture*. Academic Press, New York.

Higashiura, Y. 1989. Survival of eggs in the gypsy moth *Lymantria dispar*. I. Predation by birds. *Journal of Animal Ecology*, **58**: 403–412.

Hight, S.D. and J.J. Drea, Jr. 1991. Prospects for a classical biological control project against purple loosestrife (*Lythrum salicaria* L.). *Natural Areas Journal*, **11**: 151–157.

Higley, L.G., M.R. Zeiss, W.K. Wintersteen, and L.P. Pedigo. 1992. National pesticide policy: A call for action. *American Entomologist*, **38**: 139–146.

Hill, M.G. 1988. Analysis of the biological control of *Mythimna separata* (Lepidoptera: Noctuidae) by *Apanteles ruficrus* (Hymenoptera: Braconidae) in New Zealand. *Journal of Applied Ecology*, **25**: 197–208.

Hill, M.G. and D.J. Allan. 1986. The effects of weeds on armyworm in maize, pp. 260–263. *Proceedings of the 39th New Zealand Weed and Pest Control Conference*, Palmerston North, New Zealand, New Zealand Weed and Pest Control Society.

Hill, M.G., D.J. Allan, R.C. Henderson, and J.C. Charles. 1993. Introduction of armored scale predators and establishment of the predatory mite *Hemisarcoptes coccophagus* (Acari: Hemisarcoptidae) on latania scale, *Hemiberlesia latania* (Homoptera: Diaspididae) in kiwifruit shelter trees in New Zealand. *Bulletin of Entomological Research*, **83**: 369–376.

Hill, R.L., J.M. Gindell, C.J. Winks, J.J. Sheat, and L.M. Hayes. 1991. Establishment of gorse spider mite as a control agent for gorse, pp. 31–34. *Proceedings of the 44th New Zealand Weed and Pest Control Conference*.

Hinks, C.F. 1971. Observations on larval behaviour and avoidance of encapsulation of *Perilampus hyalinus* (Hymenoptera: Perilampidae) parasitic in *Neodiprion lecontei* (Hymenoptera: Diprionidae). *Canadian Entomologist*, **103**: 182–187.

Hirte, W.F., C. Walter, M. Grunberg, H. Sermann, and H. Adam. 1989. Selection of pathotypes of *Verticillium lecanii* for various harmful insects in glasshouses and aspects of the biotechnological spore production. *Zentrablatt für Mikrobiologie*, **144**: 405–420.

Hislop, R.G. and R.J. Prokopy. 1981a. Integrated management of phytophagous mites in Massachusetts (U.S.A.) apple orchards. 2. Influences of pesticides on the mite predator *Amblyseius fallacis* under laboratory and field conditions. *Protection Ecology*, **3**: 157–172.

Hislop, R.G. and R.J. Prokopy. 1981b. Mite predator responses to prey and predator-emitted stimuli. *Journal of Chemical Ecology*, **7**: 895–906.

Hobbs, R.J. and L.F. Huenneke. 1992. Disturbance, diversity, and invasion: implications for conservation. *Conservation Biology*, **6**: 324–337.

Hochberg, M.E. 1989. The potential role of pathogens in biological control. *Nature*, **337**(6204): 262–265.

Hochberg, M.E. and B.A. Hawkins. 1992. Refuges as a predictor of parasitoid diversity. *Science*, **255**: 973–976.

Hochberg, M.E. and J.K. Waage. 1991. A model for the biological control of *Oryctes rhinocerus* (Coleoptera: Scarabaeidae) by means of pathogens. *Journal of Applied Ecology*, **28**: 514–531.

Hochberg, M., M.P. Hassell, and R.M. May. 1990. The dynamics of host-parasitoid-pathogen interactions. *American Naturalist*, **135**: 74–94.

Hodek, I. 1970. Coccinellids and modern pest management. *BioScience*, **20**: 543–552.

Hodek, I. 1973. *Biology of the Coccinellidae.* Dr. W. Junk, N.V. Publishers, The Hague, The Netherlands.

Hodek, I., (ed.). 1986. *Ecology of Aphidophaga. Proceedings of the 2nd symposium held at Zvíkovské Podhradí, 2–8 September, 1984.* Dr. W. Junk Publishers, Dordrecht, The Netherlands.

Hodgkin, S.E. 1984. Scrub encroachment and its effects on soil fertility on Newborough Warren, Anglesey, Wales. *Biological Conservation* **29**: 99–119.

Hoelmer, K.A. and D.L. Dahlsten. 1993. Effects of malathion bait spray on *Aleyrodes spiraeoides* (Homoptera: Aleyrodidae) and its parasitoids in northern California. *Environmental Entomology*, **22**: 49–56.

Hoffman, J.D., C.M. Ignoffo, and W.A. Dickerson. 1975. *In vitro* rearing of the endoparasitic wasp, *Trichogramma pretiosum*. *Annals of the Entomological Society of America*, **68**: 335–336.

Hoffmann, J.H. and V.C. Moran. 1989. Novel graphs for depicting herbivore damage on plants: The biocontrol of *Sesbania punicea* (Fabaceae) by an introduced weevil. *Journal of Applied Ecology*, **26**: 353–360.

Hoffmann, J.H. and V.C. Moran. 1991. Biocontrol of a perennial legume, *Sesbania punicea*, using a florivorous weevil, *Trichapion lativentre*: Weed population dynamics with a scarcity of seeds. *Oecologia*, **88**: 574–576.

Hoffmann, J.H. and V.C. Moran. 1992. Oviposition patterns and the supplementary role of a seed-feeding weevil, *Rhyssomatus marginatus* (Coleoptera: Curculionidae), in the biological control of a perennial leguminous weed, *Sesbania punicea*. *Bulletin of Entomological Research*, **82**: 343–347.

Hoffmann, J.R., (ed.). 1991. Biological control of weeds in South Africa. *Agriculture, Ecosystems and Environment*, **37**: 1–255.

Hoffmann, M.P. and A.C. Frodsham. 1993. *Natural Enemies of Vegetable Insect Pests.* Cornell University, Cooperative Extension Publication, Ithaca, New York, U.S.A.

Hoffmann, M.P., L.T. Wilson, F.G. Zalom, and R.J. Hilton. 1990. Parasitism of *Heliothis zea* (Lepidoptera: Noctuidae) eggs: Effect on pest management decision rules for processing tomatoes in the Sacramento Valley of California. *Environmental Entomology*, **19**: 753–763.

Hoffmann, M.P., L.T. Wilson, F.G. Zalom, and R.J. Hilton. 1991. Dynamic sequential sampling plan for *Helicoverpa zea* (Lepidoptera: Noctuidae) eggs in processing tomatoes: Parasitism and temporal patterns. *Environmental Entomology*, **20**: 1005–1012.

Hofsvang, T. 1988. Mechanisms of host discrimination and intraspecific competition in the aphid parasitoid *Ephedrus cerasicola*. *Entomologia Experimentalis et Applicata*, **48**: 233–239.

Hokkanen, H. and D. Pimentel. 1984. New approach for selecting biological control agents. *Canadian Entomologist*, **116**: 1109–1121.

Hokkanen, H., G.B. Husberg, and M.Söderblom. 1988. Natural enemy conservation for the integrated control of the rape blossom beetle *Meligethes aeneus* F. *Annales Agricultura Fenniae*, **27**: 281–293.

Hölldobler, B. and E.O. Wilson. 1990. *The Ants.* The Belknap Press of Harvard University Press, Cambridge, Massachusetts, U.S.A.

Holling, C.S. 1959a. Some characteristics of simple types of predation and parasitism. *Canadian Entomologist*, **91**: 385–398.

Holling, C.S. 1959b. The components of predation as revealed by a study of small mammal predation of the European pine sawfly. *Canadian Entomologist*, **91**: 293–320.

Holling, C.S. 1966. The functional response of invertebrate predators to prey density. *Memoirs of the Entomological Society of Canada*, **48**: 1–86.

Holt, J.G. 1984, 1986, 1989a, 1989b. *Bergey's Manual of Systematic Bacteriology. Vol. 1 to 4.* Williams and Wilkins, Baltimore, Maryland, U.S.A.

Holt, J., D.R. Wareing, and G.A. Norton. 1992. Strategies of insecticide use to avoid resurgence of *Nilaparvata lugens* (Homoptera: Delphacidae) in tropical rice: A simulation analysis. *Journal of Economic Entomology*, **85**: 1979–1989.

Holyoak, M. 1993. New insights into testing for density dependence. *Oecologia* **93**: 435–444.

Holyoak, M. and P.H. Crowley. 1993. Avoiding erroneously high levels of detection in combinations of semi-independent tests. *Oecologia* **95**: 103–114.

Holyoak, M. and J.H. Lawton. 1993. Comments arising from a paper by Wolda and Dennis: using and interpreting the results of tests for density dependence. *Oecologia*, **95**: 435–444.

Hominick, W.M. and A.P. Reid. 1990. Perspectives on entomopathogenic nematology, pp. 327–345. *In* Gaugler, R. and H.K. Kaya, (eds.). *Entomopathogenic Nematodes in Biological Control.* CRC Press, Inc., Boca Raton, Florida, U.S.A.

Honée, G. and B. Visser. 1993. The mode of action of Bacillus thuringiensis crystal proteins. *Entomologia Experimentalis et Applicata*, **69**: 145–155.

Hope, C.A., S.A. Nicholson, and J.J. Churcher. 1990. Aerial release system for *Trichogramma minutum* Riley in plantation forests. *Memoirs of the Entomological Society of Canada*, **153**: 38–44.

Hopper, K.R., R.T. Roush, and W. Powell. 1993. Management of genetics of biological control introductions. *Annual Review of Entomology*, **38**: 27–51.

Horsfall, J.G. and E.B. Cowling, (ed.). 1980. *Plant Disease: An Advanced Treatise. Vol. 2. How Disease Develops in Populations.* Academic Press, New York.

Hörstadius, S. 1974. Linnaeus, animals and man. *Biological Journal of the Linnaean Society*, **6**: 269–275.

Hoti, S.L. and K. Balaraman. 1990. Utility of cheap carbon and nitrogen sources for the production of a mosquito-pathogenic fungus, *Lagenidium. Indian Journal of Medical Research, Section A, Infectious Diseases*, **91**: 67–69.

Houseweart, M.W., D.T. Jennings, and R.K. Lawrence. 1984. Field releases of *Trichogramma minutum* (Hymenoptera: Trichogrammatidae) for suppression of epidemic spruce budworm, *Choristoneura fumiferana* (Lepidoptera: Tortricidae), egg populations in Maine. *Canadian Entomologist*, **116**: 1357–1366.

Howard, R.W. and P.W. Flinn. 1990. Larval trails of *Cryptolestes ferrugineus* (Coleoptera: Cucujidae) as kairomonal host-finding cues for the parasitoid *Cephalonomia waterstoni* (Hymenoptera: Bethylidae). *Annals of the Entomological Society of America*, **83**: 239–245.

Howarth, F.G. 1983. Classical biocontrol: Panacea or Pandora's box. *Proceedings of the Hawaiian Entomological Society*, **24**: 239–244.

Howarth, F.G. 1985. Impacts of alien land arthropods and mollusks on native plants and animals in Hawai'i, pp. 149–179. *In* Stone, C.P. and J.M. Scott, (eds.). *Hawai'i's Terrestrial Ecosystems: Preservation and Management.* Proceedings of a symposium held 5–6 June 1984, at Hawai'i Volcanoes National Park, University of Hawaii Cooperative National Park Resources Studies Unit, 3190 Maile Way, Honolulu, Hawaii, U.S.A.

Howarth, F.G. 1991. Environmental impacts of classical biological control. *Annual Review of Entomology*, **36**: 485–509.

Hoy, M.A. 1982a. *Recent Advances in Knowledge of the Phytoseiidae.* Division of Agricultural Sciences, University of California, Special Publication #3284, Berkeley, California, U.S.A.

Hoy, M.A. 1982b. Aerial dispersal and field efficacy of a genetically improved strain of the spider mite predator *Metaseiulus occidentalis. Entomologia Experimentalis et Applicata*, **32**: 205–212.

Hoy, M.A. 1988. Biological control of arthropod pests: traditional and emerging technologies. *American Journal of Alternative Agriculture*, **3**: 63–68.

Hoy, M.A. 1992. Criteria for release of genetically-improved phytoseiids: An examination of the risks associated with release of biological control agents. *Experimental and Applied Acarology*, **14**: 393–416.

Hoy, M.A. and F.E. Cave. 1988. Guthion-resistant strain of walnut aphid parasite. *California Agriculture*, **42**(4): 4–5.

Hoy, M.A. and F.E. Cave. 1989. Toxicity of pesticides used on walnuts to a wild and azinphosmethyl-resistant strain of *Trioxys pallidus* (Hymenoptera: Aphidiidae). *Journal of Economic Entomology*, **82**: 1585–1592.

Hoy, M.A. and K.A. Standow. 1982. Inheritance of resistance to sulfur in the spider mite predator *Metaseiulus occidentalis. Entomologia Experimentalis et Applicata*, **31**: 316–323.

Hoy, M.A., P.H. Westigard, and S.C. Hoyt. 1983. Release and evaluation of laboratory-selected, pyrethroid-resistant strains of the predaceous mite *Typhlodromus occidentalis* (Acarina: Phytoseiidae) into southern Oregon pear orchards and Washington apple orchards. *Journal of Economic Entomology*, **76**: 383–388.

Hoy, M.A., F.E. Cave, R.H. Beede, J. Grant, W.H. Krueger, W.H. Olson, K.M. Spollen, W.W. Barnett, and L.C. Hendricks. 1990. Release, dispersal, and recovery of a laboratory-selected strain of the walnut aphid parasite *Trioxys pallidus* (Hymenoptera: Aphidiidae) resistant to azinphosmethyl. *Journal of Economic Entomology*, **83**: 89–96.

Hoyt, S.C. and L.E. Caltagirone. 1971. The developing programs of integrated control of pests of apples in Washington and peaches in California, pp. 395–421. *In* Huffaker, C.B., (ed.). *Biological Control.* Plenum Press, New York.

Hua, L.Z., F. Lammes, J.C. van Lenteren, P.W.T. Huisman, A. van Vianen, and O.M.B. de Ponti. 1987. The parasite-host relationship between *Encarsia formosa* Gahan (Hymenoptera, Aphelinidae) and *Trialeurodes vaporariorum* (Westwood) (Homoptera, Aleyrodidae). XXV. Influence of leaf structure on the searching activity of *Encarsia formosa*. *Journal of Applied Entomology*, **104**: 297–304.

Hubbard, S.F. and R.M. Cook. 1978. Optimal foraging by parasitoid wasps. *Journal of Animal Ecology*, **17**: 593–604.

Huber, J. 1986. Use of baculoviruses in pest management programs, pp. 181–202. *In* Granados, R.R. and B.A. Federici, (eds.). *The Biology of Baculoviruses: Volume II. Practical Application for Insect Control.* CRC Press, Inc., Boca Raton, Florida, U.S.A.

Huber, J. 1990. History of the CPGV as a biological control agent--its long way to a commercial viral pesticide, pp. 424–427. *In* Pinnock, D.E., (ed.). *Vth International Colloquium on Invertebrate Pathology and Microbial Control.* Adelaide, Australia, 20–24, August, 1990.

Huber, J. and H.G. Miltenburger. 1986. Production of pathogens. Fortschritte der Zoologie, **32**: 167–181. *In* Franz, J.M., (ed.). *Biological Plant and Health Protection, Biological Control of Plant Pests and of Vectors of Human and Animal Diseases.* Gustav Fischer Verlag, Stuttgart, Germany.

Huffaker, C.B. 1958. Experimental studies on predation: dispersion factors and predator-prey oscillations. *Hilgardia*, **27**: 343–383.

Huffaker, C.B. 1985. Biological control in integrated pest management: an entomological perspective, pp. 13–23. *In* Hoy, M.A. and D.C. Herzog, (eds.). *Biological Control in Agricultural IPM Systems.* Academic Press, Orlando, Florida, U.S.A.

Huffaker, C.B. and C.E. Kennett. 1956. Experimental studies on predation: (1) Predation and cyclamen mite populations on strawberries in California. *Hilgardia*, **26**: 191–222.

Huffaker, C.B. and C.E. Kennett. 1959. A ten-year study of vegetational changes associated with biological control of klamath weed. *Journal of Range Management*, **12**: 69–82.

Huffaker, C.B. and P.S. Messenger, (eds.). 1976. *Theory and Practice of Biological Control.* Academic Press, New York.

Huffaker, C.B., K.P. Shea, and S.G. Herman. 1963. Experimental studies on predation: Complex dispersion and levels of food in an acarine predator-prey interaction. *Hilgardia*, **34**: 305–330.

Huffaker, C.B., P.S. Messenger, and P. DeBach. 1971. The natural enemy component in natural control and the theory of biological control, pp. 16–67. *In* Huffaker, C.B., (ed.). *Biological Control.* Plenum Press, New York.

Huffaker, C.B., J. Hamai, and R.M. Nowierski. 1983. Biological control of puncturevine, *Tribulus terrestris*, in California after twenty years of activity of introduced weevils. *Entomophaga*, **28**: 387–400.

Hughes, R.D. 1962. A method for estimating the effects of mortality on aphid populations. *Journal of Animal Ecology*, **31**: 389–396.

Hughes, R.D. 1963. Population dynamics of the cabbage aphid, *Brevicoryne brassicae* (L.). *Journal of Animal Ecology*, **32**: 393–424.

Hughes, R.D., L.T. Woolcock, J.A. Roberts, and M.A. Hughes. 1987. Biological control of the spotted alfalfa aphid, *Therioaphis trifolii* F. *maculata*, on lucerne crops in Australia, by the introduced parasitic *Hymenopteran Trioxys* complanatus. *Journal of Applied Ecology*, **24**: 515–537.

Hull, L.A. and E.H. Beers. 1985. Ecological selectivity: Modifying chemical control practices to preserve natural enemies, pp. 103–122. *In* Hoy, M.A. and D.C. Herzog, (eds.). *Biological Control in Agricultural IPM Systems.* Academic Press, Orlando, Florida, U.S.A.

Hull, L.A., K.D. Hickey, and W.W. Kanour. 1983. Pesticide usage patterns and associated pest damage in commercial apple orchards of Pennsylvania. *Journal of Economic Entomology*, **76**: 577–583.

Humber, R.A. 1981. An alternative view of certain taxonomic criteria used in the Entomophthorales (Zygomycetes). *Mycotaxon*, **13**: 191–240.

Humber, R.A. 1990. Systematic and taxonomic approaches to entomophthoralean species, pp. 133–137. *In* Pinnock, D.E., (ed.). *Vth International Colloquium on Invertebrate Pathology and Microbial Control.* Adelaide, Australia, 20–24 August, 1990.

Hunter, C.D. 1992. *Suppliers of Beneficial Organisms in North America.* California Environmental Protection Agency, Department of Pesticide Regulation, 1220 N. St., P.O. Box 942871, Sacramento, California, 94271-0001, U.S.A.

Hurd, H. 1993. Reproductive disturbances induced by parasites and pathogens of insects, pp. 87–105. *In* Beckage, N.E., S.N. Thompson, and B.A. Federici, (eds.). *Parasites and Pathogens of Insects, Volume I. Parasites.* Academic Press, New York.

Hussey, N.W. 1985. History of biological control in protected culture, 1.1 Western Europe, pp. 11–22. *In* Hussey, N.W. and N. Scopes, (eds.). *Biological Pest Control, the Glasshouse Experience.* Cornell University Press, Ithaca, New York, U.S.A.

Hutchison, W.D. and D.B. Hogg. 1984. Demographic statistics for the pea aphid (Homoptera; Aphididae) in Wisconsin and a comparison with other populations. *Environmental Entomology*, **13**: 1173–1181.

Hutchison, W.D. and D.B. Hogg. 1985. Time-specific life tables for the pea aphid, *Acyrthosiphon pisum* (Harris), on alfalfa. *Researches on Population Ecology*, **27**: 231–253.

Idoine, K. and D.N. Ferro. 1990. Persistence of *Edovum puttleri* (Hymenoptera: Eulophidae) on potato plants and parasitism of *Leptinotarsa decemlineata* (Coleoptera: Chrysomelidae): effects of resource availability and weather. *Environmental Entomology*, **19**: 1732–1737.

Ignoffo, C.M. and V.H. Dropkin. 1977. Deleterious effects of the thermostable toxin of *Bacillus thuringiensis* on species of soil-inhabiting, myceliophagous, and plant-parasitic nematodes. *Journal of the Kansas Entomological Society*, **50**: 394–398.

Ignoffo, C.M. and W.F. Hink. 1971. Propagation of arthropod pathogens in living systems, pp. 541–580. *In* Burges, H.D. and Hussey, N.W., (eds.). *Microbial Control of Insects and Mites*. Academic Press, London.

Ignoffo, C.M., C. Garcia, R.W. Kapp, and W.B. Coate. 1979. An evaluation of the risks to mammals of the use of an entomopathogenic fungus, *Nomuraea rileyi*, as a microbial insecticide. *Environmental Entomology*, **8**: 354–359.

Ignoffo, C.M., C. Garcia, D.L. Hostetter, and R.E. Pinnell. 1980. Transplanting: A method of introducing an insect virus into an ecosystem. *Environmental Entomology*, **9**: 153–154.

Ignoffo, C.M., B. S. Shasha, and M. Shapiro. 1991. Sunlight ultraviolet protection of the *Heliothis* nuclear polyhedrosis virus through starch-encapsulation technology. *Journal of Invertebrate Pathology*, **57**: 134–136.

Inayatullah, C. 1983. Host selection by *Apanteles flavipes* (Cameron) (Hymenoptera: Braconidae): Influence of host and host plant. *Journal of Economic Entomology*, **76**: 1086–1087.

Infante, E.P. 1986. Clave para la identificacion de los generos y catalogo de las especies espanolas peninsulares y Balearicas de Coccinellidae (Coleoptera). *Graellsia*, **662**: 19–45.

Inoue, K., M. Osakabe, and W. Ashihara. 1987. Identification of pesticide resistant phytoseiid mite populations in citrus orchards, and on grapevines in glasshouses and vinyl-houses (Acarina: Phytoseiidae). *Japanese Journal of Applied Entomology and Zoology*, **31**: 398–403.

Isenbeck, M. and F.A. Schulz. 1986. Biological control of fireblight (*Erwinia amylovora* [Burr.] Winslow et al.) on ornamentals. II. Investigation about the mode of action of the antagonistic bacteria. *Journal of Phytopathology*, **116**: 308–314.

Ishibashi, N. and E. Kondo. 1990. Behavior of infective juveniles, pp. 139–150. *In* Gaugler, R. and H.K. Kaya, (eds.). *Entomophathogenic Nematodes in Biological Control*. CRC Press, Inc., Boca Raton, Florida, U.S.A.

Ives, A.R. and R.M. May. 1985. Competition within and between species in a patchy environment: Relations between microscopic and macroscopic models. *Journal of Theoretical Biology*, **115**: 65–92.

Ives, W.G.H. 1976. The dynamics of larch sawfly (Hymenoptera: Tenthredinidae) populations in southeastern Manitoba. *Canadian Entomologist*, **108**: 701–730.

Izawa, H., M. Osakabe, and S. Moriya. 1992. Isozyme discrimination between an imported parasitoid wasp, *Torymus sinensis* Kamijo and its sibling species, *T. beneficus* Yasumatsu et Kamijo (Hymenoptera: Torymidae), attacking *Dryocosmus kuriphilus* Yasumatsu (Hymenoptera: Cynipidae). *Japanese Journal of Applied Entomology and Zoology*, **36**: 58–60. (in Japanese)

Jackson, D.M., G.C. Brown, G.L. Nordin, and D.W. Johnson. 1992. Autodissemination of a baculovirus for management of tobacco budworms (Lepidoptera: Noctuidae) on tobacco. *Journal of Economic Entomology*, **85**: 710–719.

Jackson, T.A. 1990. Commercial development of *Serratia entomophila* as a biocontrol agent for the New Zealand grass grub, p. 15. *In* Pinnock, D.E., (ed.). *Vth International Colloquium on Invertebrate Pathology and Microbial Control*. Adelaide, Australia, 20–24 August, 1990.

Jackson, T.A. and W.M. Wouts. 1987. Delayed action of an entomophagous nematode (*Heterorhabditis* sp. (V16)) for grass grub control, pp. 33–35. Proceedings of the New Zealand Weed and Pest Control Conference. Palmerston North, New Zealand, New Zealand Weed and Pest Control Society.

Jacobs, S.E. 1951. Bacteriological control of the flour moth, *Ephestia kuehniella* Z. *Proceedings of the Society of Applied Bacteriology*, **13**: 83–91.

Jaenike, J. 1990. Host specialization in phytophagous insects. *Annual Review of Ecology and Systematics*, **21**: 243–273.

Jaenike, J and D.R. Papaj. 1992. Behavioral plasticity and patterns of host use by insects., pp. 245–264. *In* Roitberg, B.D. and M.B. Isman, (eds.). *Insect Chemical Ecology, an Evolutionary Approach*. Chapman and Hall, New York.

Jaffee, B.A. and T.M. McInnis. 1990. Effects of carbendazim on the nematophagous fungus *Hirsutella rhossiliensis* and the ring nematode. *Journal of Nematology*, **22**: 418–419.

James, D.G. 1993. Pollen, mould mites and fungi: Improvements to mass rearing of *Typhlodromus doreenae* and *Amblyseius victoriensis*. *Experimental and Applied Acarology*, **14**: 271–276.

James, D.G. 1994. The development of suppression tactics for *Biprorulus bibax* (Heteroptera: Pentatomidae) as part of an integrated pest management programme in citrus in inland south-eastern Australia. *Bulletin of Entomological Research*, **84**: 31–38.

James, D.J. 1989. Overwintering of *Amblyseius victoriensis* (Womersley) (Acarina: Phytoseiidae) in southern New South Wales. *General Applied Entomology*, **21**: 51–55.
Janssen, A., C.D. Hofker, A.R. Braun, N. Mesa, M.W. Sabelis, and A.C. Bellotti. 1990. Preselecting predatory mites for biological control: the use of an olfactometer. *Bulletin of Entomological Research*, **80**: 177–181.
Jansson, H.-B. and B. Nordbring-Hertz. 1980. Interactions between nematophagous fungi and plant parasitic nematodes: Attraction, induction of trap formation and capture. *Nematologica*, **26**: 383–389.
Jansson, R.K. 1993. Introduction of exotic entomopathogenic nematodes (Rhabditida: Heterorhabditidae and Steinernematidae) for biological control of insects: Potential and problems. *Florida Entomologist*, **76**: 82–96.
Jaques, R.P. 1990. Effectiveness of the granulosis virus of the codling moth in orchard trials in Canada, pp. 428–430. *In* Pinnock, D.E., (ed.). *Vth International Colloquium on Invertebrate Pathology and Microbial Control.* Adelaide, Australia, 20–24 August, 1990.
Jatala, P., R. Kaltenbach, and M. Bocangel. 1979. Biological control of *Meloidogyne incognita* acrita and *Globodera pallida* on potatoes. *Journal of Nematology*, **11**: 303.
Jaynes, R.A. and J.E. Elliston. 1980. Pathogenicity and canker control by mixtures of hypovirulent strains of *Endothia parasitica* in American chestnut. *Phytopathology*, **70**: 453–456.
Jeffries, P. and M.J. Jeger. 1990. The biological control of postharvest diseases of fruit. *Biocontrol News and Information*, **11**: 333–336.
Jeppson, L.R., H.H. Keifer, and E.W. Baker. 1975. *Mites Injurious to Economic Plants.* University of California Press, Berkeley, California, U.S.A.
Jepson, P.C., (ed.). 1989. *Pesticides and Non-Target Invertebrates.* Intercept, Wimborne, Dorset, U.K.
Jervis, M.A. and N.A.C. Kidd. 1986. Host-feeding strategies in hymenopteran parasitoids. *Biological Reviews*, **61**: 395–434.
Jessup, A.J. 1994. Quarantine disinfestation of "Hass" avocados against *Bactrocera tryoni* (Diptera: Tephritidae) with a hot fungicide dip followed by cold storage. *Journal of Economic Entomology*, **87**: 127–130.
Jiang, J.Q., X.D. Zhang, and D.X. Gu. 1991. A bionomical study of *Scirpophaga praelata* (Lep.: Pyralidae), a "bridging host" of *Tetrastichus schoenobii* (Hym.: Eulophidae). *Chinese Journal of Biological Control*, **7**: 13–15. (in Chinese)
Jones, D. 1985. Endocrine interaction between host (Lepidoptera) and parasite (Cheloninae: Hymenoptera): Is the host or the parasite in control? *Annals of the Entomological Society of America*, **78**: 141–148.
Jones, D. 1986. Use of parasite regulation of host endocrinology to enhance the potential of biological control. *Entomophaga*, **31**: 153–161.
Jones, D., G. Jones, R.A. Van Steenwyk, and B.D. Hammock. 1982. Effect of the parasite *Copidosoma truncatellum* on development of its host *Trichoplusia ni*. *Annals of the Entomological Society of America*, **75**: 7–11.
Jones, D., M. Snyder, and J. Granett. 1983. Can insecticides be integrated with biological control agents of *Trichoplusia ni* in celery? *Entomologia Experimentalis et Applicata*, **33**: 290–296.
Jones, R.L., W.J. Lewis, M.C. Bowman, M. Beroza, and B.A. Bierl. 1971. Host-seeking stimulant for parasite of corn earworm: Isolation, identification, and synthesis. *Science*, **173**: 842–843.
Jones, R.L., W.J. Lewis, M. Beroza, B.B. Bierl, and A.N. Sparks. 1973. Host-seeking stimulants (kairomones) for the egg parasite, *Trichogramma evanescens*. *Environmental Entomology*, **2**: 593–596.
Jones, T.H. and M.P. Hassell. 1988. Patterns of parasitism by *Trybliographa rapae*, a cynipid parasitoid of the cabbage root fly, under laboratory and field conditions. *Ecological Entomology*, **13**: 309–317.
Jones, W.A. 1988. World review of the parasitoids of the southern green stink bug, *Nezara viridula* (L.) (Heteroptera: Pentatomidae). *Annals of the Entomological Society of America*, **81**: 262–273.
Joshi, R.K. and S.K. Sharma. 1989. Augmentation and conservation of *Epiricania melanoleuca* Fletcher, for the population management of sugarcane leafhopper, *Pyrilla perpusilla* Walker, under arid conditions of Rajasthan. *Indian Sugar*, **39**(8): 625–628.
Julien, M.H. 1981. Control of aquatic *Alternanthera philoxeroides* in Australia; another success for *Agasicles hygrophila*, pp. 583–588. *Proceedings of the 5th International Symposium on Biological Control of Weeds.*
Julien, M.H. 1989. Biological control of weeds worldwide: Trends, rates of success and the future. *Biocontrol News and Information*, **10**: 299–306.
Julien, M.H., (ed.). 1992. *Biological Control of Weeds: a World Catalogue of Agents and Their Target Weeds, 3rd ed.* Commonwealth Agricultural Bureaux International, Wallingford, U.K.
Julien, M.H. and A.S. Bourne. 1988. Effects of leaf-feeding by larvae of the moth *Samea multiplicalis* Guen. (Lep., Pyralidae) on the floating weed *Salvinia molesta*. *Journal of Applied Entomology*, **106**: 518–526.
Junqueira, N.T.V. and L. Gasparotto. 1991. Controle biológico de fungos estromáticos causadores de doenças foliares em Seringueira, pp. 307–331. *In* Anon., *Controle Biológico de Doenças de Plantas.* Org. W. Bettiol., Brasilia, EMBRAPA.
Kahn, R.P., (ed.). 1989. *Plant Protection and Quarantine.* CRC Press, Inc., Boca Raton, Florida, U.S.A.

Kainoh, Y. and S. Tatsuki. 1988. Host egg kairomones essential for egg–larval parasitoid, *Ascogaster reticulatus* Watanabe (Hymenoptera: Braconidae), I. Internal and external kairomones. *Journal of Chemical Ecology*, **14**: 1475–1484.

Kainoh, Y., S. Tatsuki, H. Sugie, and Y. Tamaki. 1989. Host egg kairomones essential for egg-larval parasitoid, *Ascogaster reticulatus* Watanabe (Hymenoptera: Braconidae), II. Identification of internal kairomone. *Journal of Chemical Ecology*, **15**: 1219–1229.

Kainoh, Y., S. Tatsuki, and T. Kusano. 1990. Host moth scales: A cue for host location for *Ascogaster reticulatus* Watanabe (Hymenoptera: Braconidae). *Applied Entomology and Zoology*, **25**: 17–25.

Kakehashi, M., Y. Suzuki, and Y. Iwasa. 1984. Niche overlap of parasitoids in host-parasitoid systems: Its consequence to single versus multiple introduction controversy in biological control. *Journal of Applied Ecology*, **21**: 115–131.

Kandybin, N.V. and O.V. Smirnov. 1990. Registration requirements and safety considerations for microbial pest control agents in the U.S.S.R. and adjacent eastern European countries, pp. 19–30. *In* Laird, M., L.A. Lacey, and E.W. Davidson, (eds.). *Safety of Microbial Insecticides.* CRC Press, Inc., Boca Raton, Florida, U.S.A.

Kard, B.M.R., F.P. Hain, and W.M. Brooks. 1988. Field suppression of three white grub species (Coleoptera: Scarabaeidae) by the entomogenous nematodes *Steinernema feltiae* and *Heterorhabditis heliothidis*. *Journal of Economic Entomology*, **81**: 1033–1039.

Kareiva, P. and G.M. Odell. 1987. Swarms of predators exhibit 'preytaxis' if individual predators use area restricted search. *American Naturalist*, **130**: 233–270.

Kareiva, P. and R. Perry. 1989. Leaf overlap and the ability of ladybird beetles to search among plants. *Ecological Entomology*, **14**: 127–129.

Kareiva, P. and R. Sahakian. 1990. Tritrophic effects of a simple architectural mutation in pea plants. *Nature*, **345**: 433–434.

Kashyap, R.K., G.G. Kennedy, and R.R. Farrar, Jr. 1991. Behavioral response of *Trichogramma pretiosum* Riley and *Telenomus sphingis* (Ashmead) to trichome/methyl ketone mediated resistance in tomato. *Journal of Chemical Ecology*, **17**: 543–556.

Kassulke, R.C., K.L.S. Harley, and G.V. Maynard. 1990. Host specificity of *Acanthoscelides quadridentatus* and *A. puniceus* (Col.: Bruchidae) for biological control of *Mimosa pigra* (with preliminary data on their biology). *Entomophaga*, **35**: 85–96.

Kathirithamby, J. 1989. Review of the order Strepsiptera. *Systematic Entomology*, **14**: 41–92.

Kawakami, K. 1987. The use of an entomogenous fungus *Beauveria brongniartii*, to control the yellow-spotted longicorn beetle, *Psacothea hilaris*. Extension Bulletin, ASPAC Food and Fertilizer Technology Center for the Asian and Pacific Region (1987) No. 257, pp. 38–39. *In* Anon. *Biological Pest Control for Field Crops.* Summaries of papers presented at the International Seminar on Biological Pest Control for Field Crops. Kyushu, Japan, August–September, 1986.

Kaya, H.K. 1985. Entomogenous nematodes for insect control in IPM systems, pp. 283–302. *In* Hoy, M.A. and D.C. Herzog, (eds.). *Biological Control in Agricultural IPM Systems.* Academic Press, New York.

Kaya, H.K. 1990. Soil ecology, pp. 93–115. *In* Gaugler, R. and H.K. Kaya, (eds.). *Entomopathogenic Nematodes in Biological Control.* CRC Press, Inc., Boca Raton, Florida, U.S.A.

Kaya, H.K. 1993. Entomogenous and entomopathogenic nematodes in biological control, pp. 565–591. *In* Evans, K., D.L. Trudgill, and J.M. Webster, (eds.). *Plant Parasitic Nematodes in Temperate Agriculture.* Commonwealth Agricultural Bureaux International, Cambridge University Press, Cambridge, U.K.

Kaya, H.K. and R. Gaugler. 1993. Entomopathogenic nematodes. *Annual Review of Entomology*, **38**: 181–206.

Kaya, H.K. and P.G. Hotchkin. 1981. The nematode *Neoaplectana carpocapsae* Weiser and its effect on selected ichneumonid and braconid parasites. *Environmental Entomology*, **10**: 474–478.

Kaya, H.K., T.M. Burlando, and G.S. Thurston. 1993. Two entomopathogenic nematode species with different search strategies for insect suppression. *Environmental Entomology*, **22**: 859–864.

Keating, S.T., J.P. Burand, and J.S. Elkinton. 1989. DNA hybridization assay for detection of gypsy moth nuclear polyhedrosis virus in infected gypsy moth (*Lymantria dispar* L.) larvae. *Applied and Environmental Microbiology*, **55**: 2749–2754.

Kelleher, J.S. and M.A. Hulme, (eds.). 1984. *Biological Control Programmes Against Insects and Weeds in Canada 1969–1980.* Commonwealth Institute of *Biological Control*, Technical Communication No. 6., Commonwealth Agricultural Bureaux International, Farnham Royal, U.K., 410 pp.

Keller, M.A. 1987. Influence of leaf surfaces on movements by the hymenopterous parasitoid *Trichogramma exiguum*. *Entomologia Experimentalis et Applicata*, **43**: 55–59.

Kelly, P.M., M.R. Speight, P.H. Sterling, P.F. Entwistle, and M.L. Hirst. 1988. The potential development of a viral control agent for the brown-tail moth, *Euproctis chrysorrhoea* (L.) (Lepidoptera: Lymantriidae). *Aspects of Applied Biology*, **17**: 247–248.

Kendrick, B. 1992. *The Fifth Kingdom.* Mycologue Publications, Focus Texts, Newburyport, Massachusetts, U.S.A.

Kenmore, P.E. 1988. Conservation of natural enemies: Precept, payoff, and policy in IPM for tropical rice, p. 318. *Proceedings of the XVIII International Congress of Entomology, Vancouver, B.C., Canada, July 3–9, 1988.*

Kennett, C.E., D.L. Flaherty, and R.W. Hoffmann. 1979. Effect of wind-borne pollens on the population dynamics of *Amblyseius hibisci* (Acarina: Phytoseiidae). *Entomophaga*, **24**: 83–98.

Kenney, D.S. 1986. DeVine—The way it was developed—An industrialist's view. *Weed Science*, **34** (Supplement 1): 15–16.

Kermack, W.O. and A.G. McKendrick. 1927. A contribution to the mathematical theory of epidemics. *Proceedings of the Royal Society of London, Series A*, **115**: 700–721.

Kerns, D.L. and M.J. Gaylor. 1993. Induction of cotton aphid outbreaks by insecticides in cotton. *Crop Protection*, **12**: 387–393.

Kerry, B.R. 1988. Two microorganisms for the biological control of plant parasitic nematodes. *Proceedings of the Brighton Crop Protection* Conference, **2**: 603–607.

Kerry, B.R., D.H. Crump, and L.A. Mullen. 1980. Parasitic fungi, soil moisture and multiplication of the cereal cyst nematode, *Heterodera avenae. Nematologica*, **26**: 57–68.

Kerry, B.R., D.H. Crump, and L.A. Mullen. 1982a. Studies of the cereal cyst nematode, *Heterodera avenae* under continuous cereals, 1975–1978. II. Fungal parasitism of nematode eggs and females. *Annals of Applied Biology*, **100**: 489–499.

Kerry, B.R., D.H. Crump, and L.A. Mullen. 1982b. Natural control of the cereal cyst nematode, *Heterodera avenae* Woll., by soil fungi at three sites. *Crop Protection*, **1**: 99–109.

Kerwin, J.L. 1992. Testing the effects of microorganisms on birds, pp. 729–744. *In* Levin, M.A., R.J. Seidler, and M. Rogul, (eds.). *Microbial Ecology, Principles, Methods and Applications.* McGraw-Hill, New York.

Kerwin, J.L., D A. Dritz, and R.K. Wahino. 1990. Confirmation of the safety of *Lagenidium giganteum* (Oomycetes: Lagenidiales) to mammals. *Journal of Economic Entomology*, **83**: 374–376.

Kester, K.M. and P. Barbosa. 1992. Effects of postemergence experience on searching and landing responses of the insect parasitoid, *Cotesia congregata* (Say) (Hymenoptera: Braconidae), to plants. *Journal of Insect Behavior*, **5**: 301–320.

Kimsey, L.S. and R.M. Bohart, (eds.). 1990. *The Chrysidid Wasps of the World.* Oxford University Press, New York.

King, B.H. 1989. A test of local mate competition theory with a solitary species of parasitoid wasp, *Spalangia cameroni. Oikos*, **54**: 50–54.

King, E.G., D.F. Martin, and L.R. Miles. 1975. Advances in rearing of *Lixophaga diatraeae* (Dipt.: Tachinidae). *Entomophaga*, **20**: 307–311.

King, E.G., K.R. Hopper, and J.E. Powell. 1985. Analysis of systems for biological control of crop arthropod pests in the U.S. by augmentation of predators and parasites, pp. 201–227. *In* Hoy, M.A. and D.C. Herzog, (eds.). *Biological Control in Agricultural IPM Systems.* Academic Press, Orlando, Florida, U.S.A.

King, G.A., A.J. Daugulis, P. Faulkner, D. Bayly, and M.F.A. Goosen. 1988. Growth of baculovirus-infected insect cells in microcapsules to a high cell and virus density. *Biotechnology Letters*, **10**: 683–688.

King, J.L. 1931. The present status of the established parasites of *Popillia japonica* Newman. *Journal of Economic Entomology*, **24**: 453–462.

Kirby, W. and W. Spence. 1815. *An Introduction to Entomology.* Longman, Brown, Green and Longmans, London.

Kiritani, K. and F. Nakasuji. 1967. Estimations of the stage- specific survival rate in the insect population with overlapping stages. *Researches on Population Ecology*, **9**: 143–152.

Kiritani, K., S. Kawahara, T. Sasaba, and F. Nakasuji. 1972. Quantitative evaluation of predation by spiders on the green rice leafhopper, *Nephotettix cincticeps* Uhler, by a sight-count method. *Researches on Population Ecology*, **13**: 187–200.

Kirkland, R.L. and R.D. Goeden. 1978. An insecticidal-check study of the biological control of puncturevine (*Tribulus terrestris*) by imported weevils, *Microlarinus lareynii* and *M. lypriformis* (Col.: Curculionidae). *Environmental Entomology*, **7**: 349–354.

Klein, M.G. and R. Georgis. 1992. Persistence of control of Japanese beetle (Coleoptera: Scarabaeidae) larvae with steinernematid and heterorhabditid nematodes. *Journal of Economic Entomology*, **85**: 727–730.

Klingman, D.L. and J.R. Coulson. 1983. Guidelines for introducing foreign organisms into the United States for biological control of weeds. *Weed Science*, **30**: 661–667. Chapter 8, 9.

Klomp, H. 1958. On the synchronization of the generations of the tachinid *Carcelia obesa* Zett. (=*rutilla* B.B.) and its host *Bupalus piniarus. Zeitschrift für Angewandte Entomologie*, **42**: 210–217.

Kloot, P.M. 1983. The role of common iceplant (*Mesembryanthemum crystallinum*) in the deterioration of medic pastures. *Australian Journal of Ecology*, **8**: 301–306.

Kluge, R.L. 1991. Biological control of triffid weed *Chromolaena odorata* (Asteraceae), in South Africa. *Agriculture, Ecosystems and Environment*, **37**: 193–197.

Kluge, R.L. and P.M. Caldwell. 1992. Microsporidian diseases and biological weed control agents: to release or not to release? *Biocontrol News and Information*, **13**(3): 43N–47N.

Knutson, A.E. and F.E. Gilstrap. 1989. Direct evaluation of natural enemies of the southwestern corn borer (Lepidoptera: Pyralidae) in Texas corn. *Environmental Entomology*, **18**: 732–739.

Knutson, L., R.I. Sailer, W.L. Murphy, R.W. Carlson, and J.R. Dogger. 1990. Computerized data base on immigrant arthropods. *Annals of the Entomological Society of America*, **83**: 1–8.

Kobbe, B., J.K. Clark, and S.H. Dreistadt. 1991. *Integrated Pest Management for Citrus, 2nd ed.* University of California Press, Oakland, California, U.S.A.

Kong, J.A. and N.X. Zhang. 1986. Chemical relationship between *Amblyseius fallacis* (Garman) and its prey. *Chinese Journal of Biological Control*, **2**: 158–161. (in Chinese)

Kowalski, R. 1976. Biology of *Philonthus decorus* (Coleoptera: Staphylinidae) in relation to its role as a predator of winter moth pupae [(*Operophtera brumata*) (Lepidoptera: Geometridae)]. *Pediobiologia*, **16**: 233–242.

Kranz, J. 1981. Hyperparasitism of biotrophic fungi, pp. 327–352. *In* J.P. Blakeman, (ed.). *Microbial Ecology of the Phylloplane.* Academic Press, London.

Krebs, J.R. 1973. Behavioral aspects of predation, pp. 73–111. *In* Bateson, P.P.G. and P.H. Klopfer, (eds.). *Perspectives in Ethology.* Plenum Press, New York.

Kreiter, D. 1991. The biological characteristics of the predatory mites that prey on mites and their use in biological control. *Progrès Agricole et Viticole*, **108**: 247–262. (in French)

Kring, T.J., F.E. Gilstrap, and G.J. Michels, Jr. 1985. Role of indigenous coccinellids in regulating greenbugs (Homoptera: Aphididae) on Texas grain sorghum. *Journal of Economic Entomology*, **78**: 269–273.

Krombien, K.V., P.D. Hurd, Jr., D.R. Smith, and B.D. Burks. 1979. *Catalog of Hymenoptera in America North of Mexico (Vols. I, II, III).* Smithsonian Institute Press, Washington, D.C., U.S.A.

Kúc, J. 1981. Multiple mechanisms, reaction rates and induced resistance in plants, pp. 259–272. *In* Staples, R.C. and G.H. Toenniessen (eds.). *Plant Disease Control.* John Wiley and Sons, New York.

Kurtti, T.J. and U.G. Munderloh. 1987. Biotechnological application of invertebrate cell culture to the development of microsporidian insecticides, pp. 327–334. *In* Maramorosch, K., (ed.). *Biotechnology in Invertebrate Pathology and Cell Culture.* Academic Press, San Diego, California, U.S.A.

Lackey, B.A., A.E. Muldoon, and B.A. Jaffee. 1993. Alginate pellet formulation of *Hirsutella rhossiliensis* for biological control of plant-parasitic nematodes. *Biological Control*, **3**: 155–160.

Ladd, T.L. and P.J. McCabe. 1966. The status of *Tiphia vernalis* Rohwer, a parasite of the Japanese beetle, in southern New Jersey and southeastern Pennsylvania in 1963. *Journal of Economic Entomology*, **59**: 480.

Ladd, T.L. and P.J. McCabe. 1967. Persistence of spores of *Bacillus popilliae*, causal organism of type A milky-disease of Japanese beetle larvae in New Jersey soils. *Journal of Economic Entomology*, **60**: 493–495.

Laing, J.E. and G.M. Eden. 1990. Mass-production of *Trichogramma minutum* Riley on factitious host eggs. *Memoirs of the Entomological Society of Canada*, **153**: 10–24.

Laing, J.E. and J.E. Corrigan. 1987. Intrinsic competition between the gregarious parasite *Cotesia glomerata* and the solitary parasite *Cotesia rubecula* [Hymenoptera: Braconidae] for their host *Artogeia rapae* [Lepidoptera: Pieridae]. *Entomophaga*, **32**: 493–501.

Laing, J.E. and J. Hamai. 1976. Biological control of insect pests and weeds by imported parasites, predators, and pathogens, pp. 685–743. *In* Huffaker, C.B. and P.S. Messenger, (eds.). *Theory and Practice of Biological Control.* Academic Press, New York.

Laing, J.E. and C.B. Huffaker. 1969. Comparative studies of predation by *Phytoseiulus persimilis* Athias-Henriot and *Metaseiulus occidentalis* (Acarina: Phytoseiidae) on populations of *Tetranychus urticae* Koch (Acarina: Tetranychidae). *Researches on Population Ecology*, **11**: 105–126.

Laird, M., L.A. Lacey, and E.W. Davidson, (eds.). 1990. *Safety of Microbial Insecticides.* CRC Press, Inc., Boca Raton, Florida, U.S.A.

Lake, P.S. and D.J. O'Dowd. 1991. Red crabs in rain forest, Christmas Island: Biotic resistance to invasion by an exotic snail. *Oikos*, **62**: 25–29.

Lambert, B. and M. Peferoen. 1992. Insecticidal promise of *Bacillus thuringiensis*, facts and mysteries about a successful biopesticide. *BioScience*, **42**: 112–122.

Lambert, W.R., J.S. Bacheler, W.A. Dickerson, M.E. Roof, and R.H. Smith. Ch. 20: Insect and mite pest management in the Southeast. *In* Anon. *Cotton Insects and Mites.* The Cotton Foundation Reference Book, in press.

LaRosa, A.M., R.F. Doren, and L. Gunderson. 1992. Alien plant management in Everglades National Park: An historical perspective, pp. 47–63. *In* Stone, C. P., C.W. Smith, and J.T. Tunison, (eds.). *Alien Plant Invasions in Native Ecosystems of Hawai'i: Management and Research.* University of Hawaii Press, Honolulu, Hawaii, U.S.A.

Larsson, J.I.R. 1988. Identification of microsporidian genera (Protozoa, Microspora)—A guide with comments on the taxonomy. *Archiv für Protistenkunde*, **136**(1): 1–37.

Latgé, J.P., R.A. Hall, R.I. Cabrera, and J.C. Kerwin. 1986. Liquid fermentation of entomogenous fungi, pp. 603–606. *In* Samson, R.A., J.M. Vlak, and D. Peters, (eds.). *Fundamentals and Applied Aspects of Invertebrate Pathology.* Foundation 4th International Colloquium on Invertebrate Pathology, Wageningen, the Netherlands.

Lawrence, P.O. and B. Lanzrein. 1993. Hormonal interactions between insect endoparasites and their host insects, pp. 59–85. *In* Beckage, N.E., S.N. Thompson, and B.A. Federici, (eds.). *Parasites and Pathogens of Insects, Volume I.* Parasites. Academic Press, Inc., New York.

Lawson, F.R., R.L. Rabb, F.E. Guthrie, and T.G. Bowery. 1961. Studies of an integrated control system for hornworms on tobacco. *Journal of Economic Entomology*, **54**: 93–97.

Lawton, J.H. and M.P. Hassell. 1984. Interspecific competition in insects, pp. 451–495. *In* Huffaker, C.B., and R.L. Rabb, (eds.). *Ecological Entomology.* John Wiley and Sons, New York.

Lawton, J.H. and D.R. Strong. 1981. Community patterns and competitior in folivorous insects. *American Naturalist*, **118**: 317–338.

Le Masurier, A.D. 1991. Effect of host size on clutch size in *Cotesia glomerata. Journal of Animal Ecology*, **60**: 107–118.

Leathwick, D.M. and M.J. Winterbourn. 1984. Arthropod predation on aphids in a lucerne crop. *New Zealand Entomologist*, **8**: 75–80.

Lefkovitch, L.P. 1963. Census studies on unrestricted populations of *Lasioderma serricorne* (F.) (Coleoptera: Anobiidae). *Journal of Animal Ecology*, **32**: 221–231.

Legner, E.F. 1986a. Importation of exotic natural enemies, pp. 19–30. *In* Franz, J.M., (ed.). *Biological Plant and Health Protection: Biological Control of Plant Pests and of Vectors of Human and Animal Diseases.* International Symposium of the Akademie der Wissenschaften und der Literatur, Mainz, November 15–17th, 1984 at Mainz and Darmstadt. *Fortschritte der Zoologie*, **32**: 341 pp. Gustav Fischer Verlag, Stuttgart, Germany.

Legner, E.F. 1986b. The requirement for reassessment of interactions among dung beetles, symbovine flies and natural enemies. *Entomological Society of America, Miscellaneous Publications*, **61**: 120–131.

Legner, E.F. and G. Gordh. 1992. Lower navel orangeworm (Lepidoptera: Phycitiidae) population densities following establishment of *Goniozus legneri* (Hymenoptera: Bethylidae) in California. *Journal of Economic Entomology*, **85**: 2153–2160.

Legner, E.F. and R.D. Sjogren. 1984. Biological mosquito control furthered by advances in technology and research. *Journal of the American Mosquito Control Association*, **44**: 449–456.

Legner, E.F., R.D. Sjogren, and I.M. Hall. 1974. The biological control of medically important arthropods. *Critical Reviews in Environmental Control*, **4**: 85–113.

Leite, L.G., A.B. Filho, and W.L.A. Prada. 1990. Production of *Neoaplectana glaseri* Steiner in live and dead larvae of *Galleria mellonella* L. *Revista de Agricultura (Piracicaba)*, **65**: 225–232. (in Portuguese)

Leius, K. 1960. Attractiveness of different foods and flowers to the adults of some hymenopterous parasites. *Canadian Entomologist*, **92**: 369–376.

Leius, K. 1967. Influence of wild flowers on parasitism of tent caterpillar and codling moth. *Canadian Entomologist*, **99**: 444–446.

Lenz, C.J., A.H. McIntosh, C. Mazzacano, and U. Monderloh. 1991. Replication of *Heliothis zea* nuclear polyhedrosis virus in cloned cell lines. *Journal of Invertebrate Pathology*, **57**: 227–233.

Leonard, D.E. 1966. *Brachymeria intermedia* (Nees) (Hymenoptera: Chalcididae) established in North America. *Entomological News*, **77**: 25–27.

Leppla, N.C. and T.R. Ashley, (eds.). 1978. *Facilities for Insect Research and Production.* United States Department of Agriculture Technical Bulletin 1576.

LeRoux, E.J. 1971. Biological control attempts on pome fruit (apple and pear) in North America, 1860–1970. *Canadian Entomologist*, **103**: 963–974.

Lessells, C.M. 1985. Parasitoid foraging: Should parasitism be density dependent? *Journal of Animal Ecology*, **54**: 27–41.

Levin, S. A. 1974. Dispersion and population interactions. *American Naturalist*, **108**: 207–228.

Levin, S.A. 1976. Population dynamics in heterogeneous environments. *Annual Review of Ecology and Systematics*, **7**: 287–310.

Levy, R., M.A. Nichols, and T.W. Miller Jr. 1990. Culigel superabsorbent polymer controlled-release system: application to mosquito larvicidal bacilli, p. 107. *In* Pinnock, D.E., (ed.). *Vth International Colloquium on Invertebrate Pathology and Microbial Control.* Adelaide, Australia, 20–24, August, 1990.

Lewis, W.J. and W.R. Martin, Jr. 1990. Semiochemicals for use with parasitoids: Status and future. *Journal of Chemical Ecology*, **16**: 3067–3089.

Lewis, W.J. and K. Takasu. 1990. Use of learned odours by a parasitic wasp in accordance with host and food needs. *Nature*, **348**: 635–636.

Lewis, W.J. and B. Vinson. 1971. Suitability of certain *Heliothis* (Lepidoptera: Noctuidae) as hosts for the parasite *Cardiochiles nigriceps. Annals of the Entomological Society of America*, **64**: 970–972.

Lewis, W.J., J.W. Snow, and R.L. Jones. 1971. A pheromone trap for studying populations of *Cardiochiles nigriceps*, a parasite of *Heliothis virescens*. *Journal of Economic Entomology*, **64**: 1417–1421.

Lewis, W.J., R.L. Jones, and A.N. Sparks. 1972. A host-seeking stimulant for the egg parasite *Trichogramma evanescens*: Its source and a demonstration of its laboratory and field activity. *Annals of the Entomological Society of America*, **65**: 1087–1089.

Lewis, W.J., R.L. Jones, H.R. Gross, Jr., and D.A. Nordlund. 1976. The role of kairomones and other behavioral chemicals in host finding by parasitic insects. *Behavioral Biology*, **16**: 267–289.

Lewis, W.J., D.A. Nordlund, R.C. Gueldner, P.E.A. Teal, and J.H. Tumlinson. 1982. Kairomones and their use for management of entomophagous insects. XIII. Kairomonal activity for *Trichogramma* spp. of abdominal tips, excretion, and a synthetic sex pheromone blend of *Heliothis zea* (Boddie) moths. *Journal of Chemical Ecology*, **8**: 1323–1331.

Lewis, W.J., L.E.M. Vet, J.H. Tumlinson, J.C. van Lenteren, and D.R. Papaj. 1990. Variations in parasitoid foraging behavior: essential element of a sound biological control theory. *Environmental Entomology*, **19**: 1183–1193.

Lewis, W. J., J.H. Tumlinson, and S. Krasnoff. 1991. Chemically mediated associative learning: an important function in the foraging behavior of *Microplitis croceipes* (Cresson). *Journal of Chemical Ecology*, **17**: 1309–1325.

Li, Y.X. 1989. Study on the bionomics of *Amata pascus* (Leech)—A natural enemy of *Kuwanaspis pseudoleucaspis* (Kuwana). *Insect Knowledge*, **26**: 224–225.

Lindegren, J.E., K.A. Valero, and B.E. Mackey. 1993. Simple *in vivo* production and storage methods for *Steinernema carpocapsae* infective juveniles. *Journal of Nematology*, **25**: 193–197.

Lindow, S.E. 1985a. Foliar antagonists: Status and prospects, pp. 395–413. *In* Hoy, M.A. and D.C. Herzog, (eds.). *Biological Control in Agricultural IPM Systems*. Academic Press, Inc., Orlando, Florida, U.S.A.

Lindow, S.E. 1985b. Integrated control and the role of antibiosis in biological control of fireblight and frost injury, pp. 83–115. *In* Windels, C.E. and S.E. Lindow, (eds.). *Biological Control on the Phylloplane*. American Phytopathological Society, St. Paul, Minnesota, U.S.A.

Liu S.S. and R.D. Hughes. 1984. Effect of host age at parasitization by *Aphidius sonchi* on the development, survival, and reproduction of the sowthistle aphid, *Hyperomyzus lactucae*. *Entomologia Experimentalis et Applicata*, **36**: 239–246.

Liu, Z.C., Y.R. Sun, Z.Y. Wang, J.F. Liu, L.W. Zhang, Q.X. Zhang, K.J. Dai, and Y.G. Gao. 1985. Field release of *Trichogramma confusum* reared on artificial host eggs against sugarcane borers. *Chinese Journal of Biological Control*, **3**: 2–5. (in Chinese)

Loewenberg, J.R., T. Sullivan, and M.L. Schuster. 1959. A viral disease of *Meloidogyne incognita incognita*, the southern root-knot nematode. *Nature*, **184**: 1896.

Loke, W.H. and T.R. Ashley. 1984. Behavioral and biological responses of *Cotesia marginiventris* to kairomones of the fall armyworm, *Spodoptera frugiperda*. *Journal of Chemical Ecology*, **10**: 521–529.

Longworth, J.F. and J. Kalmakoff. 1977. Insect viruses for biological control: An ecological approach. *Intervirology*, **8**: 68–72.

Loope, L.L. 1992. An overview of problems with introduced plant species in national parks and bioshpere reserves in the United States, pp. 3–28. *In* Stone, C.P., C.W. Smith, and J.T. Tunison, (eds.). *Alien Plant Invasions in Native Ecosystems of Hawai'i: Management and Research*. University of Hawaii Press, Honolulu, Hawaii, U.S.A.

Lopez, E.R. and R.G. Van Driesche. 1989. Direct measurement of host and parasitoid recruitment for assessment of total losses due to parasitism in a continuously breeding species, the cabbage aphid *Brevicoryne brassicae* (L.) (Hemiptera: Aphidiae). *Bulletin of Entomological Research*, **79**: 47–59.

Lowery, D.T. and M.K. Sears. 1986. Stimulation of reproduction of the green peach aphid (Homoptera: Aphididae) by azinphosmethyl applied to potatoes. *Journal of Economic Entomology*, **79**: 1530–1533.

Lublinkhof, J. and L.C. Lewis. 1980. Virulence of *Nosema pyrausta* to the European corn borer, when used in combination with insecticides. *Environmental Entomology*, **9**: 67–71.

Luck, R.F. 1981. Parasitic insects introduced as biological control agents for arthropod pests, pp. 125–284. *In* Pimentel, D., (ed.). *CRC Handbook of Pest Management in Agriculture*. CRC Press, Inc., Boca Raton, Florida, U.S.A.

Luck, R.F. and D.L. Dahlsten. 1975. Natural decline of a pine needle scale (*Chionaspis pinifoliae* [Fitch]), outbreak at South Lake Tahoe, California following chionaspis cessation of adult mosquito control with malathion. *Ecology*, **56**: 893–904.

Luck, R.F. and H. Podoler. 1985. Competitive exclusion of *Aphytis lingnanensis* by A. melinus: Potential role of host size. *Ecology*, **66**: 904–913.

Luck, R.F. and N. Uygun. 1986. Host recognition and selection by Aphytis species: Response to California red, oleander, and cactus scale cover extracts. *Entomologia Experimentalis et Applicata*, **40**: 129–136.

Luck, R.F., B.M. Shepard, and P.E. Kenmore. 1988. Experimental methods for evaluating arthropod natural enemies. *Annual Review of Entomology*, **33**: 367–391.

Lunau, S., S. Stoessel, A.J. Schmidt-Peisker, and R.-U. Ehlers. 1993. Establishment of monoxenic inocula for scaling up *in vitro* cultures of the entomopathogenic nematodes *Steinernema* spp. and *Heterorhabditis* spp. *Nematologica*, **39**: 385–399.

Lüthy, P. 1986. Insect pathogenic bacteria as pest control agents, pp. 201–216. *In* Franz, J.M., (ed.). *Biological Plant and Health Protection, Biological Control of Plant Pests and of Vectors of Human and Animal Diseases.* International Symposium of the Akademie der Wissenschaften und der Literatur, Mainz, November 15–17th, 1984 at Mainz and Darmstadt. *Fortschritte der Zoologie*, **32**: 341 pp., Gustav Fischer Verlag, Stuttgart, Germany.

Lutz, G.G., J.E. Strassmann, and C.R. Hughes. 1984. Nest defense by the social wasps, *Polistes exclamans* and *P. instabilis* (Hymenoptera: Vespidae) against the parasitoid, *Elasmus polistis* (Hymenoptera: Chalcidoidea: Eulophidae). *Entomological News*, **95**: 47–50.

Lynch, J.M. 1987. Biological control within microbial communities of the rhizosphere, pp. 55–82. *In* Fletcher, M., T.R.G. Gray, and J.G. Jones, (eds.). *Ecology of Microbial Communities. Society of General Microbiology, Symposium 41.* Cambridge University Press, Cambridge, U.K.

Lynn, D.E., M. Shapiro, E.M. Dougherty, H. Rathburn, G.P. Godwin, K.M. Jeong. B.W. Belisle, and R.H. Chiarella. 1990. Gypsy moth nuclear polyhedrosis virus in cell culture: A likely commercial system for viral pesticide production, p. 12. *In* Pinnock, D.E., (ed.). *Vth International Colloquium on Invertebrate Pathology and Microbial Control.* Adelaide, Australia, 20–24, August, 1990.

MacArthur, R.H. and E.R. Pianka. 1966. On optimal use of a patchy environment. *American Naturalist*, **100**: 603–609.

MacCarter, L.E. and D.L. Gaynor. 1980. Gorse: A subject for biological control in New Zealand. *New Zealand Journal of Experimental Agriculture*, **8**: 321–330.

Macdonald, I. 1988. Invasive alien plants and nature conservation in South Africa. *African Wildlife*, **42**: 333–335.

Macdonald, I.A.W., L. Ortiz, J.E. Lawesson, and J. Bosco Nowak. 1988. The invasion of highlands in Galapagos by the red quinine-tree *Cinchona succirubra. Environmental Conservation*, **15**: 215–220.

Maceina, M.J., M.F. Cichra, R.K. Betsill, and P.W. Bettoli. 1992. Limnological changes in a large reservoir following vegetation removal by grass carp. *Journal of Freshwater Ecology*, **7**: 81–95.

Mackauer, M. 1972. Genetic aspects of insect production. *Entomophaga*, **17**: 27–48.

Mackauer, M. and L.E. Ehler, (eds.). *Critical Issues in Biological Control.* Intercept, Andover, U.K.

MacLeod, D.M. 1963. Entomophthorales infections, pp. 189–231. *In* Steinhaus, E.A., (ed.). *Insect Pathology: An Advanced Treatise, Volume 2.* Academic Press, New York.

Madden, J.L. 1968. Behavioural responses of parasites to the symbiotic fungus associated with *Sirex noctilio* F. *Nature*, **218**: 189–190.

Madeiros, J.L. 1990. The barn owl: Bermuda's unsung rat control expert. *Monthly Bulletin, Department of Agriculture, Fisheries and Parks, Bermuda*, **61**(8): 57–60.

Madsen, H. 1990. Biological methods for the control of freshwater snails. *Parasitology Today*, **6**(7): 237–241.

Madsen, J.D., J.W. Sutherland, J.A. Bloomfield, L.W. Eichler, and C.W. Boylen. 1991. The decline of native vegetation under dense Eurasian watermilfoil canopies. *Journal of Aquatic Plant Management*, **29**: 94–99.

Maeto, K. and S. Kudo. 1992. A new euphorine species of *Aridelus* (Hymenoptera, Braconidae) associated with a subsocial bug *Elasmucha putoni* (Heteroptera, Acanthosomatidae). *Japanese Journal of Entomology*, **60**: 77–84.

Maggenti, A.R. 1991. Nemata: higher classification, pp. 147–187. *In* Nickel, W.R., (ed.). *Manual of Agricultural Nematology.* Marcel Dekker, Inc., New York.

Maier, C.T. 1982. Parasitism of the apple blotch leafminer, *Phyllonorycter crataegella*, on sprayed and unsprayed apple trees in Connecticut. *Environmental Entomology*, **11**: 603–610.

Maier, C.T. 1994. Biology and impact of parasitoids of *Phyllonorycter blancardella* and *P. crataegella* (Lepidoptera: Gracillariidae) in northeastern North American apple orchards. *In* Maier, C.T., (ed.). *Integrated Management of Tentiform Leafminers, Phyllonorycter spp. (Lepidoptera: Gracillariidae), in North American Apple Orchards.* Thomas Say Publications in Entomology, Entomological Society of America, Lanham, Maryland, U.S.A.

Majchrowicz, I. and T.J. Poprawski. 1993. Effects *in vitro* of nine fungicides on growth of entomopathogenic fungi. *Biocontrol Science and Technology*, **3**: 321–336.

Malajczuk, N. 1979. Biological suppression of *Phytophthora cinnamomi* in eucalypts and avocado in Australia, pp. 635–652. *In* Schippers, B. and W. Gams, (eds.). *Soil-borne plant pathogens.* Academic Press, Inc., London.

Maltby, H.L., F.W. Stehr, R.C. Anderson, G.E. Moorehead, L.C. Barton, and J.D. Paschke. 1971. Establishment in the United States of *Anaphes flavipes*, an egg parasite of the cereal leaf beetle. *Journal of Economic Entomology*, **64**: 693–697.

Maniania, N.K. 1991. Potential of some fungal pathogens for the control of pests in the tropics. *Insect Science and Application*, **12**: 63–70.

Mankau, R. 1962. Soil fungistasis and nematophagus fungi. *Phytopathology*, **52**: 611–615.

Mankau, R. 1968. Effects of nematicides on nematode-trapping fungi associated with the citrus nematode. *Plant Disease Reporter*, **52**: 851–855.
Mankau, R. 1972. Utilization of parasites and predators in nematode pest management ecology. *Proceedings of the Tall Timbers Conference on Ecological Animal Control by Habitat Management*, **4**: 129–143.
Mankau. R. 1975. *Bacillus penetrans* n. comb. causing a virulent disease of plant-parasitic nematodes. *Journal of Invertebrate Pathology*, **26**: 333–339.
Manly, B.F.J. 1974. Estimation of stage-specific survival rates and other parameters for insect populations developing through several life stages. *Oecologia*, **15**: 277–285.
Manly, B.F.J. 1976. Extensions to Kiritani and Nakasuji's method for analyzing insect stage-frequency data. *Researches on Population Ecology*, **17**: 191–199.
Manly, B.F.J. 1977. The determination of key factors from life table data. *Oecologia*, **31**: 111–117.
Manly, B.F.J. 1989. A review of methods for the analysis of stage-frequency data. pp. 3–69. In McDonald, L.L., B.F.J. Manly, J. Lockwood, and J. Logan, (eds.). *Estimation and Analysis of Insect Populations*. Springer-Verlag, New York.
Mansour, F., D. Rosen, A. Shulov, and H.N. Plaut. 1980. Evaluation of spiders as biological control agents of *Spodoptera littoralis* larvae on apple in Israel. *Oecologica Applicata*, **1**: 225–232.
Maramorosch, K. 1987. Genetically engineered microbial and viral insecticides: safety considerations, pp. 485–492. *In* Maramorosch, K., (ed.). *Biotechnology in Invertebrate Pathology and Cell Culture*. Academic Press, New York, 511 pp.
Maramorosch, K. and K.E. Sherman, (eds.). 1985. *Viral Insecticides for Biological Control*. Academic Press, New York, 809 pp.
Marcovitch, S. 1935. Experimental evidence on the value of strip farming as a method for the natural control of injurious insects with special reference to plant lice. *Journal of Economic Entomology*, **28**: 62–70.
Margulis, L., J.O. Corliss, M. Melkonian, and D.J. Chapman. 1990. *Handbook of Protoctista*. Jones and Bartlett, Boston, Massachusetts, U.S.A.
Markin, G.P. 1970a. Foraging behavior of the Argentine ant in a California citrus grove. *Journal of Economic Entomology*, **63**: 740–744.
Markin, G.P. 1970b. The seasonal life cycle of the Argentine ant, *Iridomymrex humilis* (Hymenoptera: Formicidae), in southern California. *Annals of the Entomological Society of America*, **63**: 1238–1242.
Markin, G.P. 1982. Alien plant management by biological control, pp. 70–73. *In* Stone, C.P. and D.B Stone, (eds.). *Conservation Biology in Hawai'i*. University of Hawaii Press, Honolulu, Hawaii, U.S.A.
Markin, G.P., L.A. Dekker, I.A. Lapp, and R.F. Nagata. 1988. Distribution of the weed gorse (*Ulex europaeus* L.), a noxious weed in Hawaii. *Bulletin of the Hawaiian Botanical Society*, **27**: 110–117.
Markin, G.P., R.F. Nagata, and G. Taniguchi. 1989. Biology and behavior of the South American moth, *Cyanotricha necyria* (Felder and Rogenhofer) (Lepidoptera: Notodontidae), a potential biocontrol agent in Hawaii of the forest weed, *Passifora mollissima* (HBK) Bailey. *Proceedings of the Hawaiian Entomological Society*, **29**: 115–123.
Markin, G.P., Po-Yung Lai, and G.Y. Funasaki. 1992. Status of biological control of weeds in Hawai'i and implications for managing native ecosystems, pp. 466–482. *In* Stone, C.P., C.W. Smith, and J.T. Tunison, (eds.). *Alien Plant Invasions in Native Ecosystems of Hawai'i: Management and Research*. University of Hawaii Press, Honolulu, Hawaii, U.S.A.
Markkula, M., K. Tiittanen, M. Hamalainen, and A. Forsberg. 1979. The aphid midge *Aphidoletes aphidimyza* (Diptera: Cecidomyiidae) and its use in biological control of aphids. *Ann. Entomol. Fenn. Annales Entomologici Fenniae*, **45**: 89–98.
Markwick, N.P. 1986. Detecting variability and selecting for pesticide resistance in two species of phytoseiid mites. *Entomophaga*, **31**: 225–236.
Marten, G.G. 1990. Elimination of *Aedes albopictus* from tire piles by introducing *Macrocyclops albidus* (Copepoda, Cyclopidae). *Journal of the American Mosquito Control Association*, **6**(4): 689–693.
Martignoni, M.E. and P.J. Iwai. 1981. A catalogue of viral diseases of insects, mites and ticks, pp. 897–911. *In* Burges, H.D., (ed.). *Microbial Control of Pests and Plant Diseases, 1970–1980*. Academic Press, London.
Martin, N.A. and J.R. Dale. 1989. Monitoring greenhouse whitefly puparia and parasitism: A decision approach. *New Zealand Journal of Crop and Horticultural Science*, **17**: 115–123.
Martin, P.A.W. 1994. An iconoclastic view of *Bacillus thuringiensis* ecology. *American Entomologist*, **40**: 85–50.
Martin, W.R., Jr., D.A. Nordlund, and W.C. Nettles Jr. 1990. Response of parasitoid *Eucelatoria bryani* to selected plant material in an olfactometer. *Journal of Chemical Ecology*, **16**: 499–508.
Masutti, L., A. Battisti, N. Milani, M. Zanata, and G. Zanazzo. 1993. *In vitro* rearing of *Ooencyrtus pityocampae* [Hym., Encyrtidae], an egg parasitoid of *Thaumetopoea pityocampa* [Lep., Thaumetopoeidae]. *Entomophaga*, **38**: 327–333.
Matthews, R.E.F. 1991. *Plant Virology, 3rd ed.* Academic Press, San Diego, California, U.S.A.
Mattiacci, L., S.B. Vinson, H.J. Williams, J.R. Aldrich, and F. Bin. 1993. A long-range attractant kairomone for egg parasitoid *Trissolcus basalis*, isolated from defensive secretion of its host, *Nezara viridula*. *Journal of Chemical Ecology*, **19**: 1167–1181.

May, R.M. 1975. Biological population obeying difference equations: stable points, stable cycles and chaos. *Journal of Theoretical Biology*, **51**: 511–524.
May, R.M. 1978. Host-parasitoid systems in patchy environments: a phenomenological model. *Journal of Animal Ecology*, **47**: 833–843.
May, R.M. 1985. Regulation of population with nonoverlapping generations by microparasites: A purely chaotic system. *American Naturalist*, **125**: 573–584.
May, R.M. and M.P. Hassell. 1981. The dynamics of multiparasitoid-host interactions. *American Naturalist*, **117**: 234–261.
May, R.M. and M.P. Hassell. 1988. Population dynamics and biological control. *Philosophical Transactions of the Royal Society of London, Series B*, **318**: 129–169.
May, R. M. and G.F. Oster. 1976. Bifurcations and dynamic complexity in simple ecological models. *American Naturalist*, **110**: 573–599.
May, R.M., G.R. Conway, M.P. Hassell, and T.R.E. Southwood. 1974. Time delays, density dependence, and single-species oscillations. *Journal of Animal Ecology*, **43**: 747–770.
May, R.M., M.P. Hassell, R.M. Anderson, and D.W. Tonkyn. 1981. Density dependence in host-parasitoid models. *Journal of Animal Ecology*, **50**: 855–865.
Maynard Smith, J. 1974. *Models in Ecology*. Cambridge University Press, New York.
Maynard Smith, J. and M. Slatkin. 1973. The stability of predator-prey systems. *Ecology*, **54**: 384–391.
McBrien, H., R. Harmsen, and A. Crowder. 1983. A case of insect grazing affecting plant succession. *Ecology*, **64**: 1035–1039.
McCabe, D. and R.S. Soper. 1985. Preparation of an entomopathogenic fungal insect control agent. U.S. Patent 4,530,834.
McClain, D.C., G.C. Rock, and R.E. Stinner. 1990. Thermal requirements for development and simulation of the seasonal phenology of *Encarsia perniciosi* (Hymenoptera: Aphelinidae), a parasitoid of the San José scale (Homoptera: Diaspididae) in North Carolina orchards. *Environmental Entomology*, **19**: 1396–1402.
McClay, A.S. 1989. *Selection of Suitable Target Weeds for Classical Biological Control in Alberta*. Alberta Environmental Centre, Vegreville, AECV89-R1.
McClure, M.S. 1977. Parasitism of the scale insect *Fiorinia externa* (Homoptera: Diaspididae) by *Aspidiotiphagus citrinus* (Hymenoptera: Eulophidae) in a hemlock forest: Density dependence. *Environmental Entomology*, **6**: 551–555.
McCoid, M.J. 1991. Brown tree snake (*Boiga irregularis*) on Guam: A worst case scenario of an introduced predator. *Micronesica Supplement*, **3**: 63–69.
McCoy, C.W. 1981. Pest control by the fungus *Hirsutella thompsonii*, pp. 499–512. *In* Burges, H.D., (ed.). *Microbial Control of Pests and Plant Diseases, 1970–1980*. Academic Press, London.
McCoy, C.W. and A.M. Heimpel. 1980. Safety of the potential mycoacaricide, *Hirsutella thompsonii*, to vertebrates. *Environmental Entomology*, **9**: 47–49.
McCoy, C.W., A.J. Hill, and R.F. Kanavel. 1975. Large-scale production of the fungal pathogen *Hirsutella thompsonii* in submerged culture and its formulation for application in the field. *Entomophaga*, **20**: 229–240.
McCoy, C.W., R.A. Samson, and D.G. Boucias. 1988. Entomogenous fungi, pp. 151–236. *In* Ignoffo, C.M., (ed.). *CRC Handbook of Natural Pesticides*. Microbial Insecticides, Part A. Entomogenous Protozoa and Fungi, Vol. 5. CRC Press, Inc., Boca Raton, Florida, U.S.A.
McDaniel, S.G. and W.L. Sterling. 1982. Predation of *Heliothis virescens* (F.) eggs on cotton in east Texas. *Environmental Entomology*, **11**: 60–66.
McDonald, L.L., B.F.J. Manly, J. Lockwood, and J. Logan, (eds.). 1989. *Estimation and Analysis of Insect Populations*. Springer-Verlag, New York.
McEvoy, P. and C. Cox. 1991. Successful biological control of ragwort, *Senecio jacobaea*, by introduced insects in Oregon. *Ecological Applications*, **1**: 430–442.
McEvoy, P.B., C.S. Cox, R.R. James, and N.T. Rudd. 1990. Ecological mechanisms underlying successful biological weed control: Field experiments with ragwort *Senecio jacobaea*, pp. 55–66 *In* Delfosse, (ed.). *Proceedings of the VII International Symposium on Biological Control of Weeds*. Rome, Italy.
McEwen, P.K., M.A. Jervis, and N.A.C. Kidd. 1994. Use of a sprayed L-tryptophan solution to concentrate numbers of the green lacewing *Chrysoperla carnea* in olive tree canopy. *Entomologia Experimentalis et Applicata*, **70**: 97–99.
McGroarty, D. and B. Croft. 1975. Sampling population of *Amblyseius fallacis* (Acarina: Phytoseiidae) in the ground cover of Michigan commercial apple orchards. *Proceedings of the North Central Branch of the Entomological Society of America*, **30**: 49–52.
McGuire, M.R. and J.E. Henry. 1989. Production and partial characterization of monoclonal antibodies for detection of entomopoxvirus from *Melanoplus sanguinipes*. *Entomologia Experimentalis et Applicata*, **51**: 21–28.
McGuire, M.R., B.S. Shasha, L.C. Lewis, R.J. Bartelt, and K. Kinney. 1990. Field evaluation of granular starch formula-

tions of *Bacillus thuringiensis* against *Ostrinia nubilalis* (Lepidoptera: Pyralidae). *Journal of Economic Entomology*, **83**: 2207–2210.

McGuire, M.R., D.A. Streett, and B.S. Shasha. 1991. Evaluation of starch encapsulation for formulation of grasshopper (Orthoptera: Acrididae) entomopoxvirus. *Journal of Economic Entomology*, **84**: 1652–1656.

McKillup, S.C. and P.T. Bailey. 1990. Biological control of a pest millipede *Ommatoiulus moreleti* in south Australia using a rhabditid nematode, pp. 236. *In* Pinnock, D.E., (ed.). *Vth International Colloquium on Invertebrate Pathology and Microbial Control.* Adelaide, Australia, 20–24, August, 1990.

McKillup, S.C., P.G. Allen, and M.A. Skewes. 1988. The natural decline of an introduced species following its initial increase in abundance; an explanation for *Ommatoiulus moreletii* in Australia. *Oecologia*, **77**: 339–342.

McKnight, B.N., (ed.). 1993. *Biological Pollution: The Control and Impact of Invasive Exotic Species.* Indiana Academy of Science, Indianapolis, Indiana, U.S.A.

McMurtry, J.A. 1982. The use of phytoseiids for biological control: Progress and future prospects, pp. 23–48. *In* Hoy, M.A., (ed.). *Recent Advances in Knowledge of the Phytoseiidae.* Division of Agricultural Sciences, University of California, Special Publication #3284, Berkeley, California, U.S.A.

McMurtry, J.A. 1992. Dynamics and potential impact of "generalist" phytoseiids in agroecosystems and possibilities for establishment of exotic species. *Experimental and Applied Acarology*, **14**: 371–382.

McMurtry, J.A. and G.T. Scriven. 1964. Studies on the feeding, reproduction, and development of *Amblyseius hibisci* (Acarina: Phytoseiidae) in various food substances. *Annals of the Entomological Society of America*, **57**: 649–655.

McMurtry, J.A., E.R. Oatman, P.H. Phillips, and G.W. Wood. 1978. Establishment of *Phytoseiulus persimilis* (Acari: Phytoseiidae) in southern California. *Entomophaga*, **23**: 175–179.

McNamee, P.J., J.M. McLeod, and C.S. Holling. 1981. The structure and behavior of defoliating insect/forest systems. University of British Columbia Institute of Resource Ecology, Publication R-25, 1–89.

Meadow, R.H., W.C. Kelly, and A.M. Shelton. 1985. Evaluation of *Aphidoletes aphidimyza* (Dip.: Cecidomyiidae) for control of *Myzus persicae* (Hom.: Aphididae) in greenhouse and field experiments in the United States. *Entomophaga*, **30**: 385–392.

Meagher, R.L., Jr. and J.R. Meyer. 1990. Influence of ground cover and herbicide treatments on *Tetranychus urticae* populations in peach orchards. *Experimental and Applied Acarology*, **9**: 149–158.

Melching, J.S., K.R. Bromfield, and C.H. Kingsolver. 1983. The plant pathogen containment facility at Frederick, Maryland. *Plant Disease*, **67**: 717–722.

Mendel, Z., Y. Golan, and Z. Madar. 1984. Natural control of the eucalyptus borer, *Phoracantha semipunctata* (F.) (Coleoptera: Cerambycidae), by the Syrian woodpecker. *Bulletin of Entomological Research*, **74**: 121–127.

Merlin, M.D. and J.O. Juvik. 1992. Relationships among native and alien plants on Pacific islands with and without significant human disturbance and feral ungulates, pp. 597–624. *In* Stone, C.P., C.W. Smith, and J.T. Tunison, (eds.). *Alien Plant Invasions in Native Ecosystems of Hawai'i: Management and Research.* University of Hawaii Press, Honolulu, Hawaii, U.S.A.

Merriman, P.R., R.D. Price, and K.F. Baker. 1974. The effect of inoculation of seed with antagonists of *Rhizoctonia solani* on the growth of wheat. *Australian Journal of Agricultural Research*, **25**: 213–218.

Merritt, R.W., E.D. Walker, M.A. Wilzbach, K.W. Cummins, and W.T. Morgan. 1989. A broad evaluation of B.t.i. for black fly (Diptera: Simuliidae) control in a Michigan river: Efficacy, carryover and nontarget effects on invertebrates and fish. *Journal of the American Mosquito Control Association*, **5**: 397–415.

Messenger, P.S. 1971. Climatic limitation to biological controls. *Proceedings of the Tall Timbers Conference on Ecological Animal Control by Habitat Management*, **3**: 97–114.

Messenger, P.S., E. Biliotti, and R. van den Bosch. 1976. The importance of natural enemies in integrated control, pp. 543–563. *In* Huffaker, C. B. and P.S. Messenger, (eds.). *Theory and Practice of Biological Control.* Academic Press, Inc., New York.

Messing, R.H., L.M. Klungness, M. Purcell, and T.T.Y. Wong. 1993. Quality control parameters of mass-reared opiine parasitoids used in augmentative biological control of tephritid fruit flies in Hawaii. *Biological Control*, **3**: 140–147.

Metcalfe, J.R. 1971. Observations on the ecology of *Saccharosydne saccharivora* (Westw.) (Hom,. Delphacidae) in Jamaican sugar-cane fields. *Bulletin of Entomological Research*, **60**: 565–597.

Metcalf, R.L. 1980. Changing role of insecticides in crop protection. *Annual Review of Entomology*, **25**: 219–256.

Metcalfe, R. 1991. Of mosquitoes [sic] and coconuts. IDRC Reports **19**: 17–19.

Meyer, J.R. and C.A. Nalepa. 1991. Effect of dormant oil treatments on white peach scale (Homoptera: Diaspididae) and its overwintering parasite complex. *Journal of Entomological Science*, **26**: 27–32.

Meyerdirk, D.E. and I.M. Newell. 1979. Importation, colonization, and establishment of natural enemies on the Comstock mealybug in California. *Journal of Economic Entomology*, **72**: 70–73.

Meyerdirk, D.E., I. M. Newell, and R.W. Warkentin. 1981. Biological control of Comstock mealybug. *Journal of Economic Entomology*, **74**: 79–84.

Miao, W.C., J.G. Zhu, M.R. Zhou, W.H. Cheng, T.B. Li, and H.Lu. 1993. Development of a new mass production procedure for Beauveria bassiana conidia. *Chinese Journal of Biological Control*, **9**: 1–4. (in Chinese)

Michels, G.J., Jr. and R.W. Behle. 1992. Evaluation of sampling methods for lady beetles (Coleoptera: Coccinellidae) in grain sorghum. *Journal of Economic Entomology*, **85**: 2251–2257.

Millard, W.A. and C.B. Taylor. 1927. Antagonism of micro-organisms as the controlling factor in the inhibition of scab by green manuring. *Annals of Applied Biology*, **14**: 202–216.

Miller, J.C. 1983. Ecological relationships among parasites and the practice of biological control. *Environmental Entomology*, **12**: 620–624.

Miller, J.C. 1990. Effects of a microbial insecticide, *Bacillus thuringiensis kurstaki*, on nontarget Lepidoptera in a spruce budworm-infested forest. *Journal of Research on Lepidoptera*, **29**: 267–276.

Miller, L.A. and R.A. Bedding. 1982. Field testing of the insect parasitic nematode, *Neoaplectana bibionis* (Nematoda: Steinernematidae) against current borer moth, *Synanthedon tipuliformis* (Lep.: Sessiidae) in blackcurrants. *Entomophaga*, **27**: 109–114.

Miller, L.K., A.J. Lingg, and L.A. Bulla, Jr. 1983. Bacterial, viral, and fungal insecticides. *Science*, **219**: 715–721.

Mills, N.J. and J. Schlup. 1989. The natural enemies of *Ips typographus* in central Europe: Impact and potential use in biological control, pp. 131–146. *In* Kulhavy, D.L. and M.C. Miller, (eds.). *Potential for Biological Control of Dendroctonus and Ips Bark Beetles.* Center for Applied Studies, School of Forestry, Stephen F. Austin State University, Nacogdoches, Texas, U.S.A.

Milne, W.M. and A.L. Bishop. 1987. The role of predators and parasites in the natural regulation of lucerne aphids in eastern Australia. *Journal of Applied Ecology*, **24**: 893–905.

Milner, R.J., R.S. Soper, and G.G. Lutton. 1982. Field release of an Israeli strain of the fungus *Zoophthora radicans* (Brefeld) Batko for biological control of *Therioaphis trifolii* (Monell) f. *maculata. Journal of the Australian Entomological Society*, **21**: 113–118.

Minkenberg, O.P.J.M. 1988. Dispersal of *Liriomyza trifolii. Bulletin OEPP/EPPO*, **18**: 173–182.

Minkenberg, O.P.J.M., M. Tatar, and J.A. Rosenheim. 1992. Egg load as a major source of variability in insect foraging and oviposition behavior. *Oikos*, **65**: 134–142.

Misra, M.P., A.D. Pawar, and U.L. Srivastava. 1986. Biocontrol of sugarcane moth borers by releasing *Trichogramma* parasites at Haringar, West Champaran, Bihar. *Indian Journal Plant Protection*, **14**: 89–91.

Mitchell, D.S., T. Petr, and A.B. Viner. 1980. The water-fern *Salvinia molesta* in the Sepik River, Papua New Guinea. *Environmental Conservation*, **7**: 115–122.

Miura, T., R.M. Takahashi, and W.H. Wilder. 1984. Impact of the mosquitofish (*Gambusia affinis*) on a rice field ecosystem when used as a mosquito control agent. *Mosquito News*, **44**(4): 510–517.

Miyasono, M., S. Inagaki, M. Yamamoto, K. Ohba, T. Ishiguro, R. Takeda, and Y. Hayashi. 1994. Enhancement of δ-endotoxin activity by toxin-free spore of *Bacillus thuringiensis* against the diamondback moth, *Plutella xylostella. Journal of Invertebrate Pathology*, **63**: 111–112.

Mogi, M. and I. Miyagi. 1990. Colonization of rice fields by mosquitoes (Diptera: Culicidae) and larvivorous predators in asynchronous rice cultivation areas in the Philippines. *Journal of Medical Entomology*, **27**: 530–536.

Mohan, K.S. and G.B. Pillai. 1993. Biological control of *Oryctes rhinoceros* (L.) using an Indian isolate of *Oryctes baculovirus. Insect Science and Its Application*, **14**: 551–558.

Mohd, S. 1990. Barn owls (*Tyto alba*) for controlling rice field rats. *MAPPS Newsletter*, **14**(4): 51.

Mohyuddin, A. I. 1991. Utilization of natural enemies for the control of insect pests of sugar-cane. *Insect Science and Its Application*, **12**: 19–26.

Mohyuddin, A.I., C. Inayatullah, and E.G. King. 1981. Host selection and strain occurrence in *Apanteles flavipes* (Cameron) (Hymenoptera: Braconidae) and its bearing on biological control of graminaceous stem-borers (Lepidoptera: Pyralidae). *Bulletin of Entomological Research*, **71**: 575–581.

Moller, H., G.M. Plunkett, J.A.V. Tilley, R.J. Toft, and J.R. Beggs. 1991. Establishment of the wasp parasitoid, *Sphecophaga vesparum* (Hymenoptera: Ichneumonidae), in New Zealand. *New Zealand Journal of Zoology*, **18**: 199–208.

Montllor, C.B., E.A. Bernays, and M.L. Cornelius. 1991. Responses of two hymenopteran predators to surface chemistry of their prey: Significance for an alkaloid-sequestering caterpillar. *Journal of Chemical Ecology*, **17**: 391–399.

Mook, L.J. 1963. Birds and the spruce budworm. *In* Morris, R.F., (ed.). *The Dynamics of Epidemic Spruce Budworm Populations. Memoirs of the Entomological Society of Canada*, **31**: 268–271.

Moore, D. and C. Prior. 1993. The potential of mycoinsecticides. *Biocontrol News and Information*, **14**(2): 331N–40N.

Moore, D., P.D. Bridge, P.M. Higgins, R.P. Bateman, and C. Prior. 1993. Ultra-violet radiation damage to *Metarhizium flavoviride* conidia and the protection given by vegetable and mineral oils and chemical sunscreens. *Annals of Applied Biology*, **122**: 605–616.

Moore, J.C., D.E. Walter, and H.W. Hunt. 1988. Arthropod regulation of micro- and mesobiota in below-ground detrital food webs. *Annual Review of Entomology*, **33**: 419–439.

Moore, N.F., L.A. King, and R.D. Possee. 1987. Mini review: viruses of insects. *Insect Science and Its Application*, **8**: 275–289.

Moore, S.D. 1989. Regulation of host diapause by an insect parasitoid. *Ecological Entomology*, **14**: 93–98.

Moorehead, G.E. and H.L. Maltby. 1970. A container for releasing Anaphes flavipes from parasitized eggs of *Oulema melanopus*. *Journal of Economic Entomology*, **63**: 675–676.

Moran, V.C., S. Neser, and J.H. Hoffmann. 1986. The potential of insect herbivores for the biological control of invasive plants in South Africa, pp. 261-268. *In* Macdonald, I.A.W., F.J. Kruger, and A.A. Ferrar, (eds.). *The Ecology and Management of Biological Invasions in Southern Africa*. Proceedings of the National Synthesis Symposium on the Ecology of Biological Invasions. Oxford University Press, Cape Town, South Africa.

Moreno, D.S. and R.F. Luck. 1992. Augmentative releases of *Aphytis melinus* (Hymenoptera: Aphelinidae) to suppress California red scale (Homoptera: Diaspididae) in southern California lemon orchards. *Journal of Economic Entomology*, **85**: 1112–1119.

Morewood, W.D. 1992. Cold storage of *Phytoseiulus persimilis* (Phytoseiidae). *Experimental and Applied Acarology*, **13**: 231–236.

Morewood, W.D. and L.A. Gilkeson. 1991. Diapause induction in the thrips predator *Amblyseius cucumeris* (Acarina: Phytoseiidae) under greenhouse conditions. *Entomophaga*, **36**: 253–263.

Morgan, P.B., C.J. Jones, R.S. Patterson, and D. Milne. 1988. Use of electrophoresis for monitoring purity of laboratory colonies of exotic parasitoids (Hymenoptera: Pteromalidae). *In* Gupta, V.K., (ed.). *Advances in Parasitic Hymenoptera Research; Proceedings of the IInd Conference on the Taxonomy and Biology of Parasitic Hymenoptera.* Held at the University of Florida, Gainesville, Florida, November 19–21, 1987. E.J. Brill, Pub., Leiden, The Netherlands.

Morris, O.N. 1980. Entomopathogenic viruses: Strategies for use in forest insect pest management. *Canadian Entomologist*, **112**: 573–584.

Morris, R.F. 1959. Single-factor analysis in population dynamics. *Ecology*, **40**: 580–588.

Morris, W.F. 1992. The effects of natural enemies, competition, and host plant water availability on an aphid population. *Oecologia*, **90**: 359–365.

Morrison, R.K., W.C. Nettles, Jr., D. Ball, and S.B. Vinson. 1983. Successful oviposition by *Trichogramma pretiosum* through a synthetic membrane. *Southwestern Entomologist*, **8**: 248–251.

Morrow, B.J., Boucias, D.G., and M.A. Heath. 1989. Loss of virulence in an isolate of an entomopathogenic fungus, *Nomuraea rileyi*, after serial *in vitro* passage. *Journal of Economic Entomology*, **82**: 404–407.

Morse, J.G. and T.S. Bellows, Jr. 1986. Toxicity of major citrus pesticides to *Aphytis melinus* (Hymenoptera: Aphelinidae) and *Cryptolaemus montrouziere* (Coleoptera: Coccinellidae). *Journal of Economic Entomology*, **79**: 311–314.

Morse, J.G. and N. Zareh. 1991. Pesticide-induced hormoligosis of citrus thrips (Thysanoptera: Thripidae) fecundity. *Journal of Economic Entomology*, **84**: 1169–1174.

Morse, J.G., T.S. Bellows, Jr., L.K. Gaston, and Y. Iwata. 1987. Residual toxicity of acaricides to three beneficial species on California citrus. *Journal of Economic Entomology*, **80**: 953–960.

Moscardi, F. 1983. Utilizacão de *Baculovirus anticarsia* para o controle da lagarta da soya, *Anticarsia gemmatalis. Empresa Brasiliera de Pesquira Agropecuaria, Comunicado Tecnico No. 23.* (in Portuguese)

Moscardi, F. 1990. Development and use of soybean caterpillar baculovirus in Brazil, pp. 184–187. *In* Pinnock, D.E., (ed.). *Proceedings and Abstracts, Vth International Colloquium on Invertebrate Pathology and Microbial Control.* Adelaide, Australia, 20–24, August, 1990.

Moscardi, F., I.L.S. Bono, and F.E. Paro. 1988. Biological activities of batches of *Baculovirus anticarsia* formulated by a process developed at CNPSo-EMBRAPA. *Documentos Centro Nacional de Pesquisa de Soja, EMBRAPA No.* **36**: 33–34. (in Portuguese)

Mosjidis, J.A., R. Rodríguez-Kábana, and C.M. Owsley. 1993. Reaction to three cool-season annual legume species to *Meloidogyne arenaria* and *Heterodera glycines. Nematropica*, **23**: 35–39.

Mueller-Beilschmidt, D. and M.A. Hoy. 1987. Activity levels of genetically manipulated and wild strains of *Metaseiulus occidentalis* (Nesbitt) (Acarina: Phytoseiidae) compared as a method to assay quality. *Hilgardia*, **55**(6): 1–23.

Mulinge, S.K. and E. Griffiths. 1974. Effects of fungicides on leaf rust, berry disease, foliation and yield of coffee. *Transactions of the British Mycological Society*, **62**: 495–507.

Mulla, M.S., B.A. Federici, and H.A. Darwazeh. 1982. Larvicidal efficacy of *Bacillus thuringiensis* serotype H-14 against stagnant-water mosquitoes and its effect on nontarget organisms. *Environmental Entomology*, **11**: 788–795.

Mulla, M.S., J.D. Chaney, and J. Rodcharoen. 1990. Control of nuisance aquatic midges (Diptera: Chironomidae) with the microbial larvicide *Bacillus thuringiensis* var. *israelensis* in a man-made lake in southern California. *Bulletin of the Society for Vector Control*, **15**(2): 176–184.

Müller-Kögler, E. 1965. *Pilzkrankheiten bei Insekten.* Paul Parey, Berlin and Hamburg, Germany.

Müller-Schärer, H. 1991. The impact of root herbivory as a function of plant density and competition: Survival, growth and fecundity of *Centaurea maculosa* in field plots. *Journal of Applied Ecology*, **28**: 759–776.

Muñoz, A. and R. Murúa. 1990. Control of small mammals in a pine plantation (central Chile) by modification of the habitat of predators (*Tyto alba*, Strigiforme and *Pseudalopex* sp., Canidae). *Acta Oecologica*, **11**: 251–261.

Murakami, Y., K. Umeya, and N. Ono. 1977. A preliminary introduction and release of a parasitoid (Chalcidoidea: Torymidae) of the chestnut gall wasp, *Dryocosmus kuriphilus* Yasumatsu (Cynipidae) from China. *Japanese Journal of Applied Entomology, Zoology*, **21**: 197–203. (in Japanese)

Murakami, Y., H.-B. Ao, and C.H. Chang. 1980. Natural enemies of the chestnut gall wasp *Dryocosmus kuriphilus* in Hopei Province, China (Hymenoptera: Chalcidoidea). *Applied Entomology and Zoology*, **15**: 184–186.

Muratov, V.S., T.F. Bondarenko, V.A. Irikov, N.P. Bunyaeva, I.P. Uvarov, and N.V. Chekotina. 1990. A method of determining the quality of preparations obtained from bacteria of the group *Bacillus thuringiensis*. *Biotekhnologiya*, **6**(5): 67–68. (in Russian)

Murdie, G. and M.P. Hassell. 1973. Food distribution, searching success and predator-prey models, pp. 87–101. *In* Bartlett, M.S. and R.W. Hiorns, (eds.). *The Mathematical Theory of the Dynamics of Biological Populations*. Academic Press, Inc., London.

Murdoch, W.W. 1969. Switching in general predators: Experiments on predator specificity and stability of prey populations. *Ecological Monographs*, **39**: 335–354.

Murdoch, W.W. and A. Oaten. 1975. Predation and population stability. *Advances in Ecological Research*, **9**: 1–131.

Murdoch, W.W. and A. Stewart-Oaten. 1989. Aggregation by parasitoids and predators: Effects on equilibrium and stability. *American Naturalist*, **134**: 288–310.

Murdoch, W.W., J.D. Reeve, C.B. Huffaker, and C.E. Kennett. 1984. Biological control of olive scale and its relevance to ecological theory. *American Naturalist*, **123**: 371–392.

Murdoch, W.W., R.M. Nisbet, S.P. Blythe, W.S.C. Gurney, and J.D. Reeve. 1987. An invulnerable age class and stability in delay-differential parasitoid-host models. *American Naturalist*, **129**: 263–282.

Murdoch, W.W., C.J. Briggs, R.M. Nisbet, W.S.C. Gurney, and A. Stewart-Oaten. 1992. Aggregation and stability in metapopulation models. *American Naturalist*, **140**: 41–58.

Murray, J., E. Murray, M.S. Johnson, and B. Clarke. 1988. The extinction of Partula on Moorea. *Pacific Science*, **42**: 150–153.

Musgrove, C.H. and G.E. Carman. 1965. Argentine ant control in citrus in southern California with granular formulations of certain chlorinated hydrocarbons. *Journal of Economic Entomology*, **58**: 428–434.

Nachman, G. 1981. Temporal and spatial dynamics of an acarine predatory-prey system. *Journal of Animal Ecology*, **50**: 435–451.

Nadel, H. and R.F. Luck. 1992. Dispersal and mating structure of a parasitoid with a female-biased sex ratio: Implications for theory. *Evolutionary Ecology*, **6**: 270–278.

Nadel, H. and J.J.M. van Alphen. 1987. The role of host and host plant odours in the attraction of a parasitoid, *Epidinocarsis lopezi*, to the habitat of its host, the cassava mealybug, *Phenacoccus manihoti*. *Entomologia Experimentalis et Applicata*, **45**: 181–186.

Nafus, D.M. 1993. Movement of introduced biological control agents onto nontarget butterflies, *Hypolimnas* spp. (Lepidoptera: Nymphalidae). *Environmental Entomology*, **22**: 265–272.

Nafus, D. and I. Schreiner. 1989. Biological control activities in the Mariana Islands from 1911 to 1988. *Micronesia*, **22**: 65–106.

Napompeth, B. 1990. Use of natural enemies for controlling agricultural pests in Thailand. *In* FFTC-NARC International Seminar on "The use of parasitoids and predators to control agricultural pests." Tukuba Science City, Ibaraki-ken, 305, Japan, October 2–7, 1989.

Nappi, A.J. 1973. Parasitic encapsulation in insects, pp. 293–326. *In* Maramorosch, K. and R.E. Shope, (eds.). *Invertebrate Immunity*. Academic Press, New York.

Naser, W.L., H.G. Miltenburger, J.F. Harvey, J. Huber, and A.M. Huger. 1984. *In vitro* replication of the *Cydia pomonella* (codling moth) granulosis virus. *FEMS Microbiology Letter*, **24**: 117–121.

Navasero, R.C. and G.W. Elzen. 1989. Responses of *Microplitis croceipes* to host and nonhost plants of *Heliothis virescens* in a wind tunnel. *Entomologia Experimentalis et Applicata*, **53**: 57–63.

Nealis, V. 1985. Diapause and the seasonal ecology of the introduced parasite *Cotesia (Apanteles) rubecula* (Hymenoptera: Braconidae). *Canadian Entomologist*, **117**: 333–342.

Nealis, V.G. 1986. Responses to host kairomones and foraging behavior of the insect parasite *Cotesia rubecula* (Hymenoptera: Braconidae). *Canadian Journal of Zoology*, **64**: 2393–2398.

Nealis, V.G., K. van Frankenhuyzen, and B.L. Cadogan. 1992. Conservation of spruce budworm parasitoids following application of *Bacillus thuringiensis* var. *kurstaki* Berliner. *Canadian Entomologist*, **124**: 1085–1092.

Nechols, J.R. and R.S. Kikuchi. 1985. Host selection of the spherical mealybug (Homoptera: Pseudococcidae) by *Anagyrus indicus* (Hymenoptera: Encyrtidae): Influence of host stage on parasitoid oviposition, development, sex ratio, and survival. *Environmental Entomology*, **14**: 32–37.

Nechols, J.R., L.A. Andres, J.W. Beardsley, R.D. Goeden, and C.G. Jackson, (eds.). 1995. *Biological Control in the Western United States: Accomplishments and Benefits of Regional Project W-84, 1964–1989.* DANR Publications, University of California, Oakland, California, U.S.A.

Neel, P.L. and A.A. Will. 1978. *Grevillea chrysodendron* R. Br.: Potential weed in south Florida. *Hortscience*, **13**: 18–21.

Neil, K.A. and H.B. Specht. 1990. Field releases of *Trichogramma pretiosum* Riley (Hymenoptera: Trichogrammatidae) for suppression of corn earworm, *Heliothis zea* (Boddie) (Lepidoptera: Noctuidae), egg populations on sweet corn in Nova Scotia. *Canadian Entomologist*, **122**: 1259–1266.

Neill, W.M. 1983. The tamarisk invasion of desert riparian areas. Education Bulletin 83-4, Desert Protective Council, Spring Valley, California, U.S.A.

Nemoto, T., M. Shibuya, Y. Kuwahara, and T. Suzuki. 1987. New 2-acylcyclohexane-1,3-diones: Kairomone components against a parasitic wasp, Venturia canescens, from feces of the almond moth, Cadra cautella, the Indian meal moth, *Plodia interpunctella*. *Agric. Biol. Chem.*, **51**: 1805–1810.

Nentwig, W. 1988. Augmentation of beneficial arthropods by strip-management. 1. Succession of predacious arthropods and long-term change in the ratio of phytophagous and predacious arthropods in a meadow. *Oecologia*, **76**: 597–606.

Neser, S. and R.L. Kluge. 1986. The importance of seed-attacking agents in the biological control of invasive alien plants, pp. 285–293. *In* Macdonald, I.A.W., F.J. Kruger, and A.A. Ferrar, (eds.). *The Ecology and Management of Biological Invasions in Southern Africa, Proceedings of the National Synthesis Symposium on the Ecology of Biological Invasions.* Oxford University Press, Cape Town, South Africa.

Nettles, W.C., Jr., C.M. Wilson, and S.W. Ziser. 1980. A diet and methods for the *in vitro* rearing of the tachinid *Eucelatoria* sp. *Annals of the Entomological Society of America*, **73**: 180–184.

Nettles, W.C., Jr., R.K. Morrison, Zhong-Neng Xie, D. Ball, C.A. Shenkir, and S.B. Vinson. 1985. Effect of artificial diet media, glucose, protein hydrolyzates, and other factors on oviposition in wax eggs by *Trichogramma pretiosum*. *Entomologia Experimentalis et Applicata*, **38**: 121–129.

Neuenschwander, P. 1982. Beneficial insects caught by yellow traps used in mass-trapping of the olive fly, *Dacus oleae*. *Entomologia Experimentalis et Applicata*, **32**: 286–296.

Neuenschwander, P., F. Schulthess, and E. Madojemu. 1986. Experimental evaluation of the efficiency of *Epidinocarsis lopezi*, a parasitoid introduced into Africa against the cassava mealybug *Phenacoccus manihoti*. *Entomologia Experimentalis et Applicata*, **42**: 133–138.

Neuenschwander, P., W.N.O. Hammond, A.P. Gutierrez, A.R. Cudjoe, R. Adjakloe, J.U. Baumgärtner, and U. Regev. 1989. Impact assessment of the biological control of the cassava mealybug, *Phenacoccus manihoti* Matile-Ferrero [Hemiptera: Pseudococcidae], by the introduced parasitoid *Epidinocarsis lopezi* (De Santis) [Hymenoptera: Encyrtidae]. *Bulletin of Entomological Research*, **79**: 579–594.

New, T.R. 1992. *Insects as Predators.* New South Wales University Press, Kensington, N.S.W., Australia.

Newhook, F.J. 1957. The relationship of saprophytic antagonism to control of *Botrytis cinerea* Pers. on tomatoes. *New Zealand Journal of Science and Technology, Section A*, **38**: 473–481.

Newton, I. 1988. Monitoring of persistent pesticide residues and their effects on bird populations, pp. 33–45. *In* Harding, D.J.L., (ed.). *Britain Since "Silent Spring", An Update on the Ecological Effects of Agricultural Pesticides in the U.K.* Proceedings of a Symposium held in Cambridge, U.K., 18 March, 1988.

Newton, P.J. and W.J. Odendaal. 1990. Commercial inundative releases of *Trichogrammatoidea cryptophlebiae* [Hymen.: Trichogrammatidae] against *Cryptophlebia leucotreta* [Lep.: Tortricidae] in citrus. *Entomophaga*, **35**: 545–556.

Nguyen, K.B. and G.C. Smart, Jr. 1990. *Steinernema scapterisci* n. sp. (Rhabditida: Steinernematidae). *Journal of Nematology*, **22**: 187–199.

Nguyen, K.B. and G.C. Smart, Jr. 1991. Pathogenicity of *Steinernema scapterisci* to selected invertebrates. *Journal of Nematology*, **23**: 7–11.

Nguyen, R., J.R. Brazzel, and C. Poucher. 1983. Population density of the citrus blackfly, *Aleurocanthus woglumi* Ashby (Homoptera: Aleyrodidae), and its parasites in urban Florida in 1979–1981. *Environmental Entomology*, **12**: 878–884.

Nicholson, A.J. 1933. The balance of animal populations. *Journal of Animal Ecology*, **2**: 131–178.

Nicholson, A.J. 1954. An outline of the dynamics of animal populations. *Australian Journal of Zoology*, **2**: 9–65.

Nicholson, A.J. and V.A. Bailey. 1935. The balance of animal populations. Part I. *Proceedings of the Zoological Society of London*, **3**: 551–598.

Nicoli, G., M. Benuzzi, and N.C. Leppla. 1994. Seventh workshop of the IOBC global working group "Quality control of mass reared arthropods". Rimimi, Italy, 13–16, September, 1993.

Nilsson, C. 1985. Impact of ploughing on emergence of pollen beetle parasitoids after hibernation. *Zeitschrift für Angewandte Entomologie*, **100**: 302–308.

Nisbet, R.M. and W.S.C. Gurney. 1982. *Modeling Fluctuating Populations.* John Wiley and Sons, New York.

Nishida, T. and B. Napompeth. 1974. Trap for tephritid fruit fly parasites. *Entomophaga*, **19**: 349–352.

Noldus, L.P.J.J. 1988. Response of the egg parasitoid *Trichogramma pretiosum* to the sex pheromone of its host *Heliothis zea*. *Entomologia Experimentalis et Applicata*, **48**: 293–300.

Noldus, L.P.J.J. 1989. Semiochemicals, foraging behaviour and quality of entomophagous insects for biological control. *Journal of Applied Entomology*, **108**: 425–451.

Noldus, L.P.J.J., W.J. Lewis, and J.H. Tumlinson. 1990. Beneficial arthropod behavior mediated by airborne semiochemicals. IX. Differential response of *Trichogramma pretiosum*, an egg parasitoid of *Heliothis zea*, to various olfactory cues. *Journal of Chemical Ecology*, **16**: 3531–3544.

Norambuena, H., R. Carrillo, and M. Neira. 1986. Introduccion, establecimiento y potencial de *Apion ulicis* como antagonista de *Ulex europaeus* en el sur de Chile. *Entomophaga*, **31**: 3–10. (in Spanish)

Nordbring-Hertz, B. 1973. Peptide-induced morphogenesis in the nematode-trapping fungus *Arthrobotrys oligospora*. *Physiologica Plantarum*, **29**: 223–233.

Nordlund, D.A. 1981. Semiochemicals: A review of terminology, pp. 13–28. *In* Nordlund, D.A., R.J. Jones, and W.J. Lewis, (eds.). *Semiochemicals, Their Role in Pest Control.* John Wiley and Sons, New York.

Nordlund, D.A. and C.E. Sauls. 1981. Kairomones and their use for management of entomophagous insects. XI. Effect of host plants on kairomonal activity of frass from *Heliothis zea* larvae for the parasitoid *Microplitis croceipes*. *Journal of Chemical Ecology*, **7**: 1057–1061.

Norgaard, R.B. 1988. Economics of the cassava mealybug [*Phaenacoccus* (sic) *manihoti*; Hom.: Pseudococcidae] biological control program in Africa. *Entomophaga*, **33**: 3–6.

Norris, M. J. 1935. A feeding experiment on the adults of *Pieris rapae* Linnaeus (Lepid.: Rhop.). *Entomologist* **68**: 125–127.

North, S.G., D.J. Bullock, and M.E. Dulloo. 1994. Changes in the vegetation and reptile populations on Round Island, Mauritius, following eradication of rabbits. *Biological Conservation* **67**: 21–28.

Northrup, Z. 1914. A bacterial disease of the larvae of the June beetle, *Lachnosterna* spp. Michigan Agricultural Experiment Station Technical Bulletin 18.

Nuessly, G.S. and R.D. Goeden. 1984. Rodent predation on larvae of *Coleophora parthenica* (Lepidoptera: Coleophoridae), a moth imported for the biological control of Russian thistle. *Environmental Entomology*, **13**: 502–508.

Nuessly, G.S., A.W. Hartstack, J.A. Witz, and W.L. Sterling. 1991. Dislodgment of *Heliothis zea* (Lepidoptera: Noctuidae) eggs from cotton due to rain and wind: a predictive model. *Ecological Modeling*, **55**: 89–102.

Nunney, L. 1980. The influence of the type 3 (sigmoid) functional response upon the stability of predator-prey difference models. *Theoretical Population Biology*, **18**: 257–278.

Nuttall, M.J. 1989. *Sirex noctilio* F., sirex wood wasp (Hymenoptera: Siricidae), pp. 299–306. *In* Cameron, P.J., R.L. Hill, J. Bain, and W.P. Thomas, (eds.). *A Review of Biological Control of Invertebrate Pests and Weeds in New Zealand 1874 to 1987.* Commonwealth Agricultural Bureaux International, Wallingford, U.K.

Nyffeler, M. and G. Benz. 1987. Spiders in natural pest control: A review. *Journal of Applied Entomology*, **103**: 321–339.

Nyrop, J.P. 1988. Sequential classification of prey/predator ratios with application to European red mite (Acari: Tetranychidae) and *Typhlodromus pyri* (Acari: Phytoseiidae) in New York apple orchards. *Journal of Economic Entomology*, **81**: 14–21.

Oatman, E.R., J.A. McMurtry, F.E. Gilstrap, and V. Voth. 1977. Effect of releases of *Amblyseius californicus* on the two spotted spider mite on strawberry in southern California. *Journal of Economic Entomology*, **70**: 638–640.

Obrycki, J.J. 1986. The influence of foliar pubescence on entomophagous species, pp. 61–83. *In* Boethel, D.J. and R.D. Eikenbary, (eds.). *Interactions of Plant Resistance and Parasitoids and Predators of Insects.* Ellis Horwood Limited, John Wiley and Sons, New York.

Ochieng, R.S., G.W. Oloo, and E.O. Amboga. 1987. An artificial diet for rearing the phytoseiid mite, *Amblyseius teke* Pritchard and Baker. *Experimental and Applied Acarology*, **3**: 169–173.

Ochieng-odero, J.P.R. 1990. New strategies for quality assessment and control of insects produced in artificial rearing systems. *Insect Science and Its Application*, **11**: 133–141.

O'Donnell, M.S. and T.H. Croaker. 1975. Potential of intra-crop diversity in the control of brassica pests, pp. 101–107. *Proceedings of the 8th British Insecticide and Fungicide Conference, Brighton, U.K.*

O'Hara, J.E. 1985. Oviposition strategies in the Tachinidae, a family of beneficial parasitic flies. *Agriculture and Forestry Bulletin, University of Alberta*, **8**: 31–34.

Oien, C.T. and D.W. Ragsdale. 1993. Susceptibility of nontarget hosts to *Nosema furnacalis* (Microsporida: Nosematidae), a potential biological control agent of the European corn borer, *Ostrinia nubilalis* (Lepidoptera: Pyralidae). *Biological Control*, **3**: 323–328.

O'Neil, R.J. 1988. A model of predation by *Podisus maculiventris* (Say) on Mexican bean beetle, *Epilachna varivestis* Mulsant, in soybeans. *Canadian Entomologist*, **120**: 601–608.

O'Neil, R.J. and J.L. Stimac. 1988a. Measurement and analysis of arthropod predation on velvetbean caterpillar, *Anticarsia gemmatalis* (Lepidoptera: Noctuidae), in soybeans. *Environmental Entomology*, **17**: 821–826.

O'Neil, R.J. and J.L. Stimac. 1988b. Model of arthropod predation on velvetbean caterpillar (Lepidoptera: Noctuidae) larvae in soybean. *Environmental Entomology*, **17**: 983–987.

Oi, D.H. and M.M. Barnes. 1989. Predation by the western predatory mite (Acari: Phytoseiidae) on the Pacific spider mite (Acari: Tetranychidae) in the presence of road dust. *Environmental Entomology*, **18**: 892–896.

Oka, I.N. 1991. Success and challenges of the Indonesia National Integrated Pest Management Program in the rice-based cropping system. *Crop Protection*, **10**: 163–165.

Okuda, M.S. and K.V. Yeargan. 1988. Intra- and interspecific host discrimination in *Telenomus podisi* and *Trissolcus euschisti* (Hymenoptera: Scelionidae). *Annals of the Entomological Society of America*, **81**: 1017–1020.

Olthof, Th.H.A. and R.H. Estey. 1966. Carbon and nitrogen levels of a medium in relation to growth and nematophagous activity of *Arthrobotrys oligospora* Fres. *Nature*, **209**: 1158.

Ooi, P.A. and K.L. Heong. 1988. Operation of a brown planthopper surveillance system in the Tanjung Karang Irrigation Scheme in Malayasia. *Crop Protection*, **7**: 273–278.

Opp, S.B. and R.F. Luck. 1986. Effects of host size on selected fitness components of *Aphytis melinus* and *A. lingnanensis* (Hymenoptera: Aphelinidae). *Annals of the Entomological Society of America*, **79**: 700–704.

Orians, G.H. 1986. Site characteristics favoring invasions, pp. 133–148. *In* Mooney, H.A. and J.A. Drake, (eds.). *Ecology of Biological Invasions of North America and Hawaii.* Springer-Verlag, New York.

Osakabe, M. 1988. Relationships between food substances and developmental success in *Amblyseius sojaensis* Ehara (Acarina: Phytoseiidae). *Applied Entomology and Zoology*, **23**: 45–51.

Osborne, K.J., R.J. Powles, and P.L. Rogers. 1990. *Bacillus sphaericus* as a biocontrol agent. *Australian Journal of Biotechnology*, **4**: 205–211.

Osborne, L.S. 1981. Utility of physiological time in integrating chemical and biological control of greenhouse whitefly. *Environmental Entomology*, **10**: 885–888.

Osman, A.A. and G. Zohdi. 1976. Suppression of the spider mites on cotton with mass releases of *Amblyseius gossipi* (El-Badry). *Zeitschrift für Angewandte Entomologie*, **81**: 245–248.

Osman, G.Y., A.M. Mohamed, and K. Jamel Al-Layl. 1992. Studies on molluscicidal activity of different preparations of *Bacillus thuringiensis* as biocidal agents on *Biomphalaria alexandrina* snails as vectors of schistosomiasis (bilharziasis) in Saudi Arabia. *Anzeiger für Schädlingskunde, Pflanzenschutz, Umweltschutz*, **65**: 67–70.

Ostlie, K.R. and L.P. Pedigo. 1987. Incorporating pest survivorship into economic thresholds. *Bulletin of the Entomological Society of America*, **33**(2): 98–102.

Overmeer, W.P.J. 1985. Rearing and handling, pp. 161–170. *In* Helle, W. and M.W. Sabelis, (eds.). *Spider Mites: Their Biology, Natural Enemies and Control, Vol. 1B.* Elsevier, Amsterdam, The Netherlands.

Pacala, S.W. and M.P. Hassell. 1991. The persistence of host-parasitoid associations in patchy environments. II. Evaluation of field data. *American Naturalist*, **138**: 584–605.

Pacala, S.W., M.P. Hassell, and R.M. May. 1990. Host-parasitoid associations in patchy environments. *Nature*, **344**: 150–153.

Padidam, M. 1991. Rational deployment of *Bacillus thuringiensis* strains for control of insect pests in India. *Current Science*, **60**: 464–465.

Paine, R.T. 1966. Food web complexity and species diversity. *American Naturalist*, **100**: 65–75.

Paine, R.T. 1974. Intertidal community structure. *Oecologia*, **15**: 93–120.

Paine, R.T. 1980. Food webs: linkage, interaction strength and community infrastructure. *Journal of Animal Ecology*, **49**: 667–685.

Pak, G.A. and T.G. van Heiningen. 1985. Behavioural variations among strains of *Trichogramma* spp.: Adaptability to field-temperature conditions. *Entomologia Experimentalis et Applicata*, **38**: 3–13.

Pantone, D.J., W.A. Williams, and A.R. Maggenti. 1989. An alternative approach for evaluating the efficacy of potential biocontrol agents of weeds. 1. Inverse linear model. *Weed Science*, **37**: 771–777.

Papaj, D.R. and L.E.M. Vet. 1990. Odor learning and foraging success in the parasitoid, *Leptopilina heterotoma*. *Journal of Chemical Ecology*, **16**: 3137–3150.

Park, T. 1948. Experimental studies of interspecies competition. I. Competition between populations of the flour beetles, *Tribolium confusum* Duval and *Tribolium castaneum* Herbst. *Ecological Monographs*, **18**: 265–308.

Parker, C.A., A.D. Rovira, K.J. Moore, P.T.W. Wong, and J.F. Kollmorgen, (eds.). 1983. *Ecology and Management of Soilborne Plant Pathogens.* American Phytopathological Society, St. Paul, Minnesota, U.S.A.

Parker, P.E. 1991. Nematodes as biological control agents of weeds, pp. 58–68. *In* TeBeest, D.O., (ed.). *Microbial Control of Weeds.* Chapman and Hall, New York.

Parkman, J.P., W.G. Hudson, J.H. Frank, K.B. Nguyen, and G.C. Smart, Jr. 1993. Establishment and persistence of *Steinernema scapterisci* (Rhabditida: Steinernematidae) in field populations of *Scapteriscus* spp. mole crickets (Orthoptera: Gryllotalpidae). *Journal of Entomological Science*, **28**: 182–190.

Parrish, J.D. and S.B. Saila. 1970. Interspecific competition, predation and species diversity. *Journal of Theoretical Biology*, **27**: 207–220.

Pasqualini, E. and C. Malavolta. 1985. Possibility of natural limitation of *Panonychus ulmi* (Koch) (Acarina, Tetranychidae) on apple in Emilia-Romagna. *Bolletino dell'Istituto di Entomologia 'Guido Grandi' della Universita degli Studi di Bologna*, **39**: 221–230.

Pasteels, J.M., J.-C. Grégoire, and M. Rowell-Rahier. 1983. The chemical ecology of defense in arthropods. *Annual Review of Entomology*, **28**: 263–289.

Pathak, J.P.N., (ed.). 1993. *Insect Immunity*. Kluwer Academic Pub., Boston, Massachusetts, U.S.A.

Patterson, D.T. 1976. The history and distribution of five exotic weeds in North Carolina. *Castanea*, **41**: 177–180.

Patterson, R.S., P.G. Koehler, R.B. Morgan, and R.L. Harris, (eds.). 1981. *Status of Biological Control of Filth Flies*. USDA/SEA publication, New Orleans, Louisiana, U.S.A.

Payne, C.C. 1986. Insect pathogenic viruses as pest control agents, pp. 183–200. *In* Franz, J.M., (ed.). *Biological Plant and Health Protection: Biological Control of Plant Pests and of Vectors of Human and Animal Diseases*. International Symposium of the Akademie der Wissenschaften und der Literatur, Mainz, November 15–17th, 1984 at Mainz and Darmstadt. *Fortschritte der Zoologie*, **32**: 341 pp., Gustav Fischer Verlag, Stuttgart, Germany.

Payne, T.L., J.C. Dickens, and J.V. Richerson. 1984. Insect predator-prey coevolution via enantiomeric specificity in a kairomone-pheromone system. *Journal of Chemical Ecology*, **10**: 487–492.

Peferoen, M., H. Höfte, and W. Chungjatupornchai. 1989. Cloning and expression of *Bacillus thuringiensis* insecticidal proteins in new hosts: Applications for developing countries. *Israel Journal Entomology*, **23**: 185–188.

Peloquin, J.J. and E.G. Platzer. 1993. Control of root gnats (Sciaridae: Diptera) by *Tetradonema plicans* Hungerford (Tetradonematidae: Nematoda) produced by a novel culture method. *Journal of Invertebrate Pathology*, **62**: 79–86.

Pemberton, C.E. and H.F. Willard. 1918. A contribution to the biology of fruit-fly parasites in Hawaii. *Journal of Agricultural Research*, **15**: 419–465.

Pemberton, R.W. 1989. Insects attacking *Passiflora mollissima* and other *Passiflora* species: Field surveys in the Andes. *Proceedings of the Hawaiian Entomological Society*, **29**: 71–84.

Pemberton, R.W. and C.E. Turner. 1990. Biological control of *Senecio jacobaea* in northern California, an enduring success. *Entomophaga*, **35**: 71–77.

Pennacchio, F., S.B. Vinson, and E. Tremblay. 1993. Growth and development of *Cardiochiles nigriceps* Viereck (Hymenoptera, Braconidae) larvae and their synchronization with some changes of the hemolymph composition of their host, *Heliothis virescens* (F.) (Lepidoptera, Noctuidae). *Archives of Insect Biochemistry and Physiology*, **24**: 65–77.

Peng, G. and J.C. Sutton. 1991. Evaluation of microorganisms for biocontrol of *Botrytis cinerea* in strawberry. *Canadian Journal of Plant Pathology*, **13**: 247–257.

Peng, G., J.C. Sutton, and P.G. Kevan. 1992. Effectiveness of honey bees for applying the biocontrol agent *Gliocladium roseum* to strawberry flowers to suppress *Botrytis cinerea*. *Canadian Journal of Plant Pathology*, **14**: 117–129.

Penman, D.R., C.H. Wearing, E. Collyer, and W.P. Thomas. 1979. The role of insecticide-resistant phytoseiids in integrated mite control in New Zealand. *Recent Advances in Acarology*, **1**: 59–69.

Perera, P.A.C.R., M.P. Hassell, and H.C.J. Godfray. 1988. Population dynamics of the coconut caterpillar, *Opisina arenosella* Walker (Lepidoptera: Xyloryctidae) in Sri Lanka. *Bulletin of Entomological Research*, **78**: 479–492.

Perfecto, I. 1991. Ants (Hymenoptera: Formicidae) as natural control agents of pests in irrigated maize in Nicaragua. *Journal of Economic Entomology*, **84**: 65–70.

Perry, J.N. and L.R. Taylor. 1986. Stability of real interacting populations in space and time: implications, alternatives and the negative binomial *k*. *Journal of Animal Ecology*, **55**: 1053–1068.

Petersen, J.J. 1986. Augmentation of early season releases of filth fly (Diptera: Muscidae) parasites (Hymenoptera: Pteromalidae) with freeze-killed hosts. *Environmental Entomology*, **15**: 590–593.

Phatak, S.C., D.R. Sumner, H.D. Wells, D.K. Bell, and N.C. Glaze. 1983. Biological control of yellow nut sedge with the indigenous rust fungus *Puccinia canaliculata*. *Science*, **219**: 1446–1447.

Phillips, P. 1987. Timing *Aphytis* release in coastal citrus. *Citrograph*, **72**(7): 128–131.

Pickering, J., J.D. Dutcher, and B.S. Ekbom. 1989. An epizootic caused by *Erynia neoaphidis* and *E. radicans* (Zygomycetes, Entomophthoraceae) on *Acyrthosiphon pisum* (Hom., Aphididae) on legumes under overhead irrigation. *Journal of Applied Entomology*, **107**: 331–333.

Pickett, A.D. 1965. The influence of spray programs on the fauna of apple orchards in Nova Scotia. XIV. Supplement to II. Oystershell scale, *Lepidosaphes ulmi* (L.). *Canadian Entomologist*, **97**: 816–821.

Pickett, C.H., L.T. Wilson, and D.L. Flaherty. 1990. The role of refuges in crop protection, with reference to plantings of French prune trees in a grape agroecosystem, pp. 151–165. *In* Bostanian, N.J., L.T. Wilson, and T.J. Dennehy. *Monitoring and Integrated Management of Arthropod Pests of Small Fruit Crops*. Intercept, Andover, U.K.

Pierce, N.E., R.L. Kitching, R.C. Buckley, M.F.J. Taylor, and K.F. Benbow. 1987. The costs and benefits of cooperation between the Australian lycaenid butterfly, *Jalmenus evagoras* and its attendant ants. *Behavioral Ecology and Sociobiology*, **21**: 237–248.

Pimentel, D. 1963. Introducing parasites and predators to control native pests. *Canadian Entomologist*, **95**: 785–792.

Pimentel, D. (ed.). 1991. *CRC Handbook of Pest Management in Agriculture, 2nd edition.* CRC Press, Boca Raton, Florida, U.S.A.

Pimentel, D. 1993. Habitat factors in new pest invasions, pp. 165–181 *In* Kim, K.C. and B.A. McPheron, (eds.). *Evolution of Insect Pests, Patterns of Variation.* John Wiley and Sons, Inc. New York.

Pimentel, D., C. Glenister, S. Fast, and D. Gallahan. 1984. Environmental risks of biological pest controls. *Oikos*, **42**: 283–290.

Pimentel, D., C. Kirby, and A. Shroff. 1993. The relationship between "cosmetic standards" for foods and pesticide use, pp. 85–105. *In* Pimentel, D. and H. Lehman, (eds.). *The Pesticide Question: Environment, Economics, and Ethics.* Chapman and Hall, New York.

Pinto, J.D. and R. Stouthamer. 1994. Systematics of the *Trichogrammatidae* with emphasis on *Trichogramma*, pp. 1–36. *In* Wajnberg, E. and S.A. Hassan, (eds.). *Biological Control With Egg Parasitoids.* Commonwealth Agricultural Bureaux, Wallingford, U.K.

Pinto, J.D., D.J. Kazmer, G.R. Platner, and C.A. Sassaman. 1992. Taxonomy of the *Trichogramma minutum* complex (Hymenoptera: Trichogrammatidae): Allozymic variation and its relationship to reproductive and geographic data. *Annals of the Entomological Society of America*, **85**: 413–422.

Pivnick, K.A. 1993. Diapause initiation and pupation site selection of the braconid parasitoid *Microplitis mediator* (Haliday): A case of manipulation of host behaviour. *Canadian Entomologist*, **125**: 825–830.

Podgwaite, J.D. 1986. Effects of insect pathogens on the environment, pp. 279–287. *In* Franz, J.M., (ed.). *Biological Plant and Health Protection: Biological Control of Plant Pests and of Vectors of Human and Animal Diseases.* International Symposium of the Akademie der Wissenschaften und der Literatur, Mainz, November 15–17th, 1984 at Mainz and Darmstadt. *Fortschritte der Zoologie*, **32**, 341 pp., Gustav Fischer Verlag, Stuttgart, Germany.

Podgwaite, J.D. and R.B. Bruen. 1978. Procedures for the microbial examination of production batch preparations of the nuclear polyhedrosis virus (Baculovirus) of the gypsy moth, *Lymantria dispar*, L. *U.S. Department of Agriculture, Forest Service, General Technical Report NE-38.*

Podgwaite, J.D. and H.M. Mazzone. 1981. Development of insect viruses as pesticides: the case of the gypsy moth (*Lymantria dispar* L.) in North America. *Protection Ecology*, **3**: 219–227.

Podoler, H. and D. Rogers. 1975. A new method for the identification of key factors from life-table data. *Journal of Animal Ecology*, **44**: 85–114.

Poehling, H.-M. 1989. Selective application strategies for insecticides in agricultural crops, pp. 151–175. *In* Jepson, P.C., (ed.). *Pesticides and Non-Target Invertebrates.* Intercept, Wimborne, U.K.

Poinar, G.O., Jr. 1965. The bionomics and parasite development of *Tripius sciarae* (Bovien) (Sphaerulariidae: Aphelenchoidea), a nematode parasite of sciarid flies (Sciaridae: Diptera). *Parasitology*, **55**: 559–569.

Poinar, G.O., Jr. 1979. *Nematodes for Biological Control of Insects.* CRC Press, Inc., Boca Raton, Florida, U.S.A.

Poinar, G.O. 1986. Entomphagous nematodes, pp. 95–121. *In* Franz, J M., (ed.). *Biological Plant and Health Protection: Biological Control of Plant Pests and of Vectors of Human and Animal Diseases.* International Symposium of the Akademie der Wissenschaften und der Literatur, Mainz, November 15–17th, 1984 at Mainz and Darmstadt. *Fortschritte der Zoologie*, **32**, 341 pp., Gustav Fischer Verlag, Stuttgart, Germany.

Poinar, G.O., Jr. 1990. Taxonomy and biology of Steinernematidae and Heterorhabditidae, pp. 23–61. *In* Gaugler, R. and H.K. Kaya, (eds.). *Entomopathogenic Nematodes in Biological Control.* CRC Press, Inc., Boca Raton, Florida, U.S.A.

Pointier, J.P. and F. McCullough. 1989. Biological control of the snail hosts of *Schistosoma mansoni* in the Caribbean area using *Thiara* spp. *Acta Tropica*, **46**: 147–155.

Pointier, J.P., C. Balzan, P. Chrosciechowski, and R.N. Incani. 1991. Limiting factors in biological control of the snail intermediate hosts of *Schistosoma mansoni* in Venezuela. *Journal of Medical and Applied Malacology*, **3**: 53–67.

Pollard, E., K.H. Lakhani, and P. Rothery. 1987. The detection of density-dependence from a series of annual censuses. *Ecology*, **68**: 2046–2055.

Poolman Simons, M.T.T., B.P. Sukerkropp, L.E.M. Vet, and G. de Moed. 1992. Comparison of learning in related generalist and specialist eucoilid parasitoids. *Entomologia Experimentalis et Applicata*, **64**: 117–124.

Popov, N.A., I.A. Zabudskaja, and I.G. Burikson. 1987. The rearing of Encarsia in biolaboratories in greenhouse combines. *Zashchita Rastenii*, **6**: 33.

Port, C.M. and N.E.A. Scopes. 1981. Biological control by predatory mites (*Phytoseiulus persimilis* Athias-Henriot) of red spider mite (*Tetranychus urticae* Koch) infesting strawberries grown in "walk-in" plastic tunnels. *Plant Pathology*, **30**: 95–99.

Possee, R.D., C.J. Allen, P.F. Entwistle, L.R. Cameron, and D.H.L. Bishop. 1990. Field trials of genetically engineered baculovirus insecticides, pp. 50–60. *In* Anon. *Risk Assessment in Agricultural Biotechnology: Proceedings of the International Conference.* August, 1988, Oakland, California, U.S.A.

Potter, M.F., M.P. Jensen, and T.F. Watson. 1982. Influence of sweet bait-*Bacillus thuringiensis* var. *kurstaki* combinations on adult tobacco budworm (Lepidoptera: Noctuidae). *Journal of Economic Entomology*, **75**: 1157–1160.

Pottinger, R.P. and E.J. LeRoux. 1971. The biology and dynamics of *Lithocolletis blancardella* (Lepidoptera: Gracillariidae) on apple in Quebec. *Memoirs of the Entomological Society of Canada*, **77**: 1–437.

Powell, W. and A.F. Wright. 1988. The abilities of the aphid parasitoids *Aphidius ervi* Haliday and *A. rhopalosiphi* De Stefani Perez (Hymenoptera: Braconidae) to transfer between different known host species and the implication for the use of alternative hosts in pest control strategies. *Bulletin of Entomological Research*, **78**: 683–693.

Powell, W. and A.F. Wright. 1992. The influence of host food plants on host recognition by four aphidiine parasitoids (Hymenoptera: Braconidae). *Bulletin of Entomological Research*, **81**: 449–453.

Prasad, S.S.S.V., K.V.B.R. Tilak, and K.G. Gollakota. 1972. Role of *Bacillus thuringiensis* var. *thuringiensis* on the larval survivability and egg hatching of *Meloidogyne* spp., the causative agent of root knot disease. *Journal of Invertebrate Pathology*, **20**: 377–378.

Prasad, Y.K. 1989. The role of natural enemies in controlling Icerya purchasi in south Australia. *Entomophaga*, **34**: 391–395.

Prescott, H.W. 1960. Suppression of grasshoppers by nemestrinid parasites (Diptera). *Annals of the Entomological Society of America*, **53**: 513–521.

Price, P.W. 1970. Trail odours: Recognition by insects parasitic in cocoons. *Science*, **170**: 546–547.

Price, P.W. 1972. Methods of sampling and analysis for predictive results in the introduction of entomophagous insects. *Entomophaga*, **17**: 211–222.

Price, P.W. 1986. Ecological aspects of host plant resistance and biological control: Interactions among three trophic levels, pp. 11–30. *In* Boethel, D.J. and R.D. Eikenbary, (eds.). *Interactions of Plant Resistance and Parasitoids and Predators of Insects*. Ellis Horwood Limited, John Wiley and Sons, New York.

Price, P.W. 1990. Evaluating the role of natural enemies in latent and eruptive species: new approaches in life table construction, pp. 221–232. *In* Watt, A.D., S.R. Leather, M.D. Hunter, and N.A.C. Kidd, (eds.). *Population Dynamics of Forest Insects*. Intercept, Andover, U.K..

Price, P.W., C.E. Bouton, P. Gross, B.A. McPheron, J.N. Thompson, and A.E. Weis. 1980. Interactions among three trophic levels: influence of plants on interactions between herbivores and natural enemies. *Annual Review of Ecology and Systematics*, **11**: 41–65.

Prinsloo, H.E. 1960. Parasitiese mikro-organismes by die bruinsprinkaan *Locustana pardalina* (Walk.). *Suid-Afrikaanse Tydskrif vir Landbouwwetenskap*, **3**: 551–560.

Prokopy, R.J. and R.P. Webster. 1978. Oviposition-deterring pheromone of *Rhagoletis pomonella*, a kairomone for its parasitoid *Opius lectus*. *Journal of Chemical Ecology*, **4**: 481–494.

Prokopy, R.J., S.A. Johnson, and M.T. O'Brien. 1990. Second-stage integrated management of apple arthropod pests. *Entomologia Experimentalis et Applicata*, **54**: 9–19.

Prokrym, D.R., D.A. Andow, J.A. Ciborowski, and D.D. Sreenivasam. 1992. Suppression of *Ostrinia nubilalis* by *Trichogramma nubilale* in sweet corn. *Entomologia Experimentalis et Applicata*, **64**: 73–85.

Pschorn-Walcher, H. 1963. Historical-biogeographical conclusions from host-parasite associations in insects. *Zeitschrift für Angewandte Entomologie*, **51**: 208–214. (in German)

Purvis, G. and J.P. Curry. 1984. The influence of weeds and farmyard manure on the activity of carabidae and other ground-dwelling arthropods in a sugar beet crop. *Journal of Applied Ecology*, **21**: 271–283.

Qiu, H.G., L.F. He, D.C. Ding, J.L. Wang, Z.L. Qiu, and B.J. Shen. 1988. Effect of diet on the kairomonal activity of frass from *Ostrinia furnacalis* larvae to the parasitoid *Macrocentrus linearis*. *Contributions from Shanghai Institute of Entomology*, **8**: 77–83. (in Chinese)

Quezada, J.R. and P. DeBach. 1973. Bioecological and population studies of the cottony-cushion scale, Icerya purchasi Mask., and its natural enemies, *Rodolia cardinalis* Muls., and *Cryptochaetum iceryae* Will., in southern California. *Hilgardia*, **41**: 631–688.

Quinlan, R.J. 1990. Registration requirements and safety considerations for microbial pest control agents in the European Economic Community, pp. 11–18. *In* Laird, M., L.A. Lacey and E.W. Davidson, (eds.). *Safety of Microbial Insecticides*. CRC Press, Inc., Boca Raton, Florida, U.S.A.

Ramadan, M.M., T.T.Y. Wong, and J.W. Beardsley Jr. 1989. Insectary production of *Biosteres tryoni* (Cameron) (Hymenoptera: Braconidae), a larval parasitoid of *Ceratitis capitata* (Weidemann) (Diptera: Tephritidae). *Proceedings of the Hawaiian Entomological Society*, **29**: 41–48.

Ramakers, P.M.J. and R.A.Samson. 1984. *Aschersonia aleyrodis*, a fungal pathogen of whitefly, II. Application as a biological insecticide in glasshouses. *Zeitschrift für Angewandte Entomologie*, **97**: 1–8.

Rao, V.P. 1971. *Biological Control of Pests in Fiji*. Miscellaneous Publications, Commonwealth Institute of *Biological Control* No. 2.

Rao, V.P., M.A. Ghani, T. Sankaran, and K.C. Mathur. 1971. *A Review of Biological Control of Insects and Other Pests in South-East Asia and the Pacific Region*. Commonwealth Institute of *Biological Control* Technical Communication No. 6. Commonwealth Agricultural Bureaux, Farnham Royal, U.K.

Ratcliffe, N.A. 1982. Cellular defense reactions of insects, pp. 223–244. *In* Frank, W., (ed.). *Immune Reactions to Parasites.* Fischer-Verlag, New York.

Ratcliffe, N.A. 1993. Cellular defense responses of insects: Unresolved problems, pp. 267–304. *In* Beckage, N.E., S.N. Thompson, and B.A. Federici, (eds.). *Parasites and Pathogens of Insects, Volume I. Parasites.* Academic Press, New York.

Rath, A.C., S. Pearn, and D. Worladge. 1990. An economic analysis of production of *Metarhizium anisopliae* for control of the subterranean pasture pest *Adoryphorus couloni*, p. 13. *In* Pinnock, D.E., (ed.). *Vth International Colloquium on Invertebrate Pathology and Microbial Control.* Adelaide, Australia, 20–24, August, 1990.

Rathman, R.J., M.W. Johnson, J.A. Rosenheim, and B.E. Tabashnik. 1990. Carbamate and pyrethroid resistance in the leafminer parasitoid *Diglyphus begini* (Hymenoptera: Eulophidae). *Journal of Economic Entomology*, **83**: 2153–2158.

Raulston, J.R., S.D. Pair, J. Loera, and H.E. Cabanillas. 1992. Prepupal and pupal parasitism of *Helicoverpa zea* and *Spodoptera frugiperda* (Lepidoptera: Noctuidae) by *Steinernema* sp. in corn fields in the Lower Rio Grande Valley. *Journal of Economic Entomology*, **85**: 1666–1670.

Raupp, M.J., R.G. Van Driesche, and J.A. Davidson. 1993. *Biological Control of Insect and Mite Pests of Woody Landscape Plants: Concepts, Agents and Methods.* University of Maryland Cooperative Extension Service, College Park, Maryland, U.S.A.

Rausher, M.D. 1992. Natural selection and the evolution of plant-insect interactions, pp. 20–88. *In* Roitberg, B.D. and M.B. Isman, (eds.). *Insect Chemical Ecology, an Evolutionary Approach.* Chapman and Hall, New York.

Read, D.C. 1962. Notes on the life history of *Aleochara bilineata* (Gyll.) (Coleoptera: Staphylinidae), and on its potential value as a control agent for the cabbage maggot, *Hylemya brassicae* (Bouché) (Diptera: Anthomyiidae). *Canadian Entomologist*, **94**: 417–424.

Reardon, R.C. 1981. Alternative controls, 6.1 parasites, pp. 299–421. *In* Doane, C.C. and M.L. McManus, (eds.). *The Gypsy Moth: Research Toward Integrated Pest Management.* U.S. Forest Service Technical Bulletin No. 1584, United States Department of Agriculture, Washington, D.C., U.S.A.

Reardon, R.C. and J.D. Podgwaite. 1976. Disease-parasitoid relationships in natural populations of *Lymantria dispar* (Lep.: Lymantriidae) in the northeastern United States. *Entomophaga*, **21**: 333–341.

Redborg, K.E. 1983. A mantispid larva can preserve its spider egg prey: evidence for an aggressive allomone. *Oecologia*, **58**: 230–231.

Reddingius, J. 1971. Gambling for existence: A discussion of some theoretical problems in animal population ecology. *In Acta Theoretica Climum*, added to Acta Biotheoretica 20, Leiden: Brill, The Netherlands.

Reddingius, J. and P.J. den Boer. 1970. Simulation experiments illustrating stabilization of animal numbers by spreading of risk. *Oecologia*, **5**: 240–248.

Reddingius, J. and P.J. Den Boer. 1989. On the stabilization of animal numbers. Problems of testing. 1. Power estimates and estimation errors. *Oecologia*, **78**: 1–8.

Reed, D.K., S.D. Kindler, and T.L. Springer. 1992. Interactions of Russian wheat aphid, a hymenopterous parasitoid and resistant and susceptible slender wheatgrasses. *Entomologia Experimentalis et Applicata*, **64**· 239–246.

Reed-Larson, D.A. and J.J. Brown. 1990. Embryonic castration of the codling moth, *Cydia pomonella*, by an endoparasitoid, *Ascogaster quadridentata*. *Journal of Insect Physiology*, **36**: 111–118.

Reeve, J.D. and W.W. Murdoch. 1985. Aggregation by parasitoids in the successful control of the California red scale: a test of theory. *Journal of Animal Ecology*, **54**: 797–816.

Régnière, J. 1984. Vertical transmission of disease and population dynamics of insects with discrete generations: A model. *Journal of Theoretical Biology*, **107**: 287–301.

Reicheldefer, K.H. 1979. *Economic feasibility of a biological control technology: Using a parasitic wasp, Pediobius foveolatus, to manage Mexican bean beetle on soybeans.* Economics, Statistics and Cooperatives Service, Agricultural Economic Report No. 430. United Stated Department of Agriculture, Washington, D.C., U.S.A.

Remaudière, G. and S. Keller. 1980. Revision systematique des genres d'Entomophthoraceae a potentialite entomopathogene. *Mycotaxon*, **11**: 323–338. (in French)

Remillet, M. and C. Laumond. 1991. Sphaerularioid nematodes of importance in agriculture, pp. 967–1024. *In* Nickel, W.R., (ed.). *Manual of Agricultural Nematology.* Marcel Dekker, Inc., New York.

Renwick, J.A.A. and F.S. Chew. 1994. Oviposition behavior in Lepidoptera. *Annual Review of Entomology*, **39**: 377–400.

Rhoades, H.L. 1985. Comparison of fenamiphos and *Arthrobotrys amerospora* for controlling plant nematodes in central Florida. *Nematropica*, **15**: 1–7

Rice, M.E. and G.E. Wilde. 1988. Experimental evaluation of predators and parasitoids in suppressing greenbugs (Homoptera: Aphididae) in sorghum and wheat. *Environmental Entomology*, **17**: 836–841.

Rice, M.E. and G.E. Wilde. 1989. Antibiosis of sorghum on the convergent lady beetle (Coleoptera: Coccinellidae), a third-trophic level predator of the greenbug (Homoptera: Aphididae). *Journal of Economic Entomology*, **82**: 570–573.

Richards, O.W. and N. Waloff. 1954. Studies on the biology and population dynamics of British grasshoppers. *Anti-locust Bulletin* No. 17.

Richards, O.W., N. Waloff, and J.P. Spradberry. 1960. The measurement of mortality in an insect population in which recruitment and mortality widely overlap. *Oikos*, **11**: 306–310.

Richter, V.A. 1988. Family Nemestrinidae. Catalogue of Palaearctic Diptera 5: 171–181. Zoological Institute, Academy of Science of the U.S.S.R., 199164 Leningrad, U.S.S.R.

Ricker, W.E. 1973. Linear regressions in fishery research. *Journal of the Fisheries Research Board of Canada*, **30**: 409–434.

Ricker, W. E. 1975. A note concerning Professor Jolicoeur's comments. *Journal of the Fisheries Research Board of Canada*, **32**: 1494–1498.

Ridgway, R.L. and S.B. Vinson, (eds.). 1977. *Biological Control by Augmentation of Natural Enemies*. Plenum Press, New York.

Ridgway, R.L. and S.L. Jones. 1969. Inundative releases of *Chysopa carnea* for control of *Heliothis* on cotton. *Journal of Economic Entomology*, **62**: 177–180.

Riechert, S.E. and L. Bishop. 1990. Prey control by an assemblage of generalist predators: Spiders in garden test systems. *Ecology*, **71**: 1441–1450.

Riechert, S.E. and T. Lockley. 1984. Spiders as biological control agents. *Annual Review of Entomology*, **29**: 299–320.

Riley, C.V. 1885. Fourth report of the U.S. Entomological Commission, p. 323. *In* Scudder, S.H. (1889). *Butterflies of Eastern United States and Canada*. Cambridge University Press, U.K.

Rishbeth, J. 1963. Stump protection against Fomes annosus. III. Inoculation with *Peniophora gigantea*. *Annals of Applied Biology*, **52**: 63–77.

Rizki, R.M. and T.M. Rizki. 1990. Parasitoid virus-like particles destroy *Drosophila cellular* immunity. *Proceedings of the National Academy of Sciences of the United States of America*, **87**(21): 8388–8392.

Roberts, D.W. and S.P. Wraight. 1986. Current status on the use of insect pathogens as biological agents in agriculture: Fungi, pp. 510–513. *In* Samson, R.A., J.M. Vlak, and D. Peters, (eds.). *Fundamental and Applied Aspects of Invertebrate Pathology*. Proc. 4th International Colloquium of Invertebrate Pathology, Veldhoven, The Netherlands, August 18–22, 1986.

Robin, M.R. and W.C. Mitchell. 1987. Sticky trap for monitoring leafminers *Liriomyza sativae* and *Liriomyza trifolii* (Diptera: Agromyzidae) and their associated hymenopterous parasites in watermelon. *Journal of Economic Entomology*, **80**: 1345–1347.

Roehrdanz, R.L., D.K. Reed, and R.L. Burton. 1993. Use of polymerase chain reaction and arbitrary primers to distinguish laboratory-raised colonies of parasitic Hymenoptera. *Biological Control*, **3**: 199–206.

Rogers, C.E. 1985. Extrafloral nectar: Entomological implications. *Bulletin of the Entomological Society of America*, **31**: 15–20.

Roget, D.K. and A.D. Rovira. 1987. A review of the effect of tillage on the cereal cyst nematode. *Wheat Research Council of Australia, Workshop Report Series* No. 1: 31–35.

Rogoff, M.H. 1982. Regulatory safety data requirements for registration of microbial pesticides, pp. 645–679. *In* Kurstak, E., (ed.). *Microbial and Viral Pesticides*. Marcel Dekker, Inc., New York.

Rohani, P., H.C.J. Godfray, and M.P. Hassell. 1994. Aggregation and the dynamics of host-parasitoid systems: A discrete-generation model with within-generation redistribution. *American Naturalist*, **144**: 491–509.

Roland, J. 1990. Interaction of parasitism and predation in the decline of winter moth in Canada, pp. 289–302. *In* Watt, A.D., S.R. Leather, M.D. Hunter, and N.A.C. Kidd, (eds.). *Population Dynamics of Forest Insects*. Intercept, Andover, U.K.

Roland, J. 1994. After the decline: What maintains low winter moth density after successful biological control? *Journal of Animal Ecology*, **63**: 392–398.

Roland, J., W.G. Evans, and J.H. Myers. 1989. Manipulation of oviposition patterns of the parasitoid *Cyzenis albicans* (Tachinidae) in the field using plant extracts. *Journal of Insect Behavior*, **2**: 487–503.

Room, P.M. 1986. Biological control of floating weeds in Australia. *Biotrop Special Publication*, **24**: 51–54.

Room, P.M. 1990. Ecology of a simple plant-herbivore system: biological control of *Salvinia*. *Trends in Ecology and Evolution*, **5**(3): 74–79.

Room, P.M. and P.A. Thomas. 1985. Nitrogen and establishment of a beetle for biological control of the floating weed *Salvinia* in Papua New Guinea. *Journal of Applied Ecology*, **22**: 139–156.

Root, R.B. 1973. Organization of plant-arthropod association in simple and diverse habitats: the fauna of collards (*Brassica oleracea*). *Ecological Monographs*, **43**: 95–124.

Rose, M. 1990. Rearing and mass rearing of natural enemies, pp. 263–287 *In* Rosen, D., (ed.). *Armored Scale Insects: Their Biology, Natural Enemies and Control, Volume 4B*. Elsevier, Amsterdam, The Netherlands.

Rose, M. and P. DeBach. 1992. Biocontrol of *Parabemisia myricae* (Kuwana) (Homoptera: Aleyrodidae) in California. *Israel Journal of Entomology*, **25–26**: 73–95.

Rosen, D. and P. De Bach. 1979. *Species of Aphytis of the World (Hymenoptera: Aphelinidae).* Israel Universities Press, Jerusalem, and Junk, The Hague, The Netherlands.

Rosenheim, J.A. and M.A. Hoy. 1986. Intraspecific variation in levels of pesticide resistance in field populations of a parasitoid, *Aphytis melinus* (Hymenoptera: Aphelinidae): The role of past selection pressures. *Journal of Economic Entomology,* **79**: 1161–1173.

Rosenheim, J.A. and M.A. Hoy. 1988. Genetic improvement of a parasitoid biological control agent: Artificial selection for insecticide resistance in *Aphytis melinus* (Hymenoptera: Aphelinidae). *Journal of Economic Entomology,* **81**: 1539–1550.

Rosenheim, J.A. and D. Rosen. 1991. Foraging and oviposition decisions in the parasitoid *Aphytis lingnanensis*: Distinguishing the influences of egg load and experience. *Journal of Animal Ecology,* **60**: 873–893.

Ross, J. and A.M. Tittensor. 1986. The establishment and spread of myxomatosis and its effect on rabbit populations. *Philosophical Transactions of the Royal Society of London, Series B,* **314**: 599–606.

Rothenburger, W. and H. Sautter. 1987. Profitability of the production and sale of beneficial arthropods for biological control in greenhouse vegetable production, pp. 174–194. Schriftenreihe des Bundesministers fur Ernahrung, Landwirtschaft und Forsten, A (Angewandte Wissenschaft), German Federal Republic No. 344. *In* Anon. *Possibilities and Limitations of Biological Plant Protection.* A conference held in Bom-Rottgen, German Federal Republic, on 25–26, February, 1986, Lehrstuhl für Wirtschaftslehre des Gartenbaues, Tech. University München-Weihenstephan, 8000 Munich, German Federal Republic. (in German)

Rotheray, G.E. and P. Barbosa. 1984. Host related factors affecting oviposition behavior in *Brachymeria intermedia. Entomologia Experimentalis et Applicata,* **35**: 141–145.

Rotheray, G.E. and P. Martinat. 1984. Searching behaviour in relation to starvation of *Syrphus ribesii. Entomologia Experimentalis et Applicata,* **36**: 17–21.

Rothschild, G. 1966. A study of a natural population of *Conomelus anceps* Germar (Homoptera: Delphacidae) including observation on predation using the precipitin test. *Journal of Animal Ecology,* **35**: 413–434.

Roughgarden, J. and M. Feldman. 1975. Species packing and predator pressure. *Ecology,* **56**: 489–492.

Roush, R.T. 1990. Genetic considerations in the propagation of entomophagous species, pp. 373–387. *In* Baker, R.R. and P.E Dunn, (eds.). *New Directions in Biological Control: Alternatives for Suppressing Agricultural Pests and Diseases.* Alan R. Liss, Inc., New York, 837 pp.

Roush, R.T. and M.A. Hoy. 1981. Genetic improvement of *Metaseiulus occidentalis*: Selection with methomyl, dimethoate, and carbaryl and genetic analysis of carbaryl resistance. *Journal of Economic Entomology,* **74**: 138–141.

Royama, T. 1981. Evaluation of mortality factors in insect life table analysis. *Ecological Monographs,* **5**: 495–505.

Ruesink, W.G. 1975. Estimating time-varying survival of arthropod life stages from population density. *Ecology,* **56**: 244–247.

Russell, E.P. 1989. Enemies hypothesis: a review of the effect of vegetational diversity on predatory insects and parasitoids. *Environmental Entomology,* **18**: 590–599.

Ruth, J. and E.F. Dwumfour. 1989. Laboratory studies on the suitability of some aphid species as prey for the predatory flower bug *Anthocoris gallarum-ulmi* (DeG.) (Het., Anthocoridae). *Journal of Applied Entomology,* **108**: 321–327.

Ryan, J., M.F. Ryan, and F. McNaeidhe. 1980. The effect of interrow plant cover on populations of the cabbage root fly *Delia brassicae* (Wied.). *Journal of Applied Ecology,* **17**: 31–40.

Ryan, R.B. 1988. Evidence for mortality in addition to successful parasitism of needlemining larch casebearer (Lepidoptera: Coleophoridae) larvae by *Agathis pumila* (Ratz.) (Hymenoptera: Braconidae). *Canadian Entomologist,* **120**: 1035–1036.

Sabelis, M.W. 1981. Biological control of two-spotted spider mites using phytoseiid predators. Part 1: Modeling the predator-prey interaction at the individual level. Agricultural Research Reports 910, pp. 226–242. Centre for Agricultural Publishing and Documentation, Waginengin, The Netherlands.

Sabelis, M.W. 1992. Predatory arthropods, pp. 225–264. *In* Crawley, M.J., (ed.). *Natural Enemies: The Population Biology of Predators, Parasites, and Diseases.* Blackwell Science Publishers, London.

Sabelis, M.W. and M. Dicke. 1985. Long-range dispersal and searching behaviour, pp. 141–160. *In* Helle, W. and M.W. Sabelis, (eds.). *Spider Mites, Their Biology, Natural Enemies and Control, Vol. 1B.* Elsevier, Amsterdam, The Netherlands.

Sabelis, M.W. and W.E.M. Laane. 1986. Regional dynamics of spider-mite populations that become extinct locally because of food source depletion and predation by phytoseiid mites (Acarina: Tetranychidae, Phytoseiidae), pp. 345–376. *In* Metz, J.A.J. and O. Diekmann, (eds.). *The Dynamics of Physiologically Structured Populations.* Springer-Verlag, New York.

Sabelis, M.W. and J. van der Meer. 1986. Local dynamics of the interaction between predatory mites and two-spotted spider mites, pp. 322–344. *In* Metz, J.A.J. and O. Diekmann, (eds.). *The Dynamics of Physiologically Structured Populations.* Springer-Verlag, New York.

Sabelis, M.W., F. van Alebeek, A. Bal, J. van Bilsen, T. van Heijningen, P. Kaizer, G. Kramer, H. Snellen, R. Veenebos, and J. Vogelezang. 1983. Experimental validation of a simulation model of the interaction between *Phytoseiulus persimilis* and *Tetranychus urticae* on cucumber. *International Organization for Biological Control of Noxious Animals and Plants, Western Palearctic Regional Section, Bulletin* **6**(3): 207–229.

Saik, J.E., L.A. Lacey, and C.M. Lacey. 1990. Safety of microbial insecticides to vertebrates—domestic animals and wildlife, pp. 115–132. *In* Laird, M., L.A. Lacey, and E.W. Davidson, (eds.) *Safety of Microbial Insecticides.* CRC Press, Inc., Boca Raton, Florida, U.S.A.

Sailer, R.I. 1978. Our immigrant insect fauna. *Bulletin of the Entomological Society of America,* **24**: 3–11.

Sailer, R.I. 1983. History of insect introductions, pp. 15–38. *In* Wilson, C.L. and C.L. Graham, (eds.). *Exotic Plant Pests and North American Agriculture.* Academic Press, New York.

Saito, T. 1988. Control of *Aphis gossypii* in greenhouses by a mycoinsecticidal preparation of *Verticillium lecanii* and the effect of chemicals on the fungus. *Japanese Journal of Applied Entomology and Zoology,* **32**: 224–227.

Samuels, K.D.Z., D.E. Pinnock., and R.M. Bull. 1990. Scarabeid (sic) larvae control in sugarcane using *Metarhizium anisopliae. Journal of Invertebrate Pathology,* **55**: 135–137.

Samways, M.J. 1986. Spatial and temporal population patterns of *Aonidiella aurantii* (Maskell) (Hemiptera: Diaspididae) parasitoids (Hymenoptera: Aphelinidae and Encyrtidae) caught on yellow sticky traps in citrus. *Bulletin of Entomological Research,* **76**: 265–274.

Samways, M.J. 1988. Comparative monitoring of red scale *Aonidiella aurantii* (Mask.) (Hom., Diaspididae) and its *Aphytis* spp. (Hym., Aphelinidae) parasitoids. *Journal of Applied Entomology,* **105**: 483–489.

Samways, M.J. 1990. Ant assemblage structure and ecological management in citrus and subtropical fruit orchards in southern Africa, pp. 570–587. *In* van der Meer, R.K., K. Jaffe and A. Cedeno, (eds.). *Applied Myrmecology, a World Perspective.* Westview Press, Boulder, Colorado, U.S.A.

Samways, M.J. 1994. *Insect Conservation Biology.* Chapman and Hall, New York.

Samways, M.J., M. Nel, and A.J. Prins. 1982. Ants (Hymenoptera: Formicidae) foraging in citrus trees and attending honeydew producing Homoptera. *Phytophylactica,* **14**: 155–157.

Sanford, G.B. 1926. Some factors affecting the pathogenicity of *Actinomyces scabies. Phytopathology,* **16**: 525–547.

Santos, G.L., D. Kageler, D.E. Gardner, L.W. Cuddihy, and C.P. Stone. 1992. Herbicidal control of selected alien plant species in Hawaii Volcanoes National Park, pp. 341–375. *In* Stone, C.P., C.W. Smith, and J.T. Tunison, (eds.). *Alien Plant Invasions in Native Ecosystems of Hawai'i: Management and Research.* University of Hawaii Press, Honolulu, Hawaii, U.S.A.

Saunders, D.S. 1966. Larval diapause of maternal origin. II. The effect of photoperiod and temperature on *Nasonia vitripennis. Journal of Insect Physiology,* **12**: 569–581.

Sawyer, R.C. 1990. Monopolizing the insect trade: Biological control in the USDA, 1888–1951. *Agricultural History,* **64**(2): 271–285.

Saxena, G. and K.G. Mukerji. 1988. Biological control of nematodes, pp. 113–127. *In* Mukerji, K.G., and K.L. Garg, (eds.). *Biocontrol of Plant Diseases, Vol. 1.* CRC Press, Inc., Boca Raton, Florida, U.S.A.

Sayre, R.M. 1980. Promising organisms for biocontrol of nematodes. *Plant Disease,* **64**: 526–532.

Sayre, R.M. and M.P. Starr. 1988. Bacterial diseases and antagonisms of nematodes, pp. 69–101. *In* Poinar, G.O., Jr., and H.-B. Jansson, (eds.). *Diseases of Nematodes, Vol. I.* CRC Press, Inc., Boca Raton, Florida, U.S.A.

Sayre, R.M. and D.E. Walter. 1991. Factors affecting the efficacy of natural enemies of nematodes. *Annual Review of Phytopathology,* **29**: 149–166.

Sayre, R.M. and W.P. Wergin. 1977. Bacterial parasite of a plant nematode: Morphology and ultrastructure. *Journal of Bacteriology,* **129**: 1091–1101.

Schaefer, P.W., R.J. Dysart, R.V. Flanders, T.L. Burger, and K. Ikebe. 1983. Mexican bean beetle (Coleoptera: Coccinellidae) larval parasite *Pediobius foveolatus* (Hymenoptera: Eulophidae) from Japan: Field release in the United States. *Environmental Entomology,* **12**: 852–854.

Scherff, R.H. 1973. Control of bacterial blight of soybean by *Bdellovibrio bacteriovorus. Phytopathology,* **63**: 400–402.

Schmidt, J.M. and J.J.B. Smith. 1986. Correlations between body angles and substrate curvature in the parasitoid wasp *Trichogramma minutum*: A possible mechanism of host radius measurement. *Journal of Experimental Biology,* **125**: 271–285.

Schmidt, J.M. and J.J.B. Smith. 1987. Measurement of host curvature by the parasitoid wasp *Trichogramma minutum,* and its effect on host examination and progeny allocation. *Journal of Experimental Biology,* **129**: 151–164.

Schmidt, J.M., R.T. Cardé, and L.E.M. Vet. 1993. Host recognition by *Pimpla instigator* F. (Hymenoptera: Ichneumonidae): Preferences and learned responses. *Journal of Insect Behavior,* **6**: 1–11.

Schnathorst, W.C. and D.E. Mathre. 1966. Cross-protection in cotton with strains of *Verticillium albo-atrum. Phytopathology* **56**: 1204–1209.

Scholz, D.and C. Höller. 1992. Competition for hosts between hyperparasitoids of aphids, *Dendrocerus laticeps* and

Dendrocerus carpenteri (Hymenoptera: Megaspilidae): The benefit of interspecific host discrimination. *Journal of Insect Behavior*, **5**: 289–300.

Schonbeck, H. 1988. Biological control of aphids on wild cherry. *Allgemeine Forstzeitschrift*, **34**: 944.

Schoonhoven, L.M. 1962. Diapause and the physiology of host-parasite synchronization in *Bupalus pinarius* L. (Geometridae) and *Eucarcelia rutilla* Vill. (Tachinidae). *Archives Neerlandais de Zoologie*, **15**: 111–173.

Schreiner, I. 1989. Biological control introductions in the Caroline and Marshall islands. *Proceedings of the Hawaiian Entomological Society*, **29**: 57–69.

Schroeder, D. 1983. Biological control of weeds, pp. 41–78. *In* Fletcher, W. W., (ed.). *Recent Advances in Weed Research*. Commonwealth Agricultural Bureaux, Slough, U.K.

Schroeder, D. and R.D. Goeden. 1986. The search for arthropod natural enemies of introduced weeds for biological control—In theory and practice. *Biocontrol News and Information*, **7**: 147–155.

Schroth, M.N. and J.G. Hancock. 1985. Soil antagonists in IPM systems, pp. 415–431. *In* Hoy, M.A. and D.C. Herzog, (eds.). *Biological Control in Agricultural IPM Systems*. Academic Press, New York.

Scopes, N.E.A. 1969. The potential of *Chrysopa carnea* as a biological control agent of *Myzus persicae* on glasshouse chrysanthemums. *Annals of Applied Biology*, **64**: 433–439.

Scriber, J.M. and R.C. Lederhouse. 1992. The thermal environment as a resource dictating geographic patterns of feeding specialization of insect herbivores, pp. 429–466. *In* Hunter, M.R., T. Ohgushi, and P.W. Price (eds.). *Effects of Resource Distribution on Animal-Plant Interactions*. Academic Press, New York.

Senger, S.E. and B.D. Roitberg. 1992. Effects of parasitism by *Tomicobia tibialis* Ashmead (Hymenoptera: Pteromalidae) on reproductive parameters of female pine engravers, *Ips pini* (Say). *Canadian Entomologist*, **124**: 509–513.

Sengonca, C. and B. Frings. 1989. Enhancement of the green lacewing *Chrysoperla carnea* (Stephens), by providing artificial facilities for hibernation. *Turkiye Entomoloji Dergisi*, **13**(4): 245–250.

Sengonca, C. and N. Leisse. 1989. Enhancement of the egg parasite *Trichogramma semblidis* (Auriv.) (Hym., Trichogrammatidae) for control of both grape vine moth species in the Ahr valley. *Journal of Applied Entomology*, **107**: 41–45.

Senthamizhselvan, M. and J. Muthukrishnan. 1989. Effect of parasitization by a gregarious and a solitary parasitoid on food consumption and utilization by *Porthesia scintillans* Walker (Lepidoptera: Lymantriidae) and *Spodoptera exigua* Hübner (Lepidoptera: Noctuidae). *Parasitology Research*, **76**: 166–170.

Shadduck, J.A., S. Singer, and S. Lause. 1980. Lack of mammalian pathogenicity of entomocidal isolates of *Bacillus sphaericus*. *Environmental Entomology*, **9**: 403–407.

Shah, M.A. 1982. The influence of plant surfaces on the searching behaviour of coccinellid larvae. *Entomologia Experimentalis et Applicata*, **31**: 377–380.

Shamim, M., M. Baig, R.K. Datta, and S.K. Gupta. 1994. Development of monoclonal antibody-based sandwich ELISA for the detection of nuclear polyhedra of nuclear polyhedrosis virus infection in *Bombyx mori* L. *Journal of Invertebrate Pathology*, **63**: 151–156.

Shanks, C.H., Jr. and F. Agudelo-Silva. 1990. Field pathogenicity and persistence of heterorhabditid and steinernematid nematodes (Nematoda) infecting black vine weevil larvae (Coleoptera: Curculionidae) in cranberry bogs. *Journal of Economic Entomology*, **83**: 107–110.

Shapas, T.J., W.E. Burkholder, and G.M. Boush. 1977. Population suppression of *Trogoderma glabrum* by using pheromone luring for protozoan pathogen dissemination. *Journal of Economic Entomology*, **70**: 469–474.

Shapiro, M. 1986. *In vivo* production of baculoviruses, pp. 31–61. *In* Granados, R.R. and B.A. Federici, (eds.). *The Biology of Baculoviruses: Volume II. Practical Application for Insect Control*. CRC Press, Inc., Boca Raton, Florida, U.S.A.

Shapiro, M. 1992. Use of optical brighteners as radiation protectants for gypsy moth (Lepidoptera: Lymantriidae) nuclear polyhedrosis virus. *Journal of Economic Entomology*, **85**: 1682–1686.

Shapiro, M. and J.L. Robertson. 1990. Laboratory evaluation of dyes as ultraviolet screens for the gypsy moth (Lepidoptera: Lymantriidae) nuclear polyhedrosis virus. *Journal of Economic Entomology*, **83**: 168–172.

Shapiro, M. and J.L. Robertson. 1992. Enhancement of gypsy moth (Lepidoptera: Lymantriidae) baculovirus activity by optical brighteners. *Journal of Economic Entomology*, **85**: 1120–1124.

Sharma, R.D. 1971. Studies on the plant parasitic nematode *Tylenchorhynchus dubius*. *Mededelingen Landbouwwetenschappen Wageningen*, **71-1**: 1–154.

Sheehan, W. 1986. Response by specialist and generalist natural enemies to agroecosystem diversification: a selective review. *Environmental Entomology*, **15**: 456–461.

Sheehan, W. and A.M. Shelton. 1989. Parasitoid response to concentration of herbivore food plants: Finding and leaving plants. *Ecology*, **70**: 993–998.

Sheehan, W., F.L. Wäckers, and W.J. Lewis. 1993. Discrimination of previously searched, host-free sites by *Microplitis croceipes* (Hymenoptera: Braconidae). *Journal of Insect Behavior*, **6**: 323–331.

Shepard, M. and D.L. Dahlman. 1988. Plant-induced stresses as factors in natural enemy efficacy, pp. 363–379. *In* Heinrichs, E.A., (ed.). *Plant Stress-Insect Interactions.* John Wiley and Sons, New York.

Shepard, M., H.R. Rapusas, and D.B. Estano. 1989. Using rice straw bundles to conserve beneficial arthropod communities in ricefields. *International Rice Research News*, **14**(5): 30–31.

Sheppard, A.W., J.-P. Aeschlimann, J.-L. Sagliocco, and J. Vitou. 1991. Natural enemies and population stability of the winter-annual *Carduus pycnocephalus* L. in Mediterranean Europe. *Acta Oecologia*, **12**: 707–726.

Shetlar, D.J., P.E. Suleman, and R. Georgis. 1988. Irrigation and use of entomogenous nematodes, *Neoaplectana* spp. and *Heterorhabditis heliothidis* (Rhabditida: Steinernematidae and Heterorhabditidae), for control of Japanese beetle (Coleoptera: Scarabaeidae) grubs in turfgrass. *Journal of Economic Entomology*, **81**: 1318–1322.

Shi, G.Z., Y.N. Zhou, J.S. Zhao, T. Li, M.L. Lian, R.Q. Chang, T.X. Li, S.J. Chen, X.L. Yang, and J.Q. Niou. 1988. The techniques of protection of the *Trichogramma* population in the fields. *Colloques de l'INRA*, **43**: 581–583.

Shu, S., P.D. Swedenborg, and R.L. Jones. 1990. A kairomone for *Trichogramma nubilale* (Hymenoptera: Trichogrammatidae): Isolation, identification, and synthesis. *Journal of Chemical Ecology*, **16**: 521–529.

Siddiqi, M.R. 1986. *Tylenchida Parasites of Plants and Insects.* Commonwealth Agricultural Bureaux International, Commonwealth Institute of Parasitology, Slough, U.K.

Siegel, J.P. and J.A. Shadduck. 1990a. Clearance of *Bacillus sphaericus* and *Bacillus thuringiensis* ssp. *israelensis* from mammals. *Journal of Economic Entomology*, **83**: 347–355.

Siegel, J.P. and J.A. Shadduck. 1990b. Mammalian safety of *Bacillus sphaericus*, pp. 321–331. *In* de Barjac, J. and D.J. Sutherland, (eds.). *Bacterial Control of Mosquitoes and Black Flies: Biochemistry, Genetics and Applications of Bacillus thuringiensis israelensis and Bacillus sphaericus.* Rutgers University Press, New Brunswick, New Jersey, U.S.A.

Siegel, J.P. and J.A. Shadduck. 1990c. Mammalian safety of *Bacillus thuringiensis israelensis*, pp. 202–217. *In* de Barjac, H. and D.J. Sutherland, (eds.). *Bacterial Control of Mosquitoes and Black Flies: Biochemistry, Genetics and Applications of Bacillus thuringiensis israelensis and Bacillus sphaericus.* Rutgers University Press, New Brunswick, New Jersey, U.S.A.

Siegel, J.P. and J.A. Shadduck. 1992. Testing the effects of microbial pest control agents on mammals, pp. 745–759. *In* Levin, M.A., R.J. Seidler, and M. Rogul, (eds.). *Microbial Ecology, Principles, Methods and Applications.* McGraw-Hill, New York.

Siegel, J.P., J.V. Maddox, and W.G. Ruesink. 1988. Seasonal progress of *Nosema pyrausta* in the European corn borer, Ostrinia nubilalis. *Journal of Invertebrate Pathology*, **52**: 130–136.

Singer, S. 1987. Current status of the microbial larvicide *Bacillus sphaericus*, pp. 133–163. *In* Maramorosch, K., (ed.). *Biotechnology in Invertebrate Pathology and Cell Culture.* Academic Press, New York.

Singer, S. 1990. Introduction to the study of *Bacillus sphaericus* as a mosquito control agent, pp. 221–227 *In* de Barjac, H. and D.J. Sutherland, (eds.). *Bacterial Control of Mosquitoes and Blackflies: Biochemistry, Genetics and Applications of Bacillus thuringiensis israelensis and Bacillus sphaericus.* Rutgers University Press, New Brunswick, New Jersey, U.S.A.

Singleton, G.R. and H.I. McCallum. 1990. The potential of *Capillaria hepatica* to control mouse plagues. *Parasitology Today*, **6**(6): 190–193.

Singleton, G. and T. Redhead. 1990. Future prospects for biological control of rodents using micro- and macroparasites. *In* Quick, G.R., (ed.). *Rodents and Rice.* Report and proceedings of an expert panel meeting on rice rodent control, 10–14 September, 1990. Los Banos, Philippines

Sitaramaiah, K. and R.S. Singh. 1974. The possible effects on *Meloidogyne javanica* of phenolic compounds produced in amended soil. *Journal of Nematology*, **6**: 152.

Siven, A. and I. Chet. 1986. *Trichoderma harzianum*: An effective biocontrol agent of *Fusarium* spp., pp. 89–95. *In* Jensen, V. et al., (eds.). *Microbial Communities in Soil.* Elsevier, London.

Sjogren, R.D. and E.F. Legner. 1989. Survival of the mosquito predator, *Notonecta unifasciata* (Hemiptera: Notonectidae) embryos at low thermal gradients. *Entomophaga*, **34**: 201–208.

Slade, N.A. 1977. Statistical detection of density dependence from a series of sequential censuses. *Ecology*, **58**: 1094–1102.

Small, R.W. 1979. The effects of predatory nematodes on populations of plant parasitic nematodes in pots. *Nematologica*, **25**: 94–103.

Smart, L.E., J.H. Stevenson, and J.H.H. Walters. 1989. Development of field trial methodology to assess short-term effects of pesticides on beneficial arthropods in arable crops. *Crop Protection*, **8**: 169–180.

Smith, D. and D.F. Papacek. 1991. Studies of the predatory mite *Amblyseius victoriensis* (Acarina: Phytoseiidae) in citrus orchards in south-east Queensland: Control of *Tegolophus australis* and *Phyllocoptruta oleivora* (Acarina: Eriophyidae), effect of pesticides, alternative host plants and augmentative release. *Experimental and Applied Acarology*, **12**: 195–217.

Smith, D., D.F. Papacek, and D.A.H. Murray. 1988. The use of *Leptomastix dactylopi* Howard (Hymenoptera: Encyrtidae) to control *Planococcus citri* (Risso) (Hemiptera: Pseudococcidae) in Queensland citrus orchards. *Queensland Journal of Agricultural and Animal Science*, **45**: 157–164.

Smith, G.C. and R.C. Trout. 1994. Using Leslie matrices to determine wild rabbit population growth and the potential for control. *Journal of Applied Ecology*, **31**: 223–230.

Smith, H.D., H.L. Maltby, and E. Jimenez-Jimenez. 1964. Biological control of the citrus blackfly in Mexico. *USDA Technical Bulletin No. 1311*. United States Department of Agriculture, Washington, D.C., U.S.A.

Smith, K.A., R.W. Miller, and D.H. Simser. 1992. *Entomopathogenic Nematode Bibliography: Heterorhabditid and Steinernematid Nematodes*. Southern Cooperative Series Bulletin No. 370. Arkansas Agricultural Experiment Station, Fayetteville, Arkansas, U.S.A.

Smith, S.M., M. Hubbes, and J.R. Carrow. 1987. Ground releases of *Trichogramma minutum* Riley (Hymenoptera: Trichogrammatidae) against the spruce budworm (Lepidoptera: Tortricidae). *Canadian Entomologist*, **119**: 251–263.

Smith, S.S. 1994. Methods and timing of releases of *Trichogramma* to control lepidopterous pests, pp. 113–144. *In* Wajnberg, E. and S.A. Hassan, (eds.). *Biological Control With Egg Parasitoids*. Commonwealth Agricultural Bureaux, Wallingford, U.K.

Snyder, W.C., G.W. Wallis, and S.N. Smith. 1976. Biological control of plant pathogens, pp. 521–539. *In* Huffaker, C.B. and P.S. Messenger, (eds.). *Theory and Practice of Biological Control*. Academic Press, New York.

Solomon, M.G., M.A. Easterbrook, and J.D. Fitzgerald. 1993. Mite-management programmes based on organophosphate-resistant *Typhlodromus pyri* in U.K. apple orchards. *Crop Protection*, **12**: 249–254.

Soper, R.S. 1985. Erynia radicans as a mycoinsecticide for spruce budworm control, pp. 69–76. *In* Anon. *Proceedings of a Symposium: Microbial Control of Spruce Budworms and Gypsy Moths*. Windsor Locks, Connecticut, U.S.A. USDA Forest Service General Technical Report No. 100.

Soper, R.S., G.E. Shewell, and D. Tyrrell. 1976. *Colcondamyia auditrix* nov. sp. (Diptera: Sarcophagidae), a parasite which is attracted by the mating song of its host, *Okanagana rimosa* (Homoptera: Cicadidae). *Canadian Entomologist*, **108**: 61–68.

Sopp, P.I. 1987. Quantification of predation by polyphagous predators on *Sitobion avenae* (Homoptera: Aphididae) in winter wheat using ELISA. Ph.D. thesis, University of Southampton, U.K.

Sopp, P.I., A.T. Gillespie, and A. Palmer. 1989. Application of *Verticillium lecanii* for the control of *Aphis gossypii* by a low-volume electrostatic rotary atomizer and a high-volume hydraulic sprayer. *Entomophaga*, **34**: 417–428.

Sopp, P.I., K.D. Sunderland, J.S. Fenlon, and S.D. Wratten. 1992. An improved quantitative method for estimating invertebrate predation in the field using an enzyme-linked immunosorbent assay (ELISA). *Journal of Applied Ecology*, **29**: 295–302.

Sorensen, J.T., D.N. Kinn, and R.L. Doutt. 1983. Biological observations on *Bdella longicornis*: A predatory mite in California vineyards (Acari: Bdellidae). *Entomography*, **2**: 297–305.

Soulé, M.E. 1990. The onslaught of alien species, and other challenges in the coming decades. *Conservation Biology*, **4**: 233–239.

Southwood, T.R.E. 1978. *Ecological Methods with Particular Reference to the Study of Insect Populations, 2nd ed.* Chapman and Hall, London.

Southwood, T.R.E. and H.N. Comins. 1976. A synoptic population model. *Journal of Animal Ecology*, **45**: 949–965.

Southwood, T.R.E. and W.F. Jepson. 1962. Studies on the populations of *Oscinella frit* L. (Diptera: Chloropidae) in the oat crop. *Journal of Animal Ecology*, **31**: 481–495.

Spacie, A. 1992. Testing the effects of microbial agents on fish and crustaceans, pp. 707–728. *In* Levin, M.A., R.J. Seidler, and M. Rogul, (eds.). *Microbial Ecology, Principles, Methods and Applications*. McGraw-Hill, New York.

Sparks, A.N., J.R. Ables, and R.L. Jones. 1982. Notes on biological control of stem borers in corn, sugarcane, and rice in the People's Republic of China, pp. 193–215. *In* Anon. *Biological Control of Pests in China*. United States Department of Agriculture, Washington, D.C., U.S.A.

Speigel, Y., I. Chet, E. Cohn, S. Galper, and E. Sharon. 1988. Use of chitin for controlling plant-parasitic nematodes. III. Influence of temperature on nematicidal effect, mineralization and microbial population buildup. *Plant and Soil*, **109**: 251–256.

Speigel, Y., E. Cohn, and I. Chet. 1989. Use of chitin for controlling *Heterodera avenae* and *Tylenchulus semipenetrans*. *Journal of Nematology*, **21**: 419–422.

Speyer, E.R. 1927. An important parasite of the greenhouse whitefly. *Bulletin of Entomological Research*, **17**: 301–308.

Splittstoesser, C.M. and C.Y. Kawanishi. 1981. Insect diseases caused by bacilli without toxin mediated pathologies, pp. 189–208. *In* Davidson, E.W., (ed.). *Pathogenesis of Invertebrate Microbial Diseases*. Allanheld, Osmun, Totowa, New Jersey, U.S.A.

Stadler, B. and V. Völkl. 1991. Foraging patterns of two aphid parasitoids, *Lysiphlebus testaceipes* and *Aphidius colemani* on banana. *Entomologia Experimentalis et Applicata*, **58**: 221–229.

Stahly, D.P. and M.G. Klein. 1992. Problems with *in vitro* production of spores of *Bacillus popilliae* for use in biological control of the Japanese beetle. *Journal of Invertebrate Pathology*, **60**: 283–291.
Stam, P.A. and H. Elmosa. 1990. The role of predators and parasites in controlling populations of *Earias insulana, Heliothis armigera* and *Bemisia tabaci* on cotton in the Syrian Arab Republic. *Entomophaga*, **35**: 315–327.
Stamp, N.E. 1981. Effect of group size on parasitism in a natural population of the Baltimore checkerspot *Euphydryas phaeton. Oecologia*, **49**: 201–206.
Stamp, N.E. 1982. Behavioral interactions of parasitoids and the Baltimore checkerspot caterpillars (*Euphydryas phaeton*). *Environmental Entomology*, **11**: 100–104.
Stamp, N.E. 1984. Interactions of parasitoids and checkerspot caterpillars *Euphydryas* spp. (Nymphalidae). *Journal of Research on the Lepidoptera*, **23**: 2–18.
Stanek, E.J., S.R. Diehl, N. Dgetluck, M.E. Stokes, and R.J. Prokopy. 1987. Statistical methods for analyzing discrete responses of insects tested repeatedly. *Environmental Entomology*, **16**: 320–326.
Starnes, R.L., Chi Li Liu, and P.G. Marrone. 1993. History, use and future of microbial insecticides. *American Entomologist*, **39**(2): 83–91.
Starr, M.P. and R.M. Sayre. 1988. *Pasteuria thornei* sp. nov. and *Pasteuria penetrans sensu stricto* emend., mycelial and endospore-forming bacteria parasitic, respectively, on plant-parasitic nematodes of the genera *Pratylenchus* and *Meloidogyne. Annales de l'Institute Pasteur/Microbiology*, **139**: 11–31.
Starý, P. 1970. *Biology of Aphid Parasites (Hymenoptera: Aphidiidae) with Respect to Integrated Control.* W. Junk, N.V., The Hague, The Netherlands.
Starý, P. 1983. The perennial stinging nettle (*Urtica dioica*) as a reservoir of aphid parasitoids (Hymenoptera, Aphidiidae). *Acta Entomologica Bohemoslovaca*, **80**: 81–86.
Starý, P. and W. Völkl. 1988. Aggregations of aphid parasitoid adults (Hymenoptera, Aphidiidae). *Journal of Applied Entomology*, **105**: 270–279.
Steele, J.H. 1974. *The Structure of Marine Ecosystems.* Harvard University Press, Cambridge, Massachusetts, U.S.A.
Steinberg, S., M. Dicke, and L.E.M. Vet. 1993. Relative importance of infochemicals from first and second trophic levels in long-range host location by the larval parasitoid *Cotesia glomerata. Journal of Chemical Ecology*, **19**: 47–59.
Steinhaus, E.A. 1956. Microbial control—The emergence of an idea. A brief history of insect pathology through the nineteenth century. *Hilgardia*, **26**: 107–157.
Steinhaus, E.A., (ed.). 1963. *Insect Pathology, an Advanced Treatise, 2 Vols.* Academic Press, New York.
Sterling, W.L., A. Dean, and N.M.A. El-Salam. 1992. Economic benefits of spider (Araneae) and insect (Hemiptera: Miridae) predators of cotton leafhoppers. *Journal of Economic Entomology*, **85**: 52–57.
Stern, V.M., P.L. Adkisson, O. Beingolea, and G.A. Viktorov. 1976. Cultural controls, pp. 593–613. *In* Huffaker, C.B. and P.S. Messenger, (eds.). *Theory and Practice of Biological Control.* Academic Press, New York.
Sternlicht, M. 1973. Parasitic wasps attracted by the sex pheromone of their coccid host. *Entomophaga*, **18**: 339–342.
Stevens, L.M., A.L. Steinhauer, and T.C. Elden. 1975. Laboratory rearing of the Mexican bean beetle and the parasite *Pediobius foveolatus*, with emphasis on parasite longevity and host-parasite ratios. *Environmental Entomology*, **4**: 953–957.
Stewart, L.M.D., M. Hirst, M.L. Ferber, A.T. Merryweather, P.J. Cayley, and R.D. Possee. 1991. Construction of an improved baculovirus insecticide containing an insect-specific toxin gene. *Nature*, **352**(6330): 85–88.
Steyn, J.J. 1958. The effect of ants on citrus scales at Letaba, South Africa. *Proceedings of the 10th International Congress of Entomology*, **4**: 589–594.
Stiling, P.D. 1987. The frequency of density dependence in insect host-parasitoid systems. *Ecology*, **68**: 844–856.
Stinner, R.E., R.L. Ridgway, and R.E. Kinzer. 1974. Storage, manipulation of emergence, and estimation of numbers of *Trichogramma pretiosum. Environmental Entomology*, **3**: 505–507.
Stirling, G.R. 1984. Biological control of *Meloidogyne javanica* with *Bacillus penetrans. Phytopathology*, **74**: 55–60.
Stirling, G.R. 1991. *Biological Control of Plant Parasitic Nematodes: Progress, Problems and Prospects.* Commonwealth Agricultural Bureaux International, Wallingford, Oxon., U.K.
Stirling, G.R. and M.F. Watchel. 1980. Mass production of Bacillus penetrans for the biological control of root-knot nematodes. *Nematologica*, **26**: 308–312.
Stirling, G.R., M.V. McKenry, and R. Mankau. 1979. Biological control of root-knot nematodes (*Meloidogyne* spp.) on peach. *Phytopathology*, **69**: 806–809.
Stirling, G.R., R.D. Sharma, and J. Pery. 1990. Attachment of *Pasteuria penetrans* spores to the root knot nematode *Meloidogyne javanica* in soil and its effects on infectivity. *Nematologica*, **36**: 246–252.
Stoffolano, J.G., Jr. 1973. Maintenance of *Heterotylenchus autumnalis*, a nematode parasite of the face fly, in the laboratory. *Annals of the Entomological Society of America*, **66**: 469–471.
Stoltz, D.B. 1993. The polydnavirus life cycle, pp. 167–187. *In* Beckage, N.E., S.N. Thompson and B.A. Federici, (eds.). *Parasites and Pathogens of Insects, Volume I.* Parasites. Academic Press, New York.

Stoltz, D.B. and S.B. Vinson. 1979. Viruses and parasitism in insects. *Advances in Virus Research*, **24**: 125–171.
Storey, G.K., C.W. McCoy, K. Stenzel, and W. Andersch. 1990. Conidiation kinetics of the mycelial granules of *Metarhizium anisopliae* (BIO 1020) and its biological activity against different soil insects, pp. 320–325. *In* Pinnock, D.E., (ed.). *Vth International Colloquium on Invertebrate Pathology.* Adelaide, Australia, 20–24, August 1990.
Story, R.N. and W.H. Robinson. 1979. Biological control potential of *Taphrocerus schaefferi* (Coloeptera: Buprestidae), a leaf-miner of yellow nutsedge. *Environmental Entomology*, **8**: 1088–1091.
Stouthamer, R. 1990. Evidence for microbe-mediated parthenogenesis in Hymenoptera, pp. 417–421. *In* Pinnock, D.E., (ed.). *Vth International Colloquium on Invertebrate Pathology and Microbial Control.* Adelaide, Australia, 20–24, August, 1990.
Stowell, L.J. 1991. Submerged fermentation of biological herbicides, pp. 225–261. *In* TeBeest, D.O., (ed.). *Microbial Control of Weeds.* Chapman and Hall, New York.
Strand, M.R. and S.B. Vinson. 1982a. Behavioral response of the parasitoid *Cardiochiles nigriceps* to a kairomone. *Entomologia Experimentalis et Applicata*, **31**: 308–315.
Strand, M.R. and S.B. Vinson. 1982b. Stimulation of oviposition and successful rearing of *Telenomus heliothidis* [Hym.: Scelionidae] on nonhosts by use of a host-recognition kairomone. *Entomophaga*, **27**: 365–370.
Strand, M.R. and S.B. Vinson. 1983a. Host acceptance behavior of *Telenomus heliothidis* (Hymenoptera: Scelionidae) toward *Heliothis virescens* (Lepidoptera: Noctuidae). *Annals of the Entomological Society of America*, **76**: 781–785.
Strand, M.R. and S.B. Vinson. 1983b. Factors affecting host recognition and acceptance in the egg parasitoid *Telenomus heliothidis* (Hymenoptera: Scelionidae). *Environmental Entomology*, **12**: 1114–1119.
Strand, M.R. and S.B. Vinson. 1983c. Analysis of an egg recognition kairomone of *Telenomus heliothidis* (Hymenoptera: Scelionidae). Isolation and host function. *Journal of Chemical Ecology*, **9**: 423–432.
Strand, M.R., H.J. Williams, S.B. Vinson, and A. Mudd. 1989. Kairomonal activities of 2-acylcyclohexane-1,3 diones produced by *Ephestia kuehniella* Zeller in eliciting searching behavior by the parasitoid *Bracon hebetor* (Say). *Journal of Chemical Ecology*, **15**: 1491–1500.
Strong, D.R., J.H. Lawton, and Sir Richard Southwood. 1984. *Insects on Plants: Community Patterns and Mechanisms.* Harvard University Press, Cambridge, Massachusetts, U.S.A.
Sturm, M.M., W.L. Sterling, and A.W. Hartstack. 1990. Role of natural mortality in boll weevil (Coleoptera: Curculionidae) management programs. *Journal of Economic Entomology*, **83**: 1–7.
Sugimoto, T., Y. Shimono, Y. Hata, A. Nakai, and M. Yahara. 1988. Foraging for patchily-distributed leaf-miners by the parasitoid, *Dapsilarthra rufiventris* (Hymenoptera: Braconidae), III. Visual and acoustic cues to a close range patch-location. *Applied Entomology and Zoology*, **23**: 113–121.
Summy, K.R., F.E. Gilstrap, W.G. Hart, J.M. Caballero, and I. Saenz. 1983. Biological control of citrus blackfly (Homoptera: Aleyrodidae) in Texas. *Environmental Entomology*, **12**: 782–786.
Sundby, R.A. and G. Taksdal. 1969. Surveys of parasites of *Hylemya brassicae* (Bouché) and *H. floralis* (Fallén) (Diptera, Muscidae) in Norway. *Norsk Entomologisk Tidsskift*, **16**: 97-106.
Sunderland, K.D. 1988. Quantitative methods for detecting invertebrate predation occurring in the field. *Annals of Applied Biology*, **112**: 201–224.
Sutton, J.C. and G. Peng. 1993a. Manipulation and vectoring of biocontrol organisms to manage foliage and fruit diseases in cropping systems. *Annual Review of Phytopathology*, **31**: 473–493.
Sutton, J.C. and G. Peng. 1993b. Biocontrol of *Botrytis cinerea* in strawberry leaves. *Phytopathology*, **83**: 615–621.
Sweet, M.H. and C.W. Schaefer. 1985. Systematic status and biology of *Chauliops fallax* Scott, with a discussion of the phylogenetic relationships of the Chauliopinae (Hemiptera: Malcidae). *Annals of the Entomological Society of America*, **78**: 526–536.
Syed, A. 1985. New rearing device for *Exeristes roborator* (F.) (Hymenoptera: Ichneumonidae). *Journal of Economic Entomology*, **78**: 279–281.
Syrett, P. 1990. The biological control of heather (*Calluna vulgaris*) in New Zealand: An environmental impact assessment. DSIR Plant Protection, Christchurch, New Zealand.
Tabashnik, B.E., N.L. Cushing, N. Finson, and M.W. Johnson. 1990. Field development of resistance to *Bacillus thuringiensis* in diamondback moth (Lepidoptera: Plutellidae). *Journal of Economic Entomology*, **83**: 1671–1676.
Táborsky, V. 1992. *Small-Scale Processing of Microbial Pesticides.* FAO Agricultural Services Bulletin 96, Food and Agriculture Organization of the United Nations, Rome, Italy.
Takafuji, A. 1977. The effect of the rate of successful dispersal of a phytoseiid mite, *Phytoseiulus persimilis* Athias-Henriot (Acarina: Phytoseiidae) on the persistence in the interactive system between the predator and its prey. *Researches on Population Ecology*, **18**: 210–222.
Takafuji, A., T. Inoue, and K. Fujita. 1981. Analysis of an acarine predator-prey system in glasshouse. *First Japan/USA Symposium on Integrated Pest Management, Tsukuba (Japan), September 29–30, 1981*: 144–153.

Takafuji, A., Y. Tsuda, and T. Miki. 1983. System behaviour in predator-prey interaction, with special reference to acarine predator-prey systems. *Researches on Population Ecology*, supplement **3**: 75–92.

Takahashi, S., M. Hajika, J. Takabayashi, and M. Fukui. 1990. Oviposition stimulants in the coccoid cuticular waxes of *Aphytis yanonensis* DeBach and Rosen. *Journal of Chemical Ecology*, **16**: 1657–1665.

Takasu, K. and Y. Hirose. 1988. Host discrimination in the parasitoid *Ooencyrtus nezarae*: The role of the egg stalk as an external marker. *Entomologia Experimentalis et Applicata*, **47**: 45–48.

Talhouk, A. S. 1991. On the management of the date palm and its arthropod enemies in the Arabian Peninsula. *Journal of Applied Entomology*, **111**: 514–520.

Tamaki, G., R.L. Chauvin, and T. Hsiao. 1982. Rearing *Doryphorophaga doryphorae*, a tachinid parasite of the Colorado potato beetle, *Leptinotarsa decemlineata*. *USDA/ARS Advances in Agricultural Technology, Western Series, No. 21.*

Tanada, Y. and H.K. Kaya. 1993. *Insect Pathology*. Academic Press, San Diego, California, U.S.A.

Tanaka, T. and S.B. Vinson. 1991. Interaction of venoms with the calyx fluids of three parasitoids, *Cardiochiles nigriceps*, *Microplitis croceipes* (Hymenoptera: Braconidae), and *Campoletis sonorensis* (Hymenoptera: Ichneumonidae) in effecting a delay in the pupation of Heliothis virescens (Lepidoptera: Noctuidae). *Annals of the Entomological Society of America*, **84**: 87–92.

Tanigoshi, L.K., J. Fargerlund, J.Y. Nishio-Wong, and H.J. Griffiths. 1985. Biological control of citrus thrips, *Scirtothrips citri* (Thysanoptera: Thripidae), in southern California citrus groves. *Environmental Entomology*, **14**: 733–741.

Tatchell, G.M. and C.C. Payne. 1984. Field evaluation of a granulosis virus for control of *Pieris rapae* (Lep.: Pieridae) in the United Kingdom. *Entomophaga*, **29**: 133–144.

Tauber, C.A., J.B. Johnson, and M.J. Tauber. 1992. Larval and developmental characteristics of the endemic Hawaiian lacewing, *Anomalochrysa frater* (Neuroptera: Chrysopidae). *Annals of the Entomological Society of America*, **85**: 200–206.

Tauber, M.J. and C.A. Tauber. 1972. Geographic variation in critical photoperiod and in diapause intensity of *Chrysopa carnae* (Neuroptera). *Journal of Insect Physiology*, **18**: 25–29.

Tauber, M.J., C.A. Tauber, and S. Gardescu. 1993. Prolonged storage of *Chrysoperla carnea* (Neuroptera: Chrysopidae). *Environmental Entomology*, **22**: 843–848.

Taylor, A.D. 1988. Parasitoid competition and the dynamics of host-parasitoid models. *American Naturalist*, **132**: 417–436.

Taylor, J.N., W.R. Courtenay, Jr., and J.A. McCann. 1984. Known impacts of exotic fishes in the continental United States, pp. 322–373. *In* Courtenay, W.R., Jr. and J.R. Stauffer, Jr., (eds.). *Distribution, Biology, and Management of Exotic Fishes*. John Hopkins University Press, Baltimore, Maryland, U.S.A.

Taylor, J.S. 1932. Report on cotton insect and disease investigation. II. Notes on the American bollworm (*Heliothis obsoleta* F.) on cotton and its parasite (*Microbracon brevicornis* Wesm.). *Science Bulletin Report for Agriculture and Forestry in the Union of South Africa*, 113.

Taylor, T.H.C. 1937. *The Biological Control of an Insect in Fiji. An Account of the Coconut Leaf-Mining Beetle and its Parasite Complex*. Imperial Institute of Entomology, London.

Tayutivutikul, J. and K. Yano. 1989. Biology of insects associated with the kudzu plant, *Pueraria lobata* (Leguminosae) 1. *Chauliops fallax* (Hemiptera, Lygaeidae). *Japanese Journal of Entomology*, **57**: 831–842.

TeBeest, D.O., (ed.). 1991. *Microbial Control of Weeds*. Chapman and Hall, New York.

Templeton, G.E. 1992. Use of Colletotrichum strains as mycoherbicides, pp. 358–380. *In* Bailey, J.A. and M.J. Jeger, (eds.). *Colletotrichum: Biology, Pathology and Control*. Commonwealth Agricultural Bureaux International, Wallingford, U.K.

Templeton, G.E. and E.E. Trujillo. 1981. The use of plant pathogens in the biological control of weeds, pp. 345–350. *In* Pimentel, D. (ed.). *CRC Handbook of Pest Management in Agriculture, (vol. 2)*. CRC Press, Inc., Boca Raton, Florida, U.S.A.

Templeton, G.E., D.O. TeBeest, and R.J. Smith, Jr. 1984. Biological weed control in rice with a strain of *Colletotrichum gloeosporioides* (Penz.) Sacc. used as a mycoherbicide. *Crop Protection*, **3**: 409–422.

Terytze, K. and H. Adam. 1981. Zur Verwendung von Pheromonfallen fur die biologische Bekampfung der Kohleule (*Barathra brassicae* L.) mittels Eiparasiten der Gattung *Trichogramma* (*Trichogramma evanescens* Westw.) (Lepidoptera, Noctuidae; Hymenoptera, Trichogrammatidae). *Archieves der Phytopathologie und Pflanzenschutz*, **6**: 387–396. (in German)

Thang, M.H., O. Mochida, B. Morallo-Rejesus, and R.P. Robles. 1987. Selectivity of eight insecticides to the brown planthopper, *Nilaparvata lugens* (Stål) (Homoptera: Delphacidae), and its predator, the wolf spider, *Lycosa pseudoannulata* Boes. et Str. (Araneae: Lycosidae). *Philippine Entomologist*, **7**: 51–56.

Thiele, H.U. 1977. *Carabid Beetles in Their Environments*. Springer-Verlag. Berlin, Germany.

Thiery, I., S. Hamon, V.C. Dumanoir, and H. de Barjac. 1992. Vertebrate safety of *Clostridium bifermentans* serovar *malaysia*, a new larvicidal agent for vector control. *Journal of Economic Entomology*, **85**: 1618–1623.

Thomas, E.D., C.F. Reichelderfer, and A.M. Heimpel. 1973. The effect of soil pH on the persistence of cabbage looper nuclear polyhedrosis virus in soil. *Journal of Invertebrate Pathology*, **21**: 21–25.

Thomas, K.J. 1981. The role of aquatic weeds in changing the pattern of ecosystems in Kerala. *Environmental Conservation*, **8**: 63–66.

Thomas, M. 1990. Diversification of the arable ecosystem to control natural enemies of cereal aphids. *Game Conservancy Review*, **21**: 68–69.

Thomas, M.B., S.D. Wratten, and N.W. Sotherton. 1991. Creation of "island" habitats in farmland to manipulate populations of beneficial arthropods: predator densities and emigration. *Journal of Applied Ecology*, **28**: 906–917.

Thomas, M.B., H.J. Mitchell, and S.D. Wratten. 1992. Abiotic and biotic factors influencing the winter distribution of predatory insects. *Oecologia*, **89**: 78–84.

Thomas, P.A. and P.M. Room. 1986. Taxonomy and control of *Salvinia molesta*. *Nature*, **320**: 581–584.

Thompson, C.R. and D.H. Habeck. 1989. Host specificity and biology of the weevil *Neohydronomus affinis* [Coleoptera: Curculionidae] a biological control agent of *Pistia stratiotes*. *Entomophaga*, **34**: 299–306.

Thompson, J.N. 1986. Oviposition behaviour and searching efficiency in a natural population of a braconid parasitoid. *Journal of Animal Ecology*, **55**: 351–360.

Thompson, J.N. and O. Pellmyr. 1991. Evolution of oviposition behavior and host preference in Lepidoptera. *Annual Review of Entomology*, **36**: 65–89.

Thompson, L.C., H.M. Kulman, and R.A. Hellenthal. 1979. Parasitism of the larch sawfly by *Bessa harveyi* (Diptera: Tachinidae). *Annals of the Entomological Society of America*, **72**: 468–471.

Thompson, S.N. 1981. *Brachymeria lasus*: culture *in vitro* of a chalcid insect parasite. *Experimental Parasitology*, **52**: 414–418.

Thompson, W.R. 1924. La théorie mathematique de l'action des parasites entomophages et le facteur du hazard. *Annales de la Faculté Science de Marseille* 2, 69–89.

Thompson, W.R. and F.J. Simmonds. 1964–1965. *A Catalogue of the Parasites and Predators of Insect Pests*. Commonwealth Agricultural Bureaux, Burks, U.K.

Thoms, E.M. and W.H. Robinson. 1987. Potential of the cockroach oothecal parasite *Prosevania punctata* (Hymenoptera: Evaniidae) as a biological control agent for the oriental cockroach (Orthoptera: Blattidae). *Environmental Entomology*, **16**: 938–944.

Thomson, A.D. 1958. Interference between plant viruses. *Nature*, **181**: 1547–1548.

Thomson, S.V., D.R. Hanson, K.M. Flint, and Y.J.D. Vandenberg. 1992. Dissemination of bacteria antagonistic to *Erwinia amylovora* by honey bees. *Plant Disease*, **76**: 1052–1056.

Tillman, P.G. and J.E. Powell. 1992. Interspecific discrimination and larval competition among *Microplitis croceipes*, *Microplitis demolitor*, *Cotesia kazak* (Hym.: Braconidae), and *Hyposoter didymator* (Hym.: Ichneumonidae), parasitoids of *Heliothis virescens* (Lep.: Noctuidae). *Entomophaga*, **37**: 439–451.

Tinsley, T.W. 1979. The potential of insect pathogenic viruses as pesticide agents. *Annual Review of Entomology*, **24**: 63–87.

Tinsley, T.W. and P.F. Entwistle. 1974. The use of pathogens in the control of insect pests, pp. 115–129. *In* Price-Jones, D. and M.E. Solomon, (eds.). *Biology in Pests and Disease Control*. John Wiley and Sons, New York.

Tisdell, C. 1990. Economic impact of biological control of weeds and insects, pp. 301–316. *In* Mackauer, M., L.E. Ehler, and J. Roland, (eds.). *Critical Issues in Biological Control*. Intercept, Andover, U.K.

Tisdell, C.A., B.A. Auld, and K.M. Menz. 1984. On assessing the value of biological control of weeds. *Protection Ecology*, **6**: 169–179.

Torgersen, T.R., J.W. Thomas, R.R. Mason, and D. van Horn. 1984. Avian predators of Douglas-fir tussock moth, *Orgyia pseudotsugata* (McDunnough) (Lepidoptera: Lymantriidae) in southwestern Oregon. *Environmental Entomology*, **13**: 1018–1022.

Tothill, J.D., T.H.C. Taylor, and R.W. Paine. 1930. *The Coconut Moth in Fiji. A History of its Control by Means of Parasites*. Imperial Bureaux of Entomology (now Commonwealth Agricultural Bureaux International Institute of Entomology), London.

Townes, H. 1969. The genera of Ichneumonidae Parts 1, 2 and 3. *Memoirs of the American Entomological Institute* 11, 12, 13: 300 pp., 537 pp., and 307 pp.

Townes, H. 1988. The more important literature on parasitic Hymenoptera, pp. 491–518. *In*: *Advances in Parasitic Hymenoptera Research*. Proceedings of the II Conference on the Taxonomy and Biology of Parasitic Hymenoptera, held at the University of Florida, Gainesville, Florida, U.S.A., November 19–21, 1987, E. J. Brill, New York.

Tracewski, K.T., P.C. Johnson, and A.T. Eaton. 1984 Relative densities of predaceous Diptera (Cecidomyiidae, Chamaemyiidae. Syrphidae) and their apple aphid prey in New Hampshire, U.S.A., apple orchards. *Protection Ecology*, **6**: 199–207.

Treacy, M.F., J.H. Benedict, M.H. Walmsley, J.D. Lopez, and R.K. Morrison. 1987. Parasitism of bollworm (Lepidoptera: Noctuidae) eggs on nectaried and nectariless cotton. *Environmental Entomology*, **16**: 420–423.

Tribe, H.T. 1980. Prospects for the biological control of plant-parasitic nematodes. *Parasitology*, **81**: 619–639.

Trichilo, P.J. and L.T. Wilson. 1993. An ecosystem analysis of spider mite outbreaks: Physiological stimulation or natural enemy suppression. *Experimental and Applied Acarology*, **17**: 291–314.

Trimble, R.M. 1988. Monitoring *Pholetesor ornigis* (Hymenoptera: Braconidae), a parasite of the spotted tentiform leafminer, *Phyllonorycter blancardella* (Lepidoptera: Gracillariidae): Effect of sticky trap location on size and sex ratio of trap catches. *Environmental Entomology*, **17**: 567–571.

Tronsmo, A. and C. Dennis. 1977. The use of *Trichoderma* species to control strawberry fruit rots. *Netherlands Journal of Plant Pathology*, **83** (Suppl. 1): 449–455.

Trujillo, E.E. 1985. Biological control of Hamakua pa-makani with *Cercosporella* sp. in Hawaii, pp. 661–671. *In* Delfosse, E.S., (ed.). *Proceedings of the VI International Symposium on Biological Control of Weeds.* 19–25 August 1984, Vancouver, Canada. Agriculture Canada, Ottawa, Canada.

Trujillo, E.E., F.M. Latterell, and A.E. Rossi. 1986. *Colletotrichum gloeosporioides*, a possible biological control agent for *Clidemia hirta* in Hawaiian forests. *Plant Disease*, **70**: 974–976.

Trumble, J.T. and B. Alvarado-Rodriguez. 1990. Economics and integration of *Bacillus thuringiensis* use in vegetable production in the USA and Mexico, p. 498. *In* Pinnock, D.E., (ed.). *Vth International Colloquium on Invertebrate Pathology and Microbial Control.* Adelaide, Australia, 20–24, August, 1990.

Trumble, J.T. and B. Alvarado Rodriguez. 1993. Development and economic evaluation of an IPM program for fresh market tomato production in Mexico. *Agriculture, Ecosystems and Environment*, **43**: 267–284.

Trumble, J.T. and J.P. Morse. 1993. Economics of integrating the predaceous mite *Phytoseiulus persimilis* (Acari: Phytoseiidae) with pesticides in strawberries. *Journal of Economic Entomology*, **86**: 879–885.

Trumble, J.T., D.M. Kolodny-Hirsch, and I.P. Ting. 1993. Plant compensation for arthropod herbivory. *Annual Review of Entomology*, **38**: 93–119.

Tulisalo, U. and T. Tuovinen. 1975. The green lacewing, *Chrysopa carnea* Steph. (Neuroptera, Chrysopidae), used to control the green peach aphid, *Myzus persicae* (Sulz.), and the potato aphid, *Macrosiphum euphorbiae* Thomas (Homoptera, Aphididae), on greenhouse green peppers. *Annales Entomologici Fennici*, **41**: 94–102.

Tumlinson, J.H., T.C.J. Turlings, and W.J. Lewis. 1992. The semiochemical complexes that mediate insect parasitoid foraging. *Agricultural Zoological Reviews*, **5**: 221–252.

Turlings, T.C.J., J.H. Tumlinson, and W.J. Lewis. 1990. Exploitation of herbivore-induced plant odors by host seeking parasitic wasps. *Science*, **250**: 1251–1253.

Turlings, T.C.J., J.H. Tumlinson, F.J. Eller, and W.J. Lewis. 1991. Larval-damaged plants: Source of volatile synomones that guide the parasitoid *Cotesia marginiventris* to the micro-habitat of its hosts. *Entomologia Experimentalis et Applicata*, **58**: 75–82.

Turlings, T.C.J., P.J. McCall, H.T. Alborn, and J.H. Tumlinson. 1993. An elicitor in caterpillar oral secretions that induces corn seedlings to emit chemical signals attractive to parasitic wasps. *Journal of Chemical Ecology*, **19**: 411–425.

Turnbull, A.L. 1967. Population dynamics of exotic insects. *Bulletin of the Entomological Society of America*, **13**: 333–337.

Turnbull, A.L. and P.A. Chant. 1961. The practice and theory of biological control of insects in Canada. *Canadian Journal of Zoology*, **39**: 697–753.

U.S. Congress, Office of Technology Assessment. 1993. *Harmful Non-Indigenous Species in the United States.* U.S. Government Printing Office, Washington, D.C., U.S.A.

U.S. Department of Agriculture. 1978. *Biological Agents for Pest Control, Status and Prospectus.* Stock no. 001-000-03756-1, U.S. Government Printing Office, Washington, D.C., U.S.A.

Utida, S. 1941. Studies on experimental population of the azuki bean weevil, *Callosobruchus chinensis* (L.) IV. Analysis of density effect with respect to fecundity and fertility of eggs. *Memoirs of the College of Agriculture, Kyoto Imperial University*, **51**: 1–26.

Utida, S. 1950. On the equilibrium state of the interacting population of an insect and its parasite. *Ecology*, **31**: 165–175.

Utida, S. 1957. Cyclic fluctuations of population density intrinsic to the host-parasite system. *Ecology*, **38**: 442–449.

Utida, S. 1967. Damped oscillation of population density at equilibruim. *Researches on Population Ecology*, **9**: 1–9.

Vaeck, M., A. Reynaerts, H. Hofte, S. Jansens, M. de Beuckeleer, C. Dean, M. Zabeau, M, van Montagu, and J. Leemans. 1987. Transgenic plants protected from insect attack. *Nature*, **328** (6125): 33–37.

Vail, P.V., W. Barnett, D.C. Cowan, S. Sibbett, R. Beede, and J.S. Tebbets. 1991. Codling moth (Lepidoptera: Tortricidae) control on commercial walnuts with a granulosis virus. *Journal of Economic Entomology*, **84**: 1448–1453.

van Alfen, N.K. 1982. Biology and potential for disease control of hypovirulence of *Endothia parasitica. Annual Review of Phytopathology*, **20**: 349–362.

van Alphen, J.J.M. 1988. Patch-time allocation by insect parasitoids: Superparasitism and aggregation, pp. 215–221. *In* de Jong, G., (ed.). *Population Genetics and Evolution.* Springer-Verlag, Berlin, Germany.

van Alphen, J.J.M. and F. Galis. 1983. Patch time allocation and parasitization efficiency of *Asobara tabida* Nees, a larval parasitoid of *Drosophila*. *Journal of Animal Ecology*, **52**: 937–952.

van Alphen, J.J.M. and M.J. van Dijken. 1988. Host discrimination: The learning hypothesis revisited, pp. 35–36. *In* Anon. *Parasitoid Insects, European Workshop*, Colloques de l'INRA (1988) No. 48, Lyon, France, 7–10 September 1987.

van Alphen, J.J.M. and H.H. van Harsel. 1982. Host selection by *Asobara tabida* Nees (Bracondiae; Alysiinae), a larval parasitoid of fruit inhabiting *Drosophila* species. III. Host species selection and functional response, pp. 61–93. *In* van Alphen, J.J.M. *Foraging Behaviour of Asobara tabida, a larval parasitoid of Drosophilidae*. Ph.D. dissertation, University of Leiden, The Netherlands.

van Alphen, J.J.M. and L.E.M. Vet. 1986. An evolutionary approach to host finding and selection, pp. 23–61. *In* Waage, J.K. and D. Greathead, (eds.). *Insect Parasitoids*. Academic Press, London.

van Alphen, J.J.M., M.J. van Dijken, and J.K. Waage. 1987. A functional approach to superparasitism: Host discrimination need not be learnt. *Netherlands Journal of Zoology*, **37**: 167–179.

van Bergeijk, K.E., F. Bigler, N.K. Kaashoek, and G.A. Pak. 1989. Changes in host acceptance and host suitability as an effect of rearing *Trichogramma maidis* on a factitious host. *Entomologia Experimentalis et Applicata*, **52**: 229–238.

van den Berg, H. 1993. *Natural control of Helicoverpa armigera in smallholder crops in East Africa*. Ph.D. dissertation, Department of Entomology, Wageningen, The Netherlands.

van den Bosch, R. 1968. Comments on population dynamics of exotic insects. *Bulletin of the Entomological Society of America*, **14**: 112–115.

van den Bosch, R. 1971. Biological control of insects. *Annual Review of Ecology and Systematics*, **2**: 45–66. van den Bosch, R. and P.S. Messenger. 1973. *Biological Control*. Intext Educational Publications, New York.

van den Bosch, R., C.F. Lagace, and V.M. Stern. 1967. The interrelationship of the aphid, *Acyrthosiphon pisum*, and its parasite, *Aphidius smithi*, in a stable environment. *Ecology*, **48**: 993–1000.

van den Bosch, R., R.D. Frazer, C.S. Davis, P.S. Messenger, and R. Hom. 1970. *Trioxys pallidus*. An effective new walnut aphid parasite from Iran. *California Agriculture*, **24** (6): 8–10.

van den Bosch, R., O. Beingolea, M. Hafez, and L.A. Falcon. 1976. Biological control of insect pests of row crops, pp. 443–456. *In* Huffaker, C.B. and P.S. Messenger, (eds.). *Theory and Practice of Biological Control*. Academic Press, New York.

Van der Vecht, J. 1954. Parasitism in an outbreak of the coconut moth (*Artona cataxantha* [Hamps.]) in Java (Lep.). *Entomologische Berichten*, **15**: 122–132.

van de Vrie, M. and A. Boersma. 1970. The influence of the predaceous mite *Typhlodromus* (A.) *potentillae* (Garman) on the population development of *Panonychus ulmi* (Koch) on apple grown under various nitrogen conditions. *Entomophaga*, **15**: 291–304.

van Dijken, M.J. and J.K. Waage. 1987. Self and conspecific superparasitism by the egg parasitoid *Trichogramma evanescens*. *Entomologia Experimentalis et Applicata*, **43**: 183–192.

Van Driesche, R.G. 1983. The meaning of percent parasitism in studies of insect parasitoids. *Environmental Entomology*, **12**: 1611–1622.

Van Driesche, R.G. 1988. Field levels of encapsulation and superparasitism for *Cotesia glomerata* (L.) (Hymenoptera: Braconidae) in *Pieris rapae* (L.) (Lepidoptera: Pieridae). *Journal of the Kansas Entomological Society*, **61**: 328–331.

Van Driesche, R.G. 1989a. Extending biological control: The role of extension agents in the use of natural enemies. *The IPM Practitioner*, **11** (10): 1–3.

Van Driesche, R.G. 1989b. What every state needs—A biological control coordinator. *The IPM Practitioner*, **11** (9): 5–7.

Van Driesche, R.G. 1993. Methods for the field colonization of new biological control agents, pp. 67–86. *In* Van Driesche, R.G. and T.S. Bellows, Jr., (eds.). *Steps in Classical Arthropod Biological Control*. Thomas Say Publications in Entomology, Entomological Society of America, Lanham, Maryland, U.S.A., 88 pp.

Van Driesche, R.G. 1994. Biological control for the control of environmental pests. *Florida Entomologist*, **77**: 20–33.

Van Driesche, R.G. and T.S. Bellows, Jr. 1988. Use of host and parasitoid recruitment in quantifying losses from parasitism in insect populations with reference to *Pieris rapae* and *Cotesia glomerata*. *Ecological Entomology*, **13**: 215–222.

Van Driesche, R.G. and E. Carey. 1987. *Opportunities for Increased Use of Biological Control in Massachusetts*. Massachusetts Agricultural Experiment Station Bulletin #718, Amherst, Massachusetts, U.S.A.

Van Driesche, R.G. and D.N. Ferro. 1989. *Using biological control in Massachusetts: Cole crop Lepidoptera*. University of Massachusetts Cooperative Extension Service Publication L-597, Amherst, Massachusetts, U.S.A.

Van Driesche, R.G. and G.G. Gyrisco. 1979. Field studies of *Microctonus aethiopoides*, a parasite of the adult alfalfa weevil, *Hypera postica*, in New York. *Environmental Entomology*, **8**: 238–244.

Van Driesche, R.G. and C. Hulbert. 1984. Host acceptance and discrimination by *Comperia merceti* (Compere) (Hymenoptera: Encyrtidae) and evidence for an optimal density range for resource utilization. *Journal of Chemical Ecology*, **10**: 1399–1409.

Van Driesche, R.G. and G. Taub. 1983. Impact of parasitoids on *Phyllonorycter* leafminers infesting apple in Massachusetts, USA. *Protection Ecology*, **5**: 303–317.

Van Driesche, R.G., A. Bellotti, C.J. Herrera, and J.A. Castillo. 1987. Host feeding and ovipositor insertion as sources of mortality in the mealybug *Phenacoccus herreni* caused by two encyrtids, *Epidinocarsis diversicornis* and *Acerophagus coccois*. *Entomologia Experimentalis et Applicata*, **44**: 97–100.

Van Driesche, R.G., T.S. Bellows, Jr., D.N. Ferro, R. Hazzard, and A. Maher. 1989a. Estimating stage survival from recruitment and density data, with reference to egg mortality in the Colorado potato beetle, *Leptinotarsa decemlineata* (Say) [Coleoptera: Chrysomelidae]. *Canadian Entomologist*, **121**: 291–300.

Van Driesche, R.G., R. Prokopy, W. Coli, and T. Bellows. 1989b. *Using biological control in Massachusetts: Apple blotch leafminer.* Massachusetts Cooperative Extension Service Publication L-594, Amherst, Massachusetts, U.S.A.

Van Driesche, R.G., W. Coli, and S. Schumacher. 1990. Update: Lessons from the Massachusetts biological control initiative. *The IPM Practitioner*, **12** (4): 1–5.

Van Driesche, R.G., D.N. Ferro, E. Carey, and A. Maher. 1990. Assessing augmentative releases of parasitoids using the "recruitment method", with reference to *Edovum puttleri*, a parasitoid of the Colorado potato beetle [Coleoptera: Chrysomelidae]. *Entomophaga*, **36**: 193–204.

Van Driesche, R.G., J.S. Elkinton, and T.S. Bellows, Jr. 1994. Potential use of life tables to evaluate the impact of parasitism on population growth of the apple blotch leafminer (Lepidoptera: Gracillariidae). *In* Maier, C., (ed.). *Integrated Management of Tentiform Leafminers, Phyllonorycter (Lepidoptera: Gracillariidae) spp., in North American Apple Orchards.* Thomas Say Publications in Entomology, Entomological Society of America, Lanham, Maryland, U.S.A.

van Emdem, H.F. 1991. The role of host plant resistance in insect pest mis-management. *Bulletin of Entomological Research*, **81**: 123–126.

van Essen, F.W. and S.C. Hembree. 1980. Laboratory bioassay of *Bacillus thuringiensis israelensis* against all instars of *Aedes aegypti* and *Aedes taeniorhynchus* larvae. *Mosquito News*, **40**(3): 424–431.

van Haren, R.J.F., M.M. Steenhuis, M.W. Sabelis, and O.M.B. de Ponti. 1987. Tomato stem trichomes and dispersal success of *Phytoseiulus persimilis* relative to its prey *Tetranychus urticae*. *Experimental and Applied Acarology*, **3**: 115–121.

van Leerdam, M.B., J.W. Smith, Jr., and T.W. Fuchs. 1985. Frass-mediated, host-finding behavior of *Cotesia flavipes*, a braconid parasite of *Diatraea saccharalis* (Lepidoptera: Pyralidae). *Annals of the Entomological Society of America*, **78**: 647–650.

van Lenteren, J.C. 1986. Evaluation, mass production, quality control and release of entomophagous insects, pp. 31–56. *In* Franz, J.M., (ed.). *Biological Plant and Health Protection: Biological Control of Plant Pests and of Vectors of Human and Animal Diseases.* International Symposium of the Akademie der Wissenschaften und der Literatur, Mainz, November 15–17th, 1984 at Mainz and Darmstadt. Fortschritte der Zoologie, 32, 341 pp., Gustav Fischer Verlag, Stuttgart, Germany.

van Lenteren, J.C. 1989. Implementation and commercialization of biological control in western Europe. Proceedings and Abstracts. International Symposium of Biological Control Implementation, *North American Plant Protection Bulletin* No. 6: 50–70.

van Lenteren, J.C. 1991. Encounters with parasitized hosts: To leave or not to leave a patch. *Netherlands Journal of Zoology*, **41**: 144–157.

van Lenteren, J.C. 1995. Integrated pest management in protected crops. pp. 311–343, *In* Dent, D.R., (ed.). *Integrated Pest Management: Principles and Systems Development.* Chapman and Hall, London.

van Lenteren, J.C. and K. Bakker. 1978. Behavioural aspects of the functional responses of a parasite (*Pseudocoila bochei* Weld) to its host (*Drosophila melanogaster*). *Netherlands Journal of Zoology*, **28**: 213–233.

van Lenteren, J.C. and J. Woets. 1988. Biological and integrated pest control in greenhouses. *Annual Review of Entomology*, **33**: 239–269.

van Rensburg, P.J.J., J.D. Skinner, and R.J. van Aarde. 1987. Effects of feline panleucopaenia on the population characteristics of feral cats on Marion Island. *Journal of Applied Ecology*, **24**: 63–73.

van Riper, C. III, S.G. Van Riper, M.L. Goff, and M. Laird. 1986. The epizootiology and ecological significance of malaria in Hawaiian birds. *Ecological Monographs*, **56**: 327–344.

Van Valen, L. 1974. Predation and species diversity. *Journal of Theoretical Biology*, **44**: 19–21.

van Winkelhoff, A.J. and C.W. McCoy. 1984. Conidiation of Hirsutella thompsonii var. synnematosa in submerged culture. *Journal of Invertebrate Pathology*, **43**: 59–68.

Vandenberg, J.D. 1990. Safety of four entomopathogens for caged adult honey bees (Hymenoptera: Apidae). *Journal of Economic Entomology*, **83**: 755–759.

Vargas, M., J. Gomez, and G. Perera. 1991. Geographic expansion of *Marisa cornuarietis* and *Tarebia granifera* in the Dominican Republic. *Journal of Medical and Applied Malacology*, **3**: 69–72.

Varley, G.C. and G.R. Gradwell. 1960. Key factors in insect population studies. *Journal of Animal Ecology*, **29**: 399–401.

Varley, G.C. and G.R. Gradwell. 1968. Population models for the winter moth, pp. 132–142. *In* Southwood, T.R.E., (ed.). *Insect Abundance*. Symposium of the Royal Entomological Society of London, No. 4.

Varley, G.C. and G.R. Gradwell. 1970. Recent advances in insect population dynamics. *Annual Review of Entomology*, **15**: 1–24.

Varley, G.C. and G.R. Gradwell. 1971. The use of models and life tables in assessing the role of natural enemies, pp. 93–110. *In* C.B. Huffaker, (ed.). *Biological Control*. Plenum Press, New York.

Velu, T.S. and T. Kumaraswami. 1990. Studies on "skip row coverage" against bollworm damage and parasite emergence in cotton. *Entomon*, **15**: 69–73.

Versfeld, D.B. and B.W. van Wilgen. 1986. Impact of woody aliens on ecosystem properties, pp. 239–246. *In* Macdonald, I.A.W., F.J. Kruger, and A.A. Ferrar, (eds.). *The Ecology and Management of Biological Invasions in Southern Africa*. Proceedings of the National Synthesis Symposium on the Ecology of Biological Invasions. Oxford University Press, Cape Town, South Africa.

Vet, L.E.M. 1985. Response to kairomones by some alysiine and eucoilid parasitoid species (Hymenoptera). *Netherlands Journal of Zoology*, **35**: 486–496.

Vet, L.E.M. and K. Bakker. 1985. A comparative functional approach to the host detection behaviour of parasitic wasps. 2. A quantitative study on eight eucoilid species. *Oikos*, **44**: 487–498.

Vet, L.E.M. and M. Dicke. 1992. Ecology of infochemical use in a tritrophic level context. *Annual Review of Entomology*, **37**: 141–172.

Vet, L.E.M. and A.W. Groenewold. 1990. Semiochemicals and learning in parasitoids. *Journal of Chemical Ecology*, **16**: 3119–3135.

Vet, L.E.M., F.L. Wäckers, and M. Dicke. 1991. How to hunt for hiding hosts: the reliability-detectability problem in foraging parasitoids. *Netherlands Journal of Zoology*, **41**: 202–213.

Vickers, R.A., G.H. Rothschild, and E.L. Jones. 1985. Control of the oriental fruit moth, *Cydia molesta* (Busck) (Lepidoptera: Tortricidae), at a district level by mating disruption with synthetic female pheromone. *Bulletin of Entomological Research*, **75**: 625–634.

Vickery, W.L. and T.D. Nudds. 1984. Detection of density-dependent effects in annual duck censuses. *Ecology*, **65**: 96–104.

Viggiani, G. 1984. Bionomics of the Aphelinidae. *Annual Review of Entomology*, **29**: 257–276.

Vinson, S.B. 1975. Biochemical coevolution between parasitoids and their hosts, pp. 14–48. *In* Price, P., (ed.). *Evolutionary Strategies of Parasitic Insects and Mites*. Plenum Press, New York.

Vinson, S.B. 1976. Host selection by insect parasitoids. *Annual Review of Entomology*, **21**: 109–133.

Vinson, S.B. 1981. Habitat location, pp. 51–77. *In* Nordlund, D.A., R.J. Jones, and W.J. Lewis, (eds.). *Semiochemicals, Their Role in Pest Control*. John Wiley and Sons, New York.

Vinson, S.B. 1984a. How parasitoids locate their hosts: A case of insect espionage, pp. 325–348. *In* Lewis, T., (ed.). *Insect Communication*. Academic Press, London.

Vinson, S.B. 1984b. Parasitoid-host relationship, pp. 205–233 *In* Bell, W.J. and R.T. Cardé, (eds.). *Chemical Ecology of Insects*. Sinauer, Sunderland, Massachusetts, U.S.A.

Vinson, S.B. 1990a. Potential impact of microbial insecticides on beneficial arthropods in the terrestrial environment, pp. 43–64. *In* Laird, M., L.A. Lacey, and E.W. Davidson, (eds.). *Safety of Microbial Insecticides*. CRC Press, Inc., Boca Raton, Florida, U.S.A.

Vinson, S.B. 1990b. How parasitoids deal with the immune system of their hosts: an overview. *Archives of Insect Biochemistry and Physiology*, **13**: 3–27.

Vinson, S.B. 1990c. The interaction of two parasitoid polydnaviruses with the endocrine system of *Heliothis virescens*, pp. 195–199. *In* Pinnock, D.E., (ed.). *Vth International Colloquium on Invertebrate Pathology and Microbial Control*. 20–24 August 1990, Adelaide, Australia.

Vinson, S.B. 1991. Chemical signals used by parasitoids. *Redia*, **74**: 15–42.

Vinson, S.B. and F.S. Guillot. 1972. Host marking: Source of a substance that results in host discrimination in insect parasitoids. *Entomophaga*, **17**: 241–245.

Vinson, S.B. and G.F. Iwantsch. 1980a. Host suitability for insect parasitoids. *Annual Review of Entomology*, **25**: 397–419.

Vinson, S.B. and G.F. Iwantsch. 1980b. Host regulation by insect parasitoids. *Quarterly Review of Biology*, **55**: 143–165.

Vinson, S.B. and G.L. Piper. 1986. Source and characterization of host recognition kairomones of *Tetrastichus hagenowii*, a parasitoid of cockroach eggs. *Physiological Entomology*, **11**: 459–468.

Vinson, S.B., C.S. Barfield, and R.D. Henson. 1977. Oviposition behavior of *Bracon mellitor*, a parasitoid of the boll weevil (Anthonomus grandis). II. Associative learning. *Physiological Entomology*, **2**: 157–164.

Vinson, S.B., D.P. Harlan, and W.G. Hart. 1978. Response of the parasitoid *Microterys flavus* to the brown soft scale and its honeydew. *Environmental Entomology*, **7**: 874–878.

Vitousek, P.M. 1986. Biological invasions and ecosystem properties: Can species make a difference?, pp. 163–176. *In* Mooney, H.A. and J.A. Drake, (eds.). *Ecology of Biological Invasions of North America and Hawaii.* Springer-Verlag, New York.

Vitousek, P.M. 1990. Biological invasions and ecosystem processes: Towards an integration of population biology and ecosystem studies. *Oikos*, **57**: 7–13.

Vitousek, P.M. 1992. Effects of alien plants on native ecosystems, pp. 29–41. *In* Stone, C.P., C.W. Smith, and J.T. Tunison, (eds.). *Alien Plant Invasions in Native Ecosystems of Hawai'i: Management and Research.* University of Hawaii Press, Honolulu, Hawaii, U.S.A.

Vivrette, N.J. and C.H. Muller. 1977. Mechanism of invasion and dominance of coastal grassland by *Mesembryanthemum crystallinum. Ecological Monographs*, **47**: 301–318.

Voblyl, P.I., B.B. Kiku, L.G. Voltenko, and A.S. Abashkin. 1988. A unit for releasing *Trichogramma. Zashchita Rastenii*, **3**: 40. (in Russian)

Voegele, J.M. 1989. Biological control of *Brontispa longissima* in Western Samoa: An ecological and economic evaluation. *Agriculture, Ecosystems and Environment*, **27**: 315–329.

Völkl, W. 1992. Aphids or their parasitoids: Who actually benefits from ant-attendance? *Journal of Animal Ecology*, **61**: 273–281.

Völkl, W. 1994. The effect of ant-attendance on the foraging behaviour of the aphid parasitoid *Lysiphlebus cardui. Oikos*, **70**: 149–155.

Vorley, V.T. and S.D. Wratten. 1987. Migration of parasitoids (Hymenoptera: Braconidae) of cereal aphids (Hemiptera: Aphididae) between grassland, early-sown cereals and late-sown cereals in southern England. *Bulletin of Entomological Research*, **77**: 555–568.

Vyas, S.C. 1988. *Nontarget Effects of Agricultural Fungicides.* CRC Press, Inc., Boca Raton, Florida, U.S.A.

Waage, J.K. 1978. Arrestment responses of the parasitoid, *Nemeritis canescens*, to a contact chemical produced by its host, *Plodia interpunctella. Physiological Entomology*, **3**: 135–146.

Waage, J.K. 1979. Foraging for patchily distributed hosts by the parasitoid, *Nemeritis canescens. Journal of Animal Ecology*, **48**: 353–371.

Waage, J.K. 1983. Aggregation in field parasitoid populations: Foraging time allocation by a population of *Diadegma* (Hymenoptera, Ichneumonidae). *Ecological Entomology*, **8**: 447–453.

Waage, J.K. 1986. Family planning in parasitoids: Adaptive patterns of progeny and sex allocation, pp. 63–95. *In* Waage, J. and D. Greathead, (eds.). *Insect Parasitoids.* Academic Press, London.

Waage, J.K. 1989. The population ecology of pest-pesticide-natural enemy interactions, pp. 81–93. *In* Jepson, P.C., (ed.). *Pesticides and Non-Target Invertebrates.* Intercept, Wimborne, Dorset, U.K.

Waage, J.K. 1990. Ecological theory and the selection of biological control agents, pp. 135–157. *In* Mackauer, M. and L.E. Ehler, (eds.). *Critical Issues in Biological Control.* Intercept, Andover, U.K.

Waage, J.K. and D. Greathead, (eds.). 1986. *Insect Parasitoids.* Academic Press, London.

Waage, J.K. and D.J. Greathead. 1988. Biological control: Challenges and opportunities. *Philosophical Transactions of the Royal Society of London, Series B*, **318**: 111–128.

Waage, J.K. and M.P. Hassell. 1982. Parasitoids as biological control agents—A fundamental approach. *Parasitology*, **84**: 241–268.

Waage, J.K. and J.A. Lane. 1984. The reproductive strategy of a parasitic wasp. II. Sex allocation and local mate competition in *Trichogramma evanescens. Journal Animal Ecology*, **53**: 417–426.

Waage, J.K. and N.J. Mills. 1992. Biological control, pp. 412–430. *In* Crawley, M.J, (ed.). *Natural Enemies: The Population Biology of Predators, Parasites and Diseases.* Blackwell Scientific Publications, Oxford, U.K.

Waage, J.K., J.T. Smiley, and L.E. Gilbert. 1981. The *Passiflora* problem in Hawaii: Prospects and problems of controlling the forest weed *P. mollissima* (Passifloraceae) with heliconiine butterflies. *Entomophaga*, **26**: 275–284.

Wäckers, F.L. and W.J. Lewis. 1994. Olfactory and visual learning and their combined influence on host site location by the parasitoid *Microplitis croceipes* (Cresson). *Biological Control*, **4**: 105–112.

Walde, S.J. and W.W. Murdoch. 1988. Spatial density dependence in parasitoids. *Annual Review of Entomology*, **33**: 441–466.

Walker, H.L. 1981. *Fusarium lateritium*: A pathogen of spurred anoda (*Anoda cristata*), prickly sida (*Sida spinosa*), and velvetleaf (*Abutilon theophrasti*). *Weed Science*, **29**: 629–631.

Wallace, M.M.H. 1954. The effect of DDT and BHC on the population of the lucerne flea, *Sminthurus viridis* (L.) (Collembola), and its control by predatory mites, *Biscirus* spp. (Bdellidae). *Australian Journal of Agricultural Research*, **5**: 148–155.

Wallace, M.M.H. 1981. Tackling the lucerne flea and red-legged earth mite. *Journal of Agriculture, Western Australia*, **22**: 72–74.

Walter, G.H. 1988. Heteronomous host relationships in aphelinids—Evolutionary pathways and adaptive significance (Hymenoptera: Chalcidoidea). *Advances in Parasitic Hymenoptera Research*, **1988**: 313–326.

Walter, D.E. and D.J. O'Dowd. 1992. Leaf morphology and predators: effect of leaf domatia on the abundance of predatory mites (Acari: Phytoseiidae). *Environmental Entomology*, **21**: 478–484.

Walter, D.E., H.W. Hunt, and E.T. Elliot. 1988. Guilds or functional groups? An analysis of predatory arthropods from a shortgrass steppe soil. *Pedobiologia*, **31**: 247–260.

Walters, L.L. and E.F. Legner. 1980. Impact of desert pupfish, *Cyprinodon macularius*, and *Gambusia affinis affinis* on fauna in pond ecosystems. *Hilgardia*, **48**(3): 1–18.

Wang, Y.H. and A.P. Guttierrez. 1980. An assessment of the use of stability analyses in population ecology. *Journal of Animal Ecology*, **49**: 435–452.

Wapshere, A.J. 1974a. A strategy for evaluating the safety of organisms for biological weed control. *Annals of Applied Biology*, **77**: 201–211.

Wapshere, A.J. 1974b. Towards a science of biological control of weeds, pp. 3–12. *In* Wapshere, A.J., (ed.). *Proceedings of the Third International Symposium on Biological Control of Weeds*. Montpellier, France, 1973. Commonwealth Institute of Biological Control Miscellaneous Publication No. 8.

Wapshere, A.J. 1989. A testing sequence for reducing the rejection of potential biological control agents for weeds. *Annals of Applied Biology*, **114**: 515–526.

Wapshere, A.J. 1990. Biological control of grass weeds in Australia: an appraisal. *Plant Protection Quarterly*, **5**(2): 62–75.

Wapshere, A.J. 1992. Comparing methods of selecting effective biocontrol agents for weeds, pp. 557–560. *In* Anon. *Proceedings of the First International Weed Control Congress*. Melbourne, Australia.

Wapshere, A.J., E.S. Delfosse, and J.M. Cullen. 1989. Recent developments in biological control of weeds. *Crop Protection*, **8**: 227–250.

Wardle, A.R. and J.H. Borden. 1985. Age-dependent associative learning by *Exeristes roborator* (F.) (Hymenoptera: Ichneumonidae). *Canadian Entomologist*, **117**: 605–616.

Wardle, A.R. and J.H. Borden. 1989. Learning of an olfactory stimulus associated with a host microhabitat by *Exeristes roborator*. *Entomologia Experimentalis et Applicata*, **52**: 271–279.

Wardle, A.R. and J.H. Borden. 1990. Learning of host microhabitat form by *Exeristes roborator* (F.) (Hymenoptera: Ichneumonidae). *Journal of Insect Behavior*, **3**: 251–263.

Waterhouse, D.F. and K.R. Norris. 1987. *Biological Control, Pacific Prospects*. Inkata Press, Melbourne, Australia.

Waterhouse, G.M. 1973. Entomophthorales, pp. 219–229. *In* Ainsworth, G.C., F.K. Sparrow, and A.S. Sussman, (eds.). *The Fungi: An Advanced Treatise, Vol. 4b*. Academic Press, New York.

Watkinson, J. 1994. Global view of present and future markets for Bt products, pp. 3–7.

Watson, A.K. 1991. The classical approach with plant pathogens, pp. 3–23. *In* TeBeest, D.O., (ed.). *Microbial Control of Weeds*. Chapman and Hall, New York.

Watson, A.K. and W.E. Sackston. 1985. Plant pathogen containment (quarantine) facility at Mcdonald College. *Canadian Journal of Plant Pathology*, **7**: 177–180.

Watt, K.E.F. 1965. Community stability and the strategy of biological control. *Canadian Entomologist*, **97**: 887–895.

Way, M.J., M.E. Cammell, B. Bolton, and P. Kanagaratnam. 1989. Ants (Hymenoptera: Formicidae) as egg predators of coconut pests, especially in relation to biological control of the coconut caterpillar, *Opisina arenosella* Walker (Lepidoptera: Xyloryctidae), in Sri Lanka. *Bulletin of Entomological Research*, **79**: 219–233.

Weaver, D.B., R. Rodríguez-Kábana, and E.L. Carden. 1993. Velvetbean in rotation with soybean for management of *Heterodera glycines* and *Meloidogyne arenaria*. *Supplement to the Journal of Nematology (Annals of Applied Nematology)*, **25**: 809–813.

Webb, S.E. and A.M. Shelton. 1990. Effect of age structure on the outcome of viral epizootics in field populations of imported cabbageworm (Lepidoptera: Pieridae). *Environmental Entomology*, **19**: 111–116.

Wehling, W.F. and G.L. Piper. 1988. Efficacy diminution of the rush skeletonweed gall midge, *Cystiphora schmidti* (Diptera: Cecidomyiidae), by an indigenous parasitoid. *Pan-Pacific Entomologist*, **64**: 83–85.

Weidemann, G.J. and D.O. TeBeest. 1990. Biology of host range testing for biocontrol of weeds. *Weed Technology*, **4**: 465–470.

Weidemann, G.J. and G.E. Templeton. 1988. Efficacy and soil persistence of *Fusarium solani* f. sp. *cucurbitae* for control of Texas gourd (*Cucurbita texana*). *Plant Disease*, **72**: 36–38.

Weiss, S.A. and J.L. Vaughn. 1986. Cell culture methods for large-scale propagation of baculoviruses, pp. 63–88. *In* Granados, R.R. and B.A. Federici, (eds.). *The Biology of Baculoviruses: Volume II. Practical Application for Insect Control*. CRC Press, Inc., Boca Raton, Florida, U.S.A.

Weller, D.M. 1983. Application of fluorescent pseudomonads to control root diseases, pp. 137–140. *In* Parker, C.A. et al.

(eds.). *Ecology and Management of Soilborne Plant Pathogens.* American Phytopathological Society, St. Paul, Minnesota, U.S.A.

Welton, J.S. and M. Ladle. 1993. The experimental treatment of the blackfly *Simulium posticatum* in the Dorset Stour using the biologically produced insecticide *Bacillus thuringiensis* var. *israelensis. Journal of Applied Ecology,* **30**: 772–782.

Wermelinger, B., J.J. Oertli, and V. Delucchi. 1985. Effect of host plant nitrogen fertilization on the biology of the two-spotted spider mite, *Tetranychus urticae. Entomologia Experimentalis et Applicata,* **38**: 23–28.

Werren, J.H. 1984a. Brood size and sex ratio regulation in the parasitic wasp *Nasonia vitripennis* (Walker) (Hymenoptera: Pteromalidae). *Netherlands Journal of Zoology,* **34**: 123–143.

Werren, J.H. 1984b. A model for sex ratio selection in parasitic wasps: local mate competition and host quality effects. *Netherlands Journal of Zoology,* **34**: 81–96.

Weseloh, R.M. 1974. Host recognition by the gypsy moth larval parasitoid, *Apanteles melanoscelus. Annals of the Entomological Society of America,* **67**: 583–587.

Weseloh, R.M. 1984. Effect of exposing adults of the gypsy moth parasite *Compsilura concinnata* (Diptera: Tachinidae) to hosts on the parasite's subsequent behavior. *Canadian Entomologist,* **116**: 79–84.

Weseloh, R.M. 1986. Artificial selection for host suitability and development length of the gypsy moth (Lepidoptera: Lymantriidae) parasite, *Cotesia melanoscela* (Hymenoptera: Braconidae). *Journal of Economic Entomology,* **79**: 1212–1216.

Weseloh, R.M. 1990. Simulation of litter residence times of young gypsy moth larvae and implications for predation by ants. *Entomologia Experimentalis et Applicata,* **57**: 215–221.

Westigard, P.H. and H.R. Moffitt. 1984. Natural control of the pear psylla (Homoptera: Psyllidae): impact of mating disruption with the sex pheromone for control of the codling moth (Lepidoptera: Tortricidae). *Journal of Economic Entomology,* **77**: 1520–1523.

Whalon, M.E., B.A. Croft, and T.M. Mowry. 1982. Introduction and survival of susceptible and pyrethroid-resistant strains of *Amblyseius fallacis* (Acari: Phytoseiidae) in a Michigan apple orchard. *Environmental Entomology,* **11**: 1096–1099.

Wharton, R.A. 1993. Bionomics of the Braconidae. *Annual Review of Entomology,* **38**: 121–143.

Whistlecraft, J.W. and I.J.M. Lepard. 1989. Effect of flooding on survival of the onion fly *Delia antiqua* (Diptera: Anthomyiidae) and two parasitoids, *Aphaereta pallipes* (Hymenoptera: Braconidae) and *Aleochara bilineata* (Coleoptera: Staphylinidae). *Proceedings of the Entomological Society of Ontario,* **120**: 43–47.

Whistlecraft, J.W., C.R. Harris, J.H. Tolman, and A.D. Tomlin. 1985. Mass-rearing technique for *Aleochara bilineata* (Coleoptera: Staphylinidae). *Journal of Economic Entomology,* **78**: 995–997.

Whitcomb, W.H. 1981. The use of predators in insect control, pp. 105–123. *In* Pimentel, D., (ed.). *CRC Handbook of Pest Management in Agriculture, Vol. II.* CRC Press, Inc., Boca Raton, Florida, U.S.A.

White E.G. and C.B. Huffaker. 1969. Regulatory processes and population cyclicity in laboratory populations of *Anagasta kühniella* (Zeller) (Lepidoptera: Phycitidae). II. Parasitism, predation, competition and protective cover. *Researches on Population Ecology,* **11**: 150–185.

Whitfield, J.B. 1990. Parasitoids, polydnaviruses and endosymbiosis. *Parasitology Today,* **6**: 381–384.

Whittaker, J.B. 1973. Density regulation in a population of *Philaenus spumarius* (L.) (Homoptera: Cercopidae). *Journal of Animal Ecology,* **42**: 163–172.

Wickremasinghe, M.G.V. and H.F. van Emden. 1992. Reactions of adult female parasitoids, particularly *Aphidius rhopalosiphi,* to volatile chemical cues from the host plants of their aphid prey. *Physiological Entomology,* **17**: 297–304.

Wilding. N. 1990. Entomophthorales in pest control—Recent developments, pp. 138–141. *In* Pinnock, D.E., (ed.). *Vth International Colloquium on Invertebrate Pathology and Microbial Control.* Adelaide, Australia, 20–24 August, 1990.

Williams, D.F., (ed.). 1994. *Exotic Ants: Biology, Impact, and Control of Introduced Species.* Westview Press, Boulder, Colorado, U.S.A.

Williams, D.F. and W.A. Banks. 1987. *Pseudacteon obtusus* (Diptera: Phoridae) attacking *Solenopsis invicta* (Hymenoptera: Formicidae) in Brazil. *Psyche,* **94**: 9–13.

Williamson, M. 1991. Biocontrol risks. *Nature,* **353**(6343), p. 394.

Willis, A.J., J.E. Ash, and R.H. Groves. 1993. Combined effects of two arthropod herbivores and water stress on growth of *Hypericum* species. *Oecologia,* **96**: 517–525.

Wilson, B.W. and C.A. Garcia. 1992. Host specificity and biology of *Heteropsylla spinulosa* [Hom.: Psyllidae] introduced into Australia and Western Samoa for the biological control of *Mimosa invisa. Entomophaga,* **37**: 293–299.

Wilson, C.G. and G.J. Flanagan. 1991. Establishment of *Acanthoscelides quadridentatus* (Schaeffer) and *A. puniceus* Johnson (Coleoptera: Bruchidae) on *Mimosa pigma* in northern Australia. *Journal of the Australian Entomological Society,* **30**: 279–280.

Wilson, C.G. and R.N. Pitkethley. 1992. Botryodiplodia die-back of *Mimosa pigra*, a noxious weed in northern Australia. *Plant Pathology*, **41**: 777–779.

Wilson, C.L. 1969. Use of plant pathogens in weed control. *Annual Review of Phytopathology*, **7**: 411–434.

Wilson, C.L. and M.E. Wisniewski. 1989. Biological control of postharvest diseases of fruits and vegetables: An emerging technology. *Annual Review of Phytopathology*, **27**: 425–441.

Wilson, C.L. and M.E. Wisniewski, (eds.). 1994. *Biological Control of Postharvest Diseases, Theory and Practice*. CRC Press, Inc., Boca Raton, Florida, U.S.A.

Wilson, F. 1960. *A Review of the Biological Control of Insects and Weeds in Australia and Australian New Guinea*. Commonwealth Agricultural Bureaux Technical Communication No. 1. Lamport Gilbert and Co., Reading, U.K.

Wilson, F. 1963. *Australia as a Source of Beneficial Insects for Biological Control*. Commonwealth Institute of Biological Control Technical Communication No. 2, Commonwealth Agricultural Bureaux, Farnham Royal, U.K.

Wilson, L.T., C.H. Pickett, D.L. Flaherty, and T.A. Bates. 1989. French prune trees: Refuge for grape leafhopper parasite. *California Agriculture*, **43**(2): 7–8.

Wilson, M. and S.E. Lindow. 1993a. Interactions between the biological control agent *Pseudomonas fluorescens* A506 and *Erwinia amylovora* in pear blossoms. *Ecology and Epidemiology*, **83**: 117–123.

Wilson, M. and S.E. Lindow. 1993b. Release of recombinant microorganisms. *Annual Review of Microbiology*, **47**: 913–944.

Wilson, M. and S.E. Lindow. 1994. Ecological similarity and coexistence of epiphytic ice-nucleating (ice[1]) *Pseudomonas syringae* strains and a non-ice-nucleating (ice[2]) biological control agent. *Applied and Environmental Microbiology*, **60**: 3128–3137.

Winder, L. 1990. Predation of the cereal aphid *Sitobion avenae* by polyphagous predators on the ground. *Ecological Entomology*, **15**: 105–110.

Wiskerke, J.S.C., M. Dicke, and L.E.M. Vet. 1993. Larval parasitoid uses aggregation pheromone of adult hosts in foraging behaviour: a solution to the reliability-detectability problem. *Oecologia*, **93**: 145–148.

Witz, B.W. 1990. Antipredator mechanisms in arthropods: A twenty year literature survey. *Florida Entomologist*, **73**: 71–99.

Woets, J. 1978. Development of an introduction scheme for *Encarsia formosa* Gahan [Hymenoptera: Aphelinidae] in greenhouse tomatoes to control the greenhouse whitefly, *Trialeurodes vaporariorum* (Westwood) [Homoptera: Aleyrodidae]. *Mededelingen Faculteit Landbouwwetenschappen Rijksuniversiteit Gent*, **43**: 379–382.

Wojcik, D.P. 1989. Behavioral interaction between ants and their parasites. *Florida Entomologist*, **72**: 43–51.

Wolda, H. and B. Dennis. 1993. Density dependence tests, are they? *Oecologia*, **95**: 581–591.

Wolda, H., B. Dennis, and M.L. Taper. 1994. Density dependence tests, and largely futile comments: Answers to Holyoak and Lawton (1993) and Hanski, Woiwood and Perry (1993). *Oecologia*, **1994**: 229–234.

Wolf, F.T. 1988. Entomophthorales and their parasitism of insects. *Nova Hedwigia*, **46**: 121–142.

Wood, B.J. 1968. *Pests of Oil Palms in Malaysia and Their Control*. Incorporated Society of Planters, Kuala Lumpur.

Wood, H.A. and R.R. Granados. 1991. Genetically engineered baculoviruses as agents for pest control. *Annual Review of Microbiology*, **45**: 69–87.

Wood, H.A., P.R. Hughes, and A. Shelton. 1994. Field studies of the co-occlusion strategy with a genetically altered isolate of the *Autographa californica* nuclear polyhedrosis virus. *Environmental Entomology*, **23**: 211–219.

Wood, R.K.S. 1951. The control of diseases of lettuce by the use of antagonistic organisms. 1. Control of *Botrytis cinerea* Pers. *Annals of Applied Biology*, **38**: 203–216.

Woodburn, T.L. 1993. Host specificity testing, release and establishment of *Urophora solstitialis* (L.) (Diptera: Tephritidae), a potential biological control agent for *Carduus nutans* L., in Australia. *Biocontrol Science and Technology*, **3**: 419–426.

Woodhead, S.H., A.L. O'Leary, D.J. O'Leary, and S.C. Rabatin. 1990. Discovery, development and registration of a biocontrol agent from an industrial perspective. *Canadian Journal of Plant Pathology*, **12**: 328–331.

Woodring, J.L. and H.K. Kaya. 1988. Steinernematid and heterorhabditid nematodes: A handbook of techniques. Southern Cooperative Series Bulletin 331, Arkansas Agricultural Experimental Station, Fayetteville, Arkansas, U.S.A.

Woods, S. and J.S. Elkinton. 1987. Bimodal patterns of mortality from nuclear polyhedrosis virus in gypsy moth (*Lymantria dispar*) populations. *Journal of Invertebrate Pathology*, **50**: 151–157.

Woolhouse, M.E.J. and R. Harmsen. 1984. The mite complex on the foliage of a pesticide-free apple orchard: Population dynamics and habitat associations. *Proceedings of the Entomological Society of Ontario*, **115**: 1–11.

Wraight, S.P., D. Molloy, and H. Jamnback. 1981. Efficacy of *Bacillus sphaericus* strain 1593 against the four instars of laboratory reared and field collected *Culex pipiens pipiens* and laboratory reared *Culex salinarius*. *Canadian Entomologist*, **113**: 379–386.

Wraight, S.P., D. Molloy, and P. McCoy. 1982. A comparison of laboratory and field tests of *Bacillus sphaericus* strain 1593 and *Bacillus thuringiensis* var. *israelensis* against *Aedes stiulans* larvae (Diptera: Culicidae). *Canadian Entomologist*, **114**: 55–61.

Wratten, S.D. 1987. The effectiveness of native natural enemies, pp. 89–112. *In* Burn, A.J., T.H. Coaker, and P.C. Jepson, (eds.). *Integrated Pest Management.* Academic Press, London.

Wrensch, D.L. and M.A. Ebbert, (eds.). 1993. *Evolution and Diversity of Sex Ratio in Insects and Mites.* Chapman and Hall, New York.

Wright, J.E. 1993. Control of the boll weevil (Coleoptera: Curculionidae) with Naturalis-L: A mycoinsecticide. *Journal of Economic Entomology,* **86**: 1355–1358.

Wright, R.J., M.G. Villani, and F. Agudelo-Silva. 1988. Steinernematid and heterorhabditid nematodes for control of larval European chafers and Japanese beetles (Coleoptera: Scarabaeidae) in potted yew. *Journal of Economic Entomology,* **81**: 152–157.

Xiong, J.J., T.Y. Du, M.D. Huang, and Z.H. Deng. 1988. Preliminary field studies in citrus orchards of a phosmet-resistant strain of *Amblyseius nicholis* Ehara et Lee. *Natural Enemies of Insects,* **10**: 9–14. (in Chinese)

Xu, F.Y. and D.X. Wu. 1987. Control of bamboo scale insects by intercropping rape in the bamboo forest to attract coccinellid beetles. *Chinese Journal of Biological Control,* **5**: 117–119. (in Chinese)

Yaninek, J.S.and A.C. Bellotti. 1987. Exploration for natural enemies of cassava green mites based on agrometeorological criteria, pp. 69–75. *In* Rijks, D. and G. Mathys, (eds.). *Proceedings of the Seminar on Agrometeorology and Crop Protection in the Lowland Humid and Subhumid Tropics.* Cotonou, Benin, 7–11 July 1986, World Meteorological Organization, Geneva.

Yasem de Romero, M.G. 1986. Effect of some agricultural chemicals on the entomopathenic fungus *Verticillium lecanii* (Zimm.) Viegas. *Revista de Investigacion, Centro de Investigaciones para la Regulacion de Poblaciones de Organismos Nocivos, Argentina,* **4**: 55–62.

Yigit, A. and L. Erkilic. 1987. Studies on egg parasitoids of grape leafhopper, *Arboridia adanae* Dlab. (Hom., Cicadellidae) and their effects in the region of South Anatolia. Turkiye I. Entomoloji Kongresi Bildirileri, 13–16 Ekim 1987, Ege Universitesi, Bornova, Izmir. Bornova/Ismir, Turkey; Ege Universitesi/Ataturk Kultur Merkezi (1987) 35–42. (in Turkish)

York, G.T. 1958. Field tests with the fungus *Beauveria* sp. for control of the European corn borer. *Iowa State College Journal of Science,* **33**: 123–129.

Young, G.R. 1987. Some parasites of *Segestes decoratus* Redtenbacher (Orthoptera: Tettigoniidae) and their possible use in the biological control of tettigoniid pests of coconuts in Papua New Guinea. *Bulletin of Entomological Research,* **77**: 515–524.

Young, S.Y. 1990. Influence of sprinkler irrigation on dispersal of nuclear polyhedrosis virus from host cadavers on soybean. *Environmental Entomology,* **19**: 717–720.

Young, S.Y., III and W.C. Yearian. 1986. Formulation and application of baculoviruses, pp. 157–179. *In* Granados, R.R. and B.A. Federici, (eds.). *The Biology of Baculoviruses: Volume II. Practical Application for Insect Control.* CRC Press, Inc., Boca Raton, Florida, U.S.A.

Young, S.Y. and W.C. Yearian. 1989. Persistence and movement of nuclear polyhedrosis virus on soybean plants after death of infected *Anticarsia gemmatalis* (Lepidoptera: Noctuidae). *Environmental Entomology,* **18**: 811–815.

Zangger, A., J.-A. Lys, and W. Nentwig. 1994. Increasing the availability of food and the reproduction of *Poecilus cupreus* in a cereal field by strip-management. *Entomologia Experimentalis et Applicata,* **71**: 111–120.

Zanotto, P.M.deA., B.D. Kessing, and J.E. Maruniak. 1993. Phylogenetic interrelationships among baculoviruses: Evolutionary rates and host associations. *Journal of Invertebrate Pathology,* **62**: 147–164.

Zdárková, E. 1986. Mass rearing of the predator *Cheyletus eruditus* (Schrank) (Acarina: Cheyletidae) for biological control of acarid mites infesting stored products. *Crop Protection,* **5**: 122–124.

Zehnder, G.W., G.M. Ghidiu, and J. Speese III. 1992. Use of the occurrence of peak Colorado potato beetle (Coleoptera: Chrysomelidae) egg hatch for timing of *Bacillus thuringiensis* spray applications in potatoes. *Journal of Economic Entomology,* **85**: 281–288.

Zelazny, B., A. Lolong, and A.M. Crawford. 1990. Introduction and field comparison of baculovirus strains against *Oryctes rhinoceros* (Coloeptera: Scarabaeidae) in the Maldives. *Environmental Entomology,* **19**: 1115–1121.

Zelazny, B., A. Lolong, and B. Pattang. 1992. *Oryctes rhinoceros* (Coleoptera: Scarabaeidae) populations suppressed by a baculovirus. *Journal of Invertebrate Pathology,* **59**: 61–68.

Zgomba, M., D. Petrovic, and Z. Srdic. 1986. Mosquito larvicide impact on mayflies (Ephemeroptera) and dragonflies (Odonatoptera (sic)) in aquatic biotypes. *Proceedings of the 3rd European Congress of Entomology, Amsterdam,* **3**: 532.

Zhang, D. and D.L. Dahlman. 1989. Microplitis croceipes teratocytes cause developmental arrest of *Heliothis virescens* larvae. *Archives of Insect Biochemistry and Physiology,* **12**: 51–61.

Zhang, A. and W. Olkowski. 1989. Ageratum cover crop aids citrus biocontrol in China. *The IPM Practitioner,* **11**(9): 8–10.

Zhang, D.and D.L. Dahlman. 1989. *Microplitis croceipes* teratocytes cause developmental arrest of *Heliothis virescens* larvae. *Archives of Insect Biochemistry and Physiology,* **12**: 51–61.

Zhang, N.X. and Y.X. Li. 1989. An improved method of rearing *Amblyseius fallacis* (Acari: Phytoseiidae) with plant pollen. *Chinese Journal of Biological Control*, **5**: 149–152.

Zhang, P.G., J C. Sutton, and A.A. Hopkin. 1994. Evaluation of microorganisms for biocontrol of *Botrytis cinerea* in container-grown black spruce seedlings. *Canadian Journal of Forest Research*, **24**: 1312–1316.

Zhi-Qiang Zhang. 1992. The use of beneficial birds for biological pest control in China. *Biocontrol News and Information*, **13** (1): 11N–16N.

Zhumanov, B.Zh. 1989. Rapid method for identification of cutworm parasitoids. *Zashchita Rastenii*, **6**: 28–29. (in Russian)

Zimmermann, H.G. 1991. Biological control of mesquite, *Prosopis* spp. (Fabaceae), in South Africa. *Agriculture, Ecosystems and Environment*, **37**: 175–186.

Zimmermann, G. 1986. Insect pathogenic fungi as pest control agents, pp. 217–231. *In* Franz, J.M., (ed.). *Biological Plant and Health Protection: Biological Control of Plant Pests and of Vectors of Human and Animal Diseases.* International Symposium of the Akademie der Wissenschaften und der Literatur, Mainz, November 15–17th, 1984 at Mainz and Darmstadt. *Fortschritte der Zoologie*, **32**: 341 pp., Gustav Fischer Verlag, Stuttgart, Germany.

Zimmermann, H.G., V.C. Moran, and J.H. Hoffmann. 1986. Insect herbivores as determinants of the present distribution and abundance of invasive cacti in South Africa, pp. 269–274. *In* Macdonald, I.A.W., F.J. Kruger, and A.A. Ferrar, (eds.). *The Ecology and Management of Biological Invasions in Southern Africa.* Proceedings of the National Synthesis Symposium on the Ecology of Biological Invasions. Oxford University Press, Cape Town, South Africa.

Zuckerman, B.M., M.B. Dicklow, and N. Acosta. 1993. A strain of *Bacillus thuringiensis* for the control of plant-parasitic nematodes. *Biocontrol Science and Technology*, **3**: 41–46.

Zwölfer, H. 1971. The structure and effect of parasite complexes attacking phytophagous host insects, pp. 405–418. *In* den Boer, P.J. and G.R. Gradwell, (eds.). *Dynamics of Populations.* Centre for Agricultural Publishing and Documentation, Wageningen, The Netherlands.

Zwölfer, H. and P. Harris. 1971. Host specificity determination of insects for biological control of weeds. *Annual Review of Entomology*, **16**: 159–178.

Zwölfer, H., M.A. Ghani, and V.P. Rao. 1976. Foreign exploration and importation of natural enemies, pp. 189–207. *In* Huffaker, C.B. and P.S. Messenger, (eds.). *Theory and Practice of Biological Control.* Academic Press, New York.

Index

MIX
Papier aus verantwortungsvollen Quellen
Paper from responsible sources
FSC® C105338
FSC
www.fsc.org

If you have any concerns about our products,
you can contact us on
ProductSafety@springernature.com

In case Publisher is established outside the EU,
the EU authorized representative is:
Springer Nature Customer Service Center GmbH
Europaplatz 3, 69115 Heidelberg, Germany

Printed by Libri Plureos GmbH
in Hamburg, Germany